Haese Mathematics
specialists in mathematics publishing

Cambridge Additional Mathematics

second edition

IGCSE® (0606)

O Level (4037)

Michael Haese

Mark Humphries

Chris Sangwin

Ngoc Vo

CAMBRIDGE ADDITIONAL MATHEMATICS second edition

Michael Haese	B.Sc.(Hons.), Ph.D.
Mark Humphries	B.Sc.(Hons.)
Chris Sangwin	M.A., M.Sc., Ph.D.
Ngoc Vo	B.Ma.Sc.

Published by Haese Mathematics
152 Richmond Road, Marleston, SA 5033, AUSTRALIA
Telephone: +61 8 8210 4666, Fax: +61 8 8354 1238
Email: info@haesemathematics.com.au
Web: www.haesemathematics.com.au

National Library of Australia Card Number & ISBN 978-1-925489-48-4

First Edition 2018

Cartoon artwork by John Martin.

Artwork by Brian Houston.

Fractal artwork on the cover generated using ChaosPro, http://www.chaospro.de/

Computer software by Huda Kharrufa, Brett Laishley, Bronson Mathews, Linden May, Joshua Douglass-Molloy, Jonathan Petrinolis, and Nicole Szymanczyk.

Production work by Charlotte Frost, Rebecca Huang, Nicholas Kellett-Southby, and Bradley Steventon.

Typeset in Australia by Deanne Gallasch. Typeset in Times Roman 10.

This textbook has been endorsed by Cambridge Assessment International Education.

Printed in China by Prolong Press Limited.

Acknowledgements: The publishers acknowledge the cooperation of Oxford University Press, Australia, for the reproduction of material originally published in textbooks produced in association with Haese Mathematics.

FOREWORD

This book has been written to cover the '*Cambridge IGCSE® Additional Mathematics 0606*' and the '*Cambridge O Level Additional Mathematics 4037*' syllabuses for examination from 2020.

These syllabuses enable learners to extend the mathematics skills, knowledge, and understanding developed in the Cambridge IGCSE or O Level Mathematics courses, and use skills in the context of more advanced techniques.

The syllabuses have a Pure Mathematics only content which enables learners to acquire a suitable foundation in mathematics for further study in the subject. Knowledge of the content of the Cambridge IGCSE or O Level Mathematics syllabus (or an equivalent syllabus) is assumed.

Learners who successfully complete these courses gain lifelong skills, including:
- the further development of mathematical concepts and principles
- an ability to solve problems, present solutions logically, and interpret results.

This book is an attempt to cover, in one volume, the content outlined in the Cambridge IGCSE Additional Mathematics 0606 and Cambridge O Level Additional Mathematics 4037 syllabuses. The book can be used as a preparation for GCE Advanced level (A level) Mathematics or IB Diploma Programme Mathematics. The book has been endorsed by Cambridge Assessment International Education.

In this second edition, much of the text and questions have been rewritten. This will allow students who complete this course to progress to our A level or IB Diploma Programme textbooks, without encountering repeated material.

To reflect the principles on which the course is based, we have attempted to produce a book that embraces understanding and problem solving in order to give students different learning experiences. Review exercises appear at the end of each chapter. Answers are given at the end of the book, followed by an index.

The accompanying digital book is accessed through our Snowflake platform. It contains **Self Tutor** software (see p. 5), geometry and graphing software, demonstrations, and simulations. The digital book is a complete copy of the textbook, so students can view it on a home computer or tablet and keep the textbook at school.

The examinations for Cambridge Additional Mathematics are in the form of two papers. Many of the problems in this textbook have been written to reflect the style of the examination questions. The questions, worked solutions, and comments that appear in the book were written by the authors.

The book can be used as a scheme of work but it is expected that the teacher will choose the order of topics. Exercises in the book range from routine practice and consolidation of basic skills, to problem solving exercises that are quite demanding.

In this changing world of mathematics education, we believe that the contextual approach shown in this book will enhance the students' understanding, knowledge and appreciation of mathematics, and its universal application.

We welcome your feedback.

Email: info@haesemathematics.com.au
Web: www.haesemathematics.com.au

PMH, MH, CS, NV

ONLINE FEATURES

With the purchase of a new textbook you will gain 27 months subscription to our online product.
This subscription can be renewed for a small fee.

Access is granted through **SNOWFLAKE**, our book viewing software that can be used in your web browser or may be installed to your tablet or computer.

Students can revisit concepts taught in class and undertake their own revision and practice online.

The Snowflake icon that appears throughout the book denotes an active link on Snowflake. Simply click on the icon to access a large range of interactive features that includes:

- printable worksheets
- graphing packages
- demonstrations
- simulations
- revision games

INTERACTIVE LINK

COMPATIBILITY

For iPads, tablets, and other mobile devices, some of the interactive features may not work. However, the electronic version of the textbook can be viewed online using any of these devices.

REGISTERING

You will need to register to access the online features of this textbook.
Visit www.haesemathematics.com.au/register and follow the instructions. Once registered, you can:

- activate your electronic textbook
- use your account to make additional purchases.

To activate your electronic textbook, contact Haese Mathematics. On providing proof of purchase, your electronic textbook will be activated. **It is important that you keep your receipt as proof of purchase.**

For general queries about registering and licence keys:

- Visit our **SNOWFLAKE** help page: http://snowflake.haesemathematics.com.au/help
- Contact Haese Mathematics: info@haesemathematics.com.au

SELF TUTOR

SELF TUTOR is an exciting feature of this book.

The ◀) ***Self Tutor*** icon on each worked example denotes an active link online.

Simply click on the ◀) ***Self Tutor*** (or anywhere in the example box) to access the worked example, with a teacher's voice explaining each step necessary to reach the answer.

Play any line as often as you like. See how the basic processes come alive using movement and colour on the screen.

Ideal for students who have missed lessons or need extra help.

Example 4

◀) ***Self Tutor***

Find where the tangent to $y = x^3 - x^2 - 8x - 4$ at $(-1, 2)$ meets the curve again.

Let $f(x) = x^3 - x^2 - 8x - 4$

$\therefore\ f'(x) = 3x^2 - 2x - 8$

$\therefore\ f'(-1) = 3(-1)^2 - 2(-1) - 8 = -3$

$\therefore$ the equation of the tangent at $(-1, 2)$ is

$3x + y = 3(-1) + 2$

or $y = -3x - 1$

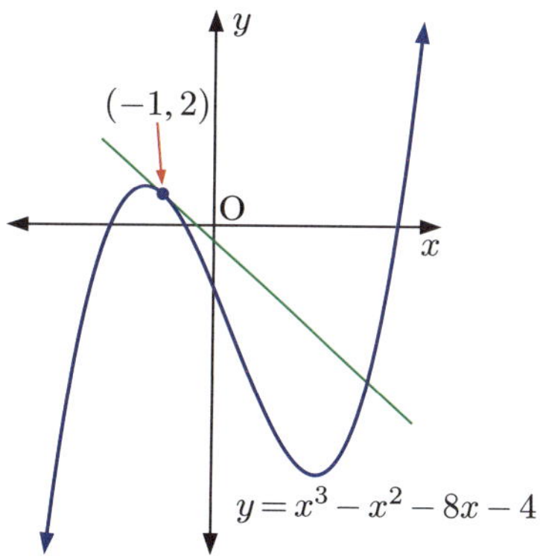

The curve meets the tangent when

$x^3 - x^2 - 8x - 4 = -3x - 1$

$\therefore\ x^3 - x^2 - 5x - 3 = 0$

$\therefore\ (x+1)^2(x-3) = 0$

{the tangent *touches* the curve at $x = -1$, so $(x+1)^2$ must be a factor}

$\therefore$ the tangent meets the curve again when $x = 3$.

Now $f(3) = (3)^3 - (3)^2 - 8(3) - 4 = -10$

$\therefore$ the tangent meets the curve again at $(3, -10)$.

See **Chapter 15, Applications of differential calculus**, page 366

SYMBOLS AND NOTATION USED IN THIS BOOK

$\mathbb{N}$	the set of natural numbers, $\{1, 2, 3, ...\}$
$\mathbb{Z}$	the set of integers, $\{0, \pm 1, \pm 2, \pm 3, ...\}$
$\mathbb{Z}^+$	the set of positive integers, $\{1, 2, 3, ...\}$
$\mathbb{Q}$	the set of rational numbers
$\mathbb{Q}^+$	the set of positive rational numbers, $\{x \in \mathbb{Q},\ x > 0\}$
$\mathbb{R}$	the set of real numbers
$\mathbb{R}^+$	the set of positive real numbers, $\{x \in \mathbb{R},\ x > 0\}$
$\{x : ....$	the set of all x such that
$\in$	is an element of
$\notin$	is not an element of
$\cup$	union
$\cap$	intersection
$a^{\frac{1}{n}},\ \sqrt[n]{a}$	a to the power of $\frac{1}{n}$, nth root of a (if $a \geqslant 0$ then $\sqrt[n]{a} \geqslant 0$)
$a^{\frac{1}{2}},\ \sqrt{a}$	a to the power $\frac{1}{2}$, square root of a (if $a \geqslant 0$ then $\sqrt{a} \geqslant 0$)
$\lvert x \rvert$	the modulus or absolute value of x, that is $\begin{cases} x \text{ for } x \geqslant 0, & x \in \mathbb{R} \\ -x \text{ for } x < 0, & x \in \mathbb{R} \end{cases}$
$\equiv$	identity or is equivalent to
$\approx$	is approximately equal to
$n!$	n factorial for $n \in \mathbb{N}$ $(0! = 1)$
$\binom{n}{r}$	the binomial coefficient $\dfrac{n!}{r!(n-r)!}$ for $n, r \in \mathbb{N},\ 0 \leqslant r \leqslant n$
$>$	is greater than
$\geq$ or $\geqslant$	is greater than or equal to
$<$	is less than
$\leq$ or $\leqslant$	is less than or equal to
u_n	the nth term of a sequence
S_n	the sum of a series $u_1 + u_2 + ... + u_n$
S_∞	the sum of an infinite series
$\sum_{i=1}^{n} a_i$	$a_1 + a_2 + ... + a_n$
f	function f
$f : x \mapsto y$	f is a function under which x is mapped to y
$f(x)$	the image of x under the function f
f^{-1}	the inverse function of the function f
$g \circ f,\ gf$	the composite function of f and g
$\lim_{x \to a} f(x)$	the limit of $f(x)$ as x tends to a

$\dfrac{dy}{dx}$	the derivative of y with respect to x
$\dfrac{d^2y}{dx^2}$	the second derivative of y with respect to x
$f'(x)$	the derivative of $f(x)$ with respect to x
$f''(x)$	the second derivative of $f(x)$ with respect to x
$\int y\,dx$	the indefinite integral of y with respect to x
$\int_a^b y\,dx$	the definite integral of y with respect to x for values of x between a and b
e	base of natural logarithms
e^x	exponential function of x
$\lg x$, $\log_{10} x$	logarithm of x to base 10
$\ln x$	natural logarithm of x
$\log_a x$	logarithm to the base a of x
sin, cos, tan, cosec, sec, cot	the circular functions
$\mathrm{A}(x,\ y)$	the point A in the plane with Cartesian coordinates x and y
AB	the line segment with endpoints A and B the distance from A to B the line containing points A and B
$\widehat{\mathrm{A}}$	the angle at A
$\mathrm{C\widehat{A}B}$	the angle between CA and AB
$\triangle$ABC	the triangle whose vertices are A, B, and C
$\mathbf{a}$	the vector $\mathbf{a}$
$\overrightarrow{\mathrm{AB}}$	the vector represented in magnitude and direction by the directed line segment from A to B
$\lvert\mathbf{a}\rvert$	the magnitude of vector $\mathbf{a}$
$\lvert\overrightarrow{\mathrm{AB}}\rvert$	the magnitude of $\overrightarrow{\mathrm{AB}}$
$\mathbf{i}$, $\mathbf{j}$	unit vectors in the directions of the Cartesian coordinate axes

TABLE OF CONTENTS

ABOUT THE AUTHORS

Michael Haese completed a BSc at the University of Adelaide, majoring in Infection and Immunity, and Applied Mathematics. He completed Honours in Applied Mathematics, and a PhD in high speed fluid flows. Michael has a keen interest in education and a desire to see mathematics come alive in the classroom through its history and relationship with other subject areas. He is passionate about girls' education and ensuring they have the same access and opportunities that boys do. His other interests are wide-ranging, including show jumping, cycling, and agriculture. He has been the principal editor for Haese Mathematics since 2008.

Mark Humphries completed a degree in Mathematical and Computer Science, and an Economics degree at the University of Adelaide. He then completed an Honours degree in Pure Mathematics. His mathematical interests include public key cryptography, elliptic curves, and number theory. Mark enjoys the challenge of piquing students' curiosity in mathematics, and encouraging students to think about mathematics in different ways. He has been working at Haese Mathematics since 2006, and is currently the writing manager.

Chris Sangwin completed a BA in Mathematics at the University of Oxford, and an MSc and PhD in Mathematics at the University of Bath. He spent thirteen years in the Mathematics Department at the University of Birmingham, and from 2000 - 2011 was seconded half time to the UK Higher Education Academy "Maths Stats and OR Network" to promote learning and teaching of university mathematics. He was awarded a National Teaching Fellowship in 2006, and is now Professor of Technology Enhanced Science Education at the University of Edinburgh.

His research interests focus on technology and mathematics education and include automatic assessment of mathematics using computer algebra, and problem solving using the Moore method and similar student-centred approaches.

Ngoc Vo completed a BMaSc at the University of Adelaide, majoring in Statistics and Applied Mathematics. Her Mathematical interests include regression analysis, Bayesian statistics, and statistical computing. Ngoc has been working at Haese Mathematics as a proof reader and writer since 2016.

1

Straight lines

Contents:

Opening problem

The graph alongside shows two straight lines.

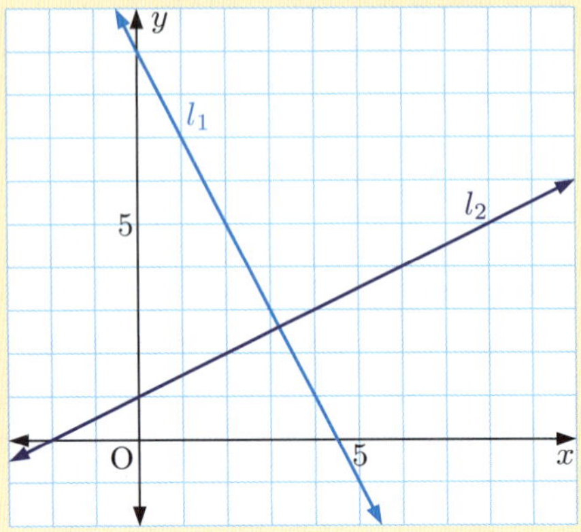

Things to think about:

a Just by looking at the graph, is it clear where the lines intersect?

b Can you find the *equation* of each line?

c How can we find the exact coordinates of the intersection point?

d Do two straight lines always meet at a single point?

From previous courses, you should be familiar with the following facts involving points and lines.

Given two points $A(x_1, y_1)$ and $B(x_2, y_2)$ on the coordinate plane:

- the **distance** between A and B is
$d = \sqrt{(x_2 - x_1)^2 + (y_2 - y_1)^2}$

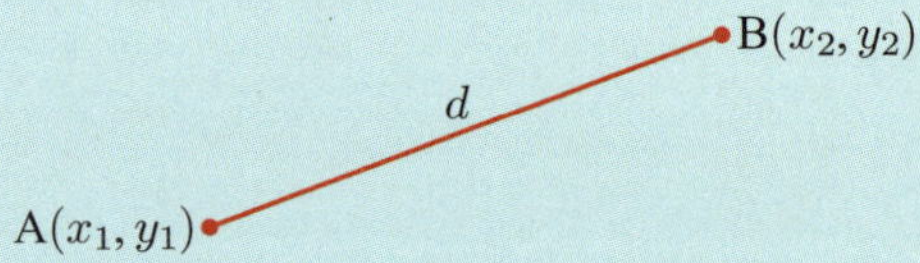

- the **midpoint** of AB is
$M\left(\frac{x_1 + x_2}{2}, \frac{y_1 + y_2}{2}\right)$

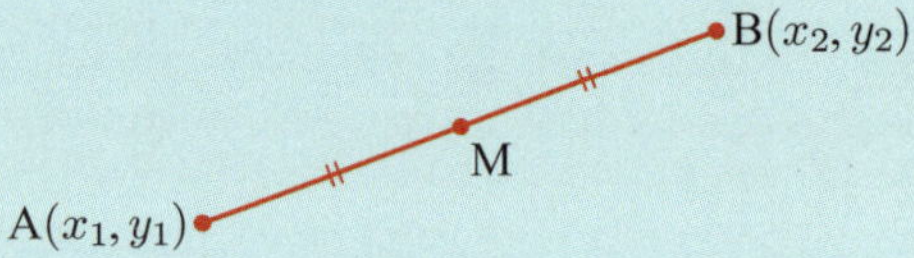

- the **gradient** of the line passing through A and B is
$\frac{y\text{-step}}{x\text{-step}} = \frac{y_2 - y_1}{x_2 - x_1}$.

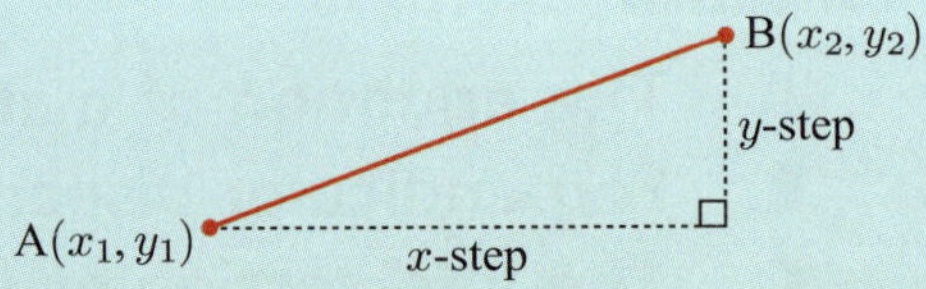

Given two lines in a plane:

- the lines are **parallel** if their gradients are **equal**

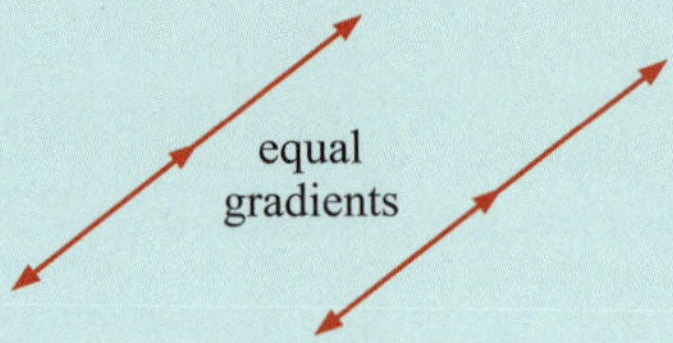

- the lines are **perpendicular** if their gradients are **negative reciprocals**.
If the gradient of one line is m, then the gradient of the other line is $-\frac{1}{m}$.

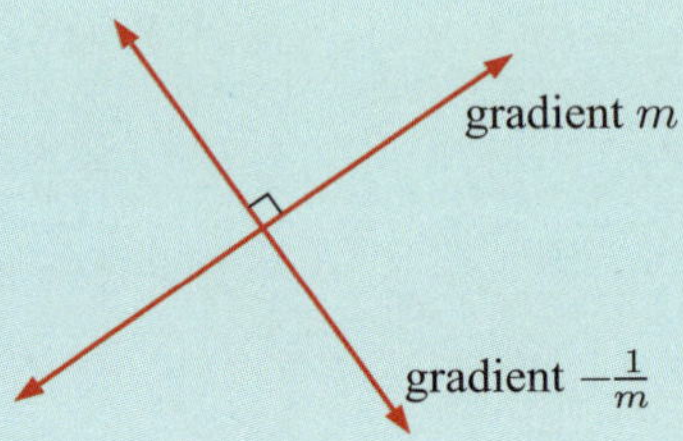

A THE EQUATION OF A STRAIGHT LINE

The **equation of a line** is an equation which connects the x and y values for every point on the line.

We can find the equation of a line if we know:
- its **gradient** and the **coordinates of any point** on the line, or
- the **coordinates of two distinct points** on the line.

Consider a line with gradient m, which passes through (x_1, y_1).

Suppose (x, y) is any point on the line.

The gradient between (x_1, y_1) and (x, y) is $\frac{y - y_1}{x - x_1}$.

$\therefore \quad \frac{y - y_1}{x - x_1} = m$

$\therefore \quad y - y_1 = m(x - x_1)$

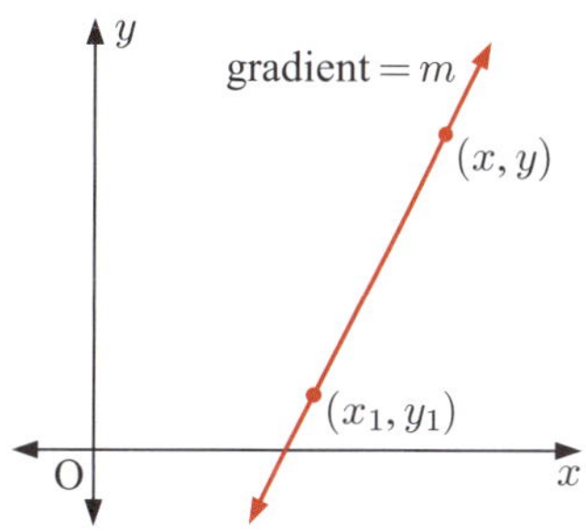

If a straight line has gradient m and passes through the point (x_1, y_1) then its equation in **gradient-point form** is $\boldsymbol{y - y_1 = m(x - x_1)}$.

We can rearrange this equation into:
- **gradient-intercept form** $y = mx + c$ where m is the gradient and c is the y-intercept of the line
- **general form** $Ax + By = D$.

Example 1 — Self Tutor

Find, in gradient-intercept form, the equation of the line through $(-1, 3)$ with gradient 5.

The equation of the line is $y - 3 = 5(x - (-1))$

$\therefore \quad y - 3 = 5x + 5$

$\therefore \quad y = 5x + 8$

Example 2 — Self Tutor

Find, in general form, the equation of the line with gradient $\frac{3}{4}$ which passes through $(5, -2)$.

The equation of the line is $y - (-2) = \frac{3}{4}(x - 5)$

$\therefore \quad 4(y + 2) = 3(x - 5)$

$\therefore \quad 4y + 8 = 3x - 15$

$\therefore \quad 3x - 4y = 23$

EXERCISE 1A

1 Find the gradient and y-intercept of the line with equation:

a $y = 3x + 5$ **b** $y = 4x - 2$ **c** $y = \frac{1}{5}x + \frac{3}{5}$

d $y = -7x - 3$ **e** $y = \dfrac{x+2}{6}$ **f** $y = \dfrac{8-5x}{3}$

2 Find the equation of the line with:

a gradient 1 and y-intercept -2 **b** gradient -1 and y-intercept 4

c gradient 2 and y-intercept 0 **d** gradient $-\frac{1}{2}$ and y-intercept 3.

3 Find, in gradient-intercept form, the equation of the line through:

a $(2, -5)$ with gradient 4 **b** $(-1, -2)$ with gradient -3

c $(7, -3)$ with gradient -5 **d** $(1, 4)$ with gradient $\frac{1}{2}$

e $(-1, 3)$ with gradient $-\frac{1}{3}$ **f** $(2, 6)$ with gradient 0.

4 Find, in general form, the equation of the line through:

a $(2, 5)$ having gradient $\frac{2}{3}$ **b** $(-1, 4)$ having gradient $\frac{3}{5}$

c $(5, 0)$ having gradient $-\frac{1}{3}$ **d** $(6, -2)$ having gradient $-\frac{2}{7}$

e $(-3, -1)$ having gradient 4 **f** $(5, -3)$ having gradient -2

g $(4, -5)$ having gradient $-3\frac{1}{2}$ **h** $(-7, -2)$ having gradient 6.

Example 3

Self Tutor

Find the equation of the line which passes through $A(-1, 5)$ and $B(2, 3)$.

The gradient of the line is $\dfrac{3-5}{2-(-1)} = -\dfrac{2}{3}$.

Using point A, the equation is

$y - 5 = -\frac{2}{3}(x - (-1))$

$\therefore\ 3(y - 5) = -2(x + 1)$

$\therefore\ 3y - 15 = -2x - 2$

$\therefore\ 2x + 3y = 13$

5 Find, in gradient-intercept form, the equation of the line which passes through:

a $A(2, 3)$ and $B(4, 8)$ **b** $A(0, 3)$ and $B(-1, 5)$ **c** $A(-1, -2)$ and $B(4, -2)$

d $C(-3, 1)$ and $D(2, 0)$ **e** $P(5, -1)$ and $Q(-1, -2)$ **f** $R(-1, -3)$ and $S(-4, -1)$.

6 Find, in general form, the equation of the line which passes through:

a $(0, 1)$ and $(3, 2)$ **b** $(1, 4)$ and $(0, -1)$ **c** $(2, -1)$ and $(-1, -4)$

d $(0, -2)$ and $(5, 2)$ **e** $(3, 2)$ and $(-1, 0)$ **f** $(-1, -1)$ and $(2, -3)$.

7 Consider the points $A(2, 5)$ and $B(-4, 2)$. Find:

a the distance between A and B **b** the midpoint of AB

c the gradient of the line which passes through A and B

d the equation of the line which passes through A and B.

Example 4 — Self Tutor

Find the equation of the line with graph:

a

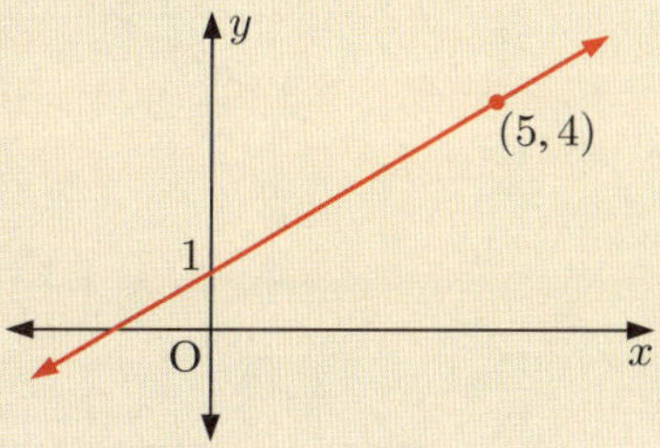

b

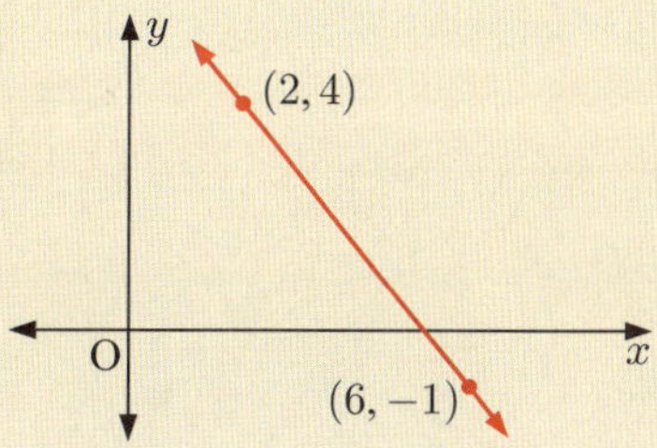

a $(0, 1)$ and $(5, 4)$ lie on the line.

$\therefore$ the gradient $m = \dfrac{4-1}{5-0} = \frac{3}{5}$

and the y-intercept $c = 1$.

The equation is $y = \frac{3}{5}x + 1$

{gradient-intercept form}

b $(2, 4)$ and $(6, -1)$ lie on the line.

$\therefore$ the gradient $m = \dfrac{-1-4}{6-2} = -\frac{5}{4}$

The equation is $y - 4 = -\frac{5}{4}(x - 2)$

$\therefore\ 4(y-4) = -5(x-2)$

$\therefore\ 4y - 16 = -5x + 10$

$\therefore\ 5x + 4y = 26$

8 Find the equation of each line:

a

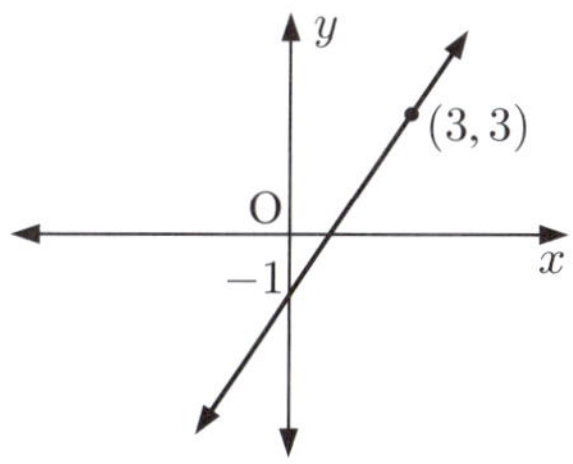

b

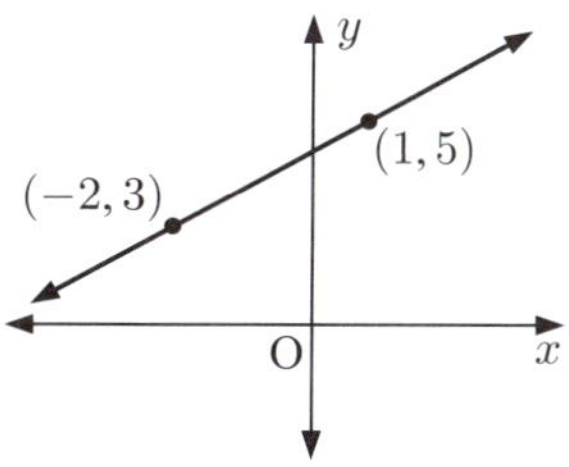

c

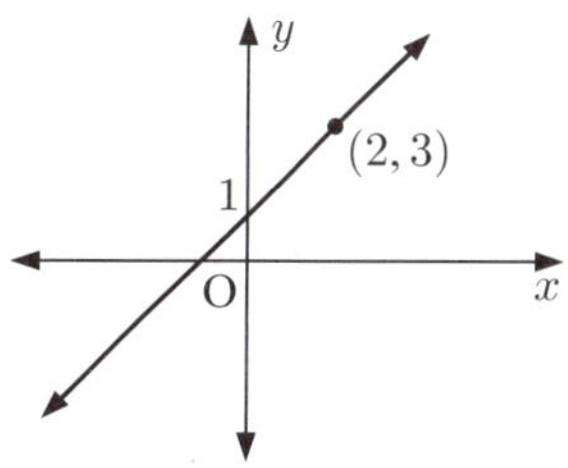

d

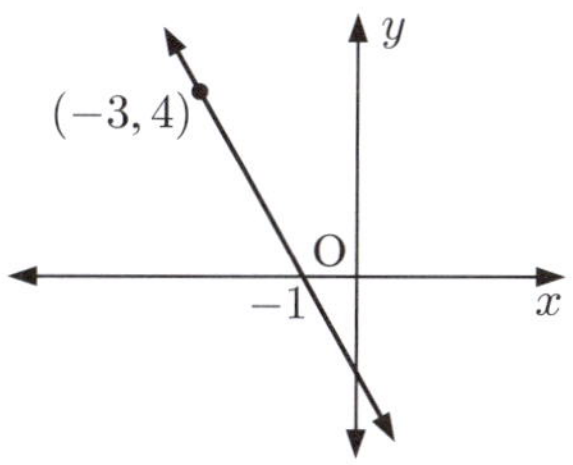

e

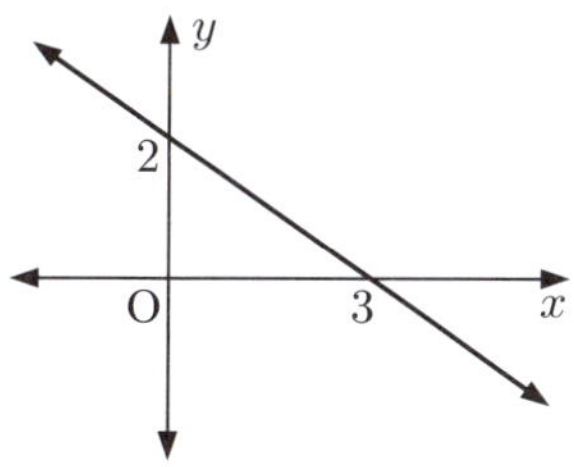

f

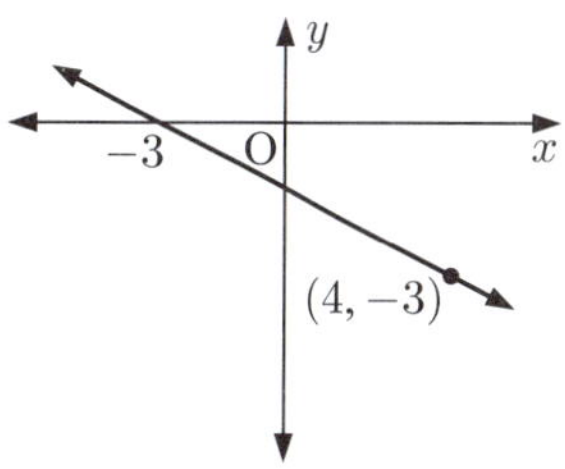

Example 5 — Self Tutor

Find the equation connecting the variables.

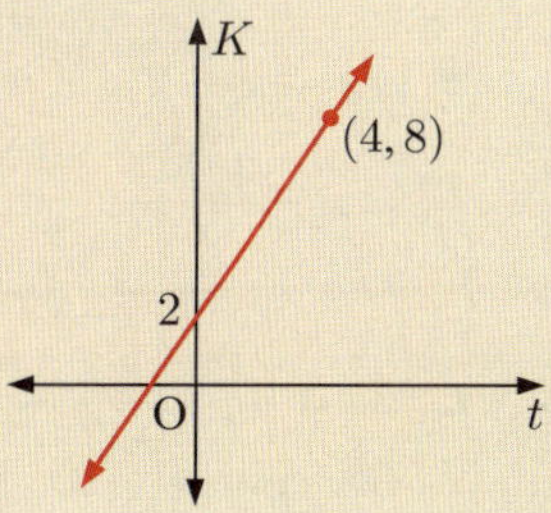

Two points on the line are $(0, 2)$ and $(4, 8)$.

$\therefore$ the gradient $m = \dfrac{8-2}{4-0} = \frac{6}{4} = \frac{3}{2}$, and the y-intercept $c = 2$.

In this case K is on the vertical axis and t is on the horizontal axis.

$\therefore$ the equation is $K = \frac{3}{2}t + 2$.

9 Find the equation connecting the variables:

a
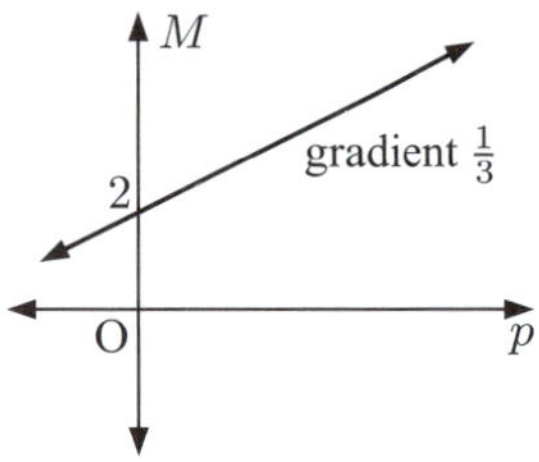

b
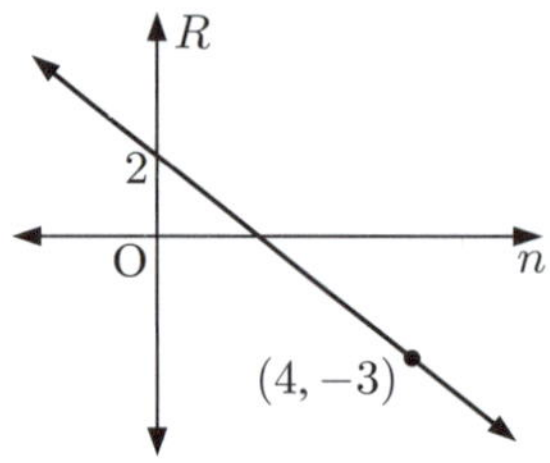

c
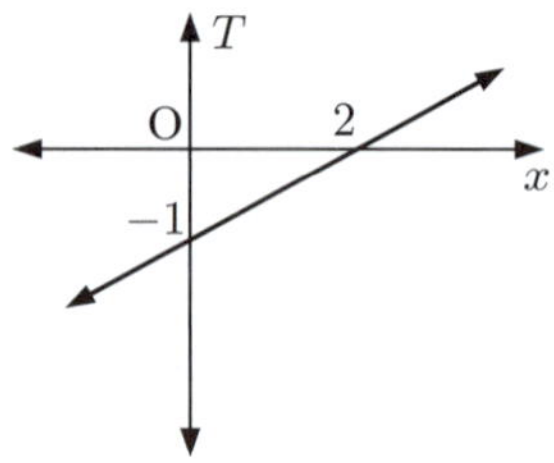

d
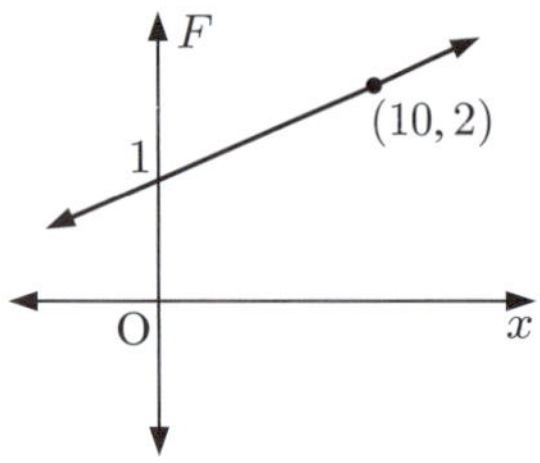

e
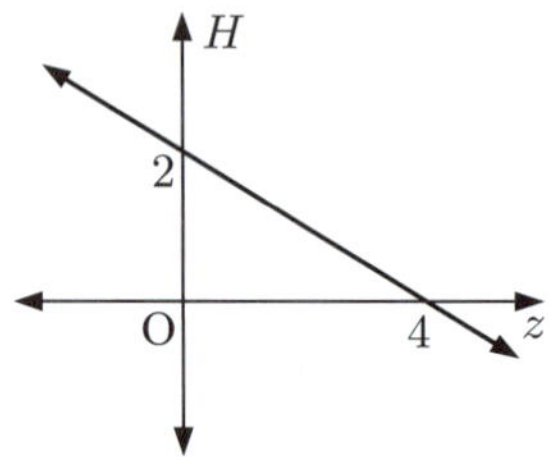

f
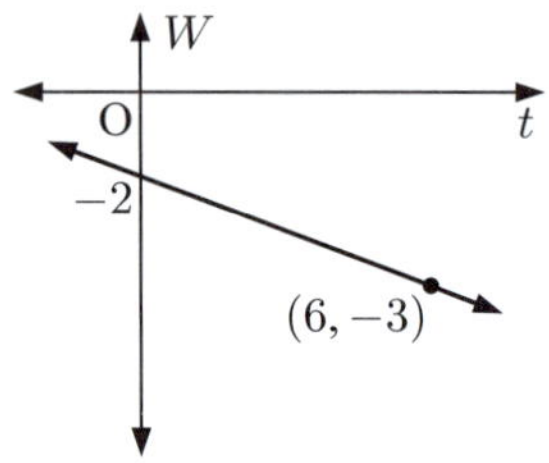

Example 6

Self Tutor

Draw the graph of:

a $y = -\frac{1}{2}x - 3$

b $3x + 5y = 30$

a The y-intercept $c = -3$

The gradient $m = \frac{-1}{2}$ ← y-step, ← x-step

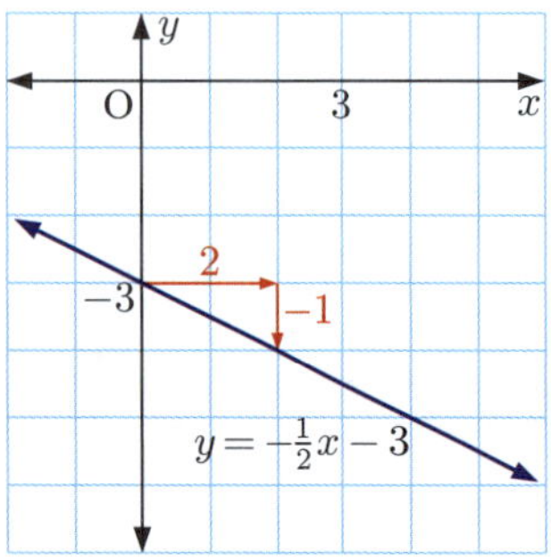

b When $x = 0$, $5y = 30$

$\therefore\ y = 6$

So, the y-intercept is 6.

When $y = 0$, $3x = 30$

$\therefore\ x = 10$

So, the x-intercept is 10.

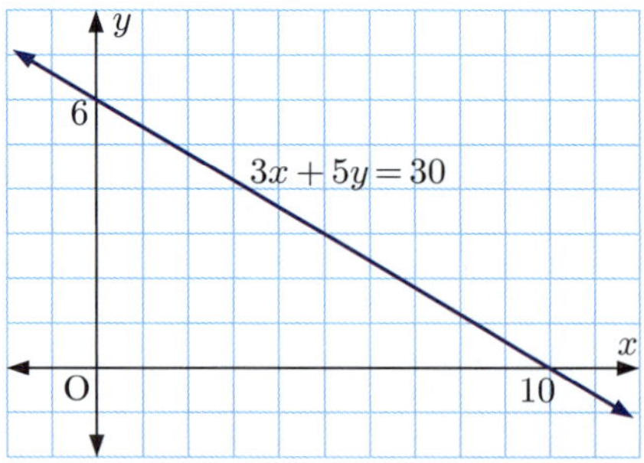

10 Draw the graph of:

a $y = x + 4$

b $y = 3x - 1$

c $y = \frac{2}{3}x + 2$

d $y = 5 - 2x$

e $y = -\frac{1}{4}x$

f $y = \dfrac{-3x - 5}{2}$

11 Draw the graph of:

a $2x + 5y = 10$

b $x - 3y = 9$

c $3x + 4y = -24$

d $5x - y = 40$

e $2x - 3y = -9$

f $5x - 4y = -12$

Example 7

Self Tutor

Consider the points A(−2, 5) and B(1, 3). A line perpendicular to AB, passes through B.

a Find the equation of the line.

b Find the coordinates of the point where the line cuts the x-axis.

a The gradient of AB $= \dfrac{3-5}{1-(-2)} = -\frac{2}{3}$

$\therefore$ the perpendicular line has gradient $\frac{3}{2}$, and passes through B(1, 3).

$\therefore$ its equation is $y - 3 = \frac{3}{2}(x-1)$

$\therefore\ 2(y-3) = 3(x-1)$

$\therefore\ 2y - 6 = 3x - 3$

$\therefore\ 3x - 2y = -3$

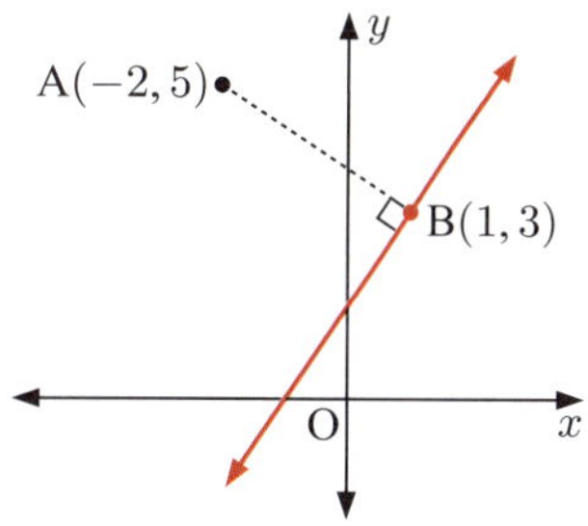

b The line cuts the x-axis when $y = 0$

$\therefore\ 3x - 2(0) = -3$

$\therefore\ x = -1$

$\therefore$ the line cuts the x-axis at (−1, 0).

12 Consider the points P(−3, −2) and Q(1, 6). A line perpendicular to PQ, passes through Q.

a Find the equation of the line.

b Find the coordinates of the point where the line cuts the x-axis.

13 Suppose A has coordinates (−7, 4) and B has coordinates (3, −2). A line parallel to AB, passes through C(5, −1).

a Find the equation of the line.

b Find the coordinates of the point where the line cuts the:

i x-axis **ii** y-axis.

14 Suppose P has coordinates (3, 8) and Q has coordinates (−5, 2). The line l is perpendicular to PQ and passes through P. l cuts the x-axis at R and the y-axis at S. Find the area of triangle ORS, where O is the origin.

15 Consider the points A(−2, −5), B(1, 7), and C(5, q). Given that BC is perpendicular to AB, find:

a q **b** the area of triangle ABC.

16 Find the shaded area.

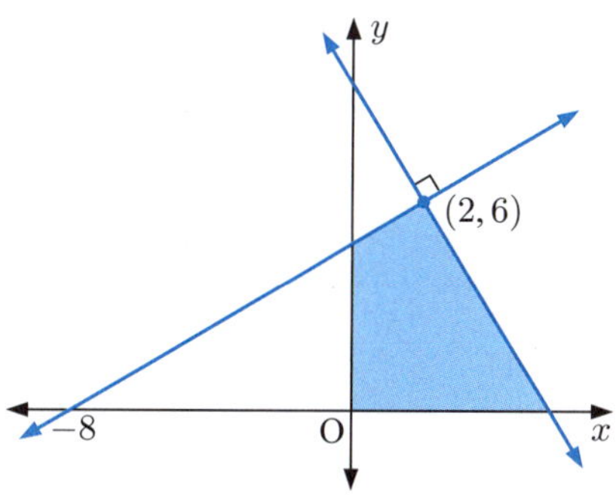

B PERPENDICULAR BISECTORS

The **perpendicular bisector** of AB is the line which is perpendicular to AB, and which passes through its midpoint M.

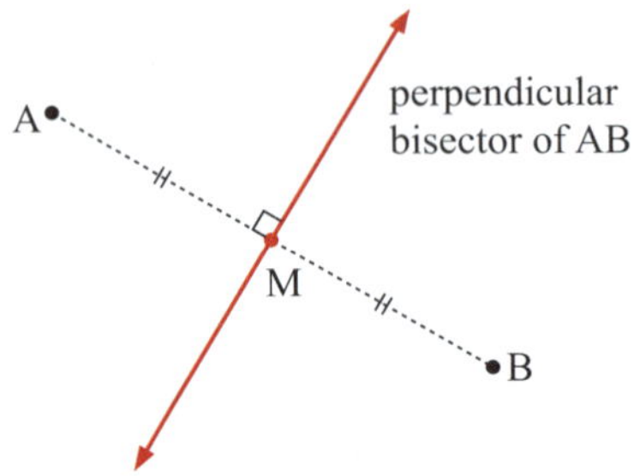

Example 8

Self Tutor

Find the equation of the perpendicular bisector of AB given A(−1, 2) and B(3, 4).

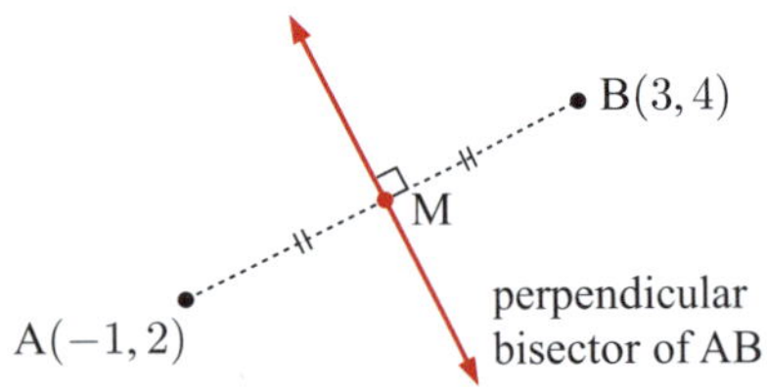

The midpoint M of AB is $\left(\frac{-1+3}{2}, \frac{2+4}{2}\right)$ or M(1, 3).

The gradient of AB is $\frac{4-2}{3-(-1)} = \frac{2}{4} = \frac{1}{2}$

$\therefore$ the gradient of the perpendicular bisector is $-\frac{2}{1}$ {the negative reciprocal of $\frac{1}{2}$}

Using M(1, 3), the equation of the perpendicular bisector is $y - 3 = -2(x-1)$

$\therefore \; y - 3 = -2x + 2$

$\therefore \; y = -2x + 5$

EXERCISE 1B

1 Consider the points P(−3, 7) and Q(1, −5). Find:

- **a** the distance between P and Q
- **b** the midpoint of PQ
- **c** the gradient of PQ
- **d** the equation of the perpendicular bisector of PQ.

2 Find the equation of the perpendicular bisector of AB given:

- **a** A(3, −3) and B(1, −1)
- **b** A(1, 3) and B(−3, 5)
- **c** A(3, 1) and B(−3, 6)
- **d** A(−1, 4) and B(2, 0)
- **e** A(2, −3) and B(−3, −3)
- **f** A(4, −2) and B(4, 4).

3 Consider the points A(6, −1) and B(−4, 5). The perpendicular bisector of AB cuts the y-axis at C.

- **a** Find the coordinates of C.
- **b** Show that C is equidistant from A and B.

4 Consider the points P(−1, 5) and Q(3, 7). The perpendicular bisector of PQ cuts the x-axis at R. Find the area of triangle PQR.

C THE INTERSECTION OF STRAIGHT LINES

To find where straight lines meet, we need to solve the equations of the lines *simultaneously*.

A set of two linear simultaneous equations can have either one, zero, or infinitely many solutions.

intersecting lines

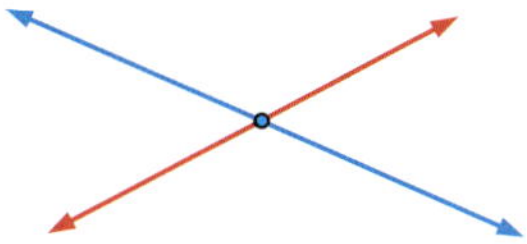

one solution

parallel lines

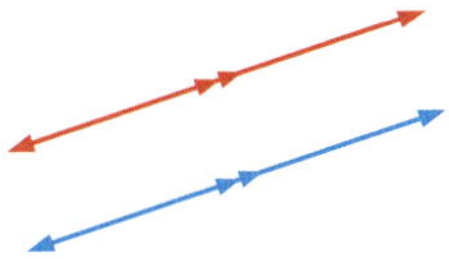

no solutions

coincident lines

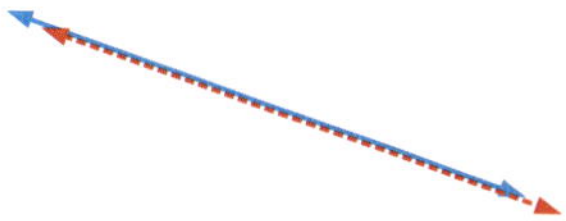

infinitely many solutions

SOLUTION BY EQUATING VALUES OF y

If both linear equations are given in gradient-intercept form $y = mx + c$, we find the solution by equating the y values.

Example 9 — Self Tutor

Find where the line $y = 2x + 3$ meets the line $y = -x + 6$.

The lines meet where $2x + 3 = -x + 6$ {equating y values}

$\therefore\ 3x = 3$

$\therefore\ x = 1$ and $y = 2(1) + 3 = 5$

The lines meet at $(1, 5)$.

SOLUTION BY SUBSTITUTION

If only one equation is in gradient-intercept form, we use it to substitute an expression for y into the other equation.

Example 10 — Self Tutor

Find where the line $y = 2x - 5$ meets the line $4x + 3y = 15$.

Substituting $y = 2x - 5$ into $4x + 3y = 15$ gives

$4x + 3(2x - 5) = 15$

$\therefore\ 4x + 6x - 15 = 15$

$\therefore\ 10x = 30$

$\therefore\ x = 3$ and $y = 2(3) - 5 = 1$

The lines meet at $(3, 1)$.

SOLUTION BY ELIMINATION

If both equations are in general form, we can use **elimination** to find the solution.

By multiplying one or both equations by (non-zero) constants as needed, we make the coefficients of x (or y) the **same size** but **opposite in sign**.

We then **add** the equations, which has the effect of eliminating one of the variables.

Example 11 — Self Tutor

Find the intersection point for the following pairs of lines:

a $2x - 3y = 5$
$5x + 3y = 9$

b $4x + 7y = 2$
$3x + 5y = -1$

a
$$\begin{aligned} 2x - 3y &= 5 \quad \text{.... (1)} \\ 5x + 3y &= 9 \quad \text{.... (2)} \end{aligned}$$

Adding, $7x = 14$

$\therefore \; x = 2$

Substituting $x = 2$ into (1) gives $2(2) - 3y = 5$

$\therefore \; -3y = 1$

$\therefore \; y = -\frac{1}{3}$

The lines meet at $\left(2, -\frac{1}{3}\right)$.

b $4x + 7y = 2$ .... (1)
$3x + 5y = -1$ .... (2)

To make the coefficients of x the same size but opposite in sign, we multiply (1) by 3 and (2) by -4.

$$\begin{aligned} 12x + 21y &= 6 \quad \{(1) \times 3\} \\ -12x - 20y &= 4 \quad \{(2) \times -4\} \end{aligned}$$

Adding, $y = 10$

Substituting $y = 10$ into (1) gives $4x + 7(10) = 2$

$\therefore \; 4x = -68$

$\therefore \; x = -17$

The lines meet at $(-17, 10)$.

EXERCISE 1C

1 Find the intersection point of each pair of lines:

a $y = 3x + 4$
$y = x - 6$

b $y = -2x + 4$
$y = 3x - 1$

c $y = 4x - 3$
$y = -2x + 5$

d $y = \frac{1}{2}x - 1$
$y = x + 3$

e $y = -4x + 1$
$y = \frac{2}{3}x - 3$

f $y = \dfrac{x + 7}{4}$
$y = -\frac{1}{2}x - 5$

2 Use the method of substitution to find the intersection point of each pair of lines:

a $y = 4x - 1$
$2x + y = 5$

b $y = 2x - 1$
$3x + 2y = 12$

c $4x + 3y = 15$
$y = 9 - 2x$

d $2x - 3y = -5$
$y = 4x + 1$

e $4x - 7y = -5$
$y = \frac{1}{2}x - 3$

f $y = -\frac{2}{3}x - 5$
$3x + 2y = 2$

3 Try to find the intersection point of the following pairs of lines. Explain your answers graphically.

a $y = -\frac{2}{3}x + 1$
$2x + 3y = 4$

b $y = \frac{1}{2}x - 4$
$x - 2y = 8$

4 Use the method of elimination to find the intersection point of each pair of lines:

a $4x + 5y = 7$
$2x - 5y = 11$

b $3x + 7y = -1$
$-3x + 4y = 23$

c $6x - 4y = 5$
$5x + 4y = -2$

d $3x + 2y = 6$
$5x - y = 23$

e $x + 6y = -10$
$3x - 5y = -7$

f $7x + 3y = 12$
$2x - 9y = 10$

g $2x - 3y = 8$
$3x + 4y = -5$

h $5x + 2y = -1$
$7x + 5y = 14$

i $4x - 9y = 2$
$7x - 12y = 5$

5 Try to find the intersection point of the following pairs of lines. Explain your answers graphically.

a $3x - 7y = -4$
$-9x + 21y = 12$

b $2x + 5y = 1$
$4x + 10y = -3$

6 Consider the **Opening Problem** on page **12**.

a Find the equations of lines l_1 and l_2.

b Hence find their intersection point.

7 Line l_1 has equation $y = 2x + 7$. Line l_2 passes through $(-7, 6)$ and $(3, 0)$.

a Find the equation of l_2.

b Find the point where l_1 and l_2 intersect.

8 Line l_1 has equation $y = -3x + 5$. Line l_2 has equation $x + 2y = 5$.

a Find the point where l_1 meets l_2.

b Line l_3 is parallel to l_1 and passes through $(-1, -2)$. Find the point where l_2 meets l_3.

9 Find the coordinates of X.

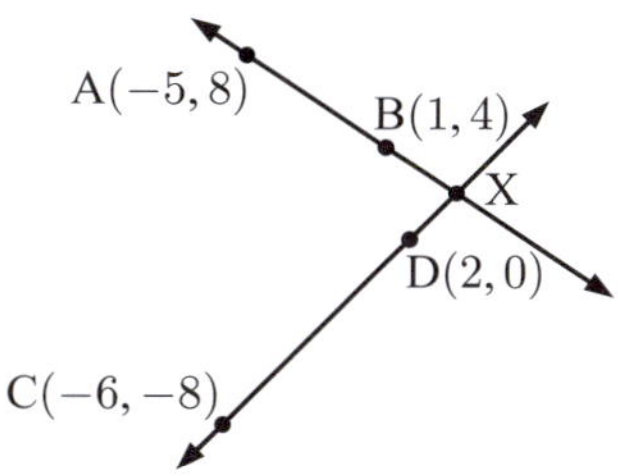

10 In the diagram alongside, a line has been drawn through C, perpendicular to AB, and meeting AB at X. Find:

a the equation of AB

b the equation of CX

c the coordinates of X.

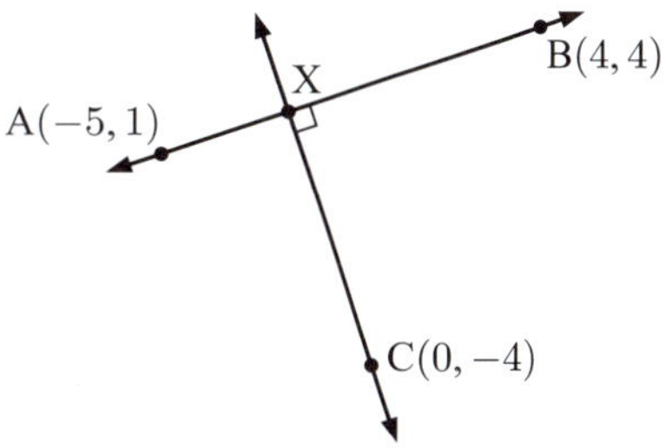

Example 12

Self Tutor

ABCD is a trapezium in which AB is parallel to DC, and $\widehat{ADC} = 90°$. Find:

a the coordinates of D

b the area of the trapezium.

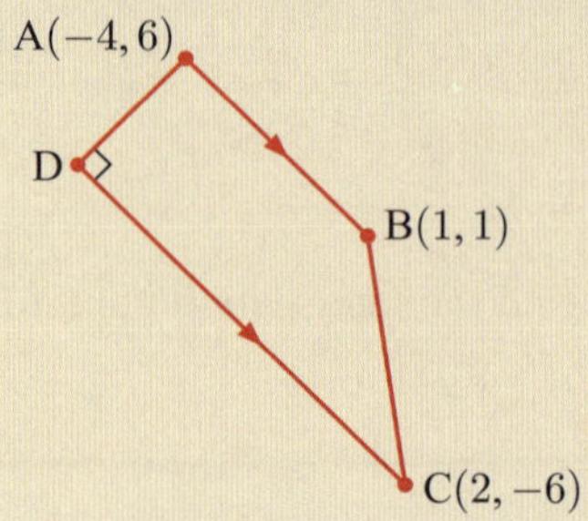

a The gradient of AB is $\dfrac{1-6}{1-(-4)} = \dfrac{-5}{5} = -1$

$\therefore$ DC also has gradient -1

$\therefore$ DC has equation $y - (-6) = -1(x - 2)$

$\therefore\ y + 6 = -x + 2$

$\therefore\ y = -x - 4$ .... (1)

AD is perpendicular to DC, so its gradient is 1

$\therefore$ AD has equation $y - 6 = 1(x - (-4))$

$\therefore\ y - 6 = x + 4$

$\therefore\ y = x + 10$ .... (2)

Equating values of y in (1) and (2), $-x - 4 = x + 10$

$\therefore\ -2x = 14$

$\therefore\ x = -7$ and $y = -(-7) - 4 = 3$

$\therefore$ D is $(-7, 3)$.

b The length of AB $= \sqrt{(1 - -4)^2 + (1-6)^2} = \sqrt{50} = 5\sqrt{2}$ units

The length of DC $= \sqrt{(2 - -7)^2 + (-6-3)^2} = \sqrt{162} = 9\sqrt{2}$ units

The length of AD $= \sqrt{(-7 - -4)^2 + (3-6)^2} = \sqrt{18} = 3\sqrt{2}$ units

$\therefore$ the area of the trapezium $= \left(\dfrac{5\sqrt{2}+9\sqrt{2}}{2}\right) \times 3\sqrt{2}$

$= 7\sqrt{2} \times 3\sqrt{2}$

$= 42$ units^2

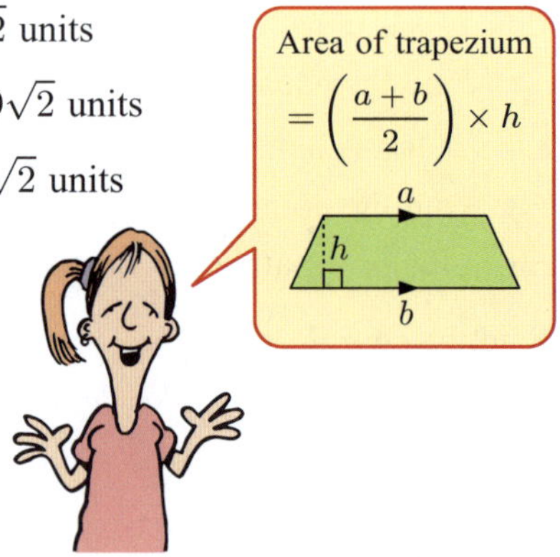

11 PQRS is a trapezium in which PQ is parallel to SR, and $\widehat{PSR} = 90°$. Find:

a the coordinates of S

b the area of the trapezium.

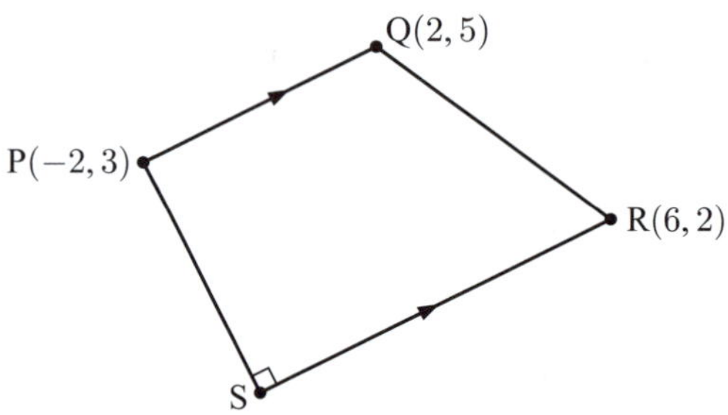

12

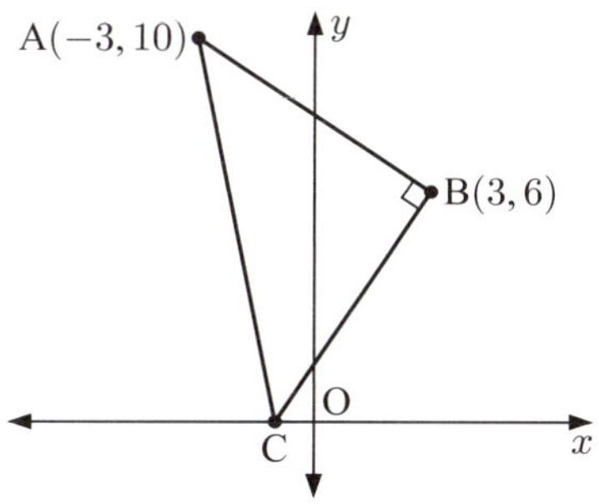

ABC is a triangle in which $A\widehat{B}C = 90°$, and C lies on the x-axis. Find:

a the coordinates of C

b the area of the triangle.

13 Consider the points J(−3, 2), K(1, −4), and L(5, 3).

a Find the equation of the line JK.

b The line through L perpendicular to JK, intersects JK at point P.

i Find the coordinates of P.

ii Show that P is the midpoint of JK.

iii Hence classify triangle JKL.

iv Find the area of triangle JKL.

14 A trapezium ABCD has vertices A(3, 0), B(−2, −5), C(−4, 1), and D. The side AD is parallel to BC, and the side CD is perpendicular to BC. Find the area of the trapezium.

15 ABC is a triangle which is right angled at B. B lies on the x-axis and AC is parallel to the x-axis. M is the midpoint of AB, and N is the midpoint of BC.

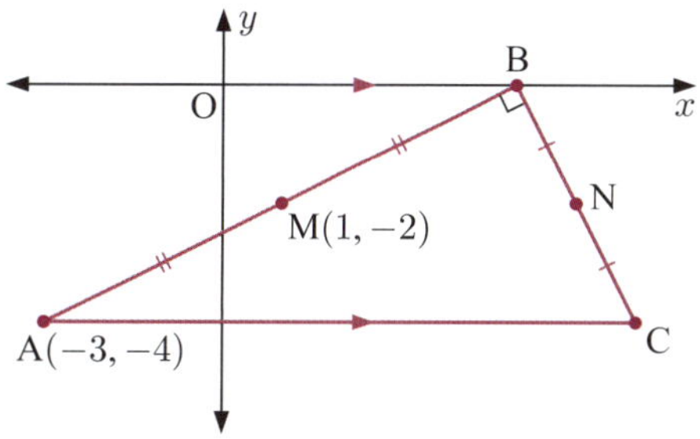

a Find the coordinates of:

i B **ii** C **iii** N.

b Show that MN is parallel to AC.

c Find the area of:

i trapezium AMNC **ii** triangle ABC.

Review set 1A

1 Consider the points A(−1, 6) and B(5, 4). Find:

a the distance between A and B

b the midpoint of AB

c the equation of the line through A and B.

2 Determine the equation of the illustrated line:

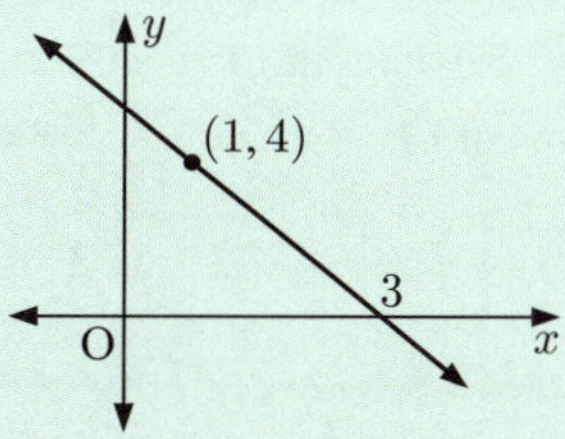

3 Explain why a vertical straight line in the plane cannot be written in gradient-intercept form $y = mx + c$.

4 Draw the graph of:

a $y = 4x - 3$ **b** $5x - 2y = 10$

5 Find the equation of the perpendicular bisector of AB given:

a A(2, 8) and B(6, −2) **b** A(−5, 1) and B(0, 4).

6 Suppose P has coordinates (−2, −3) and Q has coordinates (1, 3). A line perpendicular to PQ, passes through Q.

a Find the equation of the line.

b Find the coordinates of the point where the line cuts the x-axis.

7 Find the intersection point of each pair of lines:

a $y = 4x + 5$
$4x - 3y = 9$

b $3x + 4y = -5$
$5x - 4y = 9$

8 Line l_1 has equation $x - 2y = 5$. Line l_2 passes through (0, 3) and (3, −1).

a Find the equation of l_2. **b** Find the point where l_1 and l_2 intersect.

9 PQR is a triangle in which $\widehat{PQR} = 90°$, and R lies on the x-axis. Find:

a the coordinates of R

b the area of the triangle.

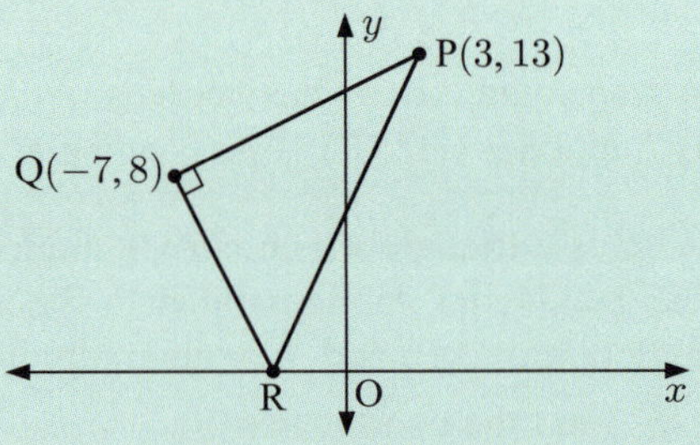

10 ABCD is a trapezium in which AB is parallel to DC, and $\widehat{ADC} = 90°$. Find:

a the coordinates of D

b the area of the trapezium.

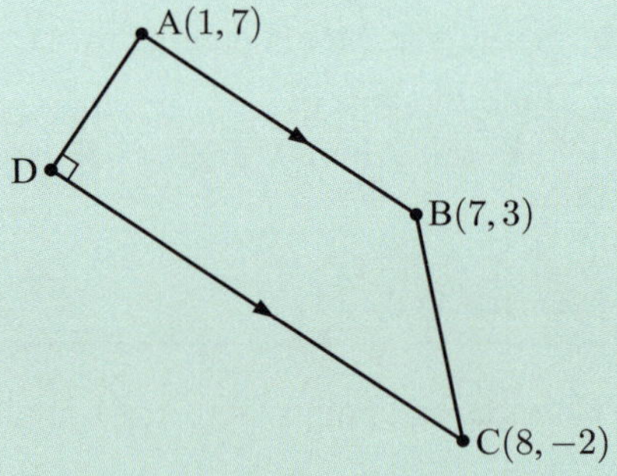

11 The points P(m, 4), Q(2, 7), R(9, n), and S(4, 3) are the vertices of a kite PQRS. The diagonals PR and QS intersect at the point T. PR is the perpendicular bisector of QS. Find:

a the coordinates of T

b the equation of the diagonal PR

c the values of m and n

d the area of the kite.

Review set 1B

1 Find, in general form, the equation of the line passing through (−5, −7) and (3, −2).

2 Given A(−2, 3) and B(4, 5), find the equation of the perpendicular bisector of AB.

3 Draw the graph of:

a $y = -\frac{5}{2}x + 3$ **b** $4x + 3y = 6$

4 Consider the points A(−3, 2), B(6, 5), and C(8, k), where BC is perpendicular to AB. Find:

a the length of AB **b** the equation of AB

c the value of k **d** the area of triangle ABC.

5 Find the equation linking the variables in each graph:

a

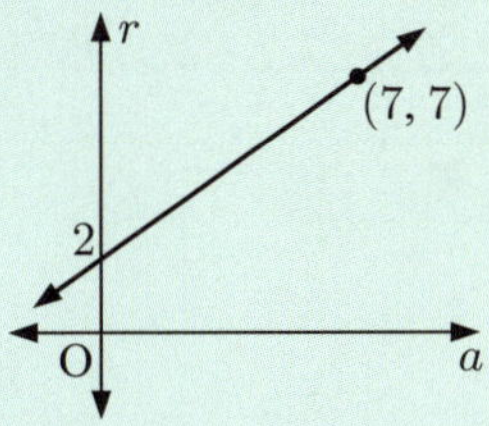

b

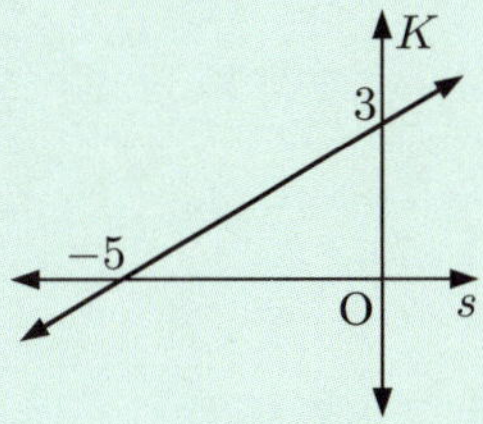

6 Find the equation of line l.

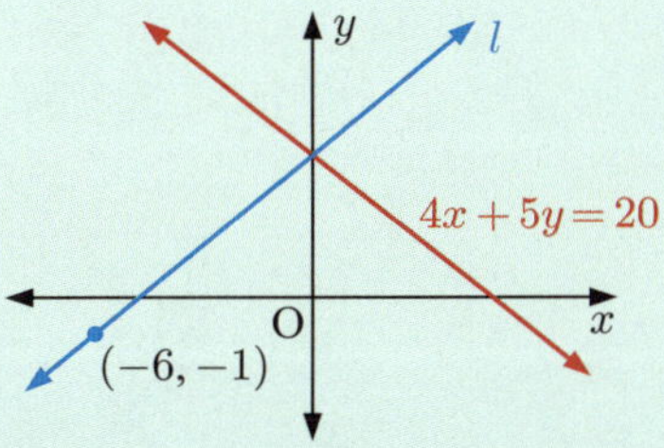

7 Find the intersection point of each pair of lines:

a $y = 3x - 7$
$y = -x + 9$

b $y = 1 - 6x$
$5x + 2y = -5$

8 Find the coordinates of X.

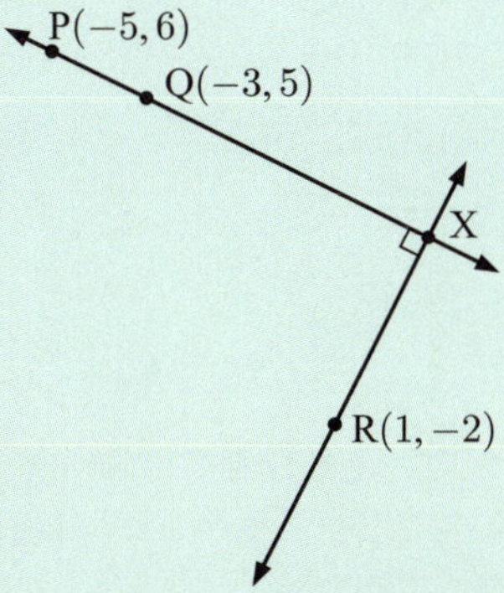

9

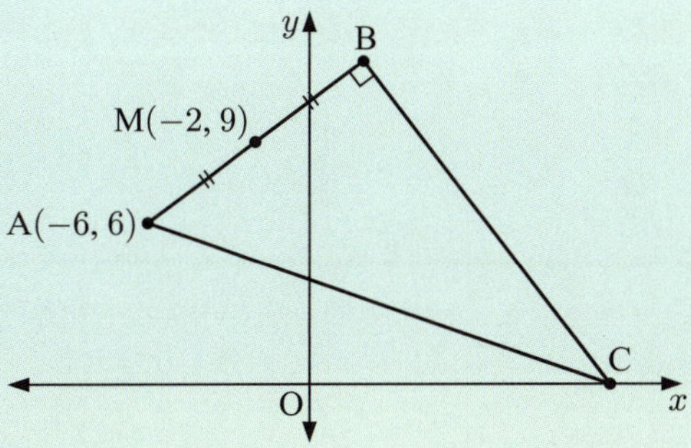

ABC is a triangle with a right angle at B. M is the midpoint of AB, and C lies on the x-axis.

a Find the coordinates of:

i B **ii** C.

b Find the area of the triangle.

10 Find the intersection point of each pair of lines:

a $2x + 3y = 11$
$4x - 5y = -33$

b $7x - 11y = 2$
$3x - 5y = -5$

11 **a** Show that ABCD is a parallelogram.

b Find:

i the equation of BX

ii the coordinates of X

iii the area of the parallelogram.

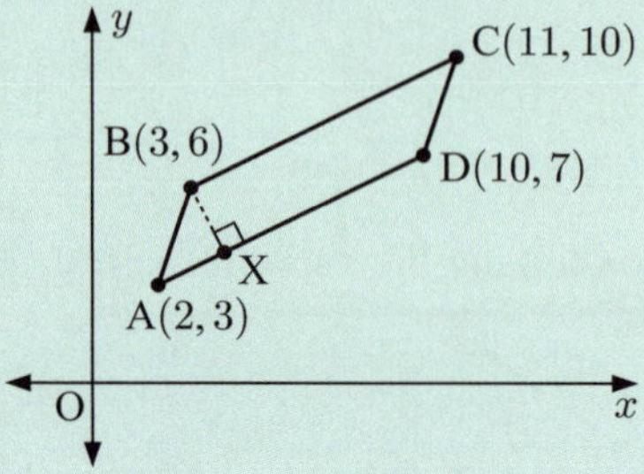

2

Functions

Contents:

Opening problem

Josephine is studying the rings of Saturn. She discovers that the rings are circular, and exist around the equator of Saturn. Their radii from the centre of the planet range from 67 000 km up to about 480 000 km. Josephine draws the rings on a set of axes with units in ten thousands of kilometres.

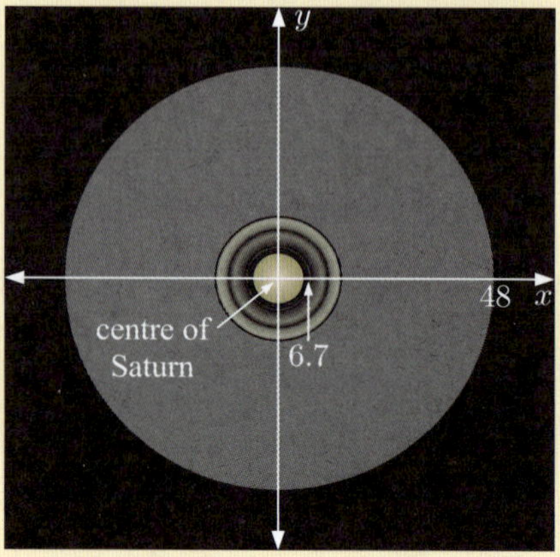

Things to think about:

- **a** How can Josephine describe the possible values that x and y can take?
- **b** The inner ring of Saturn has equation $x^2 + y^2 = 6.7^2$.
  The outer ring of Saturn has equation $x^2 + y^2 = 48^2$.
  How can Josephine describe the *region* which includes the rings of Saturn?

A RELATIONS AND FUNCTIONS

RELATIONS

A **relation** is any set of points which connects two variables.

In the **Opening Problem**, the rings of Saturn are represented as a relation between the variables x and y. In this case, the relation describes a *region*. Other relations may represent a line, a curve, or even a finite number of ordered pairs. Any relation between two variables x and y can be plotted on the Cartesian plane. For example:

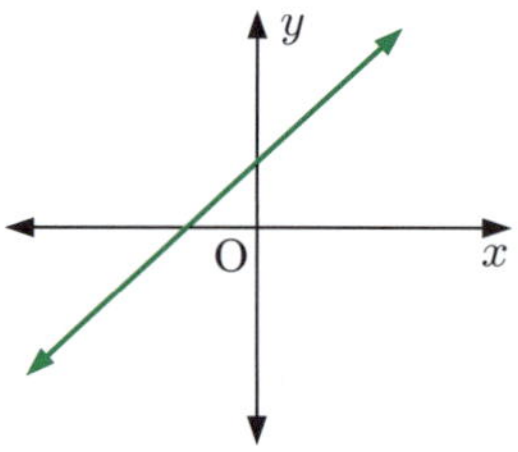

a line

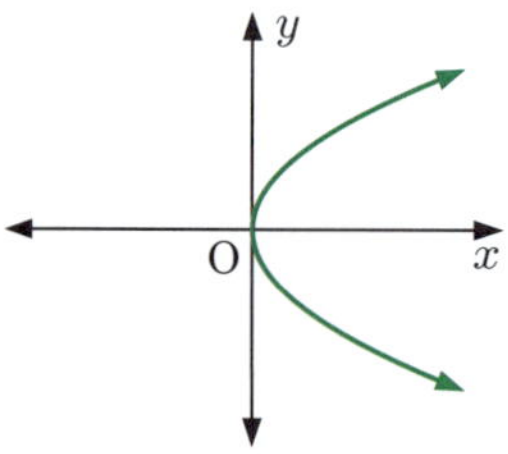

a curve

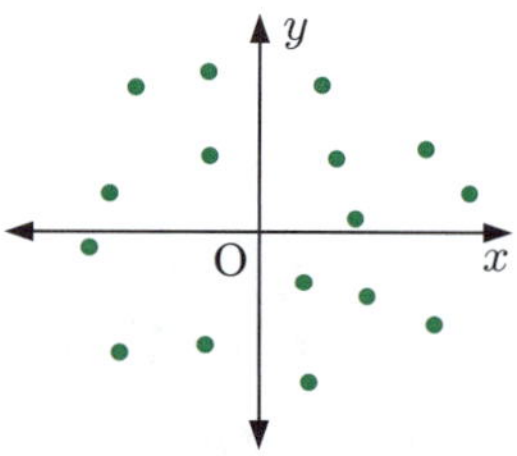

a finite set of ordered pairs

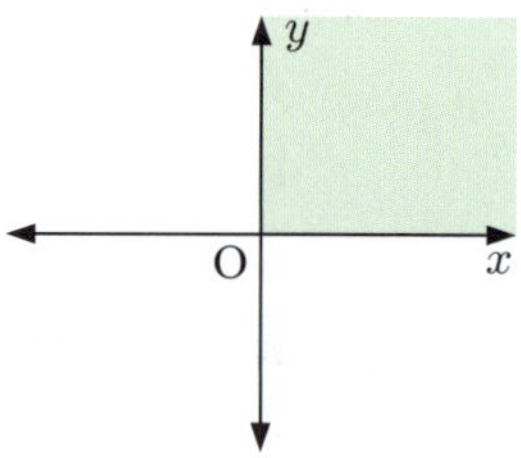

a region

In the case of lines and curves, the set of points can often be generated using an **equation**.

For a finite set of ordered pairs, we write out the set. For example: $\{(0, 1), (1, 2), (2, 2), (1, 3), (3, 3)\}$.

In the case of a region, the set of points is generated using **inequalities**.

Mapping diagrams

Another way to think of a relation is as a **mapping** from one variable to the other.

For example, consider the relation which is the set of points $\{(0, 1), (1, 2), (2, 2), (1, 3), (3, 3)\}$.

The set of possible x-values is $\{0, 1, 2, 3\}$. We call this the **domain** of the relation.

The set of possible y-values is $\{1, 2, 3\}$. We call this the **range** of the relation.

We can illustrate this relation on a mapping diagram:

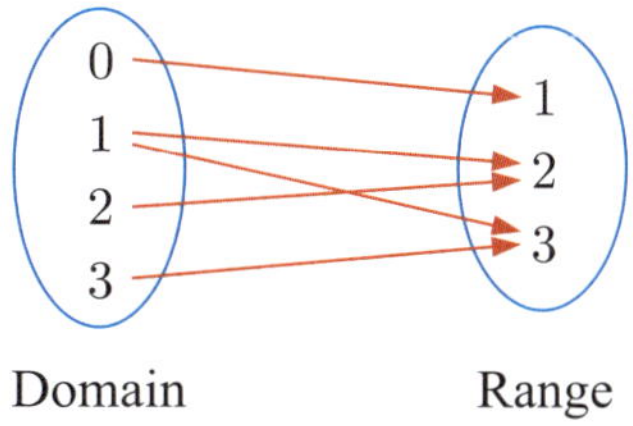

In this case, the value 1 in the domain leads to two possible values in the range, and the value 3 in the range corresponds to two possible values in the domain.

FUNCTIONS

> A **function** is a relation in which each value in the domain is mapped to exactly one value in the range.

If we remove the point $(1, 3)$ from the relation above, the mapping diagram becomes:

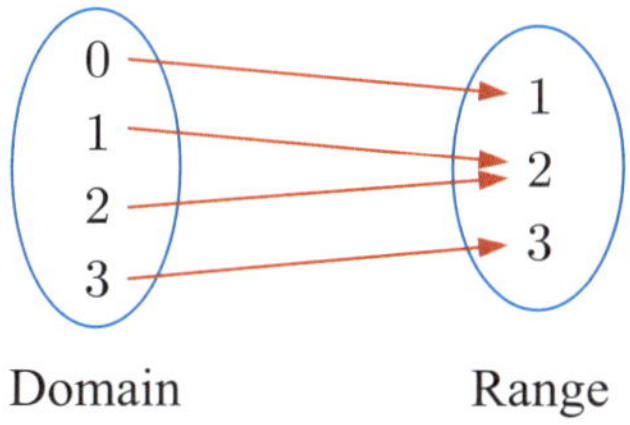

Each value in the domain is now mapped to exactly one value in the range, so this is an example of a function.

TESTING FOR FUNCTIONS

Algebraic test

> If a relation is given as an equation, and the substitution of any value of x results in exactly one value of y, then the relation is a function.

For example:

- $y = 2x + 3$ is a function, since for any value of x there is only one corresponding value of y
- $x^2 + y^2 = 1$ is not a function, since if $x = 0$ then $y = \pm 1$.

Graphical or vertical line test

> Suppose we draw all possible vertical lines on the graph of a relation.
> - If each line cuts the graph at most once, then the relation is a function.
> - If at least one line cuts the graph more than once, then the relation is *not* a function.

When we look at graphs:

- an **open circle** such as  indicates a point that *is not* included
- a **filled-in circle** such as  indicates a point that *is* included
- an **arrow head** such as  indicates that the graph continues indefinitely in that general direction, or the shape may repeat as it has done previously.

Example 1

Self Tutor

Which of the following relations are functions?

a

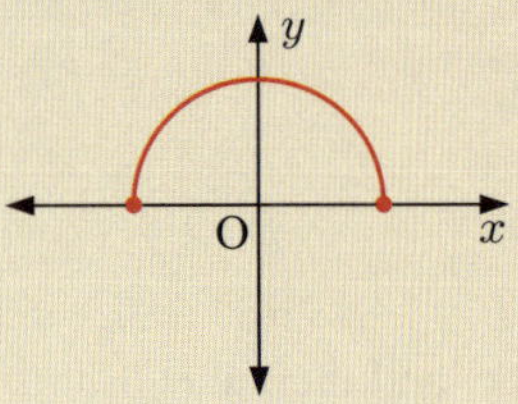

b

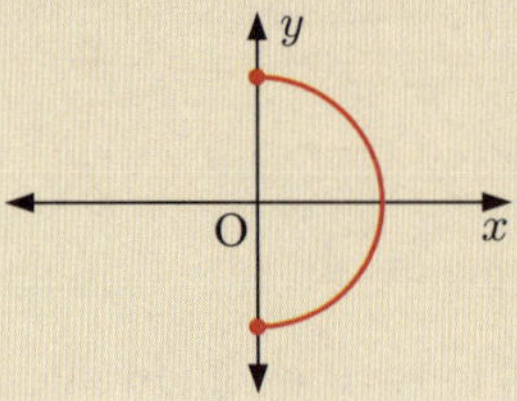

c

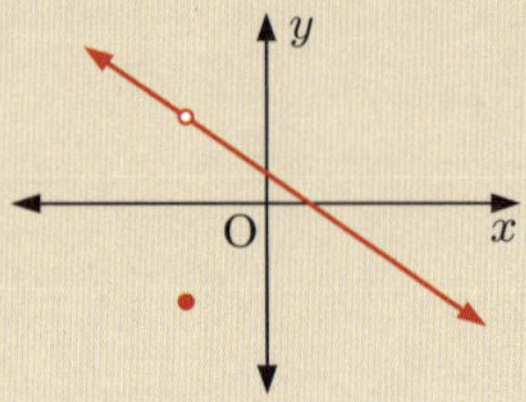

a

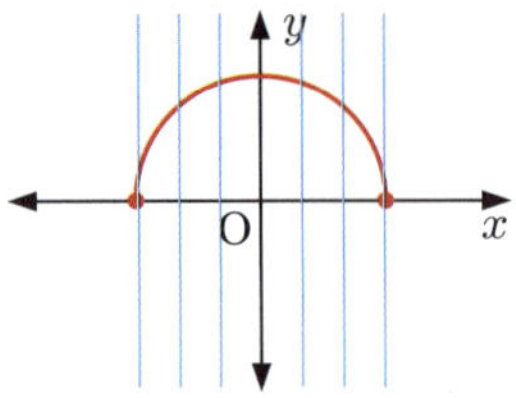

is a function

b

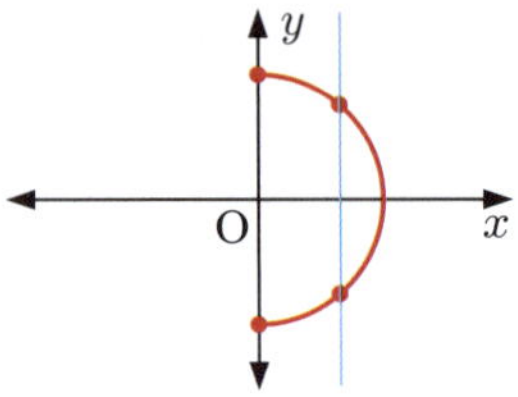

not a function

c

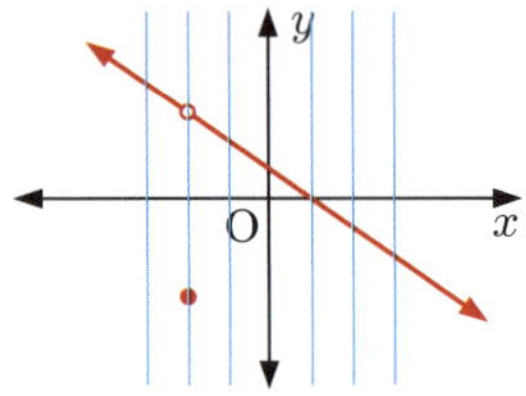

is a function

EXERCISE 2A.1

1 Draw a mapping diagram for each relation, and discuss whether the relation is a function.

- **a** $\{(1, 3), (2, 4), (3, 5), (4, 6)\}$
- **b** $\{(6, 9), (4, 3), (5, 4)\}$
- **c** $\{(1, 0), (9, 6), (8, 6), (3, 9)\}$
- **d** $\{(1, 1), (2, 2), (3, 1), (4, 2)\}$
- **e** $\{(5, 3), (2, 4), (6, 5)\}$
- **f** $\{(1, 2), (3, 3), (3, 1), (5, 1), (6, 3)\}$

2 Use the vertical line test to determine which of the following relations are functions:

a

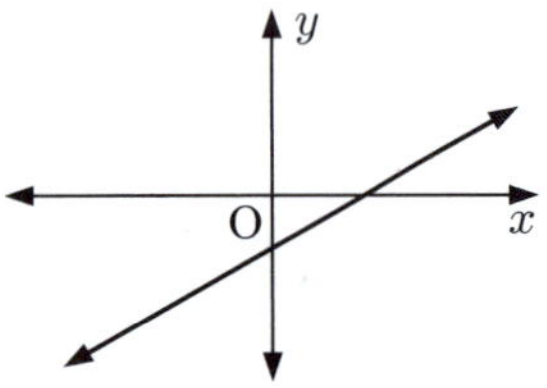

b

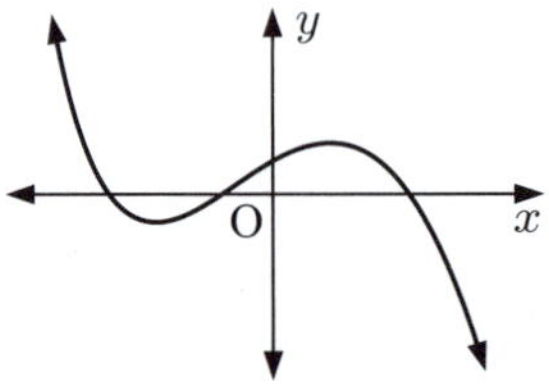

c

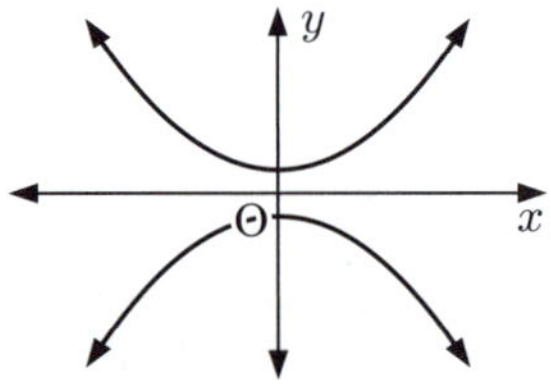

d

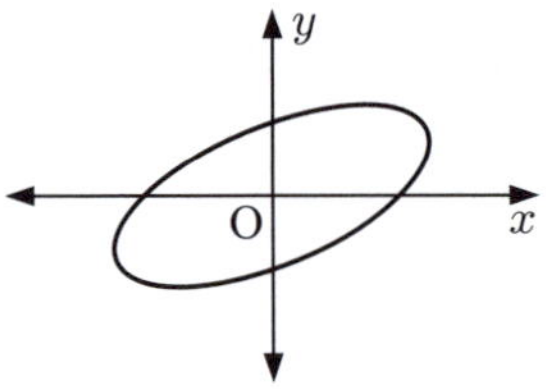

e

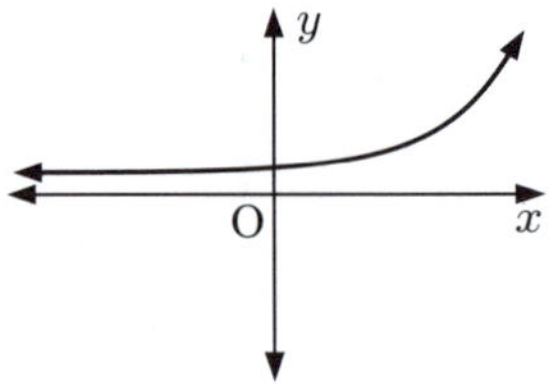

f

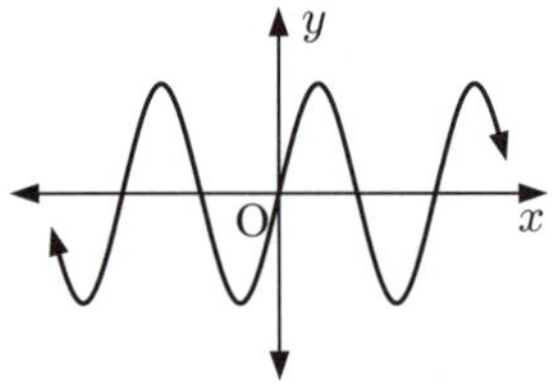

g

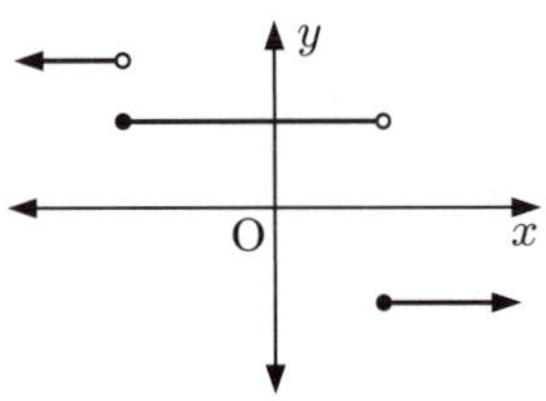

h

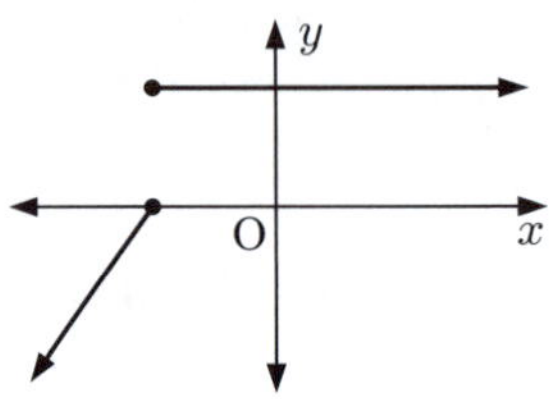

i

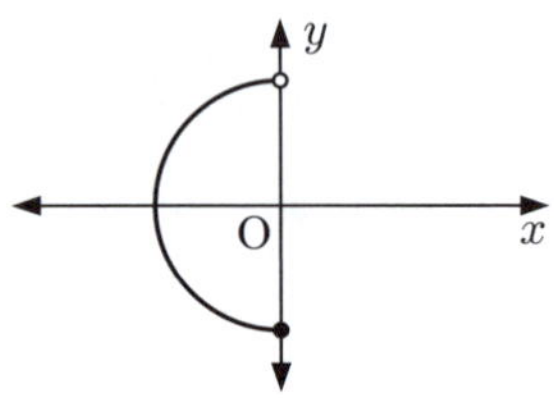

3 Give algebraic evidence to show that the relation $y^2x = 5$ is not a function.

ONE-ONE FUNCTIONS

A **one-one function** is a function in which every value in the range corresponds to exactly one value in the domain.

For example:

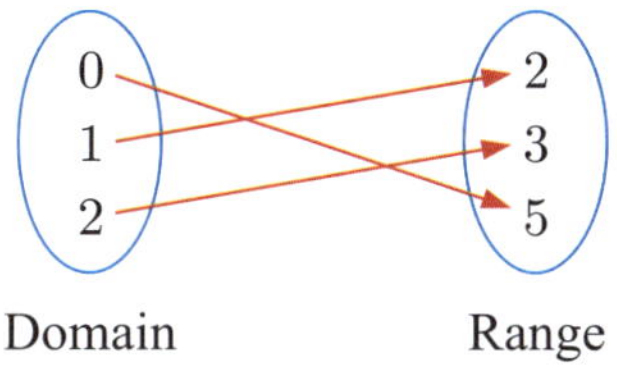

We can use the **horizontal line test** to determine whether a function is one-one:

Suppose we draw all possible horizontal lines on the graph of a function.

- If each line cuts the graph at most once, then the function is one-one.
- If at least one line cuts the graph more than once, then the function is *not* one-one.

Example 2

Self Tutor

Which of the following functions are one-one?

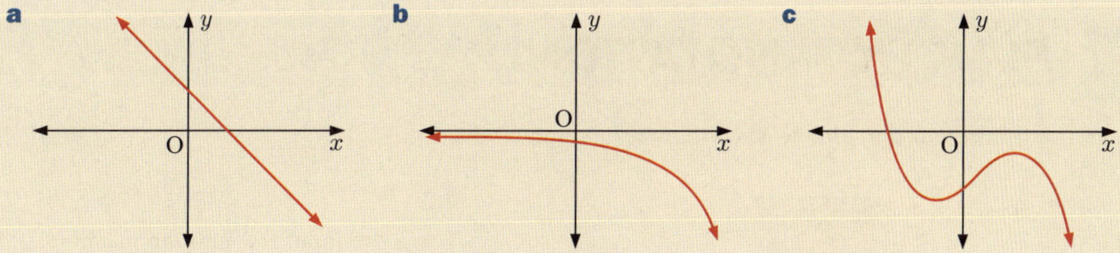

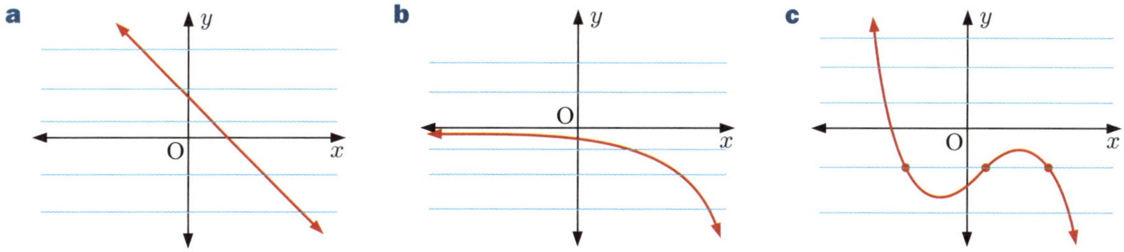

a one-one **b** one-one **c** not one-one

EXERCISE 2A.2

1 Which of the following functions are one-one?

a

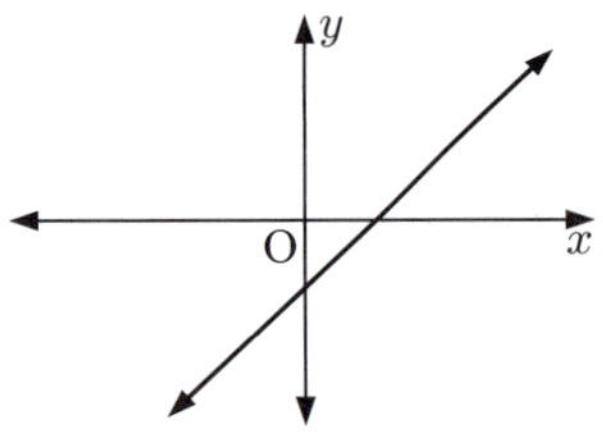

b

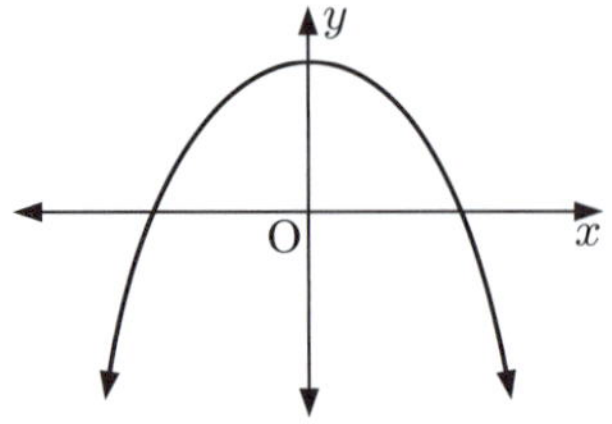

c

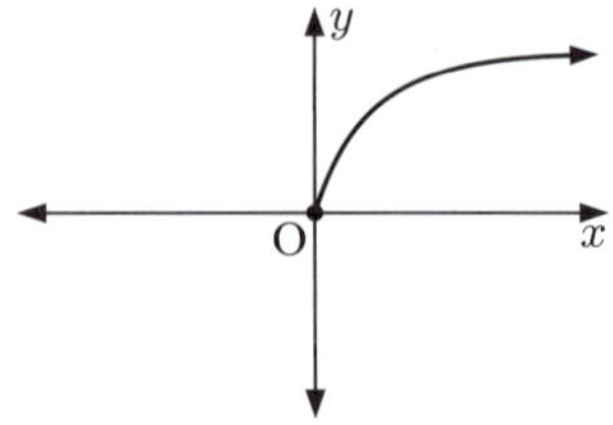

d

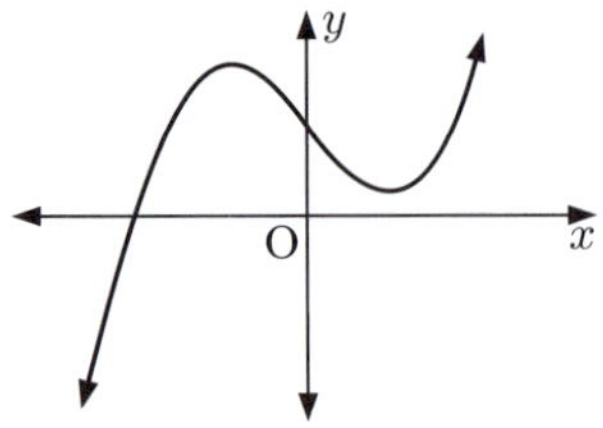

e

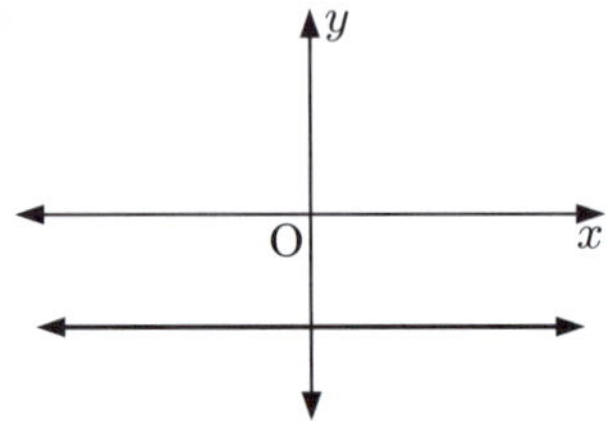

f

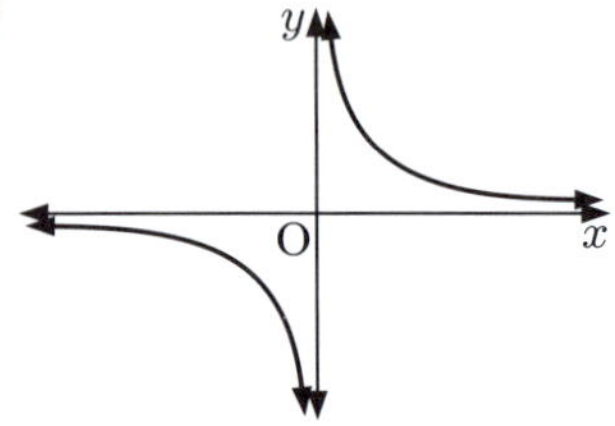

2 Determine whether the following relations are functions. If they are functions, determine whether they are one-one.

a

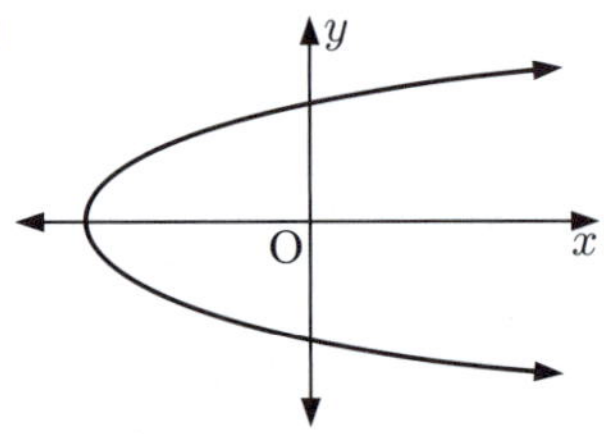

b

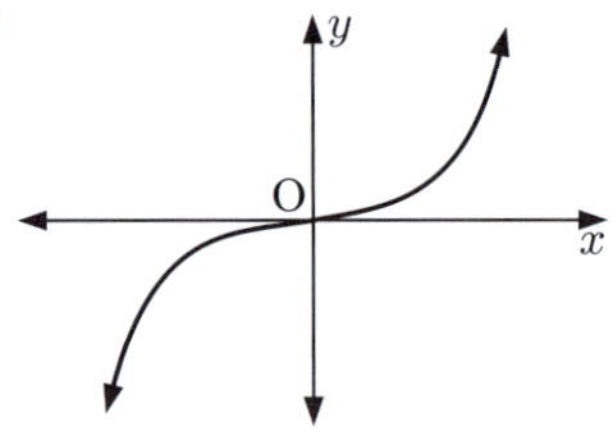

c

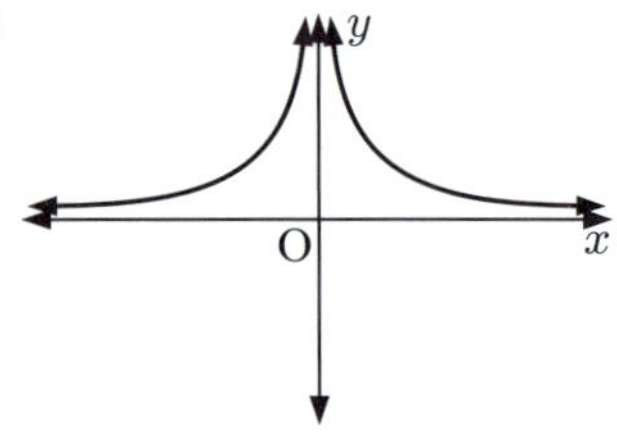

3 A relation has domain $\{1, 2, 5, 7\}$ and range $\{3, 6, 8\}$. Explain why the relation might be a function, but cannot be one-one.

B FUNCTION NOTATION

We can use a **function machine** to show how a function works.

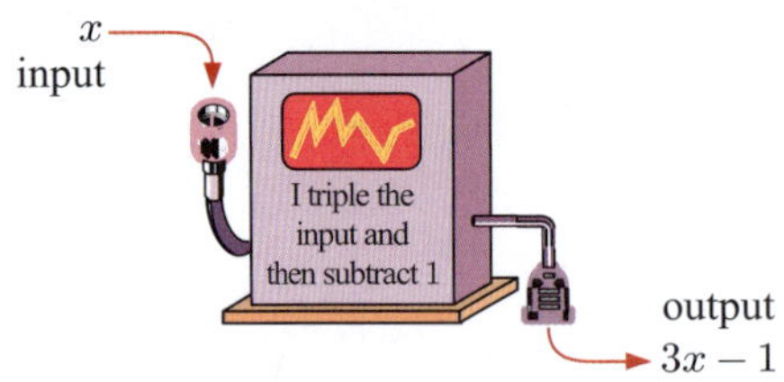

If 4 is the input fed into the machine, the output is $3(4) - 1 = 11$.

The above "machine" has been programmed to perform a particular function. If we use f to represent that particular function, we can write:

f is the function that will convert x into $3x - 1$.

So, f would convert 2 into $3(2) - 1 = 5$ and
-1 into $3(-1) - 1 = -4$.

This function can be written as:

$$\boldsymbol{f : x \mapsto 3x - 1}$$

function f such that x is converted into $3x - 1$

Two other equivalent forms we use are $f(x) = 3x - 1$ and $y = 3x - 1$.

> f is the function which converts x into $f(x)$, so we write $f : x \mapsto f(x)$.
>
> $f(x)$ is sometimes called the **function value** or **image** of x.
>
> If $y = f(x)$ then $f(x)$ is the value of y for a given value of x.

Example 3   Self Tutor

If $f : x \mapsto x^2 + 2x - 1$, find the value of:

a $f(3)$ **b** $f(-2)$

$f(x) = x^2 + 2x - 1$

a $f(3) = (3)^2 + 2(3) - 1$ {replacing x with 3}
$= 9 + 6 - 1$
$= 14$

b $f(-2) = (-2)^2 + 2(-2) - 1$ {replacing x with -2}
$= 4 - 4 - 1$
$= -1$

EXERCISE 2B

1 If $f(x) = 7 - 2x + x^2$, find the value of:

a $f(0)$ **b** $f(2)$ **c** $f(-2)$ **d** $f(-3)$ **e** $f(\frac{1}{2})$

2 If $f : x \mapsto x + \dfrac{1}{x}$, find the value of:

a $f(1)$ **b** $f(-1)$ **c** $f(2)$ **d** $f(-2)$ **e** $f(-\frac{1}{2})$

3 The graph of $y = f(x)$ is shown alongside.

a Find:

i $f(-3)$ **ii** $f(5)$

b Find the value(s) of x such that $f(x)$ equals:

i 4   **ii** 0

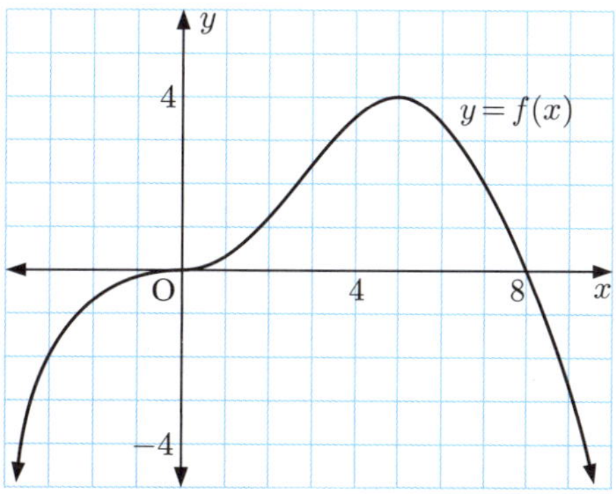

Example 4

Self Tutor

If $f(x) = x^2 + 3$, find in simplest form:

a $f(2x)$ **b** $f(x+1)$

a $f(2x) = (2x)^2 + 3$ {replacing x with $2x$}
$= 4x^2 + 3$

b $f(x+1) = (x+1)^2 + 3$ {replacing x with $x+1$}
$= x^2 + 2x + 1 + 3$
$= x^2 + 2x + 4$

4 If $f(x) = 2x + 1$, find in simplest form:

a $f(3)$ **b** $f(a)$ **c** $f(-a)$ **d** $f(a-1)$ **e** $f(x+2)$ **f** $f(2x)$

5 If $g(x) = -x^2$, find in simplest form:

a $g(1)$ **b** $g(a)$ **c** $g(a+2)$ **d** $g(2x)$ **e** $g(x-1)$ **f** $g(\sqrt{x})$

6 If $F(x) = -x^2 + 2x + 1$, find in simplest form:

a $F(3x)$ **b** $F(-x)$ **c** $F(x+1)$ **d** $F(x-1)$ **e** $F\left(\frac{1}{x}\right)$ **f** $F(x+h)$

7 Suppose $H(x) = \frac{x}{x^2 - 1}$.

a Evaluate:

i $H(2)$ **ii** $H(-3)$ **iii** $H(\frac{1}{2})$

b Find the values of x such that $H(x)$ does not exist.

c Write $H(x+1)$ in simplest form.

d For what value(s) of x does $H(x) = \frac{2}{3}$?

8 Can you write the equation of a circle using function notation? Explain your answer.

9 The altitude of a hot air balloon t minutes before landing is given by $H(t) = 32t^2$ m, $0 \leqslant t \leqslant 8$ minutes.

a Find $H(3)$ and state what this value means.

b How long before landing was the balloon 1152 m above the ground?

10 For each of the following functions, determine the values of a and b:

a $f(x) = ax + b$ where $f(-2) = 6$ and $f(3) = 2$

b $v(t) = at^2 + bt$ where $v(-1) = 8$ and $v(2) = 5$

c $g(x) = \sqrt{x + a} + bx^2$ where $g(0) = 1$ and $g(3) = 20$.

C DOMAIN AND RANGE

The **domain** of a relation is the set of values of x in the relation.
The **range** of a relation is the set of values of y in the relation.

The range is sometimes called the **image set**.

The domain and range for a finite set of points can be written as lists of values.

For relations consisting of infinitely many points, we use **interval notation** to describe the domain and range.

For example, in the relation shown in the graph, x can take any value greater than or equal to 2.

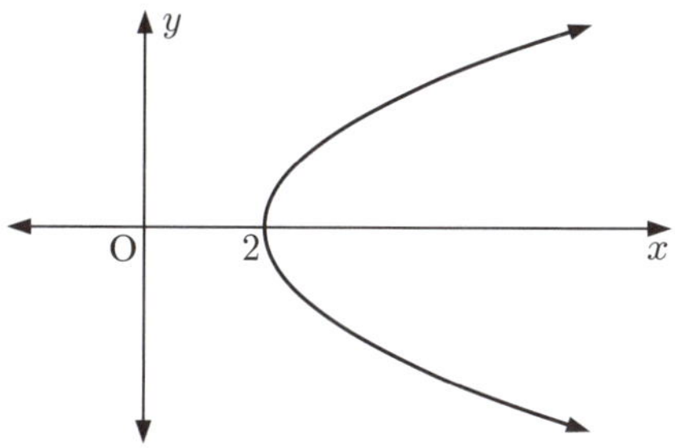

The domain of the relation is $\{x : x \geqslant 2\}$.

- $\{x$ — the set of all x
- $:$ — such that
- $x \geqslant 2\}$ — x is greater than or equal to 2

y can take any real value, so the range of the relation is

$$\{y : y \in \mathbb{R}\}$$

- $\{y$ — the set of all y
- $:$ — such that
- $y \in \mathbb{R}\}$ — y is real

$\mathbb{R}$ is the set of *all* real numbers. These are numbers which can be placed on the number line.

Example 5

Self Tutor

For each graph, state the domain and range:

a

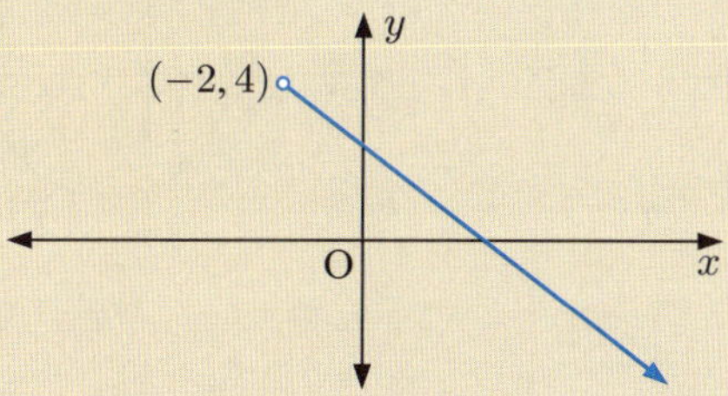

b

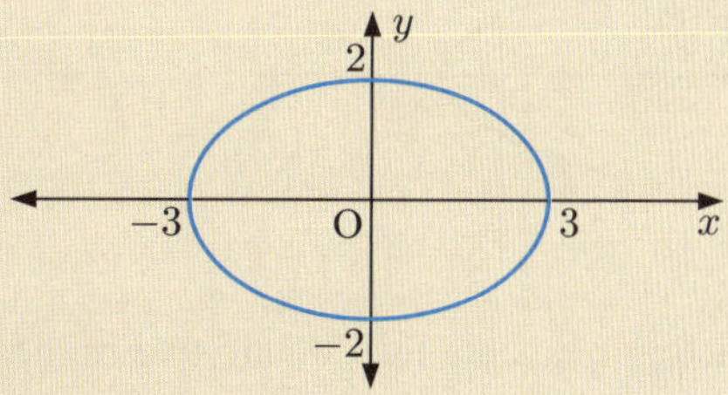

a Domain is $\{x : x > -2\}$
Range is $\{y : y < 4\}$

b Domain is $\{x : -3 \leqslant x \leqslant 3\}$
Range is $\{y : -2 \leqslant y \leqslant 2\}$

When looking at graphs to find the domain and range of a relation:

- a **solid line** ——— indicates that all points on the line are included
- a **dashed line** --------- indicates that all points on the line are *not* included
- a **shaded region** indicates that all points in the region are included.

Sometimes we use a dashed line to indicate a line which a relation gets closer and closer to, but never actually reaches. We call this line an **asymptote**.

Example 6 Self Tutor

For each graph, state the domain and range:

a

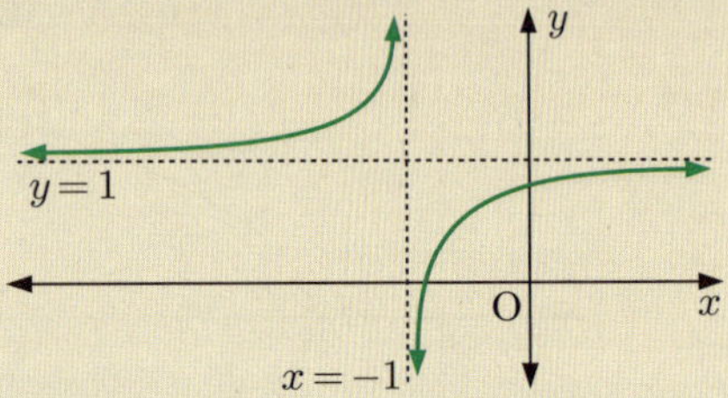

b

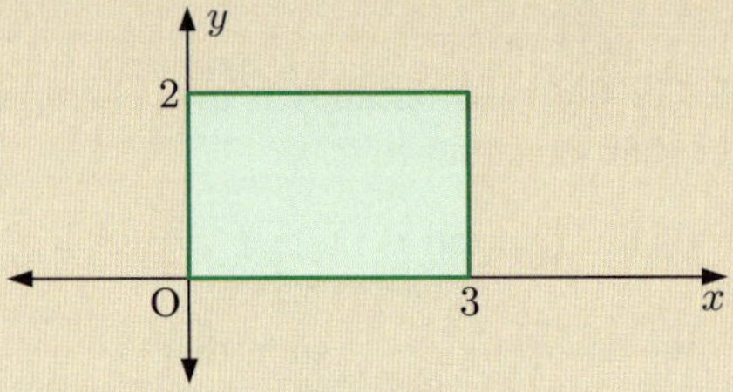

a Domain is $\{x : x \neq -1\}$
Range is $\{y : y \neq 1\}$

b Domain is $\{x : 0 \leqslant x \leqslant 3\}$
Range is $\{y : 0 \leqslant y \leqslant 2\}$

EXERCISE 2C.1

1 For each graph, state the domain and range:

a

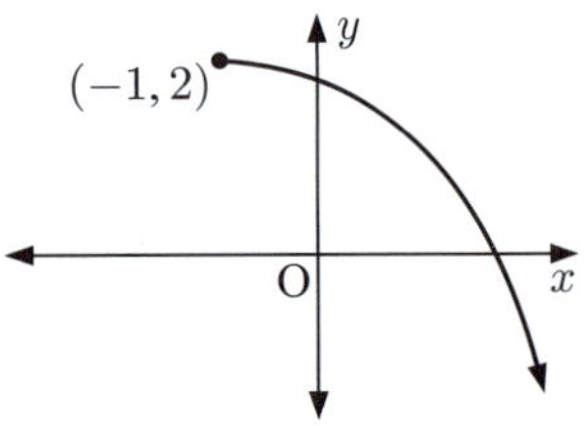

b

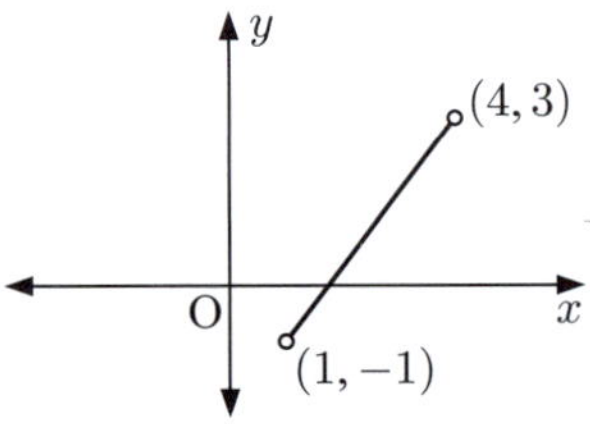

c

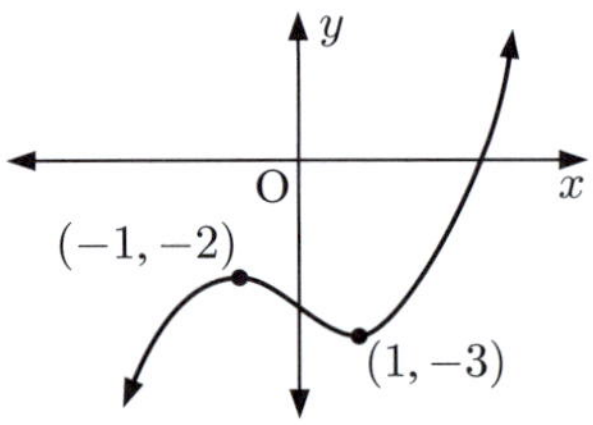

d

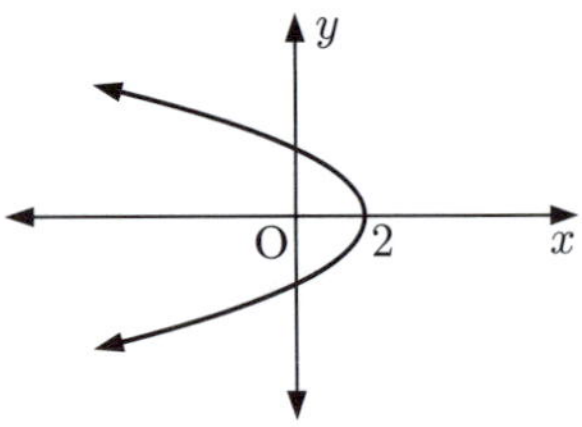

e

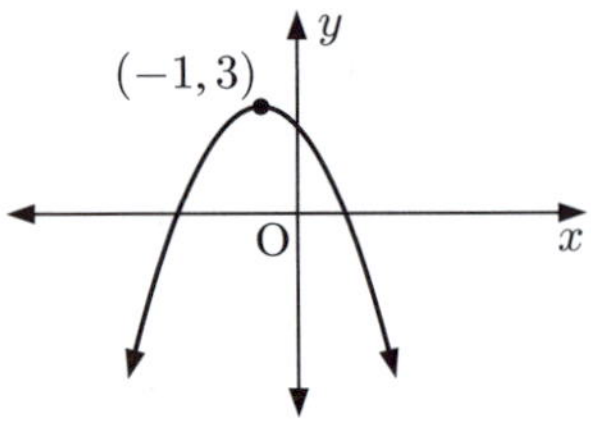

f

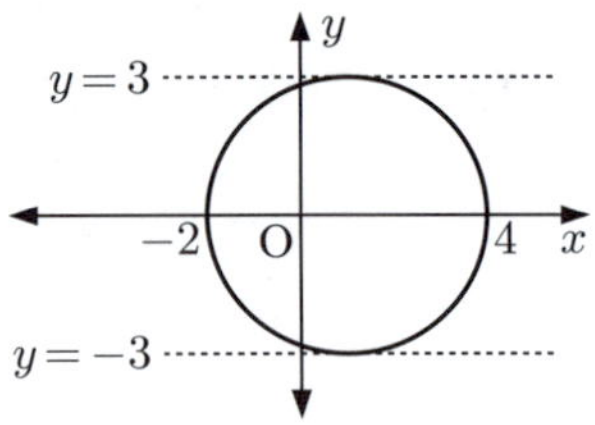

2 For each graph, state the domain and range:

a

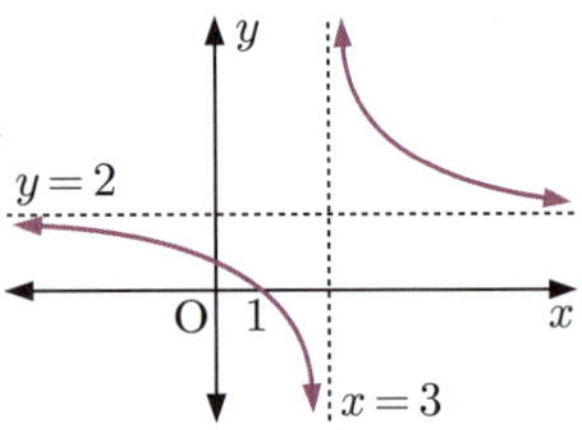

b

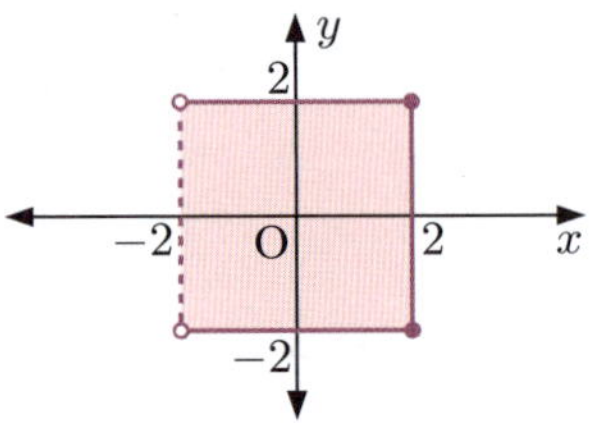

c

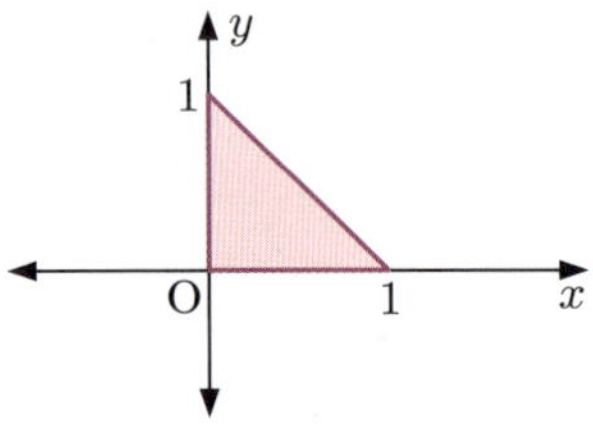

3 Every relation has a unique domain and range. Does every pair of sets D and R correspond to the domain and range of a unique relation? Give reasons for your answer.

DOMAIN AND RANGE OF FUNCTIONS

To fully describe a function, we need both a rule *and* a domain.

For example, we can specify $f(x) = 5x$ where $x \geqslant 2$.

If a domain is not specified, we use the **natural domain**. This is the largest part of $\mathbb{R}$ for which $f(x)$ is defined.

The range of a function depends on the function's formula and its domain.

The table alongside shows the natural domain and corresponding range of some common functions.

Click on the icon to obtain software for finding the domain and range of different functions.

DOMAIN AND RANGE

$f(x)$	*Natural domain*	*Range*
x	$x \in \mathbb{R}$	$y \in \mathbb{R}$
x^2	$x \in \mathbb{R}$	$y \geqslant 0$
$\sqrt{x}$	$x \geqslant 0$	$y \geqslant 0$
$\frac{1}{x}$	$x \neq 0$	$y \neq 0$
$\frac{1}{\sqrt{x}}$	$x > 0$	$y > 0$

Example 7

Self Tutor

State the domain and range of each function:

a $f(x) = \dfrac{1}{2+x}$ **b** $f(x) = \sqrt{2+x}$ **c** $f(x) = \dfrac{1}{\sqrt{2+x}}$

a $\dfrac{1}{2+x}$ is defined when $2 + x \neq 0$

$\therefore \; x \neq -2$

$\therefore$ the domain is $\{x : x \neq -2\}$.

No matter how large or small x is, $y = f(x)$ is never zero.

$\therefore$ the range is $\{y : y \neq 0\}$.

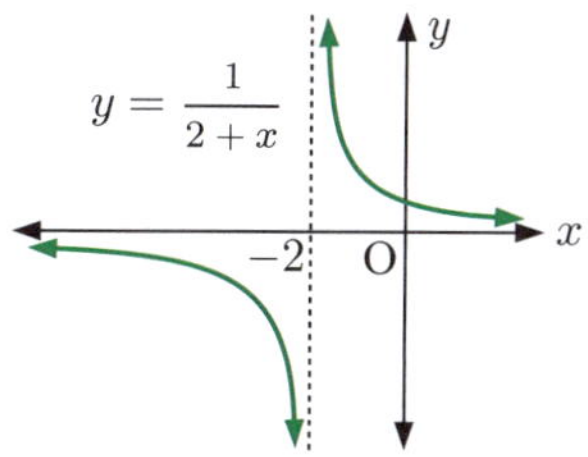

b $\sqrt{2+x}$ is defined when $2 + x \geqslant 0$

$\therefore \; x \geqslant -2$

$\therefore$ the domain is $\{x : x \geqslant -2\}$.

A square root cannot be negative.

$\therefore$ the range is $\{y : y \geqslant 0\}$.

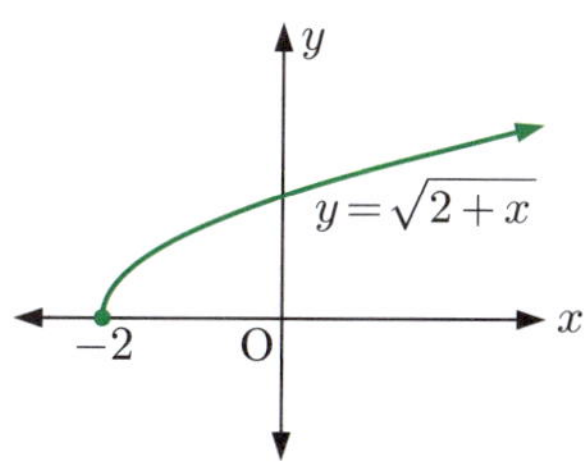

c $\dfrac{1}{\sqrt{2+x}}$ is defined when $2 + x > 0$

$\therefore \; x > -2$

$\therefore$ the domain is $\{x : x > -2\}$.

$y = f(x)$ is always positive and never zero.

$\therefore$ the range is $\{y : y > 0\}$.

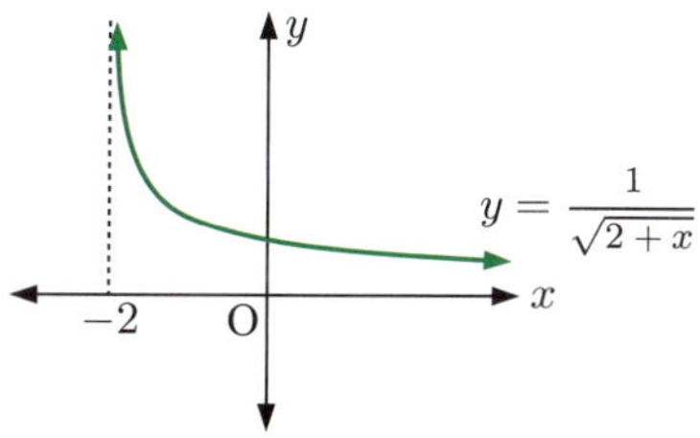

EXERCISE 2C.2

1 State the values of x for which $f(x)$ is defined, and hence state the domain of the function:

a $f(x) = \dfrac{1}{3x-5}$ **b** $f : x \mapsto \sqrt{-x}$ **c** $f(x) = \dfrac{4}{\sqrt{x+1}}$

2 Consider the function $f(x) = 2x - 1$, $a \leqslant x \leqslant b$.

a Sketch the graph of $y = f(x)$ for:

i $a = 0$, $b = 1$ **ii** $a = -2$, $b = 4$ **iii** $a = -5$, $b = 6$

b State the range of $f(x)$ for each of your sketches in **a**.

c Sketch $y = 2x - 1$ on its natural domain and find the corresponding range.

3 For each function, find the domain and range:

a $f(x) = 1$ **b** $g : x \mapsto \dfrac{3}{x}$ **c** $h(x) = 2\sqrt{x+1}$

d $y = \dfrac{1}{6x-3}$ **e** $v(t) = 9.8t + 3.1$ **f** $P(x) = \dfrac{1}{\sqrt{x-a}}$, $a \in \mathbb{R}^+$

4 Use technology to help sketch the graphs of the following functions. Find the domain and range of each.

DOMAIN AND RANGE

a $y = x^2 + 2x + 1$ **b** $f(x) = x^4 - 16$ **c** $f : x \mapsto \dfrac{1}{\sqrt{1+x^2}}$

d $f : x \mapsto \sqrt{x^2 - 1}$ **e** $y = \dfrac{x-3}{x^2 - 10x + 7}$ **f** $f(x) = 2x^2 + \dfrac{1}{x^2}$

g $f(x) = (5 + \sqrt{x})^3$ **h** $f : x \mapsto \dfrac{x}{\sqrt{3x^3 + 2}}$ **i** $y = x^5 - 5x^4 + \dfrac{1}{x^2}$

D THE MODULUS FUNCTION

The **modulus** or **absolute value** of a real number is its size, ignoring its sign.

We denote the absolute value of x by $|\,x\,|$.

For example, the modulus of 4 is 4, and the modulus of -9 is 9. We write $|\,4\,| = 4$ and $|\,-9\,| = 9$.

The relation $y = |\,x\,|$ is a function called the **modulus function**.

Example 8 — Self Tutor

Find $f(-2)$ for $f(x)$ equal to:

a $|\,x\,|$ **b** $x - |\,x\,|$ **c** $|\,x^2 - x\,|$ **d** $\left|\dfrac{8x+1}{3}\right|$

a $f(x) = |x|$

$\therefore f(-2) = |-2|$

$= 2$

b $f(x) = x - |x|$

$\therefore f(-2) = (-2) - |(-2)|$

$= -2 - 2$

$= -4$

c $f(x) = |x^2 - x|$

$\therefore f(-2) = |(-2)^2 - (-2)|$

$= |6|$

$= 6$

d $f(x) = \left|\dfrac{8x+1}{3}\right|$

$\therefore f(-2) = \left|\dfrac{8(-2)+1}{3}\right|$

$= |-5|$

$= 5$

EXERCISE 2D.1

1 Find the value of:

a $|5|$ **b** $|-5|$ **c** $|11|$ **d** $|-11|$

e $|7-3|$ **f** $|3-7|$ **g** $|2-8|$ **h** $|8-2|$

2 For any value a, does $|a| = |-a|$? Explain your answer.

3 Find:

a $|2^2 - 10|$ **b** $|15 - 3 \times 5|$ **c** $\left|\dfrac{3-1}{5+2}\right|$ **d** $\left|\dfrac{2^3}{(-3)^3}\right|$

4 Find $f(4)$ for $f(x)$ equal to:

a $|x-5|$ **b** $|10-x|$ **c** $|3x - x^2|$ **d** $\left|\dfrac{2x+1}{x-1}\right|$

5 Find $f(2)$ for $f(x)$ equal to:

a $|x|$ **b** $x|x|$ **c** $-|x - x^2|$ **d** $\dfrac{|1+3x|}{x+1}$

6 For $f(x) = |x^2 - 3x - 6|$, find:

a $f(0)$ **b** $f(5)$ **c** $f(2)$ **d** $f(-1)$

THE GRAPH OF $y = |f(x)|$

Consider the function $f(x) = -x - 1$.

The table below show the values of $f(x)$ and $|f(x)|$ for $x = -3, -2, -1, 0, 1, 2, 3$.

x	-3	-2	-1	0	1	2	3
$f(x)$	2	1	0	-1	-2	-3	-4
$\|f(x)\|$	2	1	0	1	2	3	4

We can use these values to plot $y = f(x)$ and $y = |f(x)|$ on the same set of axes.

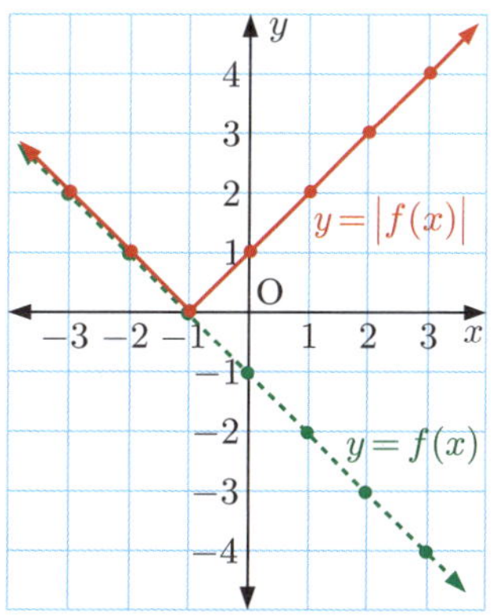

To draw the graph of $y = |f(x)|$, any parts of $y = f(x)$ that are below the x-axis are reflected in the x-axis.

Example 9 Self Tutor

For the following graphs, sketch the graph of $y = |f(x)|$:

a

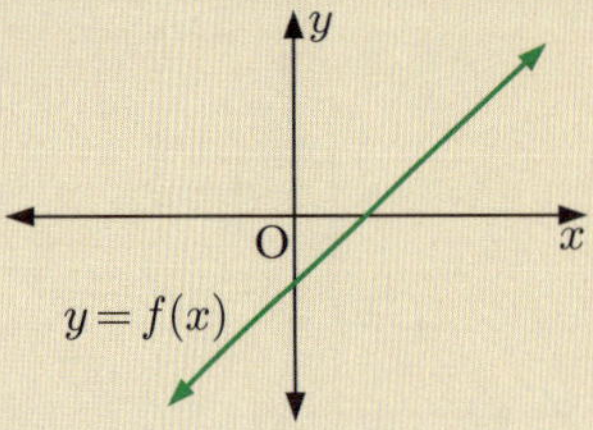

b

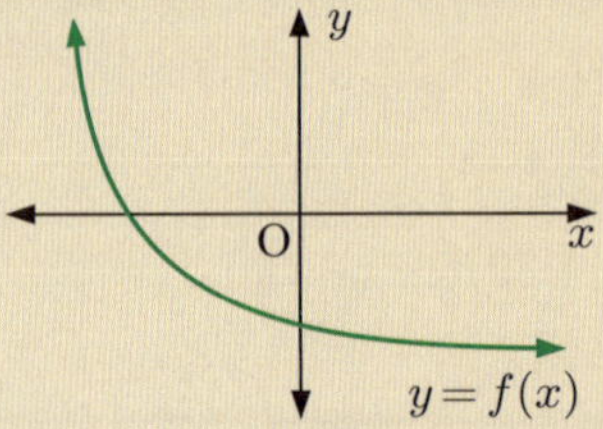

a

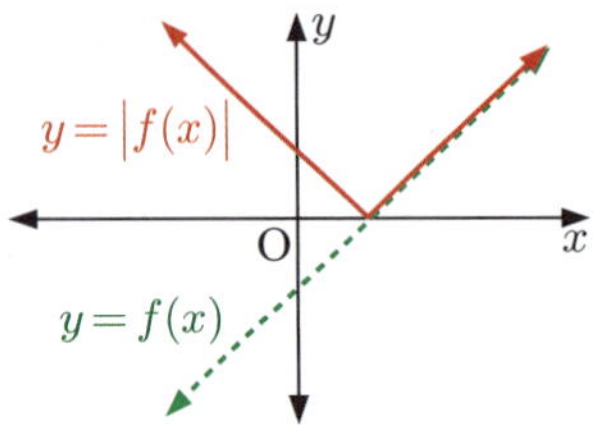

b

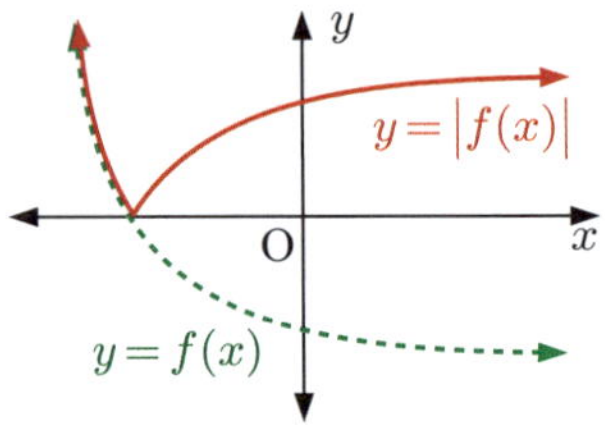

EXERCISE 2D.2

1 For the following graphs, sketch the graph of $y = |f(x)|$:

PRINTABLE GRAPHS

a

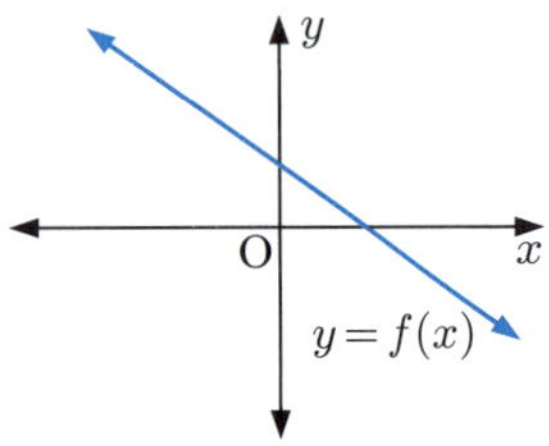

b

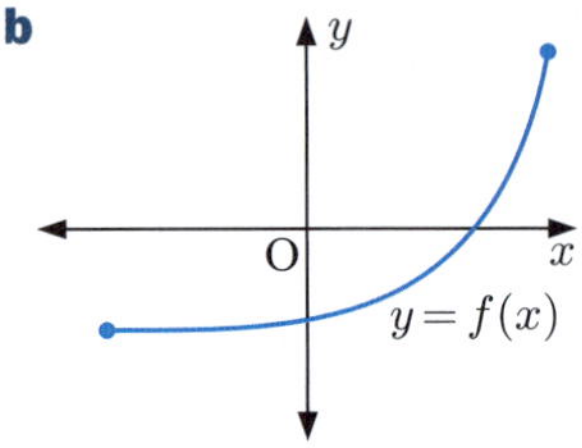

c

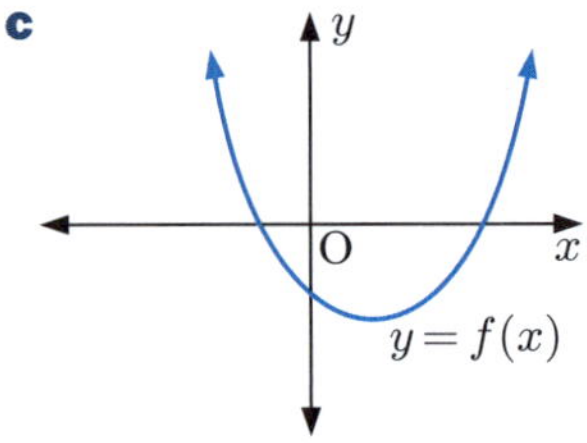

d

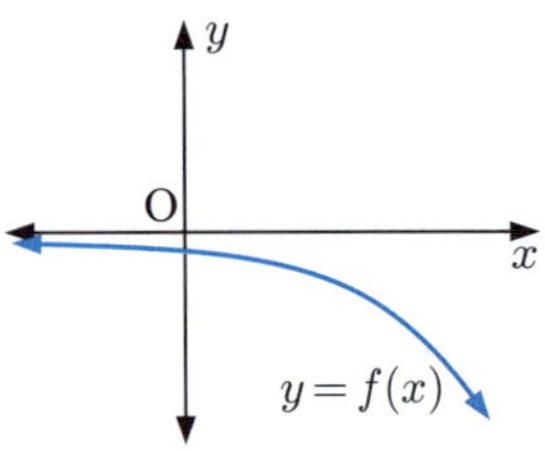

e

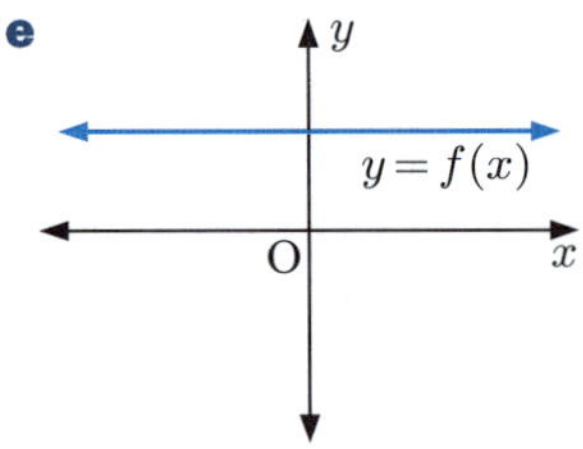

f

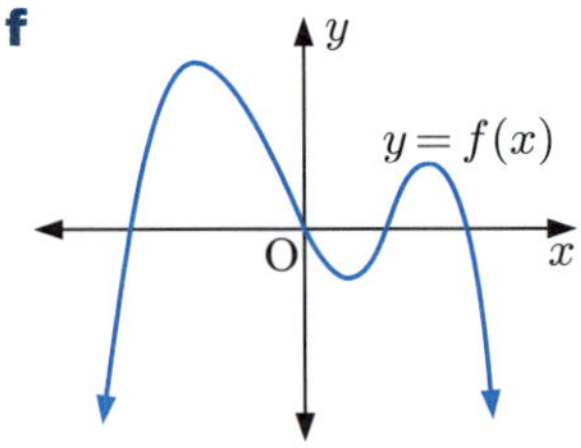

2 Which of the functions $y = |f(x)|$ in question **1** are one-one?

3 Suppose the range of $y = f(x)$ is $\{y : -6 \leqslant y \leqslant 2\}$. Write down the range of $y = |f(x)|$.

4 Determine whether the following statements are true or false:

- **a** If $y = f(x)$ is one-one, then $y = |f(x)|$ is one-one.
- **b** If $y = f(x)$ is not one-one, then $y = |f(x)|$ is not one-one.
- **c** The graphs of $y = f(x)$ and $y = |f(x)|$ always meet the x-axis at the same point(s).
- **d** The graphs of $y = f(x)$ and $y = |f(x)|$ always meet the y-axis at the same point.
- **e** $y = f(x)$ and $y = |f(x)|$ have the same domain.

Example 10

Self Tutor

Draw the graph of $y = |3x+3|$.

We first draw the graph of $y = 3x+3$.

The part of the graph that is below the x-axis is then reflected in the x-axis to produce $y = |3x+3|$.

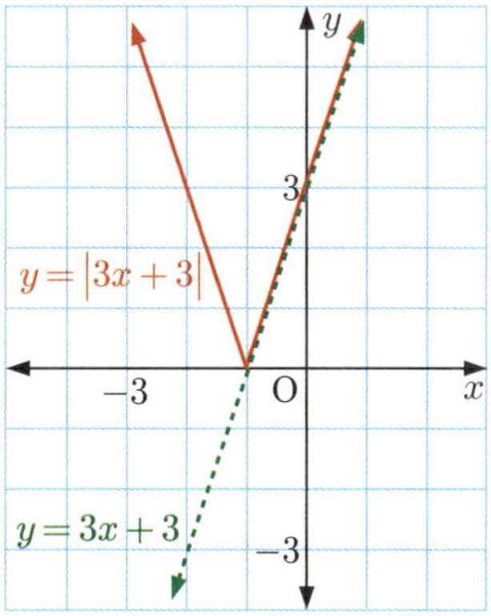

5 Draw the graph of:

a $y = |x|$ **b** $y = |x+3|$ **c** $y = |6-2x|$
d $y = |3x+1|$ **e** $y = |10-4x|$ **f** $y = \left|\frac{1}{2}x+2\right|$

E MODULUS EQUATIONS

Equations involving the modulus function, such as $|x+2| = 4$ and $|2x-1| = |x+3|$, are known as **modulus equations**.

Consider the modulus equation $|x| = 3$.

We observe that $|3| = 3$ and $|-3| = 3$, so $|x| = 3$ has two solutions: $x = 3$ and $x = -3$.

If $|x| = a$ where $a > 0$, then $x = \pm a$.
If $|x| = |b|$ then $x = \pm b$.

We use these rules to solve modulus equations.

Example 11

Self Tutor

Solve for x: **a** $|x-7| = 4$ **b** $|1+3x| = -2$

a $|x-7| = 4$

$\therefore\ x-7 = \pm 4$

$\therefore\ x-7 = 4$ **or** $x-7 = -4$

$\therefore\ x = 11$ $\quad\therefore\ x = 3$

So, $x = 11$ or 3

b $|1+3x| = -2$
has no solution as LHS is never negative.

EXERCISE 2E.1

1 Solve for x:

a $|x| = 5$ **b** $|x| = 9$ **c** $|x| = -1$ **d** $|x| = 0$

e $|x+2| = 4$ **f** $|x-3| = -1$ **g** $|6-x| = 0$ **h** $|2x-5| = 13$

i $|5x+3| = 17$ **j** $|5-7x| = 3$ **k** $|-8x-3| = 1$ **l** $|6+5x| = 11$

Example 12 Self Tutor

Solve for x:

$|3x-4| = |x+2|$

If $|3x-4| = |x+2|$, then $3x-4 = \pm(x+2)$

$\therefore\ 3x-4 = x+2$ **or** $3x-4 = -(x+2)$

$\therefore\ 2x = 6$ $\therefore\ 3x-4 = -x-2$

$\therefore\ x = 3$ $\therefore\ 4x = 2$

$\therefore\ x = \frac{1}{2}$

So, $x = 3$ or $\frac{1}{2}$.

2 Solve for x:

a $|2x+6| = |7-x|$ **b** $|3x-1| = |2x+3|$ **c** $|6-2x| = |5x+1|$

d $|x+5| = |4x-3|$ **e** $|7-2x| = |3x+4|$ **f** $|5x-2| = |2x+6|$

g $|3-6x| = |x+2|$ **h** $|4x-2| = |x-3|$ **i** $|4x+5| = |5-4x|$

j $\left|\frac{1}{2}-x\right| = \left|3x+\frac{5}{2}\right|$ **k** $|x-2| - |2x-1| = 0$ **l** $|x-4| - |3x+1| = 0$

3 Show that $|x-5| = |x-7|$ has only one solution.

SOLVING MODULUS EQUATIONS GRAPHICALLY

We can solve modulus equations graphically by graphing each side of the equation on the same set of axes. The x-coordinates of the intersection points on the graphs are the solutions to the equation.

Example 13 Self Tutor

Use graphical methods to solve:

a $|2x-1| = 3$ **b** $|x+2| = |3x-2|$

a

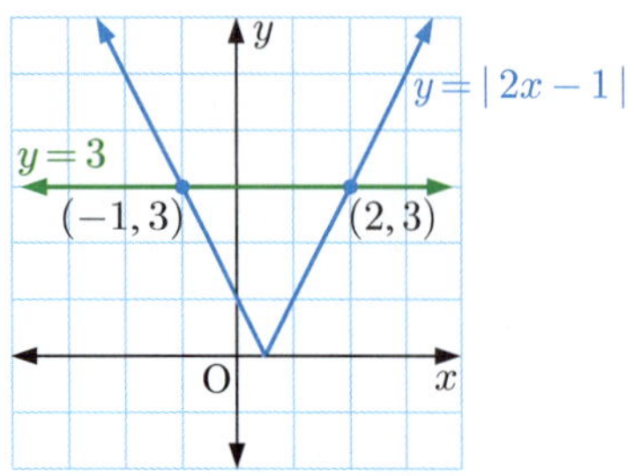

The graphs intersect at $(-1, 3)$ and $(2, 3)$.

$\therefore\ x = -1$ or 2.

b

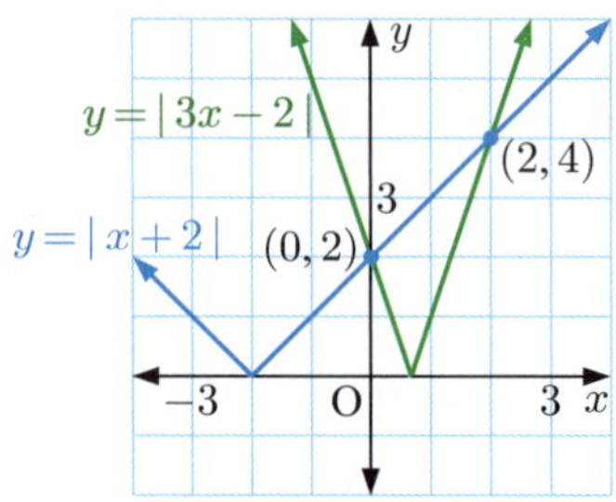

The graphs intersect at $(0, 2)$ and $(2, 4)$.

$\therefore\ x = 0$ or 2.

EXERCISE 2E.2

1 The graphs of $y = |\,2x+2\,|$, $y = 2$, and $y = 6$ are shown alongside.
Solve for x:

a $|\,2x+2\,| = 2$ **b** $|\,2x+2\,| = 6$

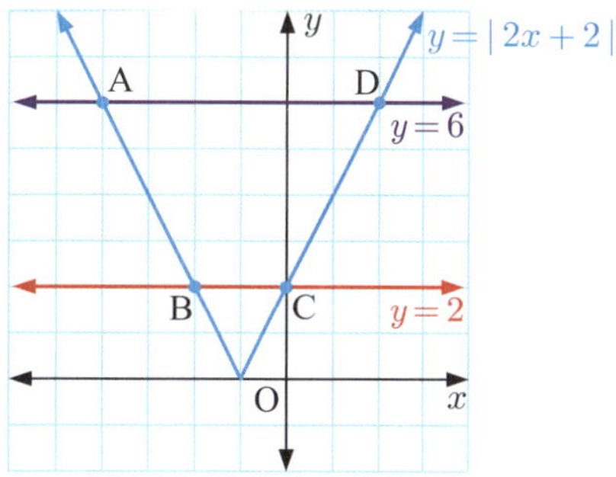

2 **a** Draw the graphs of $y = |\,4-x\,|$ and $y = 2x+1$ on the same set of axes. State the coordinates of the intersection point(s).

b Hence solve $|\,4-x\,| = 2x+1$.

3 Use graphical methods to solve:

a $|\,x-1\,| = 4$ **b** $|\,x+3\,| = 2$ **c** $|\,2x-3\,| = 5$

d $|\,3x+1\,| = -1$ **e** $|\,4-x\,| = 6$ **f** $|\,6-4x\,| = 2$

4 Use graphical methods to solve:

a $|\,x+3\,| = 2x$ **b** $|\,3x-1\,| = 4-2x$ **c** $|\,3x-3\,| = x-2$

d $|\,x+3\,| = |\,5-x|$ **e** $|\,2x+1\,| = |\,4x-7\,|$ **f** $|\,4+3x\,| = |\,x-4\,|$

5 **a** Draw the graph of $y = |\,x-3\,|$.

b Find the values of m for which the equation $|\,x-3\,| = mx$ has:

i two solutions **ii** one solution **iii** no solutions.

c Solve:

i $|\,x-3\,| = 0.5x$ **ii** $|\,x-3\,| = 2x$

F MODULUS INEQUALITIES

An **equation** is a mathematical statement that two expressions are equal.

Sometimes we have a statement that one expression is *greater than*, or else *greater than or equal to*, another. We call this an **inequality**.

$|\,2x+1\,| > 5$ is an example of a **modulus inequality**.

SOLVING MODULUS INEQUALITIES GRAPHICALLY

As with modulus equations, we can solve modulus inequalities by graphing the expressions on each side of the inequality on the same set of axes.

Example 14

Self Tutor

Solve for x:

a $|2x-6| \leqslant 4$

b $|4x-2| > |x+3|$

a The graphs meet when $x=1$ and $x=5$.
Using the graph, $|2x-6| \leqslant 4$ when $1 \leqslant x \leqslant 5$.

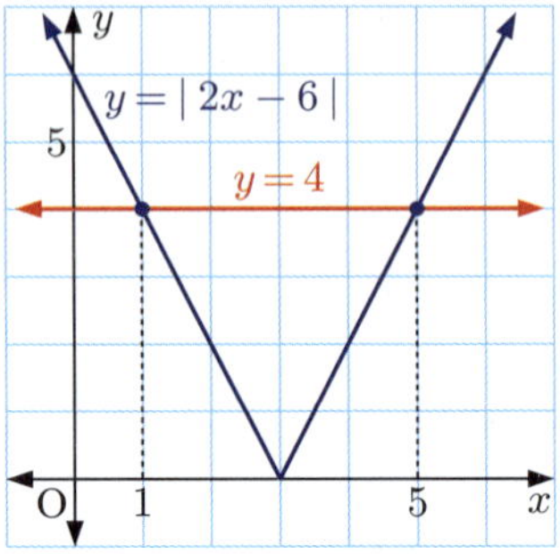

b The graphs meet where

$4x-2 = x+3$ **or** $4x-2 = -(x+3)$

$\therefore\ 3x = 5$ $\qquad \therefore\ 5x = -1$

$\therefore\ x = \frac{5}{3}$ $\qquad \therefore\ x = -\frac{1}{5}$

Using the graph, $|4x-2| > |x+3|$ when $x < -\frac{1}{5}$ *or* $x > \frac{5}{3}$.

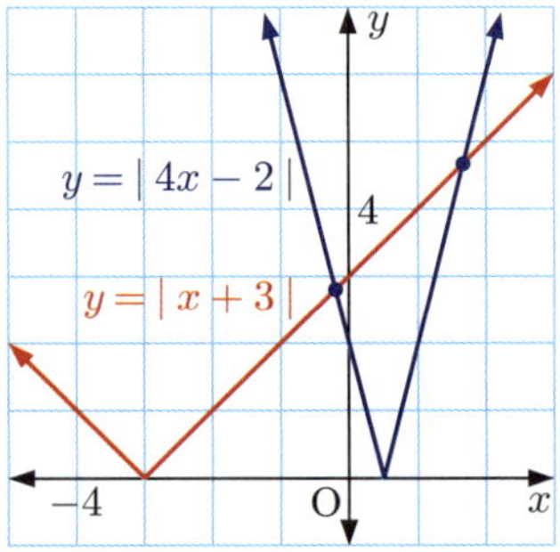

EXERCISE 2F.1

1 The graphs of $y = |x+3|$ and $y=4$ are shown alongside.

  a State the coordinates of the intersection points.

  b Hence solve:

  i $|x+3| < 4$ **ii** $|x+3| \geqslant 4$.

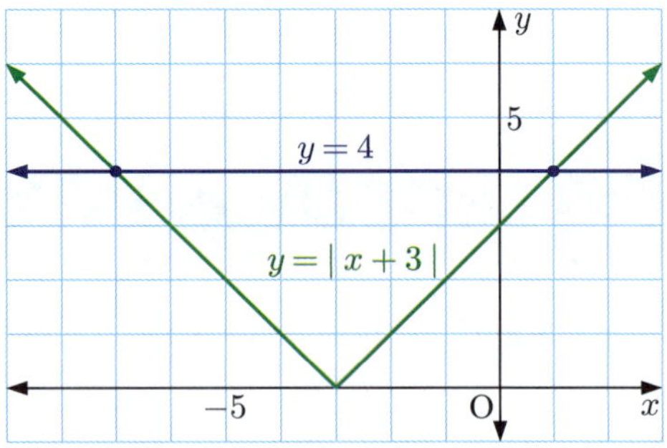

2 **a** Draw the graphs of $y = |2x-4|$ and $y = x+1$ on the same set of axes. State the coordinates of the intersection points.

  b Hence solve $|2x-4| \leqslant x+1$.

3 **a** Draw the graphs of $y = |5x-5|$ and $y = |2x+3|$ on the same set of axes.

  b Algebraically find the coordinates of the intersection points of the graphs.

  c Hence solve $|5x-5| > |2x+3|$.

4 Solve for x:

a $|x-2|>3$
b $|2x+5|\leqslant 1$
c $|3x-1|\geqslant 2x+6$
d $\left|\frac{x}{2}+1\right|<x-1$
e $|3-x|>|2x+3|$
f $|4-3x|\leqslant|3x-2|$
g $\left|\frac{x}{2}-3\right|\geqslant|2x+4|$
h $|8-x|<|3x-1|$
i $|5x+3|>\left|2-\frac{x}{3}\right|$

SOLVING MODULUS INEQUALITIES ALGEBRAICALLY

We can define the modulus of x geometrically as the distance of x from 0 on the number line.

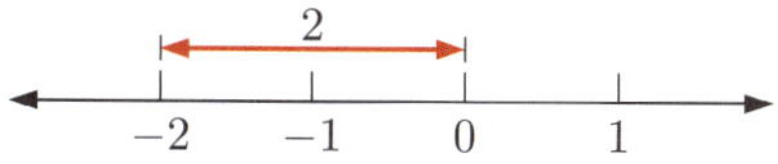

For example, -2 is 2 units from 0, so $|-2|=2$.

This definition gives us some useful rules when solving modulus inequalities algebraically.

For $c>0$:

- If $|x|<c$ then $-c<x<c$.

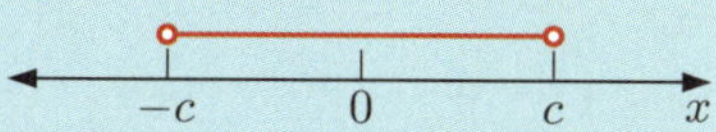

- If $|x|>c$ then $x<-c$ **or** $x>c$.

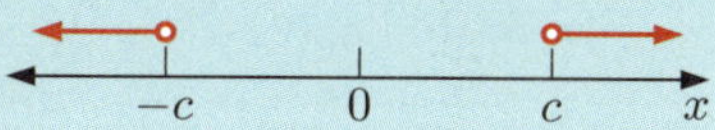

Example 15

Self Tutor

Solve for x:

a $|2x+3|\leqslant 5$
b $|7-3x|<2$

a $|2x+3|\leqslant 5$

$\therefore\ -5\leqslant 2x+3\leqslant 5$

$\therefore\ -8\leqslant 2x\leqslant 2$ {subtracting 3 from each expression}

$\therefore\ -4\leqslant x\leqslant 1$ {dividing each expression by 2}

b $|7-3x|<2$

$\therefore\ |3x-7|<2$ $\{|a|=|-a|\}$

$\therefore\ -2<3x-7<2$

$\therefore\ 5<3x<9$ {adding 7 to each expression}

$\therefore\ \frac{5}{3}<x<3$ {dividing each expression by 3}

EXERCISE 2F.2

1 Solve for x:

a $|x| < 5$ **b** $|x| \leqslant 2$ **c** $|x-2| \leqslant 3$

d $|x+5| < 4$ **e** $|2x+3| \leqslant 7$ **f** $|3x-1| < 8$

g $|6-x| < 3$ **h** $|1-2x| \leqslant 5$ **i** $|8-3x| < 7$

2 **a** Solve for x:

i $|x| < 0$ **ii** $|x| > 0$ **iii** $|x| < -2$ **iv** $|x| > -2$

b Suppose $c < 0$. Solve for x:

i $|x| < c$ **ii** $|x| > c$

Example 16 Self Tutor

Solve for x:

a $|4x-1| > 9$ **b** $|3-x| \geqslant 10$

a $|4x-1| > 9$

$\therefore\ 4x-1 > 9$ **or** $4x-1 < -9$

$\therefore\ 4x > 10$ $\quad\therefore\ 4x < -8$

$\therefore\ x > \frac{5}{2}$ $\quad\therefore\ x < -2$

So, the solution is $x > \frac{5}{2}$ or $x < -2$.

b $|3-x| \geqslant 10$

$\therefore\ |x-3| \geqslant 10$ $\quad\{|a| = |-a|\}$

$\therefore\ x-3 \geqslant 10$ **or** $x-3 \leqslant -10$

$\therefore\ x \geqslant 13$ $\quad\therefore\ x \leqslant -7$

So, the solution is $x \geqslant 13$ or $x \leqslant -7$.

3 We have seen that the statement "If $|x| < 7$ then $-7 < x < 7$" is true.
Explain what is wrong with the statement:
"If $|x| > 7$ then $-7 > x > 7$".

4 Solve for x:

a $|x| > 3$ **b** $|x| \geqslant 6$ **c** $|x| \geqslant 0$

d $|x+5| > 7$ **e** $|x-3| > 4$ **f** $|2x-6| \geqslant 4$

g $|5x-2| > 8$ **h** $|3x+5| \geqslant 3$ **i** $|4-x| > -1$

j $|10-3x| > 4$ **k** $|1-6x| \geqslant 2$ **l** $|8-2x| > 0$

5 Solve for x:

a $|2x-7| \leqslant 11$ **b** $|x+4| > 3$ **c** $|5x-3| < 7$

d $|4x-2| < -5$ **e** $|7x-3| \geqslant 12$ **f** $|11-3x| < 2$

g $|1-4x| > 9$ **h** $\left|\frac{x}{2}-5\right| < 3$ **i** $\left|\frac{x-2}{3}\right| \geqslant 4$

INEQUALITIES OF THE FORM $|ax+b| < cx+d$ AND $|ax+b| > cx+d$

In cases where there is an unknown on the right hand side, we cannot use the rules we saw in **Section F.2**. This is because the right hand side is not necessarily positive.

Instead, we consider separate intervals according to the sign of the expression within the modulus.

For example, for the inequality $|3x+1| < x+2$, we consider two separate intervals:

- if $x \geqslant -\frac{1}{3}$, $3x+1 \geqslant 0$ and so $|3x+1| = 3x+1$
- if $x < -\frac{1}{3}$, $3x+1 < 0$ and so $|3x+1| = -(3x+1)$.

We then solve the resulting linear inequalities for the separate intervals, and combine the solutions.

Example 17

Self Tutor

Solve for x:

a $|3x+1| < x+2$

b $|1-2x| > -3x+5$

a If $x \geqslant -\frac{1}{3}$, $|3x+1| = 3x+1$

$\therefore\ 3x+1 < x+2$

$\therefore\ 2x < 1$

$\therefore\ x < \frac{1}{2}$

$\therefore\ -\frac{1}{3} \leqslant x < \frac{1}{2}$ are solutions.

If $x < -\frac{1}{3}$, $|3x+1| = -(3x+1)$

$\therefore\ -(3x+1) < x+2$

$\therefore\ 4x > -3$

$\therefore\ x > -\frac{3}{4}$

$\therefore\ -\frac{3}{4} < x < -\frac{1}{3}$ are also solutions.

So, the solution is $-\frac{3}{4} < x < \frac{1}{2}$.

b If $x \leqslant \frac{1}{2}$, $|1-2x| = 1-2x$

$\therefore\ 1-2x > -3x+5$

$\therefore\ x > 4$

This is a contradiction, so there are no solutions for $x \leqslant \frac{1}{2}$.

If $x > \frac{1}{2}$, $|1-2x| = -(1-2x)$

$\therefore\ -(1-2x) > -3x+5$

$\therefore\ 5x > 6$

$\therefore\ x > \frac{6}{5}$

So, the solution is $x > \frac{6}{5}$.

EXERCISE 2F.3

1 Solve for x:

a $|2x+3| < x+6$ **b** $|3x| \leqslant 2x+10$ **c** $|4x-1| < x+2$

d $|x-5| \leqslant 2x+5$ **e** $|4x-8| < 3x-4$ **f** $|3x+2| \leqslant 5x-1$

2 Solve for x:

a $|2x-4| > x$ **b** $|4x-3| \geqslant 3x+1$ **c** $|5x+2| > 6-x$

d $|x+1| > 3x$ **e** $|6x-5| \geqslant 2x+2$ **f** $|2x+7| \geqslant 3-5x$

3 Solve for x:

a $|3x+4| > x+2$ **b** $|2x-3| \leqslant 4x+3$ **c** $|4x+4| \geqslant 5-2x$

d $|7-x| \leqslant 3x+4$ **e** $|5x+6| < 2x+4$ **f** $|3x-9| > 8x-7$

4 **a** Show that the inequality $|2x+2| < x-1$ has no solutions.

b Illustrate this result graphically.

G SIGN DIAGRAMS

In many situations we do not need to know the complete shape of a graph, but only specific information about x-intercepts, vertical asymptotes, and the *sign* of the function.

For the graph $y = f(x)$ shown, we can summarise this information in the **sign diagram** below it.

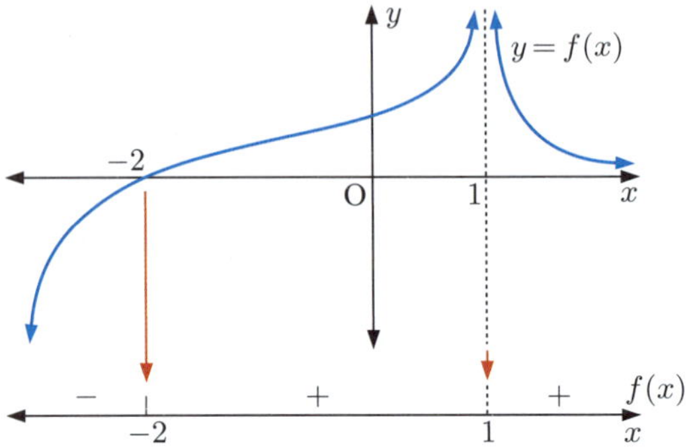

The sign diagram consists of:

- a horizontal line which represents the **x-axis**
- a tick mark at $x = -2$ which represents the x-intercept of the graph of $y = f(x)$, or a **zero** of the function
- a dotted line at $x = 1$ which represents the **vertical asymptote**
- positive and negative signs which indicate where the graph is **above** and **below** the x-axis respectively.

The **zeros** of $f(x)$ correspond to the **roots** of the equation $f(x) = 0$. The term "zeros" will not be used in Cambridge examinations.

Example 18

Self Tutor

Draw a sign diagram for each graph:

a

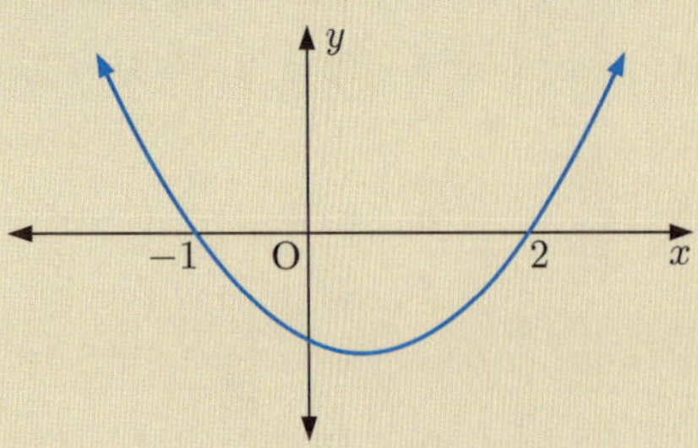

b

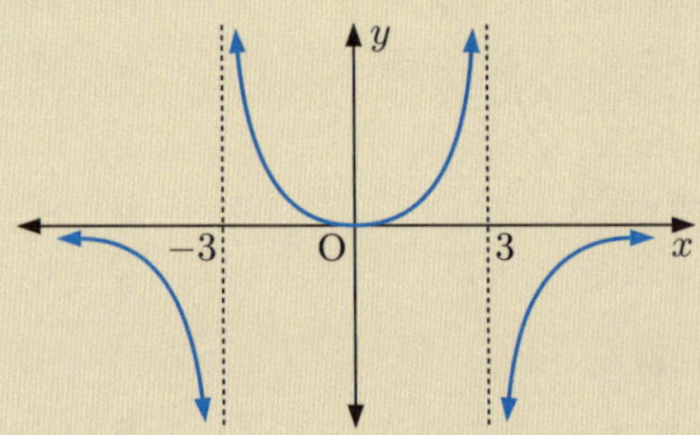

a

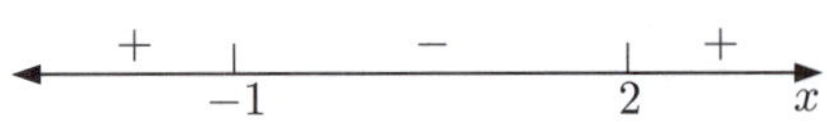

b

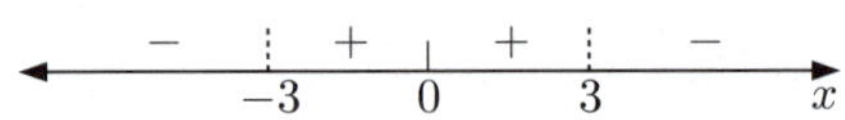

EXERCISE 2G.1

1 Draw a sign diagram for each graph:

a

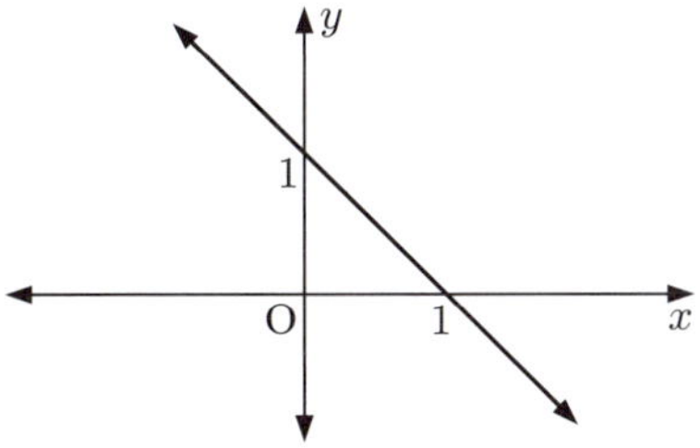

b

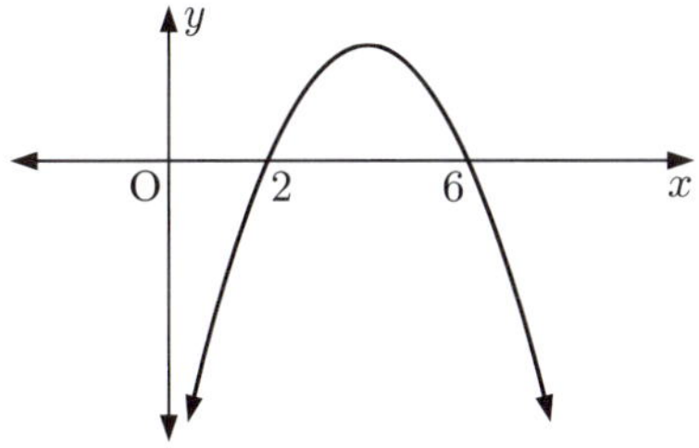

c

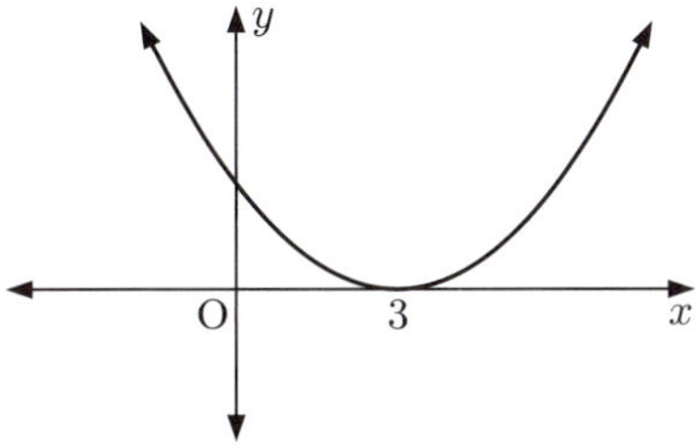

d

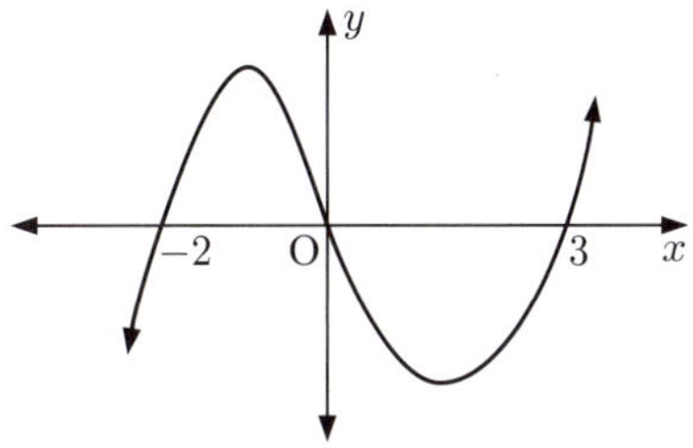

e

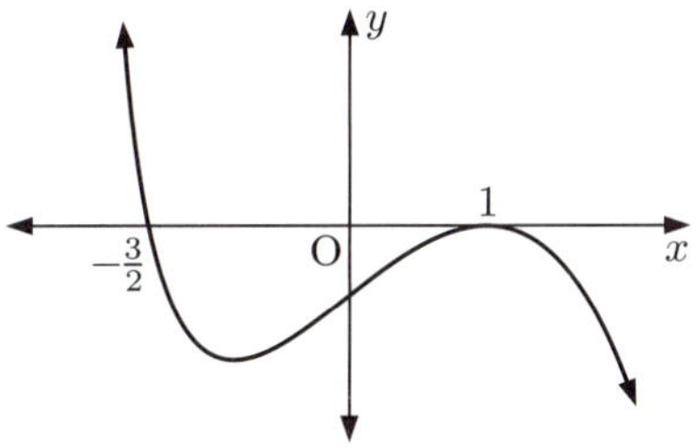

f

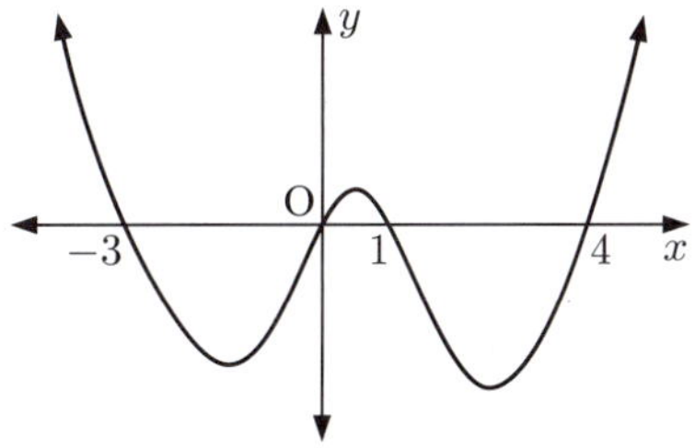

g

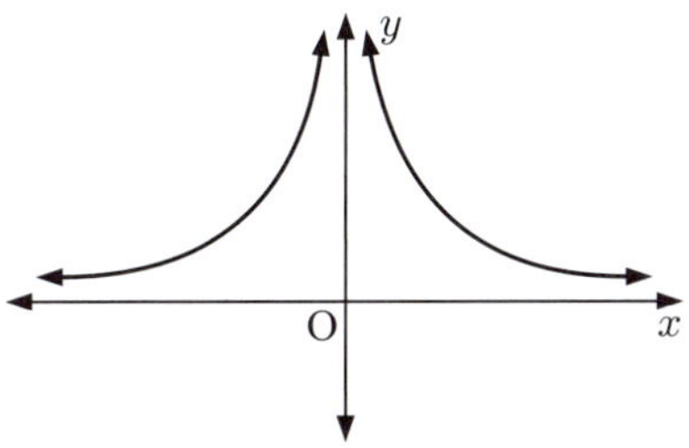

h

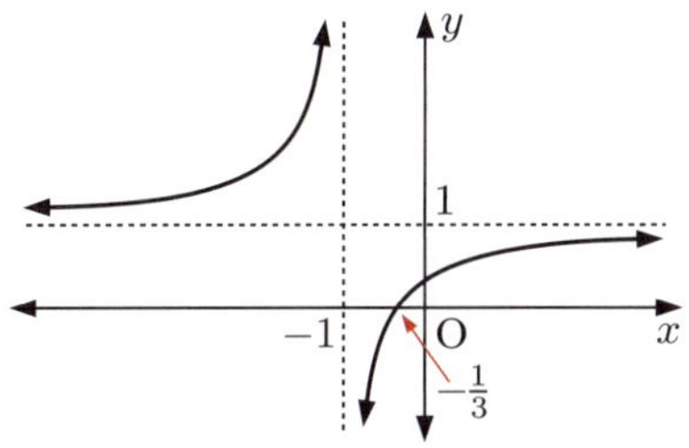

i

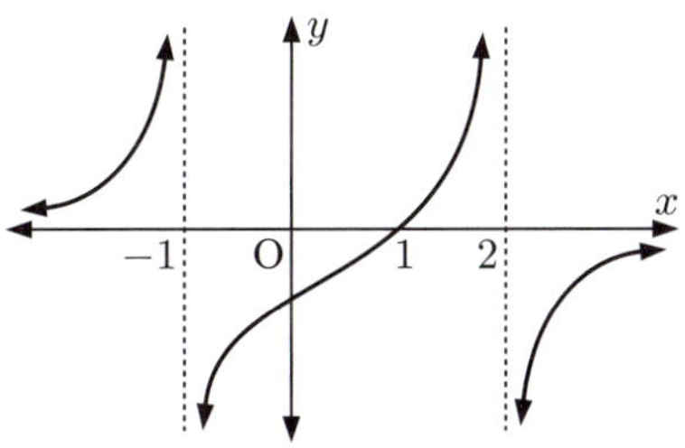

j

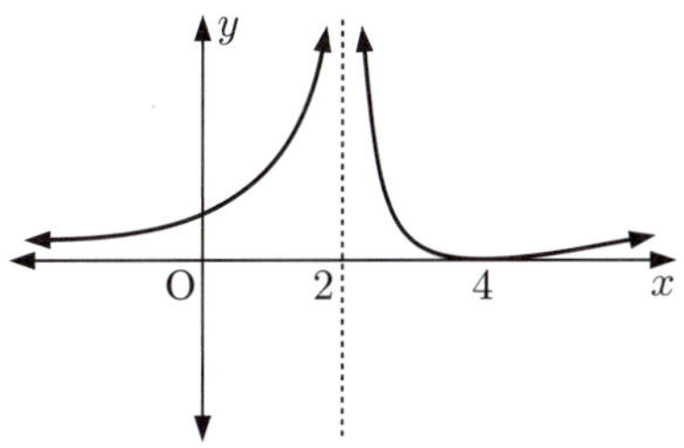

2 For each sign diagram, draw the graph of a possible function it might correspond to:

a

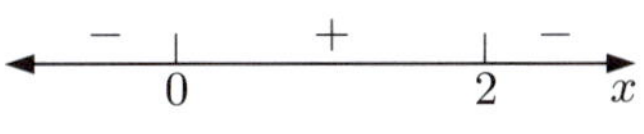

b

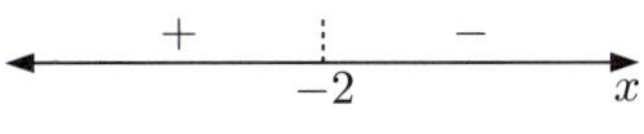

c

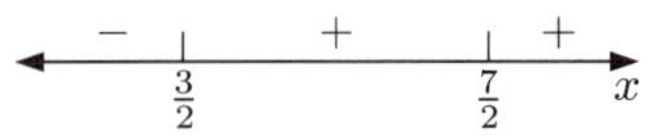

LINEAR FACTORS

Discovery 1 — Linear factors and sign diagrams

From your previous studies, you should be familiar with **linear factors**. In this Discovery, we will investigate linear factors in the context of functions and sign diagrams.

What to do:

GRAPHING PACKAGE

1 With the help of technology, match each function with its graph and sign diagram.

a $f(x) = (x-2)(x-1)$ **b** $f(x) = -(x-3)^2$ **c** $f(x) = -3(x-4)(x+2)$

d $f(x) = \frac{1}{2}(x+5)^2$ **e** $f(x) = \dfrac{4}{x+1}$ **f** $f(x) = \dfrac{x+3}{x-6}$

g $f(x) = \dfrac{2x(x-1)}{x+2}$ **h** $f(x) = \dfrac{1}{(x-3)(x+3)}$ **i** $f(x) = \dfrac{x+4}{(x-2)^2}$

A
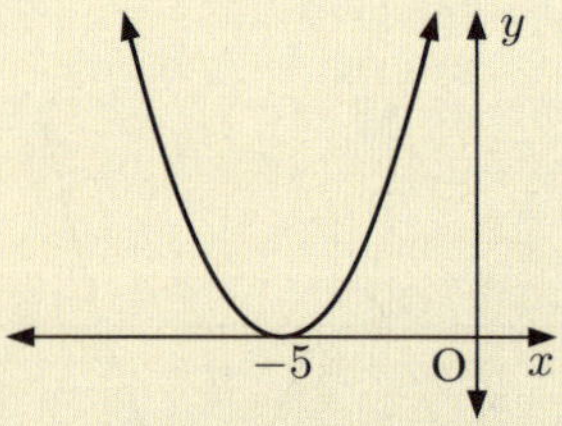

B
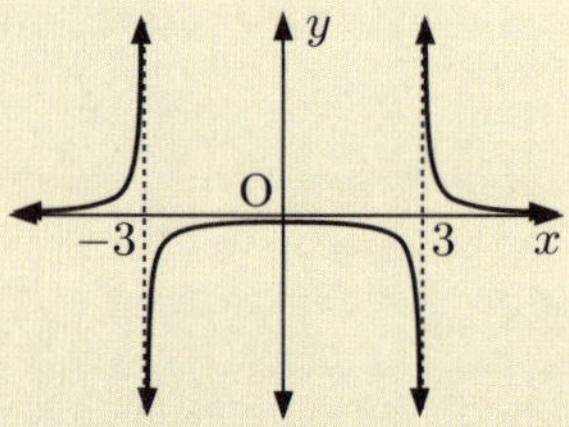

C
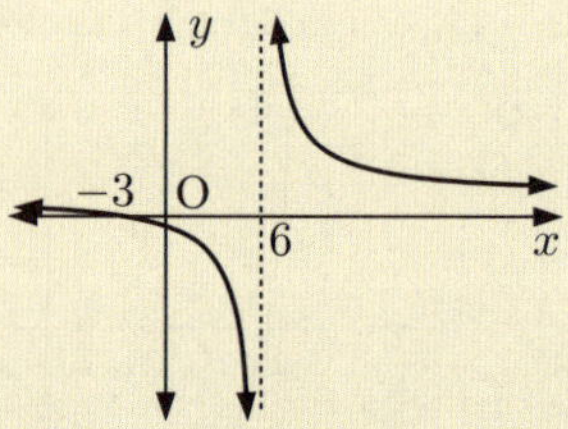

D
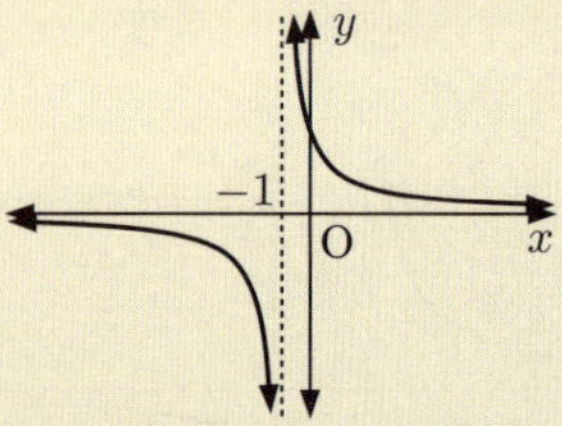

E
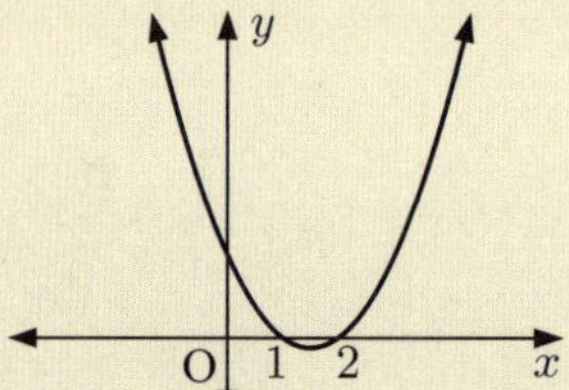

F
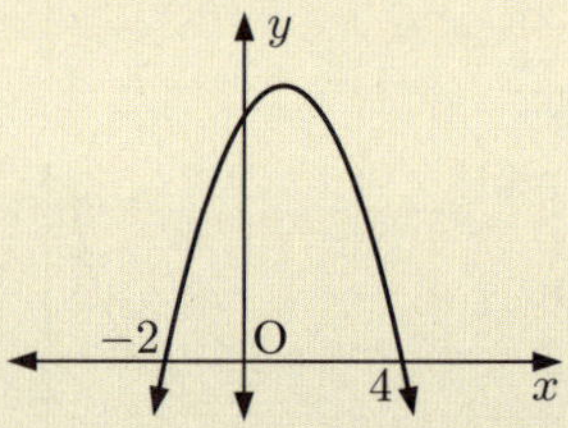

G
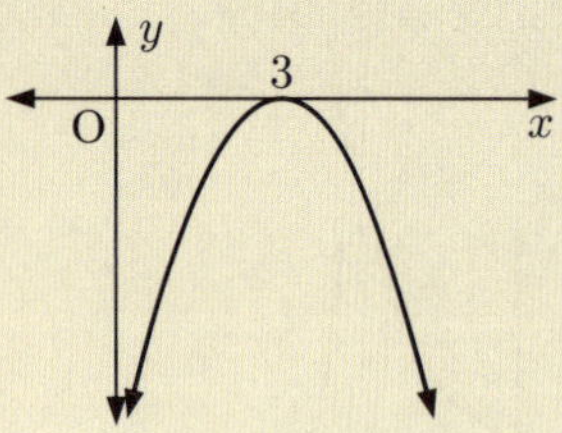

H
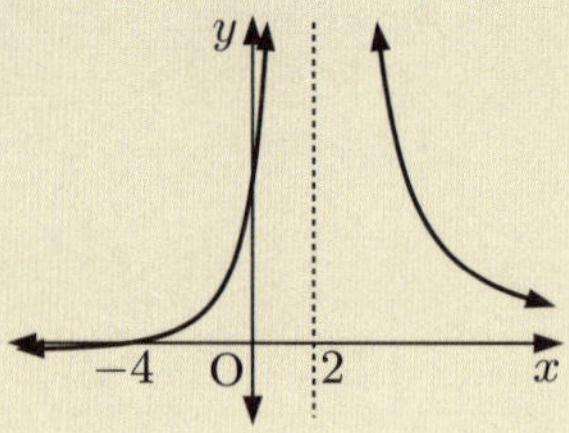

I
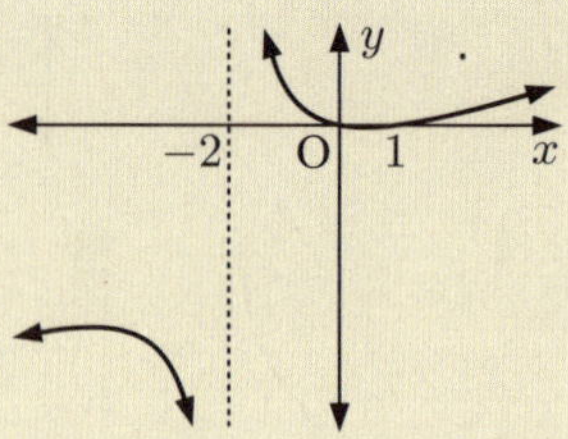

i
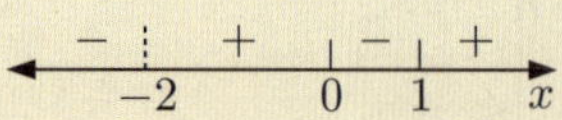

ii
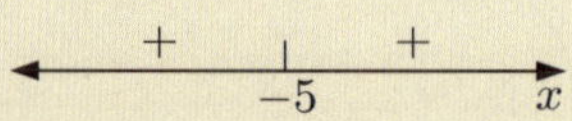

iii
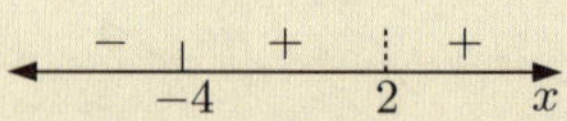

iv
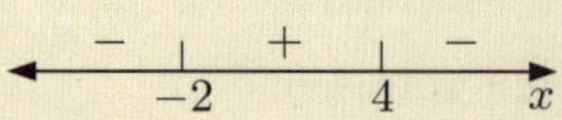

v
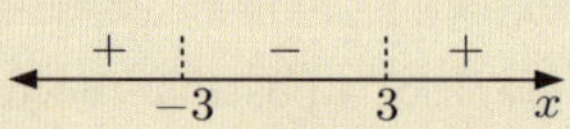

vi
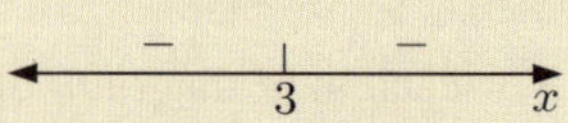

vii
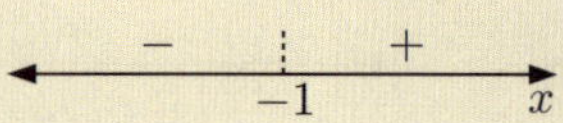

viii
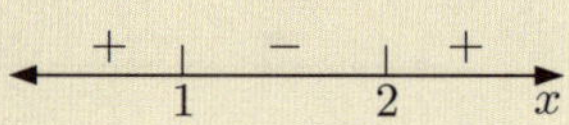

ix
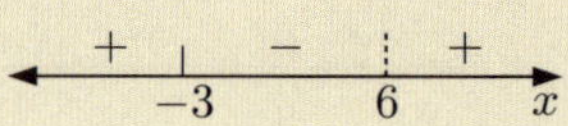

2 Look at the graphs and sign diagrams that you matched to each function in **1**.

a What do you notice about the signs on either side of each zero or vertical asymptote corresponding to:

i single linear factors **ii** squared linear factors?

b Comment on how the graphs of the functions relate to your observations in **a**.

3 Consider the function $f(x) = \dfrac{(x-a)(x-b)}{(x-c)(x-d)}$, where $a < b < c < d$.

a Find the zeros of $f(x)$.

b Find x such that $f(x)$ is undefined, and hence state the vertical asymptotes of $f(x)$.

c Determine the sign of $f(x)$ when:

i $x < a$ **ii** $a < x < b$ **iii** $b < x < c$ **iv** $c < x < d$ **v** $x > d$

d Hence draw the sign diagram of $f(x)$.

4 Consider the function $f(x) = (x+1)^n$, $n \in \mathbb{N}$.

a Based on your observations in **2**, draw the sign diagram of $f(x)$ for $n = 1$ and $n = 2$.

b Use technology to help you draw the sign diagram of $f(x)$ for $n = 3, 4, 5, 6$.

c Hence predict the sign diagram of $f(x)$ when:

i n is odd **ii** n is even.

d How will the sign diagram of $g(x) = -f(x)$ differ to that of $f(x)$?

From the **Discovery**, you should have found that:

In general, when a linear factor is raised to:
- an **odd power** there is a change of sign about the corresponding zero or asymptote
- an **even power** there is no sign change about the corresponding zero or asymptote.

Example 19

Self Tutor

Draw a sign diagram for each function:

a $f(x) = (x-3)(x-6)$ **b** $f(x) = -3(x+1)^2$

a $f(x) = (x-3)(x-6)$ has zeros 3 and 6.

We substitute any number < 3.
$f(0) = (0-3)(0-6) = 18 > 0$
so we put a $+$ sign here.

As the factors are single, the signs alternate.

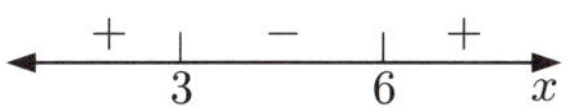

b $f(x) = -3(x+1)^2$ has zero -1.

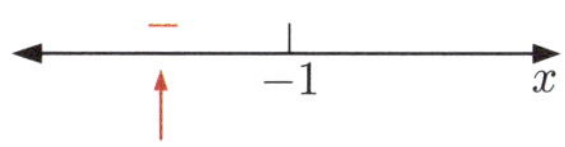

We substitute any number < -1.
$f(-2) = -3(-2+1)^2 = -3 < 0$
so we put a $-$ sign here.

As the factor is squared, the sign does not change.

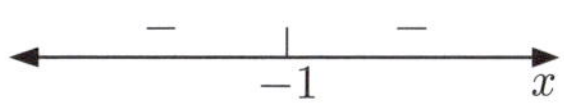

EXERCISE 2G.2

1 Draw a sign diagram for each function:

a $f(x) = (x+3)(x-1)$ **b** $f(x) = -4(x+2)^2$ **c** $f(x) = x(2x+1)$

d $f(x) = (2-x)^2(3x+4)$ **e** $f(x) = (2x-5)(x+5)^3$ **f** $f(x) = (x-6)^4$

2 Factorise each expression, and hence draw its sign diagram:

a $x^2 - 4$ **b** $x^2 + 6x + 9$ **c** $x^2 - 7x + 12$

Example 20 Self Tutor

Draw a sign diagram for $f(x) = \dfrac{x-1}{2x+1}$.

$f(x) = \dfrac{x-1}{2x+1}$ has zero 1 and vertical asymptote $x = -\frac{1}{2}$.

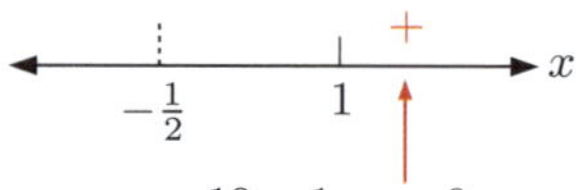

$$f(10) = \frac{10-1}{2(10)+1} = \frac{9}{21} > 0$$

Since $(x-1)$ and $(2x+1)$ are single factors, the signs alternate.

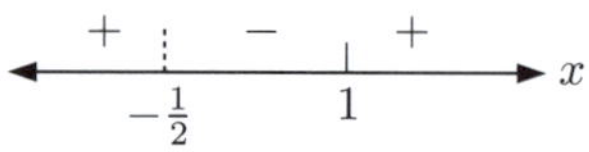

3 Draw a sign diagram for each function:

a $f(x) = \dfrac{x+3}{x-2}$ **b** $f(x) = \dfrac{1-4x}{2-x}$ **c** $f(x) = \dfrac{(x-5)^2}{x}$

d $f(x) = \dfrac{3x}{(x+1)(2x-1)}$ **e** $f(x) = \dfrac{(9-x)(9+x)}{x+10}$ **f** $f(x) = \dfrac{x+4}{x(12-6x)}$

4 Draw a sign diagram for each function:

a $f(x) = 1 + \dfrac{3}{x+1}$ **b** $f(x) = x - \dfrac{1}{x}$ **c** $f(x) = x - \dfrac{1}{x^2}$

Discussion

Is it possible to draw a sign diagram for:

- the function $f(x) = 0$
- a relation such as $x^2 = y$?

H COMPOSITE FUNCTIONS

Given $f : x \mapsto f(x)$ and $g : x \mapsto g(x)$, the **composite function** of f and g will convert x into $f(g(x))$.

$f \circ g$ or fg is used to represent the composite function of f and g. It means "f following g".

$$(f \circ g)(x) \text{ or } fg(x) = f(g(x))$$

Consider $f : x \mapsto x^2$ and $g : x \mapsto 1 - 3x$.

$f \circ g$ means that g converts x to $1 - 3x$ and then f converts $(1 - 3x)$ to $(1 - 3x)^2$.

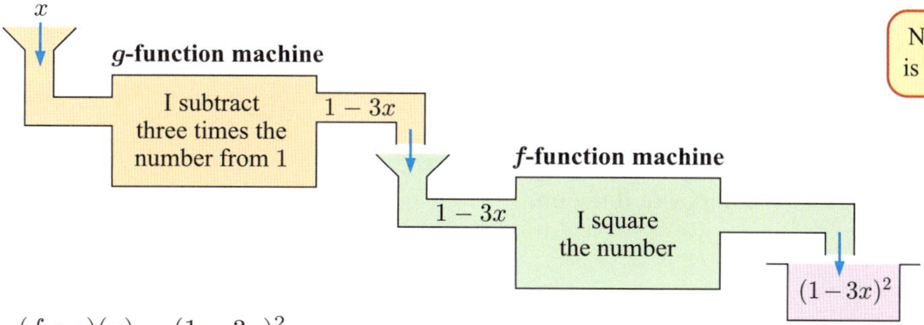

So, $(f \circ g)(x) = (1 - 3x)^2$.

Algebraically, if $f(x) = x^2$ and $g(x) = 1 - 3x$ then

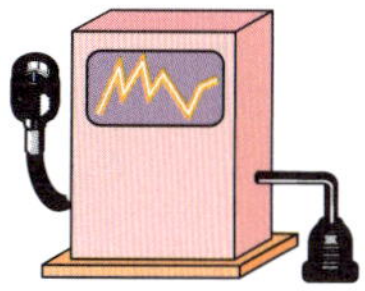

$$\begin{aligned}(f \circ g)(x) &= f(g(x)) \\ &= f(1 - 3x) && \{g \text{ operates on } x \text{ first}\} \\ &= (1 - 3x)^2 && \{f \text{ operates on } g(x) \text{ next}\}\end{aligned}$$

and
$$\begin{aligned}(g \circ f)(x) &= g(f(x)) \\ &= g(x^2) && \{f \text{ operates on } x \text{ first}\} \\ &= 1 - 3(x^2) && \{g \text{ operates on } f(x) \text{ next}\} \\ &= 1 - 3x^2\end{aligned}$$

So, $f(g(x)) \neq g(f(x))$.

In general, $(f \circ g)(x) \neq (g \circ f)(x)$.

We can also compose a function f with itself. The resulting function is $(f \circ f)(x)$ or $f^2(x)$.

In general, $(f \circ f)(x) \neq (f(x))^2$.

Example 21

Self Tutor

Given $f : x \mapsto 1 - x^2$ and $g : x \mapsto 3x - 2$, find in simplest form:

a $(f \circ g)(x)$ **b** $(g \circ f)(x)$ **c** $f^2(\frac{1}{2})$

$f(x) = 1 - x^2$ and $g(x) = 3x - 2$

a
$$\begin{aligned}&(f \circ g)(x) \\ &= f(g(x)) \\ &= f(3x - 2) \\ &= 1 - (3x - 2)^2 \\ &= 1 - 9x^2 + 12x - 4 \\ &= -9x^2 + 12x - 3\end{aligned}$$

b
$$\begin{aligned}&(g \circ f)(x) \\ &= g(f(x)) \\ &= g(1 - x^2) \\ &= 3(1 - x^2) - 2 \\ &= 3 - 3x^2 - 2 \\ &= 1 - 3x^2\end{aligned}$$

c
$$\begin{aligned}&f^2(\tfrac{1}{2}) \\ &= (f \circ f)(\tfrac{1}{2}) \\ &= f(f(\tfrac{1}{2})) \\ &= f(1 - (\tfrac{1}{2})^2) \\ &= f(\tfrac{3}{4}) \\ &= 1 - (\tfrac{3}{4})^2 \\ &= 1 - \tfrac{9}{16} \\ &= \tfrac{7}{16}\end{aligned}$$

Discussion

1 How is the domain of a composite function affected by the functions that compose it?
You may like to consider the composite function $f \circ g$ for:

a $f(x) = \sqrt{x}$ and $g(x) = x + 3$

b $f(x) = \frac{1}{x}$ and $g(x) = \sqrt{x}$

c $f(x) = \sqrt{-x}$ and $g(x) = |x|$

d $f(x) = x^2$ and $g(x) = \sqrt{x}$

2 What determines the *range* of a composite function?

EXERCISE 2H

1 Given $f : x \mapsto -2x$ and $g : x \mapsto 1 + x^2$, find in simplest form:

a $(f \circ g)(x)$ **b** $(g \circ f)(x)$ **c** $(f \circ g)(2)$ **d** $(f \circ f)(-1)$

2 Given $f(x) = 3 - x^2$ and $g(x) = 2x + 4$, find in simplest form:

a $fg(x)$ **b** $gf(x)$ **c** $g^2(\frac{1}{2})$ **d** $f^2(-\frac{1}{2})$

3 Suppose $f(x) = 9 - \sqrt{x}$ and $g(x) = x^2 + 4$.

a Find $fg(x)$ and state its domain and range.

b Find $gf(4)$.

c Find $f^2(x)$ and state its domain and range.

4 Suppose $f(x) = 1 - 2x$ and $g(x) = 3x + 5$.

a Find $f(g(x))$.

b Hence solve $(f \circ g)(x) = f(x + 3)$.

5 Suppose $f : x \mapsto 2x - x^2$ and $g : x \mapsto 1 + 3x$.

a Find in simplest form: **i** $(f \circ g)(x)$ **ii** $(g \circ f)(x)$

b Find the value(s) of x such that $(f \circ g)(x) = 3(g \circ f)(x)$.

6 For each pair of functions, find $fg(x)$ and state its domain and range:

a $f(x) = \frac{1}{x}$ and $g(x) = x - 3$

b $f(x) = \sqrt{x - 1}$ and $g(x) = |x|$

c $f(x) = -\frac{1}{x}$ and $g(x) = x^2 + 3x + 2$

7 Suppose $f(x)$ and $g(x)$ are functions. $f(x)$ has domain D_f and range R_f. $g(x)$ has domain D_g and range R_g.

a Suppose $f(x) = \sqrt{x}$ and $g(x) = -(x - 1)(x - 3)$.

i Find D_f, R_f, D_g, and R_g.

ii Find $(f \circ g)(x)$ and state its domain and range.

b In general, under what circumstance will $(f \circ g)(x)$ be defined?

8 If $f(x) = 1 - 4x^2$ and $fg(x) = -x^2 - 4x - 3$, find $g(x)$.
Hint: There are *two* possible answers.

I INVERSE FUNCTIONS

The operations of $+$ and $-$, $\times$ and $\div$, are **inverse operations** as one undoes what the other does.

For example, $x + 3 - 3 = x$ and $x \times 3 \div 3 = x$.

In a similar way, the function $y = 1 - 4x$ can be "undone" by its **inverse function** $y = \frac{1-x}{4}$.

We can think of this as two machines. If the machines are inverses then the second machine *undoes* what the first machine does. This will be true no matter what value of x enters the first machine.

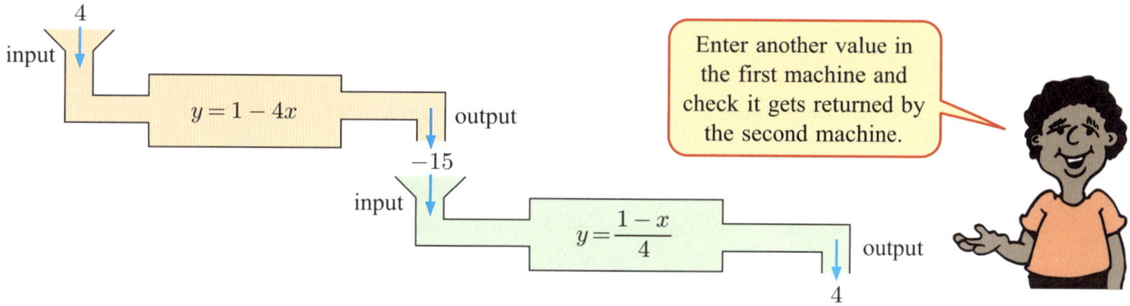

However, not all functions have an inverse function.

For example, consider the function $y = x^2$. The inputs -3 and 3 both produce an output of 9.

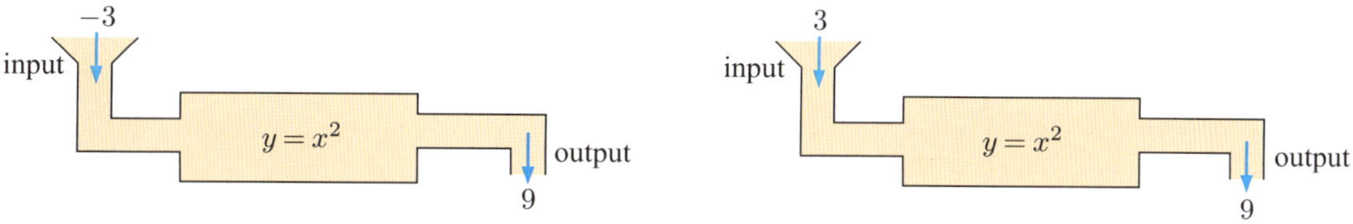

If an inverse function existed and we gave it the input 9, how would it know whether the output should be -3 or 3? It cannot answer both, since the inverse function would then fail the vertical line test.

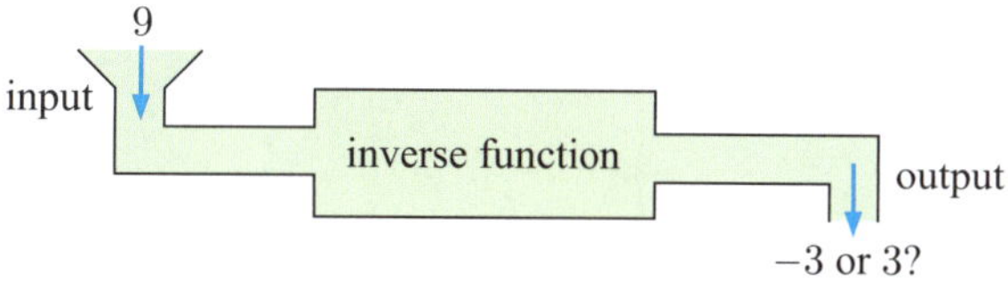

So, $y = x^2$ does not have an inverse function.

For a function to have an **inverse**, the function must be **one-one**. It must pass the horizontal line test.

If $y = f(x)$ has an **inverse function**, this new function:

- is denoted $f^{-1}(x)$
- is the reflection of $y = f(x)$ in the line $y = x$
- satisfies $(f \circ f^{-1})(x) = x$ and $(f^{-1} \circ f)(x) = x$.

The function $y = x$, defined as $f : x \mapsto x$, is the **identity function**.

Example 22

Self Tutor

If $y=f(x)$ has an inverse function, sketch $y=f^{-1}(x)$, and state the domain and range of $f(x)$ and $f^{-1}(x)$.

a

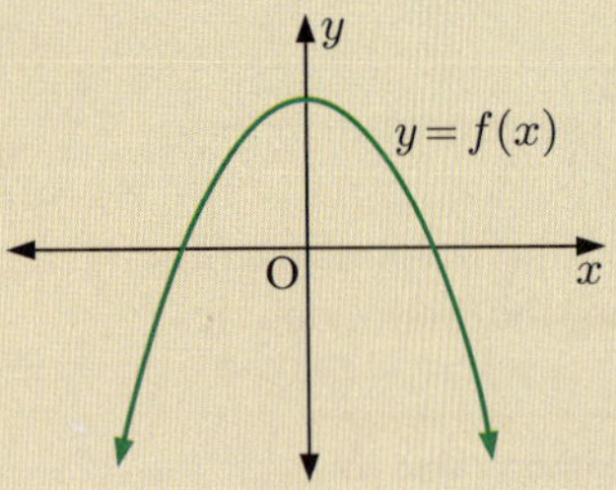

b

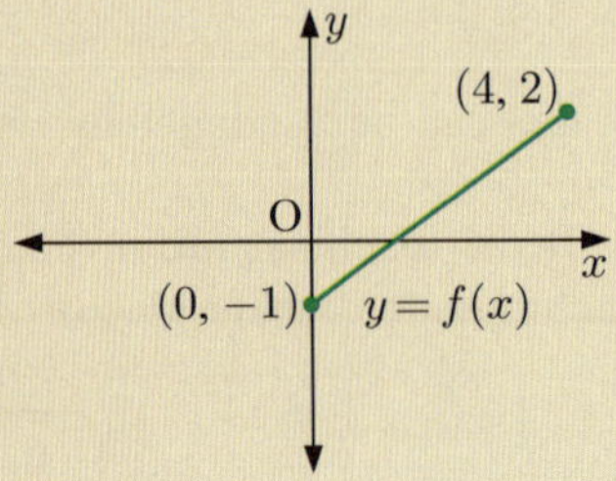

c

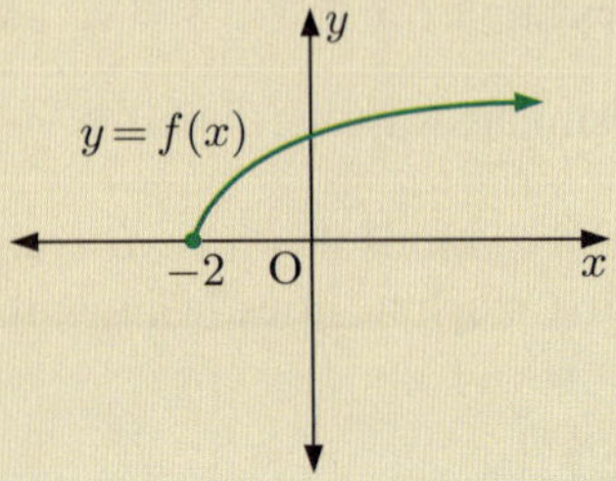

a The function fails the horizontal line test, so it is not one-one. The function does not have an inverse function.

b

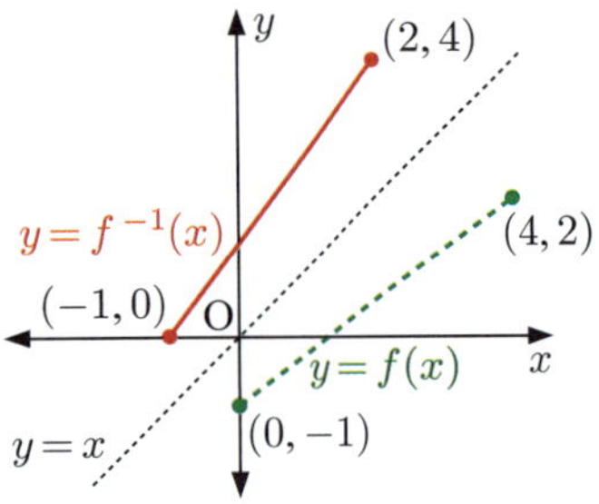

$f(x)$ has domain $\{x : 0 \leqslant x \leqslant 4\}$ and range $\{y : -1 \leqslant y \leqslant 2\}$.

$f^{-1}(x)$ has domain $\{x : -1 \leqslant x \leqslant 2\}$ and range $\{y : 0 \leqslant y \leqslant 4\}$.

c

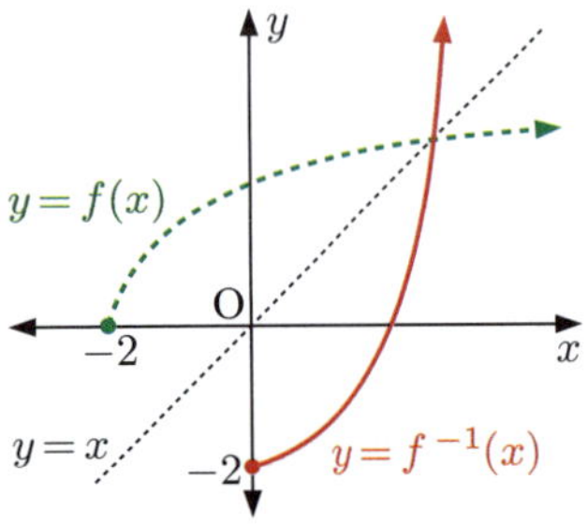

$f(x)$ has domain $\{x : x \geqslant -2\}$ and range $\{y : y \geqslant 0\}$.

$f^{-1}(x)$ has domain $\{x : x \geqslant 0\}$ and range $\{y : y \geqslant -2\}$.

$y=f^{-1}(x)$ is the reflection of $y=f(x)$ in the line $y=x$.

From **Example 22**, we can see that:

> The domain of f^{-1} is equal to the range of f.
>
> The range of f^{-1} is equal to the domain of f.

If (x, y) lies on f, then (y, x) lies on f^{-1}. Reflecting the function in the line $y=x$ has the algebraic effect of interchanging x and y. So, if the function is given as an equation, then we interchange the variables to find the equation of the inverse function.

For example, if f is given by $y=5x+2$ then f^{-1} is given by $x=5y+2$.

Example 23

Self Tutor

Consider the function $f(x) = 3x - 1$.

a On the same axes, graph f and its inverse function f^{-1}.

b Find $f^{-1}(x)$ using variable interchange.

c Check that $(f \circ f^{-1})(x) = (f^{-1} \circ f)(x) = x$.

a $f(x) = 3x - 1$ passes through $(0, -1)$ and $(1, 2)$.

$\therefore$ $f^{-1}(x)$ passes through $(-1, 0)$ and $(2, 1)$.

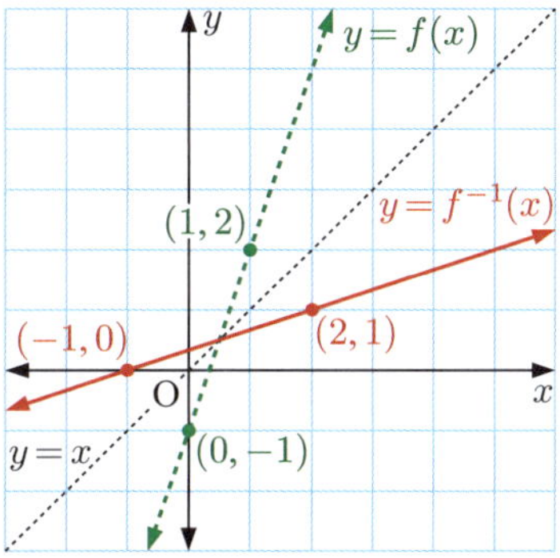

b

f is $y = 3x - 1$,

$\therefore$ f^{-1} is $x = 3y - 1$

$\therefore$ $x + 1 = 3y$

$\therefore$ $\dfrac{x+1}{3} = y$

$\therefore$ $f^{-1}(x) = \dfrac{x+1}{3}$

c

$(f \circ f^{-1})(x)$

$= f(f^{-1}(x))$

$= f\left(\dfrac{x+1}{3}\right)$

$= 3\left(\dfrac{x+1}{3}\right) - 1$

$= x$

and

$(f^{-1} \circ f)(x)$

$= f^{-1}(f(x))$

$= \dfrac{(3x-1)+1}{3}$

$= \dfrac{3x}{3}$

$= x$

SELF-INVERSE FUNCTIONS

Any function which has an inverse, and whose graph is symmetrical about the line $y = x$, is a **self-inverse function**.

If f is a self-inverse function then $f^{-1} = f$.

For example:

- The identity function $f(x) = x$ is a self-inverse function.
- The function $f(x) = \dfrac{1}{x}$, $x \neq 0$, is a self-inverse function, as $f = f^{-1}$.

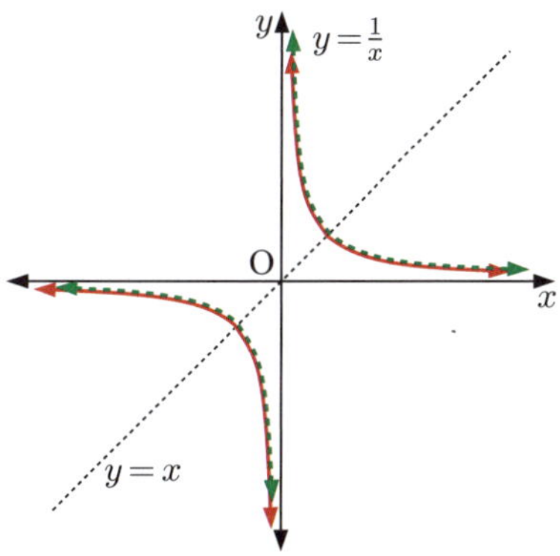

EXERCISE 2I

1 If $y = f(x)$ has an inverse function, sketch $y = f^{-1}(x)$, and state the domain and range of $f(x)$ and $f^{-1}(x)$.

PRINTABLE GRAPHS

a
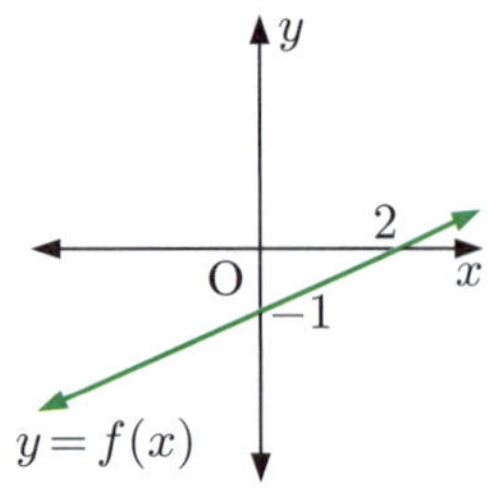

b
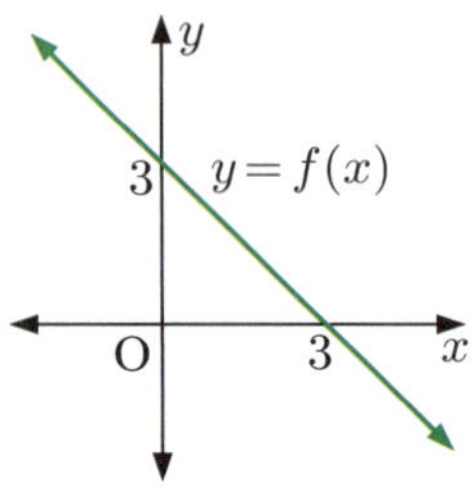

c
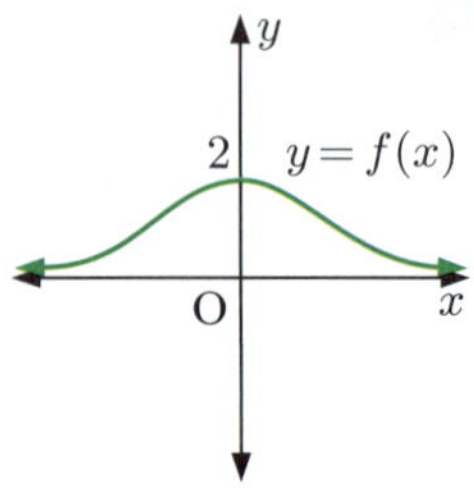

d
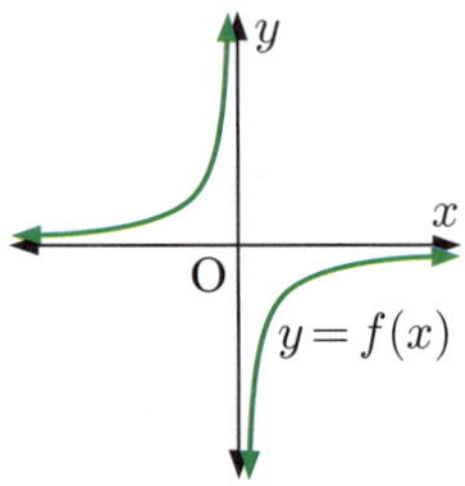

e
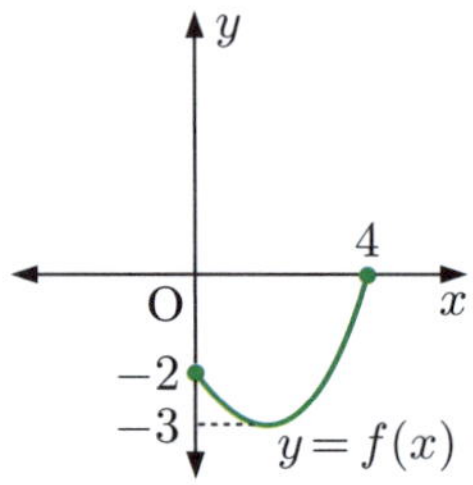

f
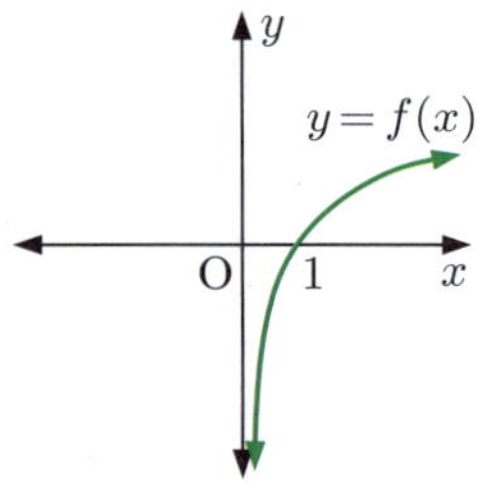

g
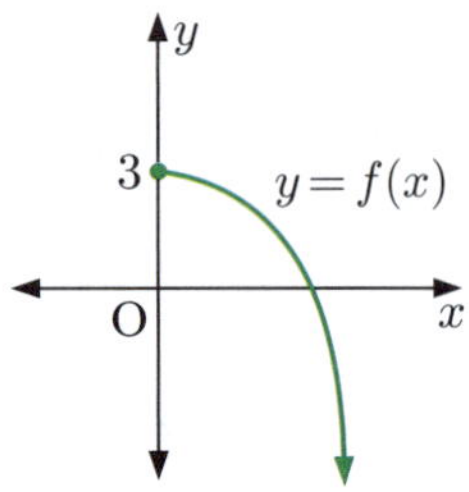

h
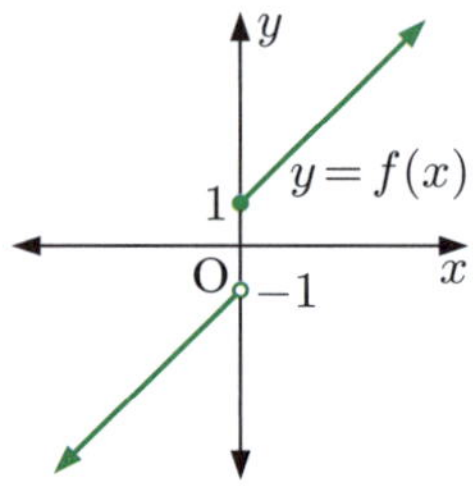

i
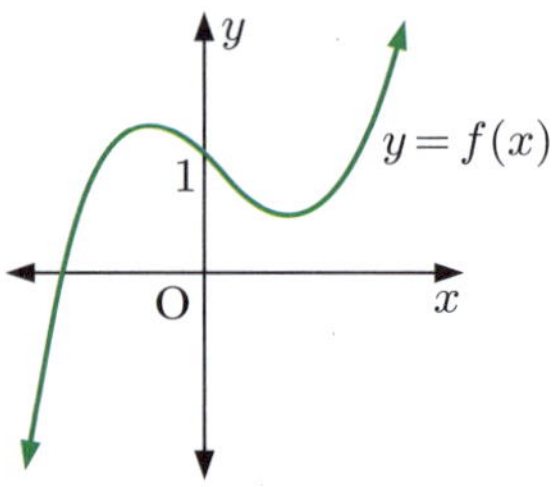

2 Which of the functions in **1** is a self-inverse function?

3 If the domain of $h(x)$ is $\{x : -1 \leqslant x \leqslant 4\}$, state the range of $h^{-1}(x)$.

4 Which of the following functions have inverses? In each of these cases, write down the inverse function.

a $\{(5, 3), (-4, -1), (1, 1)\}$ **b** $\{(-2, -3), (0, -2), (4, -3)\}$

c $\{(-5, -4), (-2, 2), (2, 5), (5, 2)\}$ **d** $\{(-4, -1), (0, 0), (3, 1), (5, 2)\}$

5 For each of the following functions f:

i Sketch $y = f(x)$, $y = f^{-1}(x)$, and $y = x$ on the same set of axes.

ii Find $f^{-1}(x)$ using variable interchange.

iii Check that $(f^{-1} \circ f)(x) = (f \circ f^{-1})(x) = x$.

a $f(x) = x - 2$ **b** $f(x) = 2x + 1$ **c** $f(x) = \frac{5 - 3x}{6}$ **d** $f(x) = \frac{1}{2}x - \frac{3}{4}$

6 For a function $f(x)$ with inverse $f^{-1}(x)$, a point $(a, f(a))$ on f is said to be **invariant** if $f(a) = f^{-1}(a)$.

a Explain why every point on a self-inverse function is invariant.

b What can be said about a point (a, a) which lies on $y = f(x)$?

7 **a** Given $f(x) = 4 - 3x$, find $(f^{-1})^{-1}(x)$.
b Explain your answer to **a** geometrically.

8 **a** Explain why the inverse of a one-one function must also be a one-one function.
b For $f(x) = 3x + 2$, verify that $f^{-1}(x)$ is a one-one function.

9 Given $f : x \mapsto \frac{m}{x}$, find the value(s) of m for which f is a self-inverse function.

10 Consider the functions $f : x \mapsto \frac{1}{2}x + 3$ and $g : x \mapsto 1 - 4x$.
a Find:
i $f(4)$ **ii** $g^{-1}(5)$ **iii** $g^{-1}f(4)$
b Solve the equation $fg^{-1}(x) = 3$.

11 Given $f : x \mapsto 3x - 1$ and $g : x \mapsto 1 - x$, show that $(f^{-1} \circ g^{-1})(x) = (g \circ f)^{-1}(x)$.

12 Verify that each function is self-inverse by:
i referring to its graph **ii** using algebra.

GRAPHING PACKAGE

a $f(x) = 3 - x$ **b** $f(x) = \frac{1}{2x}$ **c** $f(x) = \frac{1 - x}{1 + x}$

Discovery 2 — Functions and form

Click on the icon to obtain this Discovery.

FUNCTIONS AND FORM

Review set 2A

1 Draw a mapping diagram for each relation and discuss whether the relation is a function:
a $\{(4, 2), (-3, 0), (1, 5)\}$ **b** $\{(2, -6), (-5, -1), (6, -6), (-1, -1)\}$

2 Determine whether each relation is a function:

a
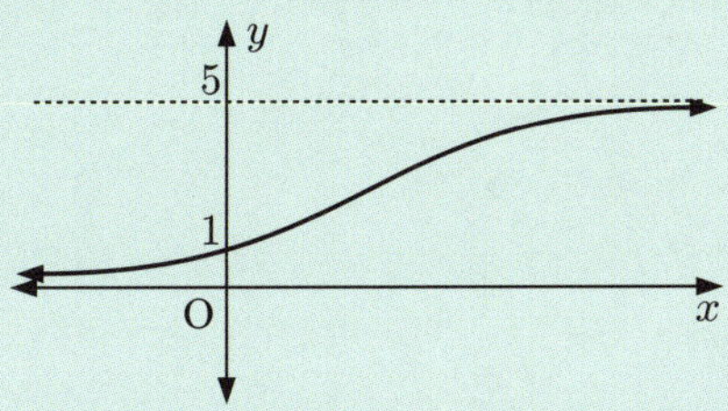

b
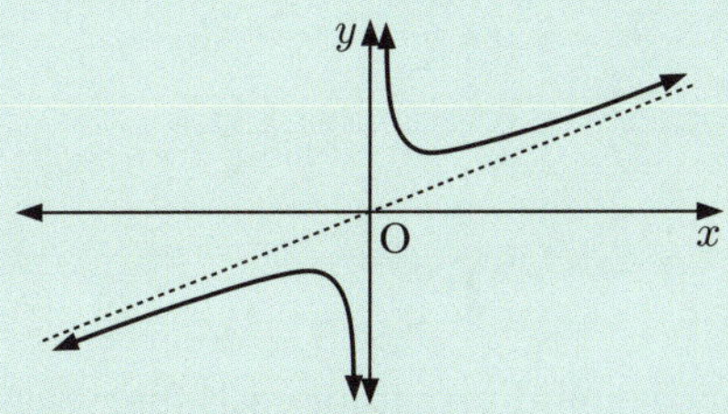

c
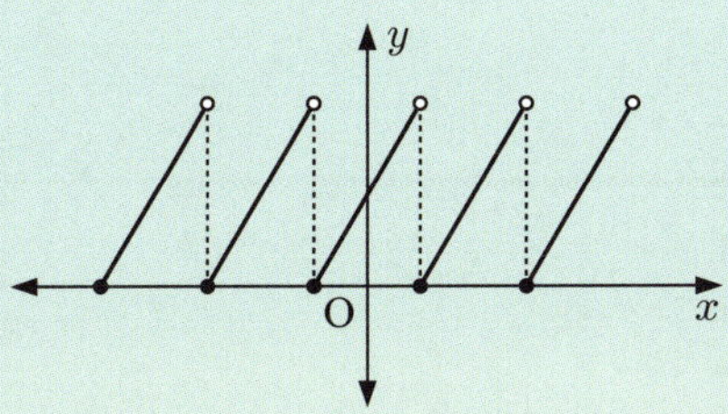

d
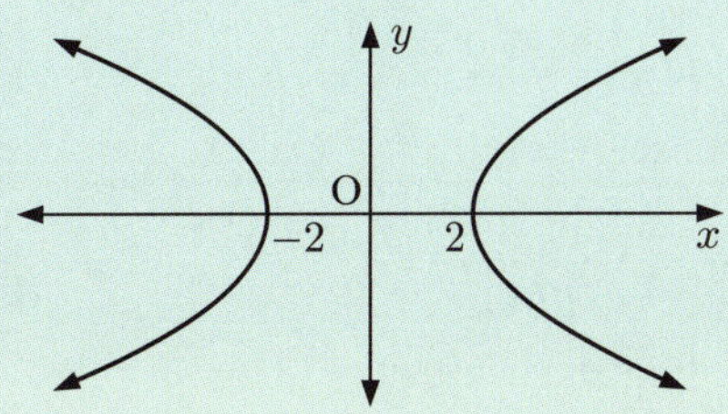

3 Given $f : x \mapsto 2x^2 + 1$, find in simplest form:

a $f(1)$ **b** $f(-3)$ **c** $f(-a)$

4 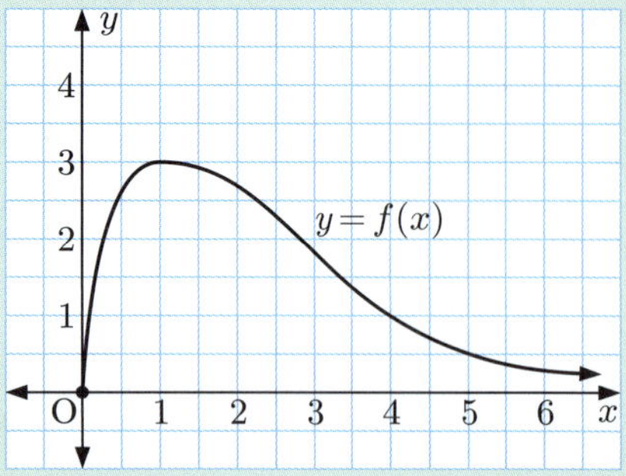

The graph of $y = f(x)$ is shown alongside. Find:

a $f(4)$

b x such that $f(x) = 3$.

5 For each of the following functions:

i find the domain and range **ii** determine whether the function is one-one.

a 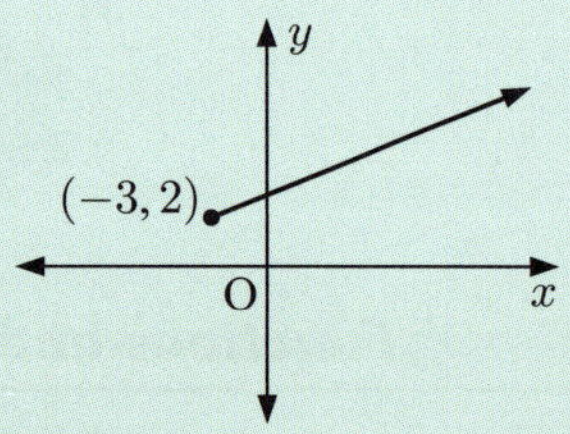

b 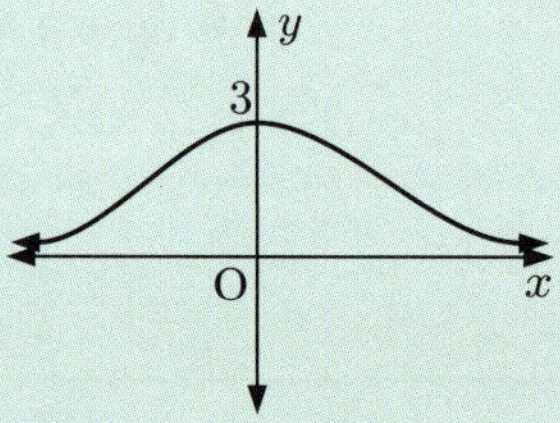

c 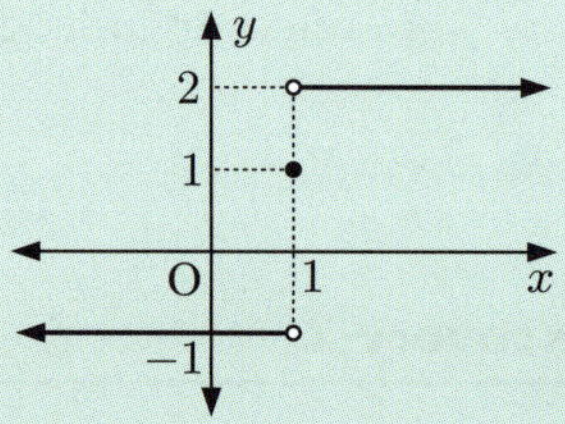

6 Draw the graph of $y = |2x - 1|$.

7 Solve for x:

a $|1 - 2x| = 11$ **b** $|5x - 1| = |9x - 13|$

8 Solve for x:

a $|3x - 2| < 5$ **b** $|2x - 5| \geqslant x + 2$

9 **a** Show algebraically that the inequality $|3x - 5| > 1 - x$ is true for all x.

b Illustrate this result graphically.

10 Draw a sign diagram for each graph:

a 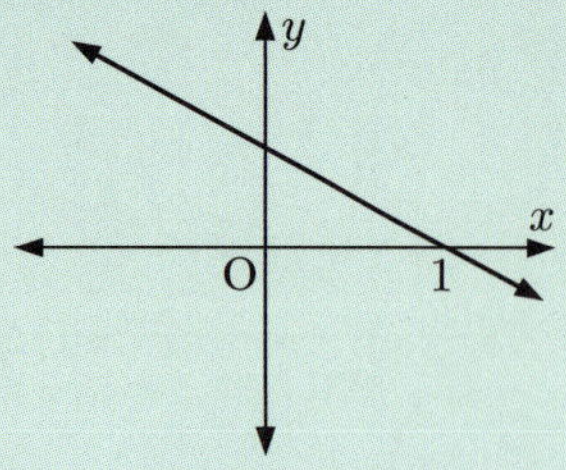

b 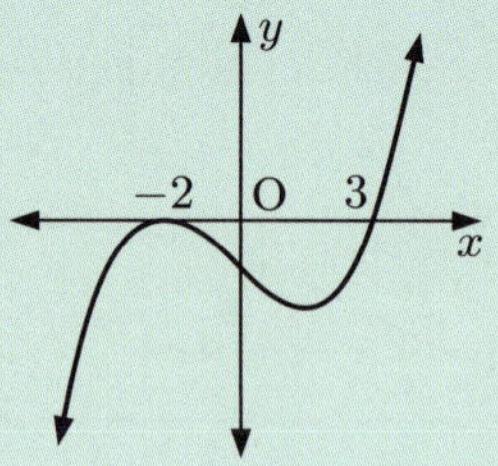

11 Draw a sign diagram for each function:

a $f(x) = x^2(x - 1)$ **b** $f(x) = \dfrac{2x + 1}{x - 4}$ **c** $f(x) = \dfrac{-(x + 2)}{(x - 3)^2}$

12 Suppose $f(x) = x^2 + 2x - 1$ and $g(x) = x - 3$.

a Find, in simplest form:

i $fg(x)$ **ii** $gf(x)$

b Find x such that $f(x) = g(x^2)$.

13 If $f(x) = 1 - 4x$ and $(f \circ g)(x) = x^2 - 7$, find $g(x)$.

14 For each of the following functions f:

i Sketch $y = f(x)$, $y = f^{-1}(x)$, and $y = x$ on the same set of axes.

ii Find $f^{-1}(x)$ using variable interchange.

a $f(x) = 5x + 10$ **b** $f(x) = \frac{4 - x}{3}$

15 Given $f : x \mapsto \frac{x}{x - a}$, $x \neq a$, for what value(s) of a is f a self-inverse function?

Review set 2B

1 Determine whether the following relations are functions. If they are functions, determine whether they are one-one.

a

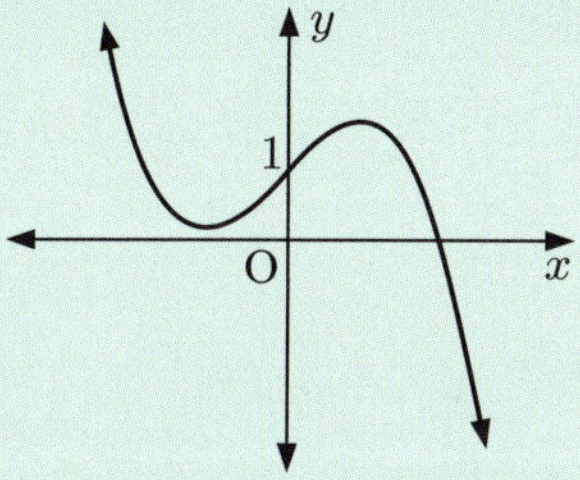

b

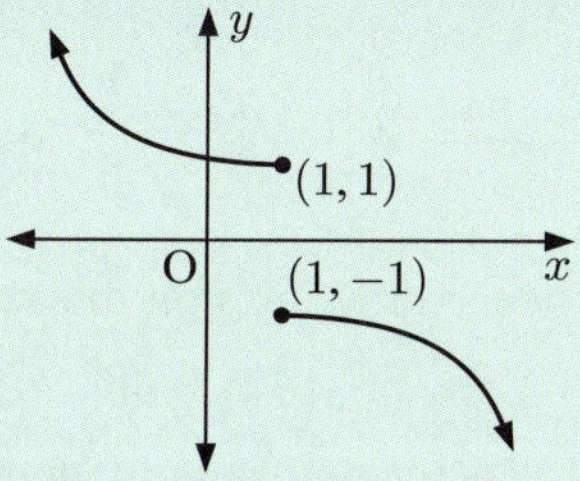

c

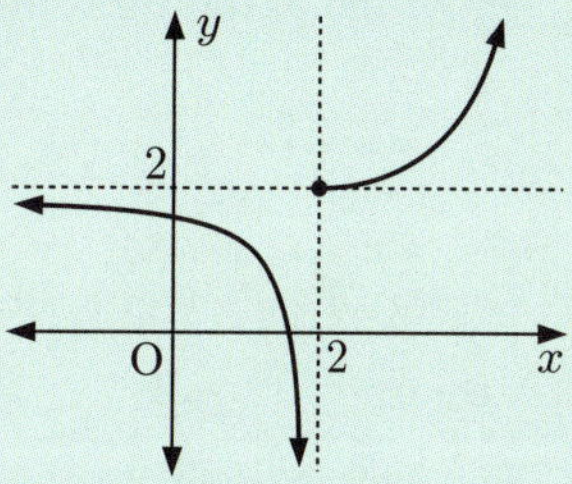

2 Give algebraic evidence to show that the relation $x^2 - y^2 = 1$ is not a function.

3 Let $f(x) = x^2 + 2$. Find, in simplest form:

a $f(x + 1)$ **b** $f(\sqrt{x} - 3)$ **c** $f(x + h)$

4 Find the domain and range for each function:

a $f(x) = \frac{2}{x + 4}$ **b** $y = \frac{-1}{\sqrt{3x - 1}}$

5 The depth of water in a tank t hours after it starts raining is given by $D(t) = \frac{0.474}{\pi}t + 1$ metres, $0 \leqslant t \leqslant 3$.

a Find $D(2)$ and state what $D(2)$ means.

b Find t when $D(t) = 1.2$ and explain what this means.

c Find the initial depth of the water.

6 If the range of $y = f(x)$ is $\{y : -4 \leqslant y \leqslant 1\}$, what is the range of $y = |f(x)|$?

7 Draw the graph of $y = |f(x)|$ for:

a

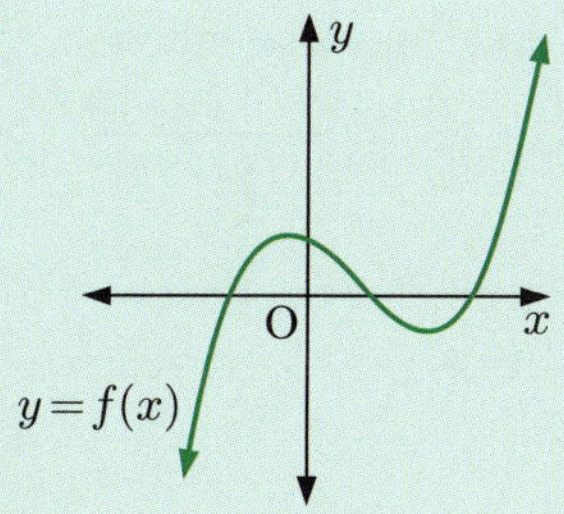

b

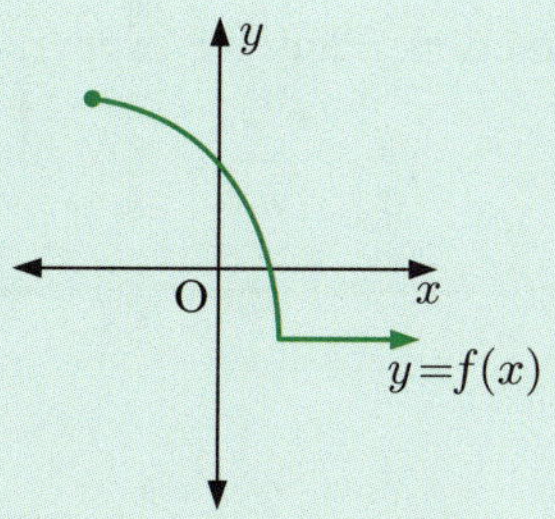

PRINTABLE GRAPHS

8 The graphs of $y = |x+4|$ and $y = |1-3x|$ are shown alongside.

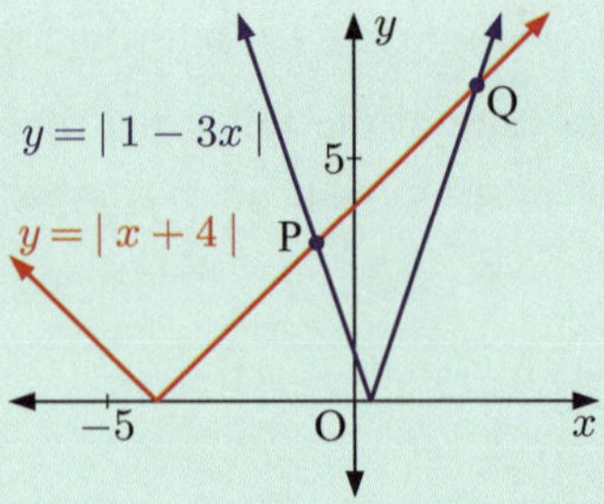

a Find the coordinates of P and Q.

b Hence solve:

i $|x+4| = |1-3x|$

ii $|x+4| < |1-3x|$

9 Solve for x:

a $\left|\frac{x}{2}+3\right| = 7$ **b** $|8-x| = 4x+1$ **c** $|2x-7| = |5x|$

10 Solve for x:

a $|3x-5| > 3$ **b** $|10-4x| \leqslant 7$ **c** $|x+4| < |4x-3|$

11 Draw the graph of a function which the sign diagram might correspond to:

a

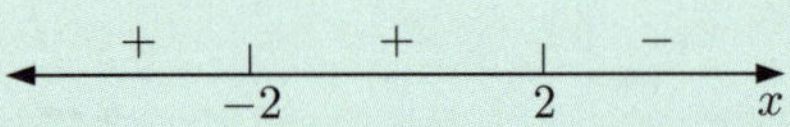

b

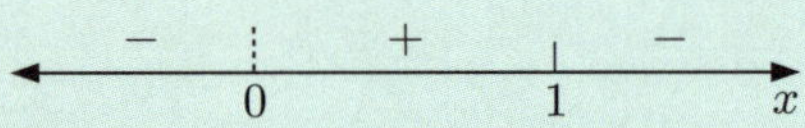

12 Factorise each expression, and hence draw its sign diagram:

a x^2-2x-3 **b** $x^2+12x+36$ **c** x^4-x^2

13 For each pair of functions, find $fg(x)$ and state its domain and range:

a $f(x) = \frac{3}{x}$ and $g(x) = |x+1|$ **b** $f(x) = \sqrt{x^2-4}$ and $g(x) = x-2$

14 If $y = f(x)$ has an inverse, sketch the graph of $y = f^{-1}(x)$. If it does not have an inverse, explain why.

a

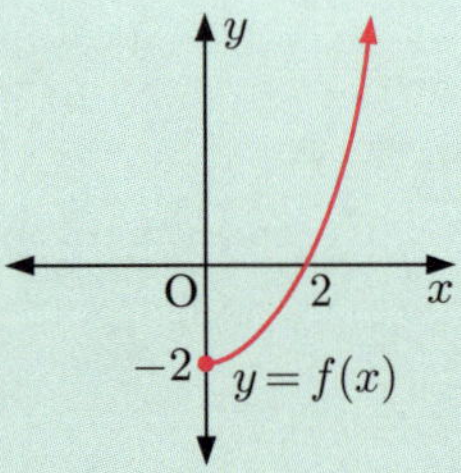

b

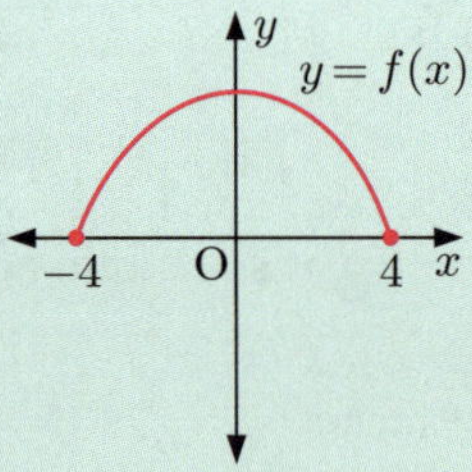

c

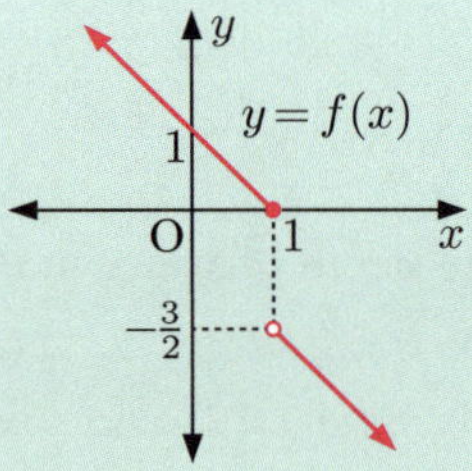

15 Suppose a function $f(x)$ has inverse function $f^{-1}(x)$.

Show that the inverse of $g(x) = k\,f(x)$ is $g^{-1}(x) = f^{-1}\left(\frac{x}{k}\right)$, provided $k \neq 0$.

Hint: Show that $(g \circ g^{-1})(x) = (g^{-1} \circ g)(x) = x$.

16 **a** Show that if a function f has the property $f^2(x) = x$, f is a self-inverse function.

b Hence show that $f : x \mapsto \frac{x}{|x|^2}$, $x \neq 0$ is a self-inverse function.

Hint: If $a, b \in \mathbb{R}$, $b \neq 0$ then $\left|\frac{a}{b}\right| = \frac{|a|}{|b|}$.

3

Quadratics

Contents:

Opening problem

In a physics experiment, a ball was filmed being thrown into the air in front of a backdrop with evenly spaced horizontal lines.

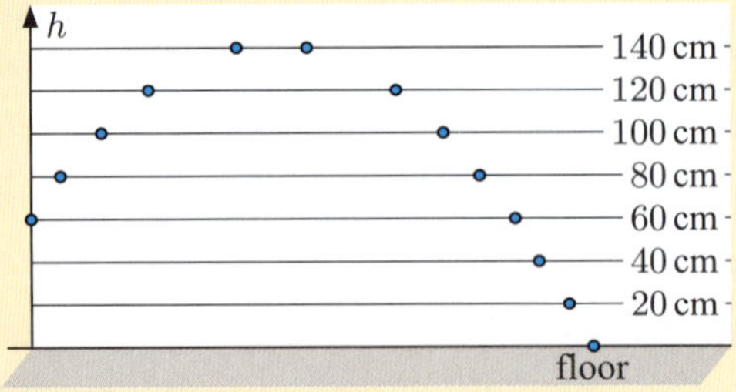

The times at which the ball was in front of a line were recorded:

Height (h cm)	60	80	100	120	140	140	120	100	80	60	40	20	0
Time (t seconds)	0	0.05	0.12	0.20	0.35	0.47	0.62	0.70	0.76	0.82	0.86	0.91	0.95

Things to think about:

- **a** Use technology to plot these points.
- **b** How long did it take for the ball to hit the floor?
- **c** What is the shape of the graph of h against t?
- **d** What formula gives the height of the ball t seconds after it has been thrown?
- **e** How can we use this formula to find the maximum height reached by the ball?

A QUADRATIC EQUATIONS

A **quadratic equation** is an equation of the form $ax^2 + bx + c = 0$ where a, b, and c are constants, $a \neq 0$.

The **roots** or **solutions** of $ax^2 + bx + c = 0$ are the values of x which satisfy the equation.

A quadratic equation may have *two*, *one*, or *zero* real solutions.

For example: $x^2 + 5x + 6 = 0$ is a quadratic equation.

If $x = -2$,
$$\begin{aligned} & x^2 + 5x + 6 \\ &= (-2)^2 + 5 \times (-2) + 6 \\ &= 4 - 10 + 6 \\ &= 0 \end{aligned}$$

If $x = -3$,
$$\begin{aligned} & x^2 + 5x + 6 \\ &= (-3)^2 + 5 \times (-3) + 6 \\ &= 9 - 15 + 6 \\ &= 0 \end{aligned}$$

Since $x = -2$ and $x = -3$ both satisfy $x^2 + 5x + 6 = 0$, they are both solutions of the equation.

METHODS FOR SOLVING QUADRATIC EQUATIONS

To solve quadratic equations we have the following methods to choose from:

- rewrite the quadratic in **factored form** and use the **Null Factor law**:

  If $ab = 0$ then $a = 0$ or $b = 0$.

- rewrite the quadratic in **completed square form** $a(x - h)^2 + k$, then rearrange $a(x - h)^2 + k = 0$ to isolate x
- use the **quadratic formula**.

SOLVING BY FACTORISATION

Step 1: If necessary, rearrange the equation so one side is zero.
Step 2: Fully factorise the other side.
Step 3: Use the Null Factor law: If $ab = 0$ then $a = 0$ or $b = 0$.
Step 4: Solve the resulting linear equations.

Example 1 Self Tutor

Solve for x:
a $3x^2 - 4x = 0$ **b** $x^2 + x = 12$

a
$$3x^2 - 4x = 0$$
$$\therefore x(3x - 4) = 0$$
$$\therefore x = 0 \text{ or } 3x - 4 = 0$$
$$\therefore x = 0 \text{ or } x = \tfrac{4}{3}$$

b
$$x^2 + x = 12$$
$$\therefore x^2 + x - 12 = 0$$
$$\therefore (x - 3)(x + 4) = 0$$
$$\therefore x = 3 \text{ or } x = -4$$

Example 2 Self Tutor

Solve for x:
a $9x^2 + 4 = 12x$ **b** $13x - 3 = 10x^2$

a
$$9x^2 + 4 = 12x$$
$$\therefore 9x^2 - 12x + 4 = 0$$
$$\therefore (3x - 2)^2 = 0$$
$$\therefore x = \tfrac{2}{3}$$

b
$$13x - 3 = 10x^2$$
$$\therefore 10x^2 - 13x + 3 = 0$$
$$\therefore (10x - 3)(x - 1) = 0$$
$$\therefore x = \tfrac{3}{10} \text{ or } x = 1$$

Caution:

- Do not be tempted to divide both sides by an expression involving x. If you do this then you may lose one of the solutions.
  For example, consider $x^2 = -2x$.

  Correct solution
  $$x^2 = -2x$$
  $$\therefore x^2 + 2x = 0$$
  $$\therefore x(x + 2) = 0$$
  $$\therefore x = 0 \text{ or } -2$$

  Incorrect solution
  $$x^2 = -2x$$
  $$\therefore \frac{x^2}{x} = \frac{-2x}{x}$$
  $$\therefore x = -2$$

- Be careful when taking square roots of both sides of an equation. If you do this then you may lose one of the solutions.
  For example, consider $x^2 = 16$.

  Correct solution
  $$x^2 = 16$$
  $$\therefore x = \pm\sqrt{16}$$
  $$\therefore x = \pm 4$$

  Incorrect solution
  $$x^2 = 16$$
  $$\therefore x = \sqrt{16}$$
  $$\therefore x = 4$$

EXERCISE 3A.1

1 Solve by factorisation:

a $3x^2 + 2x = 0$ **b** $4x^2 + x = 0$ **c** $6x^2 - 2x = 0$
d $14x^2 - 7x = 0$ **e** $x = 5x^2$ **f** $2x^2 = 3x$

2 Solve by factorisation:

a $x^2 + 2x - 15 = 0$ **b** $x^2 = 4x - 3$ **c** $x^2 + 2x - 8 = 0$
d $x + 12 = x^2$ **e** $x^2 - 13x = -42$ **f** $x^2 - 63 = 2x$

3 Solve by factorisation:

a $3x^2 - 11x - 4 = 0$ **b** $4x^2 + 4x + 1 = 0$ **c** $2x^2 + 3 = 5x$
d $1 = 4x^2 + 3x$ **e** $6x^2 = 7x + 3$ **f** $3x^2 - 10x - 8 = 0$
g $14x + 5 = -8x^2$ **h** $4x^2 + 12x + 5 = 0$ **i** $-9x^2 + 16x = 7$
j $25x + 9 = 6x^2$ **k** $5x^2 = -2x + 7$ **l** $-15x^2 + 4x = -3$

Example 3 — Self Tutor

Solve for x: $3x = 4 - \frac{1}{x}$

$$3x = 4 - \frac{1}{x}$$
$$\therefore\ 3x^2 = x\left(4 - \frac{1}{x}\right) \quad \{\text{multiplying both sides by } x\}$$
$$\therefore\ 3x^2 = 4x - 1 \quad \{\text{expanding brackets}\}$$
$$\therefore\ 3x^2 - 4x + 1 = 0 \quad \{\text{making the RHS} = 0\}$$
$$\therefore\ (3x - 1)(x - 1) = 0 \quad \{\text{factorising}\}$$
$$\therefore\ x = \tfrac{1}{3} \text{ or } 1$$

4 Solve for x:

a $(x-3)^2 - 4 = x - 5$ **b** $(x+3)(x+2) = 6$ **c** $6(x-1)^2 = 7(x+1) - 9$
d $7x - \frac{1}{x} = -6$ **e** $2x - 13 = \frac{7}{x}$ **f** $\frac{3-x}{x+1} = \frac{1}{x}$

Example 4 — Self Tutor

Solve for x: $x^4 + 5 = 6x^2$

$$x^4 + 5 = 6x^2$$
$$\therefore\ x^4 - 6x^2 + 5 = 0$$
$$\therefore\ (x^2 - 1)(x^2 - 5) = 0 \quad \{\text{compare with } a^2 - 6a + 5 = (a-1)(a-5)\}$$
$$\therefore\ x^2 = 1 \text{ or } x^2 = 5$$
$$\therefore\ x = \pm 1 \text{ or } \pm\sqrt{5}$$

5 Solve for x:

a $x^4 - 11x^2 + 28 = 0$ **b** $x^4 + 20 = 12x^2$ **c** $x^4 = x^2 + 30$

d $2x^4 + 3 = 7x^2$ **e** $4x^4 + x^2 = 3$ **f** $13x^2 + 5 = -6x^4$

g $x^6 - 7x^3 = 8$ **h** $x = 8\sqrt{x} - 15$ **i** $2\sqrt{x} = 8 - 3x$

6 Solve $\dfrac{3}{x^2} - \dfrac{7}{x} + 2 = 0$ by:

a treating it as a quadratic equation in the variable $\dfrac{1}{x}$

b first multiplying both sides by x^2.

7 To solve the equation $x - 6 = \sqrt{x}$, Dwayne performed these steps:

$$\begin{aligned} x^2 - 12x + 36 &= x \qquad \{\text{squaring both sides}\} \\ \therefore\ x^2 - 13x + 36 &= 0 \\ \therefore\ (x-4)(x-9) &= 0 \\ \therefore\ x &= 4 \text{ or } 9 \end{aligned}$$

Discuss the validity of Dwayne's solutions.

SOLVING BY "COMPLETING THE SQUARE"

As you would be aware by now, not all quadratics factorise easily.

For example, $x^2 + 6x + 3$ cannot be factorised by simple factorisation. We cannot write $x^2 + 6x + 3$ in the form $(x - p)(x - q)$ where p and q are rational.

An alternative way to solve equations like $x^2 + 6x + 3 = 0$ is by "completing the square".

In this process we write the quadratic in the **completed square form** $a(x - h)^2 + k$, from which we find the solutions.

Start with the quadratic equation in the form $ax^2 + bx + c = 0$.

Step 1: If $a \neq 1$, divide both sides by a.

Step 2: Rearrange the equation so that only the constant coefficient is on the RHS.

Step 3: Add to both sides $\left(\dfrac{\text{coefficient of } x}{2}\right)^2$.

Step 4: Factorise the LHS.

Step 5: Use the rule: If $X^2 = a$ then $X = \pm\sqrt{a}$.

Example 5 — Self Tutor

Solve exactly for x:

a $(x - 4)^2 = 3$ **b** $(x + 2)^2 = -1$

a $(x - 4)^2 = 3$

$\therefore\ x - 4 = \pm\sqrt{3}$

$\therefore\ x = 4 \pm \sqrt{3}$

b $(x + 2)^2 = -1$

has no real solutions since the square $(x + 2)^2$ cannot be negative.

Example 6

Self Tutor

Solve exactly for x: $x^2 - 2x - 4 = 0$

$$\begin{aligned} x^2 - 2x - 4 &= 0 \\ \therefore\ x^2 - 2x &= 4 && \{\text{writing the constant on the RHS}\} \\ \therefore\ x^2 - 2x + (-1)^2 &= 4 + (-1)^2 && \{\text{completing the square}\} \\ \therefore\ (x-1)^2 &= 5 && \{\text{factorising the LHS}\} \\ \therefore\ x - 1 &= \pm\sqrt{5} \\ \therefore\ x &= 1 \pm \sqrt{5} \end{aligned}$$

EXERCISE 3A.2

1 Solve exactly for x:

a $(x-2)^2 = 1$ **b** $(x+1)^2 = 5$ **c** $(x-5)^2 = -3$

d $(x+6)^2 = 2$ **e** $2(x-4)^2 = 12$ **f** $-4(x-1)^2 = -28$

g $5 + (x-2)^2 = 3$ **h** $(3x+1)^2 = 8$ **i** $(1-2x)^2 - 10 = 0$

2 Solve exactly by completing the square:

a $x^2 + 2x = 4$ **b** $x^2 - 4x + 2 = 0$ **c** $6x = x^2 - 1$

d $x^2 + 5x + 2 = 0$ **e** $x^2 + 2x + 3 = 0$ **f** $x^2 + 8x - 3 = 0$

g $x^2 = 9 - 3x$ **h** $x^2 + 6 = 4x$ **i** $x^2 - 7x = -13$

Example 7

Self Tutor

Solve exactly for x: $-2x^2 + 16x - 7 = 0$

$$\begin{aligned} -2x^2 + 16x - 7 &= 0 \\ \therefore\ x^2 - 8x + \tfrac{7}{2} &= 0 && \{\text{dividing both sides by } -2\} \\ \therefore\ x^2 - 8x &= -\tfrac{7}{2} && \{\text{putting the constant on the RHS}\} \\ \therefore\ x^2 - 8x + 4^2 &= -\tfrac{7}{2} + 4^2 && \{\text{completing the square}\} \\ \therefore\ (x-4)^2 &= \tfrac{25}{2} && \{\text{factorising the LHS}\} \\ \therefore\ x - 4 &= \pm\sqrt{\tfrac{25}{2}} \\ \therefore\ x &= 4 \pm \tfrac{5}{\sqrt{2}} \end{aligned}$$

3 Solve exactly by completing the square:

a $2x^2 + 6x + 1 = 0$ **b** $4x - 4x^2 = -5$ **c** $2x^2 - 12x + 3 = 0$

d $3x^2 - 5x + 3 = 0$ **e** $3x^2 + 9x = -2$ **f** $-5x^2 + 18x - 3 = 0$

4 Solve for x:

a $\frac{1}{x} - 3 = -2x$ **b** $x = 4 + \frac{1}{2x}$ **c** $\frac{1}{x^2} - 3 = -\frac{2}{x}$

5 Suppose $ax^2 + bx + c = 0$ where a, b, and c are constants, $a \neq 0$. Solve for x by completing the square.

Historical note — Babylonian algebra

The mathematics used by the **Babylonians** was recorded on clay tablets in cuneiform. One such tablet which has been preserved is called *Plimpton 322*, written around 1600 BC.

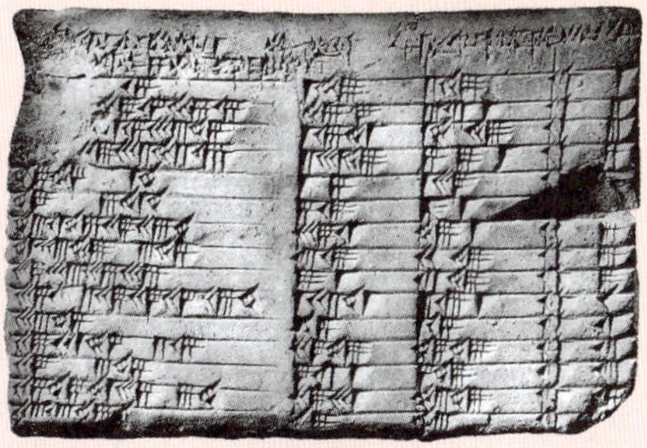

Plimpton 322

The ancient Babylonians were able to solve difficult equations using the rules we use today, such as transposing terms, multiplying both sides by like quantities to remove fractions, and factorisation.

For example, they could add $4xy$ to $(x-y)^2$ to obtain $(x+y)^2$.

However, the Babylonians did all these things without the use of letters for unknown quantities. Instead, they often used words for the unknown.

Consider the following example from about 4000 years ago:

Problem: *"I have subtracted the side of my square from the area and the result is 870. What is the side of the square?"*

Solution: Take half of 1, which is $\frac{1}{2}$, and multiply $\frac{1}{2}$ by $\frac{1}{2}$ which is $\frac{1}{4}$. Add this to 870 to get $870\frac{1}{4}$. This is the square of $29\frac{1}{2}$. Now add $\frac{1}{2}$ to $29\frac{1}{2}$ and the result is 30, the side of the square.

Using our modern symbols, the equation is $x^2 - x = 870$, and one of the solutions is $x = \sqrt{(\frac{1}{2})^2 + 870} + \frac{1}{2} = 30$.

THE QUADRATIC FORMULA

In many cases, a quadratic will not factorise easily. Since "completing the square" can be tedious, we can instead use the **quadratic formula**.

$$\text{If } ax^2 + bx + c = 0,\ a \neq 0, \text{ then } x = \frac{-b \pm \sqrt{b^2-4ac}}{2a}.$$

Proof:

If $ax^2 + bx + c = 0,\ a \neq 0$

then $x^2 + \frac{b}{a}x + \frac{c}{a} = 0$ {dividing each term by a, as $a \neq 0$}

$\therefore\ x^2 + \frac{b}{a}x = -\frac{c}{a}$

$\therefore\ x^2 + \frac{b}{a}x + \left(\frac{b}{2a}\right)^2 = -\frac{c}{a} + \left(\frac{b}{2a}\right)^2$ {completing the square on LHS}

$\therefore\ \left(x + \frac{b}{2a}\right)^2 = \frac{b^2-4ac}{4a^2}$ {factorising}

$\therefore\ x + \frac{b}{2a} = \pm\sqrt{\frac{b^2-4ac}{4a^2}}$

$\therefore\ x = \frac{-b \pm \sqrt{b^2-4ac}}{2a}$

Example 8

Self Tutor

Solve exactly for x:

a $x^2 - 4x + 2 = 0$

b $9x^2 - 6x - 4 = 0$

a $x^2 - 4x + 2 = 0$ has $a = 1,\ b = -4,\ c = 2$

$$\therefore\ x = \frac{-(-4) \pm \sqrt{(-4)^2 - 4(1)(2)}}{2(1)}$$

$$\therefore\ x = \frac{4 \pm \sqrt{16 - 8}}{2}$$

$$\therefore\ x = \frac{4 \pm \sqrt{8}}{2}$$

$$\therefore\ x = \frac{4 \pm 2\sqrt{2}}{2}$$

$$\therefore\ x = 2 \pm \sqrt{2}$$

b $9x^2 - 6x - 4 = 0$ has $a = 9,\ b = -6,\ c = -4$

$$\therefore\ x = \frac{-(-6) \pm \sqrt{(-6)^2 - 4(9)(-4)}}{2(9)}$$

$$\therefore\ x = \frac{6 \pm \sqrt{36 + 144}}{18}$$

$$\therefore\ x = \frac{6 \pm \sqrt{180}}{18}$$

$$\therefore\ x = \frac{6 \pm 6\sqrt{5}}{18}$$

$$\therefore\ x = \frac{1 \pm \sqrt{5}}{3}$$

EXERCISE 3A.3

1 Use the quadratic formula to solve exactly for x:

a $x^2 - 2x - 23 = 0$ **b** $x^2 - 4x - 24 = 0$ **c** $x^2 + 4x - 7 = 0$

d $x^2 - x - 11 = 0$ **e** $x^2 + 5x - 7 = 0$ **f** $-x^2 - x - 1 = 0$

g $-4x^2 + 3x + 2 = 0$ **h** $3x^2 - 5x + 1 = 0$ **i** $4x^2 + 3x - 9 = 0$

2 Rearrange the following equations so they are written in the form $ax^2 + bx + c = 0$, then use the quadratic formula to solve exactly for x.

a $(x-3)(x+5) = 4x - 4$ **b** $(x-2)^2 = -x + 3$ **c** $2(x+1)^2 = -3x$

d $(2x+3)(2x-5) = 4x$ **e** $-(3x+1)(x+1) = 24$ **f** $(2x-1)^2 = -3(x+2) + 7x$

g $x + \frac{1}{x} = 3$ **h** $\frac{2}{x} = \frac{2}{x^2} - 1$ **i** $1 - x = \frac{x-3}{x-2}$

B QUADRATIC INEQUALITIES

$x^2 \leqslant 1$, $x^2 + 2x - 4 < 0$, and $x^2 + 7x > 18$ are examples of quadratic inequalities.

While quadratic equations have 0, 1, or 2 solutions, quadratic inequalities may have 0, 1, or infinitely many solutions.

To solve quadratic inequalities we follow these steps:

Step 1: Make the RHS zero by shifting all terms to the LHS.
Step 2: Fully factorise the LHS.
Step 3: Draw a sign diagram for the LHS.
Step 4: Determine the set of solutions from the sign diagram.

Example 9 **Self Tutor**

Solve for x: **a** $x^2 + 10x \leqslant -9$ **b** $4x^2 > 12x - 9$

a $x^2 + 10x \leqslant -9$

$\therefore\ x^2 + 10x + 9 \leqslant 0$ {making RHS zero}

$\therefore\ (x+1)(x+9) \leqslant 0$ {factorising LHS}

Sign diagram of LHS is

$\therefore$ the solutions are $\{x : -9 \leqslant x \leqslant -1\}$.

b $4x^2 > 12x - 9$

$\therefore\ 4x^2 - 12x + 9 > 0$ {making RHS zero}

$\therefore\ (2x-3)^2 > 0$ {factorising LHS}

Sign diagram of LHS is

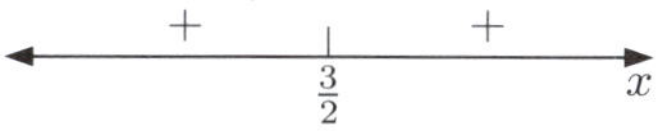

$\therefore$ the solutions are $\{x : x \neq \frac{3}{2}\}$.

EXERCISE 3B

1 Solve for x:

a $(x+1)(x-1) \leqslant 0$ **b** $(x-2)(x+4) > 0$ **c** $(2x+1)^2 \geqslant 0$

d $x^2 + x > 0$ **e** $x^2 \leqslant 2x$ **f** $x - 3x^2 < 0$

2 Solve for x:

a $x^2 - 9 < 0$ **b** $3x^2 - 12 \geqslant 0$ **c** $x^2 - 6x + 9 \leqslant 0$

d $x^2 - 3x - 10 > 0$ **e** $-x^2 + 4x + 21 \leqslant 0$ **f** $2x^2 + x - 6 \geqslant 0$

g $x^2 + 2x < 8$ **h** $3x^2 - 1 > 2x$ **i** $x(x+6) < 7$

j $4x^2 > 15 - 4x$ **k** $5x^2 > 4(1-2x)$ **l** $(x-3)^2 < x + 6$

3 In $3x^2 + 12 \square 12x$, replace $\square$ with $>$, $\geqslant$, $<$, or $\leqslant$ so that the resulting inequality has:

a no solutions **b** one solution **c** infinitely many solutions.

C THE DISCRIMINANT OF A QUADRATIC

The **discriminant** of the quadratic equation $ax^2 + bx + c = 0$ is $\Delta = b^2 - 4ac$.

The quadratic formula can therefore be written as $x = \dfrac{-b \pm \sqrt{\Delta}}{2a}$.

We can use the discriminant to determine the nature of the roots of a quadratic equation.

- If $\Delta = 0$, $x = -\dfrac{b}{2a}$ is the **only solution**, which we call a **repeated** or **double root**.
- If $\Delta > 0$, there are **two distinct real roots** $x = \dfrac{-b+\sqrt{\Delta}}{2a}$ and $x = \dfrac{-b-\sqrt{\Delta}}{2a}$.
  Additionally, if a, b, and c are rational and Δ is a **square number**, the roots are rational and can be found by factorisation.
- If $\Delta < 0$, $\sqrt{\Delta}$ is not a real number and therefore there are **no real roots**.

Example 10

Self Tutor

Use the discriminant to determine the nature of the roots of:

a $x^2 + 2x + 3 = 0$ **b** $5x^2 - x - 2 = 0$

a $x^2 + 2x + 3 = 0$ has $a = 1$, $b = 2$, $c = 3$

$\therefore \Delta = b^2 - 4ac$
$= (2)^2 - 4(1)(3)$
$= -8$

Since $\Delta < 0$, there are no real roots.

b $5x^2 - x - 2 = 0$ has $a = 5$, $b = -1$, $c = -2$

$\therefore \Delta = b^2 - 4ac$
$= (-1)^2 - 4(5)(-2)$
$= 41$

Since $\Delta > 0$, there are two distinct real roots. However, 41 is not a square number, so the roots are irrational.

EXERCISE 3C

1 Consider the quadratic equation $x^2 - 2x - 15 = 0$.

a Find the discriminant.

b Hence state the nature of the roots of the equation.

c Check your answer to **b** by solving the equation.

2 Consider the quadratic equation $4x^2 - 20x + 25 = 0$.

a Find the discriminant.

b Hence state the nature of the roots of the equation.

c Check your answer to **b** by solving the equation.

3 Use the discriminant to determine the nature of the roots of:

a $x^2 + 5x - 14 = 0$ **b** $x^2 - 7x + 2 = 0$ **c** $x^2 + 4x + 4 = 0$

d $4x^2 + 3x - 20 = 0$ **e** $2x^2 + 2x - 9 = 0$ **f** $x^2 + 2x + 2 = 0$

g $9x^2 - 12x + 16 = 0$ **h** $6x^2 - 5x + 10 = 0$ **i** $8x^2 - 2x - 3 = 0$

Example 11

Self Tutor

Consider the quadratic equation $x^2 + kx = k$.

a Find the discriminant Δ of the equation.

b Draw a sign diagram for Δ.

c Hence find the value(s) of k for which the equation has:

i a repeated root **ii** two distinct real roots **iii** no real roots.

a $\quad x^2 + kx = k$

$\therefore \; x^2 + kx - k = 0$ which has $a = 1$, $b = k$, and $c = -k$

$\therefore \; \Delta = b^2 - 4ac$

$= k^2 - 4(1)(-k)$

$= k^2 + 4k$

$= k(k + 4)$

b Δ has sign diagram:

c **i** For a repeated root, $\Delta = 0 \quad \therefore \; k = -4$ or $k = 0$.

ii For two distinct real roots, $\Delta > 0 \quad \therefore \; k < -4$ or $k > 0$.

iii For no real roots, $\Delta < 0 \quad \therefore \; -4 < k < 0$.

4 For each quadratic equation, find the discriminant in simplest form and draw its sign diagram. Hence find the value(s) of k for which the equation has:

i a repeated root **ii** two distinct real roots **iii** no real roots.

a $x^2 + 4x + k = 0$

b $kx^2 - 2x + 3 = 0$

c $x^2 + 5kx + 1 = 0$

d $kx^2 + (k + 1)x + 1 = 0$

e $x^2 + (k - 7)x + (2k - 14) = 0$

f $(k + 1)x^2 + kx + (k - 1) = 0$

5 Show that if $a > 0$ and $c < 0$ then the quadratic equation $ax^2 + bx + c = 0$ has two distinct real roots.

D QUADRATIC FUNCTIONS

A **quadratic function** is a function of the form $f(x) = ax^2 + bx + c$ where $a \neq 0$.

The graph of the simplest quadratic function $f(x) = x^2$ is shown alongside.

x	-3	-2	-1	0	1	2	3
y	9	4	1	0	1	4	9

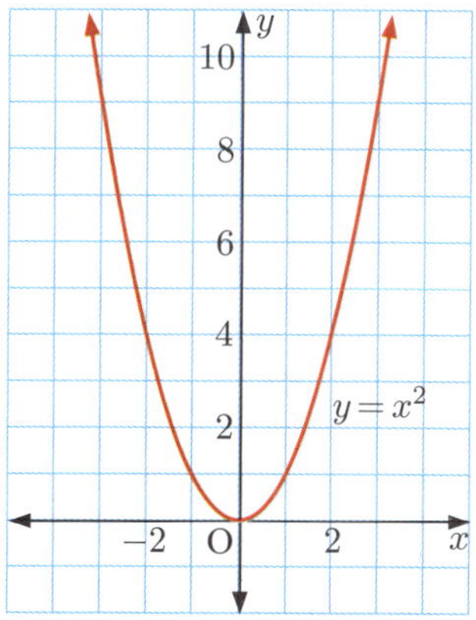

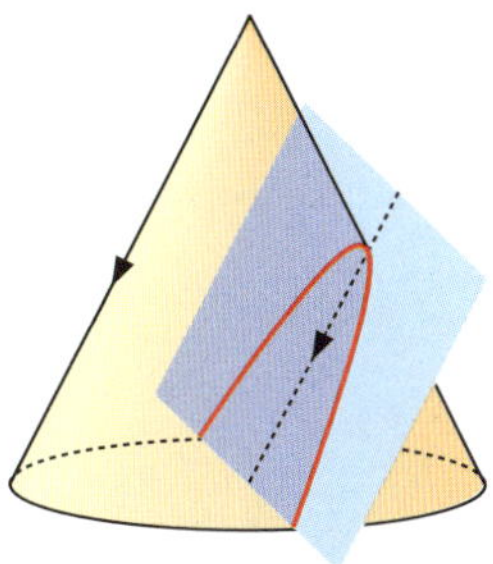

The graph of a quadratic function is called a **parabola**.

The parabola is one of the **conic sections**, the others being circles, hyperbolae, and ellipses. They are called conic sections because they can be obtained by cutting a cone with a plane. A parabola is produced by cutting the cone with a plane parallel to its slant side.

TERMINOLOGY

- The point where the parabola "turns" is called the **vertex** or **turning point**.
- A parabola is symmetric about a vertical line that passes through the vertex. This line is called the **axis of symmetry**.
- The value of y where the graph crosses the y-axis is the **y-intercept**.
- The values of x (if they exist) where the graph crosses the x-axis are called the **x-intercepts**. They correspond to the **roots** of the equation $f(x) = 0$.

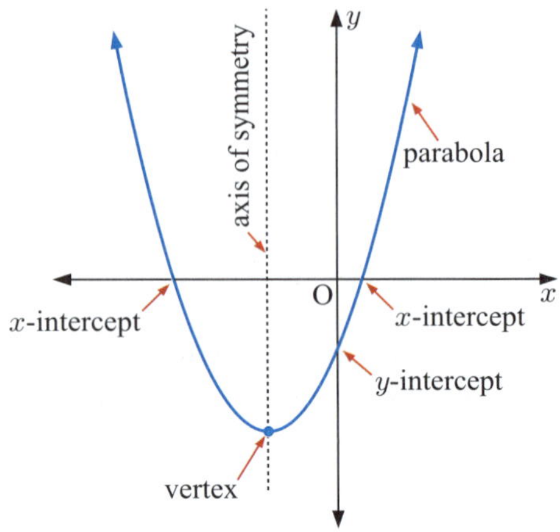

- If the parabola opens upwards, the vertex is the **minimum** and the graph is **concave upwards**.
- If the parabola opens downwards, the vertex is the **maximum** and the graph is **concave downwards**.

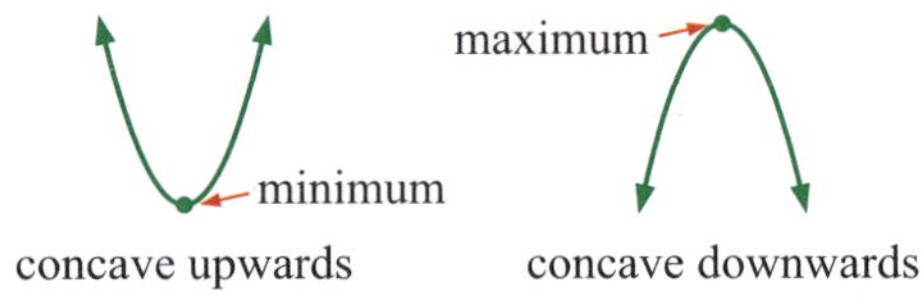

Discovery 1 — Graphing $y = a(x-p)(x-q)$

In this Discovery we consider the properties of the graph of a quadratic stated in factored form.

What to do:

GRAPHING PACKAGE

1 **a** Use technology to help you to sketch:

i $y = (x-2)(x-4)$ **ii** $y = 2(x-2)(x-4)$

iii $y = -(x-2)(x-4)$ **iv** $y = -3(x-2)(x-4)$

v $y = \frac{1}{2}(x-2)(x-4)$

b State the x-intercepts for each function in **a**.

c What is the geometrical significance of a in $y = a(x-2)(x-4)$?

2 **a** Use technology to help you to sketch:

i $y = 3(x-1)(x-4)$ **ii** $y = 3(x-3)(x-5)$ **iii** $y = 3(x+1)(x-2)$

iv $y = 3x(x+5)$ **v** $y = 3(x+2)(x+4)$ **vi** $y = 3(x-3)(x+6)$

b State the x-intercepts for each function in **a**.

c What is the geometrical significance of p and q in $y = 3(x-p)(x-q)$?

3 **a** Use technology to help you to sketch:

i $y = 3(x-1)^2$ **ii** $y = 3(x-3)^2$ **iii** $y = 3(x+2)^2$ **iv** $y = 3x^2$

b State the x-intercepts for each function in **a**.

c What is the geometrical significance of p in $y = 3(x-p)^2$?

4 Copy and complete:

a If a quadratic has the form $y = a(x-p)(x-q)$ then it ...... the x-axis at ......

b If a quadratic has the form $y = a(x-p)^2$ then it ...... the x-axis at ......

Discovery 2 — Graphing $y = a(x-h)^2 + k$

In this Discovery we consider the properties of the graph of a quadratic stated in completed square form.

What to do:

1 **a** Use technology to help you to sketch:

- **i** $y = (x-1)^2 + 3$
- **ii** $y = 2(x-1)^2 + 3$
- **iii** $y = -2(x-1)^2 + 3$
- **iv** $y = -(x-1)^2 + 3$
- **v** $y = \frac{1}{3}(x-1)^2 + 3$

GRAPHING PACKAGE

b Find the coordinates of the vertex for each function in **a**.

c What is the geometrical significance of a in $y = a(x-1)^2 + 3$?

2 **a** Use technology to help you to sketch:

- **i** $y = 3(x-1)^2 + 2$
- **ii** $y = 3(x-2)^2 + 4$
- **iii** $y = 3(x-3)^2 + 1$
- **iv** $y = 3(x+1)^2 + 4$
- **v** $y = 3(x+2)^2 - 5$
- **vi** $y = 3(x+3)^2 - 2$

b Find the coordinates of the vertex for each function in **a**.

c What is the geometrical significance of h and k in $y = 3(x-h)^2 + k$?

3 Copy and complete:

If a quadratic has the form $y = a(x-h)^2 + k$ then its vertex has coordinates ......

Summary:

Quadratic form, $a \neq 0$	*Graph*	*Facts*
$f(x) = a(x-p)(x-q)$ where $p, q \in \mathbb{R}$	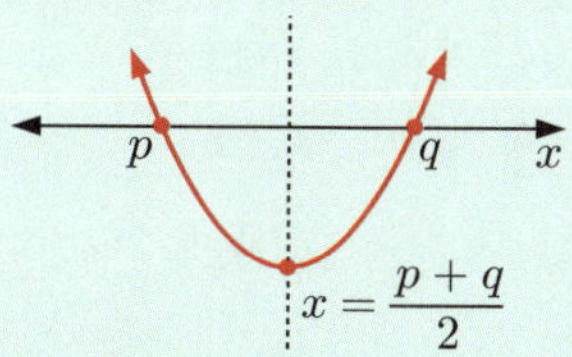	• x-intercepts are p and q. • Axis of symmetry is $x = \frac{p+q}{2}$. • Vertex has x-coordinate $\frac{p+q}{2}$.
$f(x) = a(x-h)^2$ where $h \in \mathbb{R}$	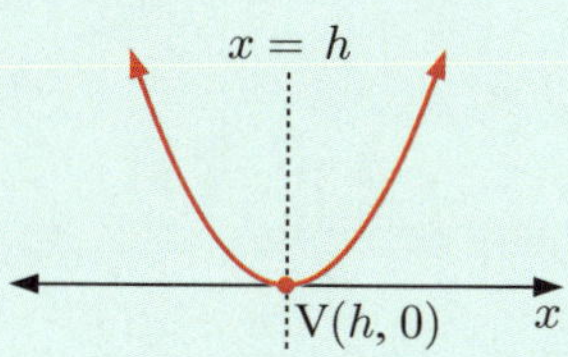	• Touches x-axis at h. • Axis of symmetry is $x = h$. • Vertex is $(h, 0)$.
$f(x) = a(x-h)^2 + k$ where $h, k \in \mathbb{R}$	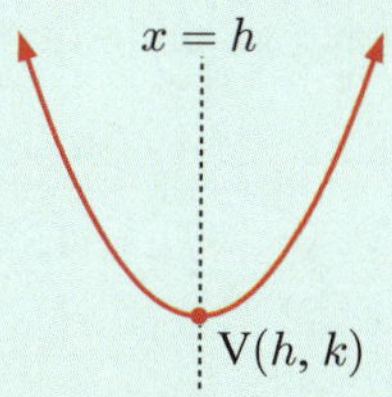	• Axis of symmetry is $x = h$. • Vertex is (h, k).

From **Discoveries 1** and **2**, you should have also found that:

For a quadratic function $f(x) = ax^2 + bx + c,\ a \neq 0$:

- The **sign** of a controls the *shape* of the graph.
  - If $a > 0$, the graph is concave upwards  .
  - If $a < 0$, the graph is concave downwards  .
- The **size** of a controls the *width* of the graph.
  - If $|a| < 1$, the graph is wider than $f(x) = x^2$.
  - If $|a| > 1$, the graph is narrower than $f(x) = x^2$.

Example 12

Self Tutor

Using axes intercepts only, sketch the graph of:

a $f(x) = 3(x-1)(x-4)$ **b** $f(x) = -3(x+1)(x-2)$ **c** $f(x) = \frac{1}{3}(x+3)^2$

a $f(x) = 3(x-1)(x-4)$ has x-intercepts 1, 4

$f(0) = 3(-1)(-4)$
$\quad = 12$

$\therefore$ the y-intercept is 12

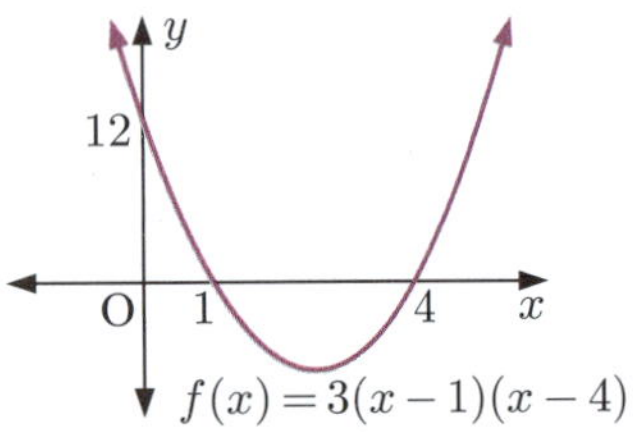

b $f(x) = -3(x+1)(x-2)$ has x-intercepts -1, 2

$f(0) = -3(1)(-2)$
$\quad = 6$

$\therefore$ the y-intercept is 6

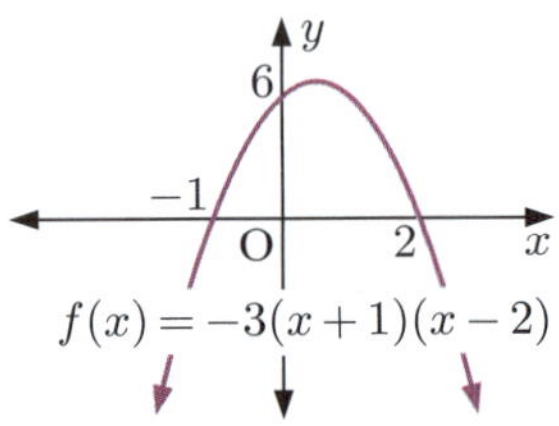

c $f(x) = \frac{1}{3}(x+3)^2$ touches the x-axis at -3

$f(0) = \frac{1}{3}(3)^2$
$\quad = 3$

$\therefore$ the y-intercept is 3

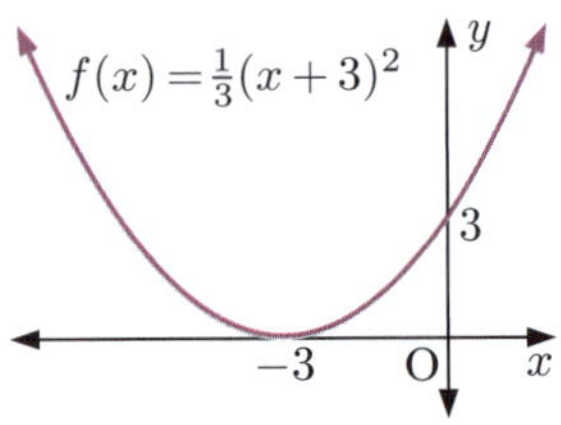

EXERCISE 3D.1

1 Using axes intercepts only, sketch the graph of:

a $f(x) = (x-3)(x+5)$ **b** $y = 2(x-3)(x+5)$

c $y = \frac{1}{2}(x+1)(x+4)$ **d** $f(x) = -(x+2)(x-2)$

e $f(x) = 3(x+5)^2$ **f** $y = -\frac{1}{4}(x-4)^2$

2 State the equation of the axis of symmetry for each graph in question **1**.

3 Match each quadratic function with its correct graph:

a $f(x) = 2(x+2)(x+6)$ **b** $f(x) = 2(x+2)(x-6)$ **c** $f(x) = -3(x+2)(x-6)$

d $f(x) = (x+2)(x+6)$ **e** $f(x) = -3(x-2)(x-6)$ **f** $f(x) = 3(x+2)(x-6)$

g $f(x) = -(x+2)(x+6)$ **h** $f(x) = -(x+2)(x-6)$ **i** $f(x) = -2(x-2)(x+6)$

A

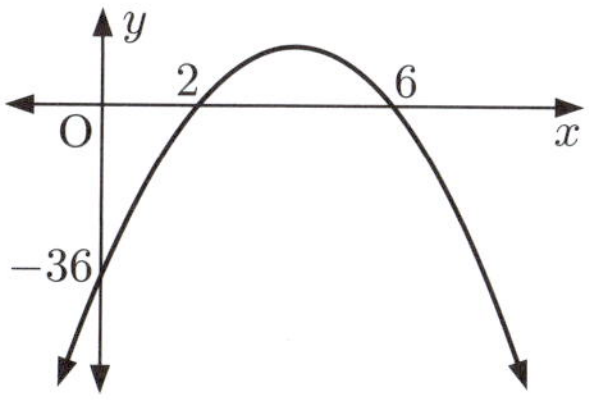

B

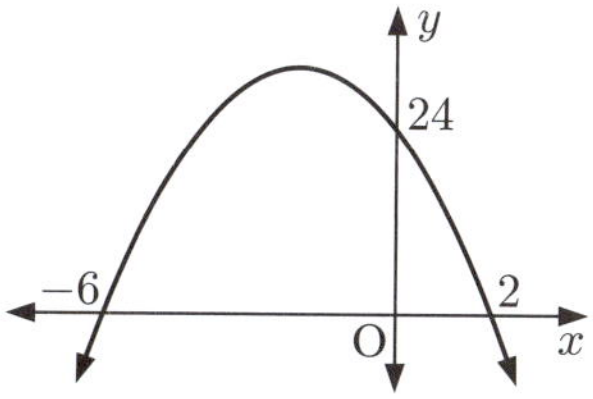

C

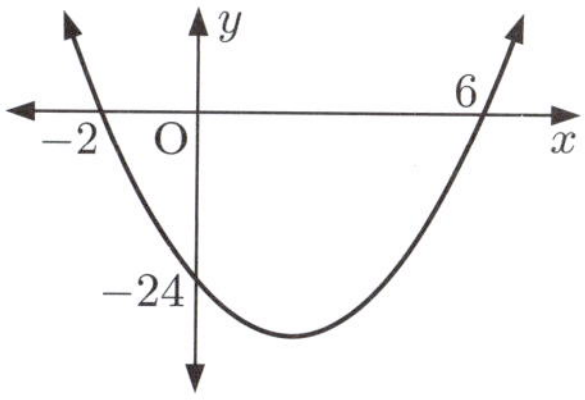

D

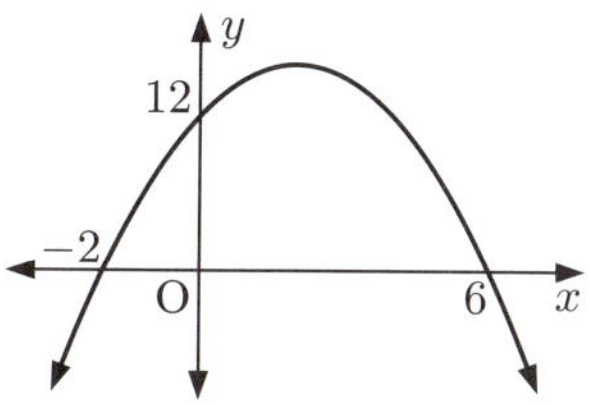

E

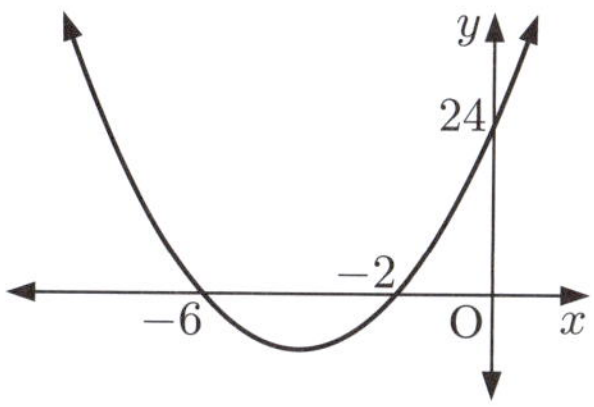

F

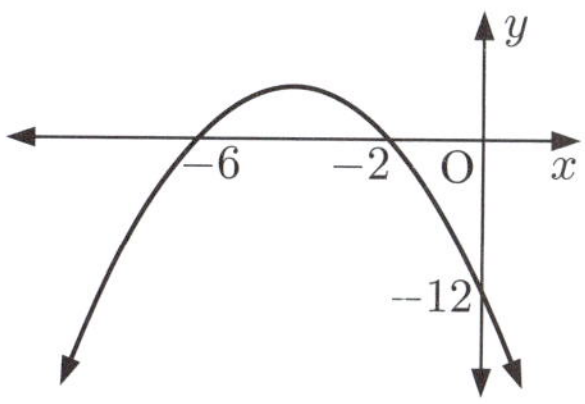

G

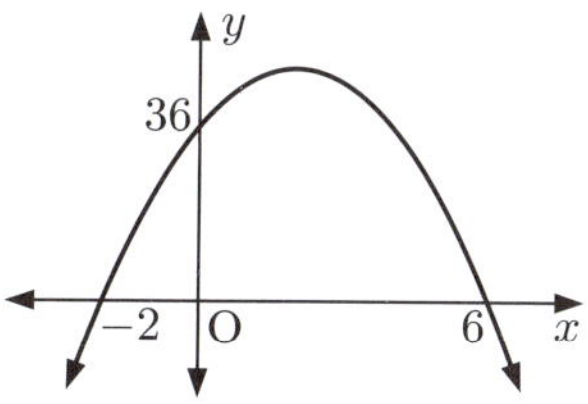

H

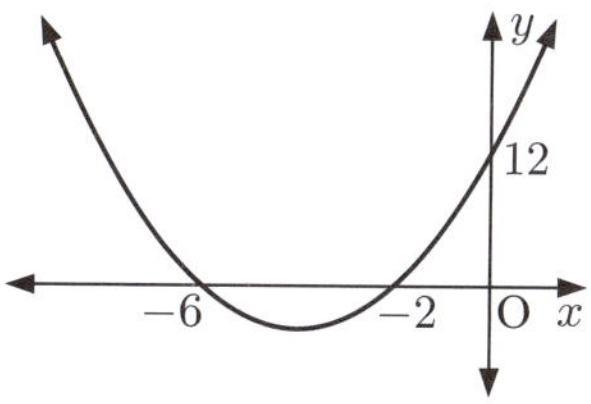

I

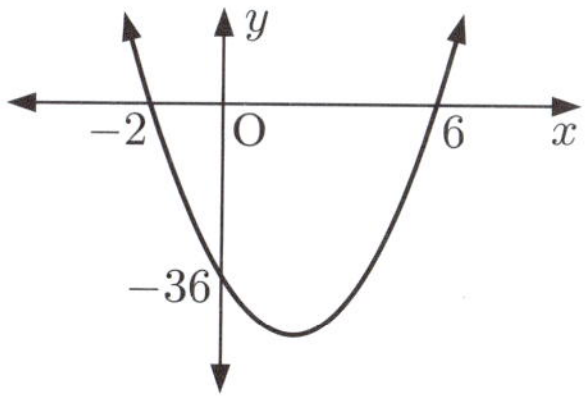

Example 13

Self Tutor

Sketch the graph of $y = |(x+2)(x-3)|$.

We first sketch $y = (x+2)(x-3)$.

$y = (x+2)(x-3)$ has x-intercepts -2 and 3, and y-intercept $2(-3) = -6$.

The part of the graph that is below the x-axis is then reflected in the x-axis to produce the graph of $y = |(x+2)(x-3)|$.

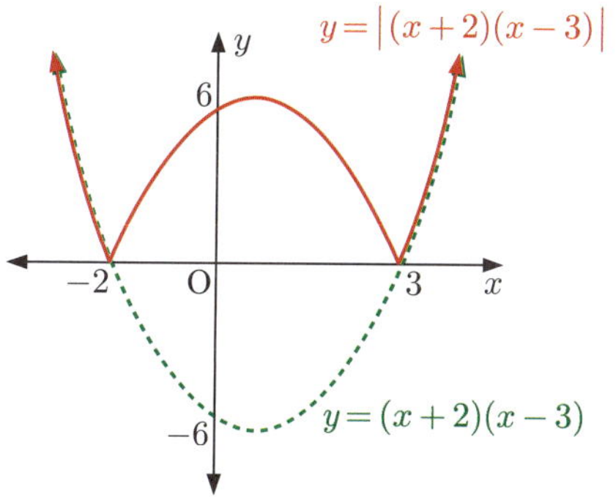

4 Sketch the graph of:

a $y = |(x+4)(x-5)|$

b $f(x) = |-(x-1)(x-6)|$

c $y = |2(x-2)(x+2)|$

d $f(x) = |-3(x+3)^2|$

Example 14

Self Tutor

Use the vertex, axis of symmetry, and y-intercept to graph $f(x) = -(x-2)^2 + 1$.

The vertex is $(2, 1)$.

The axis of symmetry is $x = 2$.

$f(0) = -(-2)^2 + 1$
$= -3$

$\therefore$ the y-intercept is -3.

$a < 0$ so the shape is 

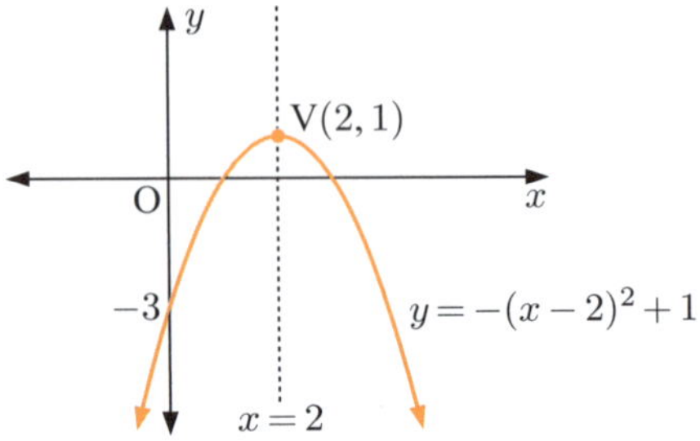

5 Use the vertex, axis of symmetry, and y-intercept to graph:

a $f(x) = (x-2)^2 + 1$ **b** $f(x) = -(x+3)^2 + 2$ **c** $f(x) = 2(x-1)^2 + 4$

d $y = \frac{1}{2}(x+5)^2 - 3$ **e** $y = -\frac{1}{2}(x-3)^2 - 1$ **f** $y = -\frac{1}{5}(x+4)^2 + 3$

6 Match each quadratic function with its correct graph:

a $f(x) = (x-1)^2 - 2$ **b** $f(x) = -(x+4)^2 + 3$ **c** $f(x) = -(x-2)^2 + 1$

d $f(x) = \frac{1}{4}(x-4)^2 - 3$ **e** $f(x) = -x^2$ **f** $f(x) = -2(x-1)^2 + 3$

g $f(x) = x^2 - 1$ **h** $f(x) = \frac{1}{2}(x+2)^2 - 4$ **i** $f(x) = 2(x+3)^2 - 1$

A

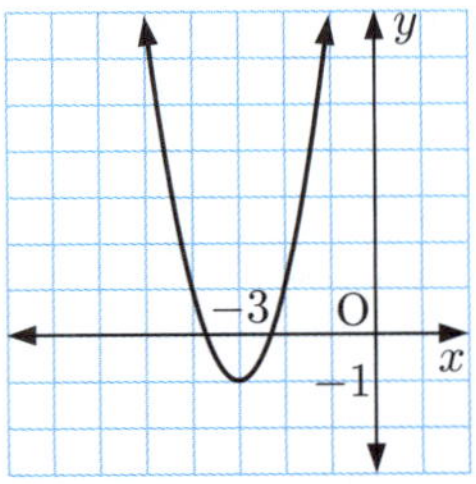

B

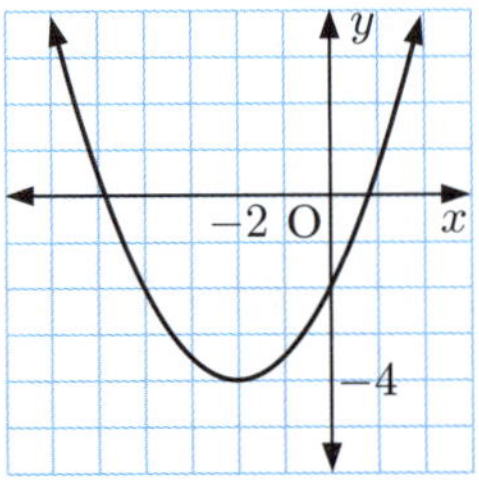

C

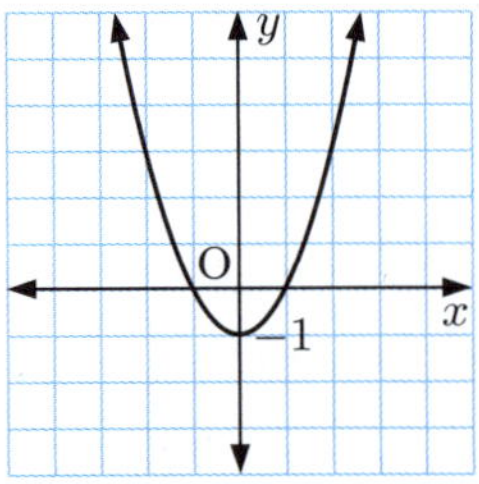

D

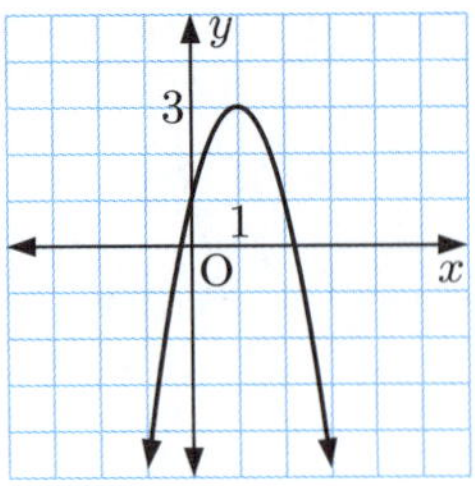

E

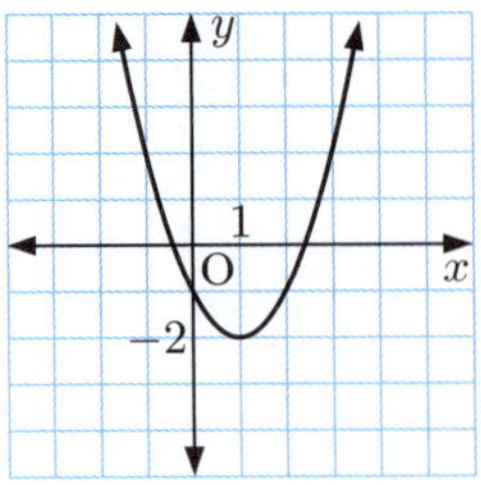

F

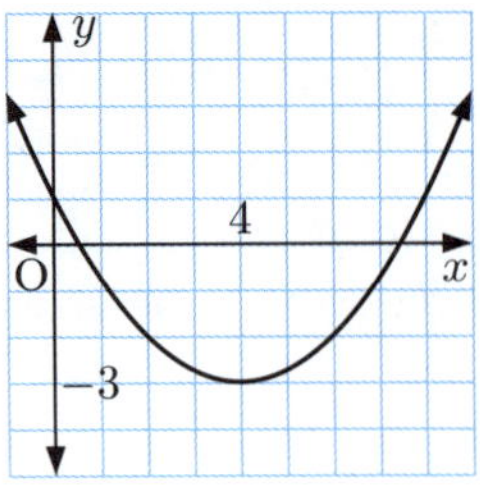

G

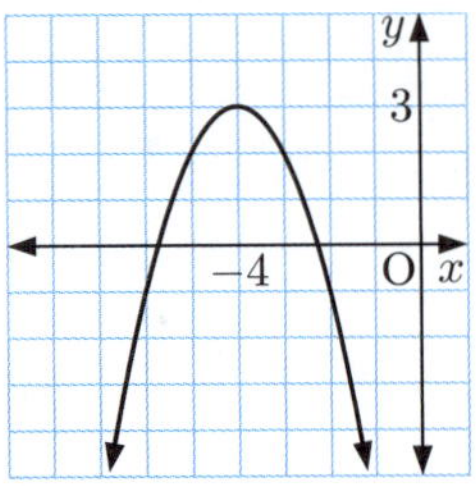

H

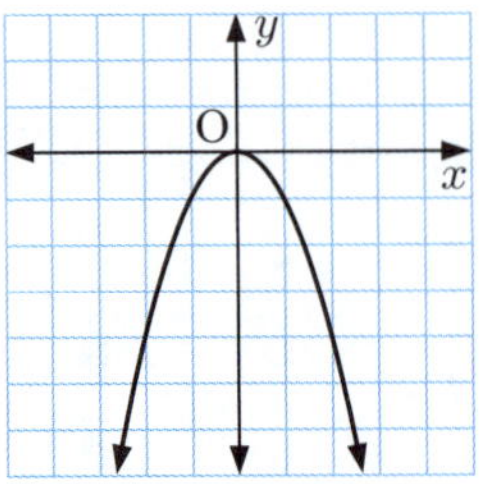

I

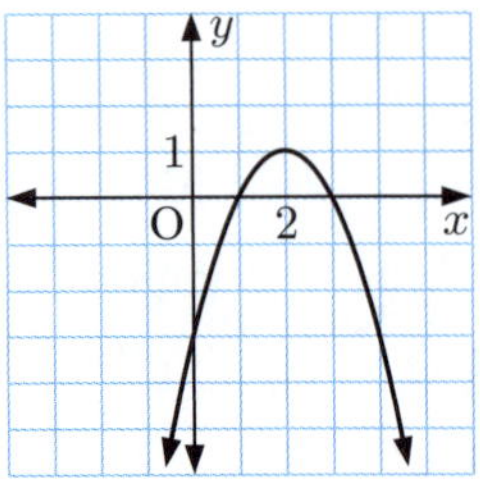

SKETCHING GRAPHS BY "COMPLETING THE SQUARE"

If we wish to graph a quadratic given in general form $f(x) = ax^2 + bx + c$, one approach is to convert it to the completed square form $f(x) = a(x-h)^2 + k$. We can then read off the coordinates of the vertex (h, k).

For example, consider $y = x^2 - 4x + 1$, for which $a = 1$.

$$y = x^2 - 4x + 1$$
$$\therefore\ y = \underbrace{x^2 - 4x + 2^2}\ \underbrace{+1 - 2^2}$$
$$\therefore\ y = (x-2)^2 \quad - 3$$

So, the axis of symmetry is $x = 2$ and the vertex is $(2, -3)$.

When $x = 0$, $y = 1$, so the y-intercept is 1.

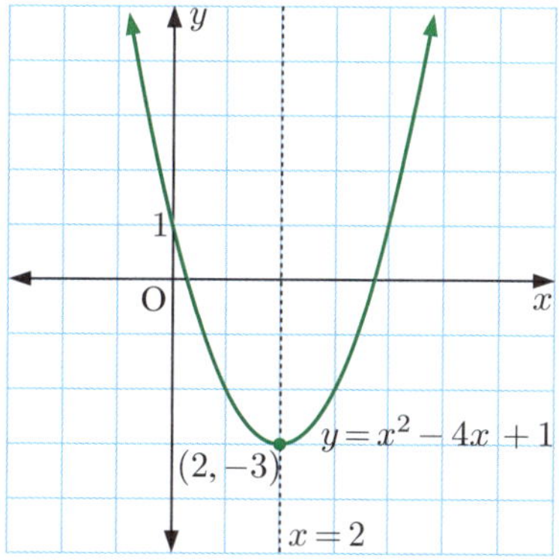

Example 15

Self Tutor

Write $f(x) = x^2 + 2x + 3$ in the form $f(x) = (x-h)^2 + k$ by "completing the square". Hence sketch the graph of $y = f(x)$.

$$f(x) = x^2 + 2x + 3$$
$$= x^2 + 2x + 1^2 + 3 - 1^2$$
$$= (x+1)^2 + 2$$

So, the axis of symmetry is $x = -1$ and the vertex is $(-1, 2)$.

$f(0) = 3$ so the y-intercept is 3.

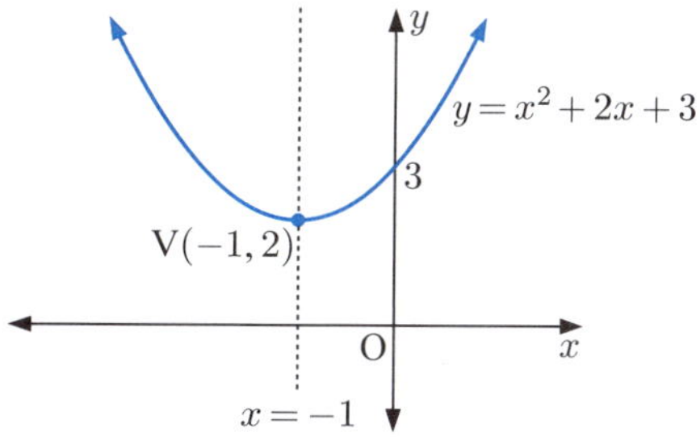

EXERCISE 3D.2

1 Consider the quadratic function $f(x) = x^2 - 4x + 7$.

a Write the function in the form $f(x) = (x-h)^2 + k$ by "completing the square".

b State the axis of symmetry and vertex of the function.

c State the y-intercept of the function.

d Hence sketch the graph of $y = f(x)$.

2 Write each quadratic function in the form $f(x) = (x-h)^2 + k$ by "completing the square". Hence sketch the graph of $y = f(x)$.

a $f(x) = x^2 - 2x + 3$ **b** $f(x) = x^2 - 6x + 1$ **c** $f(x) = x^2 - 2x$

d $f(x) = x^2 - 3x$ **e** $f(x) = x^2 + 5x - 4$ **f** $f(x) = x^2 - 3x + 1$

g $f(x) = x^2 - 8x + 2$ **h** $f(x) = x^2 + 4x - 3$ **i** $f(x) = x^2 - 7x - 2$

Example 16

Self Tutor

a Convert $f(x) = 2x^2 - 6x + 5$ to the completed square form $f(x) = a(x-h)^2 + k$.

b Hence state the coordinates of the vertex and sketch the graph of $y = f(x)$.

a
$$\begin{aligned} f(x) &= 2x^2 - 6x + 5 \\ &= 2\left[x^2 - 3x + \tfrac{5}{2}\right] && \{\text{taking out a factor of 2}\} \\ &= 2\left[x^2 - 2\left(\tfrac{3}{2}\right)x + \left(\tfrac{3}{2}\right)^2 + \tfrac{5}{2} - \left(\tfrac{3}{2}\right)^2\right] && \{\text{completing the square}\} \\ &= 2\left[\left(x - \tfrac{3}{2}\right)^2 + \tfrac{10}{4} - \tfrac{9}{4}\right] && \{\text{writing as a perfect square}\} \\ &= 2\left[\left(x - \tfrac{3}{2}\right)^2 + \tfrac{1}{4}\right] \\ &= 2\left(x - \tfrac{3}{2}\right)^2 + \tfrac{1}{2} \end{aligned}$$

b The vertex is $\left(\frac{3}{2}, \frac{1}{2}\right)$.

$f(0) = 5$ so the y-intercept is 5.

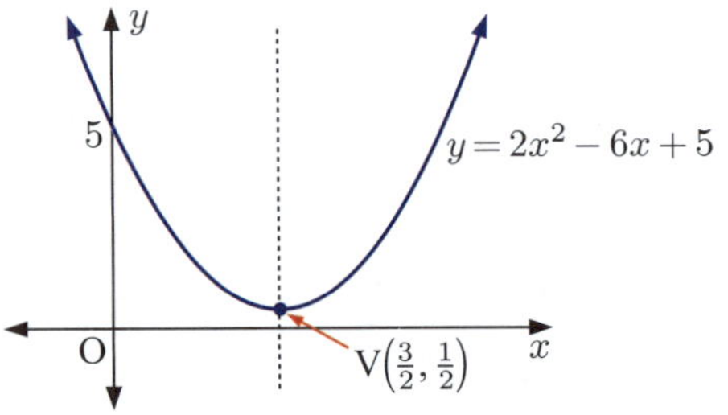

3 For each quadratic function:

  i Write the quadratic in the completed square form $f(x) = a(x-h)^2 + k$.

  ii State the coordinates of the vertex.

  iii State the y-intercept.

  iv Sketch the graph of the quadratic.

a is always the factor to be "taken out".

a $f(x) = 2x^2 - 4x + 6$ **b** $f(x) = 2x^2 + 8x - 5$ **c** $f(x) = 3x^2 + 6x + 7$

d $f(x) = 4x^2 - 16x + 1$ **e** $f(x) = -x^2 + 5x - 2$ **f** $f(x) = -2x^2 + 7x - 3$

4 By writing the quadratic function in the completed square form $f(x) = a(x-h)^2 + k$, show that the vertex of $f(x) = ax^2 + bx + c$ has x-coordinate $\dfrac{-b}{2a}$.

QUADRATIC FUNCTIONS OF THE FORM $f(x) = ax^2 + bx + c$

We now consider a method of graphing quadratics of the form $f(x) = ax^2 + bx + c$ directly, without having to first convert them to a different form.

In **Exercise 3D.2** question **4**, you should have found that the vertex of $f(x) = ax^2 + bx + c$ has x-coordinate $\dfrac{-b}{2a}$. The axis of symmetry of the function is therefore $x = \dfrac{-b}{2a}$.

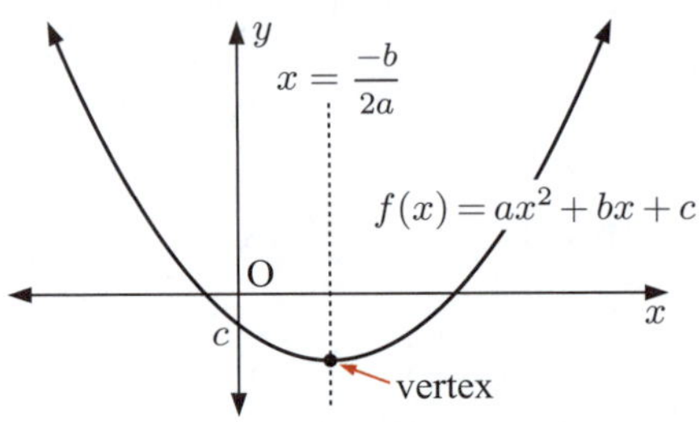

To graph a quadratic of the form $f(x) = ax^2 + bx + c$, we:

- Find the axis of symmetry $x = \dfrac{-b}{2a}$.
- Substitute this value of x to find the y-coordinate of the vertex.
- State the y-intercept c.
- Find the x-intercepts by solving $ax^2 + bx + c = 0$, either by factorisation or using the quadratic formula.

Example 17

Self Tutor

Consider the quadratic $f(x) = -x^2 - 2x + 8$.

a Find the axis of symmetry.

b Find the coordinates of the vertex.

c Find the axes intercepts.

d Hence sketch the function.

e State the range of the function.

$f(x) = -x^2 - 2x + 8$ has $a = -1$, $b = -2$, and $c = 8$.

$a < 0$, so the shape is 

a $\dfrac{-b}{2a} = \dfrac{-(-2)}{2(-1)} = -1$

The axis of symmetry is $x = -1$.

b $f(-1) = -(-1)^2 - 2(-1) + 8$
$= 9$

The vertex is $(-1, 9)$.

c The y-intercept is 8.

When $y = 0$, $-x^2 - 2x + 8 = 0$

$\therefore\ -(x^2 + 2x - 8) = 0$

$\therefore\ -(x+4)(x-2) = 0$

$\therefore\ x = -4$ or 2

$\therefore$ the x-intercepts are -4 and 2.

d 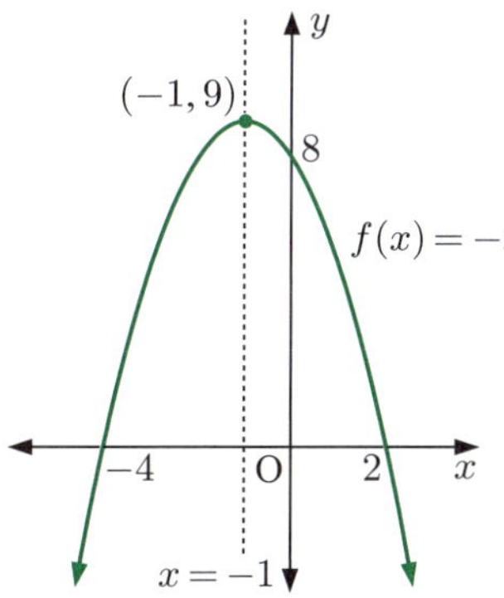

e The range is $\{y : y \leqslant 9\}$.

EXERCISE 3D.3

1 Find the coordinates of the vertex for each of the following quadratic functions:

a $f(x) = x^2 + 4x + 1$ **b** $f(x) = x^2 - 4x + 2$ **c** $f(x) = 2x^2 + 7$

d $f(x) = 3x^2 - 2x + 3$ **e** $f(x) = -3x^2 + 6$ **f** $y = -x^2 - 3x + 5$

g $y = -2x^2 - 5x + 4$ **h** $f(x) = \frac{1}{2}x^2 + 2x - 3$ **i** $y = -\frac{1}{3}x^2 + \frac{1}{6}x - 1$

2 For each quadratic function:

- **i** State the axis of symmetry.
- **ii** Find the coordinates of the vertex.
- **iii** Find the axes intercepts.
- **iv** Sketch the quadratic.
- **v** State the range.

a $f(x) = x^2 - 2x + 1$ **b** $y = x^2 - 4x + 3$

c $y = 2x^2 + x - 8$ **d** $f(x) = -x^2 - x + 8$

e $f(x) = 3x^2 + 4x - 8$ **f** $y = -2x^2 - 7x + 3$

g $y = -\frac{1}{2}x^2 + 3x + 2$ **h** $f(x) = \frac{1}{4}x^2 + 2x - 3$

3 For each of the following quadratics:

- **i** Write the quadratic in factored form and hence find the roots.
- **ii** Write the quadratic in completed square form and hence find the coordinates of the vertex.
- **iii** Sketch the quadratic, showing the details you have found.

a $f(x) = x^2 - 10x + 16$ **b** $y = x^2 + 10x + 9$ **c** $f(x) = x^2 - 14x + 45$

4 Sketch the graph of:

a $f(x) = \left|x^2 + 4x - 12\right|$ **b** $f(x) = \left|-x^2 - 3x + 10\right|$ **c** $f(x) = \left|4x^2 - 12x + 5\right|$

Example 18 — Self Tutor

Find the range of $y = x^2 - 6x - 2$ on the domain $-2 \leqslant x \leqslant 7$.

$y = x^2 - 6x - 2$ has $a = 1$, $b = -6$, and $c = -2$.

$a > 0$, so the shape is 

$\frac{-b}{2a} = \frac{-(-6)}{2(1)} = 3$

The axis of symmetry is $x = 3$.

When $x = 3$, $y = (3)^2 - 6(3) - 2$
$= -11$

$\therefore$ the vertex is $(3, -11)$.

When $x = -2$, $y = (-2)^2 - 6(-2) - 2$
$= 14$

When $x = 7$, $y = (7)^2 - 6(7) - 2$
$= 5$

So, on the domain $\{x : -2 \leqslant x \leqslant 7\}$,
the range is $\{y : -11 \leqslant y \leqslant 14\}$.

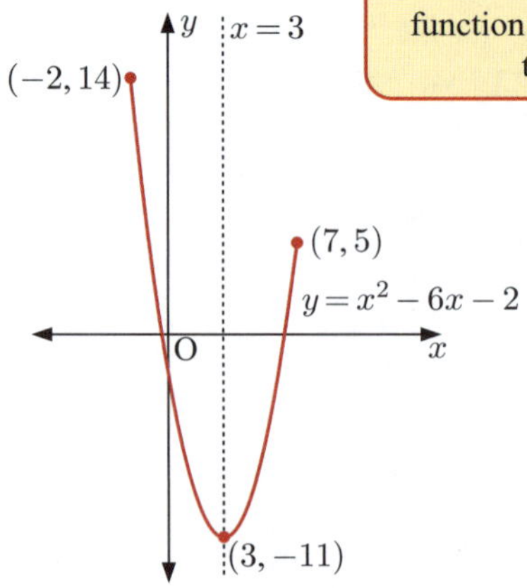

5 Find the range of:

a $f(x) = x^2 + 4x - 6$ on $-6 \leqslant x \leqslant 3$ **b** $y = -x^2 + 8x + 3$ on $0 \leqslant x \leqslant 7$

c $y = 2x^2 - 12x + 5$ on $-2 \leqslant x \leqslant 6$ **d** $f(x) = 7x - x^2$ on $-1 \leqslant x \leqslant 5$

Activity

Click on the icon to run a card game for quadratic functions.

CARD GAME

THE DISCRIMINANT AND THE QUADRATIC GRAPH

In **Section C**, we saw that the discriminant of the quadratic equation $ax^2 + bx + c = 0$ is $\Delta = b^2 - 4ac$. We used Δ to determine the number of real solutions of the equation.

We can therefore use Δ to determine how many x-intercepts a quadratic function has.

A quadratic function $f(x) = ax^2 + bx + c$ with discriminant $\Delta = b^2 - 4ac$, has three possible cases:

- $\Delta > 0$

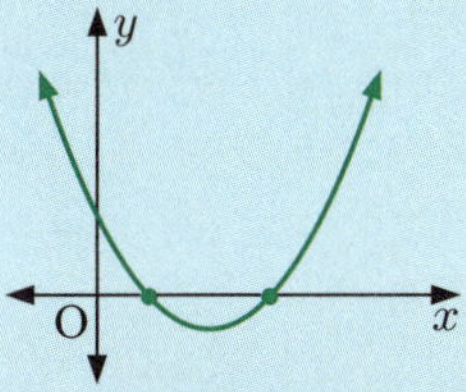

two x-intercepts

- $\Delta = 0$

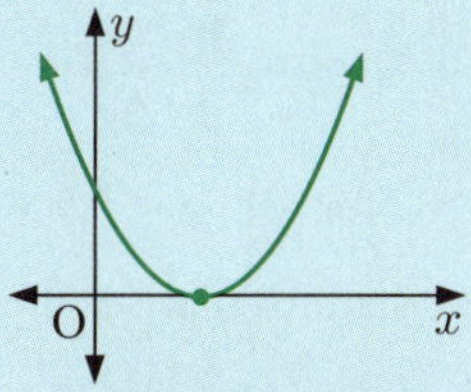

one x-intercept

- $\Delta < 0$

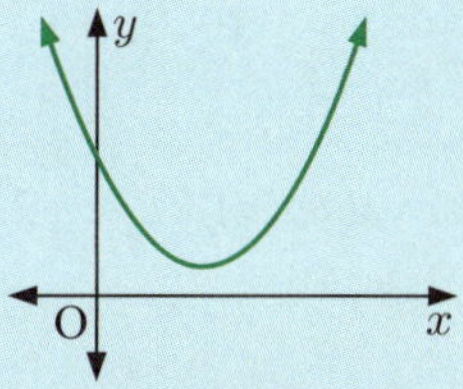

no x-intercepts

Example 19

Self Tutor

Use the discriminant to determine the relationship between the graph of each function and the x-axis:

a $f(x) = -x^2 + 3x - 3$ **b** $f(x) = 2x^2 - 4x + 1$

a $a = -1, \ b = 3, \ c = -3$

$\therefore \ \Delta = b^2 - 4ac$
$= (3)^2 - 4(-1)(-3)$
$= -3$

Since $\Delta < 0$, the graph does not cut the x-axis.
Since $a < 0$, the graph is concave downwards.

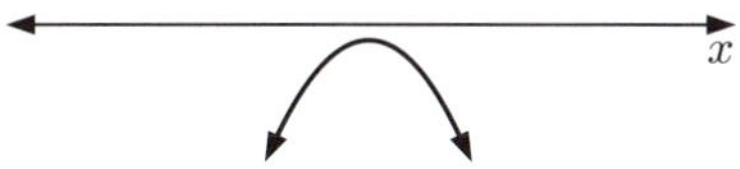

The graph lies entirely below the x-axis.

b $a = 2, \ b = -4, \ c = 1$

$\therefore \ \Delta = b^2 - 4ac$
$= (-4)^2 - 4(2)(1)$
$= 8$

Since $\Delta > 0$, the graph cuts the x-axis twice.
Since $a > 0$, the graph is concave upwards.

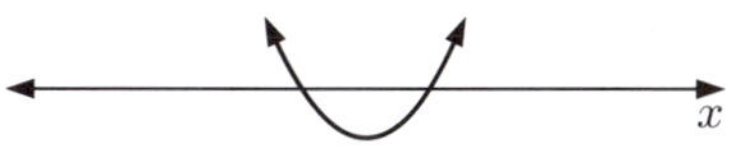

EXERCISE 3D.4

1 Use the discriminant to determine the relationship between the graph of each function and the x-axis:

a $f(x) = x^2 + 6x - 4$ **b** $f(x) = x^2 + 2x + 3$ **c** $y = x^2 + 4x + 4$

d $y = -2x^2 - x - 6$ **e** $f(x) = -x^2 + 8x + 6$ **f** $y = 3x^2 - 5x + 5$

g $f(x) = \frac{1}{2}x^2 - 3x + 4$ **h** $y = -16x^2 + 8x + 1$ **i** $f(x) = -\frac{1}{2}x^2 + 7x - 18$

2 Consider the graph of $f(x) = x^2 - 3x + 1$.

- **a** Describe the shape of the graph.
- **b** Use the discriminant to show that the graph cuts the x-axis twice.
- **c** Find the x-intercepts, rounding your answers to 2 decimal places.
- **d** State the y-intercept.
- **e** Hence sketch the function.

3 Consider the graph of $y = -2x^2 + 5x - 4$.

- **a** Use the discriminant to show that the graph does not cut the x-axis.
- **b** Does the graph lie above or below the x-axis?
- **c** Find the vertex and y-intercept.
- **d** Hence sketch the function.

4 Show that for all $x \in \mathbb{R}$:

- **a** $4x^2 - 2x + 1 > 0$
- **b** $-5x^2 + 3x - 5 < 0$
- **c** $x^2 + 4x + 5 > 0$
- **d** $3x - 1 < 3x^2$

5 For what values of k is $-2x^2 + kx - 2 < 0$ for all $x \in \mathbb{R}$?

6 Show that $x^2 - kx - (k + 2) = 0$ has two distinct real solutions for any $k \in \mathbb{R}$.

E FINDING A QUADRATIC FROM ITS GRAPH

If we are given sufficient information about a parabola, we can determine the corresponding quadratic function.

Example 20 — Self Tutor

Find the equation of the quadratic function with graph:

a

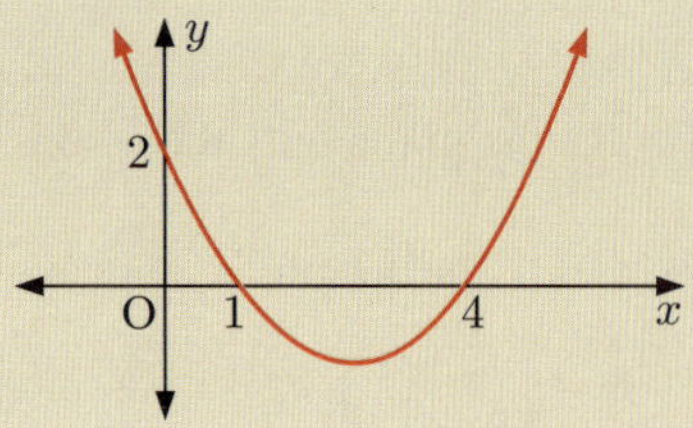

b

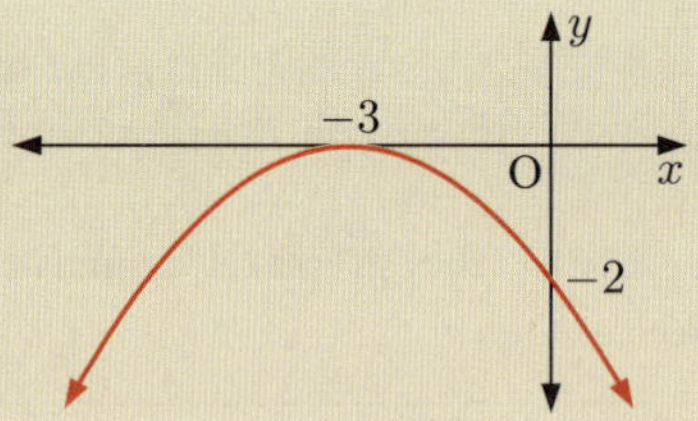

a The x-intercepts are 1 and 4.

$\therefore\ f(x) = a(x - 1)(x - 4)$

Now $f(0) = 2$

$\therefore\ a(-1)(-4) = 2$

$\therefore\ a = \frac{1}{2}$

The quadratic function is $f(x) = \frac{1}{2}(x - 1)(x - 4)$.

b The graph touches the x-axis at $x = -3$.

$\therefore\ f(x) = a(x + 3)^2$

Now $f(0) = -2$

$\therefore\ a(3)^2 = -2$

$\therefore\ a = -\frac{2}{9}$

The quadratic function is $f(x) = -\frac{2}{9}(x + 3)^2$.

Example 21

Self Tutor

Find the equation of the quadratic function with graph:

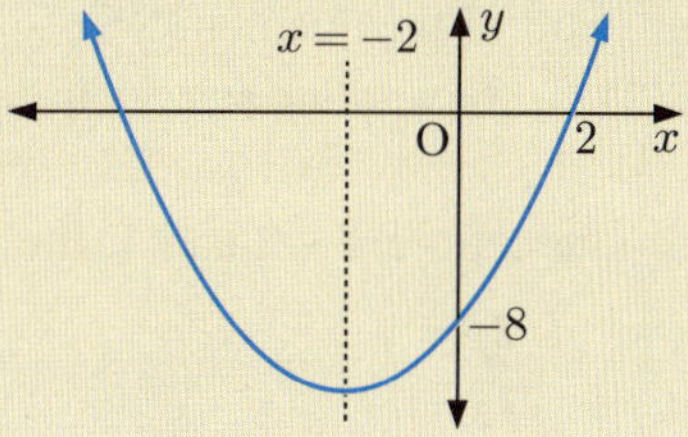

The axis of symmetry $x=-2$ lies midway between the x-intercepts.

$\therefore$ the other x-intercept is -6.

$\therefore$ the quadratic has the form $f(x)=a(x+6)(x-2)$

Now $f(0)=-8$

$\therefore$ $a(6)(-2)=-8$

$\therefore$ $a=\frac{2}{3}$

The quadratic function is $f(x)=\frac{2}{3}(x+6)(x-2)$.

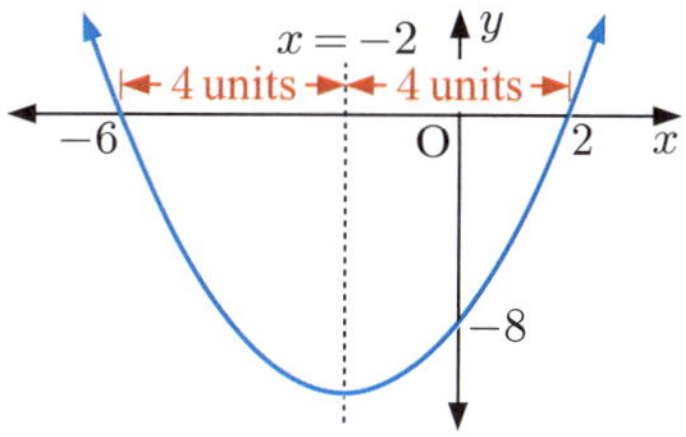

EXERCISE 3E

1 Find the equation of the quadratic function with graph:

a

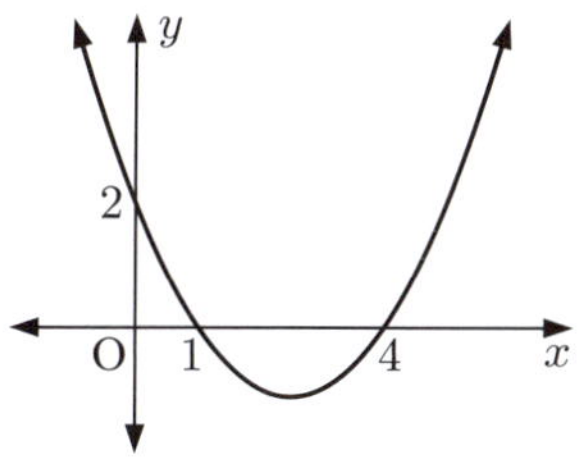

b

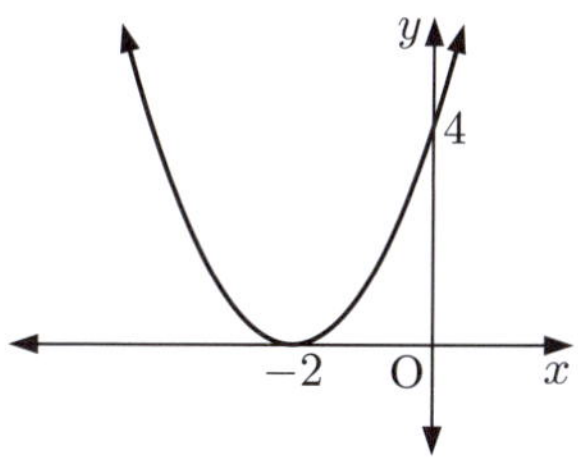

c

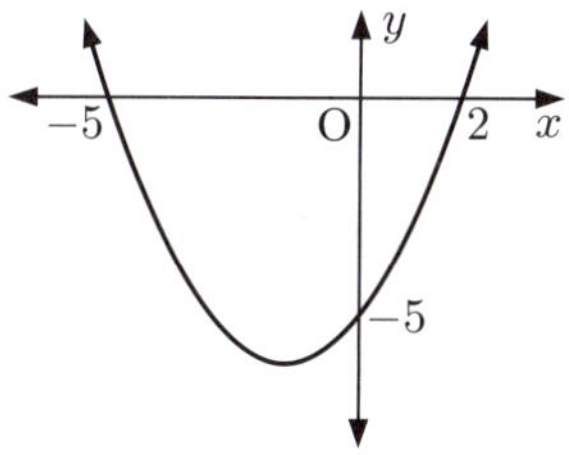

d

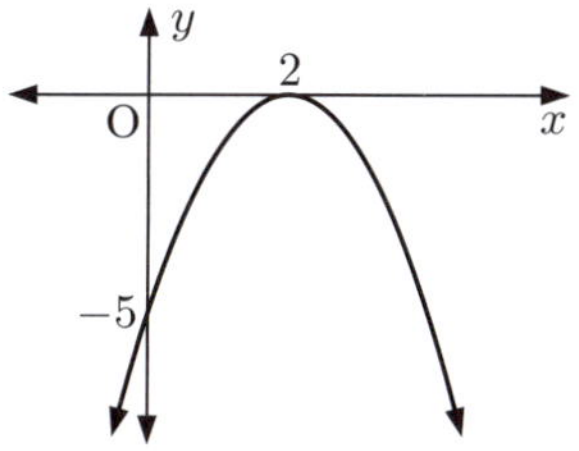

e

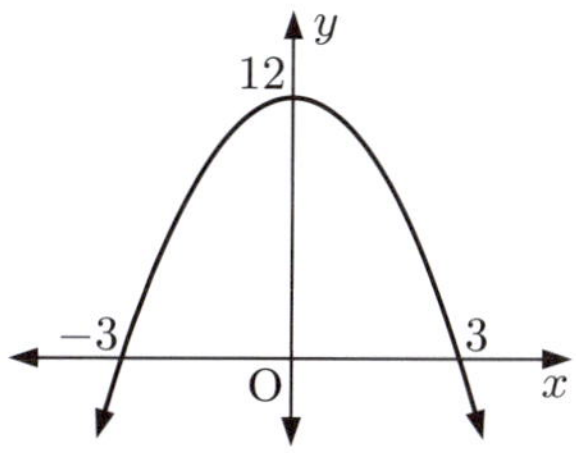

f

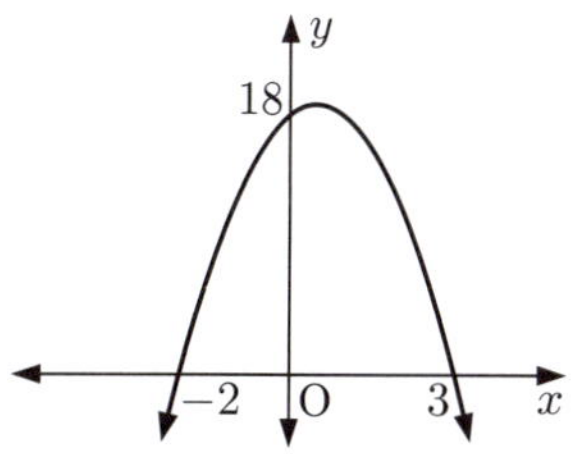

2 Find the equation of the quadratic function with graph:

a

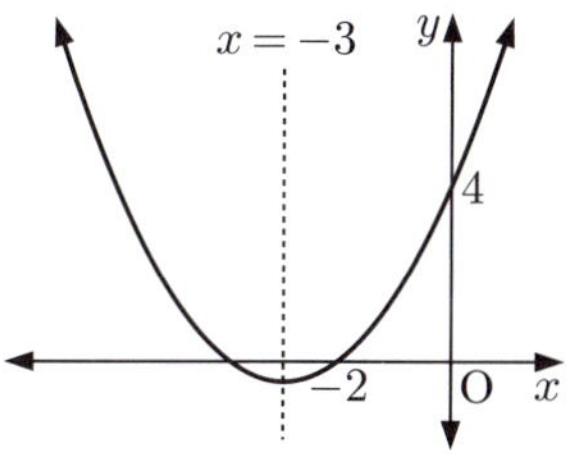

b

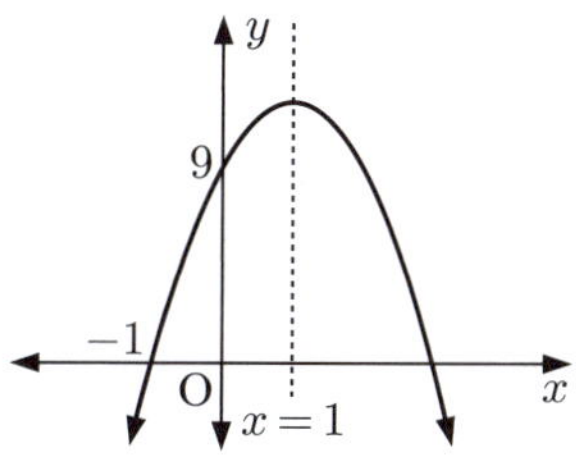

c

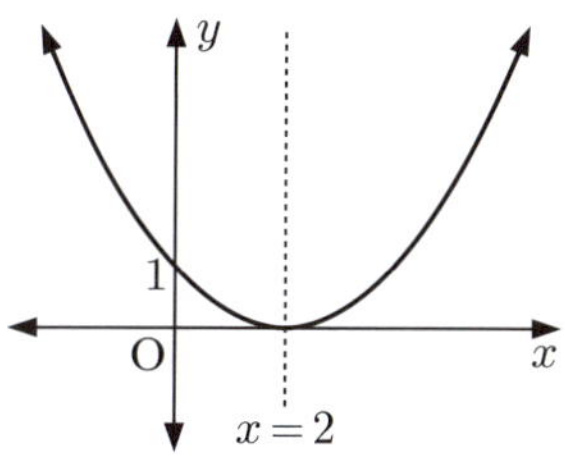

Example 22

Self Tutor

Find the equation of the quadratic whose graph cuts the x-axis at -2 and 1, and which passes through the point $(2, 8)$. Give your answer in the form $f(x) = ax^2 + bx + c$.

Since the x-intercepts are -2 and 1, the quadratic has the form $f(x) = a(x+2)(x-1)$, $a \neq 0$.

$$\begin{aligned} \text{Now } f(2) &= 8 \\ \therefore\ a(2+2)(2-1) &= 8 \\ \therefore\ a(4)(1) &= 8 \\ \therefore\ a &= 2 \end{aligned}$$

The quadratic function is $f(x) = 2(x+2)(x-1)$
$= 2(x^2 + x - 2)$
$= 2x^2 + 2x - 4$

3 Find, in the form $f(x) = ax^2 + bx + c$, the equation of the quadratic function whose graph:

- **a** cuts the x-axis at 2 and 3, and passes through $(4, -2)$
- **b** cuts the x-axis at -5 and 4, and passes through $(-1, -10)$
- **c** touches the x-axis at 1 and passes through $(3, 6)$.

4 Find, in the form $f(x) = ax^2 + bx + c$, the equation of the quadratic function whose graph:

- **a** cuts the x-axis at -1, passes through $(2, -9)$, and has axis of symmetry $x = 1$
- **b** cuts the x-axis at 3, passes through $(-1, 4)$, and has axis of symmetry $x = 2$
- **c** cuts the x-axis at -2, passes through $(1, 29)$, and has axis of symmetry $x = -4$.

Example 23

Self Tutor

Find the equation of the quadratic function with graph:

a

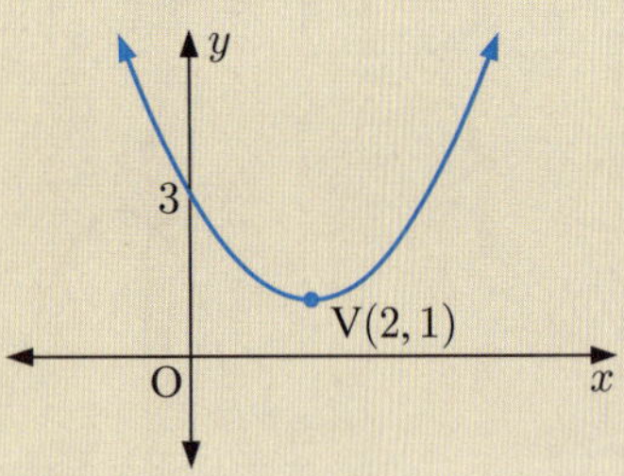

b

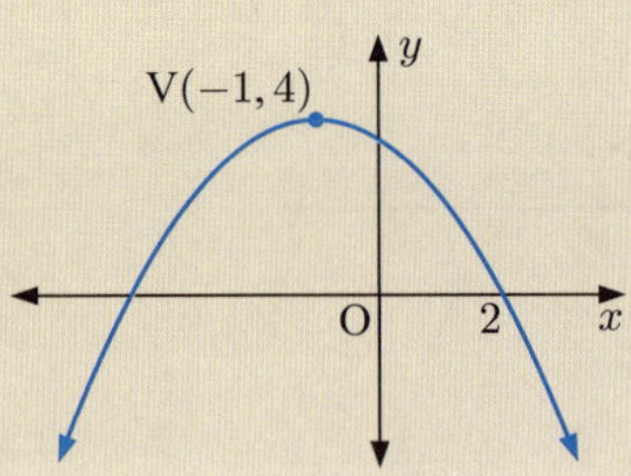

a The vertex is $(2, 1)$.

$\therefore\ f(x) = a(x-2)^2 + 1$ where $a > 0$.

Now $f(0) = 3$

$\therefore\ a(-2)^2 + 1 = 3$

$\therefore\ 4a = 2$

$\therefore\ a = \frac{1}{2}$

The quadratic function is

$f(x) = \frac{1}{2}(x-2)^2 + 1$.

b The vertex is $(-1, 4)$.

$\therefore\ f(x) = a(x+1)^2 + 4$ where $a < 0$.

Now $f(2) = 0$

$\therefore\ a(2+1)^2 + 4 = 0$

$\therefore\ 9a = -4$

$\therefore\ a = -\frac{4}{9}$

The quadratic function is

$f(x) = -\frac{4}{9}(x+1)^2 + 4$.

5 If V is the vertex, find the equation of the quadratic function with graph:

a

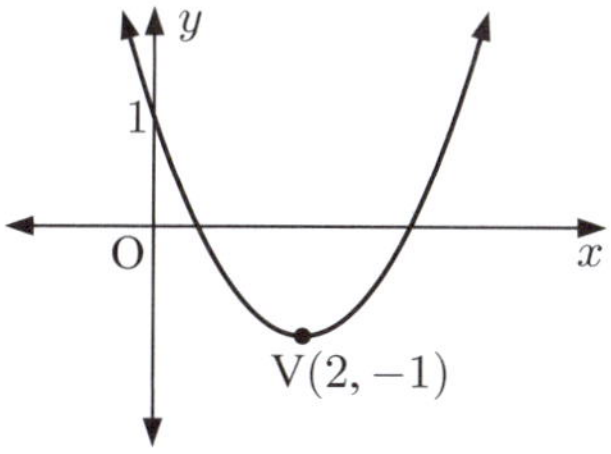

b

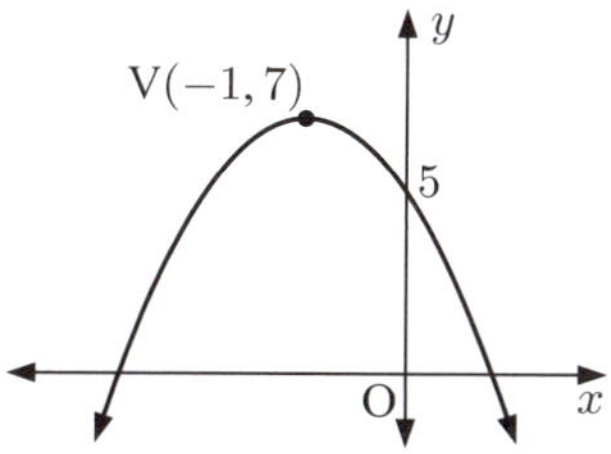

c

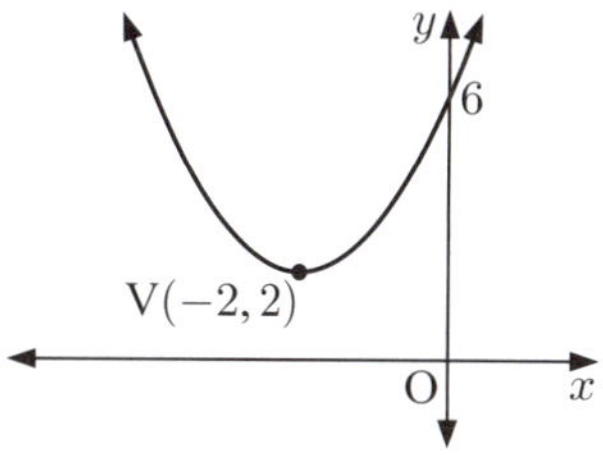

d

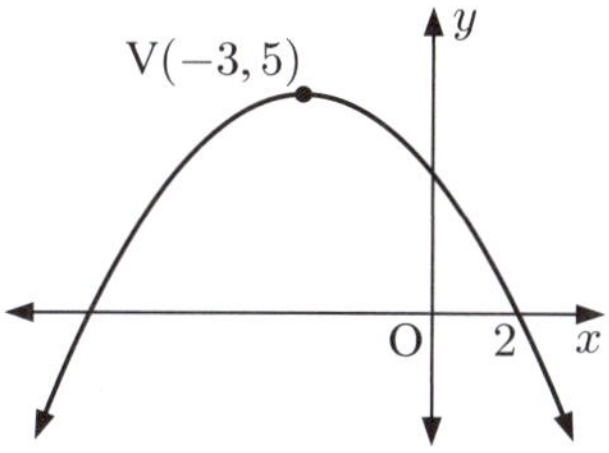

e

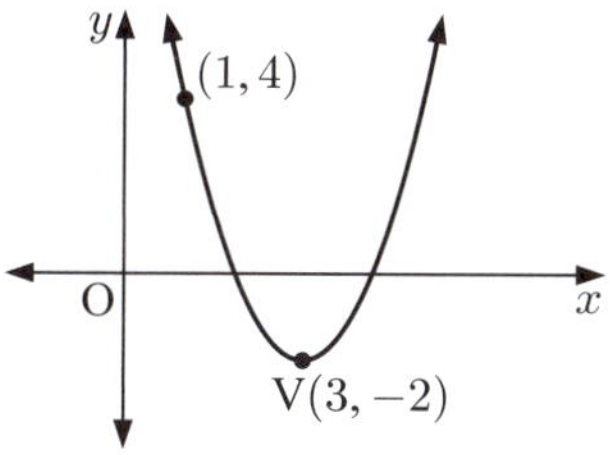

f

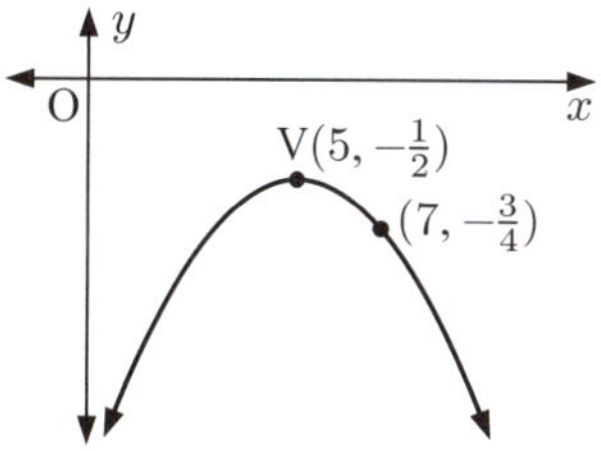

F WHERE FUNCTIONS MEET

Consider the graphs of a quadratic function $f(x)$, and a linear function $g(x)$ on the same set of axes.

There are three possible scenarios for their intersection:

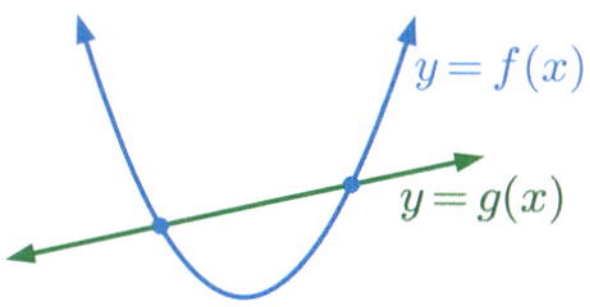

cutting
(2 points of intersection)

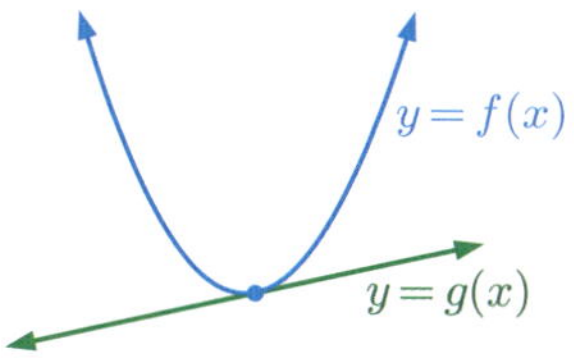

touching
(1 point of intersection)

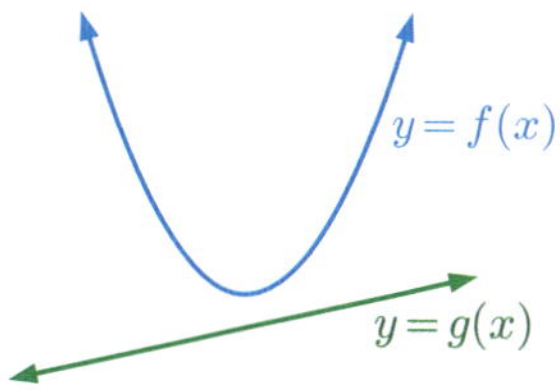

missing
(no points of intersection)

If the graphs meet, the x-coordinate(s) of the point(s) of intersection can be found by solving $f(x) = g(x)$.

Example 24

Self Tutor

Find the coordinates of the intersection points of $f(x) = x^2 + 5x + 5$ and $g(x) = -x - 3$.

$y = f(x)$ meets $y = g(x)$ where

$$f(x) = g(x)$$
$$\therefore\ x^2 + 5x + 5 = -x - 3$$
$$\therefore\ x^2 + 6x + 8 = 0 \qquad \{\text{RHS} = 0\}$$
$$\therefore\ (x+4)(x+2) = 0 \qquad \{\text{factorising}\}$$
$$\therefore\ x = -4 \text{ or } -2$$

Now $g(-4) = -(-4) - 3 = 1$ and $g(-2) = -(-2) - 3 = -1$

$\therefore$ the graphs meet at $(-4, 1)$ and $(-2, -1)$.

To find the y-coordinates of the points of intersection, substitute our solutions for x into the simpler function.

EXERCISE 3F.1

1 Find the coordinates of the point(s) of intersection of:

a $f(x) = x^2 - x + 2$ and $g(x) = x + 5$

b $y = -x^2 - 4x + 2$ and $y = 8 + x$

c $f(x) = 2x^2 + 5x + 1$ and $g(x) = 13x - 5$

d $y = -2x^2 + 9x - 15$ and $y = 3 - 3x$

Example 25 — Self Tutor

$y = -x + k$ is a tangent to $y = 3x^2 + 4x - 1$. Find k.

$y = -x + k$ meets $y = 3x^2 + 4x - 1$ where

$$3x^2 + 4x - 1 = -x + k$$

$$\therefore\ 3x^2 + 5x - (k+1) = 0$$

Since the graphs touch, this quadratic has $\Delta = 0$

$$\therefore\ (5)^2 - 4(3)(-(k+1)) = 0$$
$$\therefore\ 25 + 12(k+1) = 0$$
$$\therefore\ 25 + 12k + 12 = 0$$
$$\therefore\ 12k = -37$$
$$\therefore\ k = -\tfrac{37}{12}$$

2 For what value of c is the line $y = 4x + c$ a tangent to the parabola with equation $y = x^2 + 2x - 5$?

3 Find the values of m for which the lines $y = mx - 3$ are tangents to the curve with equation $y = -x^2 + 3x - 4$.

4 Find the gradients of the lines with y-intercept 2 that are tangents to the curve $f(x) = 2x^2 + 3x + 4$.

5 **a** For what values of c do the lines $y = x + c$ never meet the parabola with equation $y = x^2 + 3x + 2$?

b Choose one of the values of c found in part **a**. Illustrate with a sketch that these graphs never meet.

6 Consider the curve $f(x) = -x^2 - 4x - 6$ and the line $g(x) = mx + 3$. Find the values of m for which the line:

a meets the curve twice

b is a tangent to the curve

c does not meet the curve.

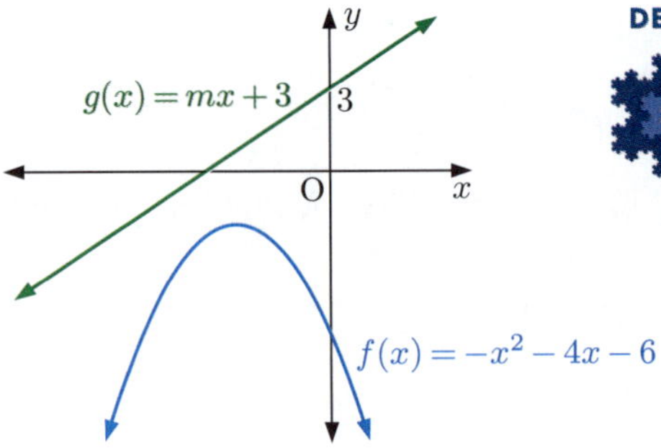

DEMO

7 Find the values of k for which the line $y = k(2x - 5)$ does not intersect the curve $y = x^2 + 2x + 9$.

8 Consider the curve $y = \frac{1}{2}x^2 - x + 2$. Find the equations of *two* tangents to the curve, which pass through the origin.

Discussion

How many possibilities are there for the intersection of *two quadratic* functions?

INTERSECTION OF A STRAIGHT LINE AND A CURVE

We can solve quadratic equations to find the intersection of straight lines and more complicated curves.

We first rearrange the equation of the line so that x or y is the subject. We then substitute this expression for x or y into the equation of the curve.

Example 26

Self Tutor

Find the points where the line $x - 3y = 4$ intersects the curve $x^2 + y^2 = 34$.

Substituting $x = 3y + 4$ into $x^2 + y^2 = 34$ gives

$$(3y+4)^2 + y^2 = 34$$
$$\therefore\ 9y^2 + 24y + 16 + y^2 = 34$$
$$\therefore\ 10y^2 + 24y - 18 = 0$$
$$\therefore\ 2(5y^2 + 12y - 9) = 0$$
$$\therefore\ 2(5y-3)(y+3) = 0$$
$$\therefore\ y = \tfrac{3}{5} \text{ or } -3$$

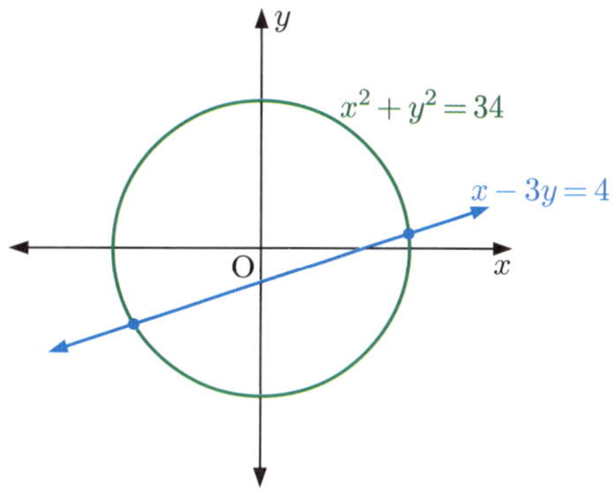

When $y = \frac{3}{5}$, $x = 3(\frac{3}{5}) + 4 = \frac{29}{5}$.

When $y = -3$, $x = 3(-3) + 4 = -5$.

$\therefore$ the line intersects the curve at $(\frac{29}{5}, \frac{3}{5})$ and $(-5, -3)$.

Example 27

Self Tutor

Find the points where the line $2x + 3y = 5$ intersects the curve $\dfrac{1}{x} - \dfrac{3}{y} = 2$.

Rearranging $2x + 3y = 5$, we find $y = \dfrac{5-2x}{3}$.

Substituting into $\dfrac{1}{x} - \dfrac{3}{y} = 2$ gives $\dfrac{1}{x} - \dfrac{3}{\frac{5-2x}{3}} = 2$

$$\therefore\ \frac{1}{x} - \frac{9}{5-2x} = 2$$
$$\therefore\ (5-2x) - 9x = 2x(5-2x) \qquad \{\times \text{ both sides by } x(5-2x)\}$$
$$\therefore\ 5 - 11x = 10x - 4x^2$$
$$\therefore\ 4x^2 - 21x + 5 = 0$$
$$\therefore\ (4x-1)(x-5) = 0$$
$$\therefore\ x = \tfrac{1}{4} \text{ or } 5$$

When $x = \frac{1}{4}$, $y = \dfrac{5 - 2(\frac{1}{4})}{3} = \frac{3}{2}$.

When $x = 5$, $y = \dfrac{5 - 2(5)}{3} = -\frac{5}{3}$.

$\therefore$ the line intersects the curve at $(\frac{1}{4}, \frac{3}{2})$ and $(5, -\frac{5}{3})$.

EXERCISE 3F.2

1 Find the points where the line $x - 2y = 3$ intersects the curve $x^2 + y^2 = 5$.

2 Find the coordinates of P and Q in the graph alongside.

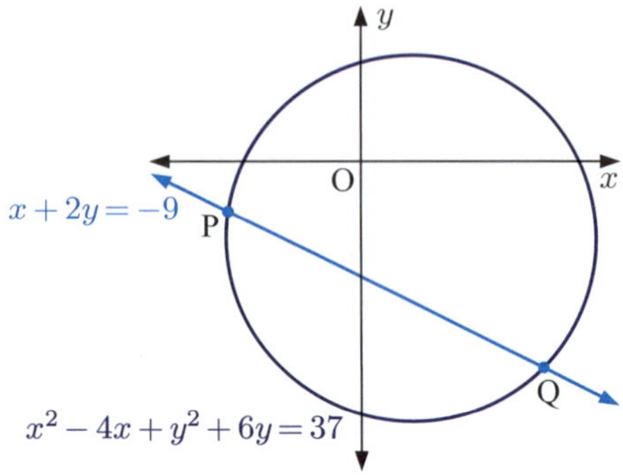

3 The line $x + y = 7$ meets the curve $x^2 + y^2 = 29$ at A and B. Find the distance between A and B.

4 The line $2x + y = 5$ meets the curve $x^2 + y^2 = 10$ at P and Q. Find the equation of the perpendicular bisector of PQ.

5 Find the points where the line $x - 2y = 4$ intersects the curve $3x^2 + y^2 + xy + 3y = 8$.

6 The line $y = 2x + 1$ meets the curve $x^2 + y^2 + xy + 16x = 29$ at P and Q. Find the distance between P and Q.

7 The line $3x + y = 1$ intersects the curve $2x^2 + y^2 + 5xy - 7x = -31$ at A and B. Find the equation of the perpendicular bisector of AB.

8 Find the points where the line:

a $x - 2y = 6$ intersects the curve $\frac{4}{x} - \frac{1}{y} = 2$

b $4x - y = 7$ intersects the curve $\frac{2}{x} + \frac{3}{y} = 1$.

9 The line $3x + 2y = 12$ intersects the curve $\frac{4}{x} + \frac{3}{y} = 3$ at P and Q. Find the midpoint of PQ.

10 For what values of k do the line $y = 3x + k$ and the curve $x^2 + 6x + y^2 = 1$ meet:

a exactly once **b** twice?

11

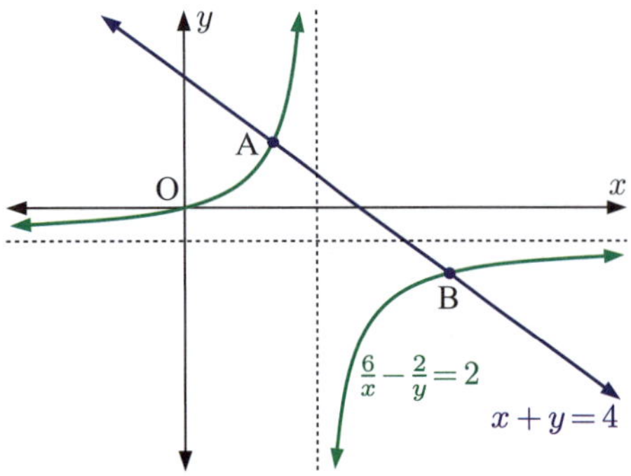

The curve $\frac{6}{x} - \frac{2}{y} = 2$ and the line $x + y = 4$ are graphed alongside.

a Find the coordinates of A and B.

b Let l_1 be the line through B perpendicular to AB. At what point does l_1 meet the curve $\frac{6}{x} - \frac{2}{y} = 2$ again?

c Let l_2 be the line through A perpendicular to AB. Does l_2 meet the curve $\frac{6}{x} - \frac{2}{y} = 2$ again?

G PROBLEM SOLVING WITH QUADRATICS

When real world problems are translated into algebraic form, a quadratic equation may result. To solve these problems, follow these steps:

Step 1: Carefully read the question to make sure you understand the problem. A sketch may be useful.

Step 2: Describe the unknown quantity using a variable such as x.

Step 3: Use the information given to write an equation involving the variable.

Step 4: Solve the equation using a suitable method.

Step 5: Check that any solutions satisfy the equation and are reasonable. For example, it may not be reasonable to have a negative solution or a non-integer solution.

Step 6: Write your answer in a sentence.

Example 28

Self Tutor

The base of an isosceles triangle is 2 cm less than its altitude. Given the area of the triangle is 60 cm^2, find its dimensions.

Let the base of the isosceles triangle be x cm.

$\therefore$ the altitude is $(x+2)$ cm

$$\therefore \tfrac{1}{2}x(x+2) = 60 \quad \{\text{equating areas}\}$$
$$\therefore x^2 + 2x = 120$$
$$\therefore x^2 + 2x - 120 = 0$$
$$\therefore (x-10)(x+12) = 0$$
$$\therefore x = 10 \quad \{\text{since } x > 0\}$$

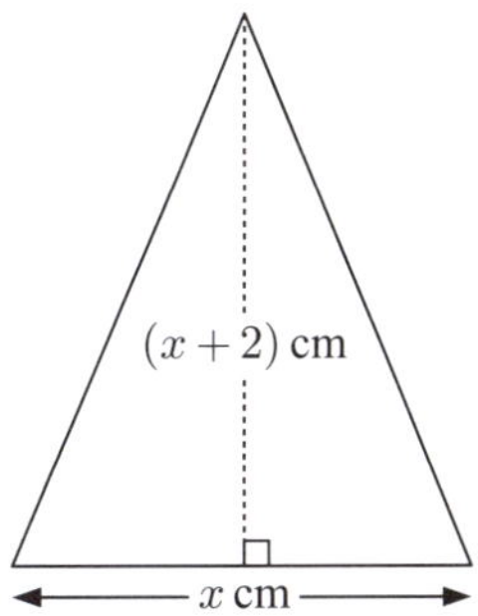

Let the other sides be y cm.

$$\therefore y^2 = 5^2 + 12^2 \quad \{\text{Pythagoras}\}$$
$$\therefore y^2 = 169$$
$$\therefore y = 13 \quad \{\text{since } y > 0\}$$

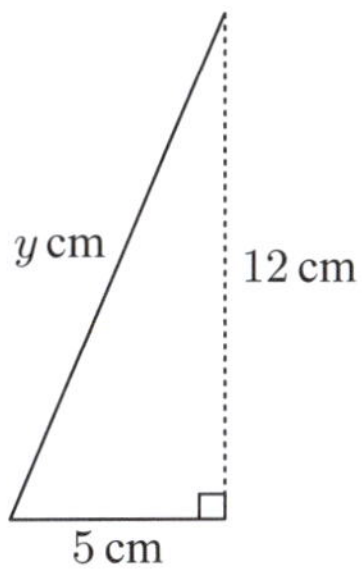

$\therefore$ the triangle has base 10 cm and equal sides 13 cm each.

EXERCISE 3G

1 When 24 is subtracted from the square of x, the result is five times x. Find x.

2 The sum of two numbers is 6, and the sum of their squares is 28. Find these numbers exactly.

3 Two numbers differ by 7, and the sum of their squares is 29. Find the numbers.

4 The base of a triangle is 4 m longer than its altitude. The area of the triangle is 70 m^2. Find the triangle's altitude.

5 A rectangular garden bed was built against an existing brick wall. 24 m of edging was used to enclose 60 m^2. Find the dimensions of the garden bed to the nearest centimetre.

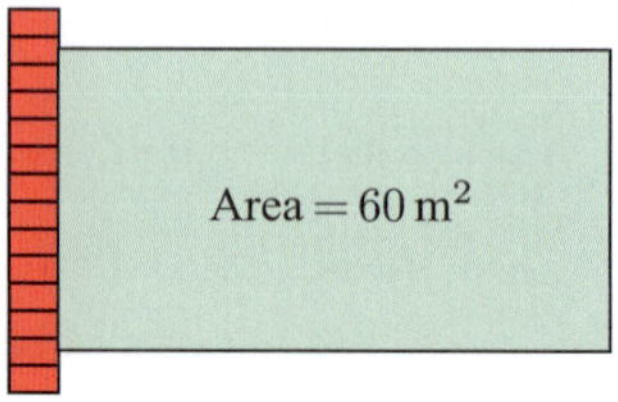

6 A right angled triangle has sides 3 cm and 8 cm respectively less than its hypotenuse. Find the length of the hypotenuse to the nearest millimetre.

7 The base of an isosceles triangle is 1 cm more than its altitude. Given the area of the triangle is 28 cm^2, find its dimensions.

8

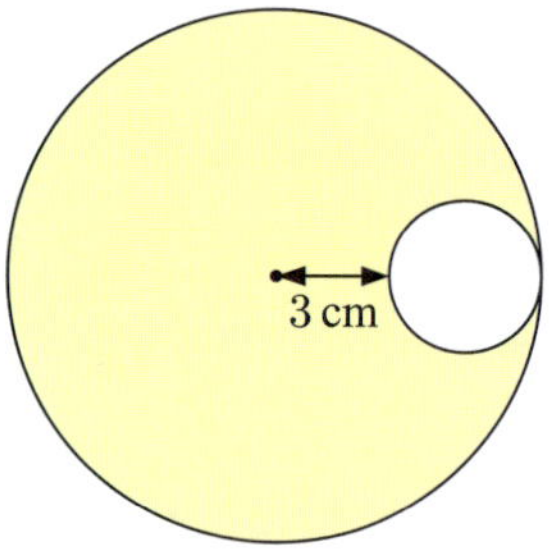

The shaded region has area 45π cm^2.
Find the radius of the large circle.

9 The sum of the first n positive integers is given by the formula $S = \dfrac{n(n+1)}{2}$.

a Find n given the sum of the first n positive integers is 300.

b The sum of the positive integers from n to $2n$ is 459. Find the value of n.

10

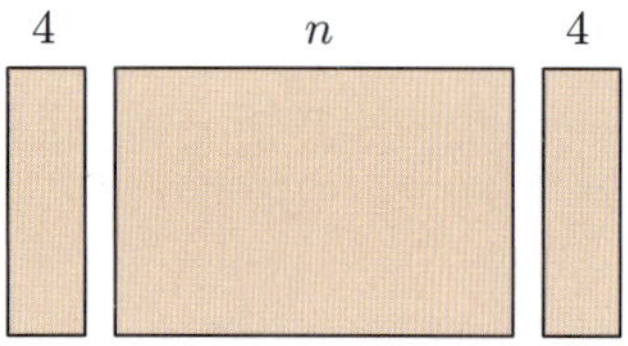

A theatre contains a central block of seats with n seats per row. Blocks on either side contain 4 seats per row. The number of rows is 5 less than the total number of seats per row. In total there are 126 seats in the theatre. Find the value of n.

11 The numerator of a fraction is 3 less than the denominator. If the numerator is increased by 6 and the denominator is increased by 5, the fraction is doubled in value. Find the original fraction.

12 A circular magnet has an inner radius of x cm, an outer radius 2 cm larger, and its depth is the same as the inner radius. The total volume of the magnet is 120π cm^3. Find x.

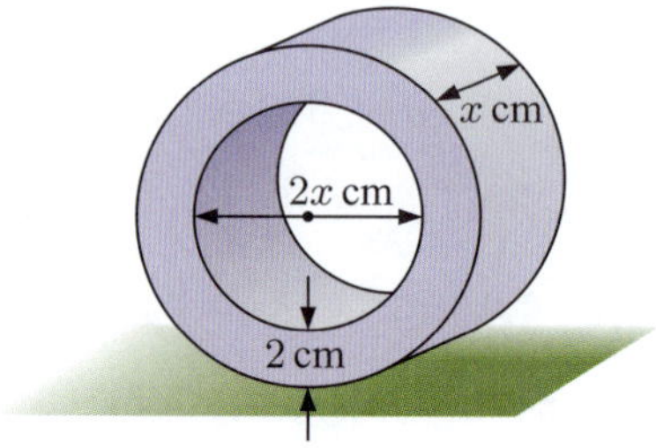

H QUADRATIC OPTIMISATION

The process of finding the **maximum** or **minimum** value of a function is called **optimisation**.

For the quadratic function $f(x) = ax^2 + bx + c$, we have seen that vertex has x-coordinate $\dfrac{-b}{2a}$.

- If $a > 0$, the **minimum** value occurs when $x = \dfrac{-b}{2a}$.

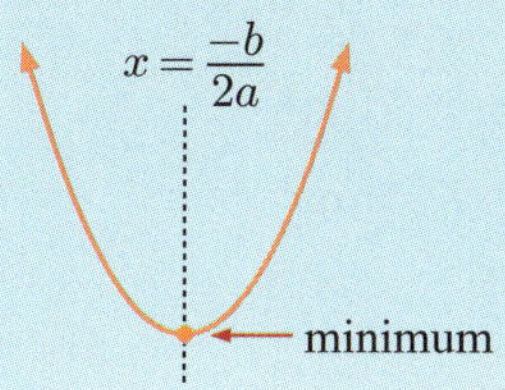

- If $a < 0$, the **maximum** value occurs when $x = \dfrac{-b}{2a}$.

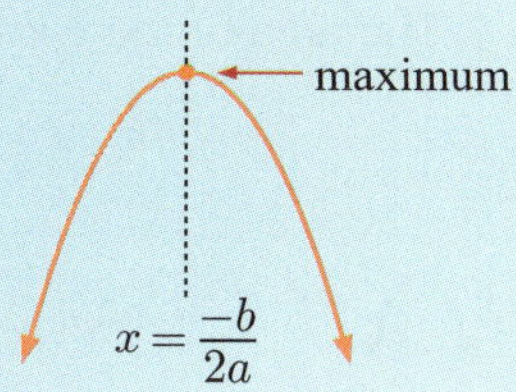

Example 29 — Self Tutor

For each quadratic function, find the value of x that maximises or minimises the function. Hence find the maximum and minimum value.

a $f(x) = x^2 - 5x + 1$ **b** $f(x) = -3x^2 - 2x + 3$

a $f(x) = x^2 - 5x + 1$ has $a = 1$, $b = -5$, and $c = 1$.
Since $a > 0$, the shape is  .

The minimum value occurs when
$x = \dfrac{-b}{2a} = \dfrac{-(-5)}{2(1)} = \frac{5}{2}$.

The minimum value is
$f(\frac{5}{2}) = (\frac{5}{2})^2 - 5(\frac{5}{2}) + 1$
$= -\frac{21}{4}$

b $f(x) = -3x^2 - 2x + 3$ has $a = -3$, $b = -2$, and $c = 3$.
Since $a < 0$, the shape is  .

The maximum value occurs when
$x = \dfrac{-b}{2a} = \dfrac{-(-2)}{2(-3)} = -\frac{1}{3}$.

The maximum value is
$f(-\frac{1}{3}) = -3(-\frac{1}{3})^2 - 2(-\frac{1}{3}) + 3$
$= \frac{10}{3}$

EXERCISE 3H

1 For each quadratic function:

i Find the value of x that maximises or minimises the function.
ii Hence find the maximum or minimum value.

a $f(x) = x^2 + 2x - 1$ **b** $y = 3x^2 - 12x + 2$ **c** $f(x) = -x^2 - 2x + 3$
d $y = 2x^2 - x + 5$ **e** $y = 5x^2 + 4x - 3$ **f** $f(x) = 3x - 2x^2$

2 The temperature in a greenhouse t hours after 7:00 pm is given by $T(t) = \frac{1}{4}t^2 - 6t + 45$ °C for $0 \leqslant t \leqslant 20$.

a At what time is the temperature lowest?
b Find the minimum temperature in the greenhouse for $0 \leqslant t \leqslant 20$.

Example 30

Self Tutor

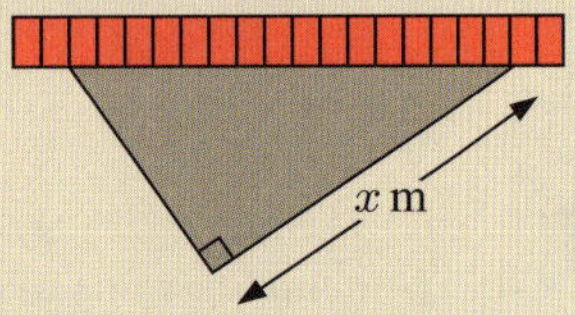

A gardener has 10 m of fencing to enclose a triangular garden bed as shown.

Show that the area of the garden bed is maximised when it is an isosceles triangle.

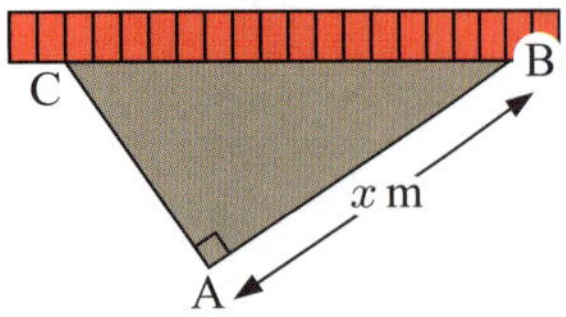

Let $\text{AB} = x$ m

$\therefore$ $\text{AC} = (10 - x)$ m

The area of the garden bed $A = \frac{1}{2} \times \text{AB} \times \text{AC}$ m^2

$= \frac{1}{2}x(10 - x)$ m^2

$= (5x - \frac{1}{2}x^2)$ m^2

Since $a < 0$, the shape is  .

$\therefore$ the maximum occurs at $x = \dfrac{-b}{2a} = \dfrac{-5}{2(-\frac{1}{2})} = 5$

$\therefore$ the area is maximised when $\text{AB} = \text{AC} = 5$ m, which is when the garden bed is an isosceles triangle.

3 A rectangular plot is enclosed by 550 m of fencing and has an area of A square metres. Show that:

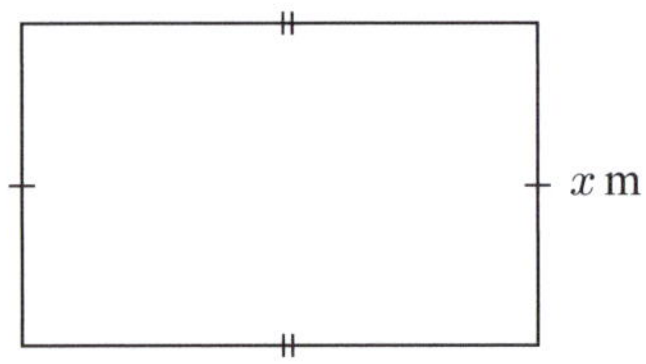

- **a** $A = 275x - x^2$ where x m is the length of one of its sides
- **b** the area is maximised if the rectangle is a square.

4 Three sides of a rectangular paddock are to be fenced, the fourth side being an existing fenceline. If 700 m of fencing is available, what dimensions should be used for the paddock to maximise its area?

5 1200 m of fencing is available to fence three identical pens as shown.

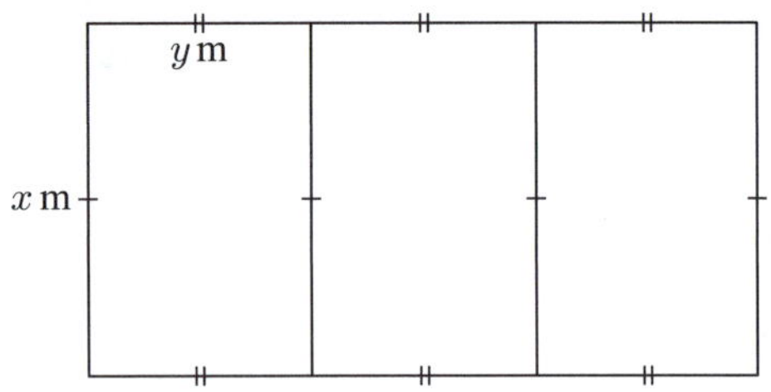

- **a** Explain why $2x + 3y = 600$.
- **b** Show that the area of each pen is given by $A = -\frac{2}{3}x^2 + 200x$ m^2.
- **c** If the area enclosed is to be maximised, what are the dimensions of each pen?

6

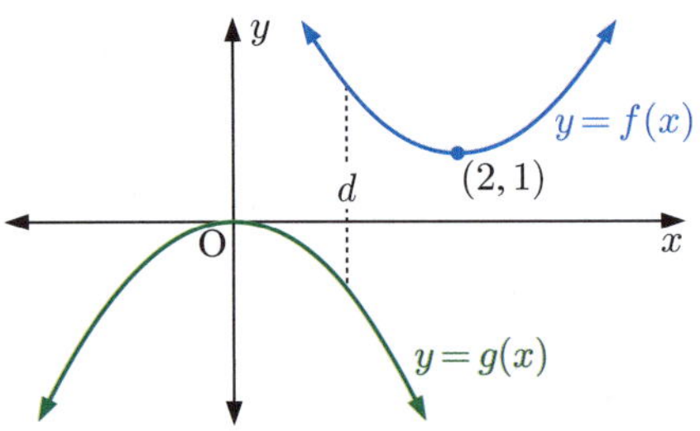

The graphs of $f(x) = (x - 2)^2 + 1$ and $g(x) = -x^2$ are shown alongside.
Find the minimum vertical separation d between the curves.

7 The total cost of producing x toasters per day is given by $C = \$\left(\frac{1}{10}x^2 + 20x + 25\right)$, and the selling price of each toaster is $45. All toasters produced each day are sold.

a Write an expression for the total profit P from producing x toasters per day.

b Find the number of toasters that should be produced to maximise the profit.

Discovery 3 — Sum and product of roots

What to do:

1 Suppose $ax^2 + bx + c = 0$ has roots p and q. Prove that $p + q = \frac{-b}{a}$ and $pq = \frac{c}{a}$.

2 Suppose $2x^2 - 5x + 1 = 0$ has roots p and q. Without finding the values of p and q, find:

a $p + q$ **b** pq **c** $p^2 + q^2$ **d** $\frac{1}{p} + \frac{1}{q}$

3 Find *all* quadratic equations whose roots are:

a one more than the roots of $2x^2 - 5x + 1 = 0$

b the squares of the roots of $2x^2 - 5x + 1 = 0$

c the reciprocals of the roots of $2x^2 - 5x + 1 = 0$.

Puzzle

Find the shortest distance between the graphs $y = -x^2 - 2x - 2$ and $y = x^2 - 6x + 7$.

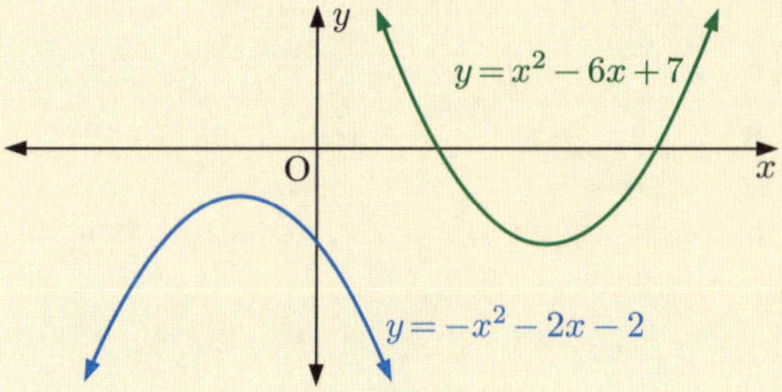

Review set 3A

1 Solve the following equations, giving exact answers:

a $-x^2 - 3x + 4 = 0$ **b** $2x^2 + x - 3 = 0$ **c** $4x^2 - 3x = 27$

d $x^4 - 15 = 2x^2$ **e** $9x^4 = 9x^2 - 2$ **f** $\sqrt{x} = 6 - 2x$

2 Solve using the quadratic formula:

a $3x^2 - 5x - 3 = 0$ **b** $-2x^2 + 4x + 3 = 0$

3 Consider the quadratic function $f(x) = \frac{1}{3}x^2 - \frac{8}{3}x + 5$.

a State the x-intercepts.

b State the equation of the axis of symmetry.

c Find the y-intercept.

d Find the coordinates of the vertex.

e Sketch the function.

f State the range of the function.

4 Solve for x:

a $x^2 - 4x - 21 < 0$ **b** $3x^2 - 2 \geqslant 5x$

5 Find the range of $f(x) = x^2 - 6x - 4$ on the domain $-1 \leqslant x \leqslant 8$.

6 Find the equation of the quadratic function with graph:

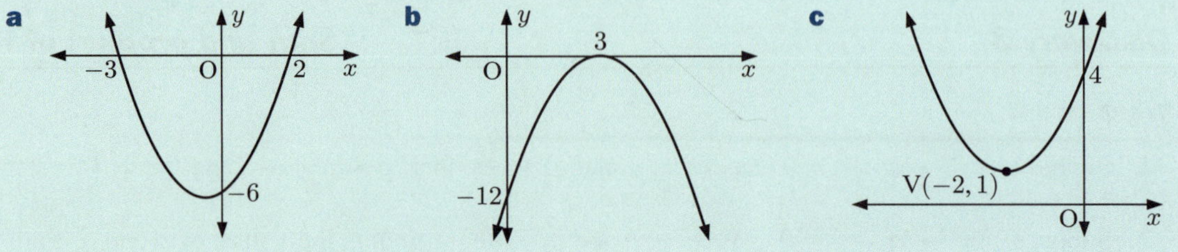

7 Find, in the form $f(x) = ax^2 + bx + c$, the equation of the quadratic whose graph:

a touches the x-axis at -2 and passes through $(1, 4)$

b has vertex $(3, -5)$ and passes through $(6, -3)$.

8 Find the values of k for which $kx^2 + kx - 2 = 0$ has:

a a repeated root **b** two distinct real roots **c** no real roots.

9 **a** Write the quadratic function $f(x) = 2x^2 + 4x - 3$ in the form $f(x) = a(x-h)^2 + k$.

b Hence sketch the graph of the function.

10 Find the point(s) of intersection of:

a $y = 2x^2 + 2x + 3$ and $y = 11 + 2x$

b $f(x) = -5x^2 - x + 1$ and $g(x) = 4x - 9$.

11 For what values of k are there *two* tangents to $f(x)$ passing through $P(0, k)$?

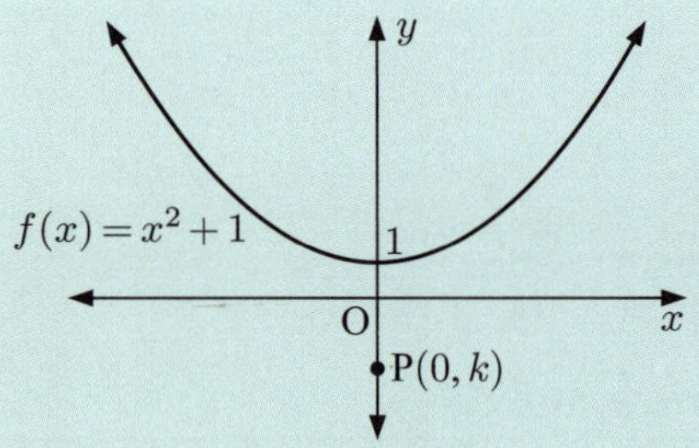

12 Find the points where the line $3x + y = 1$ intersects the curve $x^2 + y^2 = 29$.

13 The line $x + y = 5$ meets the curve $x^2 + y^2 + 3xy + 5x = 1$ at P and Q. Find the equation of the perpendicular bisector of PQ.

14 For each quadratic function:

i Find the value of x that maximises or minimises the function.

ii Find the maximum or minimum value.

a $y = 4x^2 - 8x - 3$ **b** $f(x) = -3x^2 + 7x - 1$

15 Two numbers have a sum of 4, and the sum of their reciprocals is 8. Find the numbers.

16 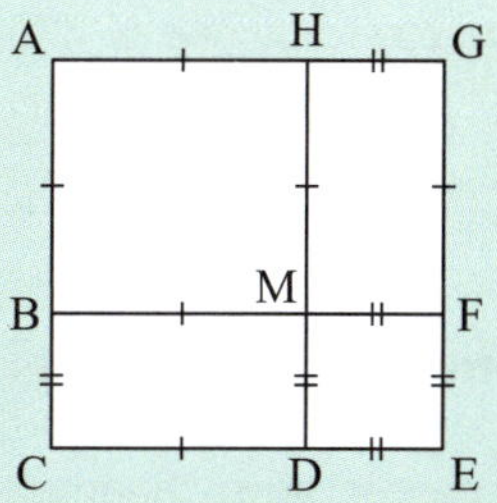

Square AGEC is divided by line segments [DH] and [BF] as shown. Given that rectangle AHDC is similar to rectangle HGFM, find the ratio of the area of AHMB to the area of MFED.

Hint: Let the ratio be $r^2 : 1$.

Review set 3B

1 Solve for x by completing the square:

a $x^2 - 6x - 5 = 0$ **b** $-x^2 + 2x + 2 = 0$ **c** $-3x^2 + 3x + 4 = 0$

2 Solve for x:

a $x^2 + 5x \leqslant 14$ **b** $2x^2 + 7x > 2(x + 6)$

3 Find the range of $y = -2x^2 + 6x + 1$ on the domain $-4 \leqslant x \leqslant 5$.

4 For each quadratic function:

i State the x-intercepts.
ii Find the y-intercept.
iii Hence sketch the function.

a $f(x) = -(x - 1)(x + 3)$ **b** $y = \frac{1}{2}(2x + 1)(x - 4)$

5 For each quadratic function:

i State the coordinates of the vertex.
ii Find the y-intercept.
iii Hence sketch the function.

a $y = 3(x + 3)^2 + 2$ **b** $f(x) = -2(x - 4)^2 + 1$

6 Find the equation of the quadratic function with graph:

a 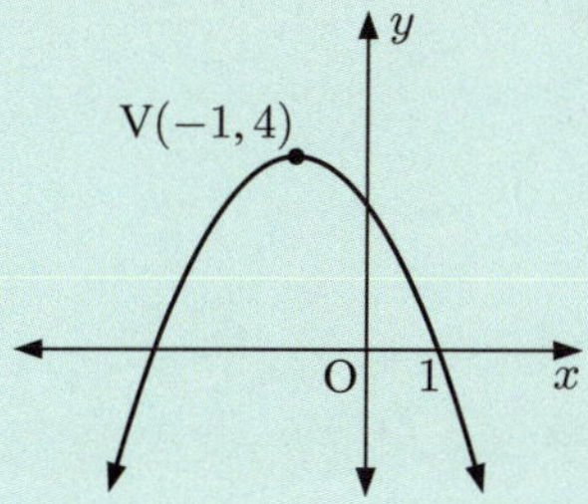

b 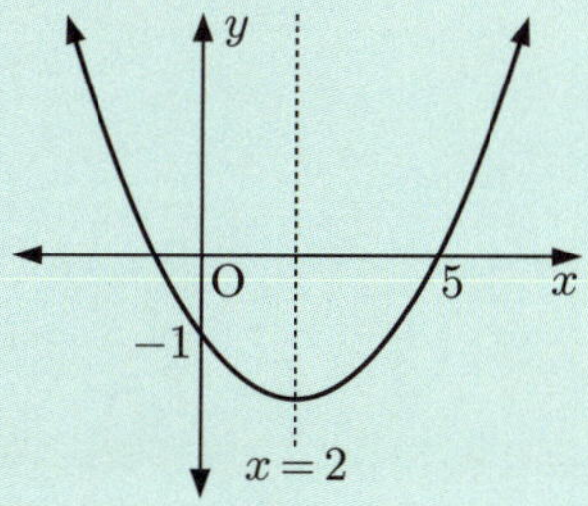

c 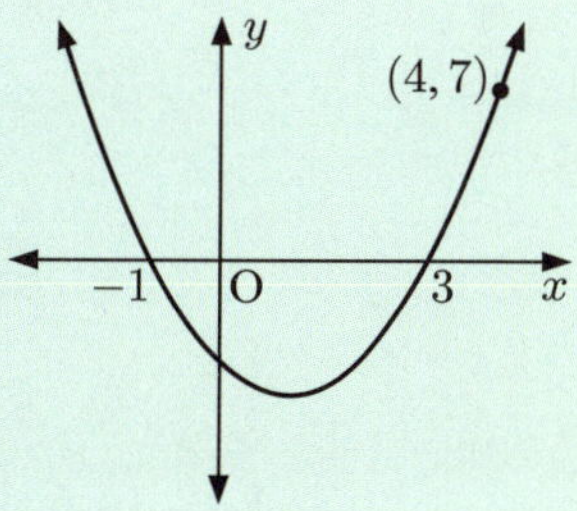

7 For what value(s) of m does the line $y = mx + 1$ *touch* the curve $y = 3x^2 - x + 3$?

8 Sketch the graph of $f(x) = \left| x^2 + x - 20 \right|$.

9 Use the discriminant to determine the relationship between the graph of each function and the x-axis:

a $f(x) = 4x^2 - 5x - 2$ **b** $f(x) = -2x^2 + 3x - 4$ **c** $y = 3x^2 - 12x + 12$

10 Find a given that $f(x) = 3x - 6$ is a tangent to $g(x) = ax^2 + 4x - 2$.

11 The hypotenuse of a right angled triangle is 10 cm long. Of the other sides, one is 3 cm shorter than the other. Find the length of the shortest side.

12 The line $x - 2y = 3$ meets the curve $x^2 + 2y^2 - 2xy + 3x = 8$ at P and Q. Find the distance between P and Q.

13 The line $4x - 3y = 2$ intersects the curve $\frac{3}{y} - \frac{1}{x} = 1$ at A and B. Find the midpoint of AB.

14

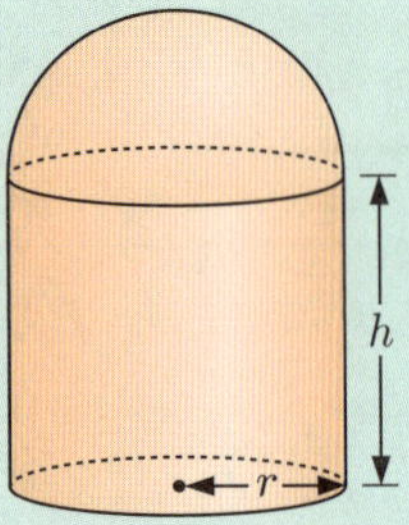

A takeaway milkshake container is cylindrical with a hemispherical lid on top.

The height h of the container is 7 cm greater than its base radius r.

The surface area of the container and lid is 96π cm^2.

Find the base radius of the container.

15 Infinitely many rectangles can be inscribed within the isosceles triangle shown alongside. Let AB $= y$ cm and BC $= x$ cm.

a Use similar triangles to write y in terms of x.

b Find the dimensions of rectangle ABCD of maximum area.

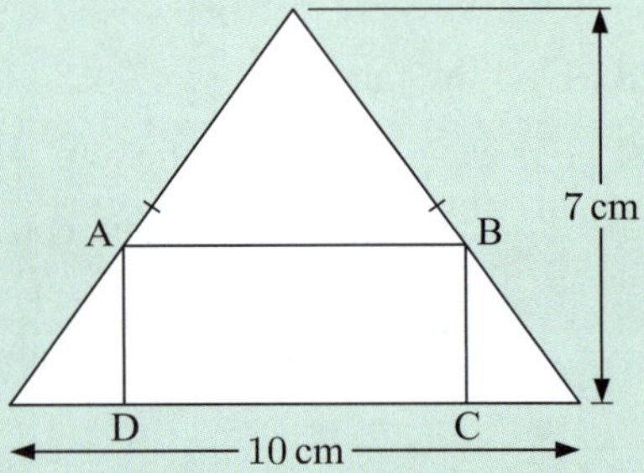

16

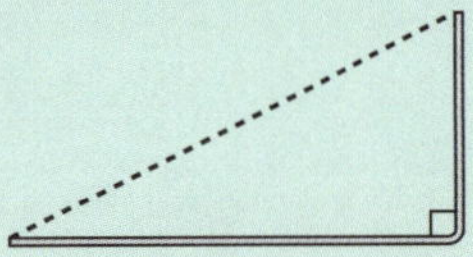

Is it possible to bend a 20 cm length of wire to form the perpendicular sides of a right angled triangle with area at least 56 cm^2?

4

Surds, indices, and exponentials

Contents:

Opening problem

A lotus plant initially covers an area of 40 cm^2. The area it covers increases by 20% each week.

Things to think about:

- **a** Does the area covered by the plant increase by a constant amount each week?
- **b** Can you explain why the area covered by the lotus plant after n weeks is given by the function $A(n) = 40 \times (1.2)^n$ cm^2?
- **c** What does the graph of $A(n)$ look like?
- **d** What area is covered by the lotus plant after 3 weeks?
- **e** Would you expect this growth in the area covered by the lotus plant to continue in the long term?

We often deal with numbers that are repeatedly multiplied together. Mathematicians use **indices**, also called **powers** or **exponents**, to represent such numbers.

A SURDS

A **radical** is any number which is written with the **radical sign** $\sqrt{\ }$.

A **surd** is a real, *irrational* radical such as $\sqrt{2}$, $\sqrt{3}$, $\sqrt{5}$, or $\sqrt{6}$. We encountered surds in **Chapter 3** in solutions to some quadratic equations. $\sqrt{9}$ is a radical, but not a surd as it simplifies to 3.

$\sqrt{a}$ is the non-negative number such that $\sqrt{a} \times \sqrt{a} = a$.

Important properties of surds are:

- $\sqrt{a}$ is never negative, so $\sqrt{a} \geqslant 0$.
- $\sqrt{a}$ is only real if $a \geqslant 0$.
- $\sqrt{ab} = \sqrt{a} \times \sqrt{b}$ for $a \geqslant 0$ and $b \geqslant 0$.
- $\sqrt{\dfrac{a}{b}} = \dfrac{\sqrt{a}}{\sqrt{b}}$ for $a \geqslant 0$ and $b > 0$.

Example 1 — Self Tutor

Write as a single surd:

a $\sqrt{5} \times \sqrt{2}$ **b** $\dfrac{\sqrt{14}}{\sqrt{7}}$

a
$$\begin{aligned} &\sqrt{5} \times \sqrt{2} \\ &= \sqrt{5 \times 2} \\ &= \sqrt{10} \end{aligned}$$

b
$$\begin{aligned} &\frac{\sqrt{14}}{\sqrt{7}} \\ &= \sqrt{\frac{14}{7}} \\ &= \sqrt{2} \end{aligned}$$

EXERCISE 4A.1

1 Write as a single surd or rational number:

a $\sqrt{5}\times\sqrt{5}$ **b** $\sqrt{2}\times\sqrt{3}$ **c** $(\sqrt{7})^2$ **d** $\sqrt{3}\times\sqrt{5}$

e $4\sqrt{3}\times\sqrt{3}$ **f** $2\sqrt{3}\times5\sqrt{3}$ **g** $(3\sqrt{5})^2$ **h** $\dfrac{\sqrt{22}}{\sqrt{2}}$

i $\dfrac{\sqrt{35}}{\sqrt{7}}$ **j** $\dfrac{\sqrt{32}}{\sqrt{8}}$ **k** $\dfrac{\sqrt{3}\times\sqrt{12}}{\sqrt{18}}$ **l** $\dfrac{\sqrt{42}}{\sqrt{7}\times\sqrt{6}}$

Example 2 — Self Tutor

Write $\sqrt{20}$ in the form $a\sqrt{b}$ where a and b are integers and a is as large as possible.

$\sqrt{20}$
$=\sqrt{4\times5}$ {4 is the largest perfect square factor of 20}
$=\sqrt{4}\times\sqrt{5}$
$=2\sqrt{5}$

2 Write in the form $a\sqrt{b}$ where a and b are integers and a is as large as possible:

a $\sqrt{12}$ **b** $\sqrt{18}$ **c** $\sqrt{27}$ **d** $\sqrt{28}$

e $\sqrt{45}$ **f** $\sqrt{32}$ **g** $\sqrt{63}$ **h** $\sqrt{75}$

i $\sqrt{125}$ **j** $\sqrt{80}$ **k** $\sqrt{132}$ **l** $\sqrt{128}$

OPERATING WITH SURDS

The rules for adding, subtracting, and multiplying by surds are the same as those for ordinary algebra.

Example 3 — Self Tutor

Simplify: **a** $\sqrt{2}+3\sqrt{2}$ **b** $7\sqrt{5}-4\sqrt{5}$

a $\sqrt{2}+3\sqrt{2}$
$=4\sqrt{2}$

b $7\sqrt{5}-4\sqrt{5}$
$=3\sqrt{5}$

Example 4 — Self Tutor

Simplify:

a $\sqrt{3}(7-\sqrt{3})$ **b** $(2+\sqrt{5})(1+3\sqrt{5})$

a $\sqrt{3}(7-\sqrt{3})$
$=\sqrt{3}\times7+\sqrt{3}\times(-\sqrt{3})$
$=7\sqrt{3}-3$

b $(2+\sqrt{5})(1+3\sqrt{5})$
$=2+2(3\sqrt{5})+\sqrt{5}(1)+\sqrt{5}(3\sqrt{5})$
$=2+6\sqrt{5}+\sqrt{5}+15$
$=17+7\sqrt{5}$

EXERCISE 4A.2

1 Simplify:

a $4\sqrt{2}+2\sqrt{2}$ **b** $2\sqrt{3}-\sqrt{3}$ **c** $\sqrt{5}-3\sqrt{5}$ **d** $\sqrt{7}+3\sqrt{7}$

e $9\sqrt{6}+2\sqrt{6}$ **f** $5\sqrt{5}-7\sqrt{5}$ **g** $-\sqrt{10}+5\sqrt{10}$ **h** $\sqrt{3}+\sqrt{3}-\sqrt{3}$

2 Simplify:

a $\sqrt{5}(2-\sqrt{5})$ **b** $\sqrt{3}(\sqrt{3}+2)$ **c** $\sqrt{7}(2+3\sqrt{7})$ **d** $\sqrt{6}(2\sqrt{6}-1)$

e $-\sqrt{2}(3+\sqrt{2})$ **f** $3\sqrt{8}(\sqrt{8}-5)$ **g** $-\sqrt{5}(\sqrt{5}-4)$ **h** $-2\sqrt{3}(5-7\sqrt{3})$

3 Simplify:

a $(2+\sqrt{3})(4+\sqrt{3})$ **b** $(1+3\sqrt{2})(5+\sqrt{2})$ **c** $(6-\sqrt{5})(2+3\sqrt{5})$

d $(\sqrt{2}+4)(1-2\sqrt{2})$ **e** $(\sqrt{10}-3)(3\sqrt{10}-1)$ **f** $(3\sqrt{7}-2)(5-2\sqrt{7})$

Example 5 — Self Tutor

Simplify:

a $(3+\sqrt{7})^2$ **b** $(5+2\sqrt{2})(5-2\sqrt{2})$

a
$$(3+\sqrt{7})^2$$
$$=3^2+2(3)(\sqrt{7})+(\sqrt{7})^2$$
$$=9+6\sqrt{7}+7$$
$$=16+6\sqrt{7}$$

b
$$(5+2\sqrt{2})(5-2\sqrt{2})$$
$$=5^2-(2\sqrt{2})^2$$
$$=25-(4\times 2)$$
$$=17$$

4 Simplify:

a $(2+\sqrt{3})^2$ **b** $(\sqrt{2}+1)^2$ **c** $(4-\sqrt{5})^2$ **d** $(\sqrt{6}-3)^2$

e $(1+2\sqrt{5})^2$ **f** $(3\sqrt{2}+1)^2$ **g** $(7-3\sqrt{3})^2$ **h** $(-\sqrt{6}-2)^2$

5 Simplify:

a $(2+\sqrt{3})(2-\sqrt{3})$ **b** $(\sqrt{5}+1)(\sqrt{5}-1)$ **c** $(\sqrt{6}-2)(\sqrt{6}+2)$

d $(3\sqrt{2}+4)(3\sqrt{2}-4)$ **e** $(5+2\sqrt{6})(5-2\sqrt{6})$ **f** $(4\sqrt{3}-7)(4\sqrt{3}+7)$

DIVISION BY SURDS

Numbers like $\frac{5}{\sqrt{3}}$ and $\frac{4}{1+\sqrt{2}}$ involve dividing by a surd.

We usually "simplify" these numbers by rewriting them without the surd in the denominator.

For any fraction of the form $\frac{b}{\sqrt{a}}$, we can remove the surd from the denominator by multiplying by $\frac{\sqrt{a}}{\sqrt{a}}$.

Since $\frac{\sqrt{a}}{\sqrt{a}}=1$, this does not change the value of the fraction.

Example 6 — Self Tutor

Write with an integer denominator:

a $\dfrac{5}{\sqrt{3}}$ **b** $\dfrac{12}{\sqrt{2}}$

a $\dfrac{5}{\sqrt{3}}$

$= \dfrac{5}{\sqrt{3}} \times \dfrac{\sqrt{3}}{\sqrt{3}}$

$= \dfrac{5\sqrt{3}}{3}$

b $\dfrac{12}{\sqrt{2}}$

$= \dfrac{12}{\sqrt{2}} \times \dfrac{\sqrt{2}}{\sqrt{2}}$

$= \dfrac{{}^{6}\cancel{12}\sqrt{2}}{\cancel{2}_{1}}$

$= 6\sqrt{2}$

For any fraction of the form $\dfrac{c}{a+\sqrt{b}}$, we can remove the surd from the denominator by multiplying by $\dfrac{a-\sqrt{b}}{a-\sqrt{b}}$.

Expressions such as $a+\sqrt{b}$ and $a-\sqrt{b}$ are known as **radical conjugates**. They are identical except for the sign in the middle.

The product of radical conjugates is rational since we have the difference between two squares. Multiplying by $\dfrac{a-\sqrt{b}}{a-\sqrt{b}}$ therefore produces a rational denominator, and the process is sometimes called **rationalising the denominator**.

Example 7 — Self Tutor

Rationalise the denominator of $\dfrac{4}{5-\sqrt{2}}$.

$$\frac{4}{5-\sqrt{2}} = \left(\frac{4}{5-\sqrt{2}}\right) \times \left(\frac{5+\sqrt{2}}{5+\sqrt{2}}\right)$$

$$= \frac{4(5+\sqrt{2})}{5^2-(\sqrt{2})^2}$$

$$= \frac{20+4\sqrt{2}}{23}$$

EXERCISE 4A.3

1 Write with an integer denominator:

a $\dfrac{1}{\sqrt{2}}$ **b** $\dfrac{3}{\sqrt{5}}$ **c** $\dfrac{7}{\sqrt{3}}$ **d** $\dfrac{11}{\sqrt{6}}$ **e** $\dfrac{9}{\sqrt{10}}$

f $\dfrac{3}{\sqrt{3}}$ **g** $\dfrac{10}{\sqrt{2}}$ **h** $\dfrac{30}{\sqrt{5}}$ **i** $\dfrac{2}{\sqrt{6}}$ **j** $\dfrac{6}{\sqrt{8}}$

k $\dfrac{\sqrt{2}}{\sqrt{5}}$ **l** $\dfrac{\sqrt{7}}{\sqrt{3}}$ **m** $\dfrac{\sqrt{5}}{3\sqrt{2}}$ **n** $\dfrac{1}{4\sqrt{6}}$ **o** $\dfrac{1}{(\sqrt{2})^3}$

2 Rationalise the denominator of:

a $\dfrac{1}{4+\sqrt{3}}$ **b** $\dfrac{1}{3-\sqrt{7}}$ **c** $\dfrac{2}{2+\sqrt{11}}$ **d** $\dfrac{5}{1-\sqrt{6}}$

e $\dfrac{7}{2-\sqrt{3}}$ **f** $\dfrac{\sqrt{3}}{\sqrt{5}+1}$ **g** $\dfrac{2+\sqrt{3}}{5+\sqrt{3}}$ **h** $\dfrac{\sqrt{7}}{2-\sqrt{7}}$

i $\dfrac{-3\sqrt{5}}{4+\sqrt{5}}$ **j** $\dfrac{4}{3-\sqrt{10}}$ **k** $\dfrac{12}{\sqrt{7}+5}$ **l** $\dfrac{\sqrt{8}+1}{\sqrt{8}-1}$

Example 8 Self Tutor

Write $\dfrac{1}{5+\sqrt{2}}$ in the form $a+b\sqrt{2}$ where $a, b \in \mathbb{Q}$.

$$\begin{aligned}\frac{1}{5+\sqrt{2}} &= \left(\frac{1}{5+\sqrt{2}}\right)\times\left(\frac{5-\sqrt{2}}{5-\sqrt{2}}\right)\\ &= \frac{5-\sqrt{2}}{25-2}\\ &= \frac{5-\sqrt{2}}{23}\\ &= \tfrac{5}{23}-\tfrac{1}{23}\sqrt{2}\end{aligned}$$

3 Write in the form $a+b\sqrt{2}$ where $a, b \in \mathbb{Q}$:

a $\dfrac{3}{\sqrt{2}-3}$ **b** $\dfrac{4}{2+\sqrt{2}}$ **c** $\dfrac{\sqrt{2}}{\sqrt{2}-5}$ **d** $\dfrac{-2\sqrt{2}}{\sqrt{2}+1}$

4 Write in the form $a+b\sqrt{3}$ where $a, b \in \mathbb{Q}$:

a $\dfrac{4}{1-\sqrt{3}}$ **b** $\dfrac{6}{\sqrt{3}+2}$ **c** $\dfrac{\sqrt{3}}{2-\sqrt{3}}$ **d** $\dfrac{1+2\sqrt{3}}{3+\sqrt{3}}$

5 **a** Suppose a, b, and c are integers, $c>0$. Show that $(a+b\sqrt{c})(a-b\sqrt{c})$ is also an integer.

b Write with an integer denominator:

i $\dfrac{1}{1+2\sqrt{3}}$ **ii** $\dfrac{\sqrt{2}}{3\sqrt{2}-5}$ **iii** $\dfrac{\sqrt{2}-1}{3-2\sqrt{2}}$

6 **a** Suppose a and b are positive integers. Show that $(\sqrt{a}+\sqrt{b})(\sqrt{a}-\sqrt{b})$ is also an integer.

b Write with an integer denominator:

i $\dfrac{1}{\sqrt{2}+\sqrt{3}}$ **ii** $\dfrac{\sqrt{3}}{\sqrt{3}-\sqrt{5}}$ **iii** $\dfrac{\sqrt{11}-\sqrt{14}}{\sqrt{11}+\sqrt{14}}$

7 Solve the equation $2x-3\sqrt{3}=1-x\sqrt{3}$. Give your solution in the form $x=a+b\sqrt{3}$, where $a, b \in \mathbb{Z}$.

8 Find the positive solution of the equation $(9+\sqrt{5})x^2+(5-2\sqrt{5})x-5=0$. Give your answer in the form $a+b\sqrt{5}$, where $a, b \in \mathbb{Q}$.

B INDICES

If n is a positive integer, then a^n is the product of n factors of a.

$$a^n = \underbrace{a \times a \times a \times a \times \ .... \ \times a}_{n \text{ factors}}$$

We say that a is the **base**, and n is the **index** or **exponent**.

base → 2^7 ← index, exponent, or power

NEGATIVE BASES

$(-1)^1 = -1$
$(-1)^2 = (-1) \times (-1) = 1$
$(-1)^3 = (-1) \times (-1) \times (-1) = -1$
$(-1)^4 = (-1) \times (-1) \times (-1) \times (-1) = 1$

$(-2)^1 = -2$
$(-2)^2 = (-2) \times (-2) = 4$
$(-2)^3 = (-2) \times (-2) \times (-2) = -8$
$(-2)^4 = (-2) \times (-2) \times (-2) \times (-2) = 16$

From the patterns above we can see that:

A **negative** base raised to an **odd** index is **negative**.
A **negative** base raised to an **even** index is **positive**.

EXERCISE 4B

1 List the first five powers of:

a 2 **b** -2 **c** 3 **d** -3 **e** 4 **f** -4

2 Copy and complete the table alongside:

n	5^n	6^n	7^n
1			
2			
3			
4			

3 Evaluate:

a $(-1)^3$ **b** $(-1)^4$ **c** $(-1)^7$ **d** $(-1)^{15}$ **e** $(-1)^{10}$ **f** -1^{10}
g $-(-1)^{10}$ **h** $(-2)^6$ **i** -2^6 **j** -2^8 **k** $-(-2)^8$ **l** $(-4)^3$

4 Use your calculator to evaluate:

a 2^8 **b** 9^4 **c** -7^6 **d** $(-7)^6$ **e** 6^5 **f** $-(-6)^5$
g 2.5^7 **h** $(-2.5)^7$ **i** -2.5^7 **j** $(0.5)^4$ **k** $-(0.5)^4$ **l** $-(-0.5)^4$

5 Use your calculator to evaluate:

a $2^4 \times 2^6$ **b** 2^{10} **c** $3^4 \times 5^4$ **d** 15^4 **e** $\left(\frac{3}{2}\right)^5$ **f** $\frac{3^5}{2^5}$
g $(4^2)^3$ **h** 4^6 **i** $\frac{5^5}{5^2}$ **j** 5^3 **k** 18^0 **l** $(0.231)^0$

Comment on your results.

6 Look for a pattern in the sequence $2^1, 2^2, 2^3, 2^4, ...$ and hence find the last digit of 2^{2019}.

Discovery 1 — Prime factorisation and HCF

The **highest common factor** (HCF) of two integers a and b is the largest integer which divides both a and b.

In this Discovery we explore the relationship between the HCF of two integers and their **prime factorisations**.

What to do:

1 Use the **prime factorisation tree** alongside to write down the prime factorisation of 72.

The prime factorisation of a number involves writing the number as a product of prime numbers.

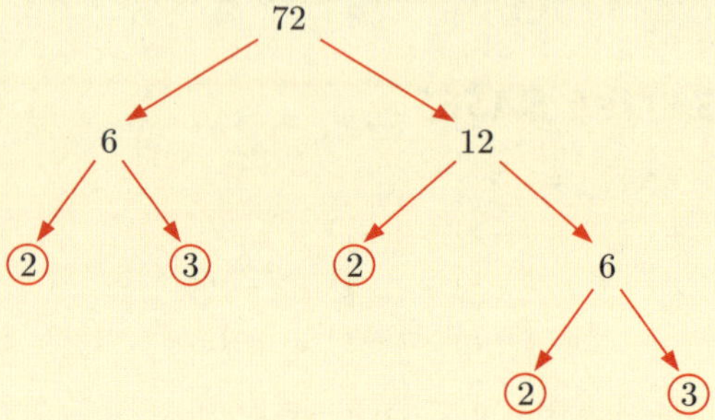

2 Find the prime factorisation for each number:

a 28 **b** 42 **c** 81 **d** 27 **e** 121 **f** 77
g 32 **h** 144 **i** 154 **j** 77 **k** 300 **l** 255

3 Find the HCF of each of the following pairs of numbers, and determine its prime factorisation:

a 28 and 42 **b** 81 and 27 **c** 121 and 77
d 32 and 144 **e** 154 and 77 **f** 300 and 255

4 Use your answers to **2** and **3** to describe the relationship between the HCF of two integers and their prime factorisations.

C INDEX LAWS

The **index laws** are a set of rules for how we work with indices. It is important that they apply for all real numbers (except 0 in some cases) and that they are consistent with one another.

The index laws for $m, n \in \mathbb{Z}$ are:

$a^m \times a^n = a^{m+n}$	To **multiply** numbers with the **same base**, keep the base and **add** the indices.
$\dfrac{a^m}{a^n} = a^{m-n}, \quad a \neq 0$	To **divide** numbers with the **same base**, keep the base and **subtract** the indices.
$(a^m)^n = a^{m \times n}$	When **raising** a **power** to a **power**, keep the base and **multiply** the indices.
$(ab)^n = a^n b^n$	The power of a product is the product of the powers.
$\left(\dfrac{a}{b}\right)^n = \dfrac{a^n}{b^n}, \quad b \neq 0$	The power of a quotient is the quotient of the powers.
$a^0 = 1, \quad a \neq 0$	Any non-zero number raised to the power of zero is 1.

$a^{-n} = \dfrac{1}{a^n}$ and $\dfrac{1}{a^{-n}} = a^n$ and in particular $a^{-1} = \dfrac{1}{a}, \quad a \neq 0.$

Example 9 Self Tutor

Simplify using the index laws:

a $2^5 \times 2^3$ **b** $\frac{7^2}{7^4}$ **c** $(x^5)^2$

a $2^5 \times 2^3$
$= 2^{5+3}$
$= 2^8$

b $\frac{7^2}{7^4}$
$= 7^{2-4}$
$= 7^{-2}$
$= \frac{1}{49}$

c $(x^5)^2$
$= x^{5\times 2}$
$= x^{10}$

EXERCISE 4C

1 Simplify using the index laws:

a $3^2 \times 3^4$ **b** $k^3 \times k^2$ **c** $\frac{5^6}{5^2}$ **d** $\frac{b}{b^3}$ **e** $(6^3)^5$ **f** $(m^4)^2$

g $x^4 \times x^3$ **h** $\frac{q^5}{q^7}$ **i** $2^n \times 2^3$ **j** $(11^2)^k$ **k** $\frac{4^x}{4^6}$ **l** $(y^z)^3$

Example 10 Self Tutor

Write as a power of 2:

a 8 **b** $\frac{1}{8}$ **c** 1 **d** $\frac{16}{2^n}$ **e** 4^k

a 8
$= 2 \times 2 \times 2$
$= 2^3$

b $\frac{1}{8}$
$= \frac{1}{2^3}$
$= 2^{-3}$

c 1
$= 2^0$

d $\frac{16}{2^n}$
$= \frac{2^4}{2^n}$
$= 2^{4-n}$

e 4^k
$= (2^2)^k$
$= 2^{2k}$

2 Write as a power of 2:

a 2 **b** $\frac{1}{2}$ **c** 16 **d** $\frac{1}{16}$

e 32 **f** $\frac{1}{32}$ **g** 128 **h** $\frac{1}{128}$

3 Write as a power of 5:

a 5 **b** $\frac{1}{5}$ **c** 25 **d** $\frac{1}{25}$

e 125 **f** $\frac{1}{125}$ **g** 1

4 Write as a power of 2:

a 2×2^n **b** 4×2^m **c** 16×2^{3x} **d** $2 \times 2^{y+1}$ **e** $2^{2k} \times 8$

f $\frac{2}{2^a}$ **g** $\frac{2^a}{2}$ **h** $\frac{2^{t+1}}{8}$ **i** $(2^{5-p})^{-2}$ **j** $\frac{8^c}{2^{6-c}}$

5 Write as a power of 5:

a $5^q \times 5$ **b** 25^r **c** 125×5^{-k} **d** $\frac{5^z}{5^{-z}}$ **e** $\frac{25}{5^y}$

6 Write as a power of 3:

a $(9^x)^2$ **b** 81×3^s **c** $\frac{3^{4n}}{81}$ **d** $\frac{9^w}{3^{w-1}}$ **e** $\frac{3^{p+2}}{27^{p-1}}$

Example 11 — Self Tutor

Write in simplest form, without brackets:

a $(3x^2)^3$ **b** $\left(-\frac{a}{b^2}\right)^5$

a $(3x^2)^3$
$= 3^3 \times (x^2)^3$
$= 27x^6$

b $\left(-\frac{a}{b^2}\right)^5$
$= \frac{(-1)^5 \times a^5}{(b^2)^5}$
$= -\frac{a^5}{b^{10}}$

7 Write without brackets:

a $(2x)^3$ **b** $(4k)^2$ **c** $(xy)^3$ **d** $(yz)^4$ **e** $\left(\frac{p}{q}\right)^3$

f $\left(\frac{m}{2}\right)^2$ **g** $\left(\frac{5}{n}\right)^3$ **h** $\left(\frac{3a}{c}\right)^2$ **i** $\left(\frac{w}{2x}\right)^4$ **j** $\left(\frac{6k}{l}\right)^0$

8 Write in simplest form, without brackets:

a $(-3x)^3$ **b** $(2y^2)^4$ **c** $(-6z^2)^2$ **d** $(-4k^2)^3$

e $\left(\frac{n}{m^2}\right)^5$ **f** $\left(\frac{u^3}{3v}\right)^2$ **g** $\left(-\frac{5w}{h}\right)^4$ **h** $\left(-\frac{7d^3}{b^2}\right)^2$

i $\frac{(2x^2y)^2}{x}$ **j** $\frac{(4a^2b)^3}{2ab^2}$ **k** $\frac{(-5a^6b^3)^2}{5b^8}$ **l** $\frac{(-2x^7y^4)^3}{4x^3y^{15}}$

Example 12 — Self Tutor

Write $\frac{x^{-2}}{y^3z^{-1}}$ without negative indices.

$x^{-2} = \frac{1}{x^2}$ and $\frac{1}{z^{-1}} = z^1$

$\therefore \frac{x^{-2}}{y^3z^{-1}} = \frac{z}{x^2y^3}$

9 Write without negative indices:

a ab^{-3} **b** $(pq)^{-2}$ **c** $(2c^{-1}d)^2$ **d** $(5xy^{-1})^2$ **e** $(3a^{-1})^{-1}$

f $\frac{p^3q^{-1}}{r^2}$ **g** $\frac{2}{a^{-3}}$ **h** $\frac{m^{-5}}{n^{-2}}$ **i** $\frac{zw^{-2}}{u^{-1}}$ **j** $\frac{2pq^{-1}}{r^{-2}}$

Example 13 — Self Tutor

Write $\dfrac{1}{3^{x-4}}$ in non-fractional form.

$$\frac{1}{3^{x-4}} = 3^{-(x-4)}$$
$$= 3^{-x+4}$$
$$= 3^{4-x}$$

10 Write in non-fractional form:

a $\dfrac{1}{x^n}$ **b** $\dfrac{1}{2^x}$ **c** $\dfrac{1}{2^{1-n}}$ **d** $\dfrac{p^2}{q^x}$ **e** $\dfrac{x^{m-1}}{x^{-2}}$

11 Write in simplest rational form:

a $\left(\frac{1}{2}\right)^3$ **b** $\left(\frac{3}{4}\right)^{-1}$ **c** $\left(\frac{5}{6}\right)^{-1}$ **d** $\left(\frac{2}{3}\right)^{-2}$

e $\left(\frac{59}{60}\right)^0$ **f** $3^2 + 3^{-2}$ **g** $\left(2\frac{1}{2}\right)^{-2}$ **h** $4^2 + 4^1 + 4^{-1}$

12 Write as the product of powers of 2, 3, and/or 5:

a $\frac{1}{12}$ **b** $\frac{1}{25}$ **c** $\frac{2}{9}$ **d** $\frac{1}{60}$

e $\frac{10}{27}$ **f** $\dfrac{9^n}{24}$ **g** $\dfrac{6^m}{45}$ **h** $\dfrac{18}{10^p}$

Example 14 — Self Tutor

Solve simultaneously: $xy = 2$
$x^3y^4 = 32$

Substituting $y = \dfrac{2}{x}$ into $x^3y^4 = 32$ gives

$$x^3\left(\frac{2}{x}\right)^4 = 32$$
$$\therefore\ x^3\left(\frac{16}{x^4}\right) = 32$$
$$\therefore\ \frac{16}{x} = 32$$
$$\therefore\ x = \tfrac{16}{32} = \tfrac{1}{2} \quad \text{and} \quad y = \frac{2}{\frac{1}{2}} = 4$$

13 Solve simultaneously:

a $xy = 5$, $x^3y^2 = 50$ **b** $xy = 3$, $x^3y^4 = \frac{27}{2}$ **c** $x^2y = -1$, $x^4y^3 = -\frac{1}{4}$

D RATIONAL INDICES

The index laws used previously can also be applied to **rational indices**, or indices which are written as a fraction.

We defined the notation a^n to mean "a multiplied together n times". Since we cannot multiply a together "half a time", the notation $a^{\frac{1}{2}}$ is an extension of this meaning. Our goal is to extend the meaning of a^n so that the index law

$$a^n a^m = a^{n+m}$$

remains true.

In this course we restrict ourselves to cases where $a > 0$. For these values of a, this law holds for all rational indices.

For $a > 0$, notice that $a^{\frac{1}{2}} \times a^{\frac{1}{2}} = a^{\frac{1}{2}+\frac{1}{2}} = a^1 = a$ {index laws}
and $\sqrt{a} \times \sqrt{a} = a$ also.

So, $\boldsymbol{a^{\frac{1}{2}} = \sqrt{a}}$ {by direct comparison}

Likewise $a^{\frac{1}{3}} \times a^{\frac{1}{3}} \times a^{\frac{1}{3}} = a^1 = a$
and $\sqrt[3]{a} \times \sqrt[3]{a} \times \sqrt[3]{a} = a$

suggests $\boldsymbol{a^{\frac{1}{3}} = \sqrt[3]{a}}$

In general, $\boldsymbol{a^{\frac{1}{n}} = \sqrt[n]{a}}$ where $\sqrt[n]{a}$ reads "the nth root of a", for $n \in \mathbb{Z}^+$.

We can now determine that $\sqrt[n]{a^m} = (a^m)^{\frac{1}{n}} = a^{\frac{m}{n}}$

$\therefore$ $\boldsymbol{a^{\frac{m}{n}} = \sqrt[n]{a^m}}$ for $a > 0$, $n \in \mathbb{Z}^+$, $m \in \mathbb{Z}$

Example 15 **Self Tutor**

Write as a single power of 2:

a $\sqrt[4]{2}$ **b** $\dfrac{1}{\sqrt[3]{2}}$ **c** $\sqrt[3]{16}$

a $\sqrt[4]{2}$
$= 2^{\frac{1}{4}}$

b $\dfrac{1}{\sqrt[3]{2}}$
$= \dfrac{1}{2^{\frac{1}{3}}}$
$= 2^{-\frac{1}{3}}$

c $\sqrt[3]{16}$
$= (2^4)^{\frac{1}{3}}$
$= 2^{4\times\frac{1}{3}}$
$= 2^{\frac{4}{3}}$

EXERCISE 4D

1 Write as a single power of 2:

a $\sqrt[3]{2}$ **b** $\frac{1}{\sqrt{2}}$ **c** $4\sqrt{2}$ **d** $(\sqrt{2})^{-4}$ **e** $\frac{1}{\sqrt[5]{2}}$

f $\frac{\sqrt{2}}{4}$ **g** $2 \times \sqrt[5]{2}$ **h** $\sqrt[6]{4}$ **i** $\frac{1}{\sqrt[3]{32}}$ **j** $\frac{8}{\sqrt[3]{2}}$

2 Write as a single power of 3:

a $\sqrt[5]{3}$ **b** $\frac{1}{\sqrt[5]{3}}$ **c** $9\sqrt{3}$ **d** $\frac{3}{\sqrt{3}}$ **e** $27\sqrt{3}$

f $\sqrt[3]{81}$ **g** $\frac{1}{\sqrt{27}}$ **h** $\frac{1}{\sqrt[3]{9}}$ **i** $27 \times \sqrt[5]{9}$ **j** $\frac{\sqrt{27}}{9}$

3 Write the following in the form a^x where a is a prime number and x is rational:

a $\sqrt[3]{25}$ **b** $\sqrt[4]{49}$ **c** $\sqrt[3]{121}$ **d** $\sqrt[5]{81}$ **e** $\sqrt[4]{125}$

f $\frac{1}{\sqrt[3]{25}}$ **g** $\frac{1}{\sqrt[4]{49}}$ **h** $\frac{1}{\sqrt[3]{121}}$ **i** $\frac{1}{\sqrt[5]{81}}$ **j** $\frac{1}{\sqrt[4]{125}}$

Example 16 — Self Tutor

Write in simplest rational form: **a** $9^{\frac{3}{2}}$ **b** $16^{-\frac{3}{4}}$

a $9^{\frac{3}{2}} = (3^2)^{\frac{3}{2}}$
$= 3^{2 \times \frac{3}{2}}$
$= 3^3$
$= 27$

b $16^{-\frac{3}{4}} = (2^4)^{-\frac{3}{4}}$
$= 2^{4 \times (-\frac{3}{4})}$
$= 2^{-3}$
$= \frac{1}{8}$

4 Write in simplest rational form:

a $4^{\frac{5}{2}}$ **b** $27^{\frac{2}{3}}$ **c** $8^{\frac{5}{3}}$ **d** $16^{\frac{5}{4}}$ **e** $81^{\frac{3}{4}}$

f $9^{-\frac{1}{2}}$ **g** $25^{-\frac{3}{2}}$ **h** $32^{-\frac{3}{5}}$ **i** $64^{-\frac{2}{3}}$ **j** $243^{-\frac{4}{5}}$

5 Use your calculator to evaluate:

a $2^{\frac{1}{3}}$ **b** $3^{\frac{5}{6}}$ **c** $5^{-\frac{1}{5}}$ **d** $9^{-\frac{3}{4}}$ **e** $\sqrt[7]{25}$

E ALGEBRAIC EXPANSION AND FACTORISATION

We can use the usual expansion laws to simplify expressions containing indices:

$$a(b+c) = ab + ac$$
$$(a+b)(c+d) = ac + ad + bc + bd$$
$$(a+b)(a-b) = a^2 - b^2$$
$$(a+b)^2 = a^2 + 2ab + b^2$$
$$(a-b)^2 = a^2 - 2ab + b^2$$

Example 17 Self Tutor

Expand and simplify:

$x^{\frac{1}{2}}(2x^{\frac{3}{2}}+2x^{\frac{1}{2}}-x^{-\frac{1}{2}})$

$$\begin{aligned}&x^{\frac{1}{2}}(2x^{\frac{3}{2}}+2x^{\frac{1}{2}}-x^{-\frac{1}{2}})\\&=x^{\frac{1}{2}}\times 2x^{\frac{3}{2}}+x^{\frac{1}{2}}\times 2x^{\frac{1}{2}}-x^{\frac{1}{2}}\times x^{-\frac{1}{2}}\\&=2x^2+2x^1-x^0\\&=2x^2+2x-1\end{aligned}$$

EXERCISE 4E

1 Expand and simplify:

a $x^2(x+2)$ **b** $3^x(1-3^x)$ **c** $x^{\frac{1}{2}}(x^{\frac{3}{2}}+x^{\frac{1}{2}})$

d $5^x(5^x+5^{-x})$ **e** $x^3(3x^2-x+x^{-1})$ **f** $x^{\frac{3}{2}}(x^{\frac{1}{2}}-x^{\frac{5}{2}}+x^{-\frac{1}{2}})$

g $2^{-x}(2^{3x}+1)$ **h** $x^{-\frac{1}{2}}(x^2+4x+x^{\frac{1}{2}})$ **i** $7^x(49^x+7^x)$

Example 18 Self Tutor

Expand and simplify: **a** $(3^x-1)(3^x+4)$ **b** $(5^x-2)(5^x+2)$

a
$$\begin{aligned}&(3^x-1)(3^x+4)\\&=3^x\times 3^x+4\times 3^x-3^x-4\\&=3^{2x}+3\times 3^x-4\\&=9^x+3^{x+1}-4\end{aligned}$$

b
$$\begin{aligned}&(5^x-2)(5^x+2)\\&=(5^x)^2-2^2\\&=5^{2x}-4\\&=25^x-4\end{aligned}$$

2 Expand and simplify:

a $(2^x-2)(2^x+1)$ **b** $(3^x-5)(3^x-1)$ **c** $(2^x-3)(2^x+3)$

d $(7^x+3)(7^x+4)$ **e** $(2^x-4)(2^x+4)$ **f** $(4^x+1)^2$

g $(8^x-3)^2$ **h** $(7-6^x)(6^x+7)$ **i** $(5^x+5)^2$

3 Expand and simplify:

a $(x^{\frac{1}{2}}+3)(x^{\frac{1}{2}}+5)$ **b** $(x^{\frac{1}{2}}-4)(x^{\frac{1}{2}}+4)$ **c** $(x^{\frac{1}{2}}-1)^2$

d $\left(x^2-\frac{1}{x}\right)^2$ **e** $(2-x^{\frac{3}{4}})(2+x^{\frac{3}{4}})$ **f** $(3x^{\frac{1}{2}}-1)(x^{\frac{1}{2}}+7)$

g $(x^{\frac{2}{3}}+x^{\frac{1}{3}})^2$ **h** $(x^{\frac{3}{2}}-x^{\frac{1}{2}})^2$ **i** $(2x^{\frac{1}{2}}-x^{-\frac{1}{2}})^2$

Example 19 Self Tutor

Factorise: **a** $2^{n+1}-2^n$ **b** $2^{n+2}+4$ **c** $3^{4n}+3^{3n}$

a
$$\begin{aligned}&2^{n+1}-2^n\\&=2^n\times 2^1-2^n\\&=2^n(2-1)\\&=2^n\end{aligned}$$

b
$$\begin{aligned}&2^{n+2}+4\\&=2^n\times 2^2+4\\&=2^n\times 4+4\\&=4(2^n+1)\end{aligned}$$

c
$$\begin{aligned}&3^{4n}+3^{3n}\\&=3^{3n}\times 3^n+3^{3n}\\&=3^{3n}(3^n+1)\end{aligned}$$

4 Factorise:

a $3^{n+3}+3^n$ **b** $5^{2n}+5^n$ **c** $2^{3n}-2^{2n}$

d $4^{n+2}+4^{n+1}$ **e** $7^{n+2}-7$ **f** $5^{n+3}+125$

g 8^n+2^n **h** 5^n-25^n **i** $9^{n+\frac{3}{2}}-27$

Example 20 Self Tutor

Factorise:

a 9^x-16 **b** $4^x+10(2^x)+25$

a 9^x-16

$=(3^x)^2-4^2$ {compare a^2-b^2}

$=(3^x+4)(3^x-4)$ $\{a^2-b^2=(a+b)(a-b)\}$

b $4^x+10(2^x)+25$

$=(2^x)^2+10(2^x)+25$ {compare $a^2+10a+25$}

$=(2^x+5)^2$ $\{a^2+10a+25=(a+5)^2\}$

5 Factorise:

a 4^x-9 **b** 25^x-9 **c** $36-4^x$

d $49-16^x$ **e** 25^x-16^x **f** $9^x+4(3^x)+4$

g $4^x-12(2^x)+36$ **h** $9^x-14(3^x)+49$ **i** $4^x+2^{x+4}+64$

6 Factorise:

a $9^x+8(3^x)+15$ **b** $4^x+5(2^x)-14$ **c** $9^x-12(3^x)+35$

d $16^x-7(4^x)-18$ **e** $49^x-9(7^x)+20$ **f** $25^x-5^{x+1}-24$

Example 21 Self Tutor

Simplify:

a $\frac{6^n}{3^n}$ **b** $\frac{4^n}{6^n}$

a $\frac{6^n}{3^n}$ **or** $\frac{6^n}{3^n}$

$=\frac{2^n\cancel{3^n}}{\cancel{3^n}_1}$ $=\left(\frac{6}{3}\right)^n$

$=2^n$ $=2^n$

b $\frac{4^n}{6^n}$ **or** $\frac{4^n}{6^n}$

$=\frac{\cancel{2^n}2^n}{\cancel{2^n}3^n}$ $=\left(\frac{4}{6}\right)^n$

$=\frac{2^n}{3^n}$ $=\left(\frac{2}{3}\right)^n$

7 Simplify:

a $\frac{12^n}{6^n}$ **b** $\frac{20^a}{2^a}$ **c** $\frac{6^b}{2^b}$ **d** $\frac{4^n}{20^n}$

e $\frac{35^x}{7^x}$ **f** $\frac{6^a}{8^a}$ **g** $\frac{5^{n+1}}{5^n}$ **h** $\frac{5^{n+1}}{5}$

Example 22 **Self Tutor**

Simplify:

a $\dfrac{3^n + 6^n}{3^n}$ **b** $\dfrac{2^{m+2} - 2^m}{2^m}$ **c** $\dfrac{2^{m+3} + 2^m}{9}$

a $\dfrac{3^n + 6^n}{3^n}$
$= \dfrac{3^n + 2^n 3^n}{3^n}$
$= \dfrac{3^n(1 + 2^n)}{3^n}$
$= 1 + 2^n$

b $\dfrac{2^{m+2} - 2^m}{2^m}$
$= \dfrac{2^m 2^2 - 2^m}{2^m}$
$= \dfrac{2^m(4 - 1)}{2^m}$
$= 3$

c $\dfrac{2^{m+3} + 2^m}{9}$
$= \dfrac{2^m 2^3 + 2^m}{9}$
$= \dfrac{2^m(8 + 1)}{9}$
$= 2^m$

8 Simplify:

a $\dfrac{6^m + 2^m}{2^m}$ **b** $\dfrac{2^n + 12^n}{2^n}$ **c** $\dfrac{8^n + 4^n}{2^n}$

d $\dfrac{12^x - 3^x}{3^x}$ **e** $\dfrac{6^n + 12^n}{1 + 2^n}$ **f** $\dfrac{5^{n+1} - 5^n}{4}$

g $\dfrac{5^{n+1} - 5^n}{5^n}$ **h** $\dfrac{4^n - 2^n}{2^n}$ **i** $\dfrac{2^n - 2^{n-1}}{2^n}$

9 Simplify:

a $2^n(n+1) + 2^n(n-1)$ **b** $3^n\left(\dfrac{n-1}{6}\right) - 3^n\left(\dfrac{n+1}{6}\right)$

F EXPONENTIAL EQUATIONS

An **exponential equation** is an equation in which the unknown occurs as part of the index or exponent.

For example: $3^x = 27$ and $20 \times 5^x = 4$ are both exponential equations.

To solve exponential equations, we look to write both sides of the equation with the **same base**. We can then **equate indices**.

If $\boldsymbol{a^x = a^k}$ then $\boldsymbol{x = k}$.

Example 23 — Self Tutor

Solve for x:

a $3^x = \frac{1}{81}$ **b** $2^{x+1} = 16$

a
$$\begin{aligned} 3^x &= \tfrac{1}{81} \\ \therefore\ 3^x &= 3^{-4} \\ \therefore\ x &= -4 \end{aligned}$$

b
$$\begin{aligned} 2^{x+1} &= 16 \\ \therefore\ 2^{x+1} &= 2^4 \\ \therefore\ x+1 &= 4 \\ \therefore\ x &= 3 \end{aligned}$$

Example 24 — Self Tutor

Solve for x:

a $9^x = \frac{1}{27}$ **b** $8^{x-3} = 32$

a
$$\begin{aligned} 9^x &= \tfrac{1}{27} \\ \therefore\ (3^2)^x &= 3^{-3} \\ \therefore\ 3^{2x} &= 3^{-3} \\ \therefore\ 2x &= -3 \\ \therefore\ x &= -\tfrac{3}{2} \end{aligned}$$

b
$$\begin{aligned} 8^{x-3} &= 32 \\ \therefore\ (2^3)^{x-3} &= 2^5 \\ \therefore\ 2^{3x-9} &= 2^5 \\ \therefore\ 3x-9 &= 5 \\ \therefore\ 3x &= 14 \\ \therefore\ x &= \tfrac{14}{3} \end{aligned}$$

EXERCISE 4F

1 Solve for x:

a $2^x = 16$ **b** $3^x = 9$ **c** $7^x = 49$ **d** $6^x = 1$

e $3^x = \frac{1}{27}$ **f** $5^x = \sqrt{5}$ **g** $2^x = 4\sqrt{2}$ **h** $5^{x+2} = 125$

i $2^{x-1} = 64$ **j** $3^{x+4} = 243$ **k** $6^{1-x} = \frac{1}{36}$ **l** $7^{2x+1} = \frac{1}{343}$

2 Solve for x:

a $4^x = 32$ **b** $27^x = 9$ **c** $25^x = \frac{1}{5}$ **d** $8^x = \frac{1}{16}$

e $81^x = 27$ **f** $4^{x+3} = 16$ **g** $27^{x-1} = 81$ **h** $25^{3x+1} = 1$

i $64^x = 32^{-x}$ **j** $\left(\frac{1}{2}\right)^{1-x} = 4$ **k** $\left(\frac{1}{5}\right)^{x+2} = 125$ **l** $\left(\frac{1}{9}\right)^{x-4} = 243$

3 Solve for x, if possible:

a $9^{x+1} = 27^{3-x}$ **b** $\left(\frac{1}{4}\right)^{2x-1} = 16^{x+4}$ **c** $4^{3x+1} = 8^{2x-2}$

4 Solve for x:

a $\dfrac{3^{2x+1}}{3^x} = 9^x$ **b** $\dfrac{25^x}{5^{x+4}} = 25^{1-x}$ **c** $\dfrac{4^x}{2^{x+2}} = \dfrac{2^{x+1}}{8^x}$

d $\dfrac{5^{2x-5}}{125^x} = \dfrac{25^{1-2x}}{5^{x+2}}$ **e** $\dfrac{4^x}{8^{2-x}} = 2^x \times 4^{x-1}$ **f** $\dfrac{9^{2x}}{27^{2-x}} = \dfrac{81^{3x+1}}{3^{1-2x}}$

5 Solve for x:

a $5 \times 2^x = 20$ **b** $3 \times 5^x = 75$ **c** $4 \times 3^{x-2} = 12$

d $5 \times 3^{x+1} = 135$ **e** $6 \times \left(\frac{1}{2}\right)^x = 96$ **f** $7 \times \left(\frac{1}{3}\right)^x = 21$

6 Solve for x:

a $2^{x^2} = 16$ **b** $3^{x^2-2x} = 27$ **c** $5^{2x^2+3x} = \frac{1}{5}$

Example 25

Self Tutor

Solve for x:

$4^x - 5(2^x) - 24 = 0$

$$4^x - 5(2^x) - 24 = 0$$
$$\therefore (2^x)^2 - 5(2^x) - 24 = 0 \quad \{\text{compare } a^2 - 5a - 24 = 0\}$$
$$\therefore (2^x - 8)(2^x + 3) = 0 \quad \{a^2 - 5a - 24 = (a-8)(a+3)\}$$
$$\therefore 2^x = 8 \text{ or } 2^x = -3$$
$$\therefore 2^x = 2^3 \quad \{2^x > 0\}$$
$$\therefore x = 3$$

7 Solve for x:

a $4^x - 3(2^x) - 4 = 0$ **b** $9^x - 3^x - 6 = 0$ **c** $4^x - 18(2^x) + 32 = 0$

d $25^x = 6(5^x) - 5$ **e** $3^{2x+1} - 10(3^x) + 3 = 0$ **f** $2^{2x+1} + 9(2^x) = -4$

8 Solve simultaneously:

a $x + y = 4$, $2^x 4^y = 32$

b $x - y = 2$, $\dfrac{9^x}{3^y} = 1$

c $xy = -6$, $8^x 4^y = \frac{1}{32}$

G EXPONENTIAL FUNCTIONS

An **exponential function** is a function in which the variable occurs as part of the index or exponent.

For example, $f(x) = 2^x$ is an exponential function.

We construct a table of values from which we graph the function:

x	-3	-2	-1	0	1	2	3
y	$\frac{1}{8}$	$\frac{1}{4}$	$\frac{1}{2}$	1	2	4	8

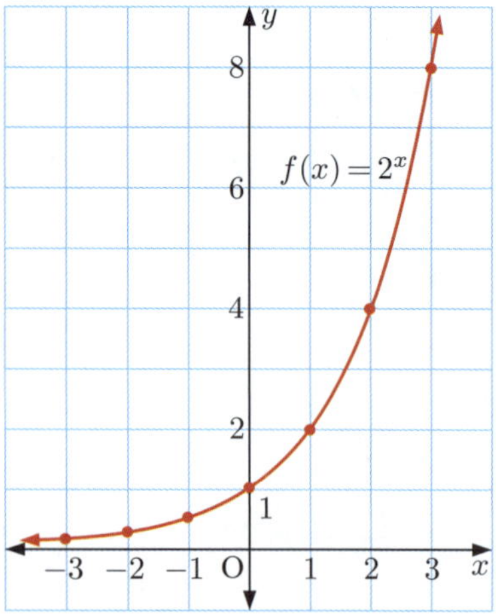

Notice that:

- the y-intercept of the function is 1
- the graph lies entirely above the x-axis
- For large negative values of x, y is very small and positive.
  We write "As $x \to -\infty$, $y \to 0^+$" which means "As x tends to minus infinity, y tends to zero from above."
  We say $y = 0$ is a **horizontal asymptote** of the function.

EXERCISE 4G.1

1 Use a table of values from $x = -3$ to $x = 3$ to help sketch each exponential function:

a $f(x) = 3^x$ **b** $f(x) = 4^x$ **c** $f(x) = \left(\frac{1}{2}\right)^x$ **d** $f(x) = \left(\frac{1}{3}\right)^x$

2 Use technology to graph the following functions on the same set of axes:

a $y = 5^x$ **b** $y = (1.8)^x$ **c** $y = (0.7)^x$ **d** $y = \left(\frac{2}{5}\right)^x$

What do you notice?

GRAPHING PACKAGE

3 Consider the graph of $y = 3^x$ alongside.

a Use the graph to estimate the value of:

i $3^{\frac{3}{2}}$ **ii** $3^{0.4}$

iii $3^{-1.9}$ **iv** $\frac{1}{\sqrt{3}}$

b Use the graph to estimate the solution to:

i $3^x = 5$ **ii** $3^x = 0.8$

iii $3^x = 4.3$ **iv** $3^x = 0.3$

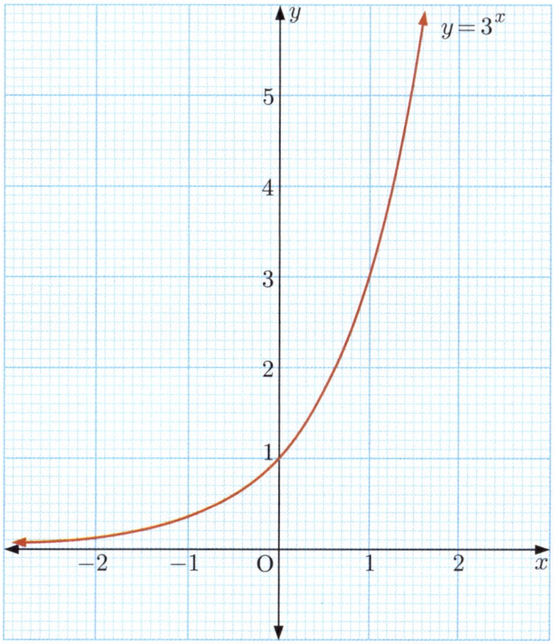

Discussion

- Which of these values are defined?
  - $(-2)^3$
  - $(-2)^{-1}$
  - $(-2)^{\frac{1}{2}}$
  - $(-2)^{\frac{1}{3}}$
  - $(-2)^{-\frac{1}{4}}$
- What would the graph of $f(x) = (-2)^x$ look like?

THE GENERAL EXPONENTIAL FUNCTION

Discovery 2 — Graphs of exponential functions

In this Discovery we examine the graphs of various families of exponential functions.

Click on the icon to run the **dynamic graphing package**, or use your **graphics calculator**.

DYNAMIC GRAPHING PACKAGE

What to do:

1 Explore the family of curves of the form $y = b^x$ where $b > 0$.
For example, consider $y = 2^x$, $y = 3^x$, $y = 5^x$, and $y = 8^x$.

a What effect does changing b have on the shape of the graph?

b What is the y-intercept of each graph?

c What is the horizontal asymptote of each graph?

2 Explore the family of curves of the form $y = 2^x + a$ where a is a constant.
For example, consider $y = 2^x$, $y = 2^x + 3$, and $y = 2^x - 1$.

- **a** What effect does changing a have on the position of the graph?
- **b** What effect does changing a have on the shape of the graph?
- **c** What is the horizontal asymptote of each graph?
- **d** What is the horizontal asymptote of $y = 2^x + a$?
- **e** What transformation is used to graph $y = 2^x + a$ from $y = 2^x$?

3 Explore the family of curves of the form $y = k \times 2^x$ where k is a constant.

- **a** Consider functions where $k > 0$, such as $y = 2^x$, $y = 3 \times 2^x$, and $y = 5 \times 2^x$. Comment on the effect k has on the graph.
- **b** Consider functions where $k < 0$, such as $y = -2^x$, $y = -3 \times 2^x$, and $y = -5 \times 2^x$. Comment on the effect k has on the graph.
- **c** What is the horizontal asymptote of $y = k \times 2^x$?

4 Explore the family of curves of the form $y = 2^{nx}$.
For example, consider $y = 2^x$, $y = 2^{3x}$, and $y = 2^{5x}$.

- **a** What effect does changing n have on the position of the graph?
- **b** What effect does changing n have on the shape of the graph?
- **c** What is the horizontal asymptote of $y = 2^{nx}$?
- **d** What transformation is used to graph $y = 2^{nx}$ from $y = 2^x$?

5 Explore the relationship between $y = b^x$ and $y = b^{-x}$ where $b > 0$.
For example, consider $y = 2^x$ and $y = 2^{-x}$.

- **a** What is the y-intercept of each graph?
- **b** What is the horizontal asymptote of each graph?
- **c** What transformation moves $y = 2^x$ to $y = 2^{-x}$?

From **Discovery 2** you should have found that:

For the general exponential function $\boldsymbol{y = k \times b^{nx} + a}$ where $b > 0$, $b \neq 1$, $k \neq 0$:

- b controls how steeply the graph increases or decreases.
- a controls the vertical translation. The horizontal asymptote is $y = a$.
- k controls the vertical stretching of the graph.
  If $k < 0$, there is a reflection in the x-axis.
- n controls the horizontal stretching of the graph.
  If $n < 0$, there is a reflection in the y-axis.

We can sketch reasonably accurate graphs of exponential functions using:

- the horizontal asymptote
- the y-intercept
- two other points, for example, when $x = 2$, $x = -2$

Example 26

Self Tutor

Sketch the graph of $y = 3 \times 2^x - 4$.
Hence state the domain and range of $y = 3 \times 2^x - 4$.

For $y = 3 \times 2^x - 4$,
the horizontal asymptote is $y = -4$.

When $x = 0$, $y = 3 \times 2^0 - 4$
$= 3 \times 1 - 4$
$= -1$

$\therefore$ the y-intercept is -1.

When $x = 2$, $y = 3 \times 2^2 - 4$
$= 3 \times 4 - 4$
$= 8$

When $x = -2$, $y = 3 \times 2^{-2} - 4$
$= 3 \times \frac{1}{4} - 4$
$= -3\frac{1}{4}$

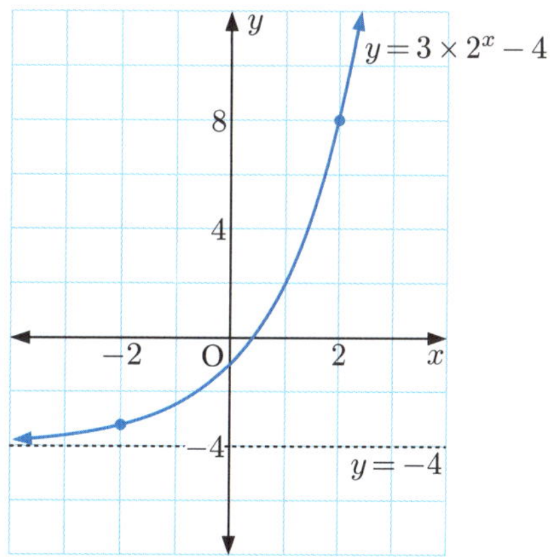

The domain is $\{x : x \in \mathbb{R}\}$. The range is $\{y : y > -4\}$.

EXERCISE 4G.2

1 Suppose $g(x) = 5^x + 2$.

a Find $g(0)$ and $g(-1)$.

b Find a such that $g(a) = 27$.

2 Draw freehand sketches of the following pairs of graphs using your observations from the previous **Discovery**:

GRAPHING PACKAGE

a $y = 2^x$ and $y = 2^x + 1$

b $y = 2^x$ and $y = 2^{3x}$

c $y = 2^x$ and $y = 2^{-x}$

d $y = 2^x$ and $y = 2(2^x)$

3 Draw freehand sketches of the following pairs of graphs:

a $y = 3^x$ and $y = 3^x - 2$

b $y = 3^x$ and $y = 3^{-x}$

c $y = 3^x$ and $y = -3^x$

d $y = 3^x$ and $y = 3^{2x}$

4 Suppose $f(x) = 2 \times 3^x$.

a Find:

i $f(0)$ **ii** $f(3)$ **iii** $f(-2)$

b State the equation of the horizontal asymptote.

c Sketch the graph of the function.

d State the domain and range of the function.

5 For each function below:

i Determine the horizontal asymptote.

ii Discuss the behaviour of y as $x \to \pm\infty$.

iii Sketch the graph of the function.

iv State the domain and range.

a $y = 2^x - 3$

b $f(x) = 5 - 3^x$

c $y = 2^{2x} + 1$

d $f(x) = 2 \times 3^x - 1$

6 Ano has invested some money in an account which pays compound interest. The value of the investment after t years is given by $A(t) = 2000 \times (1.06)^t$ dollars.

a Find the value of the investment:

i initially **ii** after 1 year **iii** after 3 years.

b By what percentage does the investment increase each year?

c Sketch the graph of $A(t)$ for $0 \leqslant t \leqslant 5$ years.

7 The population of bacteria in a culture is given by $P(t) = 10^5 \times 2^{\frac{t}{28}}$ where t is the time in minutes after the initial observation.

a Find the initial population of bacteria.

b How long does it take for the population to double?

c Find the population of bacteria after:

i 56 minutes **ii** 6 hours.

d Sketch the graph of $P(t)$ for $t \geqslant 0$.

e How long will it take for the population to reach 6.4×10^6?

8 Answer the **Opening Problem** on page **100**.

9 The mass of radioactive isotope present in a sample after t years is given by $M(t) = 4.8 \times 10^{-\frac{t}{56}}$ grams.

a Find the initial mass of radioactive isotope in the sample.

b Find the mass of radioactive isotope present after:

i 28 years **ii** 224 years

c Sketch the graph of $M(t)$ for $t \geqslant 0$.

d How long will it take for the mass of radioactive isotope in the sample to reduce to $\frac{1}{10^8}$ of its initial value?

H THE NATURAL EXPONENTIAL e^x

The diagram alongside shows a collection of graphs for exponential functions of the form $f(x) = b^x$ where $b > 0$, $b \neq 1$.

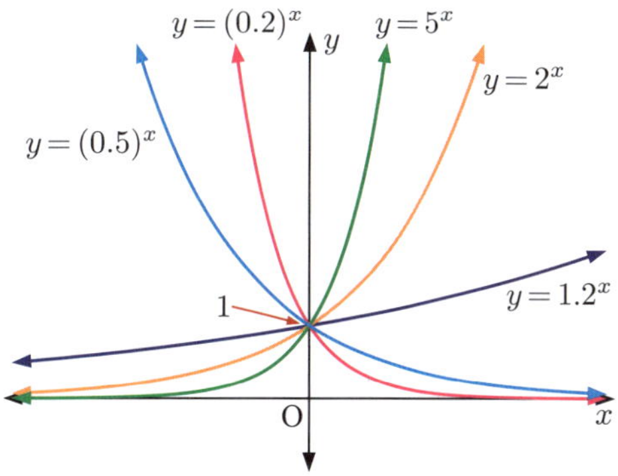

Since $b^0 = 1$ for all $b \neq 0$, each graph passes through the point $(0, 1)$.

There are an infinite number of possible choices for the base number b.

However, where exponential data is examined in science, engineering, and finance, one of the most commonly used bases is the **natural exponential** $e \approx 2.7183$.

Discovery 3 — *The natural exponential*

In **Chapter 3** we saw that the **tangent** to a curve is a line which *touches* the curve.

In this Discovery we will explore the tangent to the curve $y = b^x$, $b > 1$, at the point $(0, 1)$.

Suppose the tangent to $y = b^x$ at $(0, 1)$ cuts the x-axis at X.

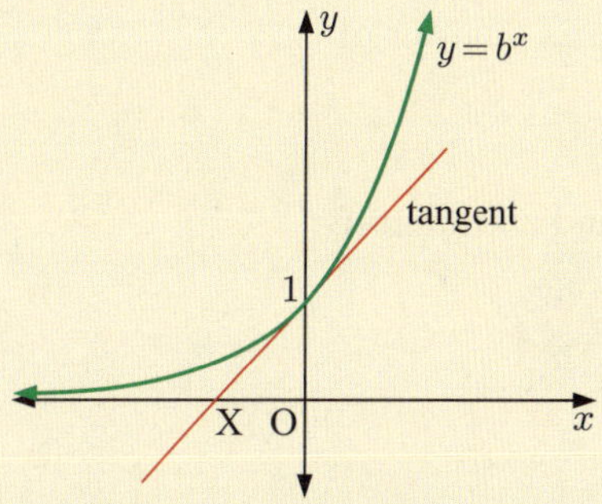

What to do:

1 Click on the icon to access the dynamic graphing package.

DYNAMIC GRAPHING PACKAGE

2 For $b = 1.5$, 2, 2.5, 3, 3.5, and 4, find the distance OX, and state whether $\text{OX} < 1$ or $\text{OX} > 1$. Record your results in a table like the one alongside.

b	OX	$\text{OX} < 1$ or $\text{OX} > 1$
1.5	2.47	$\text{OX} > 1$
2		
2.5		
3		
3.5		
4		

3 Use the slider in the software to find the value of b, correct to 2 decimal places, such that $\text{OX} = 1$.

In the **Discovery**, you should have found that OX $= 1$ when $b \approx 2.72$.

The exact value of b for which this occurs is the irrational number e, where $e \approx 2.7183....$

The tangent to $y = e^x$ at the point $(0, 1)$ has gradient 1. This property will be explored further when we study calculus.

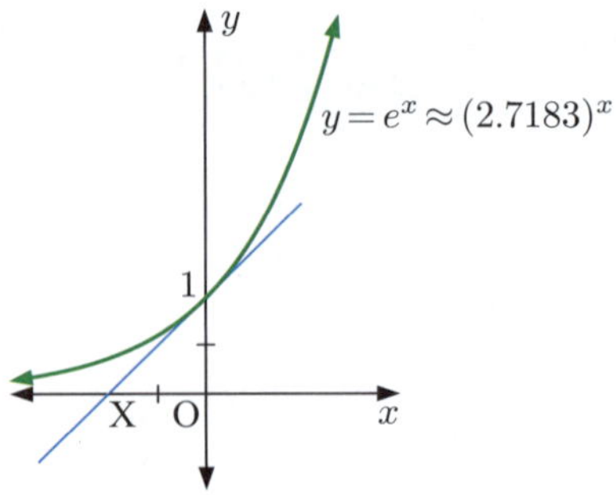

You can use your calculator to perform operations involving e.

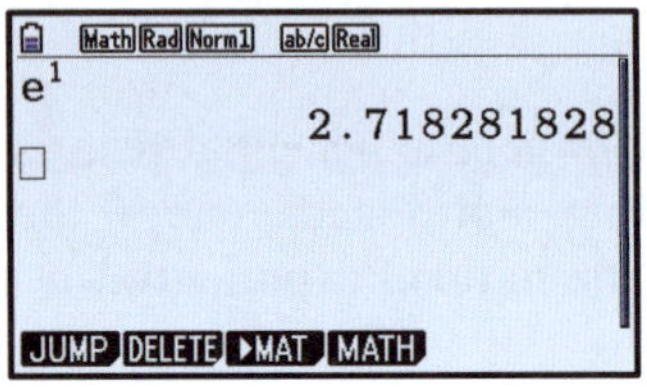

EXERCISE 4H

1 Use your calculator to evaluate, correct to 3 significant figures:

a e^2 **b** $e^4 - 1$ **c** $\frac{1}{2} \times e^5$ **d** $\frac{1}{e^3}$

2 Write as a power of e:

a $\sqrt{e}$ **b** $\frac{1}{e^5}$ **c** $\frac{1}{\sqrt[3]{e}}$ **d** $e^2\sqrt{e}$

3 Expand and simplify:

a $e(e^x + 3)$ **b** $e^x(e^x - 2)$ **c** $(e^x + 1)^2$

4 Solve for x:

a $e^x = \sqrt[3]{e}$ **b** $e^{2x} = \frac{1}{e}$ **c** $e^{\frac{x}{3}} = e\sqrt{e}$

5 Suppose $f : x \mapsto e^x$ and $g : x \mapsto 3x + 2$.

a Find $fg(x)$ and $gf(x)$.

b Solve $fg(x) = \frac{1}{e}$.

6 Graph $y = 2^x$, $y = e^x$, and $y = 3^x$ on the same set of axes.

GRAPHING PACKAGE

7 For the general exponential function $y = ke^{nx} + a$, state the:

a horizontal asymptote **b** y-intercept.

8 Sketch the following pairs of graphs on the same set of axes:

a $y = e^x$ and $y = e^x - 1$ **b** $y = e^x$ and $y = -e^x$

c $y = e^x$ and $y = e^{2x}$ **d** $y = e^x$ and $y = e^{-x}$

9 The graph of $y = 2e^{-x} + 1$ is shown alongside.

a State the equation of the horizontal asymptote.

b Find the coordinates of P.

c **i** Find k in terms of e.

ii Use your calculator to evaluate k correct to 2 decimal places.

d State the domain and range of the function.

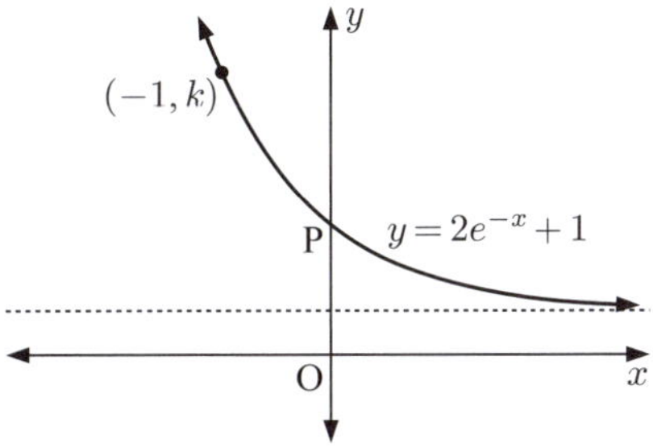

10 Consider the function $y = 3e^{x} - 5$.

a State the equation of the horizontal asymptote.

b Find the y-intercept.

c Find y when $x = 2$, giving your answer correct to 3 decimal places.

d Sketch the graph of the function.

e State the domain and range of the function.

11 For each of the following functions:

i Find the horizontal asymptote and y-intercept.

ii Sketch the graph of the function.

iii State the domain and range of the function.

a $y = e^{2x} - 3$ **b** $y = 2e^{x} + 2$ **c** $f(x) = 3e^{-x}$

12 Match each equation with the correct graph:

a $y = e^{x} + 1$

b $y = e^{2x} + 1$

c $y = 1 - e^{x}$

d $y = e^{-x} + 1$

e $y = 1 - e^{-x}$

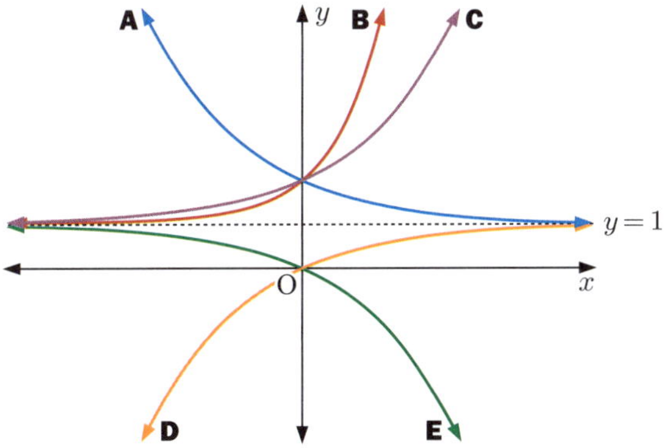

13 **a** Copy and complete the table alongside for the general exponential function $f(x) = ke^{nx} + a$.

b Under what conditions will the graph of $y = f(x)$ have an x-intercept?

Shape of $y = f(x)$

	$n > 0$	$n < 0$
$k > 0$	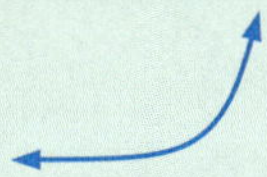 increasing	
$k < 0$		

14 Consider the function $f(x) = e^{x}$.

a On the same set of axes, sketch $y = f(x)$, $y = x$, and $y = f^{-1}(x)$.

b State the domain and range of f^{-1}.

Review set 4A

1 Simplify:

a $\sqrt{7}(4-\sqrt{7})$ **b** $(1+\sqrt{5})^2$ **c** $(3+2\sqrt{2})(3-2\sqrt{2})$

2 Write with integer denominator:

a $\frac{2}{\sqrt{3}}$ **b** $\frac{\sqrt{7}}{\sqrt{5}}$ **c** $\frac{1}{4\sqrt{7}}$

3 Rationalise the denominator:

a $\frac{1}{5-\sqrt{3}}$ **b** $\frac{\sqrt{11}}{\sqrt{7}-2}$ **c** $\frac{8+\sqrt{2}}{3-\sqrt{2}}$ **d** $\frac{4+5\sqrt{5}}{6-3\sqrt{5}}$

4 Simplify using the index laws:

a $x^3y^2 \times x^5y$ **b** $\frac{8a^2b^5}{6ab^6}$ **c** $\frac{10(c^2d)^2}{(4cd^3)^3}$

5 Let $f(x) = 3^x$.

a Write down the value of: **i** $f(4)$ **ii** $f(-1)$

b Find the value of k such that $f(x+2) = k\,f(x)$, $k \in \mathbb{Z}$.

6 Write as a single power of 2:

a $\frac{16}{8^k}$ **b** $(\sqrt{2})^{2-x} \times 4^{x+1}$

7 Evaluate:

a $27^{-\frac{1}{3}}$ **b** $4^{\frac{7}{2}}$

8 Write without negative exponents:

a mn^{-2} **b** $(mn)^{-3}$ **c** $\frac{m^2n^{-1}}{p^{-2}}$ **d** $(4m^{-1}n)^2$

9 Expand and simplify:

a $(3-e^x)^2$ **b** $(\sqrt{x}+2)(\sqrt{x}-2)$ **c** $2^{-x}(2^{2x}+2^x)$

10 Find the positive solution of the equation $(8+\sqrt{13})x^2 + (2-\sqrt{13})x - 1 = 0$.
Give your solution in the form $x = a + b\sqrt{13}$, where $a, b \in \mathbb{Q}$.

11 Solve for x:

a $2^{x-1} = 16$ **b** $27^x = \frac{1}{\sqrt{3}}$ **c** $\frac{1}{e^{2x-1}} = \sqrt[3]{e}$

12 Solve simultaneously: $xy = 6$
$x^4y^3 = 324$

13 Suppose $f(x) = 2^{-x} - 1$.

a Find $f(1)$ and $f(-2)$. **b** Find a such that $f(a) = 15$.

14 The temperature of a can of drink t minutes after it is placed in a refrigerator is given by
$T(t) = 27 \times 3^{-\frac{t}{8}}$ °C.

a Find the initial temperature of the can.

b How long will it take for the temperature to fall to 1°C?

c Sketch the graph of $T(t)$ for $t \geqslant 0$.

15 The graph of $y = 2e^{2x} - 3$ is shown alongside.

a State the equation of the horizontal asymptote.

b Find the y-intercept.

c Point P has x-coordinate -1. Find the y-coordinate of P:

i exactly **ii** to 3 decimal places.

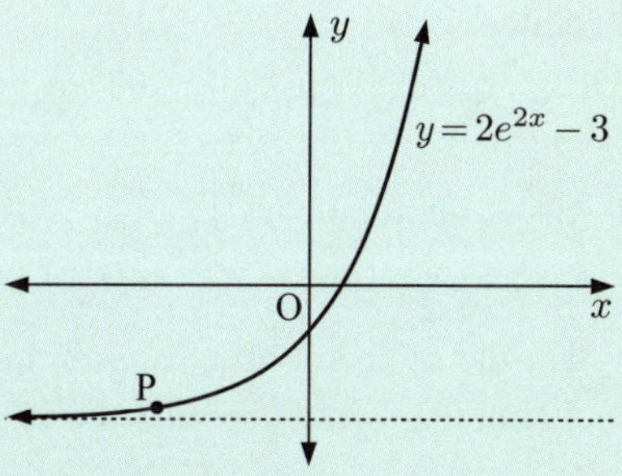

16 Consider the function $f : x \mapsto e^{-x} - 3$.

a State the range of the function. **b** Find the value of $f(0)$.

c Solve $f(x) = \frac{\sqrt{e} - 3e}{e}$.

Review set 4B

1 Simplify:

a $-\sqrt{2}(\sqrt{2} - 5)$ **b** $(\sqrt{3} - 4)^2$ **c** $(1 - 2\sqrt{7})(1 + 2\sqrt{7})$

2 Write each of the following in the form $a + b\sqrt{2}$ where $a, b \in \mathbb{Z}$:

a $(\sqrt{2} - 1)^2$ **b** $\frac{\sqrt{2} - 1}{\sqrt{2} + 1}$ **c** $\frac{1}{(\sqrt{2} + 1)^2}$ **d** $\frac{1}{3 + 2\sqrt{2}}$

3 Simplify using the laws of exponents:

a $(a^7)^3$ **b** $pq^2 \times p^3q^4$ **c** $\frac{8ab^5}{2a^4b^4}$

4 Write the following as a power of 2:

a 2×2^{-4} **b** $16 \div 2^{-3}$ **c** 8^4

5 Write the following without brackets:

a $(2m^3)^2$ **b** $\left(\frac{-a^3}{b}\right)^3$ **c** $\frac{(3x^2y)^2}{3x}$ **d** $\frac{(2a^{\frac{1}{2}}b^{\frac{1}{5}})^4}{a}$

6 Simplify $\frac{2^{x+1}}{2^{1-x}}$.

7 Write as powers of 5 in simplest form:

a 1 **b** $5\sqrt{5}$ **c** $\frac{1}{\sqrt[4]{5}}$ **d** 25^{a+3}

8 Expand and simplify:

a $x^{\frac{1}{2}}(x^{\frac{3}{2}} - x^{-\frac{1}{2}})$ **b** $(3^x + 2)(3^x - 7)$ **c** $(x^{\frac{2}{3}} - x^{\frac{1}{3}})^2$

9 Factorise:

a $5^{n+2} - 5^n$ **b** $49^x - 7^x$ **c** $4^x - 2^x - 6$

10 Simplify:

a $\frac{15^x}{5^x}$ **b** $\frac{3^{n+3} + 3^n}{3^n}$ **c** $2^n(3n + 1) + 2^n(n - 1)$

11 Solve for x:

a $8^{x+1} = \left(\frac{1}{4}\right)^x$ **b** $\dfrac{25^x}{5^{x-3}} = \dfrac{5^x}{125^{x-2}}$ **c** $\dfrac{3^{x+2}}{9^{3-x}} = \dfrac{27^{1-2x}}{3^{2x}}$

12 Solve simultaneously:
$$xy = -8$$
$$9^x 27^y = 9$$

13 On the same set of axes, sketch the graphs of $y = 3^x$ and $y = 3^x + 2$.
Label the y-intercept and the equation of the horizontal asymptote of each function.

14 In a captive breeding program for rare birds, the population P after n years is given by $P(n) = P_0 \times 2^{0.4n}$.

a Find P_0 given the population after 5 years is 68.

b Find the population after:

i 2 years **ii** 10 years.

c Sketch the graph of $P(n)$ for $n \geqslant 0$.

15 Consider the function $y = e^{2x} + 1$.

a Find the horizontal asymptote and y-intercept.

b Sketch the graph of the function.

c State the domain and range of the function.

16 Consider $f : x \mapsto 2e^{3x}$ and $g : x \mapsto x^2 + 2x - 1$.

a State the range of f.

b Find the exact value of $f(-\frac{1}{2})$. Write your answer without negative indices.

c Solve $fg(x) = \dfrac{2}{e^5}$. Write your answer in the form $x = a + b\sqrt{3}$, where $a, b \in \mathbb{Q}$.

5

Logarithms

Contents:

Opening problem

An investor buys a property for $280 000. He expects the property to increase in value by 4.8% each year.

From the previous Chapter, we know the value V of the property is given by the exponential function $V(t) = 280\,000 \times 1.048^t$ where t is the time in years after the initial investment.

Things to think about:

- **a** What does the graph of $V(t)$ look like?
- **b** The investor wants to sell the property when its value reaches $400 000. How long will the investor have to wait?
- **c** Can we write a function for the time taken t for the property to reach the value V? What will the graph of this function look like?

A LOGARITHMS IN BASE 10

Consider the exponential function $f : x \mapsto 10^x$ or $f(x) = 10^x$.

The graph of $y = f(x)$ is shown alongside, along with its inverse function f^{-1}.

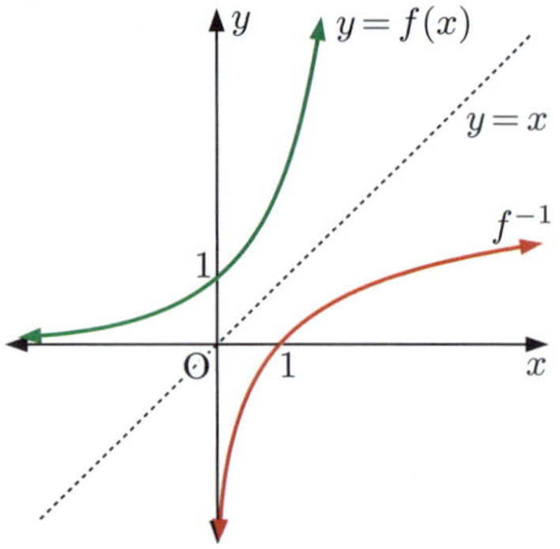

Since f is defined by $y = 10^x$,
f^{-1} is defined by $x = 10^y$.
{interchanging x and y}

y is the exponent to which the base 10 is raised in order to get x.

We write this as $y = \log_{10} x$ or $\lg x$ and say that y is the **logarithm in base 10, of x**.

Logarithms are thus defined to be the inverse of exponential functions:

If $f(x) = 10^x$ then $f^{-1}(x) = \log_{10} x$ or $\lg x$.

WORKING WITH LOGARITHMS

Many positive numbers can be easily written in the form 10^x. This allows us to quickly write down their logarithm.

For example:

$$10\,000 = 10^4 \quad \Rightarrow \quad \lg 10\,000 = 4$$
$$1000 = 10^3 \quad \Rightarrow \quad \lg 1000 = 3$$
$$100 = 10^2 \quad \Rightarrow \quad \lg 100 = 2$$
$$10 = 10^1 \quad \Rightarrow \quad \lg 10 = 1$$
$$1 = 10^0 \quad \Rightarrow \quad \lg 1 = 0$$
$$0.1 = 10^{-1} \quad \Rightarrow \quad \lg 0.1 = -1$$
$$0.01 = 10^{-2} \quad \Rightarrow \quad \lg 0.01 = -2$$
$$0.001 = 10^{-3} \quad \Rightarrow \quad \lg 0.001 = -3$$

Numbers like $\sqrt{10}$, $10\sqrt{10}$ and $\frac{1}{\sqrt[5]{10}}$ can also be written in the form 10^x:

$$\sqrt{10} = 10^{\frac{1}{2}} = 10^{0.5} \quad \therefore \ \lg\sqrt{10} = 0.5$$

$$10\sqrt{10} = 10^1 \times 10^{0.5} = 10^{1.5} \quad \therefore \ \lg 10\sqrt{10} = 1.5$$

$$\frac{1}{\sqrt[5]{10}} = 10^{-\frac{1}{5}} = 10^{-0.2} \quad \therefore \ \lg\frac{1}{\sqrt[5]{10}} = -0.2$$

In fact, all positive numbers can be written in the form 10^x.

The **logarithm in base 10** of a positive number is the power that 10 must be raised to in order to obtain the number.

$\mathbf{lg\,10^{x} = x}$ for any $x \in \mathbb{R}$.

$\lg a$ means $\log_{10} a$.
a must be positive since $10^x > 0$ for all $x \in \mathbb{R}$.

Example 1 — Self Tutor

Without using a calculator, find:

a $\lg(0.1)$ **b** $\lg(\sqrt[3]{100})$

a $\lg(0.1) = \lg 10^{-1} = -1$

b $\lg(\sqrt[3]{100}) = \lg 10^{\frac{2}{3}} = \frac{2}{3}$

The logarithms in **Example 1** can be found by hand because it is easy to write 0.1 and $\sqrt[3]{100}$ as powers of 10. To find the logarithms of other values we use a calculator.

For example, $\lg 130 \approx 2.11$
so $130 \approx 10^{2.11}$

Logarithms hence allow us to write any positive number as a power of 10. In particular:

$\mathbf{x = 10^{\lg x}}$ for any $x > 0$.

Example 2 — Self Tutor

Use your calculator to write the following in the form 10^x where x is correct to 4 decimal places:

a 9 **b** 600 **c** 0.021

a $9 = 10^{\lg 9} \approx 10^{0.9542}$

b $600 = 10^{\lg 600} \approx 10^{2.7782}$

c $0.021 = 10^{\lg 0.021} \approx 10^{-1.6778}$

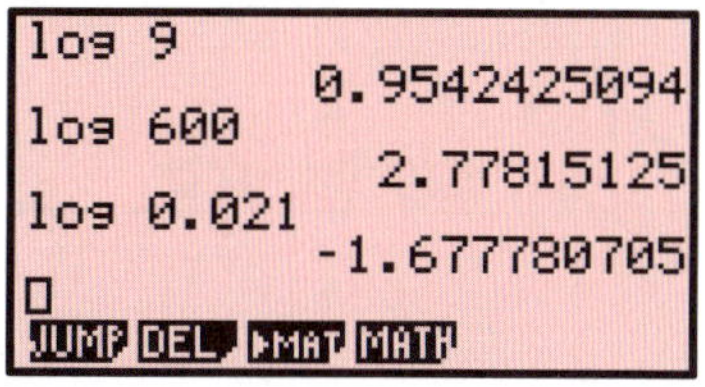

EXERCISE 5A

1 Without using a calculator, find:

a $\lg 100$ **b** $\lg(0.01)$ **c** $\lg 10\,000$ **d** $\lg 1$

e $\lg \sqrt[3]{10}$ **f** $\lg\left(\frac{1}{\sqrt{10}}\right)$ **g** $\lg(100\sqrt{10})$ **h** $\lg(10 \times \sqrt[4]{10})$

i $\lg(\sqrt{10^3})$ **j** $\lg\left(\frac{100}{\sqrt{10}}\right)$ **k** $\lg(\sqrt[4]{10^2})$ **l** $\lg\left(10 \times \frac{1}{\sqrt[4]{10}}\right)$

Check your answers using a calculator.

2 Simplify:

a $\lg 10^x$ **b** $\lg(1000 \times 10^n)$ **c** $\lg(\sqrt[n]{10^m})$ **d** $\lg\left(\frac{10^m}{\sqrt[n]{10}}\right)$

3 **a** Explain why $\lg 74$ must lie between 1 and 2.

b Use your calculator to evaluate $\lg 74$ correct to 2 decimal places.

4 Use your calculator to evaluate, correct to 3 decimal places:

a $\lg 26$ **b** $\lg 583$ **c** $\lg 5$ **d** $\lg 1425$

e $\lg(0.7)$ **f** $\lg 91$ **g** $\lg(0.03)$ **h** $\lg(-20)$

5 **a** Use your calculator to find $\lg 41$, giving your answer correct to 4 decimal places.

b Hence write 41 as a power of 10.

6 Use your calculator to write the following in the form 10^x, where x is correct to 4 decimal places:

a 5 **b** 50 **c** 500 **d** 5000 **e** 0.5

f 0.05 **g** 38 **h** 380 **i** 3800 **j** 3.8

k 0.38 **l** 0.038 **m** 0.0038

7 Copy and complete:

a $\lg x$ is positive if x is ...... **b** $\lg x$ is negative if x is ......

8 Explain why you cannot find the logarithm of a negative number.

Example 3 — Self Tutor

a Use your calculator to find: **i** $\lg 7$ **ii** $\lg 700$

b Explain why $\lg 700 = \lg 7 + 2$.

a

i $\lg 7 \approx 0.845$

ii $\lg 700 \approx 2.845$

b
$$\begin{aligned}\lg 700 &= \lg(7 \times 100)\\ &= \lg(10^{\lg 7} \times 10^2) \quad \{x = 10^{\lg x}\}\\ &= \lg 10^{\lg 7 + 2} \quad \{\text{adding indices}\}\\ &= \lg 7 + 2\end{aligned}$$

9 **a** Use your calculator to find: **i** $\lg 4$ **ii** $\lg 40$

b Explain why $\lg 40 = \lg 4 + 1$.

10 **a** Use your calculator to find: **i** $\lg 6$ **ii** $\lg(0.006)$

b Explain why $\lg(0.006) = \lg 6 - 3$.

Example 4 — Self Tutor

Find x if:

a $\lg x = 4$ **b** $\lg x \approx -1.52$

a $\lg x = 4$

$\therefore\ 10^{\lg x} = 10^4$

$\therefore\ x = 10\,000$

b $\lg x \approx -1.52$

$\therefore\ 10^{\lg x} \approx 10^{-1.52}$

$\therefore\ x \approx 0.0302$

11 Find x if:

a $\lg x = 3$ **b** $\lg x = 1$ **c** $\lg x = 0$

d $\lg x = -5$ **e** $\lg x = \frac{1}{3}$ **f** $\lg x = -\frac{3}{2}$

g $\lg x = 7$ **h** $\lg x = -\frac{1}{4}$ **i** $\lg x \approx 0.3489$

j $\lg x \approx 4.027$ **k** $\lg x \approx -0.923$ **l** $\lg x \approx -2.946$

12 Find x if:

a $\lg(x+3) = 1$ **b** $\lg(2x-1) = 0$ **c** $\lg|x| = -1$

B LOGARITHMS IN BASE a

In the previous Section we defined logarithms in base 10 as the inverse of the exponential function $f(x) = 10^x$.

If $f(x) = 10^x$ then $f^{-1}(x) = \log_{10} x$.

We can use the same principle to define logarithms in other bases:

If $f(x) = a^x$ then $f^{-1}(x) = \log_a x$.

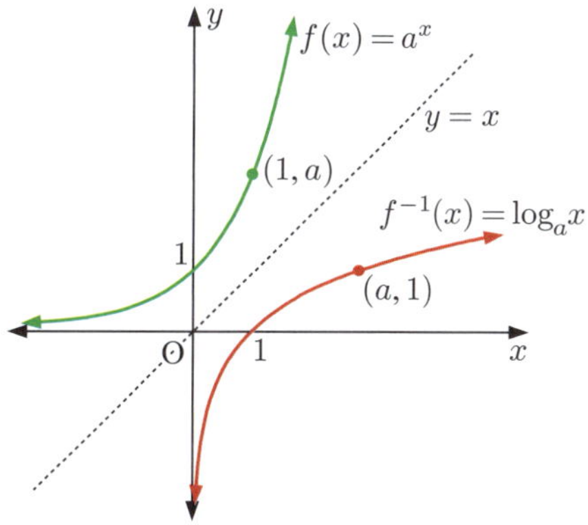

If $a > 0$, $a \neq 1$, then the **logarithm in base a** of a positive number is the power that a must be raised to in order to obtain the number.

The **logarithm in base a of b** is written $\log_a b$.

To find $\log_5 25$, we ask "What power must 5 be raised to in order to obtain 25?". We know that $5^2 = 25$, so $\log_5 25 = 2$.

$a^x = b$ and $x = \log_a b$ are *equivalent* statements.

We write: $a^x = b \Leftrightarrow x = \log_a b$

Example 5 — Self Tutor

a Write an equivalent exponential statement for $\log_2 32 = 5$.

b Write an equivalent logarithmic statement for $6^{-1} = \frac{1}{6}$.

a $\log_2 32 = 5 \Leftrightarrow 2^5 = 32$

b $6^{-1} = \frac{1}{6} \Leftrightarrow \log_6 \left(\frac{1}{6}\right) = -1$

EXERCISE 5B

1 Write an equivalent exponential statement for:

a $\log_{10} 1000 = 3$ **b** $\log_3 9 = 2$ **c** $\log_2 16 = 4$

d $\log_7 1 = 0$ **e** $\log_5 \left(\frac{1}{5}\right) = -1$ **f** $\log_2 \sqrt{2} = \frac{1}{2}$

g $\log_6 \left(\frac{1}{36}\right) = -2$ **h** $\log_3 \left(\frac{1}{\sqrt{3}}\right) = -\frac{1}{2}$ **i** $\log_4 8 = \frac{3}{2}$

2 Write an equivalent logarithmic statement for:

a $3^3 = 27$ **b** $2^3 = 8$ **c** $9^2 = 81$

d $4^1 = 4$ **e** $5^0 = 1$ **f** $2^{-3} = \frac{1}{8}$

g $7^{\frac{3}{2}} = 7\sqrt{7}$ **h** $16^{\frac{3}{4}} = 8$ **i** $9^{-\frac{1}{2}} = \frac{1}{3}$

3 Suppose $a^5 = b$, where a and b are positive. By writing an equivalent logarithmic statement, find:

a $\log_a b$ **b** $\log_b a$

Example 6 — Self Tutor

Find:

a $\log_7 49$ **b** $\log_3 \sqrt{3}$ **c** $\log_2 (0.25)$

a $\log_7 49$
$= \log_7 7^2$
$= 2$

b $\log_3 \sqrt{3}$
$= \log_3 3^{\frac{1}{2}}$
$= \frac{1}{2}$

c $\log_2 (0.25)$
$= \log_2 \left(\frac{1}{4}\right)$
$= \log_2 2^{-2}$
$= -2$

4 Find:

a $\log_{10} 100$ **b** $\log_2 8$ **c** $\log_3 \left(\frac{1}{3}\right)$
d $\log_5 \sqrt{5}$ **e** $\log_9 81$ **f** $\log_2 \left(\frac{1}{\sqrt{2}}\right)$
g $\log_3 \left(\frac{1}{243}\right)$ **h** $\log_4 2$ **i** $\log_2 1$
j $\log_6 6$ **k** $\log_2 \sqrt{32}$ **l** $\log_9 3$
m $\log_5 \sqrt[3]{5}$ **n** $\log_4 (0.5)$ **o** $\log_7 \sqrt[3]{49}$
p $\log_3 (9\sqrt{3})$ **q** $\log_5 \left(\frac{1}{\sqrt{125}}\right)$ **r** $\log_8 \left(\frac{1}{2\sqrt{2}}\right)$

Example 7 — Self Tutor

Solve for x: $\log_2 x = 7$

$$\log_2 x = 7$$
$$\therefore \quad x = 2^7$$
$$\therefore \quad x = 128$$

5 Solve for x:

a $\log_3 x = 5$ **b** $\log_2 x = -5$ **c** $\log_4 x = 2$
d $\log_9 x = -\frac{1}{2}$ **e** $\log_{16} x = \frac{3}{2}$ **f** $\log_{27} x = -\frac{2}{3}$
g $\log_2 (x + 2) = -1$ **h** $\log_3 (3x + 4) = 2$ **i** $\log_5 (2x - 1) = \frac{1}{2}$
j $\log_4 (5x + 2) = 0$ **k** $\log_2 |x - 1| = 3$ **l** $\log_3 |2x - 5| = 3$

6 Simplify:

a $\log_a (a^3)$ **b** $\log_b \left(\frac{1}{b^2}\right)$ **c** $\log_c \sqrt[3]{c}$ **d** $\log_d = \frac{1}{\sqrt{d}}$

Discussion

The following is a proof that $\log_2 3$ is irrational.

Proof: If $\log_2 3$ is rational, then $\log_2 3 = \frac{p}{q}$ where $p, q \in \mathbb{Z}$, $q \neq 0$

$$\therefore \quad 3 = 2^{\frac{p}{q}}$$
$$\therefore \quad 3^q = 2^p$$

The left hand side is always odd, and the right hand side is always even, so the statement is impossible.

Hence $\log_2 3$ must be irrational.

- Can you use similar reasoning to prove that:
  - $\log_2 5$ is irrational
  - $\log_2 6$ is irrational?
- Why can't we use this reasoning to prove that $\log_2 8$ is irrational?

C LAWS OF LOGARITHMS

Discovery — The laws of logarithms

What to do:

1 Use your calculator to find:

a $\lg 3 + \lg 4$ **b** $\lg 4 + \lg 7$ **c** $\lg 5 + \lg 9$

d $\lg 12$ **e** $\lg 28$ **f** $\lg 45$

From your answers, suggest a possible simplification for $\lg m + \lg n$.

2 Use your calculator to find:

a $\lg 8 - \lg 2$ **b** $\lg 15 - \lg 3$ **c** $\lg 3 - \lg 4$

d $\lg 4$ **e** $\lg 5$ **f** $\lg(0.75)$

From your answers, suggest a possible simplification for $\lg m - \lg n$.

3 Use your calculator to find:

a $2\lg 3$ **b** $4\lg 5$ **c** $-3\lg 7$

d $\lg(3^2)$ **e** $\lg(5^4)$ **f** $\lg(7^{-3})$

From your answers, suggest a possible simplification for $m\lg b$.

From the **Discovery**, you should have found the three important **laws of logarithms**:

- $\lg m + \lg n = \lg(mn)$ for $m, n > 0$
- $\lg m - \lg n = \lg\left(\frac{m}{n}\right)$ for $m, n > 0$
- $m\lg b = \lg(b^m)$ for $b > 0$

More generally, in any base a where $a \neq 1$, $a > 0$, we have these **laws of logarithms**:

- $\log_a m + \log_a n = \log_a(mn)$ for $m, n > 0$
- $\log_a m - \log_a n = \log_a\left(\frac{m}{n}\right)$ for $m, n > 0$
- $m\log_a b = \log_a(b^m)$ for $b > 0$

Example 8 — Self Tutor

Use the laws of logarithms to write the following as a single logarithm or as an integer:

a $\lg 6 + \lg 2$ **b** $\log_3 36 - \log_3 4$ **c** $\log_2 7 + 3$

a $\lg 6 + \lg 2$
$= \lg(6 \times 2)$
$= \lg 12$

b $\log_3 36 - \log_3 4$
$= \log_3\left(\frac{36}{4}\right)$
$= \log_3 9$
$= 2$

c $\log_2 7 + 3$
$= \log_2 7 + \log_2(2^3)$
$= \log_2(7 \times 8)$
$= \log_2 56$

Example 9

Self Tutor

Simplify by writing as a single logarithm or as a rational number:

a $2\lg 5 - 4\lg 2$ **b** $4\lg 3 - 1$ **c** $\dfrac{\lg 9}{\lg 27}$

a $2\lg 5 - 4\lg 2$
$= \lg(5^2) - \lg(2^4)$
$= \lg 25 - \lg 16$
$= \lg\left(\frac{25}{16}\right)$

b $4\lg 3 - 1$
$= \lg(3^4) - \lg(10^1)$
$= \lg 81 - \lg 10$
$= \lg\left(\frac{81}{10}\right)$

c $\dfrac{\lg 9}{\lg 27}$
$= \dfrac{\lg(3^2)}{\lg(3^3)}$
$= \dfrac{2\lg 3}{3\lg 3}$
$= \frac{2}{3}$

EXERCISE 5C

1 Write as a single logarithm or as an integer:

a $\lg 5 + \lg 3$ **b** $\lg 2 + \lg 7$ **c** $\lg 24 - \lg 6$
d $\lg 45 - \lg 5$ **e** $\lg 12 + \lg(0.5)$ **f** $\lg 25 + \lg 4$
g $\log_2 56 - \log_2 7$ **h** $\log_3 20 - \log_3 60$ **i** $\lg 3 + \lg 2 + \lg 5$
j $\lg 4 + \lg 3 - \lg 15$ **k** $\log_2 9 - \log_2 12 - \log_2 6$ **l** $\lg 6 + \lg\left(\frac{5}{3}\right) + \lg 9$

2 Write as a single logarithm:

a $\lg 4 + 2$ **b** $\lg 3 - 1$ **c** $\log_3 5 + 1$
d $\log_2 7 - 2$ **e** $3 + \lg 5$ **f** $\lg 80 - 4$
g $k + \lg m$ **h** $\log_a 11 + 2$ **i** $4 - \log_3 20$

3 Write as a single logarithm or integer:

a $3\lg 2$ **b** $-2\lg 5$ **c** $\lg 5 + 2\lg 4$
d $\frac{1}{2}\lg 36 + \lg 3$ **e** $2\lg 2 + \lg 25$ **f** $\log_3 4 - 2\log_3 6$
g $2 + \frac{1}{3}\lg 27$ **h** $\lg 14 + \frac{1}{2}\lg\left(\frac{1}{4}\right)$ **i** $1 - 2\lg 3 + 3\lg 5$

4 Simplify without using a calculator:

a $\dfrac{\lg 8}{\lg 4}$ **b** $\dfrac{\lg 25}{\lg 125}$ **c** $\dfrac{\log_5 27}{\log_5 81}$
d $\dfrac{\lg\sqrt{2}}{\lg 16}$ **e** $\dfrac{\log_3 8}{\log_3(0.5)}$ **f** $\dfrac{\lg(0.2)}{\lg 125}$

5 Find the exact value of:

a $3\lg 2 + 2\lg 5 - \frac{1}{2}\lg 4$ **b** $2\log_2 3 - \log_2 6 - \frac{1}{2}\log_2 9$
c $5\log_6 2 + 2\log_6 3 - \frac{1}{2}\log_6 16 - \log_6 12$ **d** $\dfrac{2 + 2\log_a 3}{\log_a 3a}$

6 Write:

a $\lg x^4$ in terms of $\lg x$
b $\lg 8x^3$ in terms of $\lg 2x$
c $\lg\left(\dfrac{1}{243y^5}\right)$ in terms of $\lg 3y$.

7 Suppose $x = \log_3 P$, $y = \log_3 Q$, and $z = \log_3 R$. Write in terms of x, y, and z:

a $\log_3(QR)$ **b** $\log_3(PQ^2)$ **c** $\log_3\left(\frac{PR}{Q}\right)$

d $\log_3(R^3\sqrt{P})$ **e** $\log_3\left(\frac{\sqrt[3]{Q}}{R^2}\right)$ **f** $\log_3\left(\frac{P\sqrt{R}}{Q^4}\right)$

8 Suppose $p = \log_b 2$, $q = \log_b 3$, and $r = \log_b 5$. Write in terms of p, q, and r:

a $\log_b 15$ **b** $\log_b 20$ **c** $\log_b 30$

d $\log_b\left(\frac{3}{4}\right)$ **e** $\log_b\left(\frac{\sqrt{5}}{12}\right)$ **f** $\log_b(0.009)$

9 Suppose $\log_b P = 5$ and $\log_b(P^3Q^2) = 21$. Find $\log_b Q$.

10 Suppose $\log_t(AB^3) = 15$ and $\log_t\left(\frac{A^2}{B}\right) = 9$.

a Write *two* equations connecting $\log_t A$ and $\log_t B$.

b Hence find $\log_t A$ and $\log_t B$.

c Find $\log_t\left(B^5\sqrt{A}\right)$.

d Write B in terms of t.

D LOGARITHMIC EQUATIONS

We can use the laws of logarithms to write equations in a different form. This can be particularly useful if an unknown appears as an exponent.

Since the logarithmic function is one-one, for every value of y there is only one corresponding value of x. We can therefore take the logarithm of both sides of an equation without changing the solution. However, we can only do this if both sides are positive since the domain of $\log_a x$ is $x > 0$.

Example 10 Self Tutor

Write as a logarithmic equation in base 10:

a $y = 5 \times 3^x$ **b** $P = \frac{20}{\sqrt{n}}$

a
$$y = 5 \times 3^x$$
$$\therefore \lg y = \lg(5 \times 3^x)$$
$$\therefore \lg y = \lg 5 + \lg(3^x)$$
$$\therefore \lg y = \lg 5 + x \lg 3$$

b
$$P = \frac{20}{\sqrt{n}}$$
$$\therefore \lg P = \lg\left(\frac{20}{n^{\frac{1}{2}}}\right)$$
$$\therefore \lg P = \lg 20 - \lg(n^{\frac{1}{2}})$$
$$\therefore \lg P = \lg 20 - \tfrac{1}{2}\lg n$$

Example 11

Self Tutor

Write the following equations without logarithms:

a $\lg y = x \lg 4 + \lg 3$ **b** $\log_2 M = 3\log_2 a - 5$

a
$$\lg y = x\lg 4 + \lg 3$$
$$\therefore \lg y = \lg(4^x) + \lg 3$$
$$\therefore \lg y = \lg(3 \times 4^x)$$
$$\therefore y = 3 \times 4^x$$

b
$$\log_2 M = 3\log_2 a - 5$$
$$\therefore \log_2 M = \log_2(a^3) - \log_2(2^5)$$
$$\therefore \log_2 M = \log_2\left(\frac{a^3}{32}\right)$$
$$\therefore M = \frac{a^3}{32}$$

EXERCISE 5D.1

1 Write as a logarithmic equation in base 10:

a $y = 2^x$ **b** $y = x^3$ **c** $M = d^4$
d $T = 5^x$ **e** $y = \sqrt{x}$ **f** $y = 7 \times 3^x$
g $S = \frac{9}{t}$ **h** $M = 100 \times 7^x$ **i** $T = 5\sqrt{d}$
j $F = \frac{1000}{\sqrt{n}}$ **k** $S = 200 \times 2^t$ **l** $y = \sqrt{\frac{15}{x}}$

2 Write the following equations without logarithms:

a $\lg y = x\lg 7$ **b** $\lg D = \lg x + \lg 2$ **c** $\log_a F = \log_a 5 - \log_a t$
d $\lg y = x\lg 2 + \lg 6$ **e** $\lg P = \frac{1}{2}\lg x$ **f** $\lg N = -\frac{1}{3}\lg p$
g $\lg P = 3\lg x + 1$ **h** $\lg y = x - \lg 2$ **i** $\lg y = 2\lg x - 1$
j $\log_2 T = 5\log_2 k + 1$ **k** $\log_3 P = 4\log_3 n - 2$ **l** $\log_2 y = 4x + 3$

3 Suppose $\lg y = 3\lg x - \lg 2$.

a Write y in terms of x, without using logarithms.
b Find y when: **i** $x = 2$ **ii** $x = 4$

4 Suppose $\lg y = \frac{1}{3}x + 2$.

a Write y in the form $y = a(10^{bx})$ where $a, b \in \mathbb{Q}$.
b Find y when: **i** $x = 0$ **ii** $x = 3$

5 Copy and complete:

a If there is a *power* relationship between y and x, for example $y = 5x^3$, then there is a *linear* relationship between $\lg y$ and ......
b If there is an *exponential* relationship between y and x, for example $y = 4 \times 2^x$, then there is a *linear* relationship between ...... and ......

SOLVING LOGARITHMIC EQUATIONS

Logarithmic equations can often be solved using the laws of logarithms. However, we must always check that our solutions satisfy the original equation, remembering that $\log_a x$ is only defined for $x > 0$.

Example 12

Self Tutor

Solve for x:

a $\lg(x-6)+\lg 3 = 2\lg 6$

b $\lg x + \lg(x+5) = \lg 14$

a
$$\lg(x-6)+\lg 3 = 2\lg 6$$
$$\therefore \lg(x-6) = \lg(6^2) - \lg 3$$
$$\therefore \lg(x-6) = \lg\left(\tfrac{36}{3}\right)$$
$$\therefore x-6 = 12$$
$$\therefore x = 18$$

Check: $x-6>0$, so $x>6$ ✓

b
$$\lg x + \lg(x+5) = \lg 14$$
$$\therefore \lg(x(x+5)) = \lg 14$$
$$\therefore x(x+5) = 14$$
$$\therefore x^2+5x-14 = 0$$
$$\therefore (x+7)(x-2) = 0$$
$$\therefore x = -7 \text{ or } 2$$

But $x>0$ and $x+5>0$

$\therefore$ $x=2$ is the only valid solution.

EXERCISE 5D.2

1 Solve for x:

a $\lg(x-4) = \lg 3 + \lg 7$

b $\lg(x+5) - \lg 8 = 2\lg 3$

c $\lg(2x) = 1 + \frac{1}{2}\lg 16$

d $\log_2 x = 3\log_2 5 - 6$

e $\lg x - \lg(x-4) = \lg 5$

f $\log_5(x-2) - \log_5(x+2) = \log_5 3$

g $\log_3 x - 2 = \log_3(x-1)$

h $\lg(x+2) - 1 = \lg(x-3) - \lg 12$

2 Solve for x:

a $\lg x + \lg(x+1) = \lg 30$

b $\log_5(x+9) + \log_5(x+2) = \log_5(20x)$

c $\log_7 x = \log_7 8 - \log_7(6-x)$

d $\log_6(x+4) + \log_6(x-1) = 1$

e $\lg x + \lg(2x+8) = 1$

f $\lg(x+2) + \lg(x+7) = \lg(2x+2)$

g $2\log_2 x - \log_2(8-3x) = 1$

h $\log_2 x + \log_2(2x-7) = 2$

Example 13

Self Tutor

Solve for x: $\log_x 3 + \log_x 12 = 2$

$$\log_x 3 + \log_x 12 = 2$$
$$\therefore \log_x(3\times 12) = \log_x(x^2)$$
$$\therefore 36 = x^2$$
$$\therefore x = 6 \quad \{\text{since } x>0\}$$

3 Solve for x:

a $\log_x 32 - \log_x 4 = 1$

b $\log_x 45 = 2 + \log_x 5$

c $\log_x 54 = 3 - \log_x 4$

d $2\log_x 2 - 3 = \log_x\left(\frac{1}{16}\right)$

4 Solve simultaneously:

a
$$xy = 4$$
$$\log_2 x^2 + \log_2 y^3 = 3$$

b
$$3x + y = 14$$
$$\log_3 x + \log_3(2y-1) = 3$$

E NATURAL LOGARITHMS

In **Chapter 4** we came across the **natural exponential** $e \approx 2.718\,28$.

Given the exponential function $f(x) = e^x$, the inverse function $f^{-1} = \log_e x$ is the logarithm in base e.

We use $\ln x$ to represent $\log_e x$, and call $\ln x$ the **natural logarithm** of x.

$y = \ln x$ is the reflection of $y = e^x$ in the mirror line $y = x$.

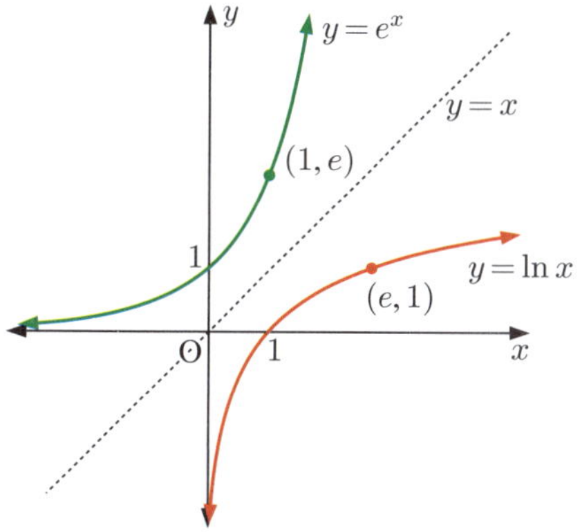

$$\ln(e^x) = x \quad \text{and} \quad e^{\ln x} = x.$$

For example:
- $\ln 1 = \ln e^0 = 0$
- $\ln e = \ln e^1 = 1$
- $\ln(e^2) = 2$
- $\ln \sqrt{e} = \ln(e^{\frac{1}{2}}) = \frac{1}{2}$
- $\ln\left(\frac{1}{e}\right) = \ln e^{-1} = -1$

As with base 10 logarithms, we can use our calculator to find natural logarithms.

For example, $\ln 50 \approx 3.91$, which means that $50 \approx e^{3.91}$.

Example 14

Self Tutor

Use your calculator to write the following in the form e^k where k is correct to 4 decimal places:

a 15 **b** 0.4

a 15
$= e^{\ln 15}$ {using $x = e^{\ln x}$}
$\approx e^{2.7081}$

b 0.4
$= e^{\ln 0.4}$
$\approx e^{-0.9163}$

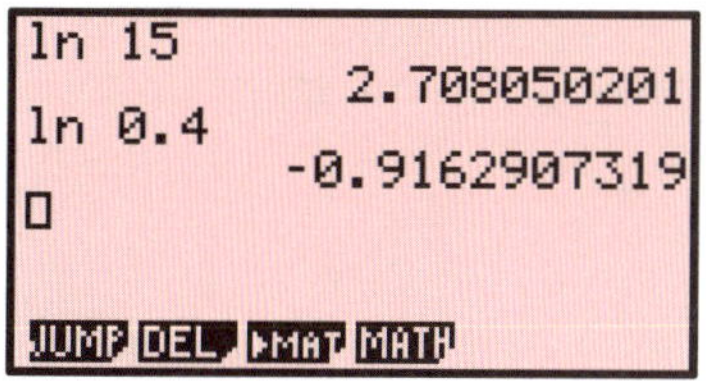

EXERCISE 5E.1

1 Without using a calculator, find:

a $\ln(e^3)$ **b** $\ln(e^5)$ **c** $\ln\left(\frac{1}{e^4}\right)$ **d** $\ln(\sqrt{e})$

e $\ln(e^k)$ **f** $\ln(e^2 \times e^k)$ **g** $\ln\left(\frac{e^a}{e^b}\right)$ **h** $\ln(e\sqrt{e})$

2 Simplify:

a $e^{\ln 5}$ **b** $e^{3\ln 2}$ **c** $e^{-\ln 4}$ **d** $e^{-2\ln 3}$

3 Use your calculator to find, correct to 3 decimal places:

a $\ln 10$ **b** $\ln 70$ **c** $\ln 2$ **d** $\ln(0.3)$ **e** $\ln 800$

4 Explain why $\ln(-5)$ cannot be found.

5 Use your calculator to write the following in the form e^k where k is correct to 4 decimal places:

a 30 **b** 7 **c** 94 **d** 0.1 **e** 500

Example 15 — Self Tutor

Find x if:

a $\ln x = 3$ **b** $\ln x = -0.2$

a $\ln x = 3$
$\therefore\ x = e^3$
$\therefore\ x \approx 20.1$

b $\ln x = -0.2$
$\therefore\ x = e^{-0.2}$
$\therefore\ x \approx 0.819$

6 Find x if:

a $\ln x = 2$ **b** $\ln x = 1$ **c** $\ln x = 0$ **d** $\ln x = -3$
e $\ln x = -6$ **f** $\ln x \approx 0.74$ **g** $\ln x \approx 1.52$ **h** $\ln x \approx -2.51$

7 Find x if:

a $\ln(x-4) = 1$ **b** $\ln(3x+1) = -1$ **c** $\ln(x^2+3) = -1$ **d** $\ln|x+2| = 0$

8 Find k such that the equation $\ln(x^2 - k) = 2$ has:

a two real solutions **b** one real solution **c** no real solutions.

9 Solve simultaneously:
$$x + y = 5$$
$$\ln\left(\frac{xy}{4}\right) = 0$$

LAWS OF NATURAL LOGARITHMS

The laws for natural logarithms are the laws for logarithms written in base e:

- $\ln m + \ln n = \ln(mn)$ for $m, n > 0$
- $\ln m - \ln n = \ln\left(\frac{m}{n}\right)$ for $m, n > 0$
- $m \ln b = \ln(b^m)$ for $b > 0$

Example 16 — Self Tutor

Use the laws of logarithms to write the following as a single logarithm:

a $\ln 2 + \ln 9$ **b** $\ln 30 - \ln 6$ **c** $\ln 4 + 1$

a $\ln 2 + \ln 9$
$= \ln(2 \times 9)$
$= \ln 18$

b $\ln 30 - \ln 6$
$= \ln\left(\frac{30}{6}\right)$
$= \ln 5$

c $\ln 4 + 1$
$= \ln 4 + \ln(e^1)$
$= \ln(4e)$

Example 17 — Self Tutor

Use the laws of logarithms to simplify:

a $3\ln 4 + 2\ln 5$ **b** $2\ln 7 - 2$

a $3\ln 4 + 2\ln 5$
$= \ln(4^3) + \ln(5^2)$
$= \ln 64 + \ln 25$
$= \ln(64 \times 25)$
$= \ln 1600$

b $2\ln 7 - 2$
$= \ln(7^2) - \ln(e^2)$
$= \ln 49 - \ln(e^2)$
$= \ln\left(\frac{49}{e^2}\right)$

EXERCISE 5E.2

1 Write as a single logarithm or integer:

a $\ln 10 + \ln 2$ **b** $\ln 10 - \ln 2$ **c** $\ln 6 + \ln 7$
d $\ln 35 - \ln 5$ **e** $\ln 9 + \ln(0.5)$ **f** $\ln 12 - \ln 22$
g $\ln 2 + \ln 5 + \ln 7$ **h** $\ln 3 + \ln 12 - \ln 6$ **i** $\ln 28 - \ln 4 - \ln 7$
j $\ln 5 + 2$ **k** $\ln 8 - 1$ **l** $4 - \ln 10$

2 Write in the form $\ln a,\ a \in \mathbb{Q}$:

a $3\ln 5$ **b** $-2\ln 4$ **c** $\frac{1}{2}\ln 25$
d $2\ln 3 + \ln 6$ **e** $4\ln 2 - 3\ln 3$ **f** $\frac{1}{3}\ln 64 + 2\ln 6$
g $\ln 2 + 2\ln 9 - \ln 8$ **h** $-\ln\left(\frac{1}{3}\right)$ **i** $-3\ln\left(\frac{1}{2}\right)$
j $2\ln\sqrt{7} - 4\ln\sqrt{5}$ **k** $-\frac{1}{2}\ln\left(\frac{1}{2}\right) + \frac{3}{2}\ln 2$ **l** $3\ln 5 - 2\ln\left(\frac{1}{3}\right) + \ln\left(\frac{1}{5}\right)$

Example 18 — Self Tutor

Show that:

a $\ln\left(\frac{1}{9}\right) = -2\ln 3$ **b** $\ln\left(\frac{e}{4}\right) = 1 - 2\ln 2$

a $\ln\left(\frac{1}{9}\right)$
$= \ln(3^{-2})$
$= -2\ln 3$

b $\ln\left(\frac{e}{4}\right) = \ln e - \ln 4$
$= \ln e^1 - \ln(2^2)$
$= 1 - 2\ln 2$

3 Show that:

a $\ln\sqrt[3]{5} = \frac{1}{3}\ln 5$ **b** $\ln\left(\frac{1}{32}\right) = -5\ln 2$
c $\ln\left(\frac{1}{\sqrt[5]{2}}\right) = -\frac{1}{5}\ln 2$ **d** $\ln\left(\frac{e^2}{8}\right) = 2 - 3\ln 2$

4 Write $\ln 27 - 5\ln\left(\frac{1}{\sqrt{3}}\right) + 3\ln\left(\frac{1}{\sqrt[3]{81}}\right) - 5$ in the form $\ln a$.

Example 19 Self Tutor

Write the following equations without logarithms:

a $\ln A = 2\ln c + 3$ **b** $\ln M = 3a - \ln 2$

a
$$\ln A = 2\ln c + 3$$
$$\therefore \ln A = \ln(c^2) + \ln(e^3)$$
$$\therefore \ln A = \ln(c^2e^3)$$
$$\therefore A = c^2e^3$$

b
$$\ln M = 3a - \ln 2$$
$$\therefore \ln M = \ln(e^{3a}) - \ln 2$$
$$\therefore \ln M = \ln\left(\frac{e^{3a}}{2}\right)$$
$$\therefore M = \tfrac{1}{2}e^{3a}$$

5 Write the following equations without logarithms:

a $\ln D = \ln x + 1$
b $\ln F = -\ln p + 2$
c $\ln P = 2x + \ln 5$
d $\ln M = 2\ln y + 3$
e $\ln B = 3t - \ln 4$
f $\ln N = -\frac{1}{3}\ln g$
g $\ln K = 5x + \ln 3$
h $\ln Q \approx 3\ln x + 2.159$
i $\ln D \approx 0.4\ln n - 0.6582$
j $\ln T \approx -x + 1.578$

F SOLVING EXPONENTIAL EQUATIONS USING LOGARITHMS

In **Chapter 4** we found solutions to simple exponential equations where we could make equal bases and then equate exponents. However, it is not always easy to make the bases the same. In these situations we use **logarithms** to find the exact solution.

Example 20 Self Tutor

Solve for x, giving your answers correct to 3 significant figures:

a $2^x = 7$ **b** $5^{3x-1} = 90$

a
$$2^x = 7$$
$$\therefore \lg(2^x) = \lg 7$$
$$\therefore x\lg 2 = \lg 7 \quad \{\lg(a^n) = n\lg a\}$$
$$\therefore x = \frac{\lg 7}{\lg 2}$$
$$\therefore x \approx 2.81$$

b
$$5^{3x-1} = 90$$
$$\therefore \lg(5^{3x-1}) = \lg 90$$
$$\therefore (3x-1)\lg 5 = \lg 90 \quad \{\lg(a^n) = n\lg a\}$$
$$\therefore 3x - 1 = \frac{\lg 90}{\lg 5}$$
$$\therefore x = \tfrac{1}{3}\left(1 + \frac{\lg 90}{\lg 5}\right)$$
$$\therefore x \approx 1.27$$

EXERCISE 5F

1 Solve for x, giving your answer correct to 3 significant figures:

a $2^x = 25$ **b** $5^x = 40$ **c** $3^x = 0.7$

d $\left(\frac{7}{2}\right)^x = 100$ **e** $\left(\frac{2}{3}\right)^x = 0.1$ **f** $(0.4)^x = 0.003$

2 Solve for x, giving your answer correct to 3 significant figures:

a $5^{2x} = 100$ **b** $2^{4x} = 75$ **c** $(0.8)^{3x} = 0.1$

d $3^{x-1} = 200$ **e** $4^{x+2} = 2.5$ **f** $6^{2x-1} = 800$

g $7^{2x+3} = 1000$ **h** $\left(3^{x+1}\right)^2 = 480$ **i** $\left(2^{x-3}\right)^{\frac{1}{2}} = 10$

Example 21 — Self Tutor

Find x exactly:

a $e^x = 12$ **b** $2e^{3x} = 40$

a $e^x = 12$

$\therefore\ x = \ln 12$

b $2e^{3x} = 40$

$\therefore\ e^{3x} = 20$

$\therefore\ 3x = \ln 20$

$\therefore\ x = \dfrac{\ln 20}{3}$

3 Find x exactly:

a $e^x = 8$ **b** $e^x = 50$ **c** $3e^x = 39$

d $e^{\frac{x}{2}} = 12$ **e** $e^{x-3} = 10$ **f** $4e^{2x} = 24$

g $e^{3x-1} = 20$ **h** $\frac{1}{2}e^{-5x} = 0.1$ **i** $6e^{-\frac{x}{2}} = 5$

4 **a** Solve $e^{2x} = 300$ exactly.

b Use your calculator to find the solution correct to 2 decimal places.

5 Solve for x, giving an exact answer:

a $4 \times 2^{-x} = 0.12$ **b** $300 \times 5^{0.1x} = 1000$ **c** $32 \times 3^{-0.25x} = 4$

6 The mass M of bacteria in a culture t hours after establishment is given by $M = 25 \times e^{0.1t}$ grams.

a Show that the time required for the mass of the culture to reach 50 grams is $10 \ln 2$ hours.

b Find the time required correct to 2 decimal places.

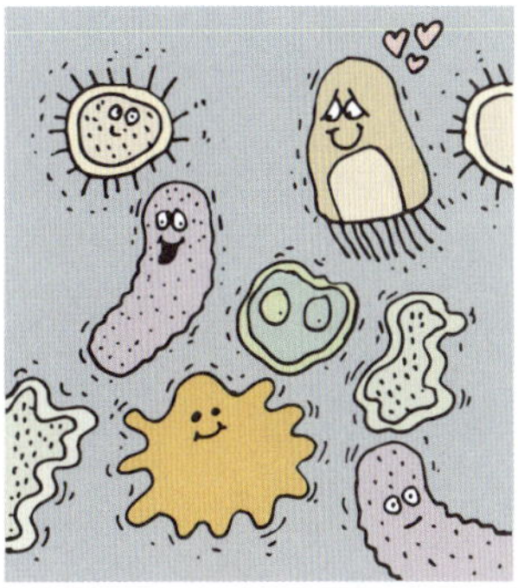

7 The weight of a radioactive uranium sample remaining after t years is given by the formula $W(t) = 50 \times 2^{-0.0002t}$ grams, $t \geqslant 0$.

a Find the initial weight of the uranium.

b Find the time required for the weight to reduce to 8 grams.

8 Answer the **Opening Problem** on page **128**.

Example 22

Self Tutor

Find the exact points of intersection of $y = e^x - 3$ and $y = 1 - 3e^{-x}$.

The functions meet where $e^x - 3 = 1 - 3e^{-x}$

GRAPHING PACKAGE

$\therefore\ e^x - 4 + 3e^{-x} = 0$

$\therefore\ e^{2x} - 4e^x + 3 = 0$ {multiplying each term by e^x}

$\therefore\ (e^x - 1)(e^x - 3) = 0$

$\therefore\ e^x = 1$ or $e^x = 3$

$\therefore\ x = \ln 1$ or $x = \ln 3$

$\therefore\ x = 0$ or $x = \ln 3$

When $x = 0$, $y = e^0 - 3 = -2$

When $x = \ln 3$, $y = e^{\ln 3} - 3 = 0$

$\therefore$ the functions meet at $(0, -2)$ and at $(\ln 3, 0)$.

9 Solve for x:

- **a** $e^{2x} = 2e^x$
- **b** $e^x = e^{-x}$
- **c** $e^{2x} - 5e^x + 6 = 0$
- **d** $e^x + 2 = 3e^{-x}$
- **e** $1 + 12e^{-x} = e^x$
- **f** $e^x + e^{-x} = 3$

10 Find algebraically the point(s) of intersection of:

- **a** $y = 2e^x + 1$ and $y = 7 - e^x$
- **b** $y = e^x$ and $y = e^{2x} - 6$
- **c** $y = 3 - e^x$ and $y = 5e^{-x} - 3$
- **d** $y = 2e^x + 5e^{-x}$ and $y = 7 + 2e^{-x}$

G THE CHANGE OF BASE RULE

A logarithm in base b can be written with a different base c using the **change of base rule**:

$$\log_b a = \frac{\log_c a}{\log_c b} \quad \text{for } a, b, c > 0 \text{ and } b, c \neq 1.$$

Proof: If $\log_b a = x$, then $b^x = a$

$\therefore\ \log_c b^x = \log_c a$ {taking logarithms in base c}

$\therefore\ x \log_c b = \log_c a$ {power law of logarithms}

$\therefore\ x = \dfrac{\log_c a}{\log_c b}$

$\therefore\ \log_b a = \dfrac{\log_c a}{\log_c b}$

We can use this rule to write logarithms in base 10 or base e. This is useful in helping us evaluate them on our calculator.

Example 23 Self Tutor

Evaluate $\log_2 9$ by:

a changing to base 10 **b** changing to base e.

a $\log_2 9 = \dfrac{\log_{10} 9}{\log_{10} 2} \approx 3.17$ **b** $\log_2 9 = \dfrac{\ln 9}{\ln 2} \approx 3.17$

The rule can also be used to solve equations involving logarithms with different bases.

Example 24 Self Tutor

Solve for x: $\log_2 x = \log_8 15$

$$\log_2 x = \log_8 15$$
$$\therefore \log_2 x = \frac{\log_2 15}{\log_2 8} \quad \{\text{writing RHS with base 2}\}$$
$$\therefore \log_2 x = \frac{\log_2 15}{3}$$
$$\therefore \log_2 x = \log_2 15^{\frac{1}{3}}$$
$$\therefore x = \sqrt[3]{15}$$

EXERCISE 5G

1 Use the rule $\log_b a = \dfrac{\log_{10} a}{\log_{10} b}$ to evaluate, correct to 3 significant figures:

a $\log_2 12$ **b** $\log_5 1250$ **c** $\log_3(0.067)$ **d** $\log_{0.4}(0.006\,984)$

2 Use the rule $\log_b a = \dfrac{\ln a}{\ln b}$ to solve for x, correct to 3 significant figures:

a $2^x = 0.051$ **b** $4^x = 213.8$ **c** $3^{2x+1} = 4.069$

3 Write:

a $\log_9 26$ in the form $a \log_3 b$, where $a, b \in \mathbb{Q}$

b $\log_2 11$ in the form $a \log_4 b$, where $a, b \in \mathbb{Z}$

c $\dfrac{6}{\log_7 25}$ in the form $a \log_5 b$, where $a, b \in \mathbb{Z}$.

4 Without using technology, show that $2^{\frac{8}{\log_3 4}} = 81$.

5 Express $\log_2 21 - (\log_a 3)(\log_2 a)$ as a single logarithm in base 2.

6 Write $(\log_{3a} 5)(1 + \log_a 3)$ as a single logarithm in base a.

7 Solve for x:

a $\log_3 x = \log_{27} 50$ **b** $\log_2 x = \log_4 13$ **c** $\log_{25} x = \log_5 7$

d $\log_3 \sqrt{x} + \log_9 x = \log_3 5$ **e** $\log_8 x^2 - \log_2 \sqrt[3]{x} = 1$ **f** $\log_4 x^3 + \log_2 \sqrt{x} = 8$

8 **a** Show that $\log_a b = \dfrac{1}{\log_b a}$.

b Solve for x:

i $\log_3 x = 4\log_x 3$ **ii** $\log_2 x - 4 = 5\log_x 2$ **iii** $2\log_4 x + 3\log_x 4 = 7$

9 Solve $2^{x-1} = 3^{2-x}$. Give your solution in the form $x = \log_a b$.

10 **a** Show that $\dfrac{1}{\frac{1}{\log_a x} + \frac{1}{\log_b x}} = \log_{ab} x$.

b Hence write $\dfrac{3}{\frac{1}{\log_2 5} + \frac{1}{\log_7 5}}$ in the form $\log_c m$ where $c, m \in \mathbb{Z}$.

H GRAPHS OF LOGARITHMIC FUNCTIONS

Consider the general exponential function $f(x) = a^x$, $a > 0$, $a \neq 1$.

For $a > 1$:

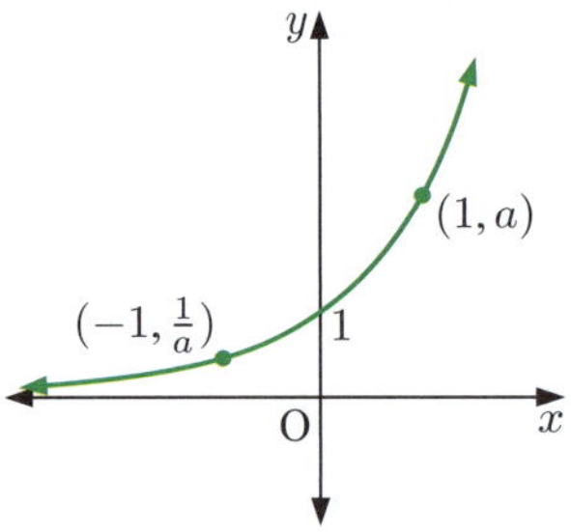

For $0 < a < 1$:

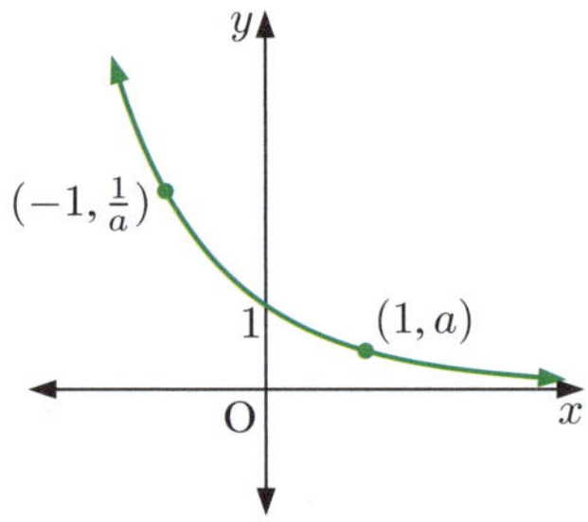

The horizontal asymptote for all of these functions is the x-axis $y = 0$.

We have seen that the inverse function of $f(x) = a^x$ is $f^{-1}(x) = \log_a x$.

Since $f^{-1}(x) = \log_a x$ is an inverse function, it is a reflection of $f(x) = a^x$ in the line $y = x$. We therefore deduce the following properties:

Function	$f(x) = a^x$	$f^{-1}(x) = \log_a x$
Domain	$\{x : x \in \mathbb{R}\}$	$\{x : x > 0\}$
Range	$\{y : y > 0\}$	$\{y : y \in \mathbb{R}\}$
Asymptote	horizontal $y = 0$	vertical $x = 0$

For $a > 1$:

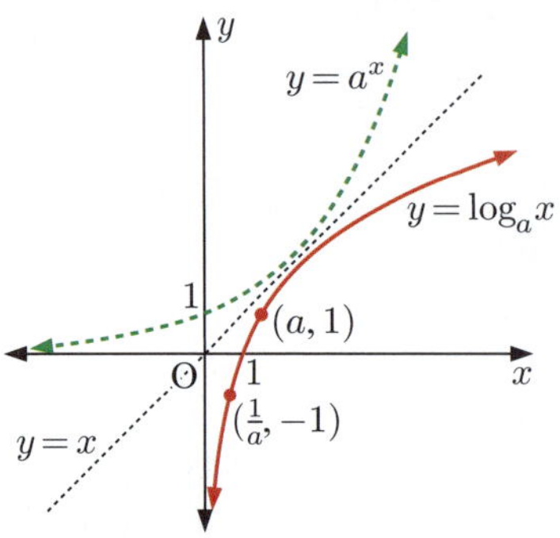

For $0 < a < 1$:

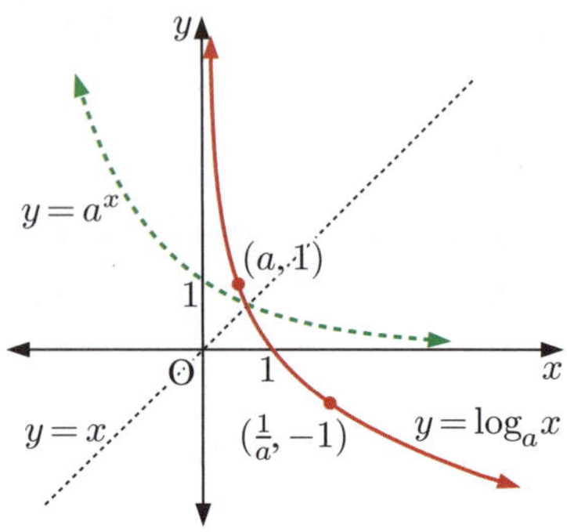

The **vertical asymptote** of $y = \log_a x$ is the y-axis $x = 0$.

Since we can only find logarithms of positive numbers, the domain of $f^{-1}(x) = \log_a x$ is $\{x : x > 0\}$.

In general, $y = \log_a(g(x))$ is defined when $g(x) > 0$.

Example 25

Self Tutor

Consider the function $f(x) = 2\log_3(x+3)$.

a Find the domain and range of f.

b Find the vertical asymptote.

c Find the axes intercepts.

d Sketch the graph of $y = f(x)$.

e Find $f^{-1}(x)$.

a $x + 3 > 0$ when $x > -3$

So, the domain is $\{x : x > -3\}$ and the range is $\{y : y \in \mathbb{R}\}$.

b The vertical asymptote is $x = -3$.

c $f(0) = 2\log_3 3 = 2(1) = 2$, so the y-intercept is 2.

$f(x) = 0$ when $2\log_3(x+3) = 0$

$\therefore \log_3(x+3) = 0$

$\therefore x + 3 = 3^0$

$\therefore x = -2$

So, the x-intercept is -2.

d

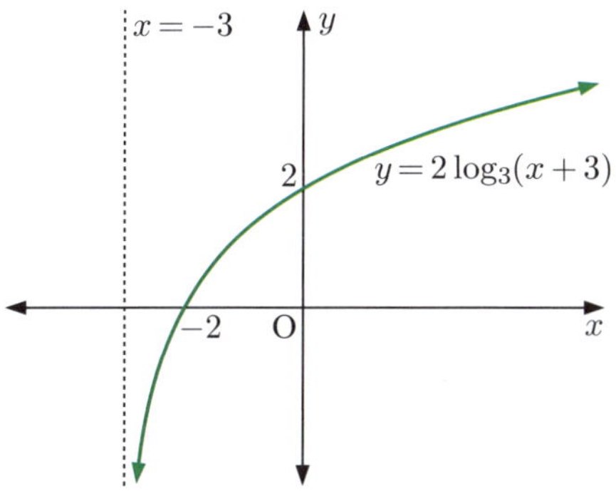

e f is $y = 2\log_3(x+3)$

$\therefore f^{-1}$ is $x = 2\log_3(y+3)$

$\therefore \frac{x}{2} = \log_3(y+3)$

$\therefore y + 3 = 3^{\frac{x}{2}}$

$\therefore y = 3^{\frac{x}{2}} - 3$

$\therefore f^{-1}(x) = 3^{\frac{x}{2}} - 3$

EXERCISE 5H

1 The graph of $f(x) = \log_2(x+4)$ is shown alongside.

a Find the equation of the vertical asymptote.

b Find the axes intercepts.

c State the domain and range of f.

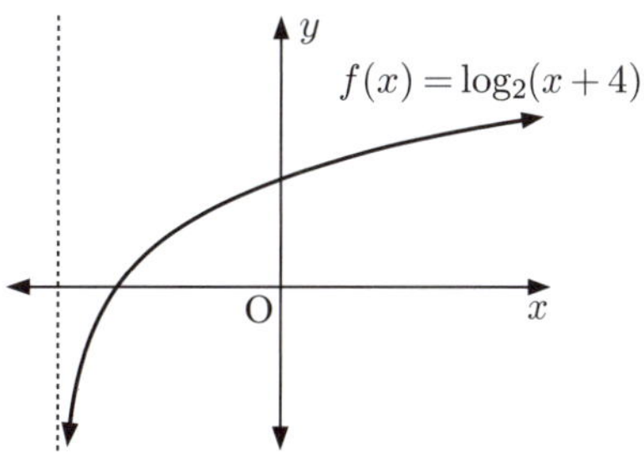

2

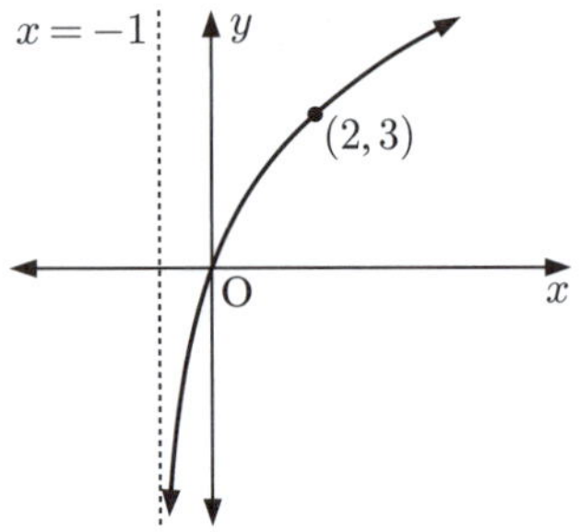

The graph alongside has the form $f(x) = k\log_3(a(x-b))$.

a Find:

i b **ii** a **iii** k

b State the domain and range of f.

c Find:

i $f(8)$ **ii** x such that $f(x) = 1$.

3 Consider the function $f(x) = \frac{1}{2}\log_2(x-3)$.

a Find the domain and range of f.

b Find the vertical asymptote.

c Find the x-intercept.

d Find x such that $f(x) = 1$.

e Sketch the graph of $y = f(x)$.

f Find $f^{-1}(x)$.

4 Consider the function $f(x) = \log_5(2x-1)$.

a Find the domain and range of f.

b Find the vertical asymptote.

c Find the x-intercept.

d Find x such that $f(x) = 1$.

e Find $f^{-1}(x)$, stating its domain and range.

f Sketch the graphs of $y = f(x)$ and $y = f^{-1}(x)$ on the same set of axes.

5 Consider the function $f(x) = 3^x + 1$.

a Find $f^{-1}(x)$.

b State the domain and range of f and f^{-1}.

c Sketch f and f^{-1} on the same set of axes.

6 Consider $f(x) = \log_2(x+3)$.

a Find:

i $f(5)$ **ii** $f(x^2)$ **iii** $f(2x-1)$

b State the domain of $f(x)$.

c Solve $f(x^2+4) = 5$.

Example 26 — *Self Tutor*

Consider the function $f : x \mapsto \ln x + 3$.

a Find the equation defining f^{-1}.

b Sketch the graphs of f and f^{-1} on the same set of axes.

c State the domain and range of f and f^{-1}.

d Find any asymptotes and axes intercepts of f and f^{-1}.

a $f(x) = \ln x + 3$

$\therefore$ f^{-1} is $x = \ln y + 3$

$\therefore$ $\ln y = x - 3$

$\therefore$ $y = e^{x-3}$

So, $f^{-1}(x) = e^{x-3}$

b

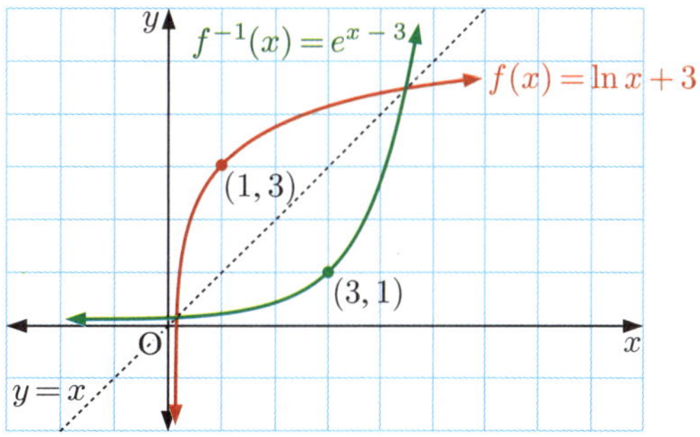

c

Function	f	f^{-1}
Domain	$x > 0$	$x \in \mathbb{R}$
Range	$y \in \mathbb{R}$	$y > 0$

d For f, the vertical asymptote is $x = 0$, and the x-intercept is e^{-3}.
For f^{-1}, the horizontal asymptote is $y = 0$, and the y-intercept is e^{-3}.

7 The graph of $y = 3\ln(x+2)$ is shown alongside.

a Find the equation of the vertical asymptote.

b Find the coordinates of P.

c Find the exact value of x when $y = 2$.

d State the domain and range of the function.

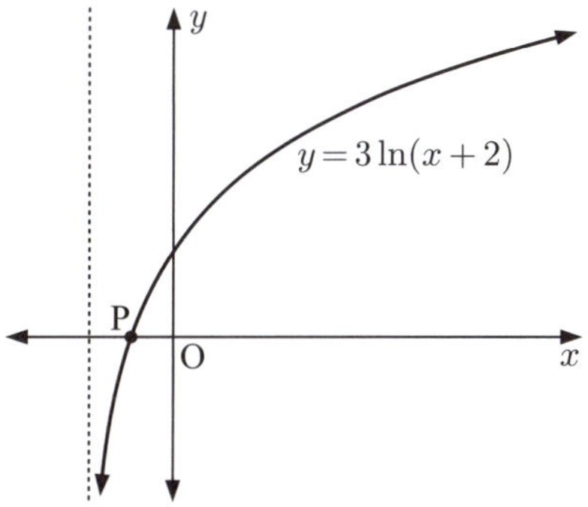

8

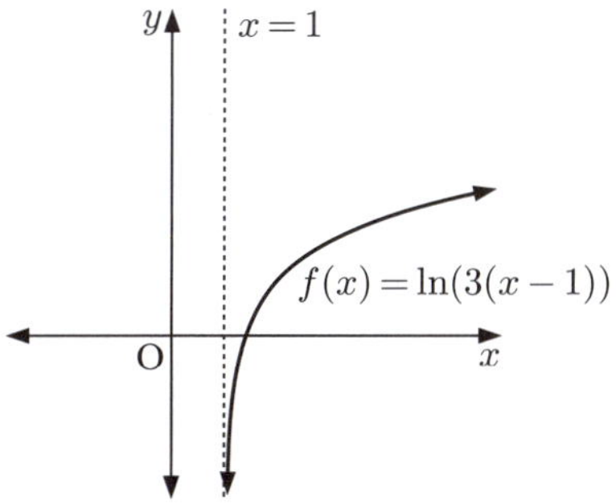

The graph of $f(x) = \ln(3(x-1))$ is shown alongside.

a State the domain and range of f.

b Find the x-intercept.

c Find $f^{-1}(x)$.

d Copy the graph, and sketch $y = f^{-1}(x)$ on the same set of axes.

e State the domain and range of f^{-1}.

PRINTABLE GRAPH

9 Match each equation with its correct graph:

a $y = 2\ln(x+1)$

b $y = \ln(x-1) - 3$

c $y = \ln(3x+3)$

d $y = \ln(x-1) + 1$

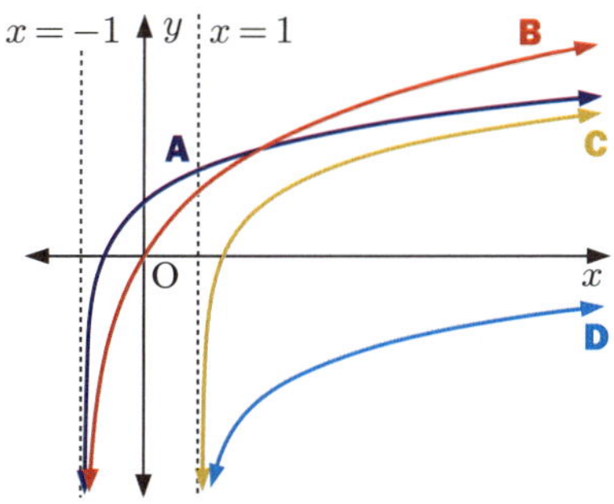

10 Consider the function $f(x) = 3\ln(2x-4)$.

a Find the domain and range of f.

b State the equation of the vertical asymptote.

c Find the x-intercept.

d Sketch the graph of the function.

11 The graphs of $f(x) = 2\ln(x-1)$ and $g(x) = \ln(x+5)$ are shown alongside.

a Find the domain of each function.

b Find the axes intercepts of each function.

c Find the exact coordinates of the intersection point of the graphs.

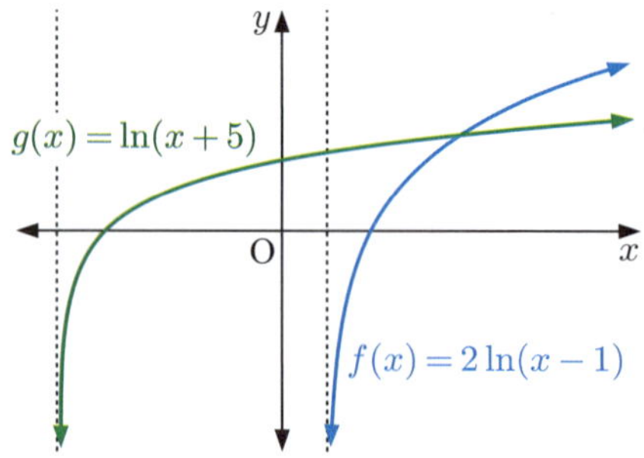

12 Suppose $f(x) = \frac{1}{3}\ln(x-1)$.

a State the domain and range of $f(x)$.

b Find $f(10)$.

c Solve $f(x) = 2$.

d Find $f^{-1}(x)$, and state its domain and range.

e Find $(f \circ f^{-1})(x)$ and $(f^{-1} \circ f)(x)$.

13 Suppose $f : x \mapsto e^{2x}$ and $g : x \mapsto 2x - 1$.

a Find: **i** $(f^{-1} \circ g)(x)$ **ii** $(g \circ f)^{-1}(x)$

b Solve $(f^{-1} \circ g)(x) = \ln 5$.

14 Consider $f : x \mapsto 10e^{-x}$ and $g : x \mapsto \ln(x-3)$.

a Find $f(1)$ and $g(6)$.

b Find the x-intercept of $g(x)$.

c Find $fg(x)$.

d Solve $f(x) = g^{-1}(x)$.

15 Let $f(x) = \ln(x+6)$ and $g(x) = x - \ln 3$.

a State the domain of $f(x)$.

b Find $f^{-1}(x)$.

c Find the axes intercepts of $f(x)$.

d Solve $fg(x) = 3$.

e Solve $gf(x) = f(x^2 - 12)$.

Activity

Click on the icon to explore the graphs of functions of the form $y = k\ln(a(x-b))$, where k, a, and b are constants.

Discuss the effect that changing k, a, and b has on the position and shape of the graph.

LOGARITHMIC GRAPHS

Review set 5A

1 Find:

a $\lg 100$

b $\lg(10 \times 10^k)$

c $\lg(\sqrt[3]{10})$

d $\lg\left(\frac{1}{100\sqrt{10}}\right)$

2 Write an equivalent logarithmic statement for:

a $3^{-4} = \frac{1}{81}$

b $8^{\frac{4}{3}} = 16$

3 Write as a single logarithm:

a $\lg 6 + \lg 9$ **b** $\log_2 42 - \log_2 6$ **c** $\lg 8 + 3$

4 Find the exact value of $\dfrac{\log_k(\frac{k}{2})}{\log_k 2 - 1}$.

5 Write as logarithmic equations:

a $P = 3 \times 7^x$ **b** $m = \dfrac{n^3}{5}$

6 Solve for x:

a $\log_2(x+5) - \log_2(x-2) = 3$ **b** $\lg x + \lg(x+15) = 2$

7 Show that $\log_3 7 \times 2\log_7 x = 2\log_3 x$.

8 Write the following equations without logarithms:

a $\lg T = 2\lg x - \lg 5$ **b** $\log_2 K = x + \log_2 3$

9 Write in the form $a\ln k$ where a and k are positive whole numbers and k is prime:

a $\ln 32$ **b** $\ln 125$ **c** $\ln 729$

10 Copy and complete:

Function	$y = \log_2 x$	$y = \ln(x+5)$
Domain		
Range		

11 If $A = \log_5 2$ and $B = \log_5 3$, write in terms of A and B:

a $\log_5 36$ **b** $\log_5 54$ **c** $\log_5(8\sqrt{3})$ **d** $\log_5(20.25)$ **e** $\log_5(0.\overline{8})$

12 Solve for x:

a $3e^x - 5 = -2e^{-x}$ **b** $2\ln x - 3\ln\left(\dfrac{1}{x}\right) = 10$

13 Solve for x, giving your answer to 2 decimal places:

a $7^x = 120$ **b** $6 \times 2^{3x} = 300$

14 Write $\dfrac{\lg 2 + \lg 4}{\log_8 10}$ in the form $(\lg k)^2$, where $k \in \mathbb{Z}$.

15 Consider $f : x \mapsto 5e^{-x} + 1$.

a State the range of f.

b Find: **i** $f^{-1}(x)$ **ii** $f^{-1}(2)$

c State the domain of f^{-1}.

d Solve $f^{-1}(x) = 0$.

e Sketch the graphs of f, f^{-1}, and $y = x$ on the same set of axes.

Review set 5B

1 Write in the form 10^x, giving x correct to 3 decimal places:

a 16 **b** 0.024 **c** 7300

2 Find the exact value of x if:

a $\lg x = \frac{1}{2}$ **b** $\log_3 x = -\frac{1}{2}$ **c** $\ln(x-3) = -2$

3 Find:

a $\ln e^6$ **b** $\ln\left(\dfrac{e^3}{e^n}\right)$ **c** $\ln(e^t \times \sqrt{e})$

4 Write the following equations without logarithms:

a $\log_2 k \approx 1.699 + x$ **b** $\log_a Q = 3\log_a P + \log_a 5$ **c** $\lg A = x\lg 2 + \lg 6$

5 Solve for x, giving exact answers:

a $3^x = 70$ **b** $20 \times 2^{2x+1} = 640$

6 Find the exact value of $\log_{12} 3 - 2\log_{12} 6$.

7 Write $\log_8 30$ in the form $a\log_2 b$, where $a, b \in \mathbb{Q}$.

8 Solve for x:

a $\log_4 x + \log_4(2x - 8) = 3$ **b** $\log_x 135 = 3 + \log_x 5$

9 Consider $f(x) = e^x$ and $g(x) = \ln(x + 4)$, $x > -4$. Find:

a $(f \circ g)(5)$ **b** $(g \circ f)(0)$

10 Write as logarithmic equations:

a $M = 5 \times 6^x$ **b** $T = \dfrac{5}{\sqrt{l}}$ **c** $G = \dfrac{4}{c}$

11 Solve exactly for x:

a $e^{2x} = 3e^x$ **b** $e^{2x} - 7e^x + 12 = 0$

12 The graph of $y = 3\ln(x + 2)$ is shown alongside.

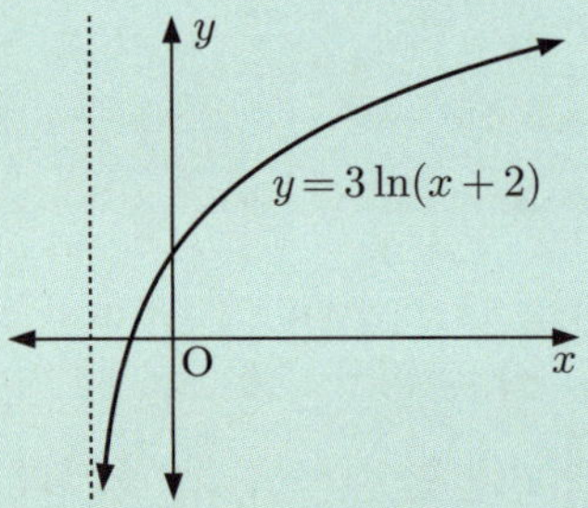

a Find the equation of the vertical asymptote.

b State the domain and range of the function.

c Find the x-intercept.

d Find the y-intercept, giving your answer in the form $\ln k$.

e Find x such that $y = 6$.

13 Consider the function $g : x \mapsto \log_3(x + 2) - 2$.

a Find the domain and range of g.

b Find any asymptotes and axes intercepts for the graph of the function.

c Find g^{-1}, and state its domain and range.

d Sketch the graphs of g, g^{-1}, and $y = x$ on the same axes.

14 Solve for x:

a $\log_2 x + \log_4 x^4 = \log_2 125$ **b** $\log_2 x = 25\log_x 2$ **c** $\log_3 x + 8\log_x 3 = 6$

15 Consider $f(x) = 5e^{2x}$ and $g(x) = \ln(x - 4)$.

a State the domain and range of g. **b** Find the axes intercepts of g.

c Find the exact solution to $fg(x) = 30$. **d** Solve $f(x) = g^{-1}(x)$.

6

Polynomials

Contents:

Opening problem

To determine whether 7 is a **factor** of 56, we divide 56 by 7. The result is exactly 8. Since there is no remainder, 7 is a factor of 56.

Things to think about:

a Can we perform a similar test for *algebraic* factors? For example, how can we determine whether $(x-3)$ is a factor of x^3-4x^2+2x+3?

b Given that $(x-3)$ is a factor of x^3-4x^2+2x+3, what does this tell us about the graph of $f(x)=x^3-4x^2+2x+3$?

We have previously studied linear, quadratic, exponential, and logarithmic functions. We now consider a more general class of functions which includes linear and quadratic. These are the **polynomials**.

A REAL POLYNOMIALS

A polynomial is a function that can be written as the sum of terms involving non-negative integer powers of the variable.

A **polynomial** is a function which can be written in the form

$P(x)=a_nx^n+a_{n-1}x^{n-1}+....+a_2x^2+a_1x+a_0$ where $a_0, a_1, ..., a_n$ are constants, $a_n \neq 0$.

We say that: x is the **variable**
a_0 is the **constant term**
a_n is the **leading coefficient**
a_r is the **coefficient of x^r** for $r=1, 2, ..., n$
n is the **degree** of the polynomial, being the highest power of the variable.

In **summation notation**, we write $P(x)=\sum_{r=0}^{n} a_rx^r$,

which reads: "the sum from $r=0$ to n, of a_rx^r".

A **real polynomial** $P(x)$ is a polynomial for which $a_r \in \mathbb{R}$, $r=0, 1, 2, ..., n$.

The low degree members of the polynomial family have special names. For these polynomials, we commonly write their coefficients as $a, b, c, ...$

Polynomial function	*Degree*	*Name*
$ax+b,\quad a\neq 0$	1	linear
$ax^2+bx+c,\quad a\neq 0$	2	quadratic
$ax^3+bx^2+cx+d,\quad a\neq 0$	3	cubic
$ax^4+bx^3+cx^2+dx+e,\quad a\neq 0$	4	quartic

EXERCISE 6A

1 State whether each function is a polynomial. If it is not, give a reason.

a $f(x)=3x^2+2$
b $g(x)=3x+\frac{1}{3}x^2$
c $h(x)=2^x-4x^2-7$
d $p(x)=-8x^5-7x^4+1$
e $q(x)=4x^3+2x^{\frac{3}{2}}$
f $r(x)=5x^2-x-5x^{-2}$

2 State the **i** degree **ii** leading coefficient **iii** constant term of:

a $f(x) = x^2 - 3x + 4$

b $g(x) = 2x^5 - x + 3$

c $p(x) = -2x^6 - x^5 + 4x^2 - x - 7$

d $q(x) = \frac{1}{2}x^4 - \frac{3}{7}x^2 + x - \frac{2}{3}$

B OPERATIONS WITH POLYNOMIALS

SCALAR MULTIPLICATION

To **multiply** a polynomial by a **scalar** (constant) we multiply each term by the scalar.

Example 1 Self Tutor

If $P(x) = x^2 + 4x - 5$ and $Q(x) = 3x^3 - 6x^2 - 2$, find:

a $2P(x)$

b $-4Q(x)$

a $2P(x) = 2(x^2 + 4x - 5)$
$= 2x^2 + 8x - 10$

b $-4Q(x) = -4(3x^3 - 6x^2 - 2)$
$= -12x^3 + 24x^2 + 8$

ADDING AND SUBTRACTING POLYNOMIALS

To add or subtract polynomials, we simply collect like terms.

Example 2 Self Tutor

Let $p(x) = x^3 + 2x^2 - 4x + 3$ and $q(x) = 5x^2 + x - 2$.

Find:

a $p(x) + q(x)$ **b** $p(x) - q(x)$ **c** $3p(x) - 2q(x)$

a $p(x) + q(x)$
$= x^3 + 2x^2 - 4x + 3$
$\quad + 5x^2 + x - 2$
$= x^3 + 7x^2 - 3x + 1$

b $p(x) - q(x)$
$= x^3 + 2x^2 - 4x + 3 - (5x^2 + x - 2)$
$= x^3 + 2x^2 - 4x + 3$
$\quad - 5x^2 - x + 2$
$= x^3 - 3x^2 - 5x + 5$

c $3p(x) - 2q(x)$
$= 3(x^3 + 2x^2 - 4x + 3) - 2(5x^2 + x - 2)$
$= 3x^3 + 6x^2 - 12x + 9$
$\quad - 10x^2 - 2x + 4$
$= 3x^3 - 4x^2 - 14x + 13$

Write "like" terms in columns to make simplification easier.

EXERCISE 6B.1

1 If $p(x) = x^2 + 3x - 2$ and $q(x) = -2x^3 + x + 4$, find:

a $3p(x)$ **b** $4q(x)$ **c** $-p(x)$
d $-3q(x)$ **e** $p(x) + q(x)$ **f** $q(x) - p(x)$

2 For each pair of functions, find: **i** $p(x) + q(x)$ **ii** $p(x) - q(x)$

a $p(x) = x^2 + 1,\quad q(x) = 2x^2 + 3x - 4$
b $p(x) = x^3 - 4x^2 + 2x + 1,\quad q(x) = 2x^3 - x^2 - x + 6$
c $p(x) = -5x^4 - x^2 + x,\quad q(x) = x^3 - 8x^2 - 2x - 4$
d $p(x) = 2x^4 - x^3 + 7x - 3,\quad q(x) = 8 - 5x - 3x^3$

3 If $p(x) = x^3 - 2x^2 - 5$ and $q(x) = 2x^2 - x + 4$, find:

a $p(x) - 2q(x)$ **b** $3p(x) + q(x)$ **c** $q(x) - 2p(x)$

POLYNOMIAL MULTIPLICATION

To multiply two polynomials, we multiply each term of the first polynomial by each term of the second polynomial, and then collect like terms.

Example 3 — Self Tutor

If $p(x) = 2x^3 - x^2 + 6$ and $q(x) = x^2 - 3x + 4$, find $p(x)q(x)$.

$$\begin{aligned} p(x)q(x) &= (2x^3 - x^2 + 6)(x^2 - 3x + 4) \\ &= 2x^3(x^2 - 3x + 4) - x^2(x^2 - 3x + 4) + 6(x^2 - 3x + 4) \\ &= 2x^5 - 6x^4 + 8x^3 \\ &\qquad\; - x^4 + 3x^3 - 4x^2 \\ &\qquad\qquad\qquad + 6x^2 - 18x + 24 \\ &= 2x^5 - 7x^4 + 11x^3 + 2x^2 - 18x + 24 \end{aligned}$$

EXERCISE 6B.2

1 Find $f(x)g(x)$ for:

a $f(x) = 3x - 1,\ g(x) = x + 2$ **b** $f(x) = 2x^2 - x - 3,\ g(x) = x - 4$
c $f(x) = 4x^3 - x^2 + 2,\ g(x) = x^2 - 5x + 4$ **d** $f(x) = -2x^3 + x + 7,\ g(x) = x^4 + x^2 - 5x$

2 For $p(x) = x^4 - 3x^3 + 4x - 1$ and $q(x) = 2x^3 - 3x^2 + 6$, find:

a $3p(x)$ **b** $-5q(x)$ **c** $2p(x) + 3q(x)$
d $4q(x) - p(x)$ **e** $p(x)q(x)$ **f** $[q(x)]^2$

3 Find:

a $(3x^2 - 4)(2x + 1)$ **b** $(x^3 + 2x + 1)(3 - x)$
c $(-x^2 + 4x - 2)(x^2 - 5)$ **d** $(x^2 - 3x - 1)(-2x^3 + 5x^2 - x)$

4 Find the:

a leading coefficient and constant term of $P(x) = (x-2)(2x^2+3x+4)$

b coefficient of x and degree of $f(x) = (x^3-2x+4)(x+3)$

c coefficient of x^2 and constant term of $Q(x) = (x+5)(x^2-3x+4)$

d degree and leading coefficient of $g(x) = (3x^2-4)(2x^3-x+3)$.

Discussion

Suppose $f(x)$ is a polynomial of degree m, and $g(x)$ is a polynomial of degree n.

What is the degree of:

- $f(x)+g(x)$
- $5f(x)$
- $[f(x)]^2$
- $f(x)g(x)$?

C ZEROS, ROOTS, AND FACTORS

A **zero** of a polynomial is a value of the variable which makes the polynomial equal to zero.

α is a **zero** of polynomial $P(x) \Leftrightarrow P(\alpha)=0$.

The **roots** of a polynomial **equation** are the solutions to the equation.

α is a **root** (or **solution**) of $P(x)=0 \Leftrightarrow P(\alpha)=0$.

The **roots** of $P(x)=0$ are the **zeros** of $P(x)$ and the x-intercepts of the graph of $y=P(x)$.

Consider
$$\begin{aligned} P(x) &= x^3+2x^2-3x-10 \\ \therefore\ P(2) &= 2^3+2(2)^2-3(2)-10 \\ &= 8+8-6-10 \\ &= 0 \end{aligned}$$

This tells us:

- 2 is a zero of $x^3+2x^2-3x-10$
- 2 is a root of $x^3+2x^2-3x-10=0$
- the graph of $y=x^3+2x^2-3x-10$ has the x-intercept 2.

FACTORISATION AND EXPANSION

Polynomials can be written in **factorised form** or **expanded form**.

For example, the **expansion of**
$$\begin{aligned} &(x-1)(x+2)(x-3) \\ &= (x^2+x-2)(x-3) \\ &= x^3-2x^2-5x+6 \end{aligned}$$

In reverse, the process of writing x^3-2x^2-5x+6 as $(x-1)(x+2)(x-3)$ is called **factorisation**.

expansion →

$(x-1)(x+2)(x-3)$ factorised form ⇄ x^3-2x^2-5x+6 expanded form

← factorisation

If $P(x) = (x-1)(x+2)(x-3)$, then $(x-1)$, $(x+2)$, and $(x-3)$ are its **linear factors**.

Likewise $P(x) = (x+3)^2(2x+3)$ has been factorised into 3 linear factors, one of which is repeated.

> $(x-\alpha)$ is a **linear factor** of the polynomial $P(x)$
> $\Leftrightarrow$ there exists a polynomial $Q(x)$ such that $P(x) = (x-\alpha)\,Q(x)$.
> If $\alpha \in \mathbb{R}$ then $(x-\alpha)$ is a **real linear factor**.

Example 4 — Self Tutor

Find the zeros of:

a $x^2 - 6x + 2$ **b** $x^3 - 5x$

a If $x^2 - 6x + 2 = 0$

then $x = \dfrac{6 \pm \sqrt{(-6)^2 - 4(1)(2)}}{2}$

$\therefore\ x = \dfrac{6 \pm \sqrt{28}}{2}$

$\therefore\ x = \dfrac{6 \pm 2\sqrt{7}}{2}$

$\therefore\ x = 3 \pm \sqrt{7}$

The zeros are $3 - \sqrt{7}$ and $3 + \sqrt{7}$.

b If $x^3 - 5x = 0$

then $x(x^2 - 5) = 0$

$\therefore\ x(x+\sqrt{5})(x-\sqrt{5}) = 0$

$\therefore\ x = 0$ or $\pm\sqrt{5}$

The zeros are $-\sqrt{5}$, 0, and $\sqrt{5}$.

EXERCISE 6C

1 Decide whether:

- **a** 4 is a zero of $x^2 - 2x - 3$
- **b** -2 is a root of $2x^3 + x^2 + 5x + 6 = 0$
- **c** -1 is a zero of $x^4 - 3x^2 + 7x + 11$
- **d** 3 is a root of $-x^4 + 2x^3 + 3x^2 - 4x + 12 = 0$

2 Find the zeros of:

- **a** $(x-2)(x+4)$
- **b** $x^2 + 6x - 16$
- **c** $3x^2 - 5x - 2$
- **d** $x(x^2 - 4)$
- **e** $x^3 - 11x$
- **f** $x^4 - 6x^2 + 8$

3 Find the roots of:

- **a** $(x+5)(x-3) = 0$
- **b** $(2x+1)(x^2-3) = 0$
- **c** $(3x-1)(x^2+x-6) = 0$
- **d** $-2x(x^2 - 2x - 2) = 0$
- **e** $x^3 = 7x$
- **f** $x^4 = 7x^2 - 10$

4 Use expansion to verify that:

- **a** $(x+1)(x-3)(x+4)$ is the factorised form of $x^3 + 2x^2 - 11x - 12$
- **b** $(2x-3)(x-1)(x+5)$ is the factorised form of $2x^3 + 5x^2 - 22x + 15$.

Example 5 — Self Tutor

Fully factorise:

a $2x^3 + 5x^2 - 3x$ **b** $x^2 + 4x - 1$

a $2x^3 + 5x^2 - 3x$
$= x(2x^2 + 5x - 3)$
$= x(2x - 1)(x + 3)$

b $x^2 + 4x - 1$ is zero when $x = \dfrac{-4 \pm \sqrt{4^2 - 4(1)(-1)}}{2}$

$\therefore\ x = \dfrac{-4 \pm \sqrt{20}}{2}$

$\therefore\ x = \dfrac{-4 \pm 2\sqrt{5}}{2}$

$\therefore\ x = -2 \pm \sqrt{5}$

$\therefore\ x^2 + 4x - 1 = (x - [-2 + \sqrt{5}])(x - [-2 - \sqrt{5}])$
$= (x + 2 - \sqrt{5})(x + 2 + \sqrt{5})$

5 Fully factorise:

a $2x^2 - 7x - 15$ **b** $x^3 - 11x^2 + 28x$ **c** $x^2 - 6x + 3$
d $x^3 + 2x^2 - 4x$ **e** $6x^3 - x^2 - 2x$ **f** $x^4 - 6x^2 + 5$

6 If $P(x) = a(x - \alpha)(x - \beta)(x - \gamma)$ then α, β, and γ are its zeros.
Verify this statement by finding $P(\alpha)$, $P(\beta)$, and $P(\gamma)$.

Example 6 — Self Tutor

Find *all* cubic polynomials with zeros:

a 1, 4, and -5 **b** $\frac{1}{2}$ and $-3 \pm \sqrt{2}$.

a 1 comes from the linear factor $(x - 1)$.
4 comes from the linear factor $(x - 4)$.
-5 comes from the linear factor $(x + 5)$.
$\therefore\ P(x) = a(x - 1)(x - 4)(x + 5),\quad a \neq 0.$

b $\frac{1}{2}$ comes from the linear factor $(2x - 1)$.
The zeros $-3 \pm \sqrt{2}$ have sum $= -3 + \sqrt{2} - 3 - \sqrt{2} = -6$
and product $= (-3 + \sqrt{2})(-3 - \sqrt{2}) = 7$
$\therefore$ they come from the quadratic factor $(x^2 + 6x + 7)$
$\therefore\ P(x) = a(2x - 1)(x^2 + 6x + 7),\quad a \neq 0.$

7 Find *all* cubic polynomials with zeros:

a 0, 3, and -4 **b** ± 1 and 5 **c** 3 and $2 \pm \sqrt{3}$ **d** $-\frac{1}{4}$ and $-1 \pm \sqrt{7}$

8 Find *all* quartic polynomials with zeros:

a ± 1 and $\pm\sqrt{2}$ **b** 2, $-\frac{1}{5}$, and $\pm\sqrt{3}$
c -3, $\frac{1}{4}$, and $1 \pm \sqrt{2}$ **d** $2 \pm \sqrt{5}$ and $-2 \pm \sqrt{7}$

D POLYNOMIAL EQUALITY

Two polynomials are **equal** if and only if they have the **same degree** *and* corresponding terms have equal coefficients.

If we know that two polynomials are **equal** then we can **equate coefficients** to find unknowns.

For example, if $2x^3+3x^2-4x+6=ax^3+bx^2+cx+d$ where $a, b, c, d \in \mathbb{R}$, then
$a=2,\quad b=3,\quad c=-4,$ and $d=6$.

Example 7

Self Tutor

Find constants a, b, and c given that:
$$6x^3+7x^2-19x+7=(2x-1)(ax^2+bx+c) \quad \text{for all } x.$$

$$\begin{aligned}6x^3+7x^2-19x+7 &= (2x-1)(ax^2+bx+c)\\ &= 2ax^3+2bx^2+2cx-ax^2-bx-c\\ &= 2ax^3+(2b-a)x^2+(2c-b)x-c\end{aligned}$$

Since this is true for all x, we equate coefficients:

$\therefore\ \underbrace{2a=6}_{x^3 \text{ terms}}\quad \underbrace{2b-a=7}_{x^2 \text{ terms}}\quad \underbrace{2c-b=-19}_{x \text{ terms}}$ and $\underbrace{7=-c}_{\text{constants}}$

$\therefore\ a=3$ and $c=-7$ and consequently $\underbrace{2b-3=7 \text{ and } -14-b=-19}$

$\therefore\ b=5$ in both equations

So, $a=3$, $b=5$, and $c=-7$.

EXERCISE 6D

1 Find constants a and b given that:

a $3x^2+5x-2=3x^2+ax+b$ for all x

b $4x^3-2x^2+3x-7=ax^3-2x^2+bx-7$ for all x

c $6x^2-x+5=2ax^2-x+(b-3)$ for all x

d $x^3+7x-6=x^3+(a-2)x^2+(a+b)x-6$ for all x.

2 Find constants a and b given that:

a $3x^3+7x^2-26x+10=(x^2+4x-2)(ax+b)$ for all x

b $4x^3+4x^2-5x-3=(2x^2+x-3)(ax+b)$ for all x.

3 Find constants a, b, and c given that:

a $x^3+x^2-22x-40=(x-5)(ax^2+bx+c)$ for all x

b $2x^3+5x^2+9=(x+3)(ax^2+bx+c)$ for all x

c $4x^3+8x^2-13x+4=(2x-1)(ax^2+bx+c)$ for all x.

4 Suppose $5x^3+kx^2-7x-6=(5x+2)(ax^2+bx+c)$ for all x. Find the values of a, b, c, and k.

5 **a** Given that $x^3 + 9x^2 + 11x - 21 = (x+3)(ax^2 + bx + c)$, find the values of a, b, and c.

b Hence fully factorise $x^3 + 9x^2 + 11x - 21$.

6 **a** Given that $4x^3 + 12x^2 + 3x - 5 = (2x-1)(px^2 + qx + r)$, find the values of p, q, and r.

b Hence find the roots of $4x^3 + 12x^2 + 3x - 5 = 0$.

7 **a** Given that $3x^3 + 10x^2 - 7x + 4 = (x+4)(ax^2 + bx + c)$, find the values of a, b, and c.

b Hence show that $3x^3 + 10x^2 - 7x + 4$ has only one real zero.

8 Fully factorise:

a $x^3 + 4x^2 - 11x - 30$ given that $(x+5)$ is a factor

b $x^3 - 12x + 16$ given that $(x-2)$ is a factor

c $2x^3 + 5x^2 - 21x - 36$ given that $(x+4)$ is a factor

d $6x^3 + 11x^2 - 46x + 24$ given that $(3x-2)$ is a factor.

9 $(x+3)$ is a factor of $x^3 + 9x^2 + kx + 27$. Find k, and fully factorise the polynomial.

10 $(2x+1)$ is a factor of $2x^3 - 5x^2 + 7x + 5$.

a Write the polynomial as the product of a linear factor and a quadratic factor.

b Explain why the polynomial cannot be written as the product of three real linear factors.

Example 8

Self Tutor

Find constants a and b given that $x^4 - 2x^3 - 3x^2 + 8x - 4 = (x^2 + ax + 2)(x^2 + bx - 2)$ for all x.
Hence fully factorise $x^4 - 2x^3 - 3x^2 + 8x - 4$.

$$\begin{aligned}
&(x^2 + ax + 2)(x^2 + bx - 2) \\
&= x^4 + bx^3 - 2x^2 \\
&\quad + ax^3 + abx^2 - 2ax \\
&\quad + 2x^2 + 2bx - 4 \\
&= x^4 + (a+b)x^3 + abx^2 + (2b - 2a)x - 4
\end{aligned}$$

Equating coefficients gives $\begin{cases} a + b = -2 & \text{.... (1)} \\ ab = -3 & \text{.... (2)} \\ 2b - 2a = 8 & \text{.... (3)} \end{cases}$

From (1), $b = -2 - a$

$\therefore$ in (3), $2(-2-a) - 2a = 8$

$\therefore -4a = 12$

$\therefore a = -3$

$\therefore b = -2 - (-3) = 1$ {using (1)}

$a = -3$ and $b = 1$ also satisfies (2)

$$\begin{aligned}
\therefore\ x^4 - 2x^3 - 3x^2 + 8x - 4 &= (x^2 - 3x + 2)(x^2 + x - 2) \\
&= (x-2)(x-1)(x+2)(x-1) \\
&= (x+2)(x-1)^2(x-2)
\end{aligned}$$

11 Find constants a and b given that
$$x^4 + x^3 - 19x^2 - 49x - 30 = (x^2 + ax - 15)(x^2 + bx + 2) \quad \text{for all } x.$$
Hence find the roots of $x^4 + x^3 - 19x^2 - 49x - 30 = 0$.

12 Find constants m and n, $m > n$, given that
$$x^4 + x^3 + x + 1 = (x^2 + mx + 1)(x^2 + nx + 1) \quad \text{for all } x.$$
Hence fully factorise $x^4 + x^3 + x + 1$.

13 Find constants p and q such that $x^4 - 2x^3 - x^2 + 2x + 10 = (x^2 + px + 5)(x^2 + qx + 2)$ for all x.
Hence show that $x^4 - 2x^3 - x^2 + 2x + 10$ has no real linear factors.

E POLYNOMIAL DIVISION

Polynomial division is a more complex operation than multiplication. We use a process similar to the long division of integers.

Consider $(3x^2 + 4x + 2)(x + 2) + 5$.

If we expand this expression we obtain $(3x^2 + 4x + 2)(x + 2) + 5 = 3x^3 + 10x^2 + 10x + 9$.

Dividing both sides by $x + 2$,

$$\frac{3x^3 + 10x^2 + 10x + 9}{x + 2} = \underbrace{3x^2 + 4x + 2}_{\text{quotient}} + \frac{5 \leftarrow \text{remainder}}{x + 2 \leftarrow \text{divisor}}$$

If $P(x)$ is divided by $ax + b$ until a constant remainder R is obtained, then

$$\frac{P(x)}{ax + b} = Q(x) + \frac{R}{ax + b}$$ where $ax + b$ is the **divisor**, $Q(x)$ is the **quotient**, and R is the **remainder**.

Notice that $P(x) = Q(x) \times (ax + b) + R$.

The following **division algorithm** explains how we perform division by a linear factor:

Step 1: What do we multiply x by to get $3x^3$?
The answer is $3x^2$, so we write $3x^2$ above the line, and we expand $3x^2(x + 2) = 3x^3 + 6x^2$ and write it underneath.

Step 2: We then subtract $3x^3 + 6x^2$ from $3x^3 + 10x^2$. The answer is $4x^2$.

Step 3: Bring down the $10x$ to obtain $4x^2 + 10x$.

$$\begin{array}{r|l} & 3x^2 + 4x + 2 \\ \hline x + 2 & 3x^3 + 10x^2 + 10x + 9 \\ & -(3x^3 + 6x^2) \\ \hline & 4x^2 + 10x \\ & -(4x^2 + 8x) \\ \hline & 2x + 9 \\ & -(2x + 4) \\ \hline & 5 \end{array}$$

Return to *Step 1* with the question:
What must we multiply x by to get $4x^2$?
The answer is $4x$, and $4x(x + 2) = 4x^2 + 8x$.
We continue the process until we are left with a constant.

We find that $\dfrac{3x^3 + 10x^2 + 10x + 9}{x + 2} = 3x^2 + 4x + 2 + \dfrac{5}{x + 2}$.

Example 9

Self Tutor

Find the quotient and remainder of $\dfrac{2x^3-5x^2+x+7}{x-3}$.

$$
\begin{array}{r|l}
 & 2x^2 + x + 4 \\
\hline
x-3 & 2x^3 - 5x^2 + x + 7 \\
 & -(2x^3 - 6x^2) \\
\hline
 & x^2 + x \\
 & -(x^2 - 3x) \\
\hline
 & 4x + 7 \\
 & -(4x - 12) \\
\hline
 & 19
\end{array}
$$

$\therefore$ the quotient is $2x^2+x+4$, and the remainder is 19.

In the long division process for integers, we only ever deal with *positive* multiples. By contrast, for polynomial division we sometimes need *negative* multiples.

Example 10

Self Tutor

Find $\dfrac{x^4-17x^2+12x-6}{x+5}$.

$$
\begin{array}{r|l}
 & x^3 - 5x^2 + 8x - 28 \\
\hline
x+5 & x^4 + 0x^3 - 17x^2 + 12x - 6 \\
 & -(x^4 + 5x^3) \\
\hline
 & -5x^3 - 17x^2 \\
 & -(-5x^3 - 25x^2) \\
\hline
 & 8x^2 + 12x \\
 & -(8x^2 + 40x) \\
\hline
 & -28x - 6 \\
 & -(-28x - 140) \\
\hline
 & 134
\end{array}
$$

The quotient is $x^3-5x^2+8x-28$, and the remainder is 134.

$\therefore \dfrac{x^4-17x^2+12x-6}{x+5} = x^3-5x^2+8x-28+\dfrac{134}{x+5}$

EXERCISE 6E

1 Find the quotient and remainder for each division, and write the result in the form $P(x) = Q(x)\,D(x) + R$ where $D(x)$ is the divisor.

a $\dfrac{x^2+5x+6}{x+4}$ **b** $\dfrac{2x^2-3x+1}{x-2}$ **c** $\dfrac{x^2-x+2}{x+3}$

d $\dfrac{x^3+3x^2+5x+11}{x+2}$ **e** $\dfrac{x^3+2x^2-7x-1}{x-1}$ **f** $\dfrac{x^3+4x^2-2x+3}{x+4}$

2 Find:

a $\dfrac{x^2-4x+5}{x-3}$ **b** $\dfrac{3x^2+x-5}{x+1}$ **c** $\dfrac{x^2+9}{x-2}$

d $\dfrac{x^3-9x^2+6x-7}{x-4}$ **e** $\dfrac{x^3+5x-4}{x+2}$ **f** $\dfrac{2x^4+5x^3+x^2+2x-4}{x+3}$

3 Perform the division, and hence factorise the cubic numerator:

a $\dfrac{x^3+2x^2-5x-6}{x-2}$ **b** $\dfrac{x^3+5x^2-8x-48}{x-3}$ **c** $\dfrac{6x^3+5x^2-44x-15}{x+3}$

Activity — Division by quadratics

As with division by linears, we can use the **division algorithm** to divide polynomials by quadratics. The division process stops when the remainder is linear.

For example, consider $\dfrac{x^4+4x^3-x+1}{x^2-x+1}$.

$$
\begin{array}{r|l}
 & \qquad\qquad x^2+5x+4 \\
x^2-x+1 & x^4+4x^3+0x^2-x+1 \\
 & -(x^4-x^3+x^2) \\ \hline
 & \qquad 5x^3-x^2-x \\
 & \qquad -(5x^3-5x^2+5x) \\ \hline
 & \qquad\qquad 4x^2-6x+1 \\
 & \qquad\qquad -(4x^2-4x+4) \\ \hline
 & \qquad\qquad\qquad -2x-3
\end{array}
$$

$\therefore\; x^4+4x^3-x+1 = (x^2+5x+4)(x^2-x+1) - 2x - 3$

What to do:

1 Find the quotient and remainder for:

a $\dfrac{x^3+2x^2+x-3}{x^2+x+1}$ **b** $\dfrac{3x^3+x-1}{x^2+1}$ **c** $\dfrac{x^4}{(x+1)^2}$ **d** $\dfrac{x^4+3x^2+x-1}{x^2-x+1}$

2 Suppose $f(x) = (x-1)(x+2)(x^2-3x+5)+15-10x$. Find the quotient and remainder when $f(x)$ is divided by x^2+x-2.

3 **a** Perform the division $\dfrac{x^4+2x^3-15x^2-68x-160}{x^2+3x+8}$.

b Hence solve $x^4-68x = 15x^2+160-2x^3$.

F THE REMAINDER THEOREM

In **Example 9** we considered the polynomial $P(x) = 2x^3 - 5x^2 + x + 7$.

When $P(x)$ was divided by $x - 3$, the remainder was 19.

Notice that $P(3) = 2(3)^3 - 5(3)^2 + 3 + 7$
$= 54 - 45 + 3 + 7$
$= 19$, which is the remainder.

This result is explained by the **Remainder theorem**:

> When a polynomial $P(x)$ is divided by $x - k$ until a constant remainder R is obtained, then $R = P(k)$.

Proof: When $P(x)$ is divided by $x - k$, we have

$$\frac{P(x)}{x-k} = Q(x) + \frac{R}{x-k} \quad \text{where } Q(x) \text{ is the quotient.}$$

$\therefore \quad P(x) = Q(x)(x - k) + R$

Letting $x = k$, $\quad P(k) = Q(k)(k - k) + R$

$\therefore \quad P(k) = Q(k) \times 0 + R$

$\therefore \quad P(k) = R$

Example 11

Self Tutor

Find the remainder when $2x^3 - x^2 + 15$ is divided by $x + 3$.

Let $P(x) = 2x^3 - x^2 + 15$.

We are dividing by $x + 3$, so we set $k = -3$.

Now $P(-3) = 2(-3)^3 - (-3)^2 + 15$
$= -54 - 9 + 15$
$= -48$

$\therefore$ when $2x^3 - x^2 + 15$ is divided by $x + 3$, the remainder is -48. {Remainder theorem}

EXERCISE 6F

1 Use the Remainder theorem to find the remainder when:

- **a** $3x^2 - 4x + 7$ is divided by $x - 2$
- **b** $x^3 + 2x^2 - 5x + 2$ is divided by $x + 4$
- **c** $2x^3 - 7x + 13$ is divided by $x - 3$
- **d** $3x^3 + 10x^2 + 10x + 9$ is divided by $x + 2$.

Example 12 — Self Tutor

Find a given that $x^2 + ax + 7$ divided by $x - 5$ has remainder 12.

Let $P(x) = x^2 + ax + 7$

Now $P(5) = 12$ {Remainder theorem}

$\therefore (5)^2 + a(5) + 7 = 12$

$\therefore 25 + 5a + 7 = 12$

$\therefore 5a = -20$

$\therefore a = -4$

2 Find a given that:

a $x^2 + 5x + a$ divided by $x + 1$ has remainder -6

b $-2x^2 + ax + 8$ divided by $x - 3$ has remainder 2

c $x^3 + ax^2 + 3x - 1$ divided by $x + 4$ has remainder 3.

Example 13 — Self Tutor

When $2x^3 + 2x^2 + ax + b$ is divided by $x + 3$, the remainder is -11.
When the same polynomial is divided by $x - 2$, the remainder is 9.
Find a and b.

Let $P(x) = 2x^3 + 2x^2 + ax + b$

Now $P(-3) = -11$ and $P(2) = 9$ {Remainder theorem}

So, $2(-3)^3 + 2(-3)^2 + a(-3) + b = -11$

$\therefore -54 + 18 - 3a + b = -11$

$\therefore -3a + b = 25$ .... (1)

and $2(2)^3 + 2(2)^2 + a(2) + b = 9$

$\therefore 16 + 8 + 2a + b = 9$

$\therefore 2a + b = -15$ .... (2)

Solving simultaneously: $3a - b = -25$ $\{-1 \times (1)\}$

$2a + b = -15$ $\{(2)\}$

Adding, $5a = -40$

$\therefore a = -8$

Substituting $a = -8$ in (2) gives $2(-8) + b = -15$

$\therefore b = 1$

3 Find a and b given that:

a $x^2 + ax + b$ has remainder 4 when divided by $x - 3$, and remainder 19 when divided by $x + 2$

b $x^3 + 4x^2 + ax + b$ has remainder 20 when divided by $x - 2$, and remainder 6 when divided by $x + 5$.

4 When the polynomial $p(x) = x^3 + ax^2 + bx + 3$ is divided by $x + 3$, the remainder is 9.

a Given that $p(1) = 1$, find a and b.

b Find the remainder when $p(x)$ is divided by $x - 4$.

5 **a** Find, in terms of m, the remainder when $x^3 + 2mx^2 - m^2x - 2$ is divided by $x - 2$.

b Hence find the values of m for which the remainder is positive.

6 **a** Let $P(x) = 3x^3 + 2x^2 - 8x - 15$.

i Use polynomial division to find the remainder when $P(x)$ is divided by $2x + 1$.

ii Evaluate $P(-\frac{1}{2})$ and comment on your answer.

b Suppose a polynomial $P(x)$ is divided by $2x + 1$ until a constant remainder R is obtained. Use $P(x) = Q(x)(2x + 1) + R$ to explain why $R = P(-\frac{1}{2})$.

c Find the remainder when:

i $2x^2 - 3x + 5$ is divided by $2x + 1$ **ii** $3x^4 - 2x^3 - x$ is divided by $3x + 1$

iii $-2x^3 + 4x^2 - 1$ is divided by $2x - 1$.

G THE FACTOR THEOREM

For any polynomial $P(x)$, k is a zero of $P(x)$ $\Leftrightarrow$ $(x - k)$ is a factor of $P(x)$.

Proof:

k is a zero of $P(x) \Leftrightarrow P(k) = 0$ {definition of a zero}

$\Leftrightarrow R = 0$ {Remainder theorem}

$\Leftrightarrow P(x) = Q(x)(x - k)$ {division algorithm}

$\Leftrightarrow (x - k)$ is a factor of $P(x)$ {definition of a factor}

The **Factor theorem** says that if 3 is a zero of $P(x)$ then $(x - 3)$ is a factor of $P(x)$, and vice versa.

We can use the Factor theorem to determine whether $(x - k)$ is a factor of a polynomial, without having to perform polynomial division.

Example 14 Self Tutor

Determine whether:

a $(x - 2)$ is a factor of $x^3 + 3x^2 - 13x + 6$ **b** $(x + 3)$ is a factor of $x^3 - 8x + 7$.

a Let $P(x) = x^3 + 3x^2 - 13x + 6$

$\therefore$ $P(2) = (2)^3 + 3(2)^2 - 13(2) + 6$
$= 8 + 12 - 26 + 6$
$= 0$

Since $P(2) = 0$, $(x - 2)$ is a factor of $x^3 + 3x^2 - 13x + 6$. {Factor theorem}

b Let $P(x) = x^3 - 8x + 7$

$\therefore$ $P(-3) = (-3)^3 - 8(-3) + 7$
$= -27 + 24 + 7$
$= 4$

Since $P(-3) \neq 0$, $(x + 3)$ is *not* a factor of $x^3 - 8x + 7$. {Factor theorem}

EXERCISE 6G

1 Use the Factor theorem to determine whether:

a $(x-1)$ is a factor of $4x^3 - 7x^2 + 5x - 2$
b $(x-3)$ is a factor of $x^4 - x^3 - 4x^2 - 15$
c $(x+2)$ is a factor of $3x^3 + 5x^2 - 6x - 8$
d $(x+4)$ is a factor of $2x^3 + 6x^2 + 4x + 16$.

2 **a** Find c given that $(x+1)$ is a factor of $5x^3 - 3x^2 + cx + 10$.
b Find c given that $(x-3)$ is a factor of $x^4 - 2x^3 + cx^2 - 4x + 3$.
c Find b given that $(x+2)$ is a factor of $x^6 + bx^5 - 2x^3 - 5x + 6$.

Example 15 — Self Tutor

$(2x-1)$ is a factor of $f(x) = 4x^3 - 4x^2 + ax + b$. When $f(x)$ is divided by $x-1$, the remainder is -1. Find a and b.

Since $(2x-1)$ is a factor of $f(x)$, $f(\frac{1}{2}) = 0$

$\therefore\ 4(\frac{1}{2})^3 - 4(\frac{1}{2})^2 + a(\frac{1}{2}) + b = 0$

$\therefore\ \frac{1}{2}a + b = \frac{1}{2}$ .... (1)

Also, $f(1) = -1$ {Remainder theorem}

$\therefore\ 4(1)^3 - 4(1)^2 + a(1) + b = -1$

$\therefore\ a + b = -1$ .... (2)

Solving simultaneously:

$-a - 2b = -1$ $\{-2 \times (1)\}$
$a + b = -1$ $\{(2)\}$

Adding, $-b = -2$

$\therefore\ b = 2$ and $a = -3$

If $(2x - 1)$ is a factor of $f(x)$, then $f(\frac{1}{2}) = 0$.

3 **a** Find a and b given that $(x-1)$ and $(x+2)$ are factors of $ax^3 - 4x^2 - 7x + b$.
b Find p and q given that $(x+1)$ and $(x-3)$ are factors of $px^4 - 5x^3 - 5x^2 + qx + 9$.
c Find m and n given that $(2x+1)$ and $(x-2)$ are factors of $mx^3 - 9x^2 + nx + 6$.

4 $(x-2)$ is a factor of $f(x) = x^3 + ax^2 - 11x + b$. When $f(x)$ is divided by $x+1$, the remainder is 15. Find a and b.

5 $(2x-3)$ is a factor of $f(x) = 2x^3 + ax^2 + bx + 6$. When $f(x)$ is divided by $x-2$, the remainder is 4. Find a and b.

6 $(x-3)$ is a factor of $P(x) = 3x^3 + kx^2 - 5x + 6$.

a Find k.
b Write $P(x)$ in the form $P(x) = (x-3)(ax^2 + bx + c)$.
c Find all solutions to $P(x) = 0$.

7 Consider the cubic polynomial $f(x) = x^3 - 2x^2 - 23x + 60$.

a Show that $(x-3)$ is a factor of $f(x)$.
b Use polynomial division to help write $f(x)$ in the form $f(x) = (x-3)Q(x)$, where $Q(x)$ is a quadratic polynomial.
c Hence fully factorise $f(x)$.

8 Consider $P(x) = x^3 - a^3$ where a is real.

- **a** Find $P(a)$ and explain the significance of this result.
- **b** Factorise $x^3 - a^3$ as the product of a real linear and a quadratic factor.

9 Consider $P(x) = x^3 + a^3$, where a is real.

- **a** Find $P(-a)$ and explain the significance of this result.
- **b** Factorise $x^3 + a^3$ as the product of a real linear and a quadratic factor.

10 Find $n > 0$ such that $(x - 2)$ is a factor of $x^{2n} - 9x^n - 14x + 36$.

H CUBIC EQUATIONS

We have seen that a quadratic with roots α and β has the form

$$(x-\alpha)(x-\beta) = x^2 - (\alpha+\beta)x + \alpha\beta.$$

If we perform a similar expansion for a cubic, we find that

$$(x-\alpha)(x-\beta)(x-\gamma) = x^3 - (\alpha+\beta+\gamma)x^2 + (\alpha\beta+\beta\gamma+\gamma\alpha)x - \alpha\beta\gamma.$$

Notice that the *product of the roots* $\alpha\beta\gamma$ is the negative of the constant term.

If you think a cubic equation has integer roots, you can try to find them by factorising the constant term.

Example 16 — Self Tutor

Solve for x: $x^3 - 31x - 30 = 0$.

Let $P(x) = x^3 - 31x - 30$.

The constant term is -30, so the product of the roots is 30.

Since $30 = 5 \times 3 \times 2 \times 1$, likely integer roots are $\pm1, \pm2, \pm3, \pm5$. They could also be ±6 since $2 \times 3 = 6$, and so on.

Now $P(1) = -60$, so 1 is not a root.

But $P(-1) = 0$, so -1 is a root, and $(x+1)$ is a factor of $P(x)$.

The coefficient of x^3 is $1 \times 1 = 1$

The constant term is $1 \times -30 = -30$

So, $P(x) = x^3 + 0x^2 - 31x - 30 = (x+1)(x^2 + bx - 30)$
$= x^3 + (b+1)x^2 + (b-30)x - 30$

Equating x^2 terms: $b + 1 = 0$
$\therefore\ b = -1$

Hence $P(x) = (x+1)(x^2 - x - 30)$
$= (x+1)(x+5)(x-6)$

$\therefore$ the solutions are $x = -1$, $x = -5$, and $x = 6$.

EXERCISE 6H

1 Solve for x:

a $x^3 - 6x^2 + 11x - 6 = 0$ **b** $x^3 - 3x^2 + 4 = 0$ **c** $x^3 + 2x^2 - x - 2 = 0$

d $x^3 - 6x^2 + 5x + 12 = 0$ **e** $x^3 + 5x^2 - 16x - 80 = 0$ **f** $x^3 + 13x^2 + 55x + 75 = 0$

2 Solve for x:

a $x^3 + 2x^2 - 11 = 9x + 7$ **b** $(x+4)(x-2)(x-6) = 18 - 9x$

c $x(x-1)(x-7) = -24$

3 Solve for x:

a $2x^3 - 6x^2 - 8x + 24 = 0$

b $2x^3 - 2x^2 - 48x - 72 = 0$

c $3x^3 - 24x^2 - 15x + 252 = 0$

Discussion

Consider the general cubic $p(x) = ax^3 + bx^2 + cx + d$, $a, b, c, d \in \mathbb{R}$.

What happens to $p(x)$ if x gets:

- very large and positive
- very large and negative?

What does this tell you about the number of solutions that $p(x) = 0$ may have?

I GRAPHS OF CUBIC FUNCTIONS

A **cubic function** has the form $f(x) = ax^3 + bx^2 + cx + d$ where $a \neq 0$ and a, b, c, and d are constants.

Discovery — Cubic functions

To discover the shape of different cubic functions, you can either use the graphing package or your graphics calculator.

GRAPHING PACKAGE

What to do:

1 **a** Use technology to help sketch:

i $y = x^3$, $y = 2x^3$, $y = \frac{1}{2}x^3$, $y = \frac{1}{3}x^3$ **ii** $y = x^3$ and $y = -x^3$

iii $y = -x^3$, $y = -2x^3$, $y = -\frac{1}{2}x^3$, $y = -\frac{1}{10}x^3$

b Discuss the geometrical significance of a in $y = ax^3$. Comment on both the sign and the size of a.

2 **a** Use technology to help sketch:

$y = x(x+1)(x-2)$, $y = 2(x+3)(x-1)(x-2)$,

$y = 2x(x+2)(x-1)$, $y = -3(x-2)(x+1)(x+4)$.

b Discuss the geometrical significance of α, β and γ for the cubic $y = a(x-\alpha)(x-\beta)(x-\gamma)$.

3 **a** Use technology to help sketch:

$$y = (x-1)(x+1)(x+3), \qquad y = 2(x-1)(x+1)(x+3),$$
$$y = \tfrac{1}{2}(x-1)(x+1)(x+3), \qquad y = -2(x-1)(x+1)(x+3).$$

b Discuss the geometrical significance of a in $y = a(x-\alpha)(x-\beta)(x-\gamma)$.

4 **a** Use technology to help sketch:

$$y = (x-2)^2(x+1), \qquad y = (x+1)^2(x-3), \qquad y = 2(x-3)^2(x-1),$$
$$y = -x(x-2)^2, \qquad y = -2(x+1)(x-2)^2.$$

b Discuss the geometrical significance of α and β for the cubic $y = a(x-\alpha)^2(x-\beta)$.

5 **a** Predict the geometrical significance of a and α for the cubic $y = a(x-\alpha)^3$.

b Check your prediction is correct by sketching:

$$y = 2(x-3)^3, \quad y = -1(x+2)^3, \quad y = 3(x+1)^3.$$

You should have discovered that:

- If $a > 0$, the graph's shape is 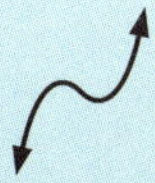 or , if $a < 0$ it is 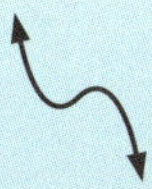 or 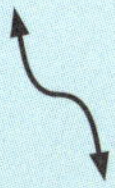.
- For a cubic function of the form $y = a(x-\alpha)(x-\beta)(x-\gamma)$, the graph has x-intercepts α, β, and γ, and the graph crosses over or **cuts** the x-axis at these points.
- For a cubic function of the form $y = a(x-\alpha)^2(x-\beta)$, the graph **touches** the x-axis at α and **cuts** it at β.
- For a cubic function of the form $y = a(x-\alpha)^3$, the graph cuts the x-axis at α. The curve changes shape at that point.

Example 17 — Self Tutor

Use axes intercepts to sketch the graph of:

a $f(x) = \frac{1}{2}(x+2)(x-1)(x-2)$ **b** $f(x) = 2x(x-2)^2$

a $f(x) = \frac{1}{2}(x+2)(x-1)(x-2)$ has x-intercepts -2, 1, and 2.

$f(0) = \frac{1}{2}(2)(-1)(-2) = 2$

$\therefore$ the y-intercept is 2.

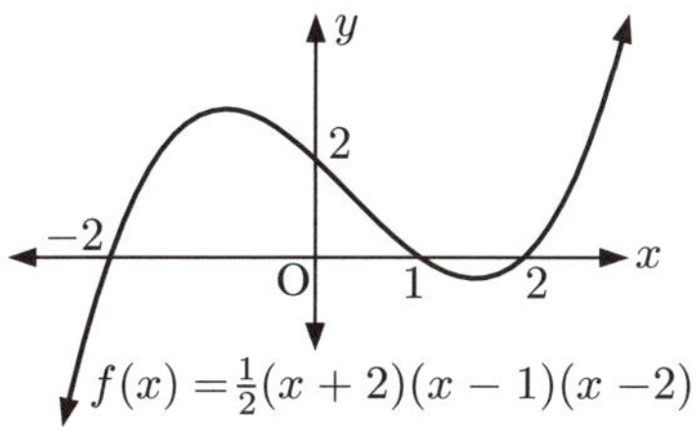

b $f(x) = 2x(x-2)^2$ cuts the x-axis when $x = 0$ and touches the x-axis when $x = 2$.

$f(0) = 2(0)(-2)^2 = 0$

$\therefore$ the y-intercept is 0.

$a = 2 > 0$, so the graph has shape .

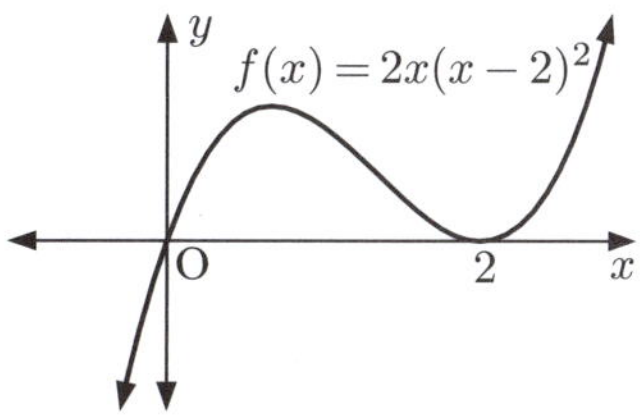

EXERCISE 6I

1 Use axes intercepts to sketch the graph of:

a $y = (x+1)(x-2)(x-3)$
b $y = x(x+4)(x-5)$
c $y = 2(x-4)(x-1)(x+2)$
d $y = -(x+2)(x-1)(x-5)$
e $y = \frac{1}{2}x(x-4)(x+3)$
f $y = 4(x+6)(x-1)(x-3)$
g $y = -3(x+4)(x+2)(x+1)$
h $y = -\frac{1}{4}(x+6)(x+3)(x-2)$
i $y = 2(2x-1)(x+3)(x-4)$
j $y = -\frac{1}{3}(x+4)(x+1)(x-\frac{3}{2})$

2 Use axes intercepts to sketch the graph of:

a $y = (x+1)^2(x-2)$
b $y = -(x-3)^2(x+4)$
c $y = 2x^2(x-3)$
d $y = -\frac{1}{4}(x-2)^2(x+1)$
e $y = -4(x+2)^2(3x-1)$
f $y = -3(x+1)^2(x-\frac{2}{3})$

3 Use axes intercepts to sketch the graph of:

a $y = (x+4)^3$
b $y = -2(x-2)^3$
c $y = -(x+1)^3$
d $y = 3(x-1)^3$

Example 18

Self Tutor

Sketch the graph of $y = |(x+4)(x+1)(x-1)|$.

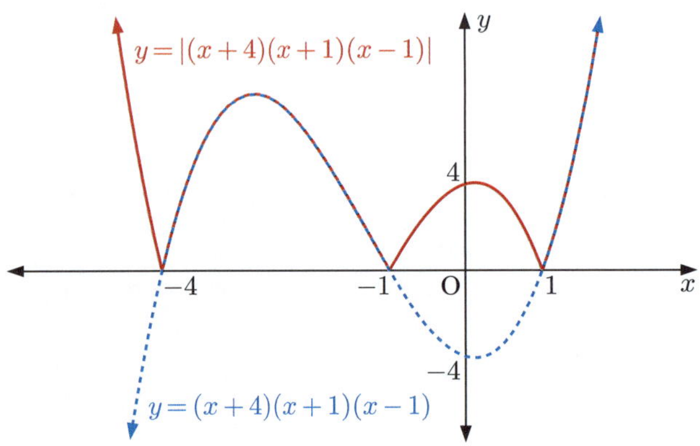

To sketch the graph of $y = |f(x)|$, any parts of $y = f(x)$ that are below the x-axis are reflected in the x-axis.

4 Sketch the graph of:

a $y = |(x+3)(x-1)(x-2)|$
b $y = |2(x-2)(x-5)(x+2)|$
c $y = \left|\frac{1}{2}(x+2)^2(x-3)\right|$
d $y = |-3x(x+1)(x-5)|$
e $y = \left|4(x+4)(x+2)(x-\frac{3}{2})\right|$
f $y = \left|-\frac{1}{3}(3x-2)(x+4)(x+1)\right|$

Example 19

Self Tutor

a Sketch the graph of $y = 2(x+4)(x+2)(x-3)$.

b Hence solve $2(x+4)(x+2)(x-3) \leqslant 0$.

a $y = 2(x+4)(x+2)(x-3)$ has x-intercepts -4, -2, and 3.

When $x = 0$, $y = 2(4)(2)(-3) = -48$

$\therefore$ the y-intercept is -48.

b $2(x+4)(x+2)(x-3) \leqslant 0$ when $x \leqslant -4$ or $-2 \leqslant x \leqslant 3$.

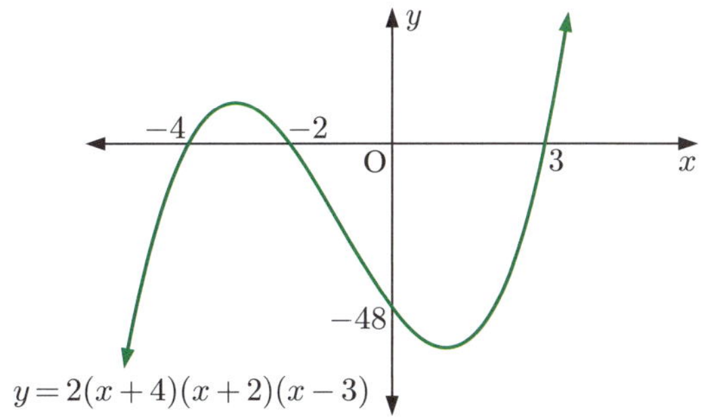

5 Use the graph shown to solve:

a $(x+1)(x-2)(x-5) < 0$

b $(x+1)(x-2)(x-5) \geqslant 0$

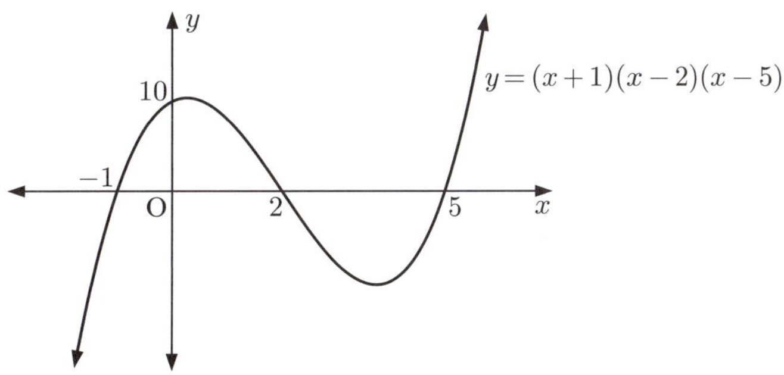

6 **a** Sketch the graph of $y = 2(x+1)(x-4)(x-6)$.

b Hence solve $2(x+1)(x-4)(x-6) > 0$.

7 Solve for x:

a $(x-3)(x-5)(x+1) \geqslant 0$

b $-\frac{1}{2}(x+3)(x-2)(x-6) < 0$

c $(x-1)(x+4)(x+\frac{1}{2}) < 0$

d $-3(x+3)^2(2x-3) > 0$

8 **a** Sketch the graph of $y = (x-2)(x-6)(x+3)$.

b Hence solve $x^3 + 36 > 5x^2 + 12x$.

9 **a** Sketch the graph of $y = x(x-2)(x+1)$.

b Solve algebraically the equation $x(x-2)(x+1) = -2$.

c Hence solve $x(x-2)(x+1) < -2$.

10 **a** Sketch the graph of $y = (x+2)(x-3)(x-5)$.

b Solve $(x+2)(x-3)(x-5) > 0$.

c Solve algebraically the equation $(x+2)(x-3)(x-5) = 24$.

d Hence solve $(x+2)(x-3)(x-5) \leqslant 24$.

Review set 6A

1 For the polynomial function $P(x) = 3x^5 + 5x^4 - x^3 + 6x - 4$, state the:

a degree **b** leading coefficient

c constant term **d** coefficient of x^3.

2 Given $p(x) = 5x^2 - x + 4$ and $q(x) = 3x^2 + 7x - 1$, find:

a $p(x) + q(x)$ **b** $2p(x) - q(x)$ **c** $p(x)q(x)$

3 Find the quotient and remainder of:

a $\dfrac{2x^2 + 11x + 18}{x + 3}$ **b** $\dfrac{x^3 - 6x^2 + 10x - 9}{x - 2}$

4 Decide whether:

a 5 is a zero of $x^2 - 3x + 10$ **b** -1 is a root of $x^4 - 3x^2 + 4x + 6 = 0$.

5 Find the zeros of:

a $3x^2 + 2x - 8$ **b** $x^2 + 8x + 11$

6 Fully factorise:

a $x^2 - 3x - 28$ **b** $x^2 + 2x - 6$

7 Suppose $6x^3 - 13x^2 + kx + 15 = (2x - 5)(ax^2 + bx + c)$. Find the values of a, b, c, and k.

8 **a** Given that $x^3 + x^2 - 3x + 9 = (x + 3)(ax^2 + bx + c)$, find the values of a, b, and c.

b Show that $x^3 + x^2 - 3x + 9$ has only one real zero.

9 Use the Remainder theorem to find the remainder when:

a $x^3 - 4x^2 + 5x - 1$ is divided by $x - 2$

b $2x^3 + 6x^2 - 7x + 12$ is divided by $x + 5$.

10 Use the Factor theorem to determine whether:

a $(x + 1)$ is a factor of $2x^4 - 9x^2 - 6x - 1$

b $(x - 3)$ is a factor of $x^4 - 2x^3 - 4x^2 + 5x - 6$.

11 $2x^2 + kx - 5$ has remainder 3 when divided by $x + 4$. Find k.

12 $ax^3 + 5x^2 - x + b$ has remainder 7 when divided by $x - 1$, and remainder -11 when divided by $x + 2$. Find a and b.

13 Find c given that $(x - 2)$ is a factor of $x^5 - 2x^4 + cx^3 - 7x^2 + 5x - 6$.

14 $(x - 4)$ is a factor of $f(x) = x^3 + 2x^2 + ax + b$. When $f(x)$ is divided by $x + 2$, the remainder is 18.

a Find a and b. **b** Find all zeros of $f(x)$.

15 Consider the cubic $f(x) = x^3 + x^2 - 17x + 15$.

a Show that $(x + 5)$ is a factor of $f(x)$.

b Use polynomial division to help write $f(x)$ in the form $f(x) = (x + 5)Q(x)$, where $Q(x)$ is a quadratic polynomial.

c By factorising $Q(x)$, write $f(x)$ in fully factorised form.

16 Solve for x: $x^3 - x^2 - 17x - 15 = 0$

17 Sketch the graph of:

a $y = 2(x+3)(x+1)(x-6)$ **b** $y = -\frac{1}{4}(x-2)(x-4)(x+3)$

18 **a** Sketch the graph of $y = \frac{1}{3}(x+1)(x-3)(x-7)$.

b Hence solve $(x+1)(x-3)(x-7) > 0$.

Review set 6B

1 Find the coefficient of x^2 and the degree of $(2x^2 - x + 1)(3x^2 - 5)$.

2 For $f(x) = x^3 - 4x + 5$ and $g(x) = x^2 - 6x + 1$, find:

a $f(x) + g(x)$ **b** $f(x) - g(x)$ **c** $f(x)g(x)$

3 Carry out the following divisions:

a $\dfrac{x^3 + 5}{x + 1}$ **b** $\dfrac{x^4 - 6x^3 + 9x^2 - 22}{x - 4}$

4 Find:

a the zeros of $x^3 - 9x^2 + 20x$ **b** the roots of $(4x - 3)(x^2 + x - 5) = 0$.

5 Use expansion to verify that $(x-2)(x+3)(x-6)$ is the factorised form of $x^3 - 5x^2 - 12x + 36$.

6 **a** Find *all* cubic polynomials with zeros $\frac{1}{4}$, $1 \pm \sqrt{5}$.

b Find *all* quartic polynomials with zeros -4, $\frac{1}{2}$, $2 \pm \sqrt{7}$.

7 If $f(x) = x^3 - 3x^2 - 9x + b$ has $(x-k)^2$ as a factor, show that there are two possible values of k. For each of these two values of k, find the corresponding value for b, and hence solve $f(x) = 0$.

8 Find the remainder when:

a $x^3 - 5x^2 + 9$ is divided by $x - 2$ **b** $4x^3 + 7x - 11$ is divided by $2x - 1$.

9 When $f(x) = 2x^3 - x^2 + ax - 4$ is divided by $x - 3$, the remainder is 56.

a Find a. **b** Find the remainder when $f(x)$ is divided by $x + 1$.

10 **a** Find, in terms of k, the remainder when $2x^3 - 3kx^2 + k^2x - 12$ is divided by $x - 3$.

b Hence find the values of k for which the remainder is negative.

11 **a** Use the Factor theorem to show that $(x-2)$ is a factor of $x^3 - 13x + 18$.

b Write $x^3 - 13x + 18$ in the form $(x-2)(ax^2 + bx + c)$ where $a, b, c \in \mathbb{Z}$.

c Find the real roots of $x^3 + 18 = 13x$.

12 $(x-2)$ and $(x+3)$ are factors of $ax^3 - 3x^2 - 11x + b$. Find a and b.

13 $(2x-1)$ is a factor of $f(x) = 2x^3 - 9x^2 + ax + b$. When $f(x)$ is divided by $x - 1$, the remainder is -15.

a Find a and b. **b** Write $f(x)$ as a product of linear factors.

14 Suppose $f(x) = x^4 - 5x^3 + ax^2 + 4x - 8$.

a Given $(x+1)$ is a factor of $f(x)$, find a.

b Show that $(x-2)$ is also a factor of $f(x)$.

c Show that $f(x) = (x+1)(x-2)(x^2-4x+4)$ by expanding the right hand side.

d Hence write $f(x)$ in fully factorised form.

15 Solve for x: $2x^3 - 2x^2 - 28x + 48 = 0$

16 Sketch the graph of:

a $y = |(x-3)(x-6)(x+1)|$

b $y = -2(x+2)^3$

17 **a** Sketch the graph of $y = x(x+4)(x+5)$.

b Solve $x(x+4)(x+5) < 0$.

c Solve algebraically the equation $x(x+4)(x+5) = -12$.

d Hence solve $x(x+4)(x+5) \geqslant -12$.

7

Transforming relationships to straight line form

Contents:

Opening problem

This table shows the value $V of Doug's father's car t years after purchase.

t (years)	1	3	5	8	10
V (dollars)	42 900	28 000	18 500	9800	6400

Doug is trying to find an equation connecting V and t. When he plots the values on a graph, the result is a curve:

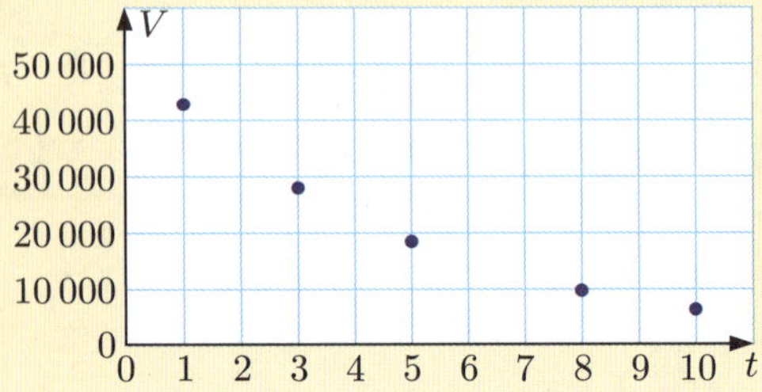

Doug's father suggests that he plot $\lg V$ against t. When Doug does this, the result is a straight line:

t	1	3	5	8	10
$\lg V$	4.63	4.45	4.27	3.99	3.81

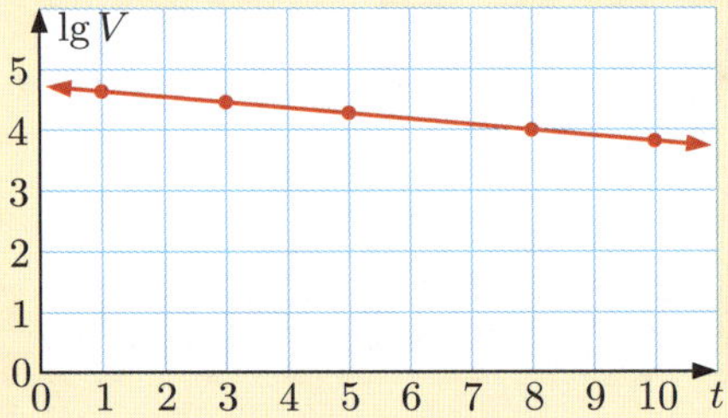

Things to think about:

a Is it easier to find the equation of a curve or a straight line?

b What does the linear relationship between $\lg V$ and t tell us about the relationship between V and t?

c How can Doug use the equation of the straight line to find a formula for V in terms of t?

If two variables are not linearly related, the graph of their relationship will not be a straight line.

However, it is sometimes possible to *transform* one or both variables so we can plot a linear relationship. The equation of this straight line graph enables us to describe the relationship between the original variables.

A TRANSFORMING RELATIONSHIPS TO STRAIGHT LINE FORM

Consider the relationship $y = 2x^2 + 1$.

x and y are not linearly related, but x^2 and y *are* linearly related since $y = 2(x^2) + 1$.

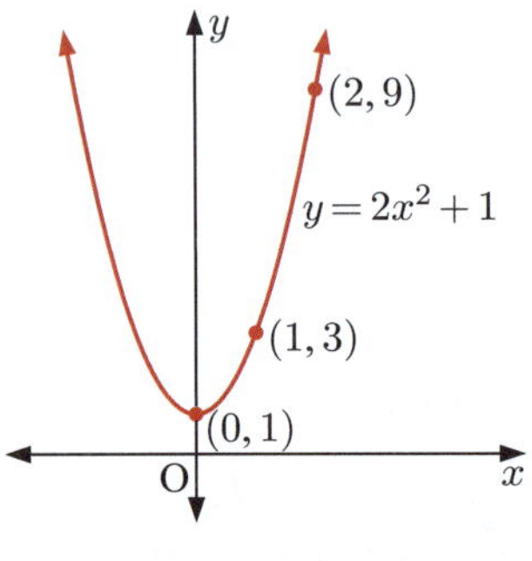

We can use a table of values to plot y against x^2:

x	-2	-1	0	1	2
x^2	4	1	0	1	4
y	9	3	1	3	9

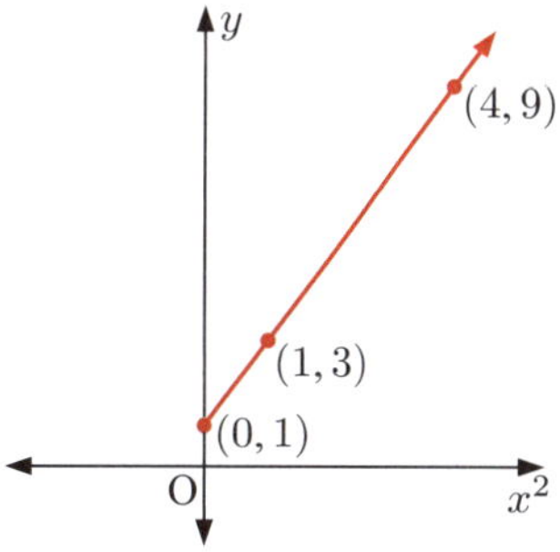

The graph of y against x^2 is a straight line with gradient 2 and y-intercept 1.

The line terminates at $(0,\ 1)$, since $x^2 \geqslant 0$ for all x.

Click on the icon to view a demonstration of how the two graphs are related.

DEMO

Example 1

Self Tutor

Find y in terms of x:

a

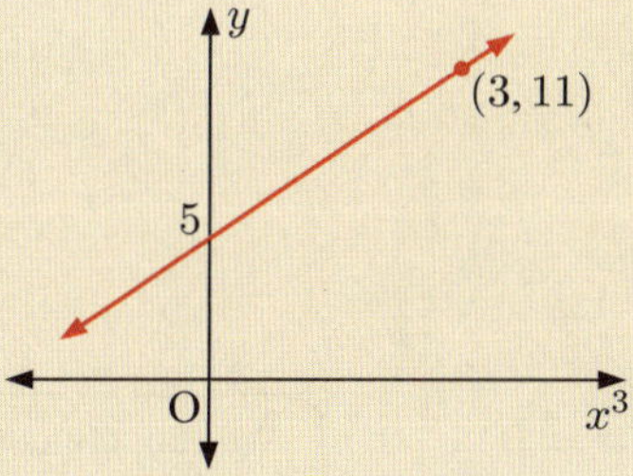

b

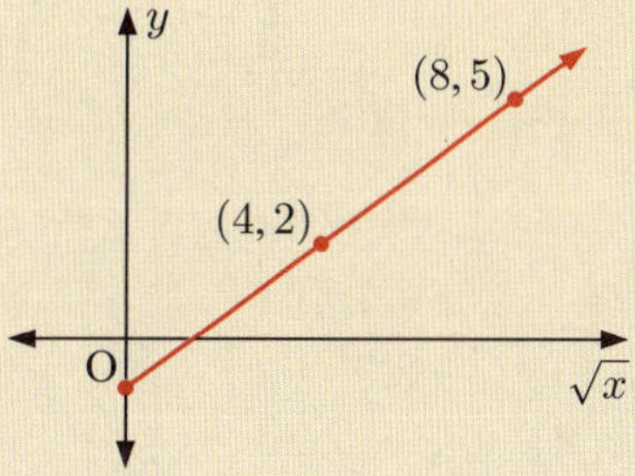

a The graph of y against x^3 is linear.

The gradient is $\dfrac{11-5}{3-0} = 2$ and the y-intercept is 5.

$\therefore$ the equation is $y = 2x^3 + 5$.

b The graph of y against $\sqrt{x}$ is linear.

The gradient is $\dfrac{5-2}{8-4} = \dfrac{3}{4}$.

$\therefore$ the equation is

$$y - 2 = \tfrac{3}{4}(\sqrt{x} - 4)$$

$$\therefore\ y - 2 = \tfrac{3}{4}\sqrt{x} - 3$$

$$\therefore\ y = \tfrac{3}{4}\sqrt{x} - 1,\ \ x \geqslant 0$$

EXERCISE 7A

1 Find y in terms of x:

a

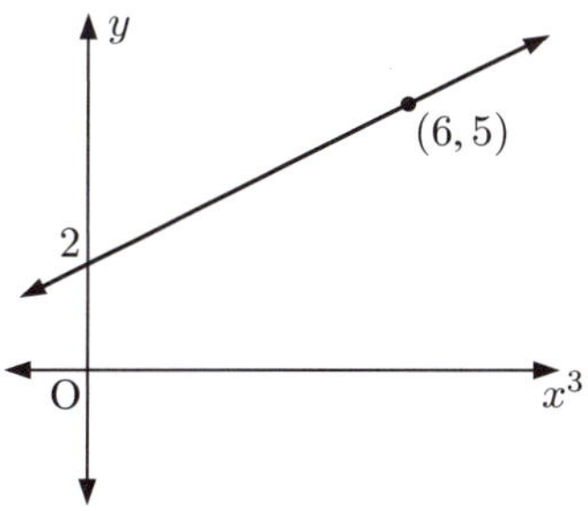

b

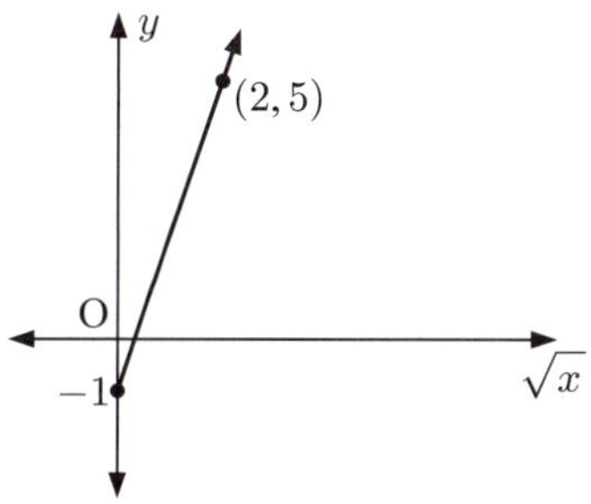

c

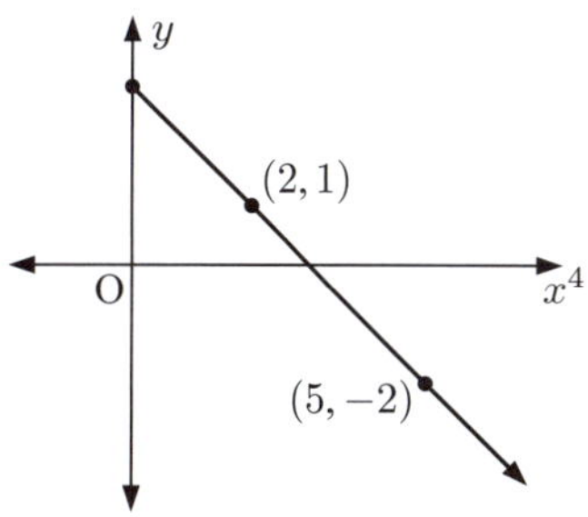

d

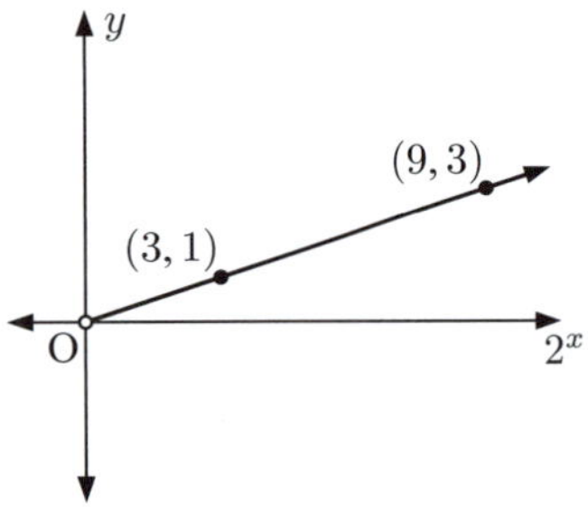

e

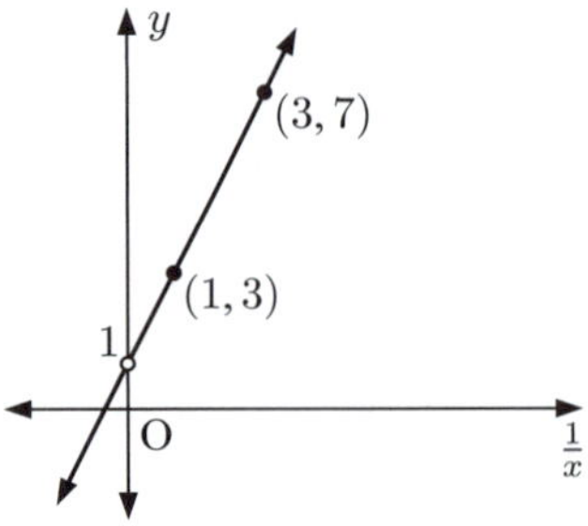

f

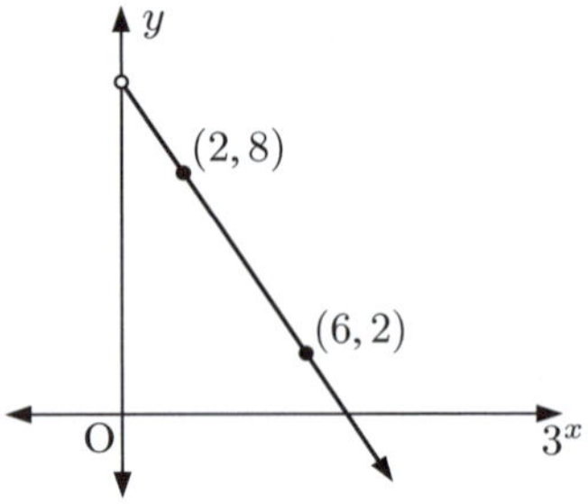

Example 2 — Self Tutor

a Find y in terms of x.

b Find y when $x = 4$.

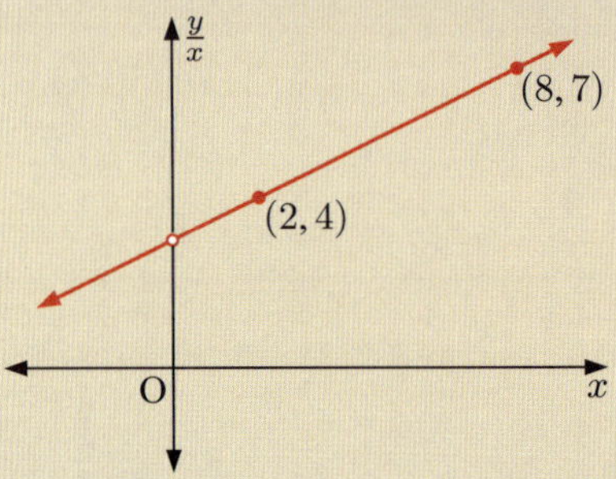

a The graph of $\frac{y}{x}$ against x is linear.

The gradient is $\frac{7-4}{8-2} = \frac{1}{2}$.

Using $(2, 4)$, the equation is $\frac{y}{x} - 4 = \frac{1}{2}(x - 2)$ where $x \neq 0$

$$\therefore \frac{y}{x} - 4 = \frac{1}{2}x - 1$$

$$\therefore \frac{y}{x} = \frac{1}{2}x + 3$$

$$\therefore y = \tfrac{1}{2}x^2 + 3x, \quad x \neq 0$$

b When $x = 4$, $y = \frac{1}{2}(4)^2 + 3(4)$

$= 20$

2 For each of the following relations:

i Write y in terms of x.

ii Find the value of y when $x = 3$.

a

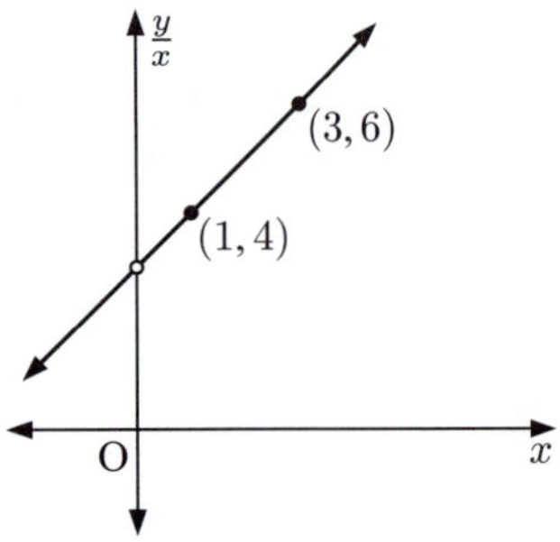

b

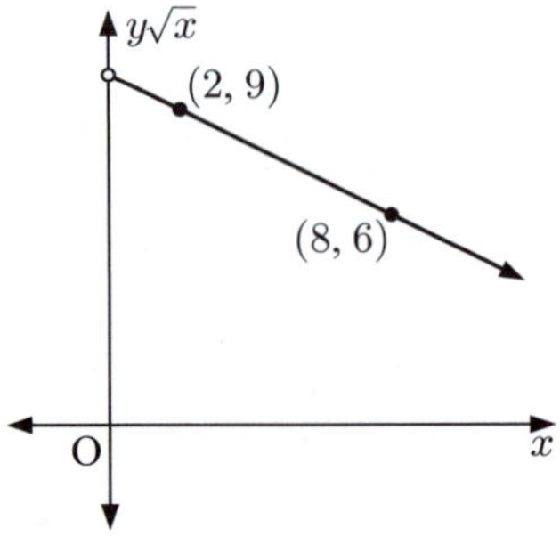

c

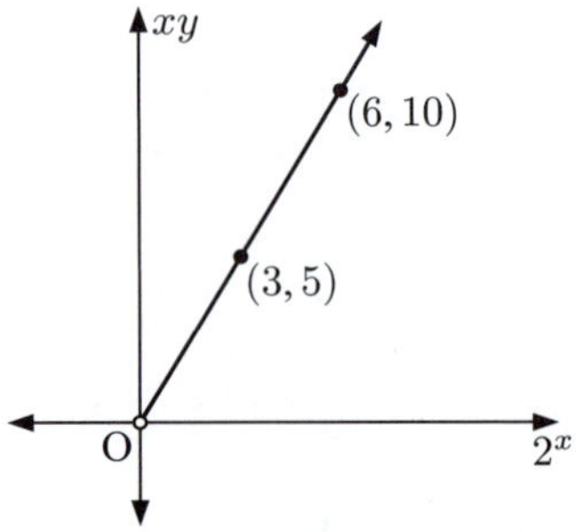

d

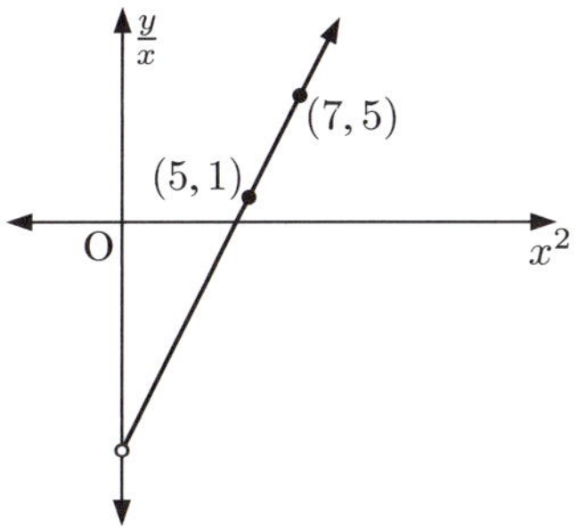

e

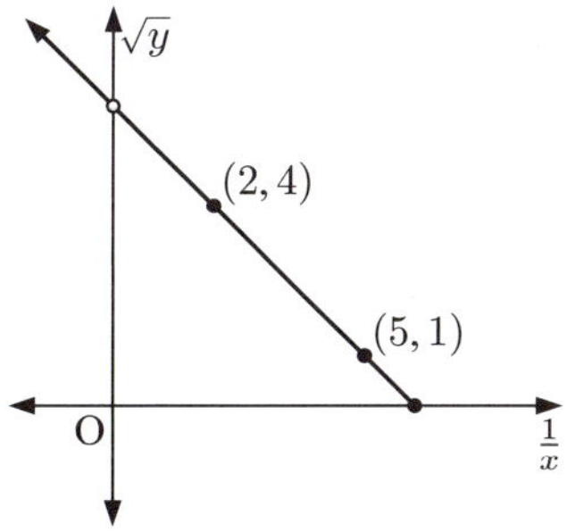

f

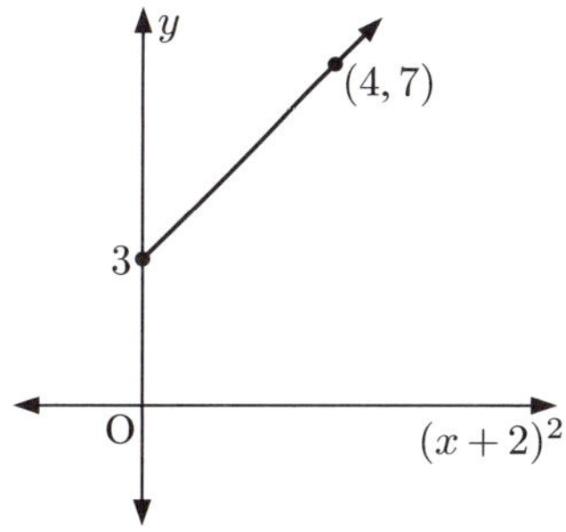

Example 3

Self Tutor

Write y in terms of x, giving your answer in the form $y = a \times 10^{bx}$ where $a, b \in \mathbb{Q}$.

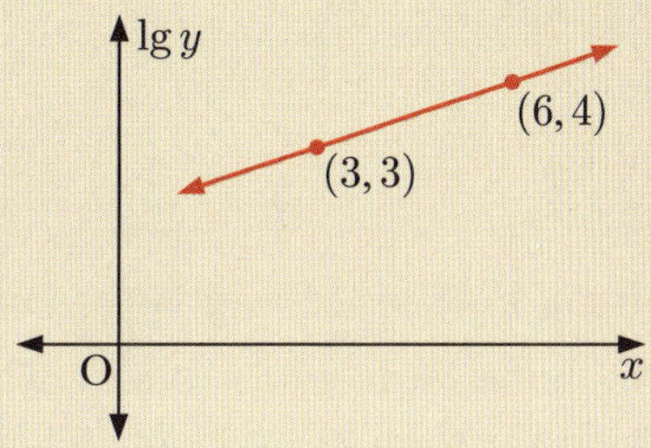

The graph of $\lg y$ against x is linear.

The gradient is $\frac{4-3}{6-3} = \frac{1}{3}$.

$\therefore$ the equation is $\lg y - 3 = \frac{1}{3}(x-3)$

$$\lg y - 3 = \tfrac{1}{3}x - 1$$

$$\therefore \lg y = \tfrac{1}{3}x + 2$$

$$\therefore y = 10^{\frac{1}{3}x+2} \quad \{\text{if } \lg p = q \text{ then } p = 10^q\}$$

$$\therefore y = 10^{\frac{1}{3}x} \times 10^2$$

$$\therefore y = 100 \times 10^{\frac{1}{3}x}$$

In **Chapter 5**, we saw that a linear relationship between $\lg y$ and x indicates an *exponential* relationship between y and x.

3 **a** Find $\lg y$ in terms of x.

b Write y in terms of x, giving your answer in the form $y = a \times 10^{bx}$ where $a, b \in \mathbb{Q}$.

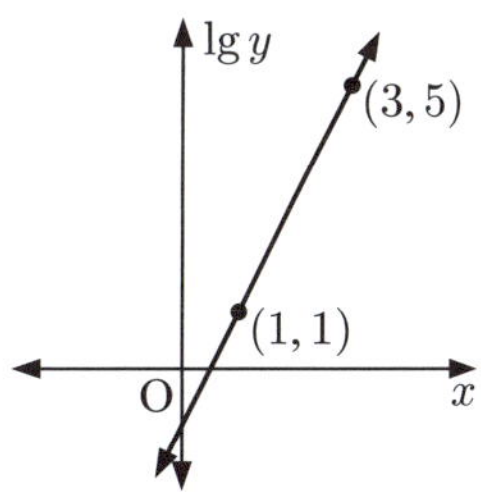

4 Write y in terms of x, giving your answer in the form $y = a \times 10^{bx}$ where $a, b \in \mathbb{Q}$.

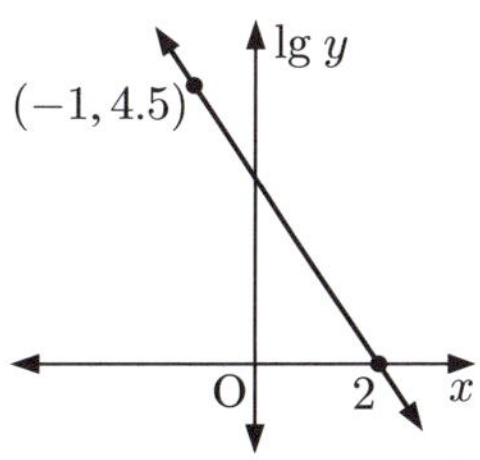

5 Write y in terms of x, giving your answer in the form $y = a \times b^x$ where $a, b \in \mathbb{Q}$.

a

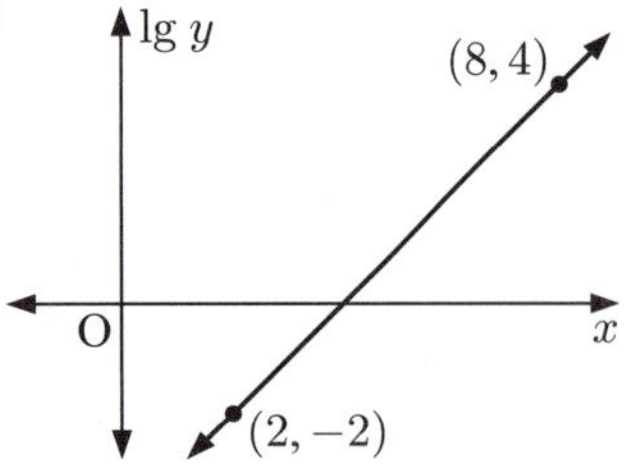

b

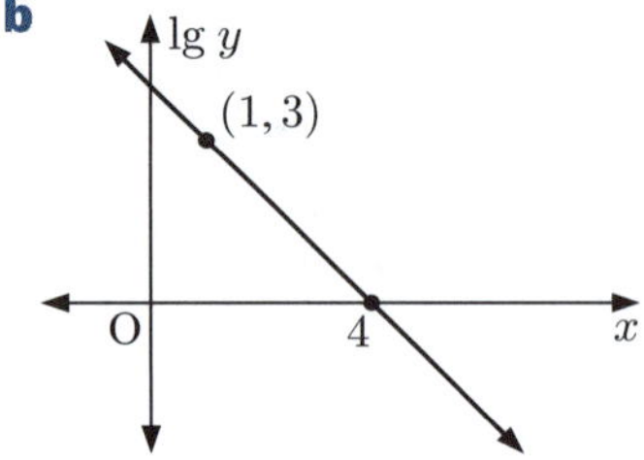

c

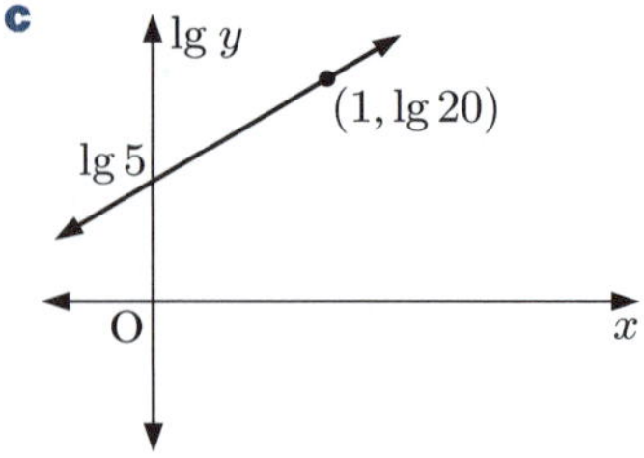

6 **a** Write y in terms of x, giving your answer in the form $y = a \times 10^{bx}$ where $a, b \in \mathbb{Q}$.

b Hence find y when $x = 6$.

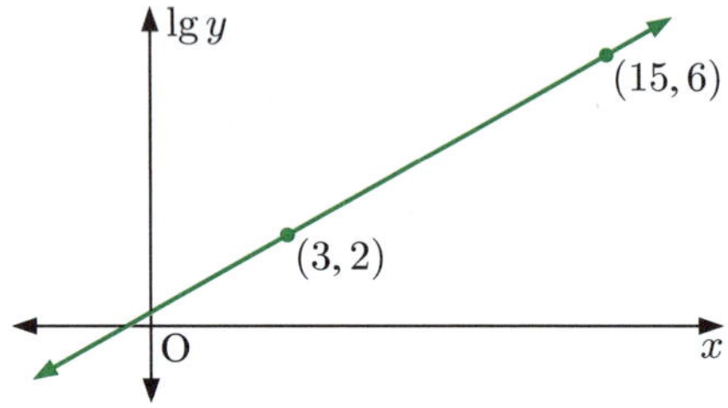

7 **a** Write y in terms of x, giving your answer:

i in the form $y = a \times e^{kx}$ where a and k are exact

ii in the form $y = A \times b^x$ where A and b are rounded to 3 significant figures.

b Find y when $x = 2$.

c Find x when $y = 300$.

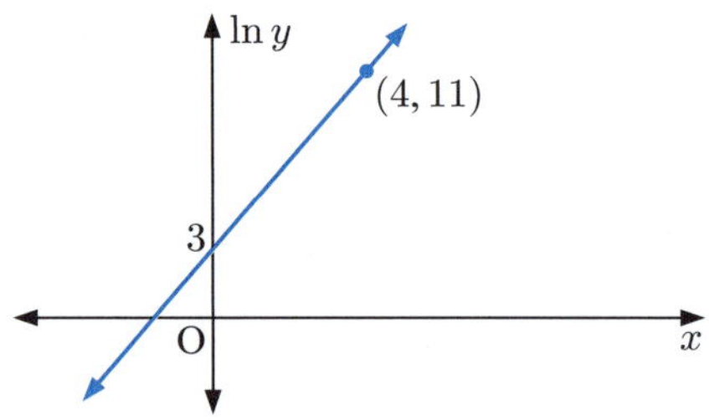

8 Write y in terms of x. Give your answer in the form $y = A \times b^x$, where A and b are rounded to 3 significant figures.

a

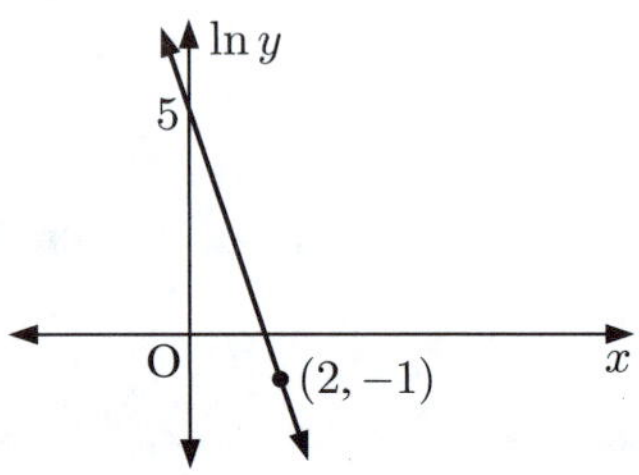

b

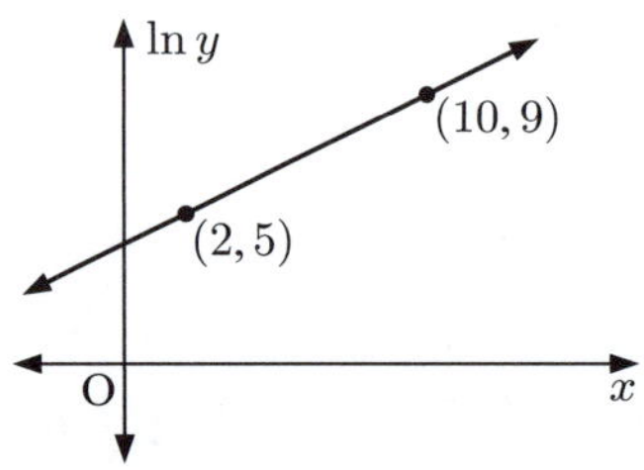

c

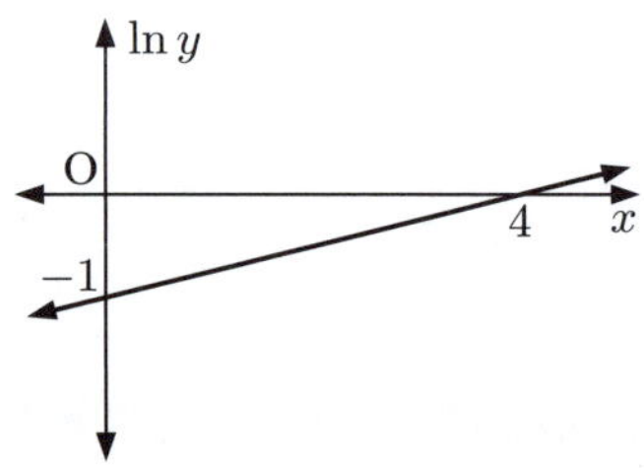

d

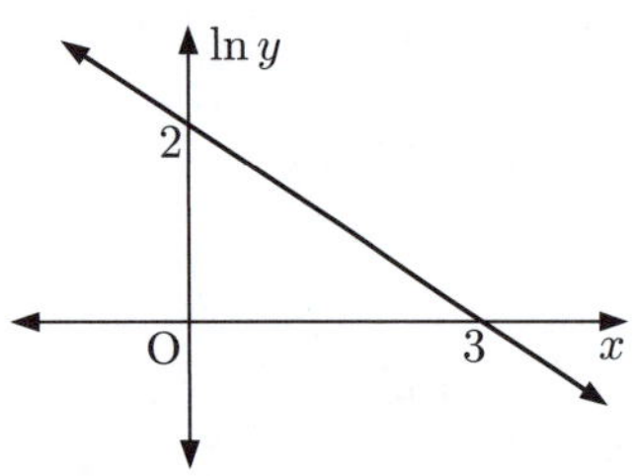

e

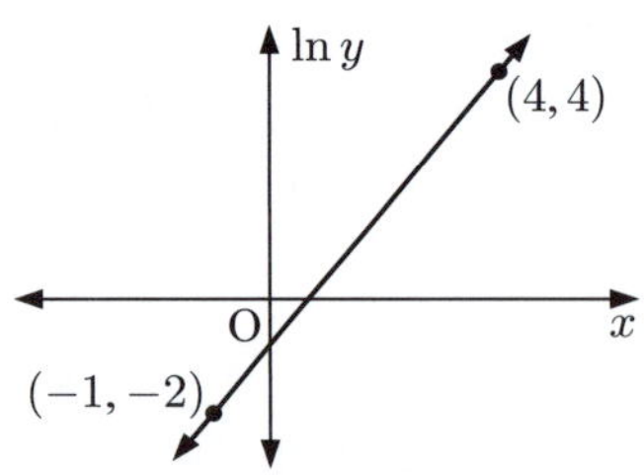

f

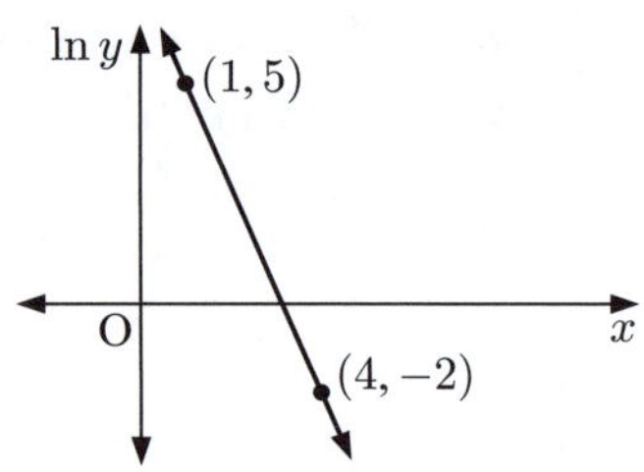

9 Two variables x and y are connected by the relationship $y = A \times b^{x^2}$.

a Find A and b.

b Find y when $x = 0.5$.

c Find the positive value of x such that $y = 0.1$.

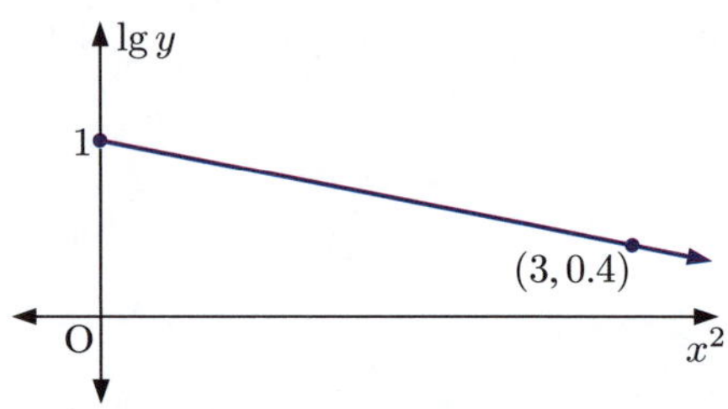

10 Two variables x and y are connected by the relationship $y = A \times b^{\sqrt{x}}$.

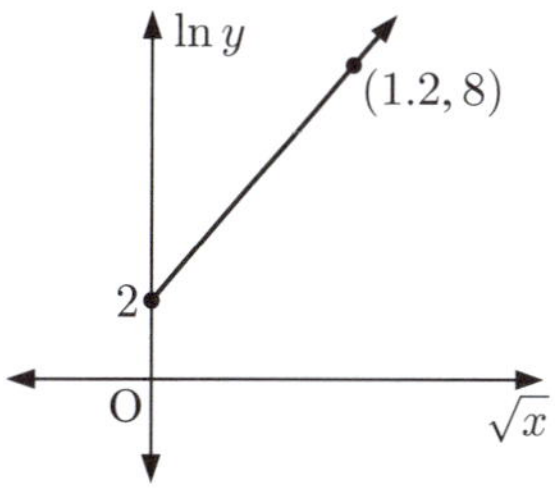

- **a** Find the exact values of A and b.
- **b** Find y when $x = 4$.
- **c** Find x such that $y = e^3$.

Example 4 — Self Tutor

Write y in terms of x, giving your answer in the form $y = a \times x^b$ where $a, b \in \mathbb{Q}$.

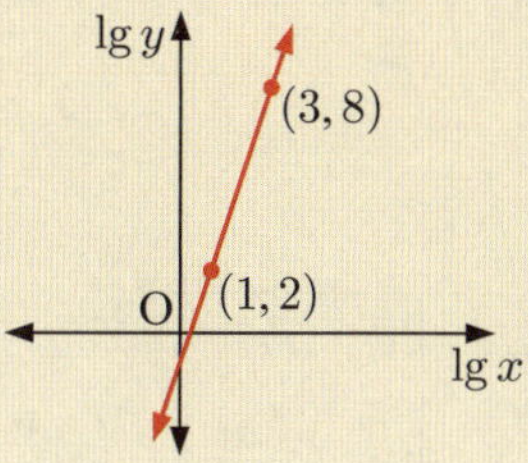

The graph of $\lg y$ against $\lg x$ is linear.

The gradient is $\dfrac{8-2}{3-1} = 3$.

$\therefore$ the equation is $\lg y - 2 = 3(\lg x - 1)$

$\therefore \lg y - 2 = 3\lg x - 3$

$\therefore \lg y = 3\lg x - 1$

$\therefore \lg y = \lg(x^3) - \lg 10$

$\therefore \lg y = \lg\left(\dfrac{x^3}{10}\right)$

$\therefore y = \frac{1}{10} \times x^3$

11 Consider the graph alongside.

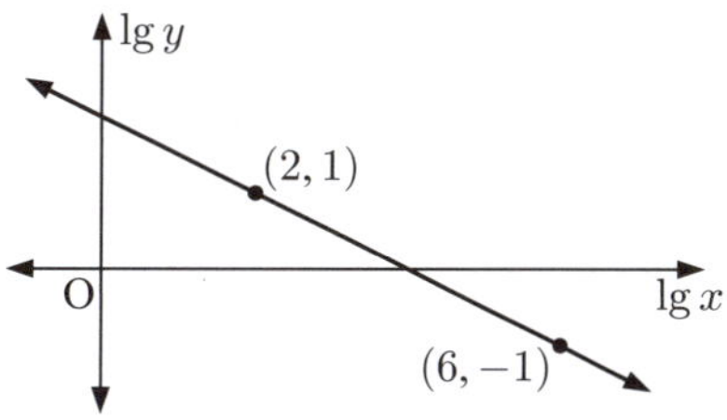

- **a** Write the equation of the line in the form $\lg y = m \lg x + c$.
- **b** Hence write y in terms of x.

12 Write y in terms of x:

a

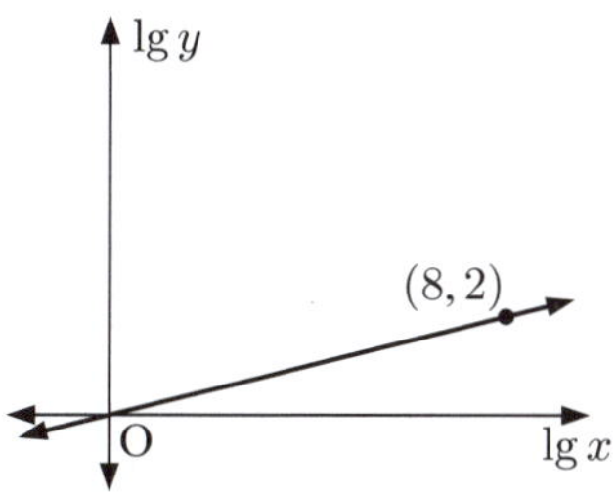

b

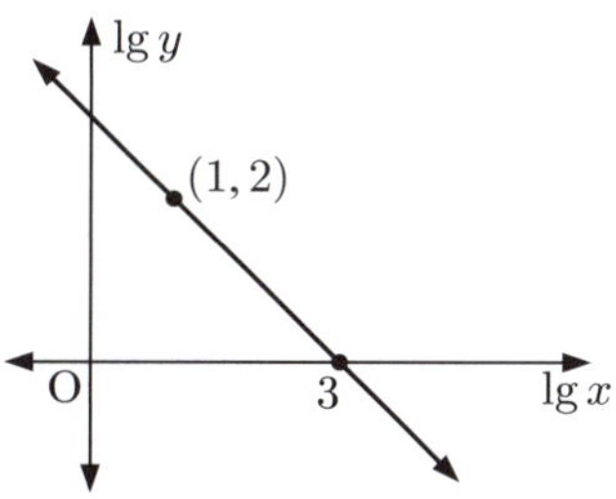

c

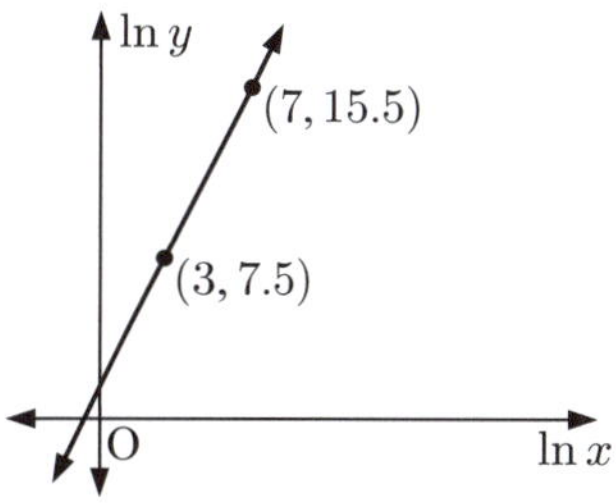

13 **a** Write K in terms of t.

b Hence find K when $t = 9$.

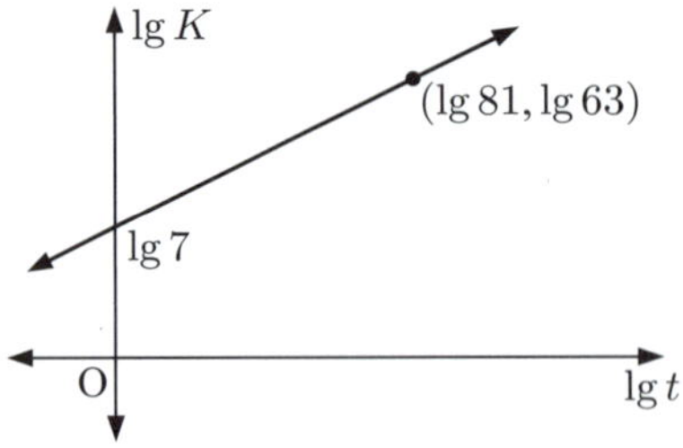

14

Graph A

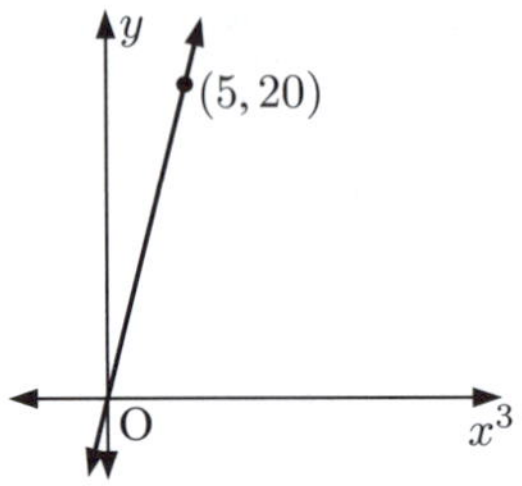

Graph B

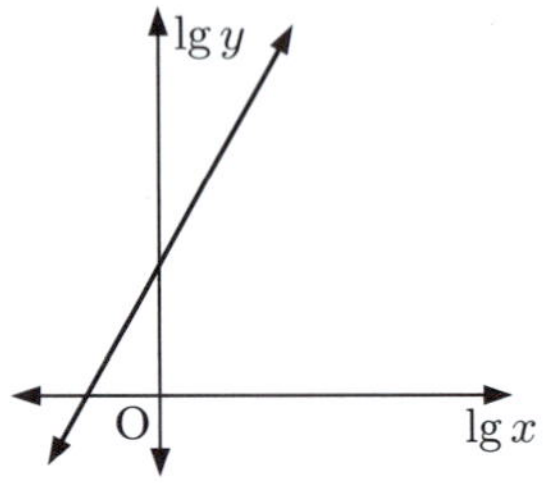

The linear relationship illustrated in *Graph A* can also be plotted as a straight line in *Graph B*. For the straight line in *Graph B*, find the:

a gradient **b** intercept on the vertical axis.

B FINDING RELATIONSHIPS FROM DATA

We have seen how a transformation of variables may allow us to display a non-linear relationship using a straight line graph. It is particularly useful to do this if we are trying to use a function to model data.

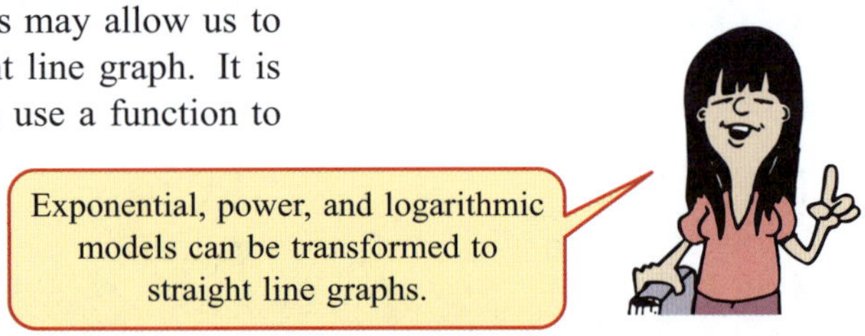

Case Study — Exponential growth and decay

Logarithms are particularly important in science. Many physical processes are modelled accurately by exponential laws.

For example, the United Nations published the following data on world population:

Year	*Population P* (in billions)	ln P
1750	0.79	−0.236
1800	0.98	−0.0202
1850	1.26	0.231
1900	1.65	0.501
1910	1.75	0.560
1920	1.86	0.621
1930	2.07	0.728
1940	2.30	0.833

Year	*Population P* (in billions)	ln P
1950	2.52	0.924
1960	3.02	1.11
1970	3.70	1.31
1980	4.44	1.49
1990	5.27	1.66
1999	5.98	1.79
2000	6.06	1.80
2010	6.79	1.92

The population data is presented on the graph below:

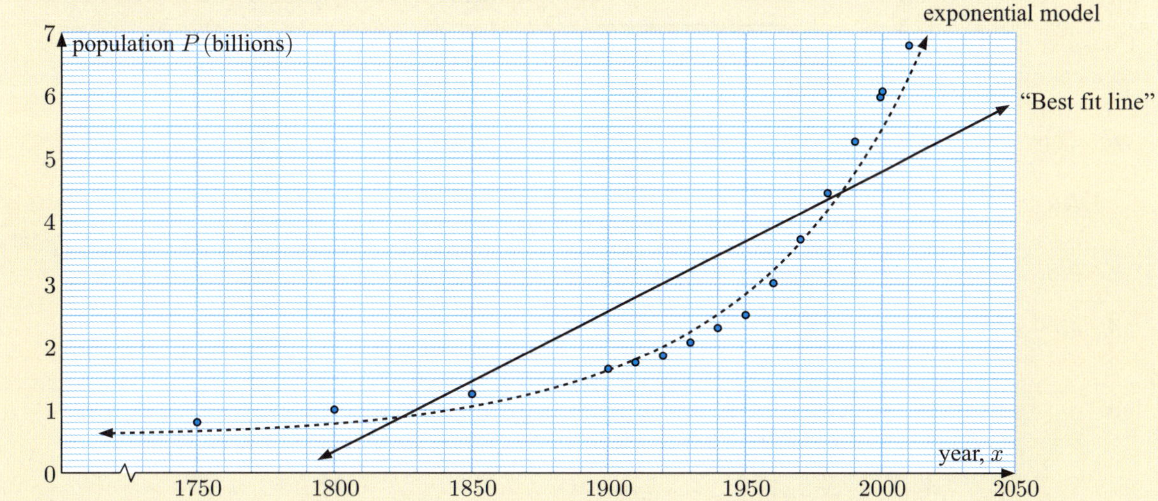

The "best fit line" $P = 0.0222x - 39.6$ does not fit the data very well, and is clearly not a good model for population over time. Instead, we try to fit an exponential curve of the form $P = ae^{mx}$.

Taking the natural logarithm of both sides, we have $\ln P = mx + \ln a$, which is the equation of a straight line.

We now plot $\ln P$ against x:

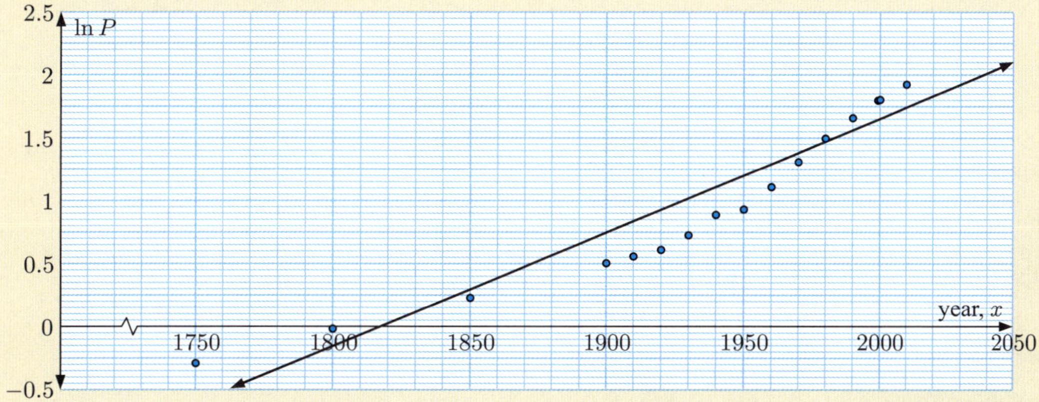

The equation of this "best fit line" is $\ln P = -15.5 + 0.008\,55x$.

$\therefore$ the data can be modelled by $P = e^{-15.5 + 0.008\,55x}$.

This exponential model is shown as a dashed line on the original graph. This is not a perfect fit either, but it is a considerable improvement on the original straight line graph.

Example 5

Self Tutor

Consider this table of data connecting x and y:

x	1	2	3	4
y	3.5	10	22.5	44

a Copy and complete:

x^2				
$\frac{y}{x}$				

b Plot $\frac{y}{x}$ against x^2, and draw a straight line through the points.

c Find y in terms of x.

a

x^2	1	4	9	16
$\frac{y}{x}$	3.5	5	7.5	11

b

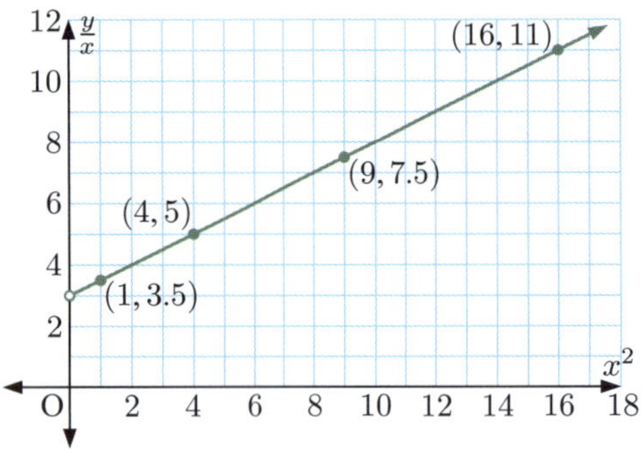

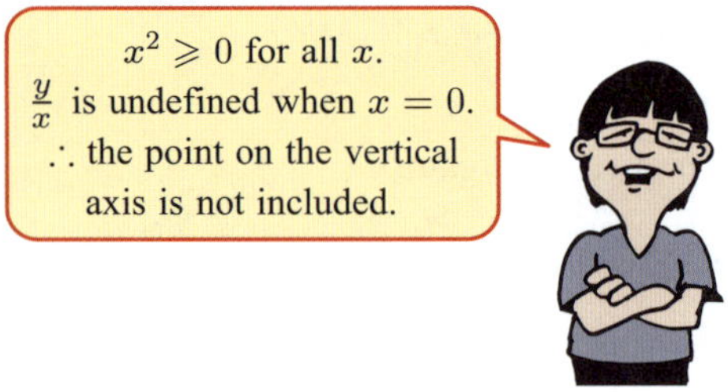

c The graph of $\frac{y}{x}$ against x^2 is linear.

Using the points (4, 5) and (16, 11), the gradient is $\frac{11-5}{16-4} = \frac{1}{2}$

$\therefore$ the equation is $\frac{y}{x} - 5 = \frac{1}{2}(x^2 - 4)$ where $x \neq 0$

$$\therefore \quad \frac{y}{x} = \frac{1}{2}x^2 + 3$$

$$\therefore \quad y = \tfrac{1}{2}x^3 + 3x, \quad x \neq 0$$

EXERCISE 7B

1 Consider this table of data connecting x and y:

x	1	2	3	4
y	2	11	26	47

a Copy and complete the following table:

x^2				
y				

b Plot y against x^2, and draw a straight line through the points.

c Find y in terms of x.

2 This table shows experimental data values for x and y:

x	1	2	3	4
y	9	9.90	10.97	12

a Copy and complete the following table:

x				
$y\sqrt{x}$				

b Plot $y\sqrt{x}$ against x, and draw a straight line through the points.

c Find y in terms of x.

d Hence find y when $x = 16$.

3 This table shows experimental values for x and y.

x	1	2	3	5
y	-1	0	0.11	0.12

It is known that x and y are related by the equation $y = \frac{a}{x} + \frac{b}{x^2}$, where a and b are constants.

a Copy and complete the following table:

$\frac{1}{x}$				
xy				

b Plot xy against $\frac{1}{x}$, and draw a straight line through the points.

c Hence find a and b.

d Find y when $x = 10$.

e Find x such that $y = \frac{2}{25}$.

4 This table shows a set of data pairs (x, y):

x	2	4	6	8
y	5.24	5	5.45	6.12

a Copy and complete:

$x\sqrt{x}$				
$y\sqrt{x}$				

b Plot $y\sqrt{x}$ against $x\sqrt{x}$, and draw a straight line through the points.

c Find y in terms of x.

d Hence find y when $x = 9$.

5 The mass of bacteria in a culture is measured each day for 5 days.

t (days)	1	2	3	4	5
M (grams)	3.98	6.31	10	15.85	25.12

a Copy and complete:

t	1	2			
$\lg M$					

b Plot $\lg M$ against t, and draw a straight line through the points.

c Find M in terms of t.

d Find the original mass of the bacteria.

Example 6

Self Tutor

This table shows experimental data values for x and y:

By plotting a suitable straight line graph, show that y and x are related by the equation $y = ax + \frac{b}{x}$, where a and b are constants.

x	1	2	3	4
y	14	10	10	11

If $y = ax + \frac{b}{x}$, then

$xy = ax^2 + b$

$\therefore$ if y and x are related in this way, then we should observe a linear relationship between xy and x^2.

x^2	1	4	9	16
xy	14	20	30	44

The graph of xy against x^2 is linear.

Using points (1, 14) and (4, 20),

the gradient is $\frac{20-14}{4-1} = 2$.

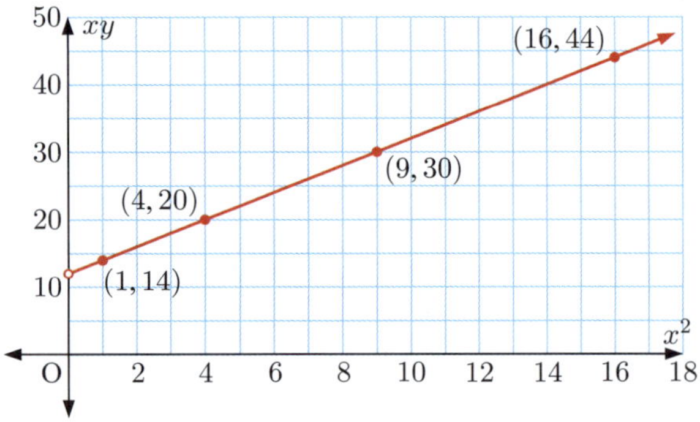

$\therefore$ the equation is $xy - 14 = 2(x^2 - 1)$

$\therefore$ $xy - 14 = 2x^2 - 2$

$\therefore$ $xy = 2x^2 + 12$

$\therefore$ $y = 2x + \frac{12}{x}$ where $x \neq 0$ $\{a = 2, b = 12\}$

6 This table shows experimental values of x and y:

x	1	2	3	4
y	1	26	99	244

It is known that x and y are related by the equation $y = ax^3 + bx$, where a and b are constants.

a A straight line graph is to be drawn to represent this information. If $\frac{y}{x}$ is plotted on the vertical axis, which variable should be plotted on the horizontal axis?

b Draw the straight line graph.

c Find the values of a and b.

d Hence find y when $x = 5$.

7 This table shows experimental values of x and y:

x	1	2	3	4
y	4	1.17	0.36	0

By plotting a suitable straight line graph, show that x and y are related by an equation of the form $y = \frac{a}{x} + \frac{b}{\sqrt{x}}$, where a and b are constants.

8 A stone is dropped from the top of an 80 m high cliff. This table shows the distance the stone has fallen at various times.

Time (t s)	1	1.7	2	2.7
Distance (D m)	4.9	14.16	19.6	35.72

a By plotting a suitable straight line graph, show that t and D are related by the equation $D = a \times t^b$, where a and b are constants.

b How far has the stone fallen after 3 seconds?

c How long does the stone take to hit the water?

Research — Logarithmic scales in science

If the data for a variable ranges over many orders of magnitude, it can be difficult to compare or represent it on a graph.

For example, the Richter scale for earthquake measurement uses logarithms in base 10. So, there is a linear relationship between the *magnitude* of an earthquake, and the logarithm of its shaking amplitude.

An earthquake measuring 6.0 on the Richter scale has a shaking amplitude $10^{6-4} = 100$ times larger than one that measures 4.0.

Research some other scientific scales that use logarithms to compress very large ranges into manageable values. You may like to consider:

- the decibel scale for the loudness of sound
- the stellar magnitude scale for brightness of stars
- the pH scale for acidity and alkalinity
- counting f-stops for ratios of photographic exposure.

Review set 7A

1 Find y in terms of x:

a

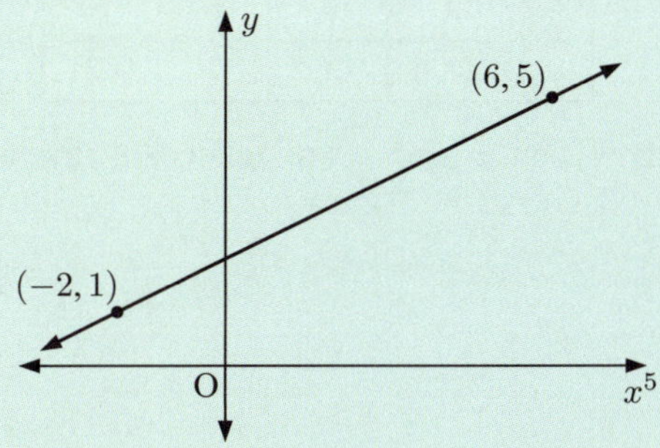

b

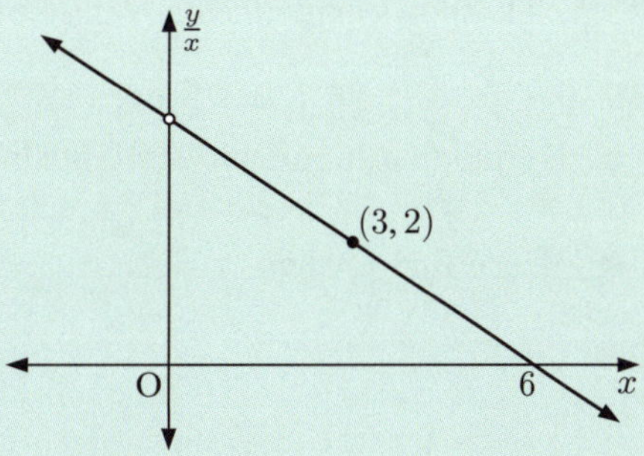

2 **a** Use the information in the graph to write y in terms of x.

b Hence find y when $x = 9$.

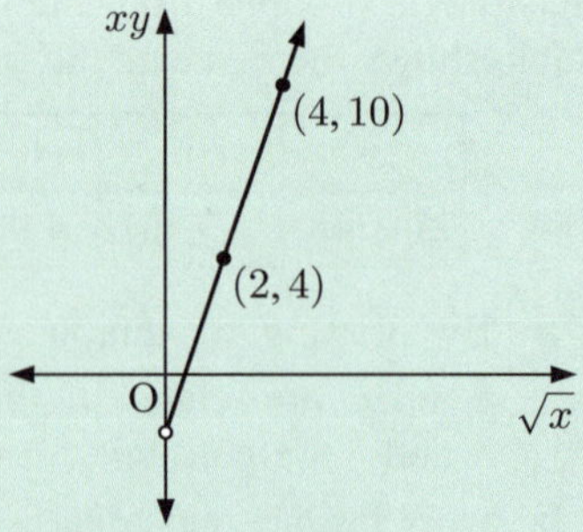

3

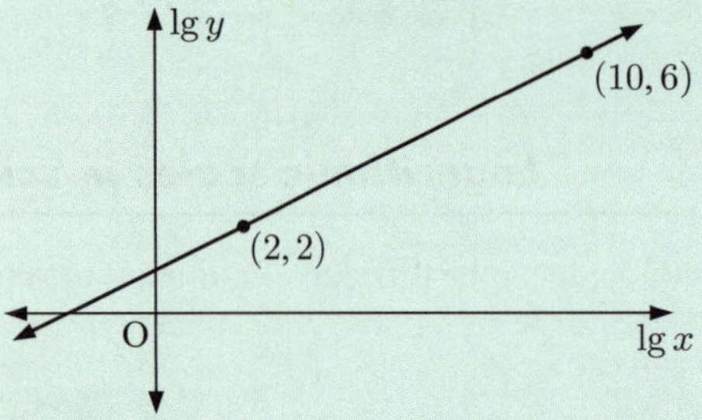

Consider the graph alongside.

a Write an equation for the line in the form $\lg y = m \lg x + c$.

b Hence write y in terms of x.

4 Suppose $y = Ae^{bx}$, where A and b are constants. The straight line graph obtained when $\ln y$ is graphed against x passes through $(1.5, 2.9)$ and $(4, 4.4)$.

a Find A and b.

b Find y when $x = 5$.

c Find x such that $y = 50$.

5 This table shows experimental values of x and y:

x	1	2	3	4
y	8	7.5	11.33	17.75

a Copy and complete:

x^3				
xy				

b Plot xy against x^3, and draw a straight line through the points.

c Find y in terms of x.

d Hence find y when $x = 7$.

6 This table shows experimental values of x and y:

x	2	4	6	8
y	21.54	4.64	1	0.21

a By plotting a suitable straight line graph, show that x and y are related by the equation $y = a \times b^x$, where a and b are constants.

b Hence find y when $x = 1$.

Review set 7B

1 **a** Write y in terms of x.

b Hence find y when $x = 8$.

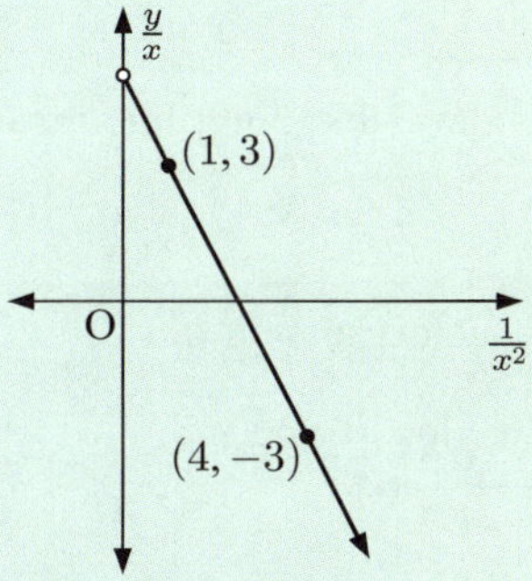

2 **a** Write y in terms of x. Give your answer in the form $y = A \times b^x$, where A and b are rounded to 3 significant figures.

b Hence find y when $x = 3$.

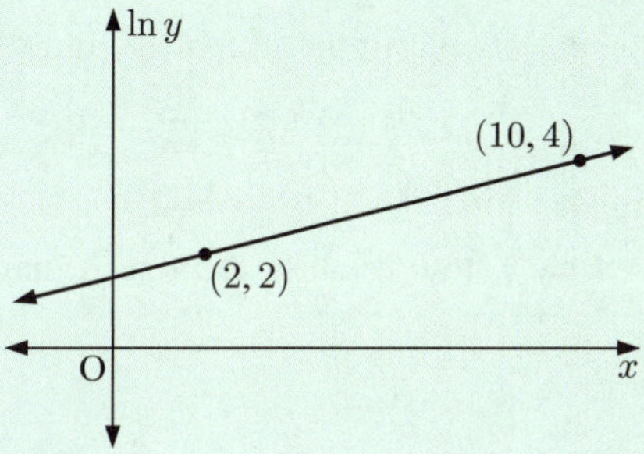

3 Consider this table of data connecting x and y:

x	1	2	3	4
y	2.5	5.29	8.13	11

a Copy and complete:

$\sqrt{x}$				
$\frac{y}{\sqrt{x}}$				

b Plot $\frac{y}{\sqrt{x}}$ against $\sqrt{x}$, and draw a straight line through the points.

c Hence write y in terms of x.

4 Suppose x and y are related by the equation $y = \frac{a}{\sqrt{x}} + b\sqrt{x}$, where a and b are constants.

a A straight line graph is drawn to represent this information. If $\frac{y}{\sqrt{x}}$ is plotted on the vertical axis, which variable is plotted on the horizontal axis?

b The straight line graph has gradient 6, and passes through $(2, 15)$.

i Find a and b.

ii Find y when $x = 9$.

iii Find x when $y = 9$.

5 This table shows experimental data values for x and y:

x	1	2	3	4
y	1.6	0.4	0.178	0.1

a Plot $\lg y$ against $\lg x$, and draw a straight line through the points.

b Find y in terms of x.

c Find y when $x = 6$.

d Find the positive value of x such that $y = 0.5$.

6 Consider the table of values connecting P and m:

m	1	2	3	4
P	-2.5	0.5	3.5	7.25

a By plotting a suitable straight line graph, show that P and m are related by an equation of the form $P = am^2 + \frac{b}{m}$, where a and b are constants.

b Find P when $m = 5$.

c Find *all* values of m such that $P = 3.5$.

8

The unit circle and radian measure

Contents:

Opening problem

In the circle alongside, suppose we measure the angle AOB using the ratio of the arc length AB it subtends, to the radius of the circle. We say the angle is $\frac{10}{5} = 2$ *radians*.

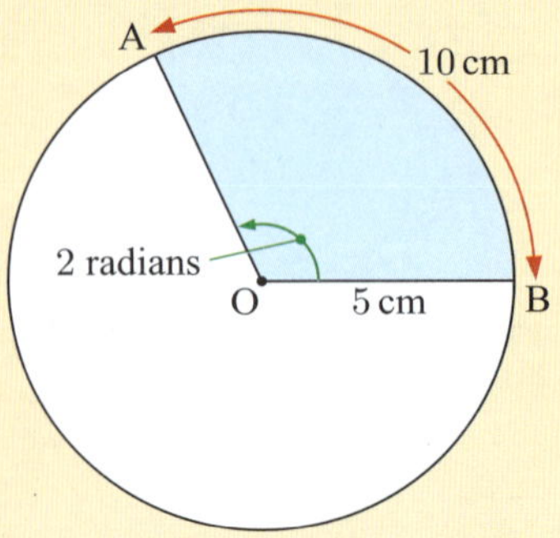

Things to think about:

- **a** Does the angle 2 radians always correspond to the same angle in degrees, or does it depend on the radius of the circle?
- **b** How many radians are in a full circle?
- **c** How could you find the size of angle AOB in degrees?
- **d** Can you find the perimeter and area of the shaded sector?

A RADIAN MEASURE

DEGREE MEASUREMENT OF ANGLES

We have seen previously that one full revolution makes an angle of $360°$, and the angle on a straight line is $180°$.

One **degree**, $1°$, is $\frac{1}{360}$th of one full revolution.

This measure of angle is commonly used by surveyors and architects.

RADIAN MEASUREMENT OF ANGLES

An angle is said to have a measure of one **radian**, 1^c, if it is subtended at the centre of a circle by an arc equal in length to the radius.

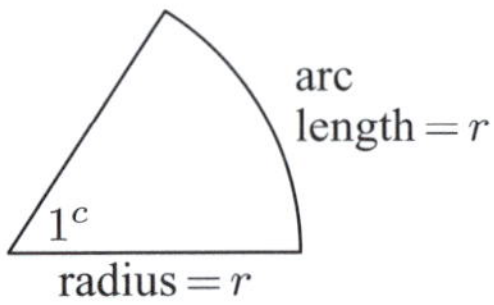

The symbol "*c*" indicates radian measure, but is usually omitted. By contrast, the degree symbol is *always* used when the measure of an angle is given in degrees.

From the diagram below, it can be seen that 1^c is slightly smaller than $60°$. In fact, $1^c \approx 57.3°$.

The word "radian" is an abbreviation of "radial angle".

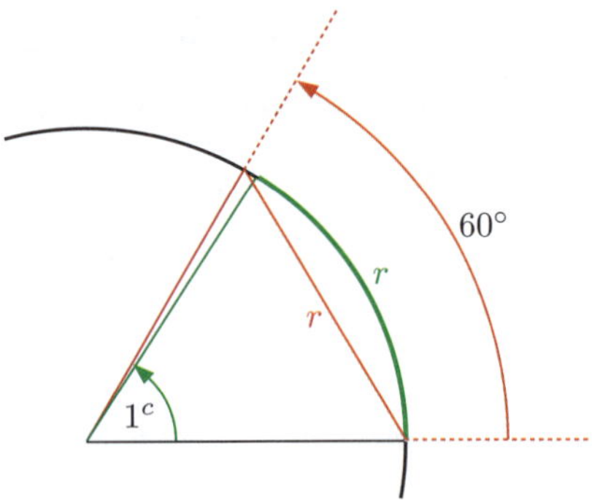

Discussion

A *dimensional unit* is a unit which corresponds to a physical dimension such as length, mass, or time.

Is a radian a dimensional or dimensionless unit?

DEGREE-RADIAN CONVERSIONS

If the radius of a circle is r, then an arc of length πr, or half the circumference, will subtend an angle of π radians.

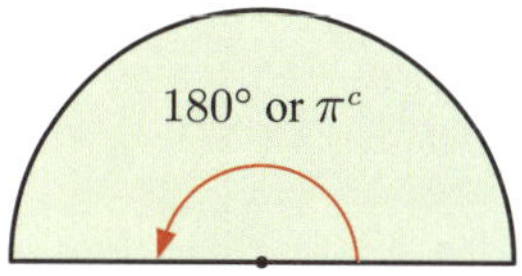

Therefore, π radians $= 180°$.

So, $1^c = \left(\frac{180}{\pi}\right)^\circ \approx 57.3°$ and $1° = \left(\frac{\pi}{180}\right)^c \approx 0.0175^c$.

To convert from degrees to radians, we multiply by $\frac{\pi}{180}$.

To convert from radians to degrees, we multiply by $\frac{180}{\pi}$.

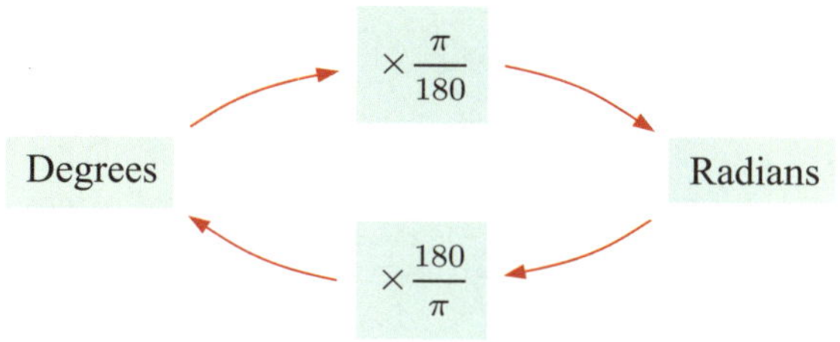

Example 1 — Self Tutor

Convert 30° to radians in terms of π.

$30° = \left(30 \times \frac{\pi}{180}\right)$ radians

$= \frac{\pi}{6}$ radians

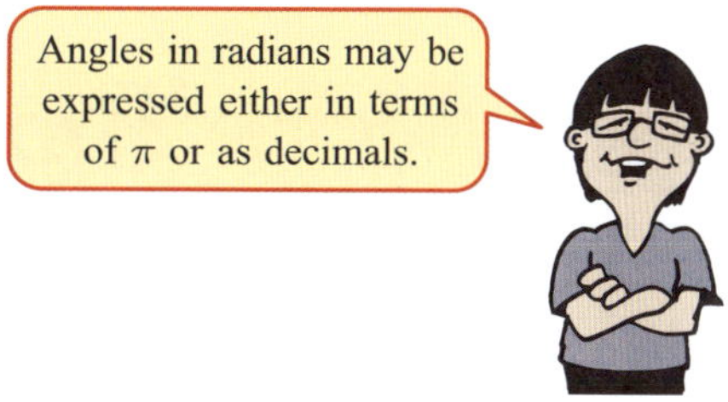

Example 2 — Self Tutor

Convert 73.2° to radians.

73.2°

$= \left(73.2 \times \frac{\pi}{180}\right)$ radians

≈ 1.28 radians

EXERCISE 8A

1 Convert to radians, in terms of π:

a 180° **b** 90° **c** 60° **d** 20° **e** 45°

f 10° **g** 3° **h** 36° **i** 360° **j** 720°

k 270° **l** 150° **m** 72° **n** 120° **o** 100°

2 Convert to radians, correct to 3 significant figures:

a 21.5° **b** 106.1° **c** 302.3° **d** 184.2° **e** 237.5°

Example 3 Self Tutor

Convert the following radian measures to degrees:

a $\frac{5\pi}{4}$ **b** 0.24 radians

a $\frac{5\pi}{4}$

$= \left(\frac{5\pi}{4} \times \frac{180}{\pi}\right)^\circ$

$= 225^\circ$

b 0.24 radians

$= \left(0.24 \times \frac{180}{\pi}\right)^\circ$

$\approx 13.8^\circ$

3 Convert the following radian measures to degrees:

a $\frac{\pi}{2}$ **b** $\frac{\pi}{4}$ **c** $\frac{2\pi}{3}$ **d** $\frac{\pi}{10}$ **e** $\frac{5\pi}{6}$

f $\frac{2\pi}{5}$ **g** 2π **h** $\frac{\pi}{20}$ **i** $\frac{11\pi}{6}$ **j** $\frac{7\pi}{20}$

4 Convert the following radian measures to degrees. Give your answers correct to 2 decimal places.

a 0.5 **b** 3 **c** 0.76 **d** 1.29 **e** 4.13

5 Copy and complete, giving your answers in terms of π:

a

Degrees	0	45	90	135	180	225	270	315	360
Radians									

b

Degrees	0	30	60	90	120	150	180	210	240	270	300	330	360
Radians													

B ARC LENGTH AND SECTOR AREA

The diagram alongside illustrates terms relating to the parts of a circle.

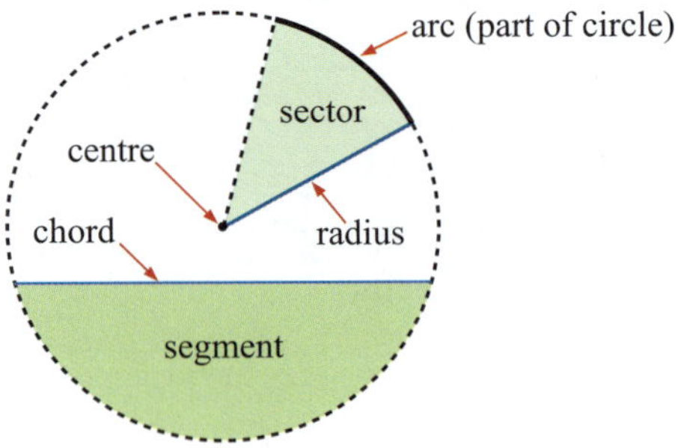

An arc, sector, or segment is described as:

- **minor** if it involves less than half the circle
- **major** if it involves more than half the circle.

For example:

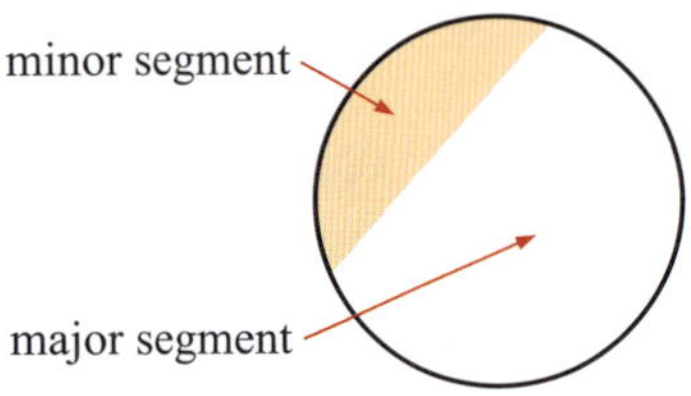

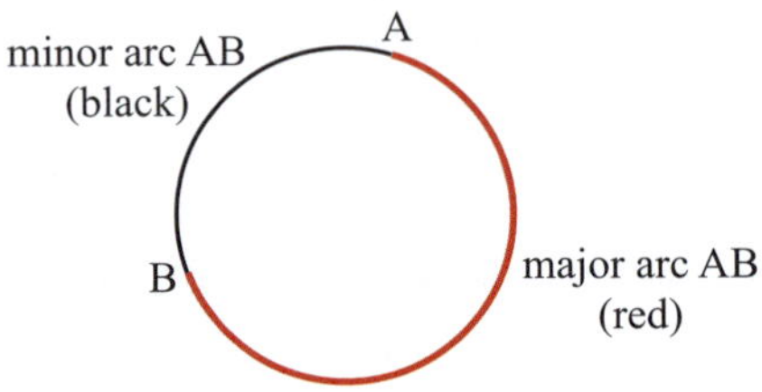

ARC LENGTH

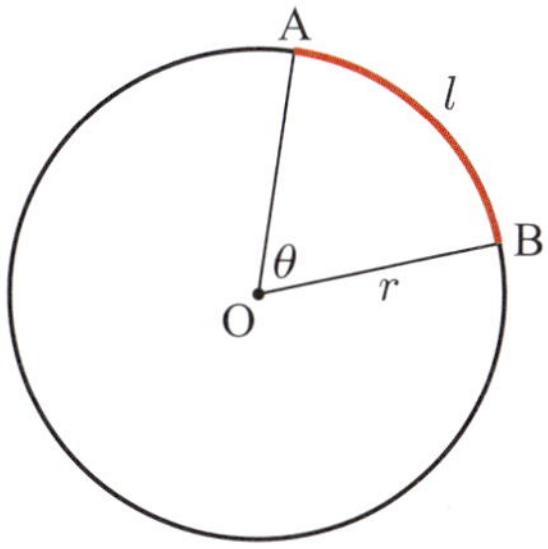

In the diagram, the **arc length** AB is l.

Suppose θ is measured in radians.

There are 2π radians in a circle, so

$$\text{arc length} = \frac{\theta}{2\pi} \times \text{circumference}$$

$$\therefore\ l = \frac{\theta}{2\pi} \times 2\pi r$$

$$\therefore\ l = \theta r$$

For θ in **radians**, arc length $\boldsymbol{l = \theta r}$.

For θ in **degrees**, arc length $\boldsymbol{l = \frac{\theta}{360} \times 2\pi r}$.

AREA OF SECTOR

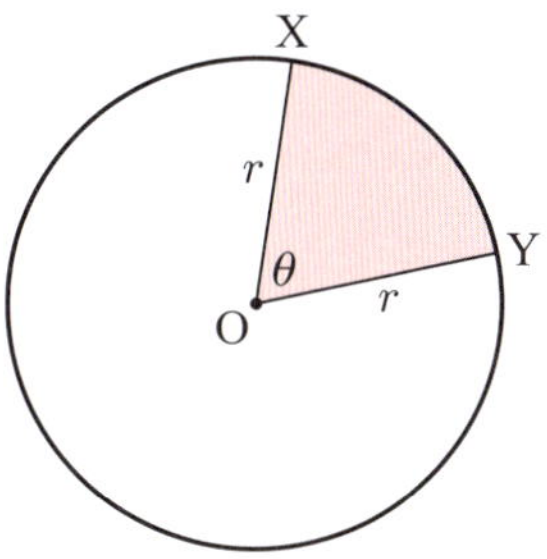

In the diagram, the area of minor sector XOY is shaded.

Suppose θ is measured in radians.

$$\text{area of sector} = \frac{\theta}{2\pi} \times \text{area of circle}$$

$$\therefore\ A = \frac{\theta}{2\pi} \times \pi r^2$$

$$\therefore\ A = \tfrac{1}{2}\theta r^2$$

For θ in **radians**, area of sector $\boldsymbol{A = \frac{1}{2}\theta r^2}$.

For θ in **degrees**, area of sector $\boldsymbol{A = \frac{\theta}{360} \times \pi r^2}$.

Example 4 — Self Tutor

A sector has radius 10 cm and angle 2 radians. Find its:

a arc length **b** area

a arc length $= \theta r$
$= 2 \times 10$
$= 20$ cm

b area $= \frac{1}{2}\theta r^2$
$= \frac{1}{2} \times 2 \times 10^2$
$= 100\ \text{cm}^2$

EXERCISE 8B

1 Find the arc length of each sector:

a

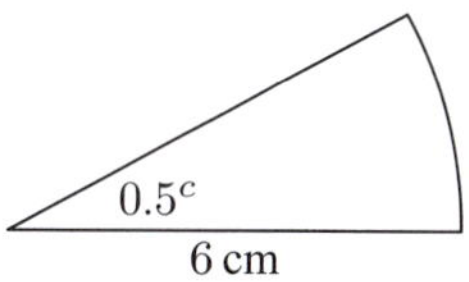

b

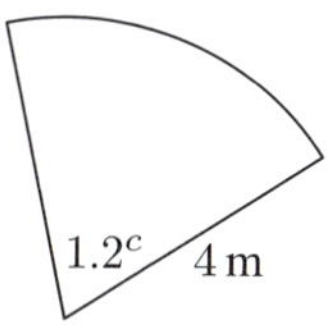

c

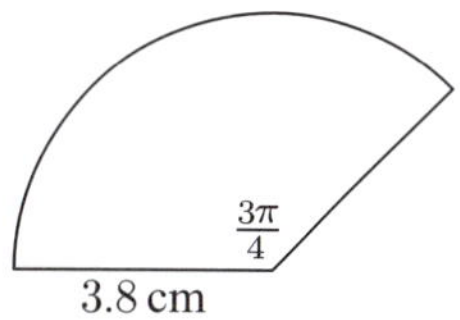

2 Find the area of each sector:

a

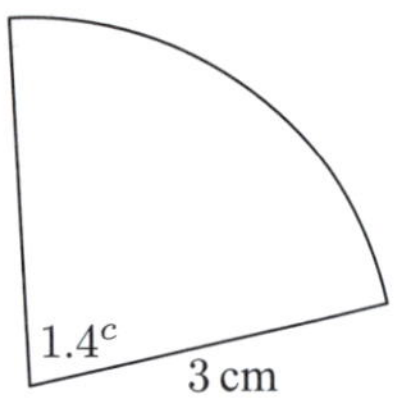

b

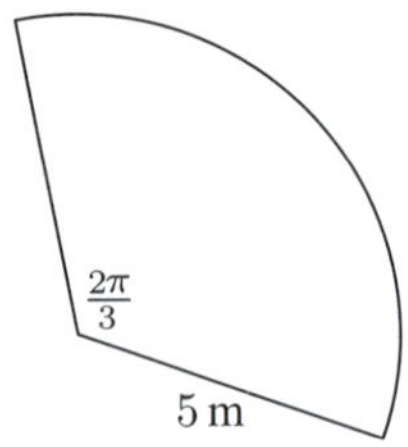

c

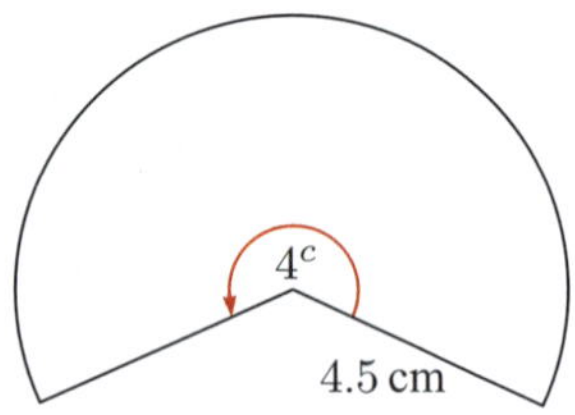

3 For the sector alongside, find the:

a arc length

b perimeter

c area.

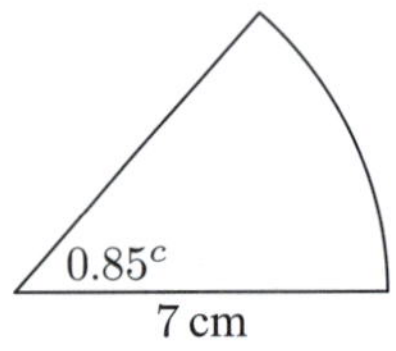

Example 5 — Self Tutor

A sector has radius 4 m and arc length 3 m. Find its:

a angle **b** area.

a $l = \theta r$ {θ in radians}

$\therefore \theta = \frac{l}{r} = \frac{3}{4} = 0.75$ radians

b $\text{area} = \frac{1}{2}\theta r^2$

$= \frac{1}{2} \times 0.75 \times 4^2$

$= 6 \text{ m}^2$

4 A sector has radius 5 cm and arc length 6 cm. Find its:

a angle **b** area.

5 Find θ (in radians) and hence find the area of the figure:

a

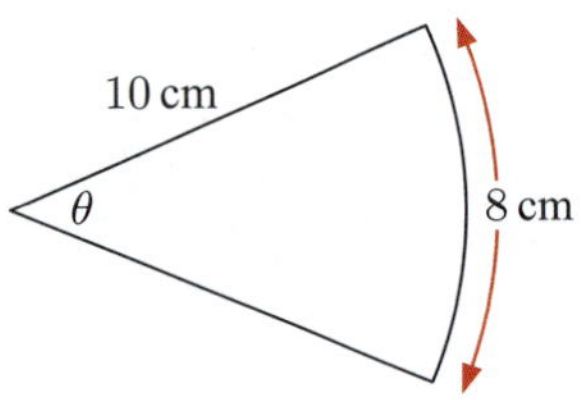

b

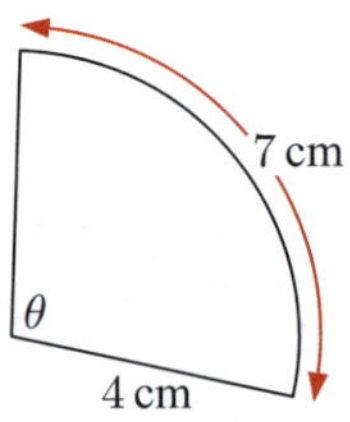

c

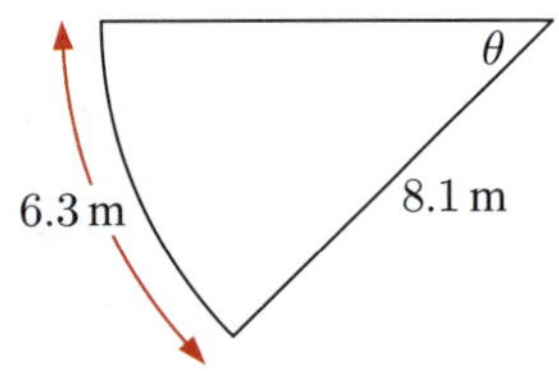

6 A sector has angle 75° and arc length 8 m. Find its:

a radius **b** area.

7 A sector has angle 2.1^c and area 30 cm^2. Find its:

a radius **b** arc length **c** perimeter.

8 **a** Find the area of sector ABC.

b Use trigonometry to find the area of triangle ABC.

c Hence find the area of the shaded region.

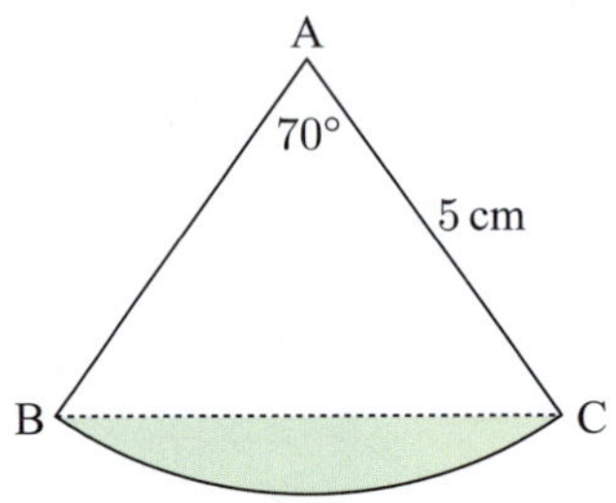

9

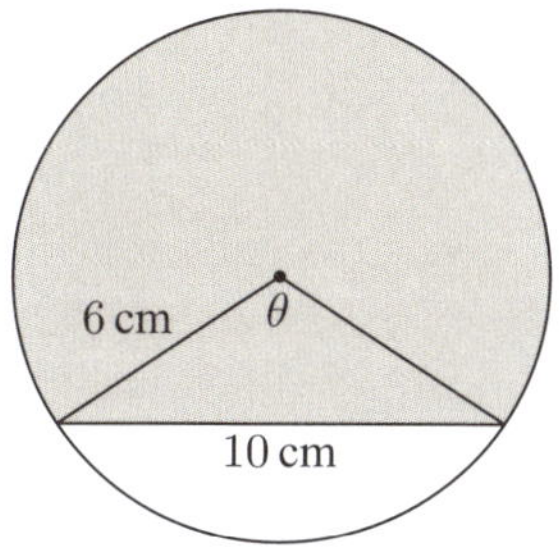

Find:

a the value of θ in:

  i degrees **ii** radians

b the perimeter of the shaded region

c the area of the shaded region.

10 The touching circles alongside have centres P and Q. Each circle has radius 4 cm. The line segment QR is tangential to the circle with centre P.

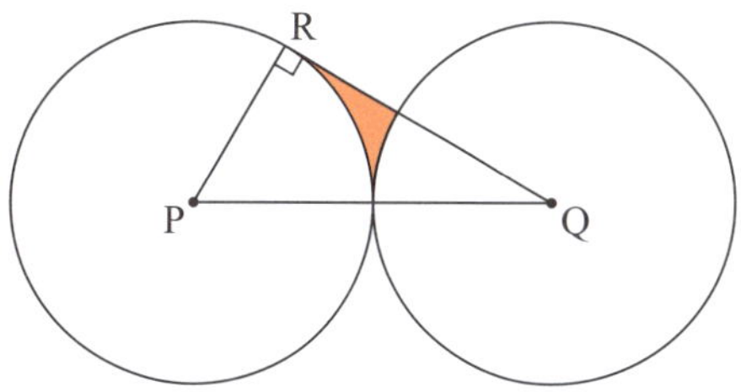

a Find $P\widehat{Q}R$ and $Q\widehat{P}R$.

b Find the perimeter and area of the shaded region.

C THE UNIT CIRCLE AND THE TRIGONOMETRIC RATIOS

The **unit circle** is the circle with centre $(0, 0)$ and radius 1 unit.

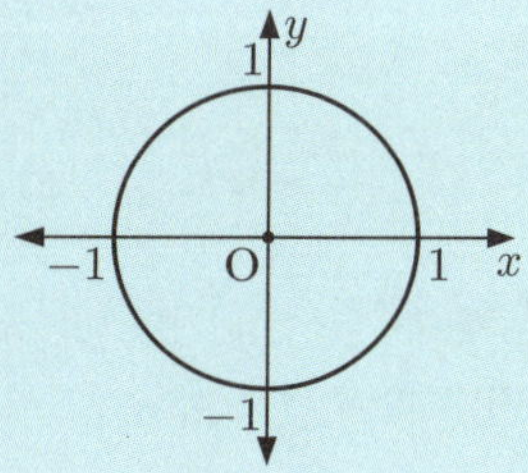

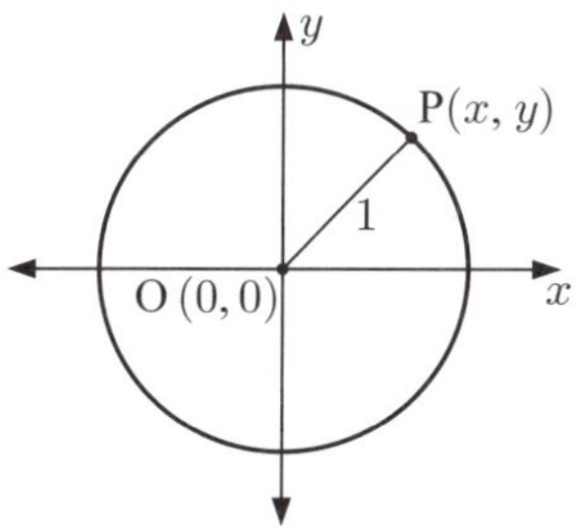

Suppose $P(x, y)$ is any point on the unit circle.

Since $OP = 1$,

$\sqrt{(x-0)^2 + (y-0)^2} = 1$ {distance formula}

$\therefore\ x^2 + y^2 = 1^2$

The equation of the **unit circle** is $x^2 + y^2 = 1$.

ANGLE MEASUREMENT

Suppose P lies anywhere on the unit circle, and A is $(1, 0)$. Let θ be the angle measured from OA on the positive x-axis.

θ is **positive** for anticlockwise rotations and **negative** for clockwise rotations.

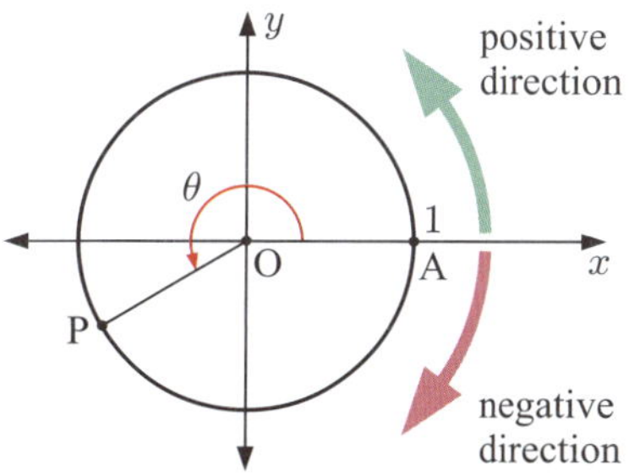

For example: $\theta = 225° = \frac{5\pi}{4}$

$\phi = -135° = -\frac{3\pi}{4}$

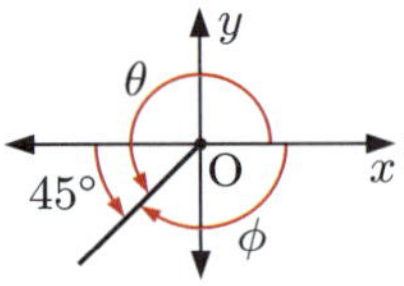

DEFINITION OF SINE AND COSINE

Consider a point P(a, b) which lies on the unit circle in the first quadrant. Suppose OP makes an angle θ with the x-axis as shown.

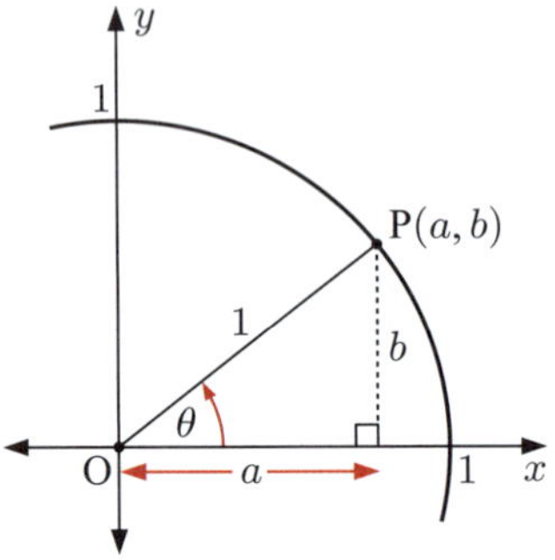

Using right angled triangle trigonometry:

$$\cos\theta = \frac{\text{ADJ}}{\text{HYP}} = \frac{a}{1} = a$$

$$\sin\theta = \frac{\text{OPP}}{\text{HYP}} = \frac{b}{1} = b$$

$$\tan\theta = \frac{\text{OPP}}{\text{ADJ}} = \frac{b}{a} = \frac{\sin\theta}{\cos\theta}$$

More generally:

If P is any point on the unit circle such that OP makes an angle θ measured from the positive x-axis:

- $\boldsymbol{\cos\theta}$ is the x-coordinate of P
- $\boldsymbol{\sin\theta}$ is the y-coordinate of P.

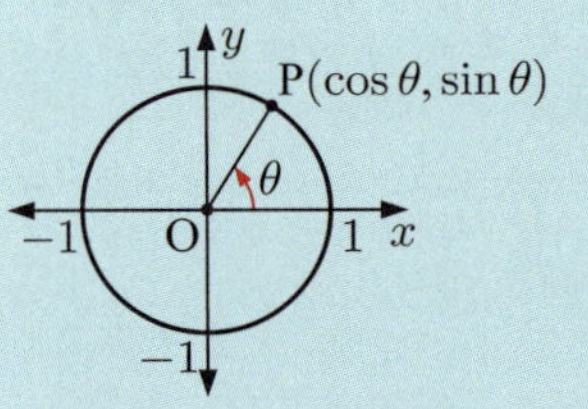

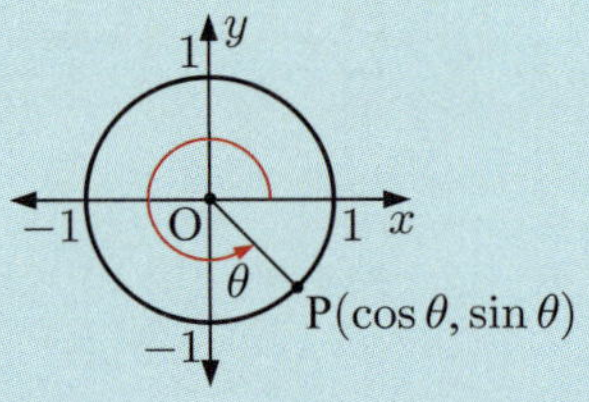

We can use these definitions to find the coordinates of any point on the unit circle with given angle θ measured from the positive x-axis.

For example:

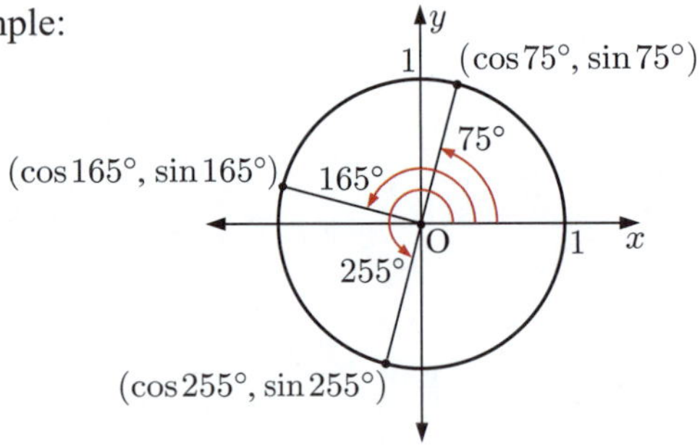

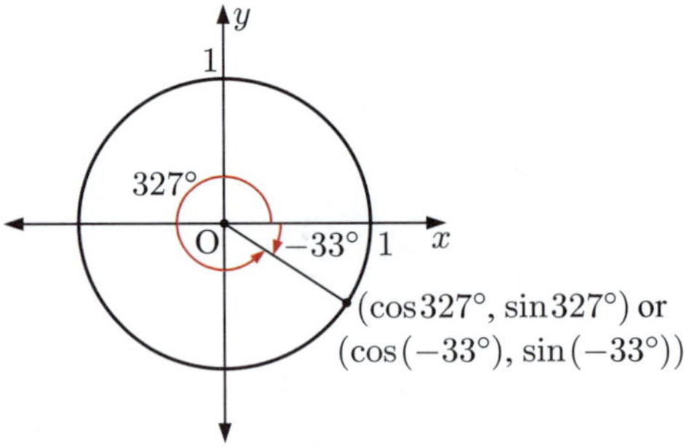

Since the unit circle has equation $x^2 + y^2 = 1$, $(\cos\theta)^2 + (\sin\theta)^2 = 1$ for all θ.

We commonly write this as $\boldsymbol{\cos^2\theta + \sin^2\theta = 1}$.

$\cos^2\theta + \sin^2\theta = 1$ is called the **Pythagorean identity**.

For all points on the unit circle, $-1 \leqslant x \leqslant 1$ and $-1 \leqslant y \leqslant 1$.

We therefore conclude that:

$\boldsymbol{-1 \leqslant \cos\theta \leqslant 1}$ and $\boldsymbol{-1 \leqslant \sin\theta \leqslant 1}$ for all θ.

DEFINITION OF TANGENT

Suppose we extend OP to meet the tangent from A(1, 0).

We let the intersection between these lines be point Q.

Note that as P moves, so does Q.

The position of Q relative to A is defined as the **tangent function**.

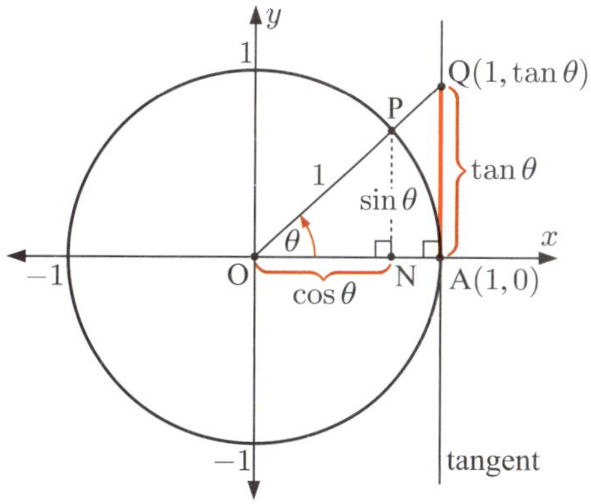

Notice that triangles ONP and OAQ are equiangular and therefore similar.

Consequently $\frac{AQ}{OA} = \frac{NP}{ON}$ and hence $\frac{AQ}{1} = \frac{\sin\theta}{\cos\theta}$.

Under the definition that $AQ = \tan\theta$, $\boldsymbol{\tan\theta = \frac{\sin\theta}{\cos\theta}}$.

Notice that the gradient of $OP = \frac{y\text{-step}}{x\text{-step}} = \frac{\sin\theta}{\cos\theta} = \tan\theta$.

For any point P on the unit circle, $\tan\theta$ is the **gradient** of OP.

Discovery 1 — The trigonometric ratios

In this Discovery we explore the signs of the trigonometric ratios in each quadrant of the unit circle.

What to do:

1 Click on the icon to run the software.
Drag the point P slowly around the circle.
Note the *sign* of each trigonometric ratio in each quadrant.

THE UNIT CIRCLE

Quadrant	$\cos\theta$	$\sin\theta$	$\tan\theta$
1	positive		
2			
3			
4			

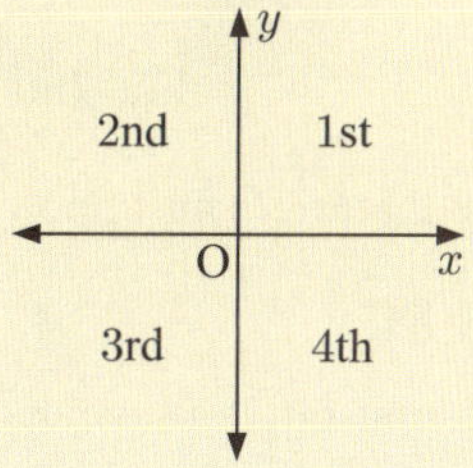

2 Hence write down the trigonometric ratios which are *positive* in each quadrant.

From the **Discovery** you should have found that:

- $\sin\theta$, $\cos\theta$, and $\tan\theta$ are positive in quadrant 1
- only $\sin\theta$ is positive in quadrant 2
- only $\tan\theta$ is positive in quadrant 3
- only $\cos\theta$ is positive in quadrant 4.

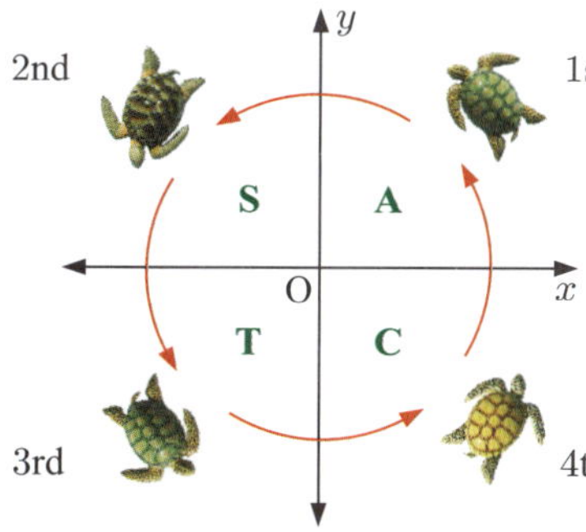

We can use a letter to show which trigonometric ratios are positive in each quadrant. The A stands for *all* of the ratios.

You might like to remember them using

All **S**illy **T**urtles **C**rawl.

Example 6

Self Tutor

Use a unit circle diagram to find the values of $\cos(-270°)$ and $\sin(-270°)$.

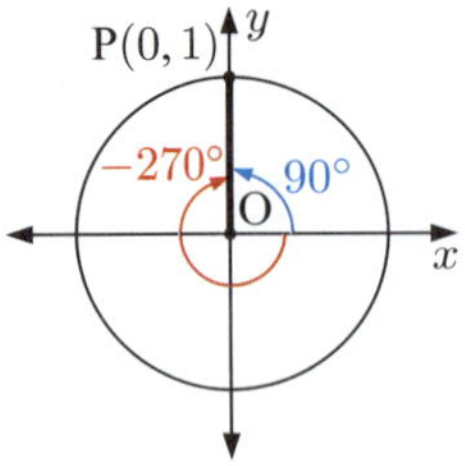

$\cos(-270°) = 0$ {the x-coordinate of P}
$\sin(-270°) = 1$ {the y-coordinate of P}

PERIODICITY OF TRIGONOMETRIC RATIOS

Since there are 2π radians in a full revolution, if we add any integer multiple of 2π to θ (in radians) then the position of P on the unit circle is unchanged.

> For θ in radians and $k \in \mathbb{Z}$,
> $\mathbf{\cos(\theta + 2k\pi) = \cos\theta}$ and $\mathbf{\sin(\theta + 2k\pi) = \sin\theta}$.

We notice that for any point $(\cos\theta, \sin\theta)$ on the unit circle, the point directly opposite is $(-\cos\theta, -\sin\theta)$

$\therefore \quad \cos(\theta + \pi) = -\cos\theta$

$\sin(\theta + \pi) = -\sin\theta$

and $\tan(\theta + \pi) = \dfrac{-\sin\theta}{-\cos\theta} = \dfrac{\sin\theta}{\cos\theta} = \tan\theta$

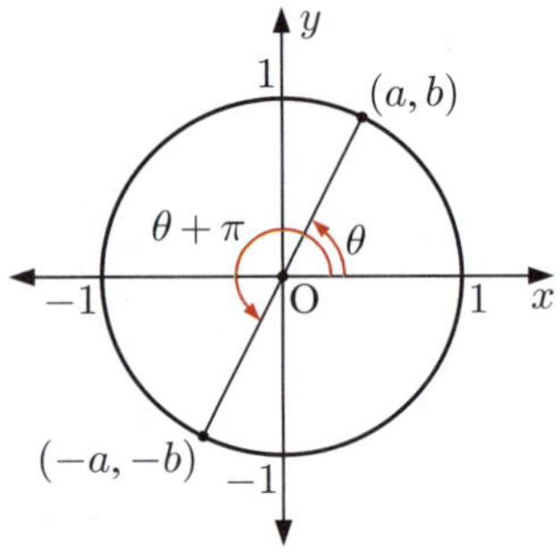

> For θ in radians and $k \in \mathbb{Z}$, $\mathbf{\tan(\theta + k\pi) = \tan\theta}$.

This **periodic** feature is an important property of the trigonometric ratios.

EXERCISE 8C

1 For each unit circle illustrated:

- **i** state the exact coordinates of points P, Q, and R in terms of sine and cosine
- **ii** use your calculator to find the coordinates of P, Q, and R correct to 3 significant figures.

a

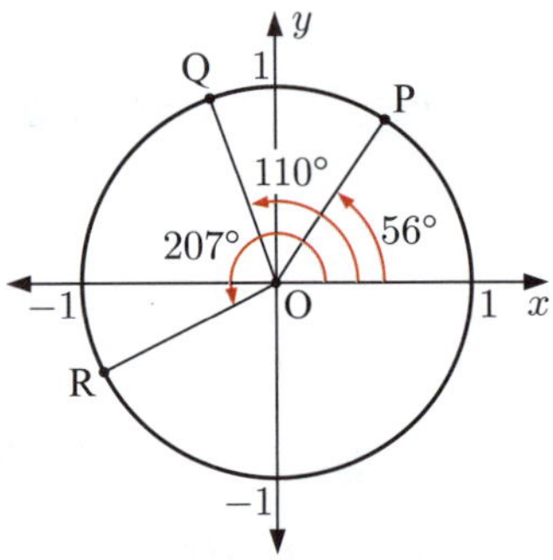

b

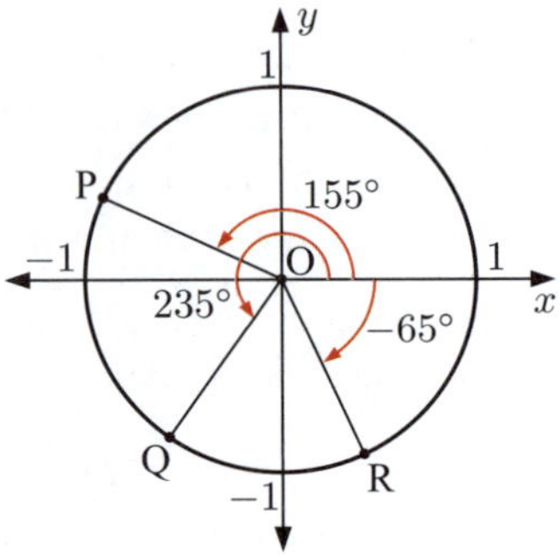

2 **a** Copy and complete:

Quadrant	*Degree measure*	*Radian measure*	$\cos\theta$	$\sin\theta$	$\tan\theta$
1	$0° < \theta < 90°$	$0 < \theta < \frac{\pi}{2}$	positive	positive	
2					
3					
4					

b In which quadrants are the following true?

i $\cos\theta$ is positive.

ii $\cos\theta$ is negative.

iii $\cos\theta$ and $\sin\theta$ are both negative.

iv $\cos\theta$ is negative and $\sin\theta$ is positive.

3 With the aid of a unit circle, complete the table alongside:

θ (degrees)	0°	90°	180°	270°	360°	450°
θ (radians)						
sine						
cosine						
tangent						

4 **a** Use your calculator to evaluate: **i** $\frac{1}{\sqrt{2}}$ **ii** $\frac{\sqrt{3}}{2}$ **iii** $\sqrt{3}$ **iv** $\frac{1}{\sqrt{3}}$

b Copy and complete the following table. If necessary, use your calculator to evaluate the trigonometric ratios, then **a** to write them exactly.

θ (degrees)	30°	45°	60°	135°	150°	240°	315°
θ (radians)							
sine							
cosine							
tangent							

5 **a** Use your calculator to evaluate:

i $\sin 70°$ **ii** $\sin 110°$ **iii** $\sin 20°$ **iv** $\sin 160°$

v $\sin 50°$ **vi** $\sin 130°$ **vii** $\sin 35°$ **viii** $\sin 145°$

b Use the results from **a** to copy and complete:

$\sin(180° - \theta) = ...$

c Justify your answer using the diagram alongside:

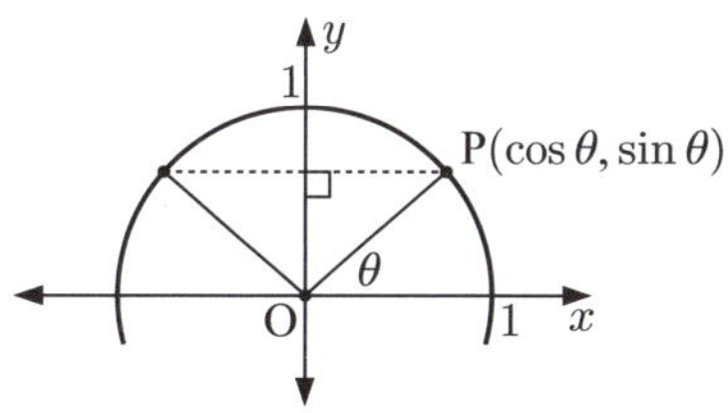

d Find the obtuse angle with the same sine as:

i 10° **ii** 62° **iii** $\frac{\pi}{3}$ **iv** $\frac{\pi}{6}$

6 **a** Use your calculator to evaluate:

i $\cos 40°$ **ii** $\cos 140°$ **iii** $\cos 170°$ **iv** $\cos 10°$

v $\cos 30°$ **vi** $\cos 150°$ **vii** $\cos 65°$ **viii** $\cos 115°$

b Use the results from **a** to copy and complete:

$\cos(180° - \theta) = ...$

c Justify your answer using the diagram alongside:

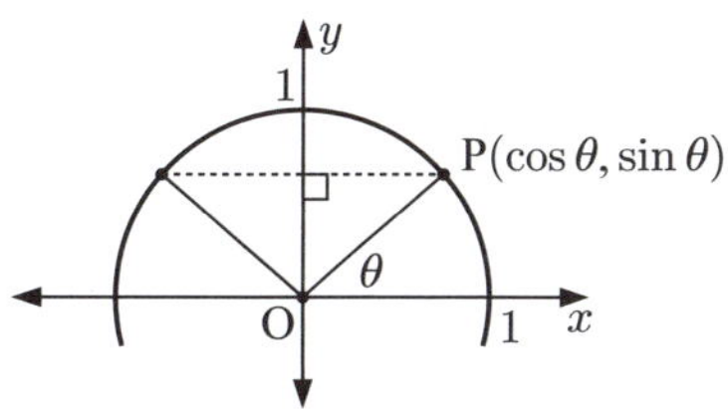

d Find the obtuse angle which has the negative cosine of:

i 20° **ii** 84° **iii** $\frac{\pi}{5}$ **iv** $\frac{2\pi}{5}$

7 Without using your calculator, evaluate:

a $\sin 108°$ given $\sin 72° \approx 0.9511$
b $\sin 34°$ given $\sin 146° \approx 0.5592$
c $\cos 99°$ given $\cos 81° \approx 0.156$
d $\cos 49°$ given $\cos 131° \approx -0.656$
e $\sin 27°$ given $\sin 153° \approx 0.454$
f $\cos 166°$ given $\cos 14° \approx 0.970$

8 **a** Copy and complete:

θ (radians)	$\sin\theta$	$\sin(-\theta)$	$\cos\theta$	$\cos(-\theta)$
0.62				
1.403				
4.283				
5.901				
-2.42				

b What trigonometric formulae can be deduced from your results in **a**?

c Justify your answer using the diagram alongside:

d Hence explain why:

i $\cos(2\pi - \theta) = \cos\theta$

ii $\sin(2\pi - \theta) = -\sin\theta$

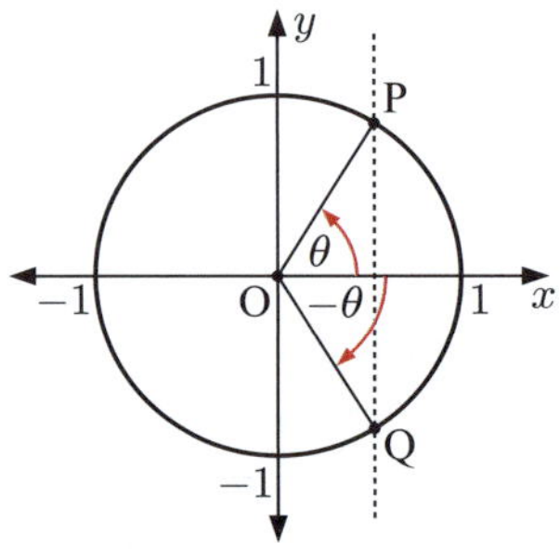

D THE MULTIPLES OF $\frac{\pi}{6}$ AND $\frac{\pi}{4}$

Angles which are multiples of $\frac{\pi}{6}$ and $\frac{\pi}{4}$ occur frequently, so it is important for us to write their trigonometric ratios exactly.

MULTIPLES OF $\frac{\pi}{4}$ OR 45°

Suppose $\mathrm{B\widehat{O}P} = 45°$. Angle OPB also measures 45°, so triangle OBP is isosceles.

Letting $\mathrm{OB} = \mathrm{BP} = a$,

$$a^2 + a^2 = 1^2 \quad \{\text{Pythagoras}\}$$
$$\therefore\ 2a^2 = 1$$
$$\therefore\ a^2 = \tfrac{1}{2}$$
$$\therefore\ a = \tfrac{1}{\sqrt{2}} \quad \{\text{as } a > 0\}$$

So, P is $\left(\frac{1}{\sqrt{2}}, \frac{1}{\sqrt{2}}\right)$ where $\frac{1}{\sqrt{2}} \approx 0.707$.

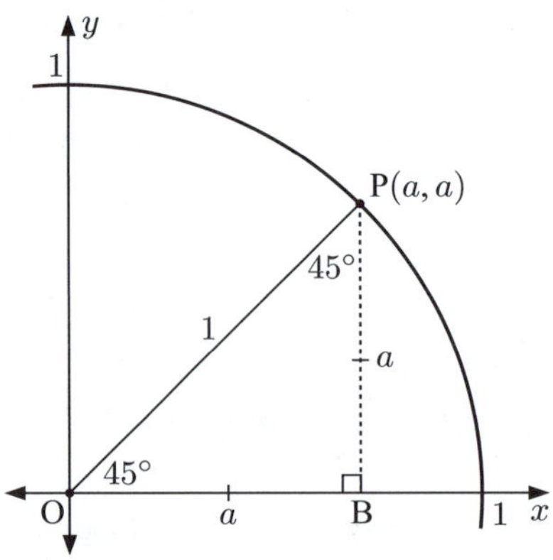

$$\cos\frac{\pi}{4} = \frac{1}{\sqrt{2}} \quad \text{and} \quad \sin\frac{\pi}{4} = \frac{1}{\sqrt{2}}$$

You should remember these values. If you forget, draw a right angled isosceles triangle with equal sides of length 1.

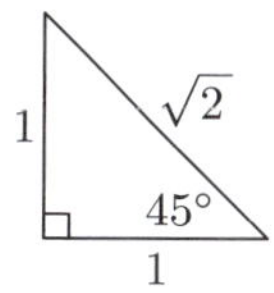

For **multiples of $\frac{\pi}{4}$**, we have:

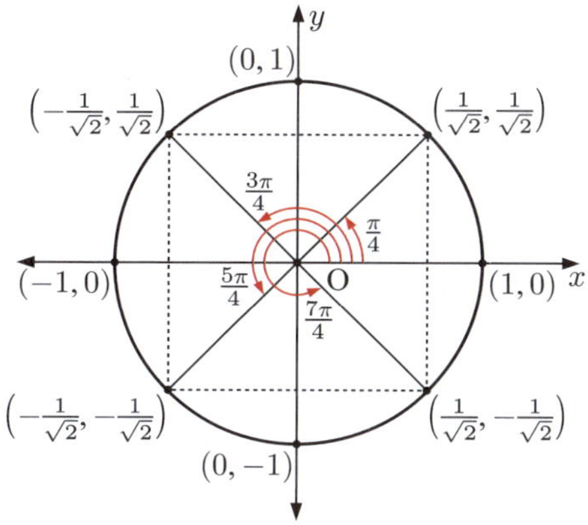

MULTIPLES OF $\frac{\pi}{6}$ OR 30°

Suppose $\widehat{AOP} = 60°$.

Since OA = OP, triangle OAP is isosceles.

The remaining angles are therefore also 60°, and so triangle AOP is equilateral.

The altitude PN bisects base OA, so $ON = \frac{1}{2}$.

If P is $(\frac{1}{2}, k)$, then $(\frac{1}{2})^2 + k^2 = 1^2$ {Pythagoras}

$\therefore\ k^2 = \frac{3}{4}$

$\therefore\ k = \frac{\sqrt{3}}{2}$ {as $k > 0$}

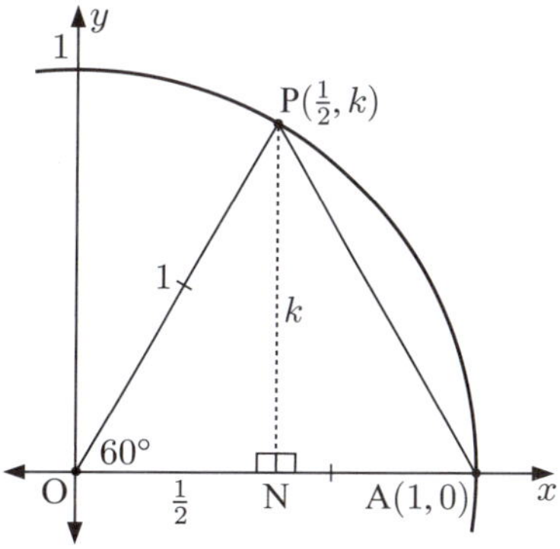

So, P is $(\frac{1}{2}, \frac{\sqrt{3}}{2})$ where $\frac{\sqrt{3}}{2} \approx 0.866$.

$$\cos\frac{\pi}{3} = \frac{1}{2} \quad \text{and} \quad \sin\frac{\pi}{3} = \frac{\sqrt{3}}{2}$$

Now $\widehat{NPO} = \frac{\pi}{6} = 30°$.

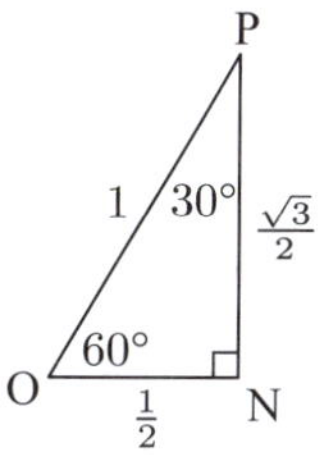

Hence $$\cos\frac{\pi}{6} = \frac{\sqrt{3}}{2} \quad \text{and} \quad \sin\frac{\pi}{6} = \frac{1}{2}$$

You should remember these values. If you forget, draw an equilateral triangle with side length 2.

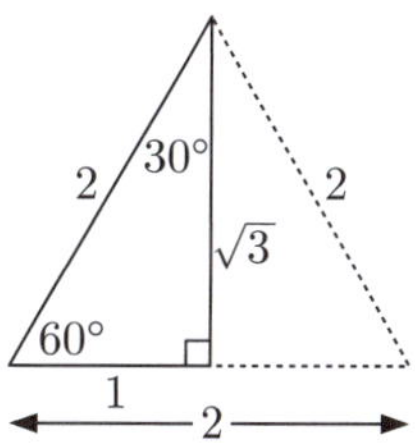

For **multiples of $\frac{\pi}{6}$**, we have:

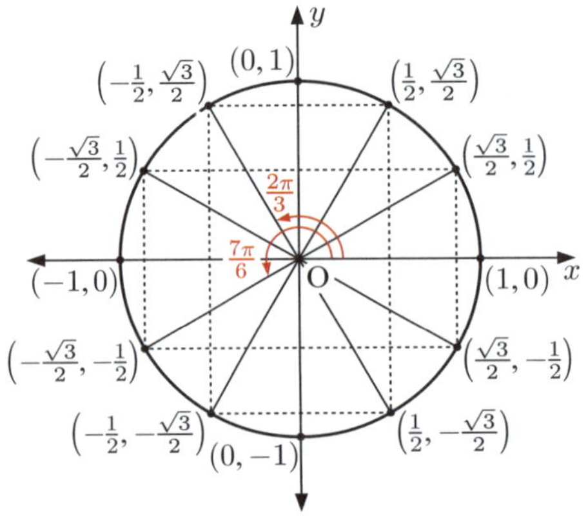

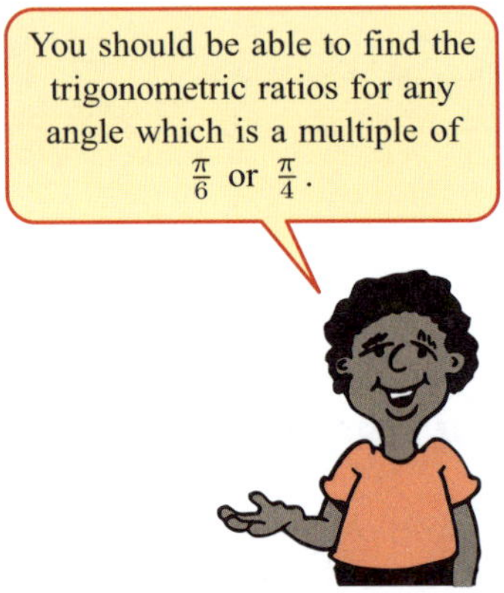

Summary

- For **multiples of $\frac{\pi}{2}$**, the coordinates of the points on the unit circle involve 0 and ± 1.
- For *other* **multiples of $\frac{\pi}{4}$**, the coordinates involve $\pm\frac{1}{\sqrt{2}}$.
- For *other* **multiples of $\frac{\pi}{6}$**, the coordinates involve $\pm\frac{1}{2}$ and $\pm\frac{\sqrt{3}}{2}$.
- The signs of the coordinates are determined by which quadrant the angle is in.

Example 7

Self Tutor

Find the exact values of $\sin\alpha$, $\cos\alpha$, and $\tan\alpha$ for: **a** $\alpha = \frac{5\pi}{4}$ **b** $\alpha = \frac{2\pi}{3}$

a $\frac{5\pi}{4}$ is a multiple of $\frac{\pi}{4}$.
The angle lies in quadrant 3,
so only $\tan\frac{5\pi}{4}$ is positive.

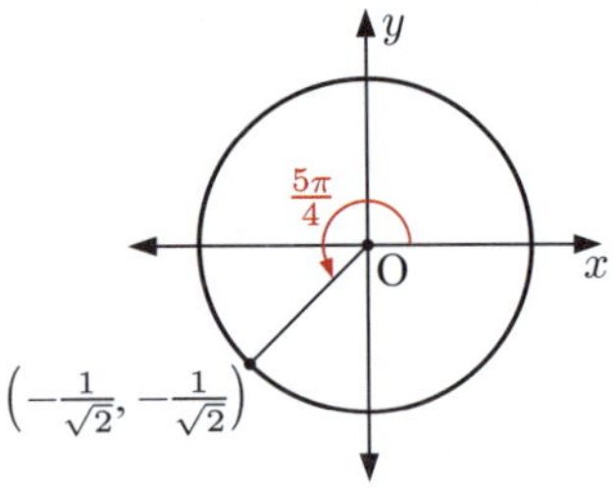

$\sin\frac{5\pi}{4} = -\frac{1}{\sqrt{2}}$

$\cos\frac{5\pi}{4} = -\frac{1}{\sqrt{2}}$

$\tan\frac{5\pi}{4} = 1$

b $\frac{2\pi}{3}$ is a multiple of $\frac{\pi}{6}$.
The angle lies in quadrant 2,
so only $\sin\frac{2\pi}{3}$ is positive.

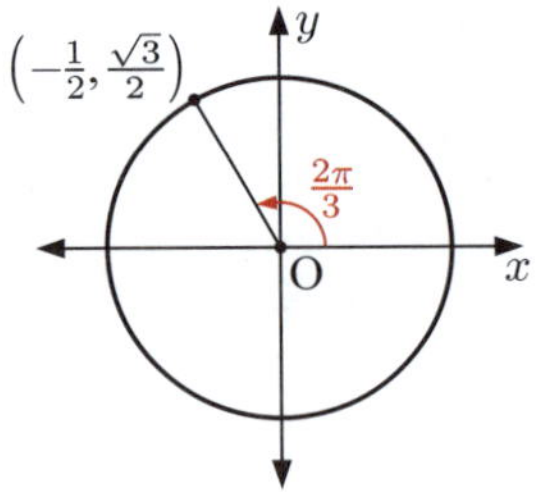

$\sin\frac{2\pi}{3} = \frac{\sqrt{3}}{2}$

$\cos\frac{2\pi}{3} = -\frac{1}{2}$

$\tan\frac{2\pi}{3} = \frac{\frac{\sqrt{3}}{2}}{-\frac{1}{2}} = -\sqrt{3}$

EXERCISE 8D

1 Use a unit circle diagram to find exact values for $\sin\theta$, $\cos\theta$, and $\tan\theta$, for θ equal to:

a $\frac{3\pi}{4}$ **b** $\frac{\pi}{2}$ **c** $\frac{7\pi}{4}$ **d** $\frac{3\pi}{2}$ **e** $-\frac{\pi}{4}$

2 Use a unit circle diagram to find exact values for $\sin\beta$, $\cos\beta$, and $\tan\beta$, for β equal to:

a $\frac{\pi}{3}$ **b** $\frac{4\pi}{3}$ **c** $\frac{5\pi}{6}$ **d** $\frac{7\pi}{6}$ **e** $-\frac{\pi}{3}$

3 Find the exact values of:

a $\cos 150°$, $\sin 150°$, and $\tan 150°$ **b** $\cos(-135°)$, $\sin(-135°)$, and $\tan(-135°)$

4 **a** Find the exact values of $\cos 270°$ and $\sin 270°$.

b What can you say about $\tan 270°$?

Example 8 — Self Tutor

Without using a calculator, evaluate $12\cos\frac{\pi}{3}\sin\frac{5\pi}{6}$.

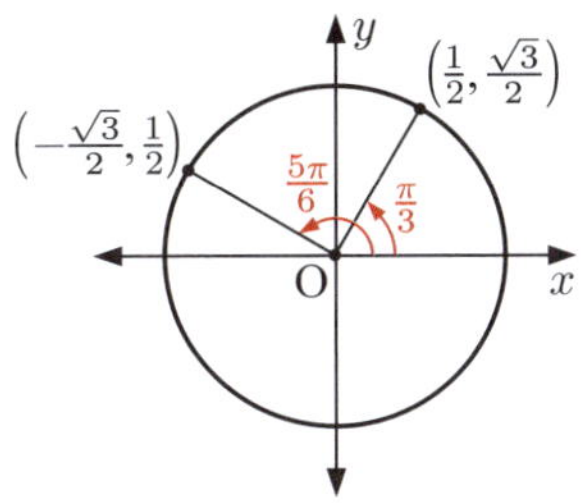

$\cos\frac{\pi}{3} = \frac{1}{2}$ and $\sin\frac{5\pi}{6} = \frac{1}{2}$

$\therefore\ 12\cos\frac{\pi}{3}\sin\frac{5\pi}{6} = 12\left(\frac{1}{2}\right)\left(\frac{1}{2}\right)$
$= 3$

5 Without using a calculator, evaluate:

a $\sin^2 30°$ **b** $\cos 30° \sin 60°$

c $8\sin 45° \cos 45°$ **d** $1 - \sin^2\left(\frac{\pi}{3}\right)$

e $\cos^2\left(\frac{3\pi}{4}\right) - 2$ **f** $\cos^2\left(\frac{\pi}{6}\right) + \tan\frac{\pi}{4}$

g $\cos\frac{7\pi}{6} - \sin\frac{2\pi}{3}$ **h** $5 - 2\sin^2\left(\frac{\pi}{2}\right)$

i $\sin^2\left(\frac{5\pi}{6}\right) - \cos^2\left(\frac{5\pi}{6}\right)$ **j** $\tan^2\left(\frac{2\pi}{3}\right) - \sin^2\left(\frac{\pi}{6}\right)$

k $2\tan\frac{7\pi}{4} - \sin\pi$ **l** $\dfrac{\tan 30°}{\cos 210°}$

Check your answers using your calculator.

Example 9 — Self Tutor

Find all angles $0 \leqslant \theta \leqslant 2\pi$ with a cosine of $\frac{1}{2}$.

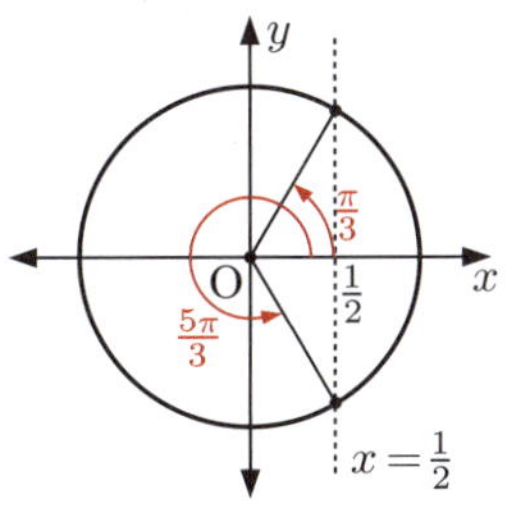

Since the cosine is $\frac{1}{2}$, we draw the vertical line $x = \frac{1}{2}$.

Because $\frac{1}{2}$ is involved, we know the required angles are multiples of $\frac{\pi}{6}$.

They are $\frac{\pi}{3}$ and $\frac{5\pi}{3}$.

6 Find all angles between $0°$ and $360°$ with:

a a sine of $\frac{1}{2}$ **b** a sine of $\frac{\sqrt{3}}{2}$ **c** a cosine of $\frac{1}{\sqrt{2}}$

d a cosine of $-\frac{1}{2}$ **e** a cosine of $-\frac{1}{\sqrt{2}}$ **f** a sine of $-\frac{\sqrt{3}}{2}$

7 Find all angles between 0 and 2π (inclusive) which have:

a a tangent of 1 **b** a tangent of -1 **c** a tangent of $\sqrt{3}$

d a tangent of 0 **e** a tangent of $\frac{1}{\sqrt{3}}$ **f** a tangent of $-\sqrt{3}$

8 Find all angles between 0 and 4π with:

a a cosine of $\frac{\sqrt{3}}{2}$ **b** a sine of $-\frac{1}{2}$ **c** a sine of -1

9 Find *all* values of θ for which $\tan\theta$ is:

a zero **b** undefined.

10 Find the angle between 0 and 2π which has:

a a cosine of $\frac{1}{2}$ and a sine of $\frac{\sqrt{3}}{2}$

b a cosine of $\frac{1}{\sqrt{2}}$ and a sine of $-\frac{1}{\sqrt{2}}$

c a sine of 1 and a cosine of 0

d a sine of $-\frac{1}{2}$ and a cosine of $-\frac{\sqrt{3}}{2}$

e a tangent of $\sqrt{3}$ and a cosine of $-\frac{1}{2}$.

E FINDING TRIGONOMETRIC RATIOS

We can use the identity $\cos^2\theta + \sin^2\theta = 1$ to find unknown trigonometric ratios.

Example 10 — Self Tutor

If $\cos\theta = \frac{3}{4}$, find the possible values of $\sin\theta$. Illustrate your answers.

$$\cos^2\theta + \sin^2\theta = 1$$

$$\therefore \left(\tfrac{3}{4}\right)^2 + \sin^2\theta = 1$$

$$\therefore \sin^2\theta = \tfrac{7}{16}$$

$$\therefore \sin\theta = \pm\tfrac{\sqrt{7}}{4}$$

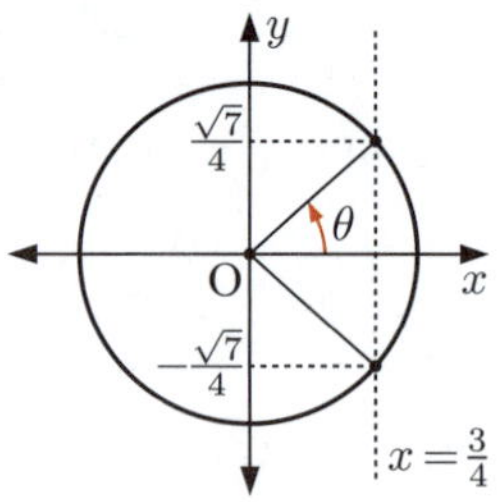

EXERCISE 8E

1 Find the possible values of $\sin\theta$ given:

a $\cos\theta = \frac{1}{2}$ **b** $\cos\theta = \frac{2}{3}$ **c** $\cos\theta = 1$ **d** $\cos\theta = -\frac{3}{5}$

2 Find the possible values of $\cos\theta$ given:

a $\sin\theta = \frac{1}{3}$ **b** $\sin\theta = -\frac{2}{5}$ **c** $\sin\theta = 0$ **d** $\sin\theta = -1$

Example 11 — Self Tutor

If $\sin\theta = -\frac{1}{3}$ and $\frac{3\pi}{2} < \theta < 2\pi$, find exact values of $\cos\theta$ and $\tan\theta$.

Now $\cos^2\theta + \sin^2\theta = 1$

$\therefore \cos^2\theta + \frac{1}{9} = 1$

$\therefore \cos^2\theta = \frac{8}{9}$

$\therefore \cos\theta = \pm\frac{2\sqrt{2}}{3}$

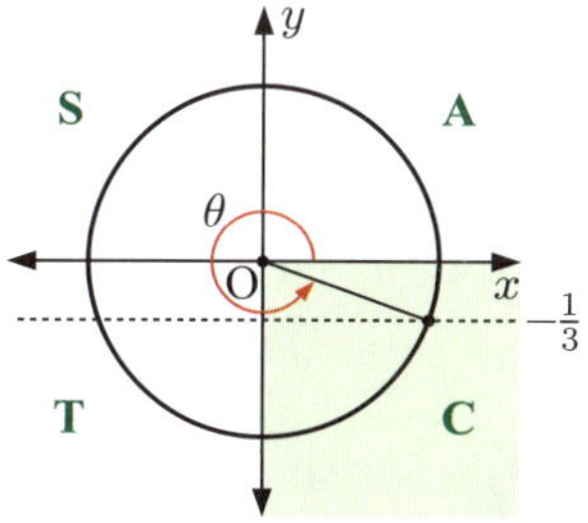

But $\frac{3\pi}{2} < \theta < 2\pi$, so θ lies in quadrant 4.

$\therefore \cos\theta$ is positive.

$\therefore \cos\theta = \frac{2\sqrt{2}}{3}$ and $\tan\theta = \frac{\sin\theta}{\cos\theta} = \frac{-\frac{1}{3}}{\frac{2\sqrt{2}}{3}} = -\frac{1}{2\sqrt{2}}$

3 Find the exact value of:

a $\sin\theta$ if $\cos\theta = \frac{1}{4}$ and $0 < \theta < \frac{\pi}{2}$

b $\cos\theta$ if $\sin\theta = \frac{5}{6}$ and $\frac{\pi}{2} < \theta < \pi$

c $\cos\theta$ if $\sin\theta = -\frac{4}{5}$ and $\frac{3\pi}{2} < \theta < 2\pi$

d $\sin\theta$ if $\cos\theta = -\frac{2}{7}$ and $\pi < \theta < \frac{3\pi}{2}$.

4 Find the exact value of $\tan\theta$ given that:

a $\sin\theta = \frac{2}{3}$ and $\frac{\pi}{2} < \theta < \pi$

b $\cos\theta = \frac{2}{5}$ and $\frac{3\pi}{2} < \theta < 2\pi$

c $\sin\theta = -\frac{1}{\sqrt{5}}$ and $\pi < \theta < \frac{3\pi}{2}$

d $\cos\theta = -\frac{4}{7}$ and $\frac{\pi}{2} < \theta < \pi$.

Example 12 — Self Tutor

If $\tan\theta = 3$ and $\pi < \theta < \frac{3\pi}{2}$, find the exact values of $\cos\theta$ and $\sin\theta$.

$\tan\theta = \frac{\sin\theta}{\cos\theta} = 3$

$\therefore \sin\theta = 3\cos\theta$

Now $\cos^2\theta + \sin^2\theta = 1$

$\therefore \cos^2\theta + (3\cos\theta)^2 = 1$

$\therefore \cos^2\theta + 9\cos^2\theta = 1$

$\therefore 10\cos^2\theta = 1$

$\therefore \cos\theta = \pm\frac{1}{\sqrt{10}}$

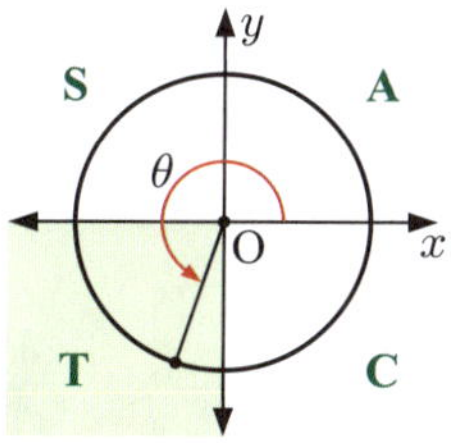

But $\pi < \theta < \frac{3\pi}{2}$, so θ lies in quadrant 3.

$\therefore \cos\theta$ and $\sin\theta$ are both negative.

$\therefore \cos\theta = -\frac{1}{\sqrt{10}}$ and $\sin\theta = -\frac{3}{\sqrt{10}}$.

5 Find the exact values of $\cos x$ and $\sin x$, given that:

a $\tan x = 2$ and $0 < x < \frac{\pi}{2}$

b $\tan x = -\frac{3}{4}$ and $\frac{\pi}{2} < x < \pi$

c $\tan x = \frac{\sqrt{3}}{2}$ and $\pi < x < \frac{3\pi}{2}$

d $\tan x = -\frac{1}{3}$ and $\frac{3\pi}{2} < x < 2\pi$.

6 Suppose $\tan\theta = m$ where $\frac{\pi}{2} < \theta < \pi$.

a Is m positive or negative? Explain your answer.

b Write expressions for $\cos\theta$ and $\sin\theta$ in terms of m.

F FINDING ANGLES

From **Exercise 8C** you should have discovered that:

For θ in radians:
- $\sin(\pi - \theta) = \sin\theta$
- $\cos(\pi - \theta) = -\cos\theta$
- $\cos(2\pi - \theta) = \cos\theta$

We need results such as these, and also the periodicity of the trigonometric ratios, to find angles which have a particular sine, cosine, or tangent.

Example 13

Self Tutor

Find the two angles θ on the unit circle, with $0 \leqslant \theta \leqslant 2\pi$, such that:

a $\cos\theta = \frac{1}{4}$ **b** $\sin\theta = \frac{5}{8}$ **c** $\tan\theta = \frac{3}{5}$

a $\cos^{-1}\left(\frac{1}{4}\right) \approx 1.32$

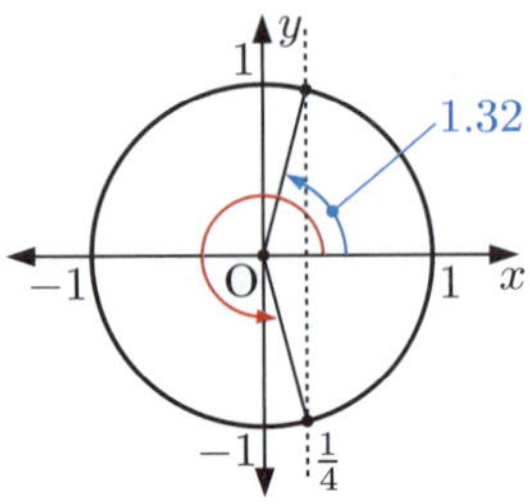

$\therefore\ \theta \approx 1.32$ or $2\pi - 1.32$

$\therefore\ \theta \approx 1.32$ or 4.96

b $\sin^{-1}\left(\frac{5}{8}\right) \approx 0.675$

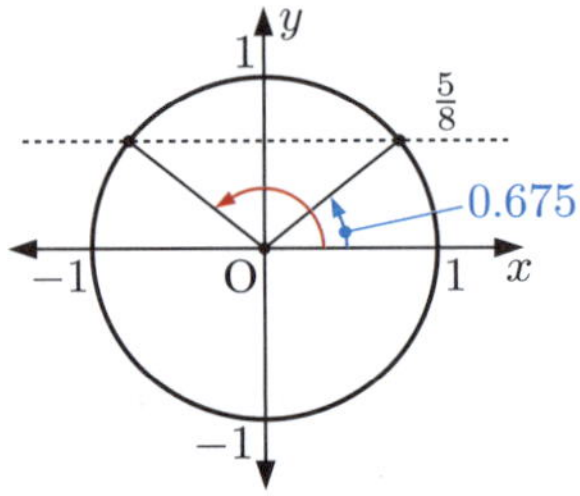

$\therefore\ \theta \approx 0.675$ or $\pi - 0.675$

$\therefore\ \theta \approx 0.675$ or 2.47

c $\tan^{-1}\left(\frac{3}{5}\right) \approx 0.540$

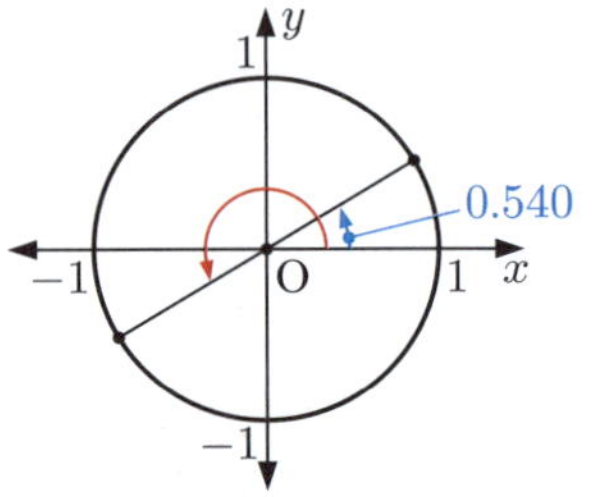

$\therefore\ \theta \approx 0.540$ or $\pi + 0.540$

$\therefore\ \theta \approx 0.540$ or 3.68

If $\cos\theta$, $\sin\theta$, or $\tan\theta$ is positive, your calculator will give the solution for θ in the domain $0 < \theta < \frac{\pi}{2}$.

The blue arrow shows the angle that your calculator gives.

EXERCISE 8F

1 Find two angles θ on the unit circle, with $0 \leqslant \theta \leqslant 2\pi$, such that:

a $\cos\theta = \frac{1}{3}$ **b** $\sin\theta = \frac{3}{4}$ **c** $\tan\theta = 3$

d $\sin\theta = \frac{2}{3}$ **e** $\tan\theta = 0.8$ **f** $\cos\theta = 0.2$

g $\tan\theta = 1.52$ **h** $\cos\theta = 0.91$ **i** $\sin\theta = 0.64$

Example 14

Self Tutor

Find two angles θ on the unit circle, with $0 \leqslant \theta \leqslant 2\pi$, such that:

a $\sin\theta = -0.7$ **b** $\cos\theta = -\frac{3}{5}$ **c** $\tan\theta = -1.5$

a $\sin^{-1}(-0.7) \approx -0.775$

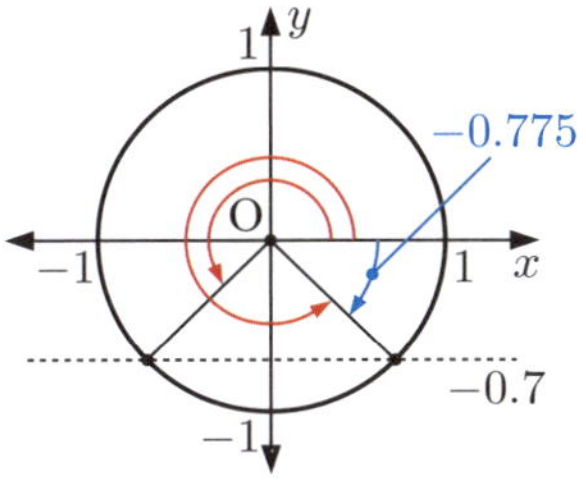

But $0 \leqslant \theta \leqslant 2\pi$

$\therefore\ \theta \approx \pi + 0.775$ or $2\pi - 0.775$

$\therefore\ \theta \approx 3.92$ or 5.51

b $\cos^{-1}\left(-\frac{3}{5}\right) \approx 2.21$

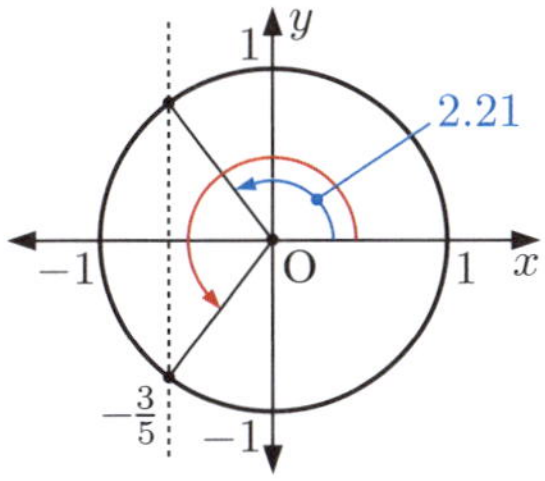

But $0 \leqslant \theta \leqslant 2\pi$

$\therefore\ \theta \approx 2.21$ or $2\pi - 2.21$

$\therefore\ \theta \approx 2.21$ or 4.07

c $\tan^{-1}(-1.5) \approx -0.983$

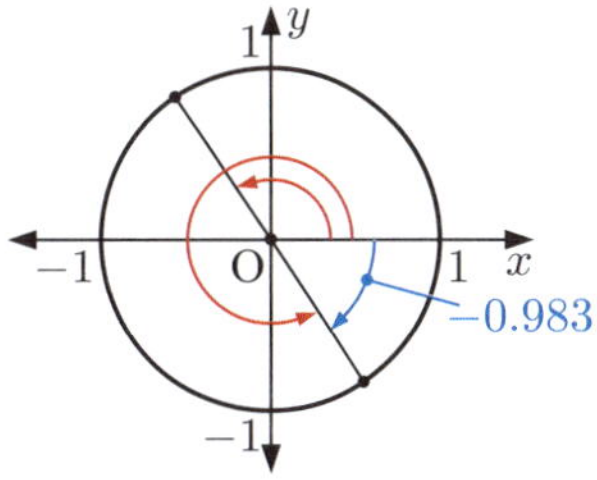

But $0 \leqslant \theta \leqslant 2\pi$

$\therefore\ \theta \approx \pi - 0.983$ or $2\pi - 0.983$

$\therefore\ \theta \approx 2.16$ or 5.30

If $\sin\theta$ or $\tan\theta$ is negative, your calculator will give the solution for θ in the domain $-\frac{\pi}{2} < \theta < 0$.

2 Find two angles θ on the unit circle, with $0 \leqslant \theta \leqslant 2\pi$, such that:

a $\sin\theta = -0.6$ **b** $\cos\theta = -\frac{2}{7}$ **c** $\tan\theta = -0.8$

d $\sin\theta = -\frac{2}{5}$ **e** $\tan\theta = -4$ **f** $\cos\theta = -0.3$

g $\tan\theta = -\sqrt{2}$ **h** $\cos\theta = -0.29$ **i** $\sin\theta = -\frac{10}{13}$

3 Find two angles θ on the unit circle, with $0° \leqslant \theta \leqslant 360°$, such that:

a $\cos\theta = \frac{1}{2}$ **b** $\sin\theta = 0.7$ **c** $\tan\theta = \sqrt{3}$

d $\sin\theta = \frac{1}{6}$ **e** $\cos\theta = 0$ **f** $\sin\theta = -\frac{\sqrt{3}}{2}$

g $\cos\theta = -0.3$ **h** $\tan\theta = -2.1$ **i** $\sin\theta = -\frac{9}{10}$

Discovery 2 — Parametric equations

We usually write functions in the form $y = f(x)$. For example, the functions $y = 3x + 7$, $y = x^2 - 6x + 8$, and $y = \sin x$ are written this way.

However, it is sometimes useful to express **both** x and y in terms of another variable t, called the **parameter**. In this case we say we have **parametric equations**.

The use of parametric equations is not required for the syllabus.

What to do:

PARAMETRIC PLOTTER

1 **a** Use the graphing package to plot $\{(x, y) : x = \cos t,\ \ y = \sin t,\ \ 0° \leqslant t \leqslant 360°\}$.
Use the same scale on both axes.

b Describe the resulting graph. Is it the graph of a function?

c Evaluate $x^2 + y^2$. Hence determine the equation of this graph in terms of x and y only.

2 Use the graphing package to plot:

a $\{(x, y) : x = 2\cos t,\ \ y = \sin 2t,\ \ 0° \leqslant t \leqslant 360°\}$

b $\{(x, y) : x = 2\cos t,\ \ y = 2\sin 3t,\ \ 0° \leqslant t \leqslant 360°\}$

c $\{(x, y) : x = 2\cos t,\ \ y = \cos t - \sin t,\ \ 0° \leqslant t \leqslant 360°\}$

d $\{(x, y) : x = \cos^2 t + \sin 2t,\ \ y = \cos t,\ \ 0° \leqslant t \leqslant 360°\}$

e $\{(x, y) : x = \cos^3 t,\ \ y = \sin t,\ \ 0° \leqslant t \leqslant 360°\}$

G RECIPROCAL TRIGONOMETRIC RATIOS

The **reciprocals** of the trigonometric ratios $\cos\theta$, $\sin\theta$, and $\tan\theta$ are given the names secant θ, cosecant θ, and cotangent θ respectively. These reciprocal trigonometric ratios are written as $\sec\theta$, $\text{cosec}\,\theta$, and $\cot\theta$.

- $\sec\theta = \dfrac{1}{\cos\theta}$
- $\text{cosec}\,\theta = \dfrac{1}{\sin\theta}$
- $\cot\theta = \dfrac{1}{\tan\theta} = \dfrac{\cos\theta}{\sin\theta}$

Historical note — The reciprocal trigonometric functions

Before logarithms and modern calculators, the calculation of trigonometric ratios was time consuming. To avoid calculating the same ratio multiple times, tables of values were written out. However, if a trigonometric function appeared in the denominator of a fraction, the division would still be time consuming.

The main purpose of the reciprocal trigonometric functions was so that tables for these functions would also be created, and the user would simply multiply by the reciprocal function.

With the invention of logarithms came quicker methods for performing such calculations, and the reciprocal trigonometric functions are now less used.

Example 15 — Self Tutor

Without using a calculator, find:

a $\sec\frac{\pi}{3}$ **b** $\text{cosec}\,\frac{5\pi}{4}$ **c** $\cot\frac{\pi}{6}$

a $\sec\frac{\pi}{3} = \dfrac{1}{\cos\frac{\pi}{3}} = \dfrac{1}{\frac{1}{2}} = 2$

b $\text{cosec}\,\frac{5\pi}{4} = \dfrac{1}{\sin\frac{5\pi}{4}} = \dfrac{1}{-\frac{1}{\sqrt{2}}} = -\sqrt{2}$

c $\cot\frac{\pi}{6} = \dfrac{1}{\tan\frac{\pi}{6}} = \dfrac{1}{\frac{1}{\sqrt{3}}} = \sqrt{3}$

EXERCISE 8G

1 Without using a calculator, find:

a $\sec\frac{\pi}{6}$ **b** $\operatorname{cosec}\frac{\pi}{4}$ **c** $\cot\frac{\pi}{3}$ **d** $\operatorname{cosec}\frac{5\pi}{6}$

e $\sec\pi$ **f** $\cot\frac{7\pi}{6}$ **g** $\operatorname{cosec}\frac{5\pi}{3}$ **h** $\cot\left(-\frac{\pi}{4}\right)$

2 Find $\cos\theta$, $\sin\theta$, $\tan\theta$, $\sec\theta$, $\operatorname{cosec}\theta$, and $\cot\theta$ for:

a $\theta = \frac{2\pi}{3}$ **b** $\theta = \frac{3\pi}{4}$ **c** $\theta = \frac{11\pi}{6}$

3 Without using a calculator, find $\operatorname{cosec}x$, $\sec x$, and $\cot x$ given:

a $\sin x = \frac{4}{5}$, $0 \leqslant x \leqslant \frac{\pi}{2}$ **b** $\cos x = \frac{1}{4}$, $\frac{3\pi}{2} < x < 2\pi$

4 Find the other *five* trigonometric ratios if:

a $\cos\theta = \frac{1}{3}$ and $\frac{3\pi}{2} < \theta < 2\pi$ **b** $\sin x = -\frac{3}{4}$ and $\pi < x < \frac{3\pi}{2}$

c $\sec x = 1\frac{1}{2}$ and $0 < x < \frac{\pi}{2}$ **d** $\operatorname{cosec}\theta = 3$ and $\frac{\pi}{2} < \theta < \pi$

e $\tan\beta = \frac{2}{5}$ and $\pi < \beta < \frac{3\pi}{2}$ **f** $\cot\theta = 4$ and $\pi < \theta < \frac{3\pi}{2}$

5 Find all angles between $0°$ and $360°$ which have:

a a secant of 2 **b** a cosecant of $\frac{2}{\sqrt{3}}$ **c** a cotangent of $-\sqrt{3}$.

6 Find *all* values of θ for which:

a $\operatorname{cosec}\theta$ is undefined **b** $\sec\theta$ is undefined

c $\cot\theta$ is zero **d** $\cot\theta$ is undefined.

Discussion

What range of possible values can $\sec\theta$, $\operatorname{cosec}\theta$, and $\cot\theta$ take?

Review set 8A

1 Convert to radians, in terms of π:

a 18° **b** 50° **c** 540° **d** 200°

2 Find the arc length of each sector:

a

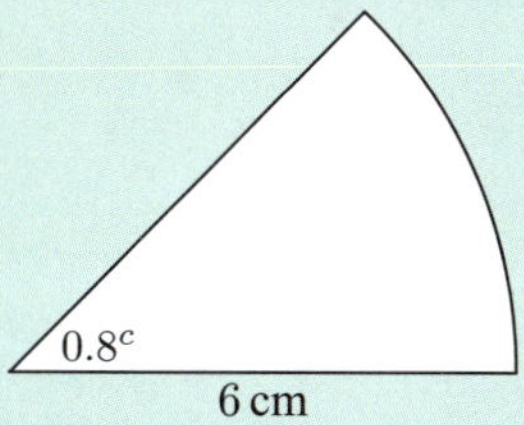

b

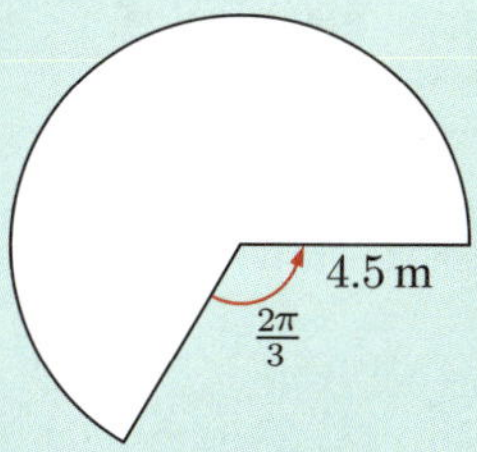

3 Find:

a θ

b the area of the sector.

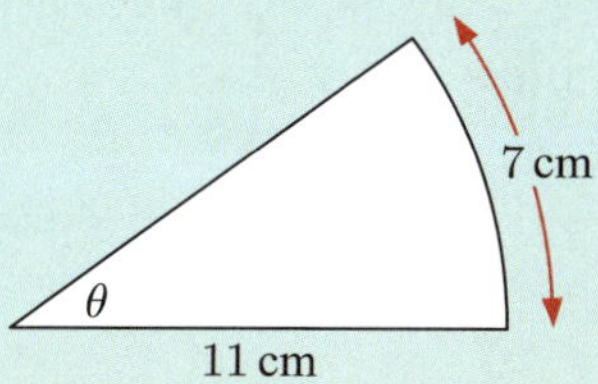

4 The two circles alongside each have centre O. Find:

a the area of the shaded region

b the perimeter of the shaded region.

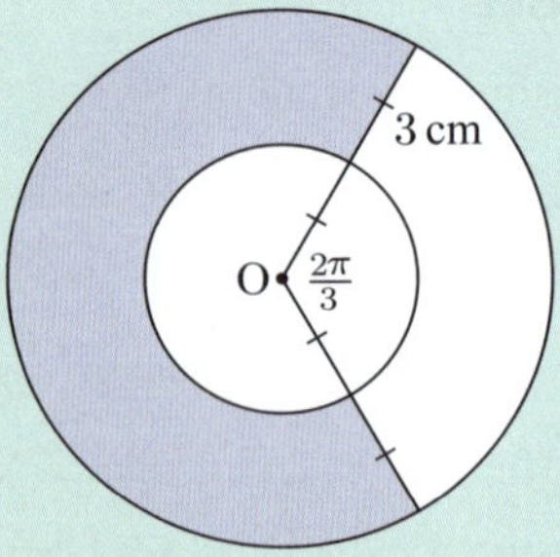

5 The circle alongside has radius 5 cm. Find:

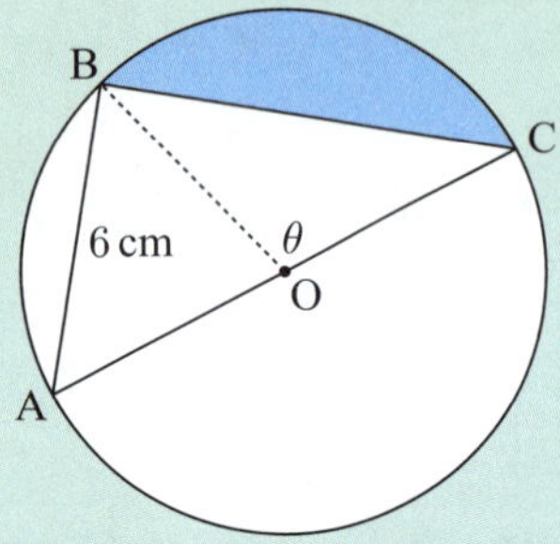

a the length of BC

b θ

c the area of the shaded region.

6 Find the acute angle that has the same:

a sine as $123°$ **b** cosine as $\frac{7\pi}{4}$ **c** tangent as $216°$.

7 Find the exact values of $\sin\theta$, $\cos\theta$, and $\tan\theta$ for θ equal to:

a $\frac{\pi}{3}$ **b** π **c** $\frac{5\pi}{4}$

8 Find the angle between 0 and 2π which has:

a a cosine of $-\frac{1}{2}$ and a sine of $\frac{\sqrt{3}}{2}$

b a tangent of $\frac{1}{\sqrt{3}}$ and a cosine of $-\frac{\sqrt{3}}{2}$.

9 If $\sin\theta = \frac{3}{8}$, find the possible values of $\cos\theta$.

10 Evaluate:

a $6\cos\frac{\pi}{3}\sin\frac{\pi}{3}$ **b** $\tan^2\left(\frac{3\pi}{4}\right) - 2$ **c** $\sin^2\left(\frac{\pi}{6}\right) - \cos^2\left(\frac{\pi}{3}\right)$

11 Given $\cos x = -\frac{1}{3}$ and $\frac{\pi}{2} < x < \pi$, find $\sin x$ and $\tan x$.

12 Suppose $\sin\theta = -\frac{2}{\sqrt{5}}$ and $\frac{3\pi}{2} < \theta < 2\pi$. Find the exact value of $\tan\theta$.

13 Find all angles between $0°$ and $360°$ which have:

a a cosine of $-\frac{\sqrt{3}}{2}$ **b** a secant of $\sqrt{2}$ **c** a cotangent of $-\frac{1}{\sqrt{3}}$

14 Find two angles on the unit circle, with $0 \leqslant \theta \leqslant 2\pi$, such that:

a $\cos\theta = \frac{1}{5}$ **b** $\tan\theta = 0.35$ **c** $\sin\theta = \frac{2}{7}$

15 If $\sin x = -\frac{1}{4}$ and $\pi < x < \frac{3\pi}{2}$, find the other *five* trigonometric ratios exactly.

Review set 8B

1 Convert the following radian measures to degrees:

a $\frac{2\pi}{9}$ **b** $\frac{7\pi}{10}$ **c** 0.82^c **d** 1.93^c

2 Find the area of each sector:

a

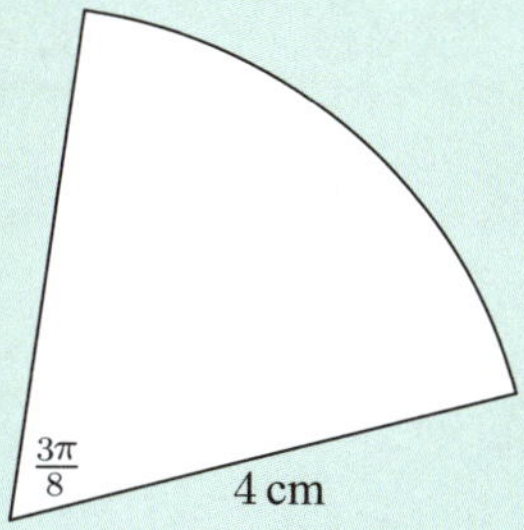

b

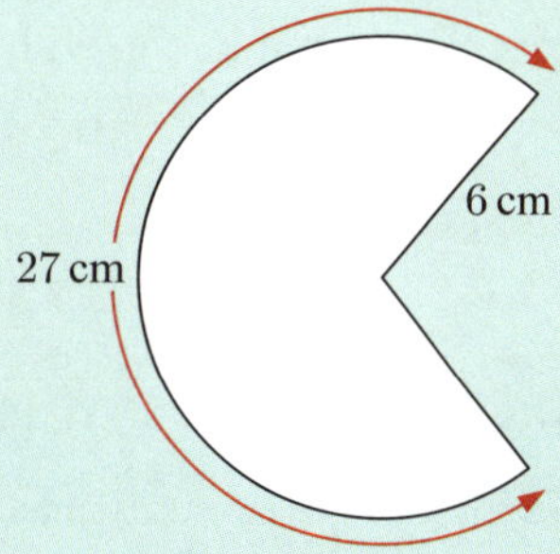

3

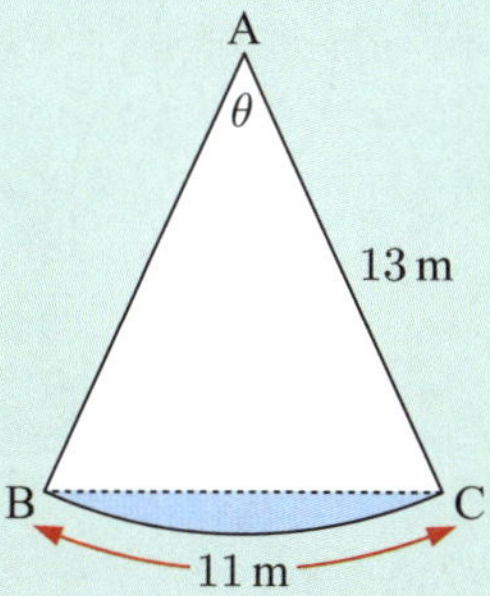

a Explain, without calculation, why θ must be less than 1 radian.

b Find θ correct to 3 significant figures.

c Find the area of sector ABC.

d Find the area of the shaded region.

4 The circle alongside has centre O and radius 5 cm. XY and XZ are tangents to the circle.

a Find the length of XY.

b Find the perimeter and area of the shaded region.

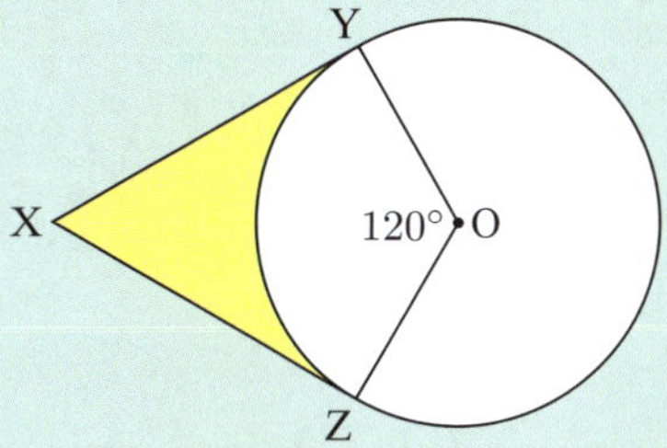

5 **a** State the coordinates of P exactly.

b Find the coordinates of P correct to 3 significant figures.

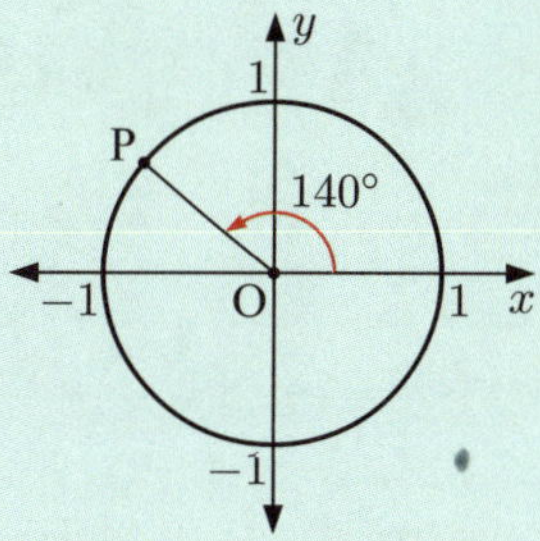

6 Use a unit circle diagram to find $\cos\frac{\pi}{2}$ and $\sin\frac{\pi}{2}$.

7 Show that $\tan\frac{\pi}{3} + \tan\frac{7\pi}{6} = \frac{4\sqrt{3}}{3}$.

8 Without using a calculator, evaluate:

a $\tan^2 120° - \sin^2 60°$ **b** $\cos\frac{7\pi}{6} - \sin\frac{\pi}{4}$ **c** $2\sin\frac{2\pi}{3} - \tan\left(-\frac{\pi}{4}\right)$

9 Find all angles between 0 and 4π with:

a a cosine of $-\frac{1}{2}$ **b** a tangent of -1.

10 If $\cos\theta = -\frac{1}{6}$, find the possible values of $\sin\theta$.

11 If $\tan\theta = -\frac{1}{2}$ and $\frac{\pi}{2} < \theta < \pi$, find the exact values of $\cos\theta$ and $\sin\theta$.

12 Find two angles on the unit circle, with $0 \leqslant \theta \leqslant 2\pi$, such that:

a $\sin\theta = -0.3$ **b** $\cos\theta = -\frac{6}{7}$ **c** $\tan\theta = -0.14$

13 Without using a calculator, find:

a $\operatorname{cosec}\frac{\pi}{2}$ **b** $\sec\frac{4\pi}{3}$ **c** $\cos\left(-\frac{\pi}{3}\right)$

14 If $\sec\alpha = -3\frac{1}{3}$ and $0 < \alpha < \pi$, find the other *five* trigonometric ratios exactly.

15 Three circles with radius r are drawn as shown, each with its centre on the circumference of the other two circles. A, B, and C are the centres of the three circles.

Prove that an expression for the area of the shaded region is $A = \frac{r^2}{2}(\pi - \sqrt{3})$.

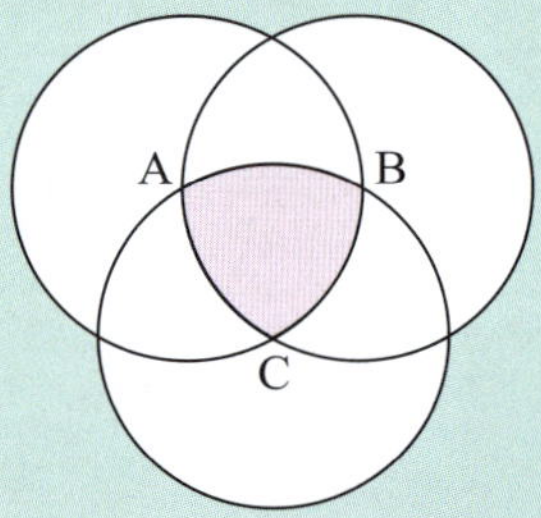

9

Trigonometric functions and equations

Contents:

Opening problem

A Ferris wheel rotates at a constant speed. The wheel's radius is 10 m and the bottom of the wheel is 2 m above ground level. From a point in front of the wheel, Andrew is watching a green light on the perimeter of the wheel. Andrew notices that the green light moves in a circle. He estimates how high the light is above ground level at two second intervals, and draws a scatter diagram of his results.

Things to think about:

a What will Andrew's scatter diagram look like?

b What function can be used to model the data?

c How could this function be used to find:

  i the light's position at any point in time

  ii the times when the light is at its maximum and minimum heights?

d What part of the function indicates the time for one full revolution of the wheel?

DEMO

A PERIODIC BEHAVIOUR

Periodic phenomena occur all the time in the physical world. Their behaviour repeats again and again over time.

We see periodic behaviour in:

- seasonal variations in our climate
- variations in average maximum and minimum monthly temperatures
- the number of daylight hours at a particular location
- tidal variations in the depth of water in a harbour
- the phases of the moon
- animal populations.

TERMINOLOGY USED TO DESCRIBE PERIODICITY

A **periodic function** is one which repeats itself over and over in a horizontal direction, in intervals of the same length.

The **period** of a periodic function is the length of one repetition or cycle.

$f(x)$ is a periodic function with period p if $\boldsymbol{f(x+p)=f(x)}$ for all x, and p is the smallest positive value for this to be true.

For example:

- A **cycloid** is a periodic function. It is the curve traced out by a point on a circle as the circle rolls across a flat surface in a straight line.

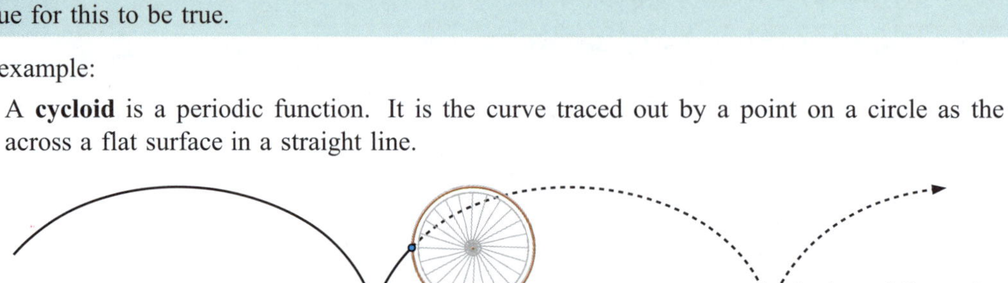

DEMO

- This function is periodic with period 5.

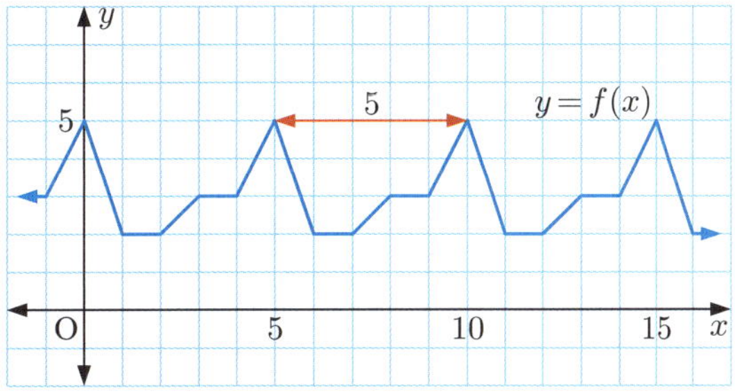

WAVES

In this course we are mainly concerned with periodic phenomena which show a wave pattern.

A wave oscillates about a horizontal line called the **principal axis** or **mean line**. The equation of the principal axis is $y = \dfrac{\textbf{max} + \textbf{min}}{2}$ where max is the **maximum value** at the top of a crest, and min is the **minimum value** at the bottom of a trough.

The **amplitude** is the distance between a maximum (or minimum) value and the principal axis.

$$\textbf{amplitude} = \frac{\textbf{max} - \textbf{min}}{2}$$

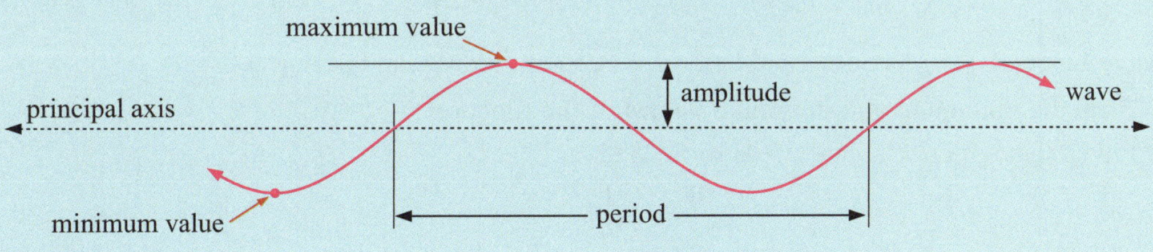

EXERCISE 9A

1 Which of these graphs show periodic behaviour?

a

b

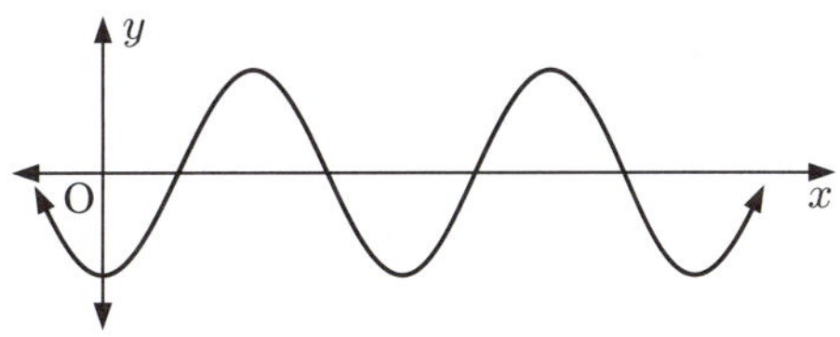

c

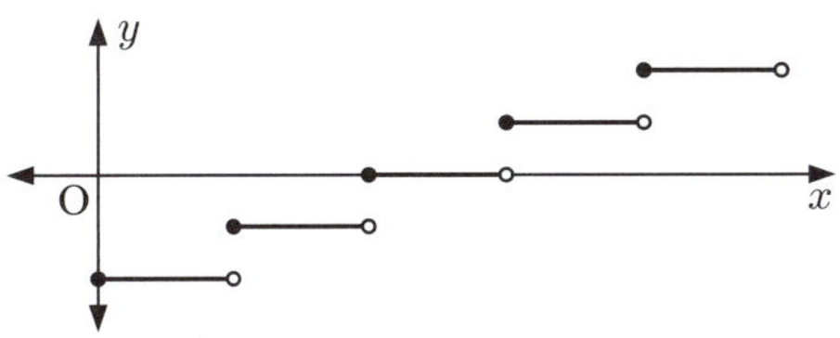

d

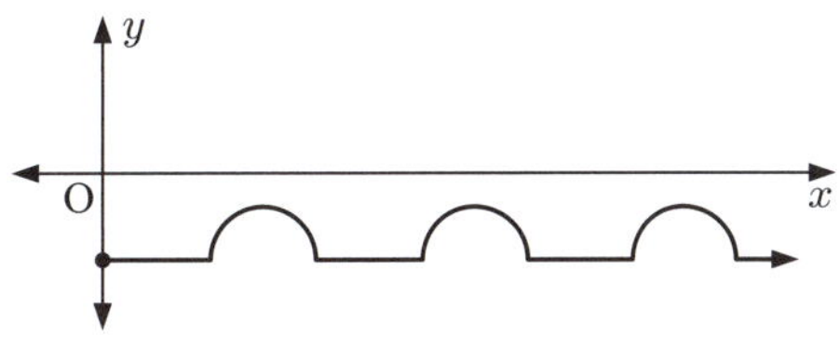

e

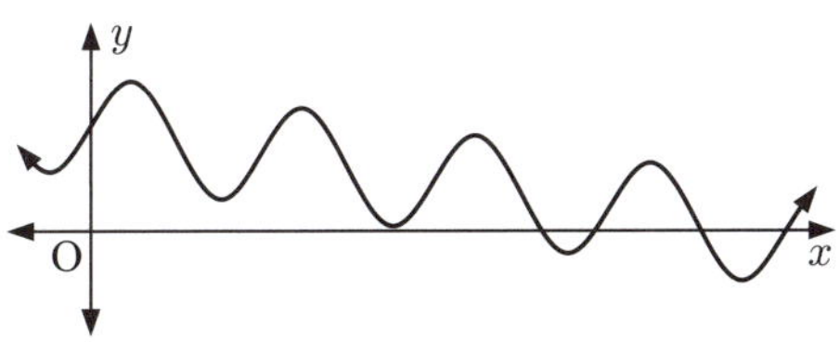

f

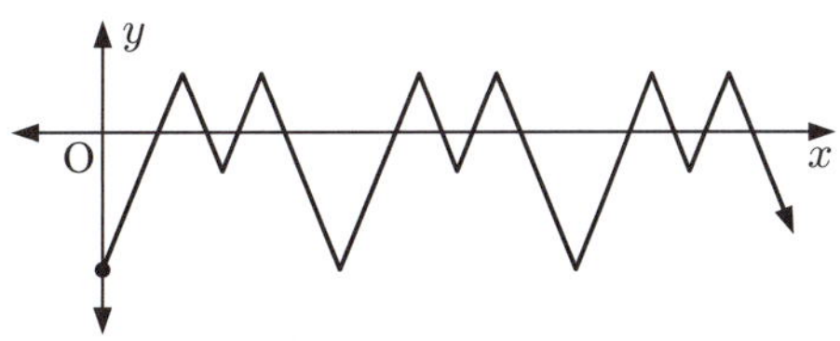

2 Find the period of the following periodic functions:

a

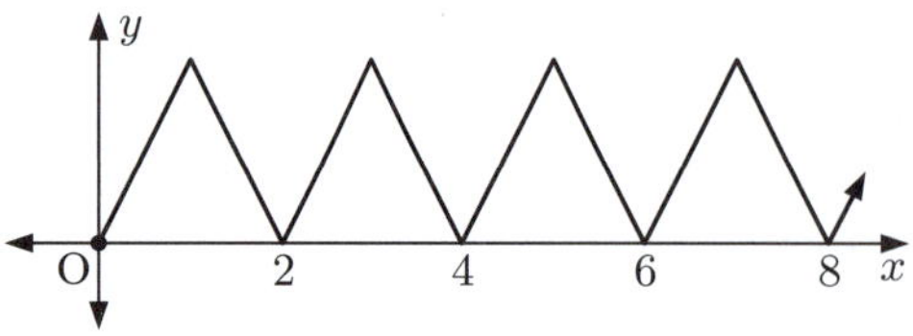

b

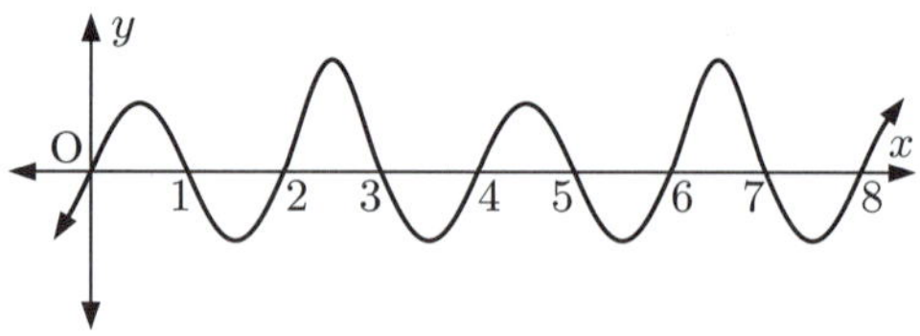

3 For the following waves, find:

i the period **ii** the equation of the principal axis **iii** the amplitude.

a

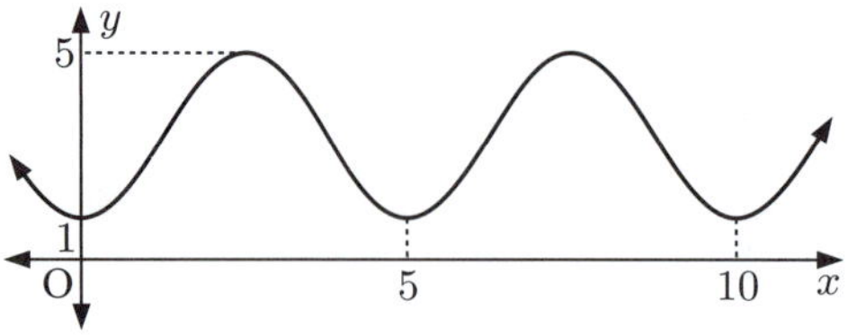

b

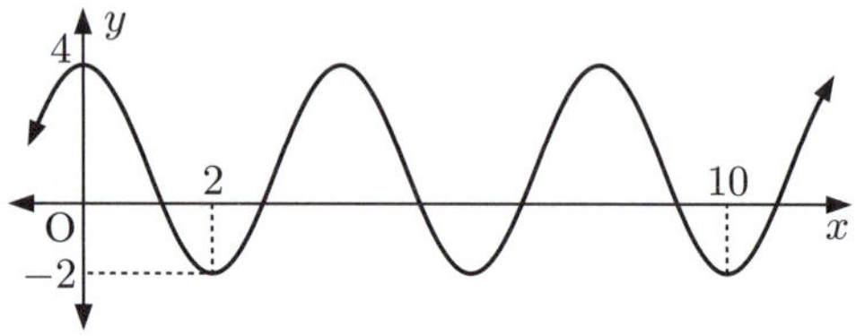

c

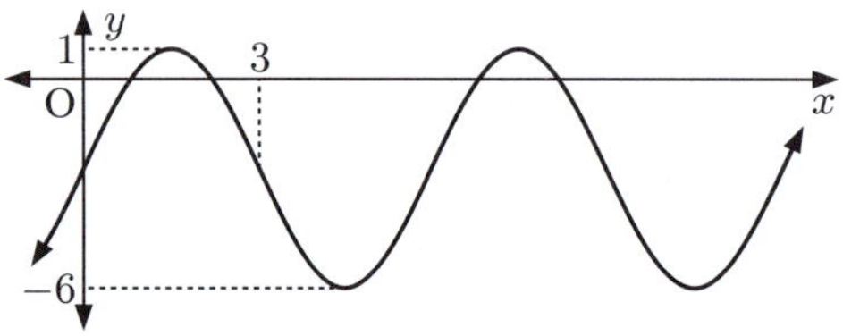

d

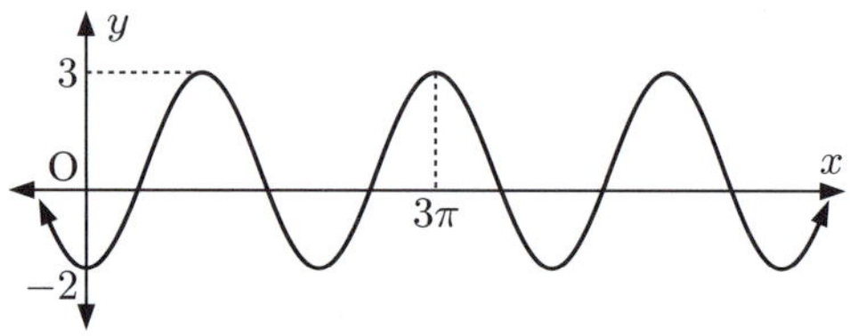

4 A wave function $f(x)$ has principal axis $y = 3$, amplitude 4, and period 5.

a State the maximum and minimum values of the function.

b Given that $f(1) = 2$ and $f(4) = 4$, find:

i $f(9)$ **ii** $f(21)$.

5 Suppose f is a periodic function with period p, and g is a periodic function with period q.

a State the period of fg.

b Is $f + g$ necessarily a periodic function? Explain your answer.

B THE SINE FUNCTION

In previous studies of trigonometry we have only considered static situations where an angle is fixed. However, when an object moves around a circle, the situation is dynamic. The angle θ between the radius OP and the positive x-axis continually changes with time.

Consider again the **Opening Problem** in which a Ferris wheel of radius 10 m revolves at constant speed. We let P represent the green light on the wheel.

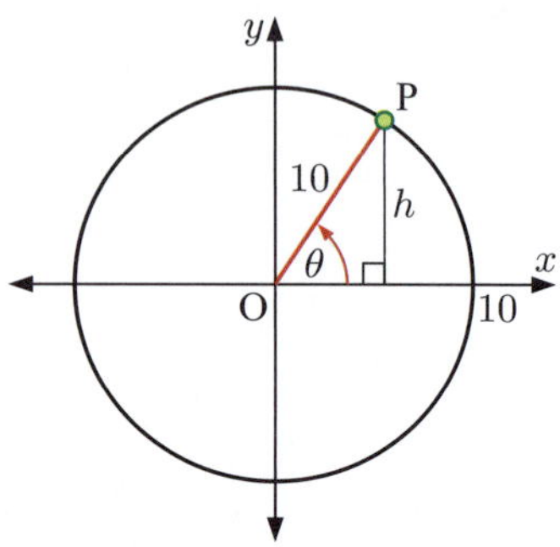

The height of P relative to the x-axis can be determined using right angled triangle trigonometry:

$$\sin\theta = \frac{h}{10}$$

$$\therefore\ h = 10\sin\theta$$

As time goes by, θ changes and so does h.

So, we can write h as a function of θ, or alternatively we can write h as a function of time t.

Suppose the Ferris wheel observed by Andrew takes 100 seconds for a full revolution. The graph below shows the height of the light above or below the principal axis against the time in seconds.

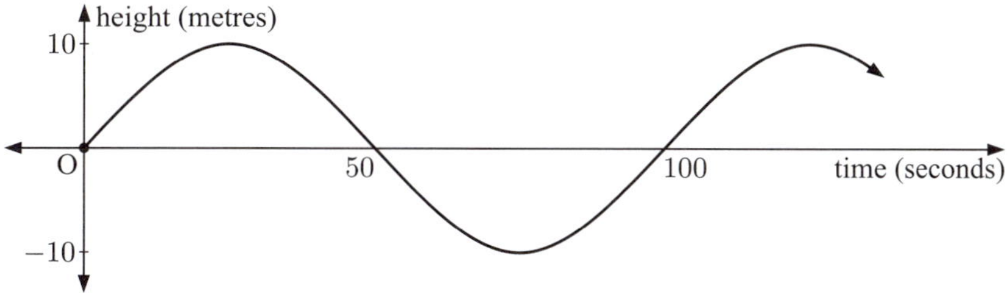

DEMO

We observe that the amplitude is 10 metres and the period is 100 seconds.

THE BASIC SINE CURVE

Suppose point P moves around the unit circle so the angle OP makes with the positive horizontal axis is x. In this case P has coordinates $(\cos x, \sin x)$.

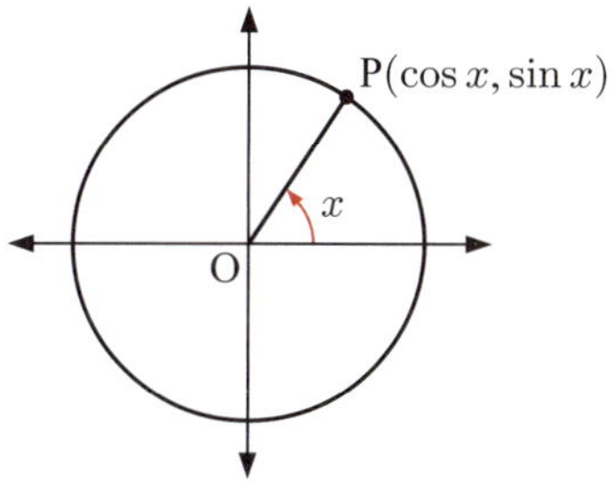

If we project the values of $\sin x$ from the unit circle to a set of axes alongside, we can obtain the graph of $y = \sin x$.

Notice that x on the unit circle diagram is an *angle*, and becomes the horizontal coordinate of the **sine function**.

Unless indicated otherwise, you should assume that x is measured in radians. Degrees are only included on this graph for the sake of completeness.

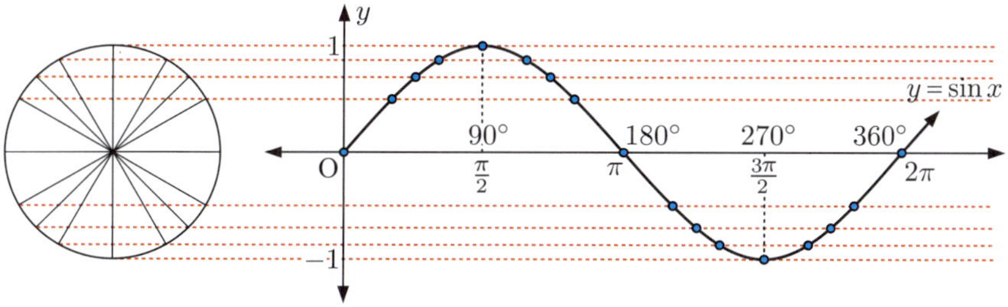

Click on the icon to generate the sine function for yourself.

SINE FUNCTION

You should observe that the sine function can be continued beyond $0 \leqslant x \leqslant 2\pi$ in either direction.

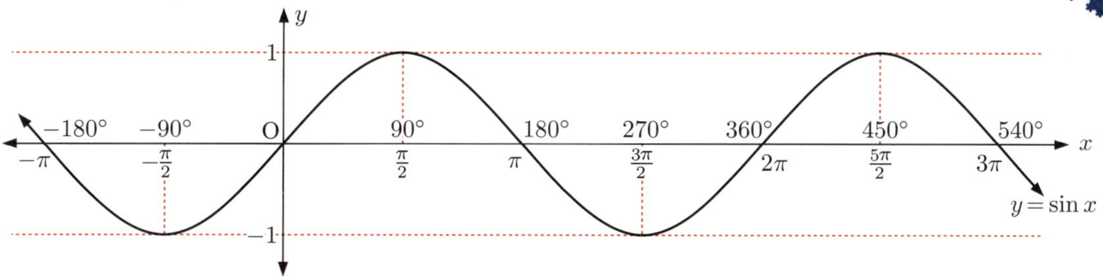

The unit circle repeats itself after one full revolution, so its **period** is 2π.

The **maximum** value is 1 and the **minimum** value is -1.

The **principal axis** is $y = 0$ and the **amplitude** is 1.

Discovery 1 — Transformations of the sine curve

In this Discovery we consider different transformations of the sine curve $y = \sin x$. Our aim is to understand each component of the **general sine function** $y = a\sin bx + c$, $a > 0$, $b > 0$.

PART 1: The family $y = a\sin x$, $a > 0$

Click on the icon to explore the family $y = a\sin x$, $a > 0$.

DYNAMIC SINE FUNCTION

What to do:

1 Use the slider to vary the value of a. Observe the changes to the graph of the function.

2 Use the software to help complete the table:

a	*Function*	*Maximum*	*Minimum*	*Period*	*Amplitude*
1	$y = \sin x$	1	-1	2π	1
2	$y = 2\sin x$				
3	$y = 3\sin x$				
0.5	$y = 0.5\sin x$				
a	$y = a\sin x$				

3 How does a affect the function $y = a\sin x$?

PART 2: The family $y = \sin bx$, $b > 0$

Click on the icon to explore the family $y = \sin bx$, $b > 0$.

What to do:

1 Use the slider to vary the value of b. Observe the changes to the graph of the function.

2 Use the software to help complete the table:

b	*Function*	*Maximum*	*Minimum*	*Period*	*Amplitude*
1	$y = \sin x$	1	-1	2π	1
2	$y = \sin 2x$				
3	$y = \sin 3x$				
$\frac{1}{2}$	$y = \sin \frac{x}{2}$				
b	$y = \sin bx$				

3 How does b affect the function $y = \sin bx$?

PART 3: The family $y = \sin x + c$

Click on the icon to explore the family $y = \sin x + c$.

What to do:

1 Use the slider to vary the value of c. Observe the changes to the graph of the function.

2 Use the software to help complete the table:

c	*Function*	*Maximum*	*Minimum*	*Period*	*Amplitude*
0	$y = \sin x$	1	-1	2π	1
3	$y = \sin x + 3$				
-2	$y = \sin x - 2$				
c	$y = \sin x + c$				

3 How does c affect the function $y = \sin x + c$?

THE GENERAL SINE FUNCTION

The **general sine function** is

$$\boldsymbol{y = a\sin bx + c} \quad \text{where } a > 0,\ b > 0.$$

- a affects **amplitude**
- b affects **period**
- c affects **vertical translation**

The **principal axis** of the general sine function is $\boldsymbol{y = c}$.

The **period** of the general sine function is $\boldsymbol{\dfrac{2\pi}{b}}$.

The **amplitude** of the general sine function is $\boldsymbol{a}$.

Example 1

Self Tutor

Without using technology, sketch the following graphs for $0 \leqslant x \leqslant 4\pi$:

a $y = 2\sin x$ **b** $y = \sin 2x$ **c** $y = \sin x - 1$

a The amplitude is 2 and the period is 2π.

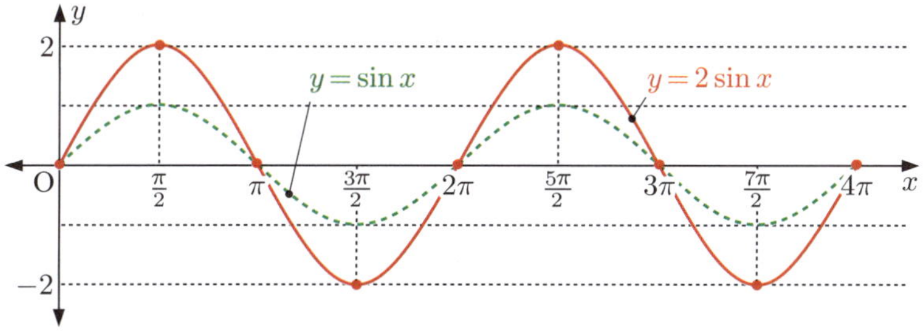

b The period is $\frac{2\pi}{2} = \pi$.

$\therefore$ the maximum values are π units apart.

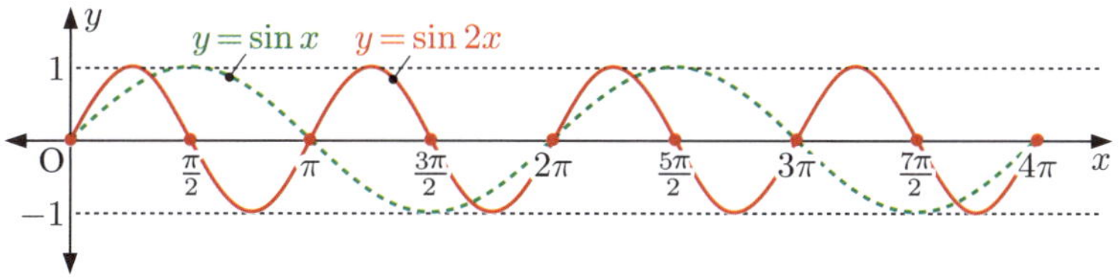

c This is a vertical translation of $y = \sin x$ downwards by 1 unit.
The principal axis is now $y = -1$.

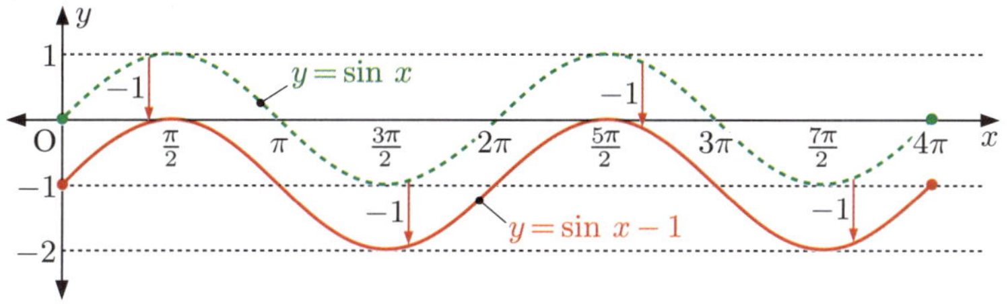

EXERCISE 9B

1 State the amplitude of:

a $y = 2\sin x$ **b** $y = \frac{1}{2}\sin x - 3$ **c** $y = 4\sin 2x$

2 State the principal axis of:

a $y = \sin x + 3$ **b** $y = 2\sin x$ **c** $y = \frac{1}{3}\sin x - 4$

3 Find the period of:

a $y = \sin 3x$ **b** $y = 4\sin x$ **c** $y = \frac{1}{2}\sin\frac{x}{3}$

4 Find the value of b given that the function $y = \sin bx$, $b > 0$, has period:

a $\frac{2\pi}{3}$ **b** $\frac{2\pi}{5}$ **c** $\frac{\pi}{3}$ **d** $\frac{\pi}{2}$ **e** $\frac{4\pi}{3}$

5 Without using technology, sketch the following graphs for $0 \leqslant x \leqslant 4\pi$:

GRAPHING PACKAGE

a $y = 3\sin x$ **b** $y = 4\sin x$ **c** $y = \frac{1}{2}\sin x$

d $y = \sin 3x$ **e** $y = \sin 4x$ **f** $y = \sin\frac{x}{2}$

g $y = \sin x + 2$ **h** $y = \sin x - 3$ **i** $y = \sin x - \frac{2}{3}$

Check your answers using technology.

6 Sketch the following graphs for $0° \leqslant x \leqslant 720°$:

a $y = 5\sin x$ **b** $y = \frac{1}{3}\sin x$

c $y = \sin 2x$ **d** $y = \sin\frac{2x}{3}$

e $y = \sin x + 1$ **f** $y = \sin x - 2$

If x is measured in degrees, the period of $y = a\sin bx + c$ is $\frac{360°}{b}$.

Example 2

Self Tutor

Without using technology, sketch $y = 2\sin 3x + 1$ for $0 \leqslant x \leqslant 2\pi$.

Starting with $y = \sin x$, we:

- double the amplitude to produce $y = 2\sin x$, then
- divide the period by 3 to produce $y = 2\sin 3x$, then
- translate the graph 1 unit upwards to produce $y = 2\sin 3x + 1$, so the principal axis is now $y = 1$.

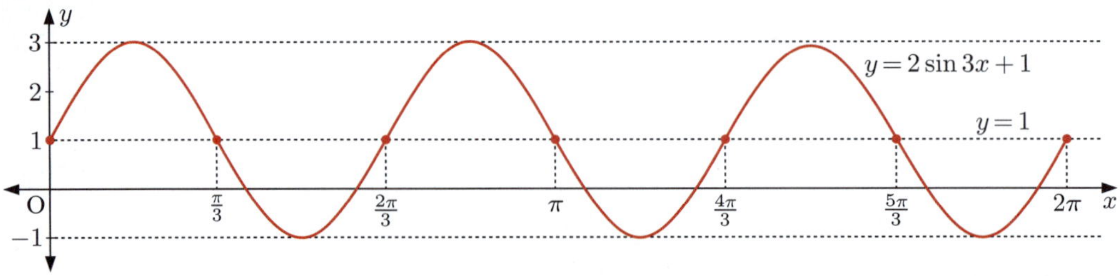

7 Without using technology, sketch the following graphs for $0 \leqslant x \leqslant 2\pi$:

a $y = 3\sin x - 1$ **b** $y = 2\sin 3x$ **c** $y = \sin 2x + 3$

d $y = 3\sin 2x - 1$ **e** $y = 5\sin 2x + 3$ **f** $y = 4\sin 3x - 2$

Check your answers using technology.

8 Find a, b, and c given that the function $y = a \sin bx + c$, $a > 0$, $b > 0$, has:

a amplitude 3, period 2π, and principal axis $y = 0$

b amplitude 2, period $\frac{2\pi}{5}$, and principal axis $y = 6$

c amplitude 5, period $\frac{2\pi}{3}$, and principal axis $y = -2$

d amplitude $\frac{1}{4}$, period $\frac{12\pi}{5}$, and principal axis $y = 7$.

9 Find the range of:

a $y = \sin x + 4$ **b** $y = 3 \sin x - 1$ **c** $y = 2 \sin 3x - 5$

Example 3 — Self Tutor

Find the equation of this sine function.

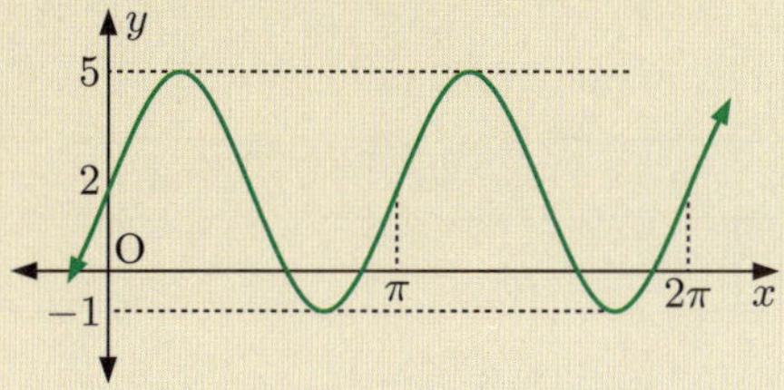

The amplitude is 3, so $a = 3$.

The period is π, so $\frac{2\pi}{b} = \pi$ and $\therefore b = 2$.

The principal axis is $y = 2$, so $c = 2$.

The equation of the function is $y = 3 \sin 2x + 2$.

10 Find the equation of each sine function:

a

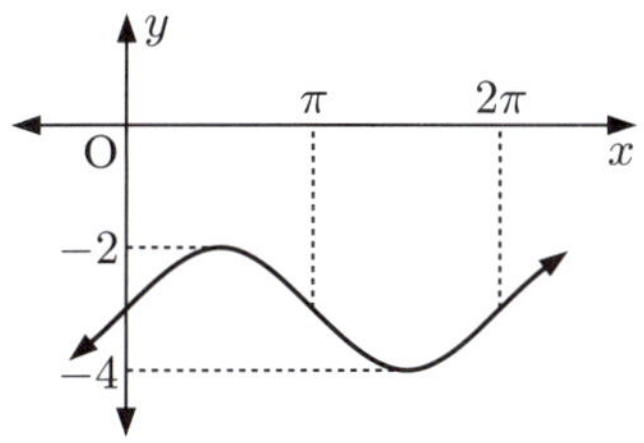

b

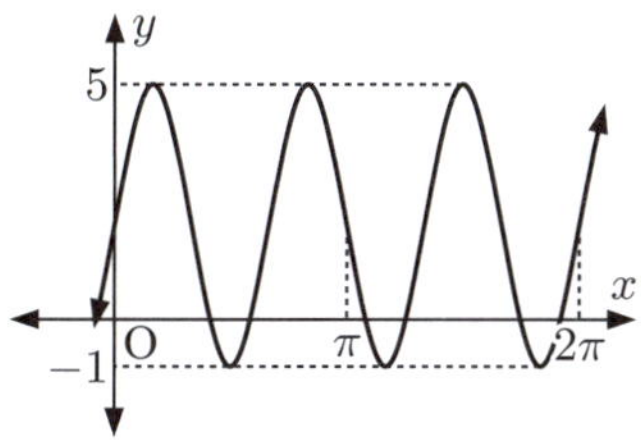

c

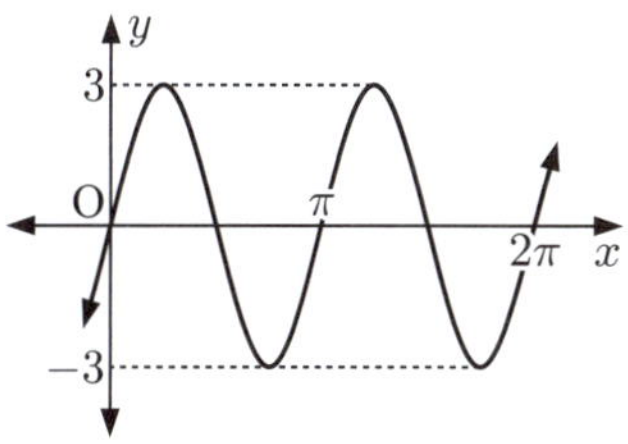

11 Find m and n given the following graph of the function $y = m \sin x + n$.

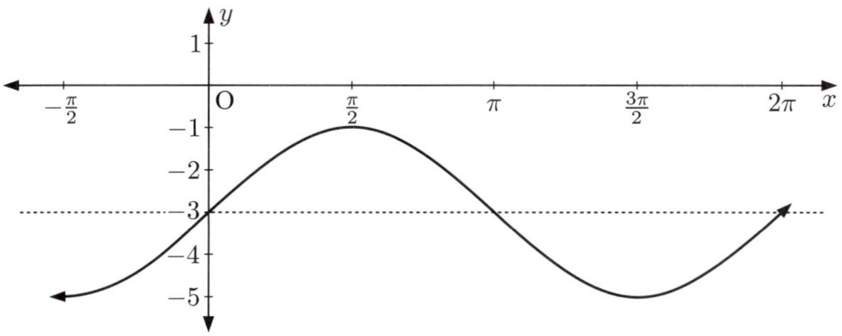

12 Sketch the following graphs for $0° \leqslant x \leqslant 360°$:

a $y = 3\sin 2x$ **b** $y = 2\sin x - 3$ **c** $y = 4\sin 3x - 1$

13 On the same set of axes, sketch for $0 \leqslant x \leqslant 2\pi$:

a $y = \sin x$ and $y = |\sin x|$ **b** $y = 3\sin 2x$ and $y = |3\sin 2x|$

Discovery 2 Modelling using sine functions

When patterns of variation can be identified and quantified using a formula or equation, predictions may be made about behaviour in the future. Examples of this include tidal movement which can be predicted many months ahead, and the date of a future full moon.

What to do:

1 Consider the mean monthly maximum temperature for Cape Town, South Africa:

Month	Oct	Nov	Dec	Jan	Feb	Mar	Apr	May	Jun	Jul	Aug	Sep
Temperature T (°C)	$21\frac{1}{2}$	24	26	28	27	$25\frac{1}{2}$	22	$18\frac{1}{2}$	16	15	16	18

The graph over a two year period is shown below:

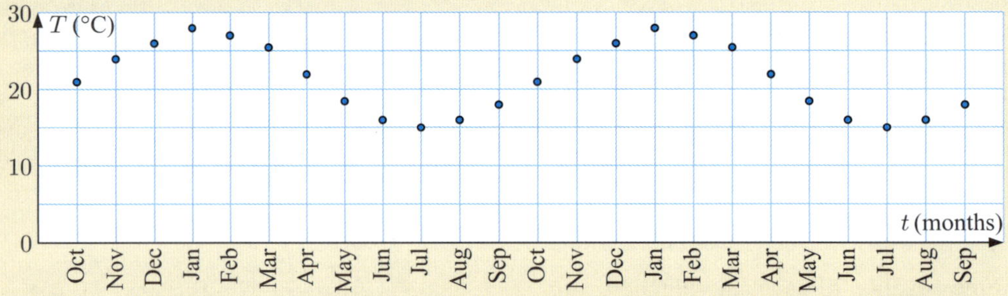

We attempt to model this data using a general sine function of the form $T = a\sin bt + c$, where t is the number of months after October of the first year.

a State the period of the function. Hence show that $b = \frac{\pi}{6}$.

b Use the amplitude to show that $a \approx 6.5$.

c Use the principal axis to show that $c \approx 21.5$.

d Superimpose the model $T \approx 6.5\sin\left(\frac{\pi}{6}t\right) + 21.5$ on the original data to confirm its accuracy.

2 Some of the largest tides in the world are observed in Canada's Bay of Fundy. The difference between high and low tides is 14 metres, and the average time difference between high tides is about 12.4 hours.

Suppose the mean tide occurs at midnight.

a Find a sine model for the height of the tide H in terms of the time t.

b Sketch the graph of the model over one period.

C THE COSINE FUNCTION

We return to the Ferris wheel from the **Opening Problem** and consider the **horizontal displacement** d of the light.

Now $\cos\theta = \dfrac{d}{10}$ so $d = 10\cos\theta$.

The graph being generated over time is therefore a **cosine function**.

DEMO

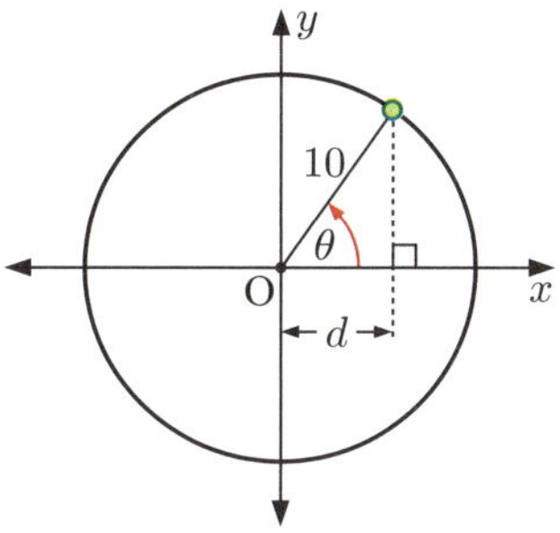

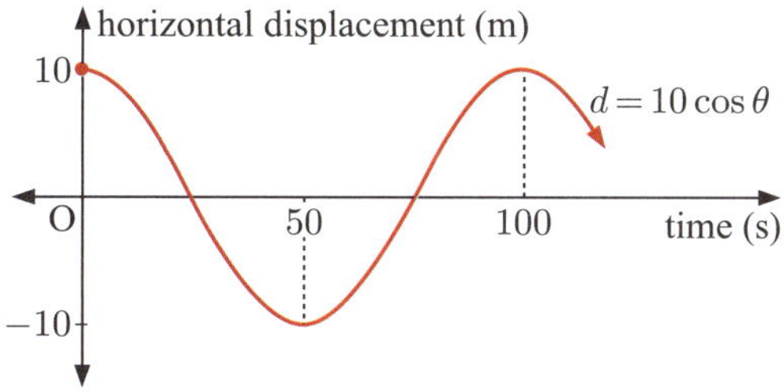

Use the graphing package to graph $y = \cos x$ and $y = \sin x$ on the same set of axes.

GRAPHING PACKAGE

Like the sine curve $y = \sin x$, the cosine curve $y = \cos x$ has **period** 2π and **amplitude** 1. Its **range** is $-1 \leqslant y \leqslant 1$.

You should observe that $y = \cos x$ and $y = \sin x$ are identical in shape, but the cosine function is $\frac{\pi}{2}$ units left of the sine function.

Use the graphing package to graph $y = \cos x$ and $y = \sin\left(x + \frac{\pi}{2}\right)$ on the same set of axes.

You should observe that for all real x,

$$\cos x = \sin\left(x + \tfrac{\pi}{2}\right).$$

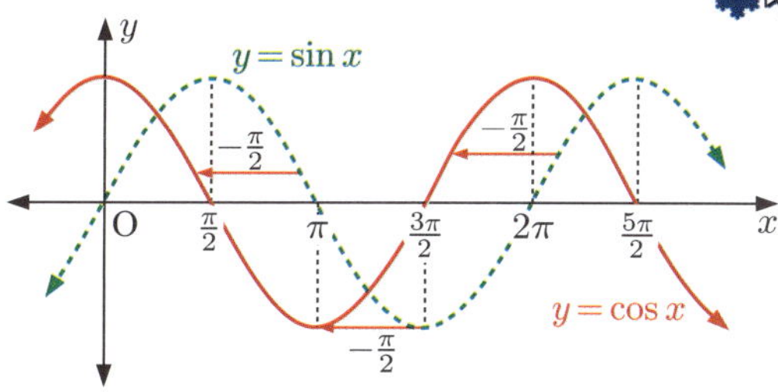

THE GENERAL COSINE FUNCTION

The **general cosine function** is $y = a\cos bx + c$ where $a > 0$, $b > 0$.

Since the cosine function is a horizontal translation of the sine function, the constants a, b, and c have the same effects as for the general sine function. Click on the icon to check this.

DYNAMIC COSINE FUNCTION

The **principal axis** of the general cosine function is $\boldsymbol{y = c}$.

The **period** of the general cosine function is $\dfrac{\mathbf{2\pi}}{\boldsymbol{b}}$.

The **amplitude** of the general cosine function is $\boldsymbol{a}$.

$y = a\cos bx + c$, $a > 0$ has a maximum when $x = 0$.

Example 4

Self Tutor

Without using technology, sketch the graph of $y = 3\cos 2x$ for $0 \leqslant x \leqslant 2\pi$.

$a = 3$, so the amplitude is 3.

$b = 2$, so the period is $\frac{2\pi}{b} = \frac{2\pi}{2} = \pi$.

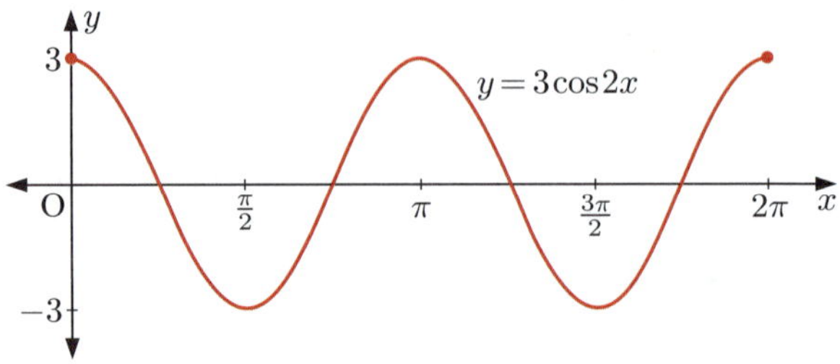

EXERCISE 9C

1 Given the graph of $y = \cos x$, sketch the following graphs for $0 \leqslant x \leqslant 2\pi$:

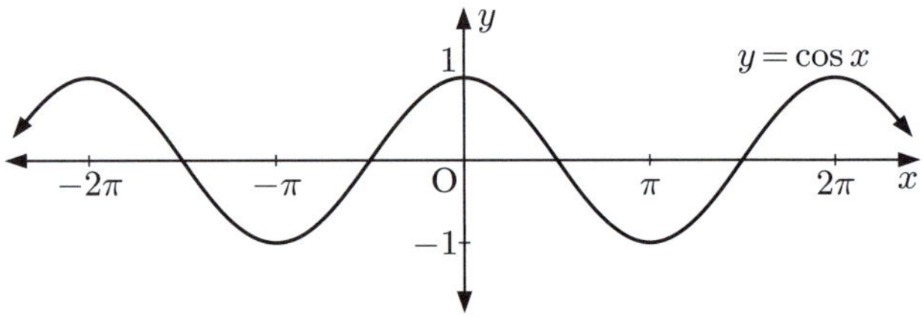

a $y = 3\cos x$ **b** $y = 5\cos x$ **c** $y = \cos 2x$

d $y = \cos \frac{x}{2}$ **e** $y = \cos x + 2$ **f** $y = \cos x - 1$

g $y = 2\cos 2x$ **h** $y = \cos 3x + 1$ **i** $y = 4\cos x + 10$

j $y = 2\cos 3x + 4$ **k** $y = 4\cos 2x - 2$ **l** $y = 3\cos 2x + 5$

2 Find a, b, and c given that the function $y = a\cos bx + c$, $a > 0$, $b > 0$, has:

a amplitude 4, period $\frac{2\pi}{3}$, and principal axis $y = -1$

b amplitude 3, period $\frac{2\pi}{5}$, and principal axis $y = 3$

c amplitude $\frac{1}{6}$, period $\frac{8\pi}{3}$, and principal axis $y = -4$.

3 Find the maximum and minimum values of:

a $y = 5\cos x$ **b** $y = 2\cos 3x + 1$ **c** $y = 3\cos 4x - 7$

4 Find the cosine function shown in the graph:

a

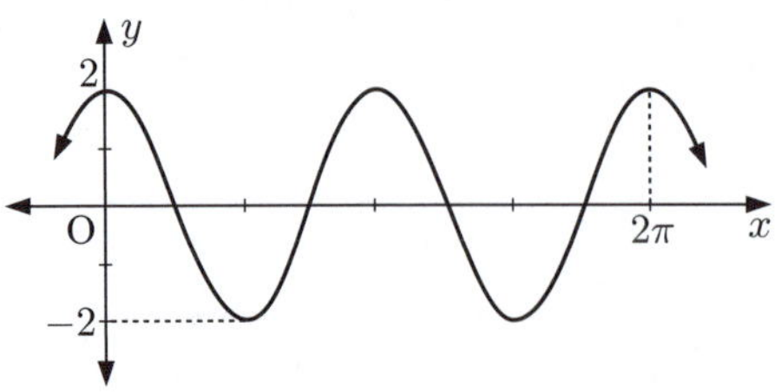

b

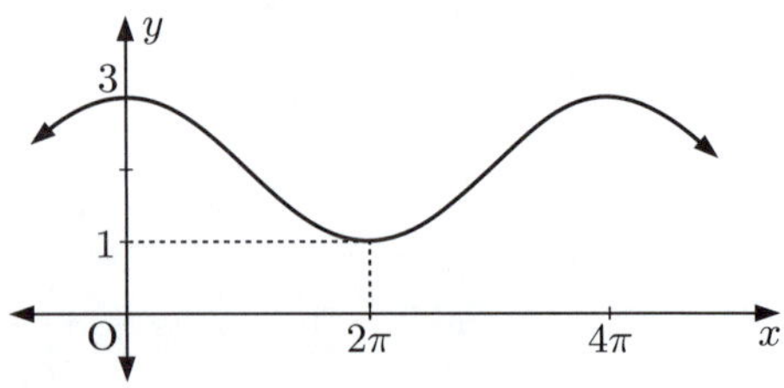

5 The function $y = a\cos bx + c$, $a > 0$, $b > 0$, has amplitude 5, period $\frac{3\pi}{2}$, and principal axis $y = 1$.

a Find the values of a, b, and c. **b** Sketch the function for $0 \leqslant x \leqslant 3\pi$.

6 Sketch the graphs of the following for $0° \leqslant x \leqslant 360°$:

a $y = \cos x + 1$ **b** $y = 3\cos x$ **c** $y = 2\cos 3x$

7 The graph shown has the form $y = a\cos bx + c$ where $a > 0$, $b > 0$.

a Find the values of a, b, and c.

b Sketch the reflection of the function in the x-axis.

c Write down the equation of the reflection in **b**.

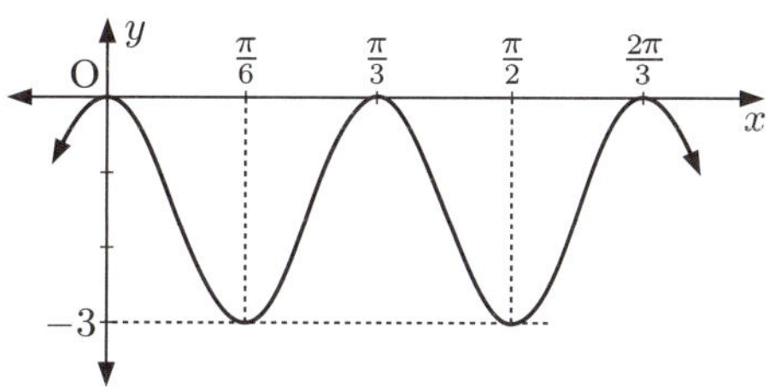

D THE TANGENT FUNCTION

We have seen that if $P(\cos\theta, \sin\theta)$ is a point which is free to move around the unit circle, and if OP is extended to meet the tangent at $A(1, 0)$, the intersection between these lines occurs at $Q(1, \tan\theta)$.

This allows us to define the **tangent function**

$$\tan\theta = \frac{\sin\theta}{\cos\theta}.$$

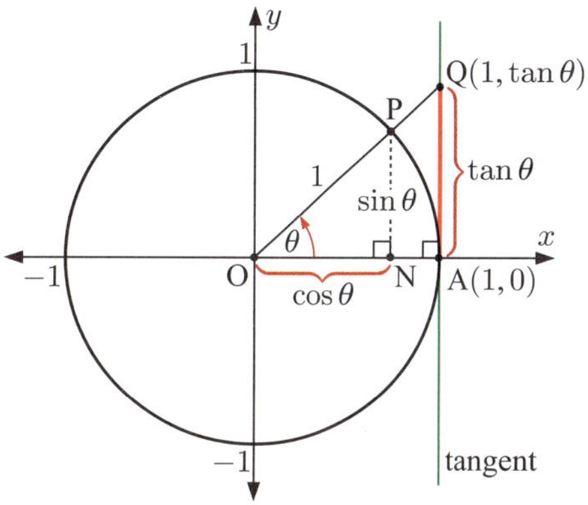

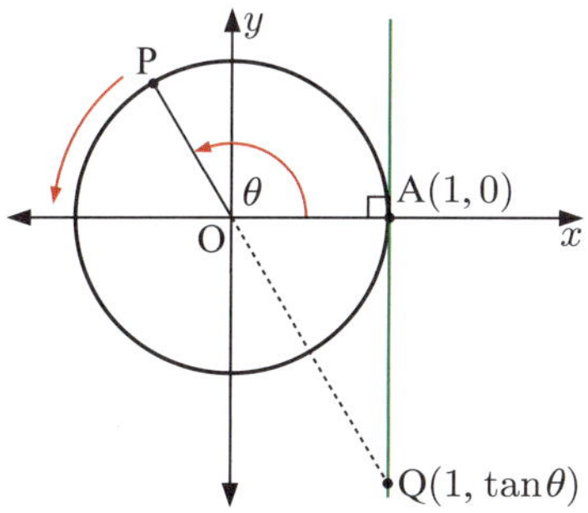

For θ in quadrant 2, $\sin\theta$ is positive and $\cos\theta$ is negative, so $\tan\theta = \dfrac{\sin\theta}{\cos\theta}$ is negative.

As before, OP is extended to meet the tangent at A at the point $Q(1, \tan\theta)$. We see that Q is below the x-axis.

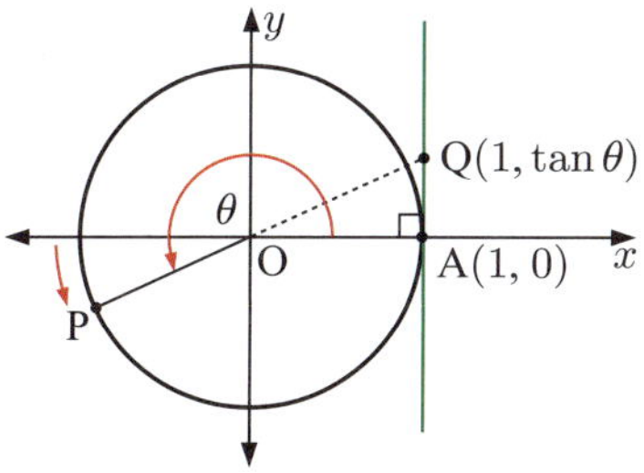

For θ in quadrant 3, $\sin\theta$ and $\cos\theta$ are both negative, and so $\tan\theta$ is positive. We see that Q is back above the x-axis.

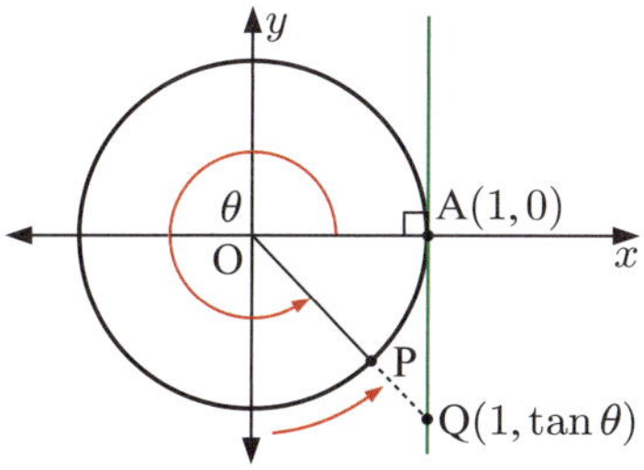

For θ in quadrant 4, $\sin\theta$ is negative and $\cos\theta$ is positive. $\tan\theta$ is again negative. We see that Q is below the x-axis.

Discussion

What happens to $\tan\theta$ when P is at $(0, 1)$ and $(0, -1)$?

THE GRAPH OF $y = \tan x$

Since $\tan x = \dfrac{\sin x}{\cos x}$, $\tan x$ will be undefined whenever $\cos x = 0$.

The zeros of the function $y = \cos x$ correspond to vertical asymptotes of the function $y = \tan x$.

The graph of $y = \tan x$ is shown alongside.

DEMO

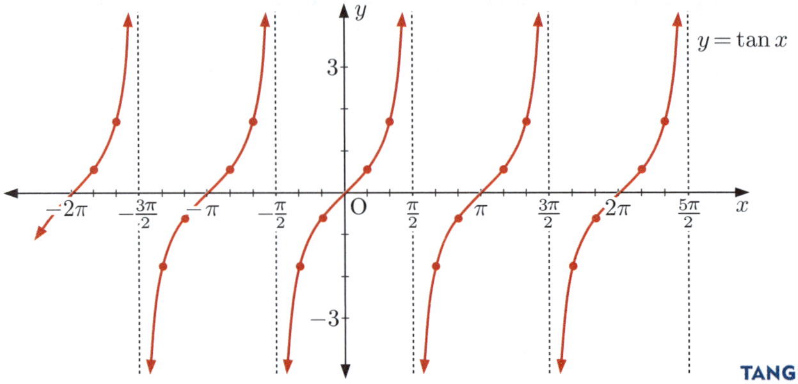

We observe that $y = \tan x$ has:

- **period** π
- **range** $y \in \mathbb{R}$
- **vertical asymptotes** $x = \frac{\pi}{2} + k\pi$ for all $k \in \mathbb{Z}$.

TANGENT FUNCTION

Click on the icon to explore how the tangent function is produced from the unit circle.

THE GENERAL TANGENT FUNCTION

The **general tangent function** is $y = a \tan bx + c$, $a > 0$, $b > 0$.

- The **principal axis** is $y = c$.
- The **period** is $\dfrac{\pi}{b}$.
- The **amplitude** is undefined.

DYNAMIC TANGENT FUNCTION

Click on the icon to explore the properties of this function.

Example 5

Self Tutor

Without using technology, sketch the graph of $y = \tan 3x$ for $-\pi \leqslant x \leqslant \pi$.

Since $b = 3$, the period is $\frac{\pi}{3}$.

The vertical asymptotes are $x = \pm\frac{\pi}{6}$, $x = \pm\frac{\pi}{2}$, $x = \pm\frac{5\pi}{6}$.

The x-axis intercepts are $0, \pm\frac{\pi}{3}, \pm\frac{2\pi}{3}, \pm\pi$.

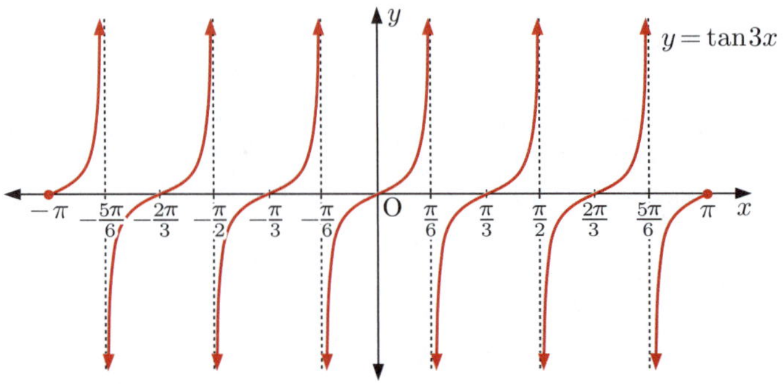

Discussion

- Discuss how to find the x-intercepts of $y = \tan x$.
- How can we simplify $\tan(x - \pi)$?
- How many solutions does the equation $\tan x = 2$ have?

EXERCISE 9D

1 State the period of:

a $y = \tan 2x$ **b** $y = 3\tan 4x$ **c** $y = \tan \frac{x}{6} - 3$

2 Sketch the following functions for $-\pi \leqslant x \leqslant \pi$:

a $y = 2\tan x$ **b** $y = \tan 2x$ **c** $y = \tan x + 2$

d $y = \tan x - 1$ **e** $y = \tan 3x + 1$ **f** $y = 2\tan \frac{x}{2} + 2$

Use technology to check your answers.

GRAPHING PACKAGE

3 Find b and c given that the function $y = \tan bx + c$, $b > 0$, has:

a period $\frac{2\pi}{3}$ and principal axis $y = 2$

b period $\frac{\pi}{2}$ and principal axis $y = -3$.

4 Find m and n given the following graph of the function $y = \tan mt + n$.

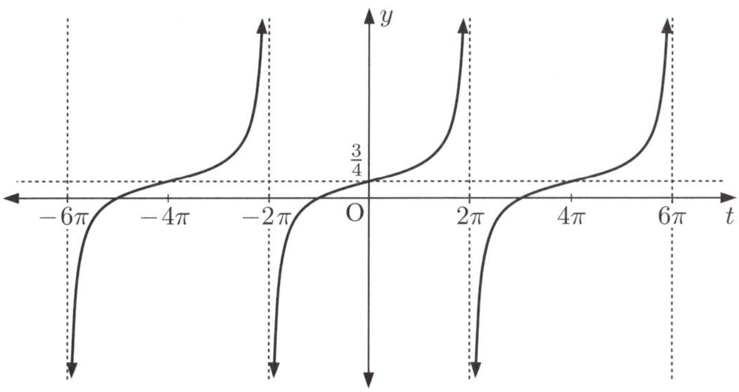

5 Suppose $y = a\tan bx + c$ has period $\frac{\pi}{3}$ and passes through the points $(0, 2)$ and $(\frac{\pi}{4}, -2)$. Find the values of a, b, and c.

Activity

Click on the icon to run a card game for trigonometric functions.

CARD GAME

E TRIGONOMETRIC EQUATIONS

Linear equations such as $2x + 3 = 11$ have exactly one solution. Quadratic equations of the form $ax^2 + bx + c = 0$, $a \neq 0$ have at most two real solutions.

Trigonometric equations generally have infinitely many solutions unless a restricted domain such as $0 \leqslant x \leqslant 3\pi$ is given.

For example, suppose that Andrew in the **Opening Problem** wants to know when the green light will be 16 metres above the ground. To find out, he will need to solve a trigonometric equation. If the wheel keeps rotating, the equation will have infinitely many solutions. Andrew may therefore specify that he is interested in the *first* time the green light is 16 metres above the ground, or look for solutions in a particular time period.

GRAPHICAL SOLUTION OF TRIGONOMETRIC EQUATIONS

Accurate graphs of trigonometric functions can be used to estimate solutions to trigonometric equations.

Example 6 — Self Tutor

Solve $\cos x = 0.4$ for $0 \leqslant x \leqslant 10$ radians using the graph of $y = \cos x$.

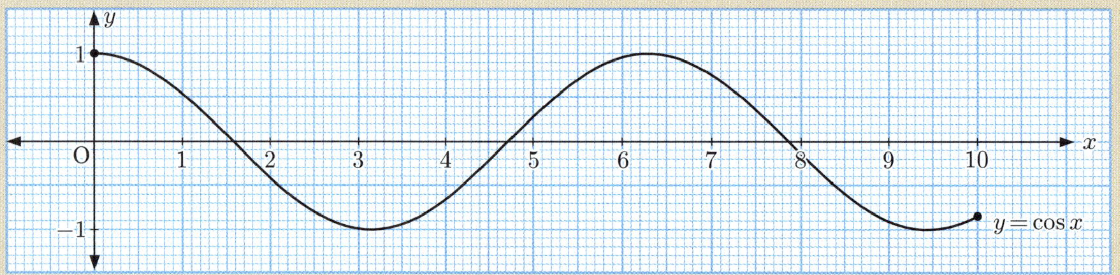

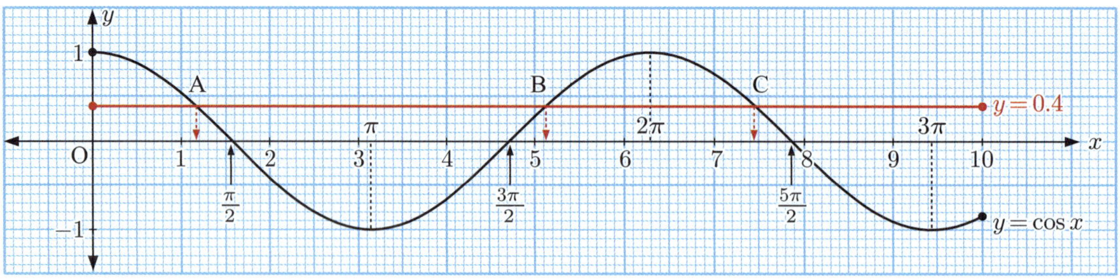

On the domain $0 \leqslant x \leqslant 10$, $y = 0.4$ meets $y = \cos x$ at A, B, and C.

$\therefore$ the solutions of $\cos x = 0.4$ for $0 \leqslant x \leqslant 10$ radians are $x \approx 1.2$, 5.1, and 7.4.

EXERCISE 9E.1

1

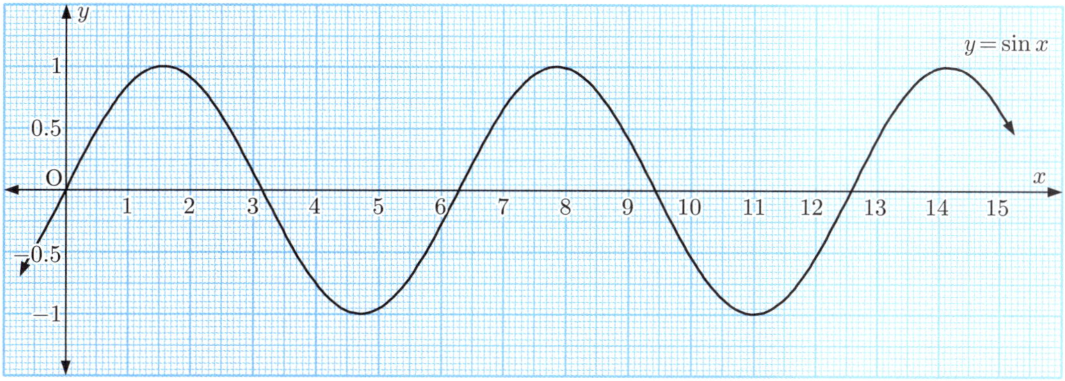

Use the graph of $y = \sin x$ to find, correct to 1 decimal place, the solutions of:

a $\sin x = 0.3$ for $0 \leqslant x \leqslant 15$

b $\sin x = -0.4$ for $5 \leqslant x \leqslant 15$.

2

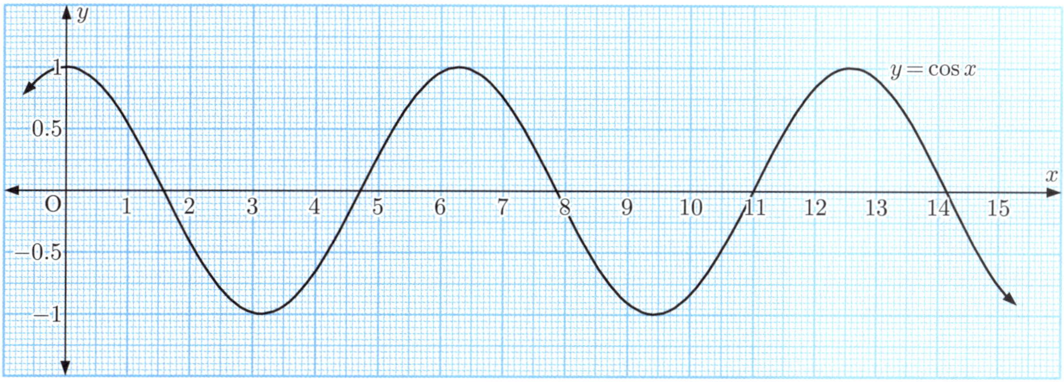

Use the graph of $y = \cos x$ to find, correct to 1 decimal place, the solutions of:

a $\cos x = 0.4$ for $0 \leqslant x \leqslant 10$

b $\cos x = -0.3$ for $4 \leqslant x \leqslant 12$.

3

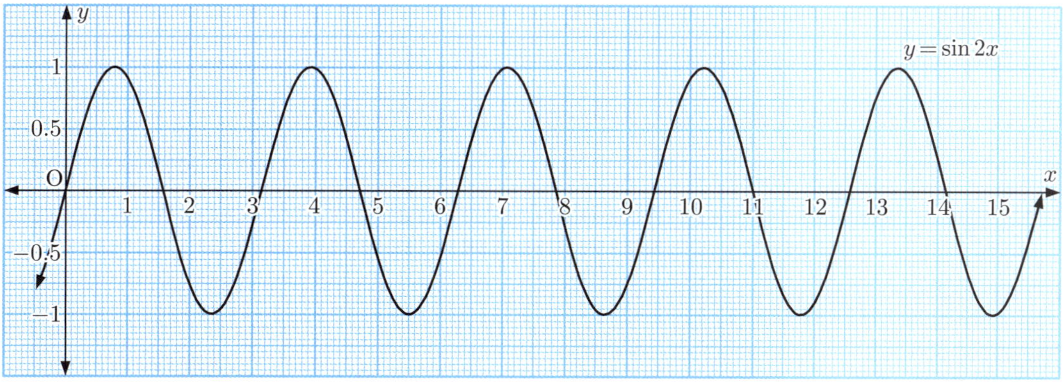

Use the graph of $y = \sin 2x$ to find, correct to 1 decimal place, the solutions of:

a $\sin 2x = 0.7$ for $0 \leqslant x \leqslant 15$

b $\sin 2x = -0.3$ for $0 \leqslant x \leqslant 15$.

4 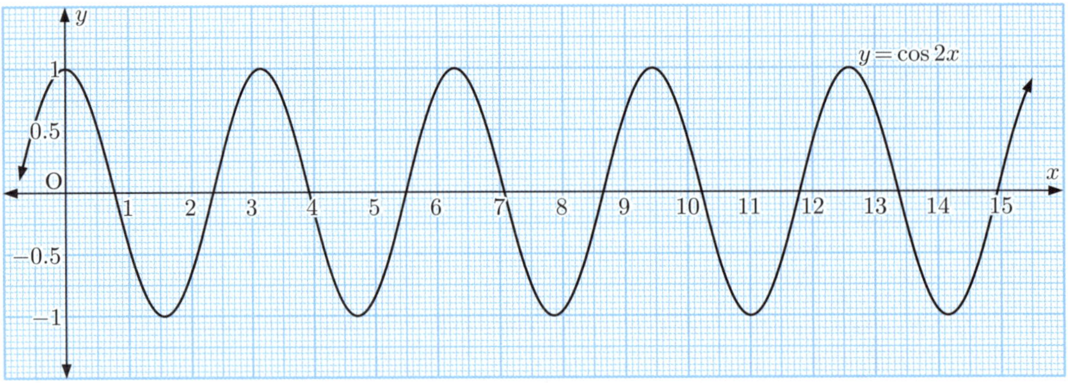

Use the graph of $y = \cos 2x$ to find, correct to 1 decimal place, the solutions of:

a $\cos 2x = 0.8$ for $0 \leqslant x \leqslant 15$ **b** $\cos 2x = -0.2$ for $0 \leqslant x \leqslant 15$.

5 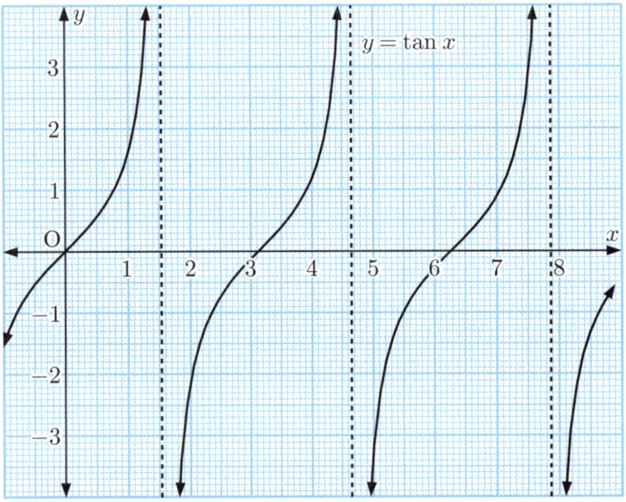

a Use the graph of $y = \tan x$ to estimate: **i** $\tan 1$ **ii** $\tan 2.3$
Check your answers with a calculator.

b Use the graph to find, correct to 1 decimal place, the solutions of:
i $\tan x = 2$ for $0 \leqslant x \leqslant 8$ **ii** $\tan x = -1.4$ for $2 \leqslant x \leqslant 7$.

SOLVING TRIGONOMETRIC EQUATIONS USING ALGEBRA

Using a graph we get approximate decimal or **numerical** solutions to trigonometric equations.

Sometimes exact solutions are needed in terms of π, and these arise when the solutions are multiples of $\frac{\pi}{6}$ or $\frac{\pi}{4}$. Exact solutions obtained using algebra are called **analytical** solutions.

We use the periodicity of the trigonometric functions to give us all solutions in the required domain.

For example, consider $\sin x = 1$. We know from the unit circle that a solution is $x = \frac{\pi}{2}$. However, since the period of $\sin x$ is 2π, there are infinitely many solutions spaced 2π apart.

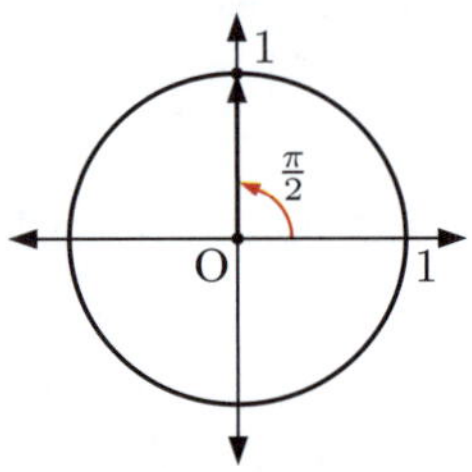

Hence $x = \frac{\pi}{2} + k2\pi$ is a solution for any $k \in \mathbb{Z}$.

In this course we will be solving equations on a finite domain. This means there will be a finite number of solutions.

Reminder:

You should remember the trigonometric ratios for angles which are multiples of $\frac{\pi}{4}$ or $\frac{\pi}{6}$.

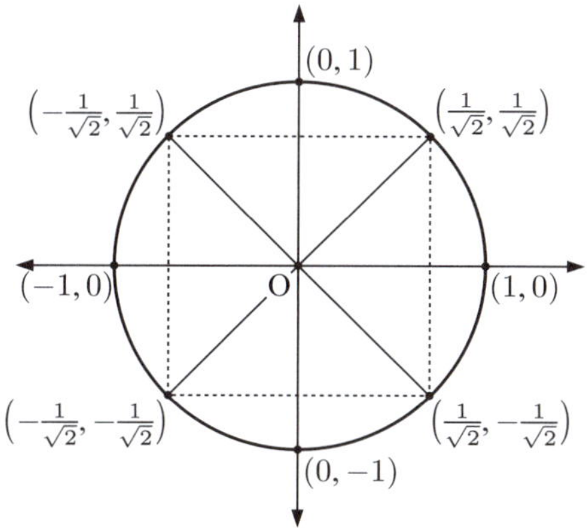

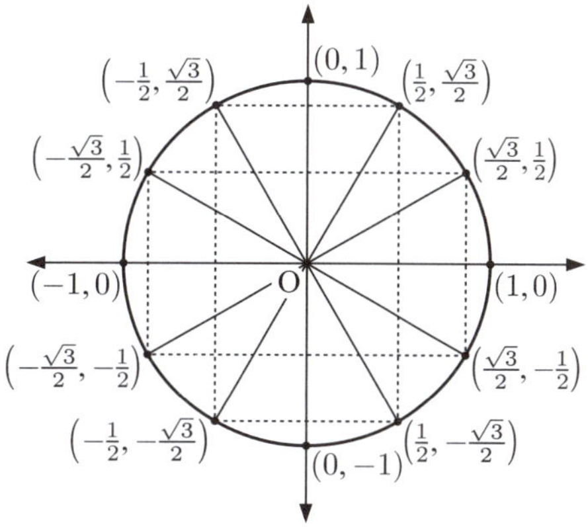

Example 7 — Self Tutor

Solve for x:

a $2\sin x - \sqrt{3} = 0$ for $0 \leqslant x \leqslant \pi$

b $\tan x + 1 = 0$ for $0 < x < 4\pi$

a $2\sin x - \sqrt{3} = 0$

$\therefore \sin x = \frac{\sqrt{3}}{2}$

There are two points on the unit circle with sine $\frac{\sqrt{3}}{2}$.

They correspond to angles $\frac{\pi}{3}$ and $\frac{2\pi}{3}$.

These are the only solutions on the domain $0 \leqslant x \leqslant \pi$, so $x = \frac{\pi}{3}$ or $\frac{2\pi}{3}$.

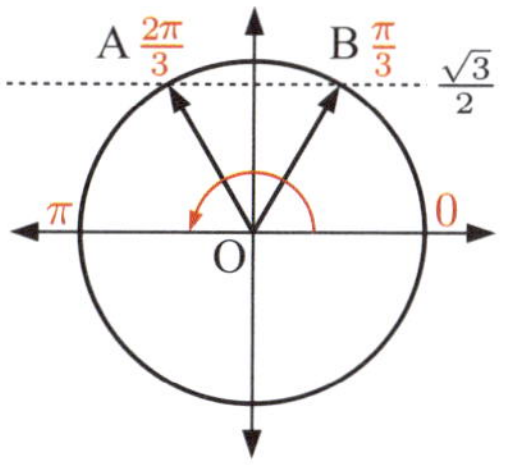

b $\tan x + 1 = 0$

$\therefore \tan x = -1$

There are two points on the unit circle with tangent -1.

They correspond to angles $\frac{3\pi}{4}$ and $\frac{7\pi}{4}$.

For the domain $0 < x < 4\pi$ we have 4 solutions: $x = \frac{3\pi}{4}, \frac{7\pi}{4}, \frac{11\pi}{4}$, or $\frac{15\pi}{4}$.

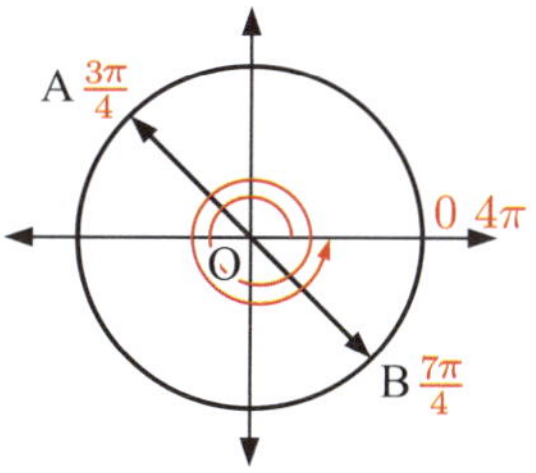

EXERCISE 9E.2

1 Solve for x on the domain $0 \leqslant x \leqslant 4\pi$:

a $2\cos x - 1 = 0$ **b** $\sqrt{2}\sin x = -1$ **c** $\tan x = \sqrt{3}$

2 Solve for x on the domain $-2\pi \leqslant x \leqslant 2\pi$:

a $2\sin x - \sqrt{3} = 0$ **b** $\sqrt{2}\cos x - 1 = 0$ **c** $\tan x - 1 = 0$

Example 8

Self Tutor

Solve exactly for $0 \leqslant x \leqslant 3\pi$: **a** $\sin x = \frac{1}{2}$ **b** $\sin 2x = \frac{1}{2}$

The equations both have the form $\sin\theta = \frac{1}{2}$.

There are two points on the unit circle with sine $\frac{1}{2}$.

They correspond to angles $\frac{\pi}{6}$ and $\frac{5\pi}{6}$.

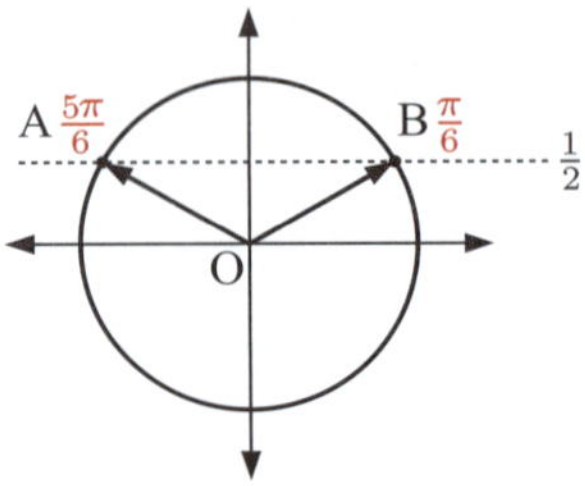

a In this case θ is simply x, so we have the domain $0 \leqslant x \leqslant 3\pi$.
The solutions for this domain are
$x = \frac{\pi}{6}, \frac{5\pi}{6}, \frac{13\pi}{6}, \text{ or } \frac{17\pi}{6}$.

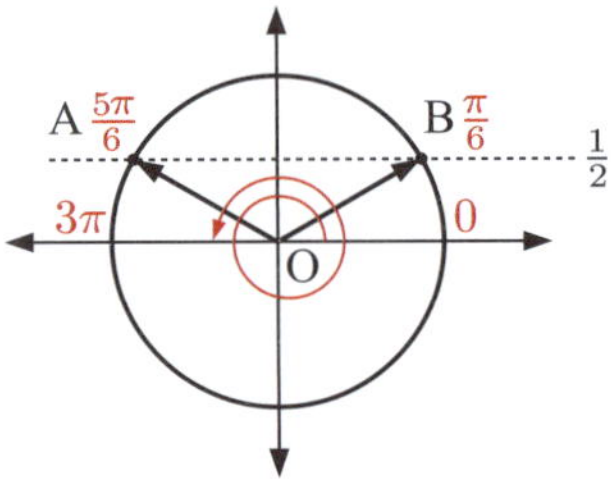

b In this case θ is $2x$.
If $0 \leqslant x \leqslant 3\pi$ then $0 \leqslant 2x \leqslant 6\pi$.

$\therefore\ 2x = \frac{\pi}{6}, \frac{5\pi}{6}, \frac{13\pi}{6}, \frac{17\pi}{6}, \frac{25\pi}{6}, \text{ or } \frac{29\pi}{6}$

$\therefore\ x = \frac{\pi}{12}, \frac{5\pi}{12}, \frac{13\pi}{12}, \frac{17\pi}{12}, \frac{25\pi}{12}, \text{ or } \frac{29\pi}{12}$

3 Solve exactly for $0 \leqslant x \leqslant 3\pi$:

a $\cos x = -\frac{1}{2}$ **b** $\cos 2x = -\frac{1}{2}$

4 Solve exactly for $0 \leqslant x \leqslant 2\pi$:

a $\sin x = -\frac{1}{\sqrt{2}}$ **b** $\sin 3x = -\frac{1}{\sqrt{2}}$

5 Solve exactly for $0° \leqslant x \leqslant 360°$:

a $\tan x = \frac{1}{\sqrt{3}}$ **b** $\tan\frac{x}{2} = \frac{1}{\sqrt{3}}$

6 Find the exact solutions of:

a $\cos 2x = -\frac{1}{2},\ 0 \leqslant x \leqslant 2\pi$ **b** $2\sin\frac{x}{2} - 1 = 0,\ -360° \leqslant x \leqslant 360°$

c $\tan 3x + \sqrt{3} = 0,\ 0 \leqslant x \leqslant \pi$ **d** $3\cos 2x + 3 = 0,\ 0 \leqslant x \leqslant 3\pi$

e $4\cos 3x + 2 = 0,\ -180° \leqslant x \leqslant 180°$ **f** $5\tan 2x - 5 = 0,\ 0 \leqslant x \leqslant 2\pi$

7 Find the coordinates of A, B, C, and D:

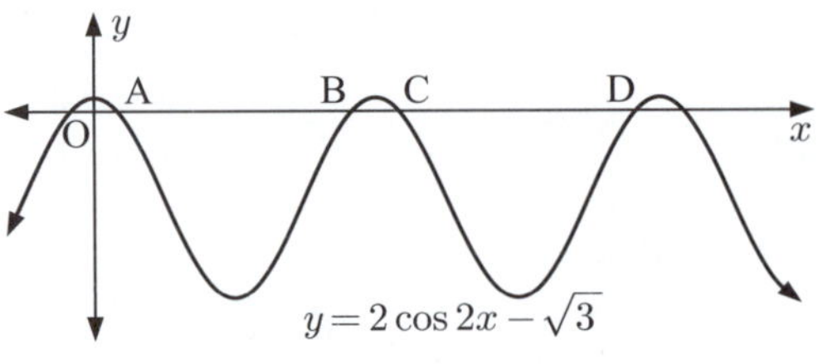

8 **a** Sketch the graph of $y = 2\sin\frac{5x}{4} + 1$ for $0° \leqslant x \leqslant 360°$.

b Find the coordinates of the points where the function cuts the x-axis.

Example 9

Self Tutor

Solve $2\cos\left(x+\frac{\pi}{2}\right)=\sqrt{3}$ for $0 \leqslant x \leqslant 2\pi$.

$2\cos\left(x+\frac{\pi}{2}\right)=\sqrt{3}$

$\therefore \cos\left(x+\frac{\pi}{2}\right)=\frac{\sqrt{3}}{2}$

If $0 \leqslant x \leqslant 2\pi$, then $\frac{\pi}{2} \leqslant x+\frac{\pi}{2} \leqslant \frac{5\pi}{2}$

$\therefore x+\frac{\pi}{2}=\frac{11\pi}{6}$ or $\frac{13\pi}{6}$

$\therefore x=\frac{4\pi}{3}$ or $\frac{5\pi}{3}$

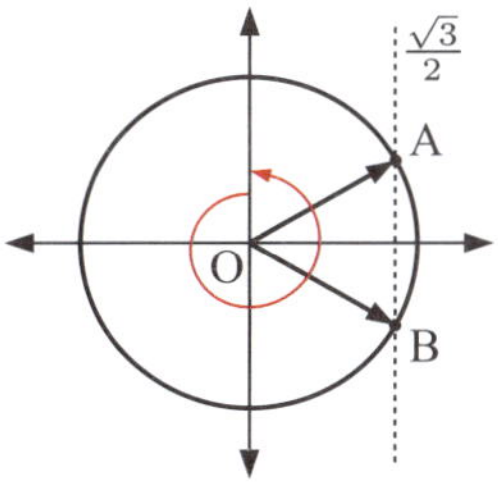

9 Solve:

- **a** $\sin\left(x+\frac{\pi}{3}\right)=\frac{1}{\sqrt{2}}$ for $0 \leqslant x \leqslant 2\pi$
- **b** $\tan\left(x-\frac{\pi}{2}\right)=-1$ for $0 \leqslant x \leqslant 2\pi$
- **c** $2\cos(x-\pi)=-1$ for $-2\pi \leqslant x \leqslant 2\pi$
- **d** $\sin\left(2x+\frac{\pi}{2}\right)=\frac{\sqrt{3}}{2}$ for $0 \leqslant x \leqslant \pi$
- **e** $\cos\left(3x-\frac{\pi}{3}\right)=-1$ for $0 \leqslant x \leqslant \pi$
- **f** $-\sqrt{3}\tan(2x-\pi)=1$ for $-\pi \leqslant x \leqslant \pi$

Example 10

Self Tutor

Find the exact solutions of $\sqrt{3}\sin x=\cos x$ for $0° \leqslant x \leqslant 360°$.

$\sqrt{3}\sin x=\cos x$

$\therefore \frac{\sin x}{\cos x}=\frac{1}{\sqrt{3}}$ {dividing both sides by $\sqrt{3}\cos x$}

$\therefore \tan x=\frac{1}{\sqrt{3}}$

$\therefore x=30°$ or $210°$

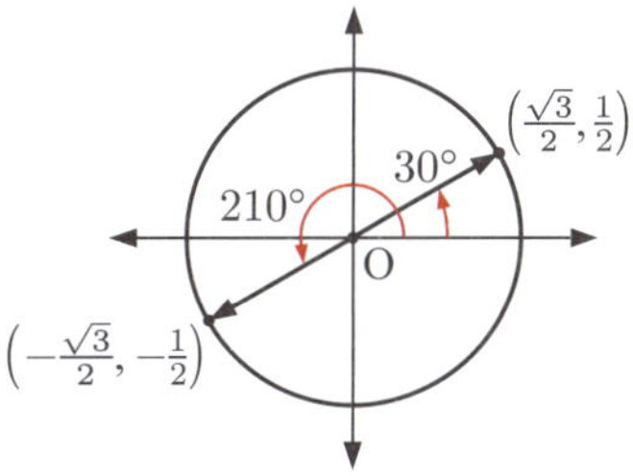

10 Solve for $0 \leqslant x \leqslant 2\pi$:

- **a** $\sin x-\cos x=0$
- **b** $\sin x=-\cos x$
- **c** $\sin 3x=\cos 3x$
- **d** $\sin 2x=\sqrt{3}\cos 2x$

Check your answers using the graphing package.

GRAPHING PACKAGE

11 **a** If $\tan^2 x=3$, state the possible values of $\tan x$.

b Hence solve $\tan^2 x=3$ for $0 \leqslant x \leqslant 2\pi$.

12 Solve:

- **a** $\sin^2 x=\frac{1}{4}$ for $0 \leqslant x \leqslant 2\pi$
- **b** $\cos^2 x=\frac{1}{2}$ for $0° \leqslant x \leqslant 360°$
- **c** $\tan^2 x=1$ for $0 \leqslant x \leqslant 4\pi$
- **d** $3\sin^2 x=3$ for $-2\pi \leqslant x \leqslant 2\pi$
- **e** $\tan^2 2x=\frac{1}{3}$ for $0° \leqslant x \leqslant 180°$
- **f** $\cos^2\left(x+\frac{\pi}{2}\right)=\frac{3}{4}$ for $-\pi \leqslant x \leqslant \pi$
- **g** $|\sin x|=\frac{\sqrt{3}}{2}$ for $0 \leqslant x \leqslant 2\pi$
- **h** $|\tan 3x|-2=-1$ for $0 \leqslant x \leqslant \pi$

13 **a** If $\sec x=2$, find the value of $\cos x$.

b Hence solve $\sec x=2$ for $0 \leqslant x \leqslant 2\pi$.

14 Solve for $-\pi \leqslant x \leqslant \pi$:

a $\operatorname{cosec} x = -2$ **b** $\sec x = \sqrt{2}$ **c** $\cot x = \sqrt{3}$

d $\sec 2x = -1$ **e** $\sqrt{3}\operatorname{cosec} 2x = 2$ **f** $\cot\left(x - \frac{\pi}{4}\right) = -\frac{1}{\sqrt{3}}$

15 Solve for $0 \leqslant x \leqslant 2\pi$:

a $\sin x = \operatorname{cosec} x$ **b** $4\cos x = \sec x$ **c** $3\tan x - \cot x = 0$

SOLVING TRIGONOMETRIC EQUATIONS USING A CALCULATOR

If the solutions of trigonometric equations are not multiples of $\frac{\pi}{6}$ or $\frac{\pi}{4}$, then we must use the $\cos^{-1}$, $\sin^{-1}$, or $\tan^{-1}$ function on our calculator to solve the equation.

We use the unit circle and periodicity to find all of the solutions within the required domain.

When solving trigonometric equations using your calculator, you must make sure your calculator is correctly set to either *degree* or *radian* mode.

Example 11 — Self Tutor

Solve for x, giving your answers correct to 2 decimal places:

a $\cos x = 0.35$ for $0 \leqslant x \leqslant 2\pi$ **b** $\tan 2x = -\frac{1}{3}$ for $-\pi \leqslant x \leqslant \pi$

a $\cos^{-1}(0.35) \approx 1.21$

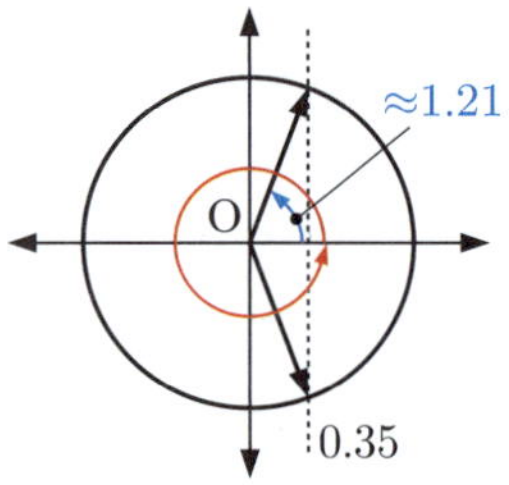

$\therefore\ x \approx 1.21$ or $2\pi - 1.21$

$\therefore\ x \approx 1.21$ or 5.07

b $\tan^{-1}\left(-\frac{1}{3}\right) \approx -0.32$

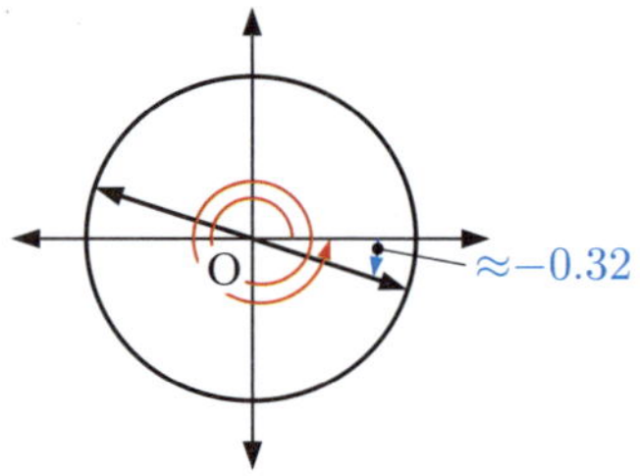

If $-\pi \leqslant x \leqslant \pi$ then $-2\pi \leqslant 2x \leqslant 2\pi$

$\therefore\ 2x \approx -3.46,\ -0.32,\ 2.82,$ or 5.96

($-3.46 = -0.32 - \pi$, $2.82 = -0.32 + \pi$, $5.96 = -0.32 + 2\pi$)

$\therefore\ x \approx -1.73,\ -0.16,\ 1.41,$ or 2.98

EXERCISE 9E.3

1 Solve for $0 \leqslant x \leqslant 2\pi$, giving your answers correct to 2 decimal places:

a $\sin x = 0.8$ **b** $\cos x = \frac{1}{4}$ **c** $\tan x = 0.7$

d $\cos x = -0.23$ **e** $\sin x = -\frac{2}{3}$ **f** $\tan x = -1.5$

2 Solve for x, giving your answers correct to 2 decimal places:

a $\sin x = 0.18$ on $0 \leqslant x \leqslant 4\pi$ **b** $\cos x = 0.63$ on $-2\pi \leqslant x \leqslant 2\pi$

c $\tan x = -0.6$ on $0 \leqslant x \leqslant 3\pi$ **d** $\sin 2x = 0.45$ on $0 \leqslant x \leqslant 2\pi$

e $\tan(x+1) = 0.3$ on $-\pi \leqslant x \leqslant \pi$ **f** $\cos 3x = -0.55$ on $0 \leqslant x \leqslant 5$

g $3\sin\left(x - \frac{\pi}{2}\right) = 2$ on $0 \leqslant x \leqslant 2\pi$ **h** $5\cos\left(2x - \frac{\pi}{4}\right) = -3$ on $0 \leqslant x \leqslant \pi$

3 Solve for x, giving your answers correct to 1 decimal place:

a $3\sin x + 1 = 0$ on $0° \leqslant x \leqslant 360°$
b $5\cos 2x - 1 = 0$ on $0° \leqslant x \leqslant 180°$
c $4\tan 3x = 5$ on $-180° \leqslant x \leqslant 180°$
d $6\sin 4x = -1$ on $0° \leqslant x \leqslant 180°$
e $4\cos 3x = -3$ on $0° \leqslant x \leqslant 180°$
f $5\tan(2x - 60°) = 2$ on $0° \leqslant x \leqslant 360°$

4 Solve for $0 \leqslant x \leqslant 2\pi$, giving your answers correct to 2 decimal places:

a $\operatorname{cosec} x = 3$
b $\sec x = -5$
c $\cot x = 2$
d $\sec 2x = \frac{9}{2}$
e $\cot\left(x + \frac{\pi}{2}\right) = \frac{3}{4}$
f $\operatorname{cosec}(2x - 1) = \frac{3}{2}$

F TRIGONOMETRIC RELATIONSHIPS

SIMPLIFYING TRIGONOMETRIC EXPRESSIONS

For any given angle θ, $\sin\theta$ and $\cos\theta$ are real numbers. $\tan\theta$ is also real whenever it is defined. The algebra of trigonometry is therefore identical to the algebra of real numbers.

An expression like $3\cos\theta + 4\cos\theta$ can be compared with $3x + 4x$, so $3\cos\theta + 4\cos\theta = 7\cos\theta$.

Example 12 — Self Tutor

Simplify:

a $2\sin\theta + 6\sin\theta$
b $4\cos\alpha - 9\cos\alpha$

a $2\sin\theta + 6\sin\theta = 8\sin\theta$
{compare with $2x + 6x = 8x$}

b $4\cos\alpha - 9\cos\alpha = -5\cos\alpha$
{compare with $4x - 9x = -5x$}

To simplify more complicated trigonometric expressions, we can use **trigonometric identities**. These are rules that are true for all values of the angle.

We have already seen the definition $\boldsymbol{\tan\theta = \dfrac{\sin\theta}{\cos\theta}}$

and the **Pythagorean identity** $\boldsymbol{\sin^2\theta + \cos^2\theta = 1}$.

The Pythagorean identity can be rearranged into the forms $\sin^2\theta = 1 - \cos^2\theta$
and $\cos^2\theta = 1 - \sin^2\theta$.

Example 13 — Self Tutor

Simplify:

a $\cos^2\theta - 1$
b $\tan\theta \times \cos\theta$

a
$$\begin{aligned}&\cos^2\theta - 1\\ &= (1 - \sin^2\theta) - 1\\ &= -\sin^2\theta\end{aligned}$$

b
$$\begin{aligned}&\tan\theta \times \cos\theta\\ &= \frac{\sin\theta}{\cos\theta} \times \cos\theta\\ &= \sin\theta\end{aligned}$$

We can use $\sin^2\theta+\cos^2\theta=1$ and $\tan\theta=\dfrac{\sin\theta}{\cos\theta}$ to establish two more important identities:

- $\tan^2\theta+1=\sec^2\theta$
- $1+\cot^2\theta=\operatorname{cosec}^2\theta$

Proof:

- $\tan^2\theta+1=\dfrac{\sin^2\theta}{\cos^2\theta}+\dfrac{\cos^2\theta}{\cos^2\theta}$
$=\dfrac{\sin^2\theta+\cos^2\theta}{\cos^2\theta}$
$=\dfrac{1}{\cos^2\theta}$
$=\sec^2\theta$

- $1+\cot^2\theta=\dfrac{\sin^2\theta}{\sin^2\theta}+\dfrac{\cos^2\theta}{\sin^2\theta}$
$=\dfrac{\sin^2\theta+\cos^2\theta}{\sin^2\theta}$
$=\dfrac{1}{\sin^2\theta}$
$=\operatorname{cosec}^2\theta$

Example 14 — Self Tutor

Simplify:

a $3\tan^2\theta+3$ **b** $\cos^2\theta(1+\cot^2\theta)$

a $3\tan^2\theta+3$
$=3(\tan^2\theta+1)$
$=3\sec^2\theta$

b $\cos^2\theta(1+\cot^2\theta)$
$=(1-\sin^2\theta)\operatorname{cosec}^2\theta$
$=\operatorname{cosec}^2\theta-1$
$=\cot^2\theta$ $\{$as $1+\cot^2\theta=\operatorname{cosec}^2\theta\}$

EXERCISE 9F.1

1 Simplify:

a $3\sin\theta+5\sin\theta$ **b** $\tan\theta+4\tan\theta$ **c** $7\cos\theta-5\cos\theta$
d $2\tan\theta-6\tan\theta$ **e** $\sin^2\theta+3\sin^2\theta$ **f** $4\cos^2\theta-7\cos^2\theta$

2 Simplify:

a $4\sin^2\theta+4\cos^2\theta$ **b** $-3\sin^2\theta-3\cos^2\theta$ **c** $\cos^2\theta+3+\sin^2\theta$
d $2-2\cos^2\theta$ **e** $5-5\sin^2 A$ **f** $\sin^2\theta-1$
g $4\cos^2 A-4$ **h** $\dfrac{1-\sin^2\theta}{\cos\theta}$ **i** $\dfrac{\cos^2\theta-1}{\sin^2\theta}$

3 Simplify:

a $5\tan\theta-\dfrac{\sin\theta}{\cos\theta}$ **b** $\dfrac{\sin^3\alpha}{\cos^3\alpha}$ **c** $\dfrac{\tan x}{\sin x}$
d $\sin x+3\cos x\tan x$ **e** $2\sec\theta\sin\theta$ **f** $\cot\alpha\tan\alpha$
g $\dfrac{\operatorname{cosec}\theta}{\sec\theta}$ **h** $\dfrac{\tan A+\cot A}{\operatorname{cosec} A}$ **i** $\dfrac{2\sin x\cot x+3\cos x}{\cot x}$

4 Simplify:

a $5\tan^2\theta+5$ **b** $4+4\cot^2 A$ **c** $\sin^2\theta(1+\cot^2\theta)$
d $\sec^2\theta-1$ **e** $\dfrac{\tan^2\alpha+1}{\operatorname{cosec}^2\alpha}$ **f** $\dfrac{\operatorname{cosec}^2 A-\cot^2 A}{\sec A}$

5 Show that:

a $(\cos\theta + \sin\theta)^2 = 1 + 2\cos\theta\sin\theta$

b $\sin\theta\tan\theta - \sec\theta = -\cos\theta$

c $(1 - \sin\theta)(1 + \operatorname{cosec}\theta) = \cos\theta\cot\theta$

d $\cos^4\theta + \sin^2\theta + \sin^2\theta\cos^2\theta = 1$

e $(\operatorname{cosec}\theta + \cot\theta)(1 - \cos\theta) = \sin\theta$

f $\dfrac{1}{1 + \operatorname{cosec}\theta} + \dfrac{1}{1 - \operatorname{cosec}\theta} = -2\tan^2\theta$

FACTORISING TRIGONOMETRIC EXPRESSIONS

Example 15 — Self Tutor

Factorise:

a $\tan^2\beta - \sin^2\beta$

b $\cos^2\theta - 6\cos\theta + 8$

a $\tan^2\beta - \sin^2\beta$
$= (\tan\beta + \sin\beta)(\tan\beta - \sin\beta)$ {compare with $a^2 - b^2 = (a+b)(a-b)$}

b $\cos^2\theta - 6\cos\theta + 8$
$= (\cos\theta - 4)(\cos\theta - 2)$ {compare with $x^2 - 6x + 8 = (x-4)(x-2)$}

Example 16 — Self Tutor

Simplify: $\dfrac{\sin\alpha + 1}{\sin^2\alpha - 1}$

$$\frac{\sin\alpha + 1}{\sin^2\alpha - 1} = \frac{\cancel{\sin\alpha + 1}^{\,1}}{(\cancel{\sin\alpha + 1})(\sin\alpha - 1)}$$

$$= \frac{1}{\sin\alpha - 1}$$

EXERCISE 9F.2

1 Factorise:

a $\cos^2\theta - \sin^2\theta$

b $4 - \tan^2 A$

c $\sin^2\theta - 1$

d $2\cos^2\theta + \cos\theta$

e $\sin^2\alpha - 5\sin\alpha$

f $4\sin^2\phi - \cos^2\phi$

g $\tan^2\theta - \tan\theta - 2$

h $\cos^2 A + 5\cos A + 4$

i $\sin^2\beta - 3\sin\beta - 10$

j $2\cos^2\theta - \cos\theta - 6$

k $3\tan^2\phi + \tan\phi - 2$

l $6\sin^2 A + 11\sin A - 10$

2 Factorise:

a $1 - \sin^2\theta - \cos\theta$

b $\sin\theta - \cos^2\theta - 1$

c $\sec^2\theta + 2\tan\theta$

3 Simplify:

a $\dfrac{1 - \cos^2\theta}{1 - \cos\theta}$

b $\dfrac{\cos^2 A - \sin^2 A}{\cos A + \sin A}$

c $\dfrac{9 - \tan^2\phi}{3 - \tan\phi}$

d $\dfrac{\sin^2\theta + 2\sin\theta + 1}{\sin\theta + 1}$

e $\dfrac{\cos^2\phi + \cos\phi - 2}{1 - \cos\phi}$

f $\dfrac{\sec^2\theta - 2}{\tan\theta + 1}$

4 Show that $\dfrac{\cos A}{1-\tan A} - \dfrac{\sin A}{\cot A - 1} = \cos A + \sin A$.

5 Write $3\cos x \cot x - 2\cot x + 3\cos x - 2$ in the form $\dfrac{(a\cos x - b)(\cos x + \sin x)}{\sin x}$, where $a, b \in \mathbb{Z}$.

USING SIMPLIFICATION AND FACTORISATION TO SOLVE TRIGONOMETRIC EQUATIONS

Example 17 — Self Tutor

a Show that $\dfrac{\sec\theta}{\sec\theta - \cos\theta} = \operatorname{cosec}^2\theta$.

b Hence solve $\dfrac{\sec\theta}{\sec\theta - \cos\theta} = 2$ for $0 \leqslant x \leqslant 2\pi$.

a $$\begin{aligned}\frac{\sec\theta}{\sec\theta - \cos\theta} &= \frac{\sec\theta}{\sec\theta - \cos\theta} \times \frac{\cos\theta}{\cos\theta} \\ &= \frac{1}{1-\cos^2\theta} \\ &= \frac{1}{\sin^2\theta} \\ &= \operatorname{cosec}^2\theta\end{aligned}$$

b $$\frac{\sec\theta}{\sec\theta - \cos\theta} = 2$$

$\therefore\ \operatorname{cosec}^2\theta = 2$ {from **a**}

$\therefore\ \sin^2\theta = \frac{1}{2}$

$\therefore\ \sin\theta = \pm\frac{1}{\sqrt{2}}$

$\therefore\ \theta = \frac{\pi}{4}, \frac{3\pi}{4}, \frac{5\pi}{4}$, or $\frac{7\pi}{4}$

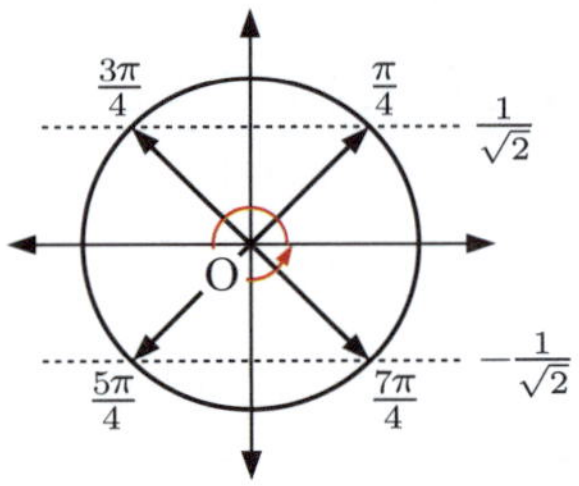

EXERCISE 9F.3

1 **a** Show that $(1+\cos\theta)(\operatorname{cosec}\theta - \cot\theta) = \sin\theta$.

b Hence solve $(1+\cos\theta)(\operatorname{cosec}\theta - \cot\theta) = -\frac{1}{2}$ for $0 \leqslant \theta \leqslant 2\pi$.

2 **a** Show that $(\sec\theta - \tan\theta)(1+\operatorname{cosec}\theta) = \cot\theta$.

b Hence solve $(\sec\theta - \tan\theta)(1+\operatorname{cosec}\theta) = \sqrt{3}$ for $0° \leqslant \theta \leqslant 360°$.

3 **a** Show that $\dfrac{\operatorname{cosec} A}{\operatorname{cosec} A - \sin A} = \sec^2 A$.

b Hence solve $\dfrac{\operatorname{cosec} A}{\operatorname{cosec} A - \sin A} = \frac{4}{3}$ for $0 \leqslant A \leqslant 4\pi$.

4 **a** Show that $\tan\theta(\operatorname{cosec}\theta - \sin\theta) = \cos\theta$.

b Hence solve $\tan\theta(\operatorname{cosec}\theta - \sin\theta) = \sin\theta - \cos\theta$ for $0 \leqslant \theta \leqslant 2\pi$. Give your answers correct to 3 significant figures.

5 **a** Show that $\frac{\sin x}{\cot x} + \cos x = \sec x$.

b Hence solve $\left|\frac{\sin x}{\cot x} + \cos x\right| = \frac{2}{\sqrt{3}}$ for $0 \leqslant x \leqslant 2\pi$.

Example 18 — Self Tutor

Solve $2\sin^2 x - \sin x - 1 = 0$ for $0 < x < 2\pi$.

$2\sin^2 x - \sin x - 1 = 0$

$\therefore (2\sin x + 1)(\sin x - 1) = 0$ {compare with $2a^2 - a - 1 = (2a+1)(a-1)$}

$\therefore \sin x = -\frac{1}{2}$ or 1

$\sin x = -\frac{1}{2}$ when $x = \frac{7\pi}{6}$ or $\frac{11\pi}{6}$

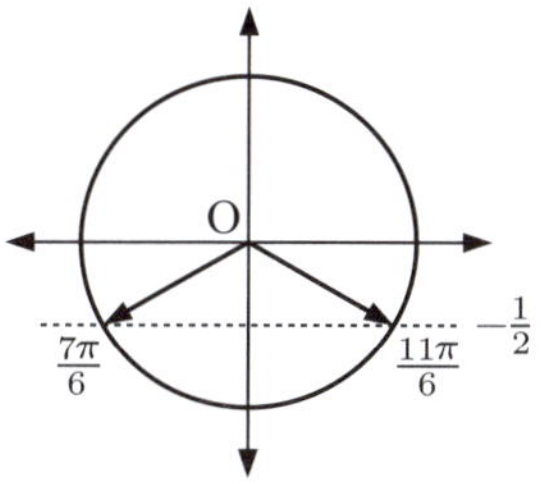

$\sin x = 1$ when $x = \frac{\pi}{2}$

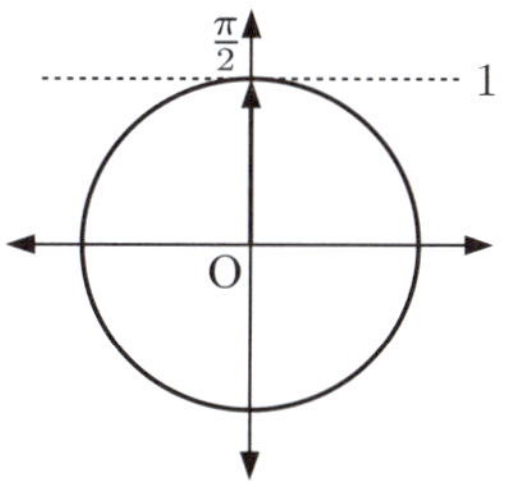

The solutions are $x = \frac{\pi}{2}, \frac{7\pi}{6}$, or $\frac{11\pi}{6}$.

6 **a** Factorise $2\cos^2 x - \cos x$.

b Hence solve $2\cos^2 x - \cos x = 0$ for $0 < x < 2\pi$.

7 **a** Factorise $2\sin^2\theta + \sin\theta - 1$.

b Hence solve $2\sin^2\theta + \sin\theta = 1$ for $0° \leqslant \theta \leqslant 360°$.

8 Solve for $0 \leqslant x \leqslant 2\pi$:

a $\tan^2 x - \tan x = 0$ **b** $2\cos^2 x + 3\cos x + 1 = 0$ **c** $\sin^2 2x - \sin 2x = 0$

d $2\cos^2 x - 7\cos x + 3 = 0$ **e** $3\sin^2 x - 2\sin x = 5$ **f** $\tan^2 x - 2\tan x - 3 = 0$

9 Solve for $0° \leqslant \theta \leqslant 360°$:

a $\cos\theta - \sin^2\theta + 1 = 0$ **b** $\cos^2\theta + \sin\theta = -1$ **c** $2\cos\theta + 3 = 3\sin^2\theta$

d $\sec^2\theta - 2\tan\theta = 0$ **e** $\cot^2 2\theta + \operatorname{cosec} 2\theta = 1$ **f** $\tan^2\theta + \sec\theta - 5 = 0$

10 Solve for $0 \leqslant x \leqslant 2\pi$, giving answers correct to 2 decimal places where appropriate:

a $3\sin^2 x - 2\sin x - 1 = 0$ **b** $\cos^2 x - 3\sin^2 x = 7\cos x - 1$

c $\tan^2 x - 9\sec x = 4 - \sec^2 x$

11 **a** Factorise $\tan^4 y - 4\tan^2 y + 3$ into two quadratic factors.

b Hence solve $\tan^4 y - 4\tan^2 y + 3 = 0$ for $0 < y < 4\pi$.

12 **a** Write $2\sqrt{3}\sin x - \cot x = 2\cos x - \sqrt{3}$ in the form $(a\sin x + b)(c\sin x - \cos x) = 0$ where $a,\ b,\ c \in \mathbb{R}$.

b Hence solve $2\sqrt{3}\sin x - \cot x = 2\cos x - \sqrt{3}$ for $0° < x < 360°$.

Review set 9A

1 Which of these graphs show periodic behaviour?

a

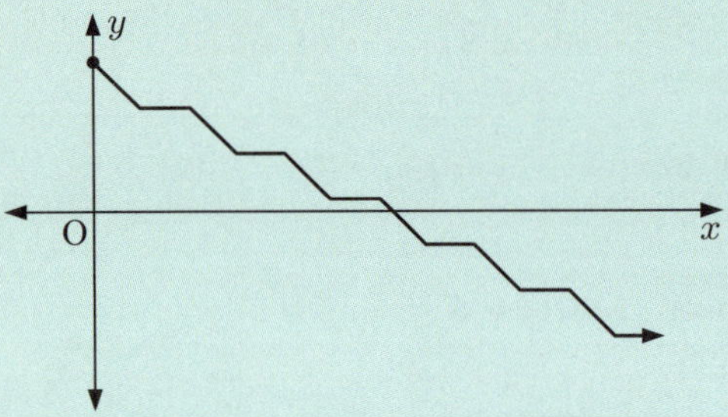

b

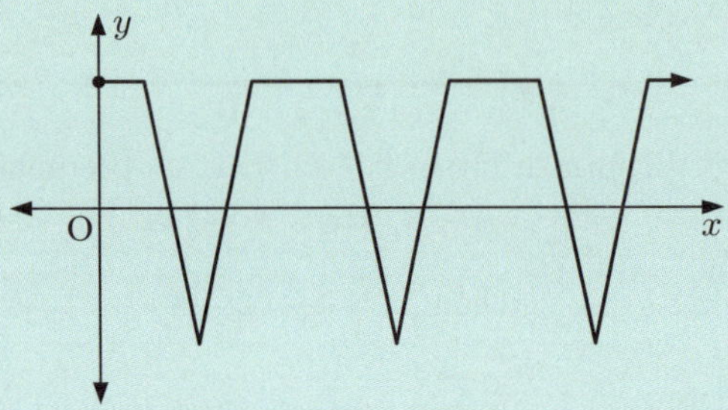

2 State the minimum and maximum values of:

a $y = 3 + \sin x$ **b** $y = 2\cos 3x$ **c** $y = 3\sin 2x$ **d** $y = \cos 4x - 1$

3 State the period of:

a $y = 4\sin x$ **b** $y = 2\cos 4x$ **c** $y = 4\cos 2x + 4$ **d** $y = 2\tan\frac{3x}{8}$

4 Draw each of the following graphs for $0 \leqslant x \leqslant 2\pi$:

a $y = 5\sin x$ **b** $y = \cos 3x - 1$ **c** $y = \tan 2x + 4$ **d** $y = \tan\frac{x}{2} - 1$

5 Complete the table:

Function	*Period*	*Amplitude*	*Domain*	*Range*
$y = 3\sin 2x + 1$				
$y = \tan 2x$				
$y = 2\cos 3x - 3$				

6 Find the cosine function represented in the graph.

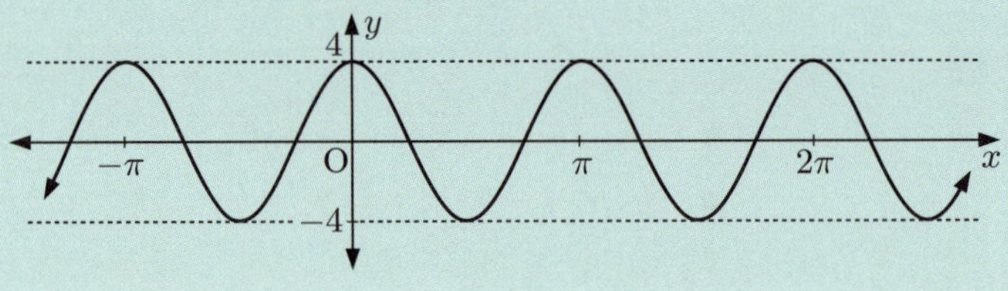

7 On the same set of axes, graph $y = 2\cos x$ and $y = |2\cos x|$ for $0 \leqslant x \leqslant 2\pi$.

8 Solve exactly:

a $2\sin x = -1$ for $0 \leqslant x \leqslant 4\pi$ **b** $\sqrt{2}\sin x - 1 = 0$ for $-2\pi \leqslant x \leqslant 2\pi$

c $2\sin 3x + \sqrt{3} = 0$ for $0 \leqslant x \leqslant 2\pi$ **d** $\sqrt{2}\cos x - 1 = 0$ for $0 \leqslant x \leqslant 4\pi$

9 Find exact solutions for $-\pi \leqslant x \leqslant \pi$:

a $\tan 2x = -\sqrt{3}$ **b** $\sin\left(x - \frac{\pi}{6}\right) = -\frac{1}{\sqrt{2}}$ **c** $\cos^2 x = \frac{1}{4}$

10 Solve for $0° < x < 180°$:

a $\operatorname{cosec} x = 2$ **b** $\sec x = -\sqrt{2}$ **c** $\cot 2x = -1$

11 Solve for $0 \leqslant x \leqslant 2\pi$, giving your answers correct to 2 decimal places:

a $\sin x = 0.72$ **b** $\cos x = -\frac{4}{7}$ **c** $3\tan x - 8 = 0$

12 Simplify:

a $5\cos\theta + 4\cos\theta$ **b** $\dfrac{2 - 2\cos^2\theta}{\sin\theta}$ **c** $\dfrac{\cot\theta}{\operatorname{cosec}\theta}$

13 Show that:

a $\dfrac{\cos\theta - \sec\theta}{\tan\theta} = -\sin\theta$ **b** $(1 + \tan\theta)\left(\dfrac{\sin\theta}{\sin\theta + \cos\theta}\right) = \tan\theta$

14 **a** Show that $\sin^2 x + \cos^2 x + \tan^2 x = \sec^2 x$.

b Hence solve $\sin^2 x + \cos^2 x + \tan^2 x = 1$ for $0 \leqslant x \leqslant 4\pi$.

15 Factorise:

a $9 - \sin^2\theta$ **b** $\cos^2\theta - 3\cos\theta + 2$ **c** $2\tan^2\theta + 7\tan\theta - 4$

16 Solve for $0 \leqslant x \leqslant 2\pi$:

a $2\sin^2 x - 3\sin x + 1 = 0$ **b** $2\cos^2 2x - 5\cos 2x = 3$ **c** $\tan^2 x - 3\tan x = 4$

Review set 9B

1 For the following waves, find:

i the period **ii** the equation of the principal axis **iii** the amplitude.

a

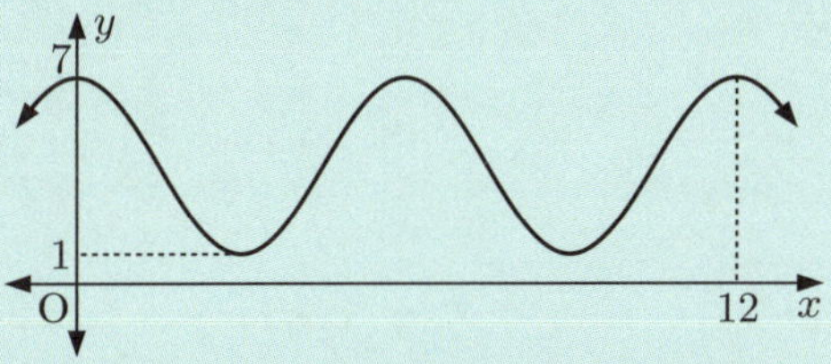

b

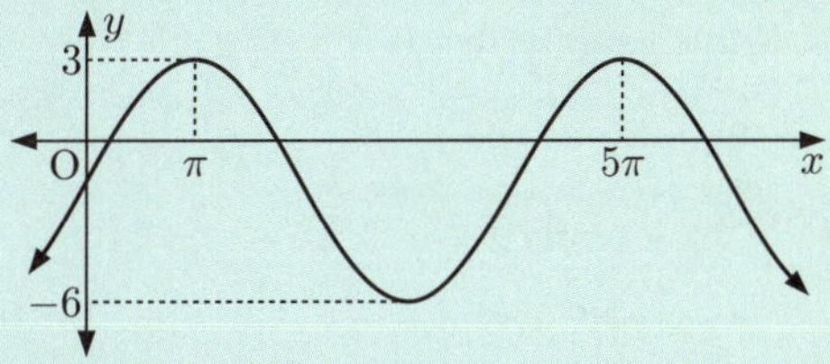

2 Find b given that the function $y = \sin bx$, $b > 0$ has period:

a $\frac{\pi}{3}$ **b** 4π **c** 6

3 State the minimum and maximum values of:

a $y = 2\sin x - 3$ **b** $y = 3\cos x + 1$ **c** $y = 4\cos 2x + 9$

4 On the same set of axes, for the domain $0 \leqslant x \leqslant 2\pi$, sketch:

a $y = \cos x$ and $y = \cos x - 3$ **b** $y = \tan x$ and $y = 2\tan x$

c $y = \cos x$ and $y = \cos 2x + 1$ **d** $y = \sin x$ and $y = 3\sin x + 1$

5 The function $y = a\sin bx + c$, $a > 0$, $b > 0$, has amplitude 2, period $\frac{\pi}{3}$, and principal axis $y = -2$.

a Find the values of a, b, and c. **b** Sketch the function for $0 \leqslant x \leqslant \pi$.

6 Consider the function $y = 2\tan x$.

a State a function which has the same shape, but has principal axis $y = 2$.

b Draw $y = 2\tan x$ and your function from **a** on the same set of axes, for $-2\pi \leqslant x \leqslant 2\pi$.

7 **a** Find m and n given the following graph of the function $y = 2\sin mx + n$:

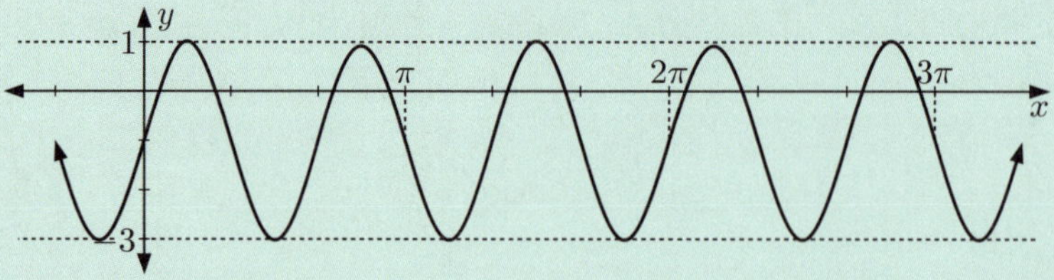

b Find the coordinates of the points on $0 \leqslant x \leqslant 2\pi$ where the function crosses the x-axis.

8 Consider $y = \sin\frac{x}{3}$ on the domain $-7 \leqslant x \leqslant 7$. Use the graph to solve, correct to 1 decimal place:

a $\sin\frac{x}{3} = -0.9$ **b** $\sin\frac{x}{3} = \frac{1}{4}$

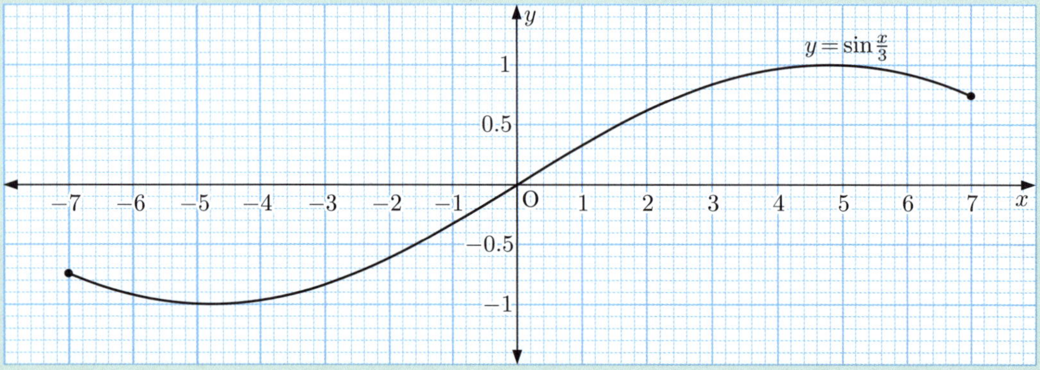

9 Solve exactly for $0 \leqslant x \leqslant 2\pi$:

a $\sin x = -\frac{\sqrt{3}}{2}$ **b** $\cos 2x = \frac{1}{\sqrt{2}}$ **c** $\tan\left(x - \frac{\pi}{2}\right) = -\sqrt{3}$

10 Solve exactly for $-180° < x < 180°$:

a $\text{cosec}^2 x = 2$ **b** $\tan x = \cot x$ **c** $\cos 2x + \sqrt{3}\sin 2x = 0$

11 Solve for $0° \leqslant A \leqslant 360°$, giving your answers correct to 1 decimal place:

a $\cos A = \frac{2}{5}$ **b** $9\sin A + 2 = 0$ **c** $\cot 2A = 4$

12 Simplify:

a $6 - 6\sin^2\phi$ **b** $\dfrac{\cot\theta + \tan\theta}{\sec\theta}$ **c** $\dfrac{\sec^2\alpha - 1}{\text{cosec}^2\alpha - 1}$

13 **a** Show that $(\sin\theta + \sin^2\theta)(\sec\theta - \tan\theta) = \sin\theta\cos\theta$.

b Hence solve $(\sin\theta + \sin^2\theta)(\sec\theta - \tan\theta) = \cos\theta$ for $0 \leqslant \theta \leqslant 4\pi$.

14 Solve for $0 \leqslant x \leqslant 2\pi$:

a $2\sin^2 x - 7\sin x = 4$ **b** $2\cos 2x - 2 = \sin^2 2x$ **c** $2\sec^2 x - \tan x - 5 = 0$

15 Solve for $0 \leqslant y \leqslant 2\pi$:

a $|\cot y| = \sqrt{3}$ **b** $7\sin^2 y - 1 = 5\cos^2 y + \sin y$

16 **a** Factorise $8\sin^4 x - 10\sin^2 x + 3$ into two quadratic factors.

b Hence solve $8\sin^4 x - 10\sin^2 x + 3 = 0$ for $0° < x < 360°$.

10

Counting and the binomial expansion

Contents:

Opening problem

At a traditional ball, there are 48 gentlemen and 48 ladies. Each gentleman shakes hands with every other gentleman, and bows to every lady. Each lady curtsies to every other person.

Things to think about:

a How many bows are made?

b How many handshakes are made? How can this problem be solved using a polygon?

c How many curtsies are made?

d Five ladies are asked to line up on the dance floor to display their beautiful gowns. In how many different orders can they stand?

The **Opening Problem** is an example of a **counting** problem.

In this Chapter we learn how to solve counting problems without having to list and count the possibilities one by one. To do this we will examine:

- the product principle
- counting permutations
- counting combinations.

A THE PRODUCT PRINCIPLE

A brook runs through the city park. Three paths lead from the park entrance to the bridge over the brook, and two paths lead on from the far side of the bridge to the statue commemorating the city's founder.

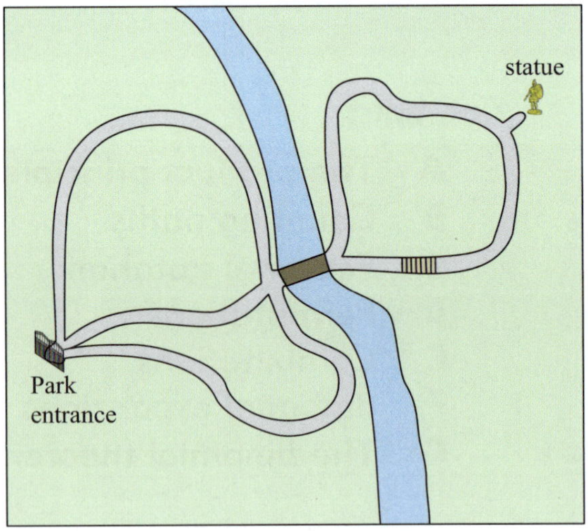

For any of the three paths to the bridge, there are two possible paths onwards to the statue.

The total number of possible paths to the statue is therefore $2+2+2=3\times 2=6$.

Notice that this is the *product* of the 3 paths to the bridge and the 2 paths onwards to the statue.

THE PRODUCT PRINCIPLE

If there are m different ways of performing an operation, and for each of these there are n different ways of performing a second **independent** operation, then there are mn different ways of performing the two operations in succession.

The product principle can be extended to three or more successive independent operations.

Example 1 — Self Tutor

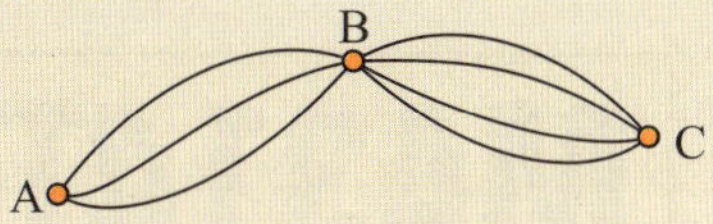

Three towns A, B, and C are connected by the road network shown. How many different ways are there to get from A to C via B?

For each of the 3 roads from A to B, there are 4 ways of getting from B to C.
Using the product principle, there are $3 \times 4 = 12$ different ways of getting from A to C via B.

EXERCISE 10A

1 Find the number of possible paths from X to Y in each network:

a

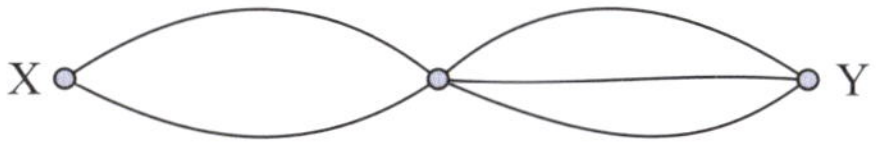

b

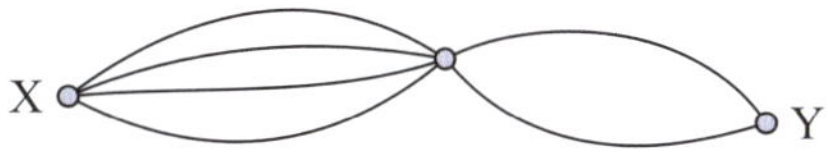

2 The diagram shows the possible pathways for a bus service between four towns P, Q, R, and S.
How many different routes are possible from:

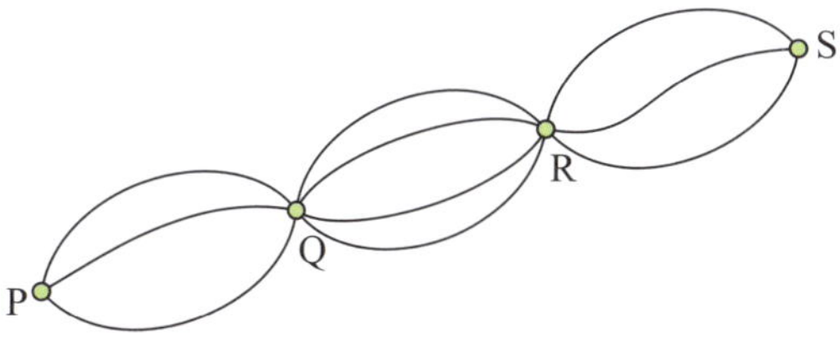

a Q to S via R

b S to P via R and Q?

3 Look at the menu alongside.
How many different combinations are there for a person to order an entreé, a main course, and a dessert?

4

The code for this lock consists of five letters of the alphabet.
How many different codes are possible?

5 In how many different ways can the top two positions be filled in a badminton competition of 6 teams?

6 How many 2-digit numbers can be formed using the digits 1, 2, 3, and 4:

a as often as desired **b** at most once each?

7 How many different alpha-numeric car registration plates can be made if the first two places and last two places are letters of the English alphabet, and the remaining places are 3 digits from 0 to 9?

B COUNTING PATHS

The diagram alongside shows a system of paths in a park.

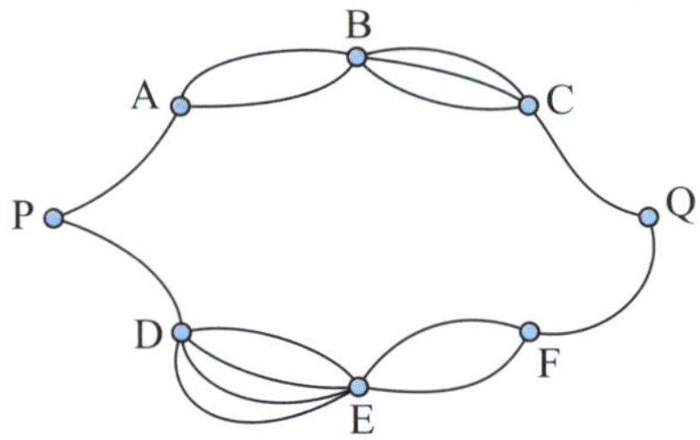

From A to C there are $2 \times 3 = 6$ routes.

From D to F there are $4 \times 2 = 8$ routes.

So, in total there are $6 + 8 = 14$ different routes from P to Q.

Notice that:
- When going from A to C, we go from A to B **and** then from B to C. We **multiply** the possibilities from these two steps.
- When going from P to Q, we can first go from P to A **or** P to D. We **add** the possibilities from each of these possible first steps.

The word **and** suggests *multiplying* the possibilities.
The word **or** suggests *adding* the possibilities.

Example 2 — Self Tutor

How many different paths lead from P to Q?

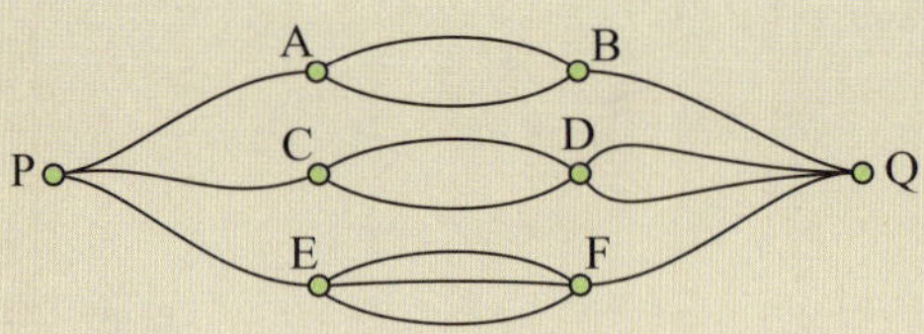

From P we could go to A **or** C **or** E.

From A there are 2 paths to Q.

From C there are $2 \times 2 = 4$ paths to Q.

From E there are 3 paths to Q.

$\therefore$ in total there are $2 + 4 + 3 = 9$ different paths from P to Q.

EXERCISE 10B

1 How many different paths lead from A to B?

a

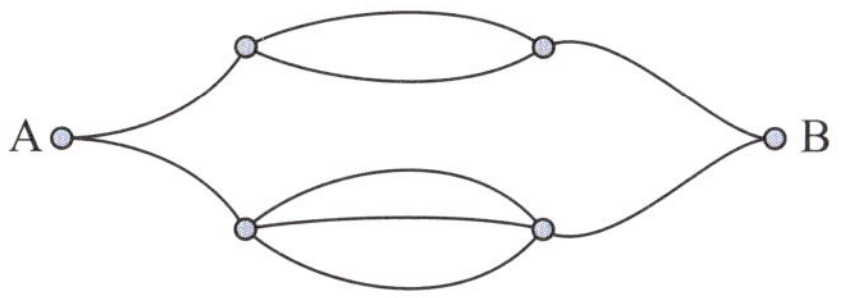

b

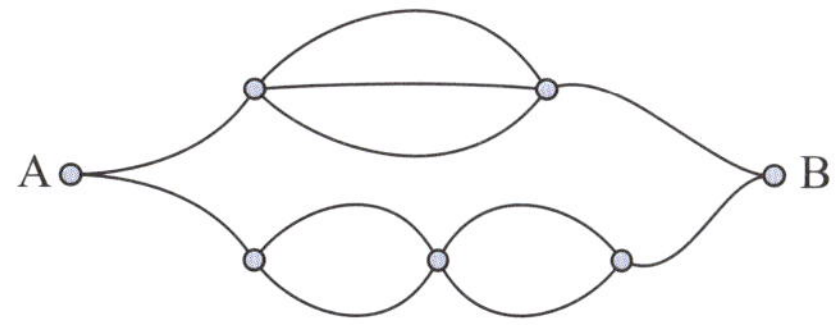

c

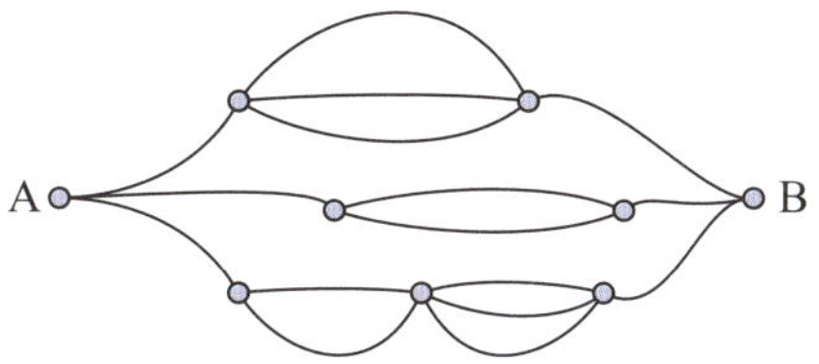

d

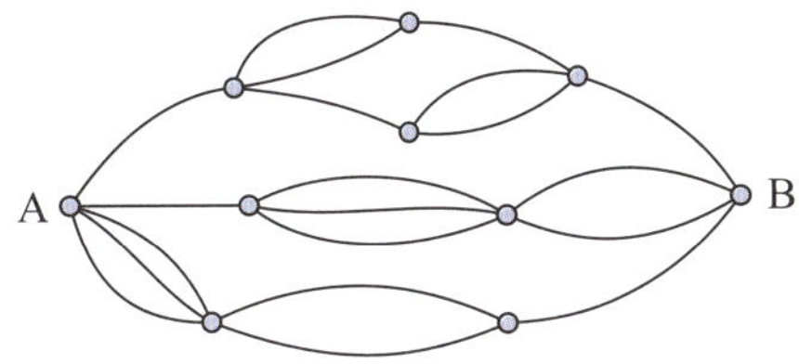

2 Deanne is driving from town A to visit her sister, who lives in town B. She has a choice of routes which could take her through other towns P and Q, on the way.
Given that Deanne will not return to a town already visited, how many different routes could she take?

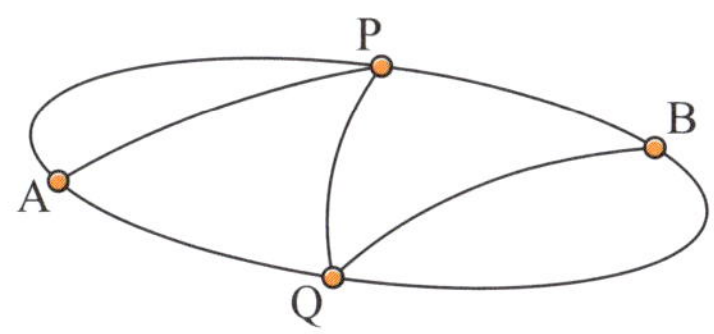

3

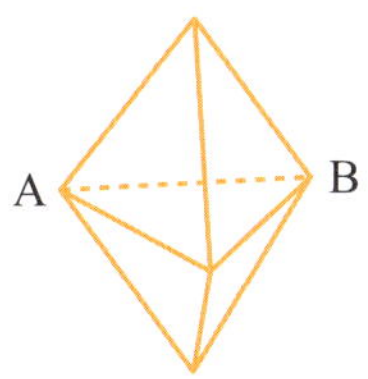

The edges of the 3-dimensional shape shown are created using straws of equal length.
How many different paths from A to B have length:

a 2 straws **b** 3 straws?

C FACTORIAL NOTATION

In problems involving counting, products of consecutive positive integers are common.

For example, $6 \times 5 \times 4$ or $5 \times 4 \times 3 \times 2 \times 1$.

For convenience, we use **factorial numbers** to represent the products of consecutive positive integers.

> For $n \geqslant 1$, $n!$ is the product of the first n positive integers.
>
> $$\boldsymbol{n! = n(n-1)(n-2)(n-3) \ldots . \times 3 \times 2 \times 1}$$

For example, the product $5 \times 4 \times 3 \times 2 \times 1 = 5!$

An alternative **recursive definition** of factorial numbers is $n! = n \times (n-1)!$ for $n \geqslant 1$.

For example, we can write $6! = 6 \times 5!$

Under this rule we notice that $1! = 1 \times 0!$

We therefore define $0! = 1$

Example 3 — Self Tutor

Evaluate:

a $3!$ **b** $\frac{6!}{4!}$ **c** $\frac{5!}{3! \times 2!}$

a $3! = 3 \times 2 \times 1 = 6$

b $\frac{6!}{4!} = \frac{6 \times 5 \times \cancel{4 \times 3 \times 2 \times 1}}{\cancel{4 \times 3 \times 2 \times 1}} = 6 \times 5 = 30$

c $\frac{5!}{3! \times 2!} = \frac{5 \times 4 \times \cancel{3 \times 2 \times 1}}{\cancel{3 \times 2 \times 1} \times 2 \times 1} = 10$

EXERCISE 10C

1 Evaluate:

a $2!$ **b** $5!$ **c** $6!$ **d** $10!$

2 Evaluate without using a calculator:

a $\frac{7!}{6!}$ **b** $\frac{5!}{2!}$ **c** $\frac{8!}{6!}$ **d** $\frac{6!}{3!}$

e $\frac{3!}{5!}$ **f** $\frac{4!}{7!}$ **g** $\frac{6!}{4! \times 2!}$ **h** $\frac{10!}{8! \times 2!}$

3 Simplify:

a $\frac{(n+1)!}{n!}$ **b** $\frac{n!}{2!(n-2)!}$ **c** $\frac{(n+3)!}{n!\,3!}$

Example 4 — Self Tutor

Express in factorial form:

a $9 \times 8 \times 7$ **b** $\frac{11 \times 10 \times 9 \times 8}{4 \times 3 \times 2 \times 1}$

a $9 \times 8 \times 7 = \frac{9 \times 8 \times 7 \times 6 \times 5 \times 4 \times 3 \times 2 \times 1}{6 \times 5 \times 4 \times 3 \times 2 \times 1} = \frac{9!}{6!}$

b $\frac{11 \times 10 \times 9 \times 8}{4 \times 3 \times 2 \times 1} = \frac{11 \times 10 \times 9 \times 8 \times 7 \times 6 \times 5 \times 4 \times 3 \times 2 \times 1}{4 \times 3 \times 2 \times 1 \times 7 \times 6 \times 5 \times 4 \times 3 \times 2 \times 1} = \frac{11!}{4! \times 7!}$

4 Express in factorial form:

a $4 \times 3 \times 2 \times 1$ **b** $7 \times 6 \times 5 \times 4 \times 3 \times 2 \times 1$ **c** $8 \times 7 \times 6$

d $15 \times 14 \times 13 \times 12$ **e** $\frac{9 \times 8 \times 7}{3 \times 2 \times 1}$ **f** $\frac{13 \times 12 \times 11 \times 10}{4 \times 3 \times 2 \times 1}$

5 A football competition is organised between 10 teams. In how many ways can the top 4 places be filled? Write your answer in factorial form.

D PERMUTATIONS

A **permutation** of a group of symbols is *any arrangement* of those symbols in a *definite order*.

For example, consider the three symbols A, B, and C.

A permutation which uses all of these symbols is BCA. We say they have been taken "three at a time".

The set {ABC, ACB, BAC, BCA, CAB, CBA} contains all of the different permutations on the symbols A, B, and C taken three at a time.

Example 5 — Self Tutor

List the permutations on the symbols □, △, and ○ when they are taken:

a one at a time **b** two at a time **c** three at a time.

a {□, △, ○}

b {□△, □○, △□, △○, ○□, ○△}

c {□△○, □○△, △□○, △○□, ○□△, ○△□}

When we have lots of symbols, listing the complete set of permutations becomes extremely time-consuming. However, there is a quicker way to count permutations.

Suppose we want to find the number of different permutations on the symbols A, B, C, D, E, F, and G, taken 3 at a time.

There are 3 positions to fill:

1st	2nd	3rd

In the 1st position we could have any of the 7 symbols.

7		
1st	2nd	3rd

In the 2nd position we could have any of the remaining 6 symbols.

7	6	
1st	2nd	3rd

In the 3rd position we could have any of the remaining 5 symbols.

7	6	5
1st	2nd	3rd

So, the total number of permutations $= 7 \times 6 \times 5$ {product principle}

$$= \frac{7 \times 6 \times 5 \times 4 \times 3 \times 2 \times 1}{4 \times 3 \times 2 \times 1}$$

$$= \frac{7!}{4!} \quad \text{or} \quad \frac{7!}{(7-3)!}$$

The number of **permutations** on n distinct symbols taken r at a time is:

$$\underbrace{n \times (n-1) \times (n-2) \times \dots \times (n-r+1)}_{r \text{ of these}} = \frac{n!}{(n-r)!}$$

For the three symbols in **Example 5**:

- the number of permutations when taken 1 at a time is $\dfrac{3!}{(3-1)!} = \dfrac{3!}{2!} = 3$
- the number of permutations when taken 2 at a time is $\dfrac{3!}{(3-2)!} = \dfrac{3!}{1!} = 6$
- the number of permutations when taken 3 at a time is $\dfrac{3!}{(3-3)!} = \dfrac{3!}{0!} = 6.$

Example 6 — Self Tutor

A netball association runs a tournament with 12 teams. In how many different ways could the top 4 positions be filled on the competition ladder?

Any of the 12 teams could fill the "top" position.
Any of the remaining 11 teams could fill the 2nd position.
Any of the remaining 10 teams could fill the 3rd position.
Any of the remaining 9 teams could fill the 4th position.

12	11	10	9
1st	2nd	3rd	4th

$$\begin{aligned}\text{The total number of permutations} &= 12 \times 11 \times 10 \times 9\\ &= \frac{12!}{8!}\\ &= 11\,880\end{aligned}$$

The top 4 positions could be filled in 11 880 ways.

EXERCISE 10D

1 List the permutations on the symbols A and B taken:

a one at a time **b** two at a time.

2 List the permutations on the symbols P, Q, R, and S taken:

a one at a time **b** two at a time **c** three at a time **d** four at a time.

3 Find the number of permutations there are on the symbols:

a A, B, C, D, E, and F taken 3 at a time **b** P, Q, R, S, T, U, and V taken 5 at a time
c A to Z taken 15 at a time.

4 A hockey competition consists of 9 teams. In how many different ways can the top two positions be filled?

5 A gridiron competition is organised between 8 teams. In how many ways is it possible to fill the top four places on the competition ladder?

6 How many 3-digit numbers can be formed using the digits 2, 3, 4, and 5:

a as often as desired **b** at most once each?

7 There are 16 students in a debating club. In how many different ways can they select:

a a captain and vice captain **b** a captain, vice captain, and treasurer?

Example 7

Self Tutor

A 4-digit number is formed using the digits 1, 2, 3, 5, 8, and 9 at most once each.
How many such numbers can be formed if:

a there are no other restrictions

b the number must be greater than 6000

c the first and last digits must be even?

a There are 6 digits taken 4 at a time

$\therefore \quad \frac{6!}{(6-4)!} = \frac{6!}{2!} = 360$ numbers can be formed.

b

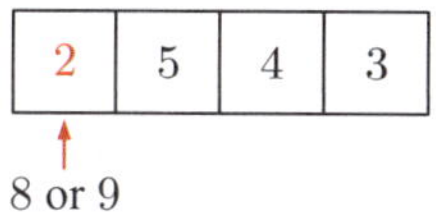

The first digit must be 8 or 9.
There are 5 remaining digits which we take 3 at a time.

$\therefore \quad 2 \times \frac{5!}{(5-3)!} = 2 \times \frac{5!}{2!} = 120$ numbers can be formed.

c

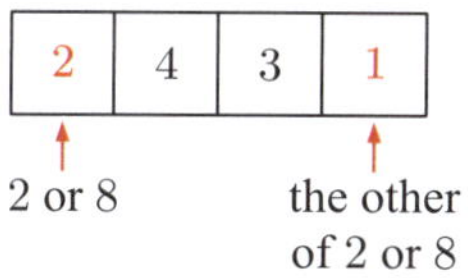

The first digit must be 2 or 8, and the other of these must be the last digit.
For the rest we have 4 remaining digits which we take 2 at a time.

$\therefore \quad 2 \times 1 \times \frac{4!}{(4-2)!} = 2 \times 1 \times \frac{4!}{2!} = 24$ numbers can be formed.

8 Georgie has 6 coloured ribbons: red, orange, yellow, green, blue, and purple. In how many ways can she arrange them on a flagpole if:

a there are no restrictions

b green must be at the top

c red and yellow must fill the bottom two places?

9 Consider the letters P, Q, R, S, and T on 5 alphabet blocks. These blocks are placed in a row. How many permutations are there if:

a there are no restrictions

b the row must end in RQ

c the row must begin with S and end with P?

10 A 4-digit number is made up using the digits 1, 2, 3, 4, 5, and 6 at most once each. How many different numbers can be formed if:

a there are no other restrictions

b the number must be divisible by 5

c the number must start with 6 and end with 2?

11 A 3-digit number is made up using the digits 0, 1, 2, 3, 4, 5, 6, and 7 at most once each. The number cannot start with 0. How many different numbers can be formed if:

a there are no other restrictions

b the number must end in 2, 4, or 6

c the number must end in 0

d the number must be even?

Example 8

Self Tutor

Twins Olivier and Philipe are inseparable. In any photo they always have to stand together.

A photo is taken with 8 people in a line, including the twins. In how many ways could the people be ordered?

Olivier and Philipe can be ordered in 2! ways.

Since the twins must be together, they can now be considered as one person in a group of 7.

The group of 7 can be ordered in 7! ways, so the total number of orderings $= 2! \times 7!$

$= 10\,080$

12 In how many ways can 4 different books be arranged on a shelf if:

a there are no restrictions
b books A and B must be together?

13 8 students stand in a line for a game of tug of war. In how many ways can this be done if:

a Tanky Tony has to be at the back
b Bernard and Eric cannot be next to one another?

14 Sprinters A, B, C, and D are members of a 4×100 m relay team. In how many ways can the sprinters be ordered if:

a there are no restrictions
b B must sprint last
c D must not sprint last
d C must sprint either first or second
e A must be followed directly by B?

15 3 boys and 3 girls are to sit in a row. How many ways can this be done if:

a there are no restrictions
b there is a girl at each end
c boys and girls must alternate
d all the boys sit together?

Discovery 1 — Knockout tournament

A knockout tournament begins with 64 players randomly drawn into 32 pairs.

In each round of the tournament, the players in each pair play one another. The loser is "knocked out" and the winner advances to the next round, being placed in a new pair according to a fixed tree.

All players "knocked out" in the same round are given equal final placing.

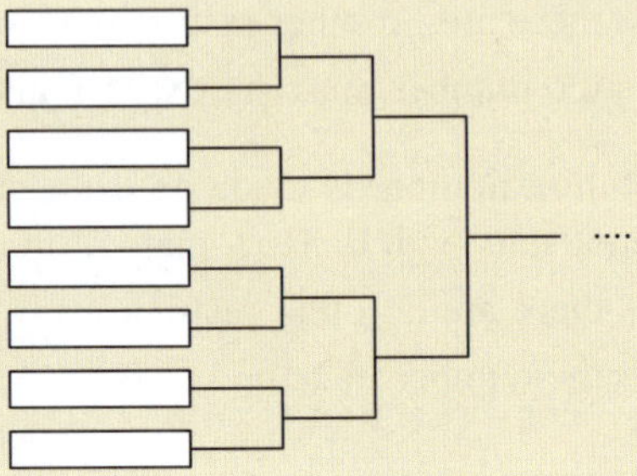

What to do:

1 In how many ways can the competition determine the:

a winner **b** runner-up **c** winner *and* runner-up?

2 **a** What placing in the tournament are the losing semi-finalists given?

b List the possible final placings a player could obtain.

c In how many ways can the placings of *all* players be selected?

3 The draw has been made for a knockout tournament with n rounds. The players reaching the last k rounds will be awarded ranking points according to the round they reach, with extra points for the winner.

Write a formula for the number of ways in which the points can be distributed.

E COMBINATIONS

A **combination** is a selection of objects *without* regard to order.

For example, there are 10 different possible teams of 3 people that can be selected from P, Q, R, S, and T:

PQR PQS PQT PRS PRT PST
QRS QRT QST
RST

There are 10 combinations in total.

Now given the five people P, Q, R, S, and T, we know that there are $5 \times 4 \times 3 = 60$ permutations for taking three of them at a time. So why is this 6 times larger than the number of combinations?

The answer is that for the combinations, order is not important. Selecting P, Q, and R results in the same team as selecting Q, R, and P. For each of the 10 possible combinations, there are $3! = 6$ ways of ordering the members of the team.

In general, when choosing r objects from n objects,

$$\begin{aligned}\text{number of combinations} &= \text{number of permutations} \div \text{number of ways } r \text{ objects can be ordered}\\ &= \frac{n!}{(n-r)!} \div r!\\ &= \frac{n!}{r!(n-r)!}\end{aligned}$$

This value is written as $\binom{n}{r}$ or C_r^n.

The number of **combinations** on n distinct symbols taken r at a time is $\binom{n}{r} = \dfrac{n!}{r!(n-r)!}$.

Example 9

Self Tutor

Use the formula $\binom{n}{r} = \dfrac{n!}{r!(n-r)!}$ to evaluate:

a $\binom{6}{2}$ **b** $\binom{9}{5}$

a $$\binom{6}{2} = \frac{6!}{2!(6-2)!} = \frac{6!}{2! \times 4!} = \frac{6 \times 5 \cancel{\times 4 \times 3 \times 2 \times 1}}{2 \times 1 \cancel{\times 4 \times 3 \times 2 \times 1}} = \frac{30}{2} = 15$$

b $$\binom{9}{5} = \frac{9!}{5!(9-5)!} = \frac{9!}{5! \times 4!} = \frac{9 \times 8 \times 7 \times 6 \cancel{\times 5 \times 4 \times 3 \times 2 \times 1}}{\cancel{5 \times 4 \times 3 \times 2 \times 1} \times 4 \times 3 \times 2 \times 1} = \frac{3024}{24} = 126$$

EXERCISE 10E

1 Evaluate:

a $\binom{3}{2}$ **b** $\binom{6}{4}$ **c** $\binom{5}{1}$ **d** $\binom{7}{2}$

Use your calculator to check your answers.

2 **a** List the different teams of 2 that can be chosen from a squad of 5 players named P, Q, R, S, and T.

b Show that $\binom{5}{2}$ gives the total number of teams.

3 Show that:

a $\binom{n}{0} = \binom{n}{n} = 1$ for all $n \in \mathbb{Z}^+$

b $\binom{n}{r} = \binom{n}{n-r}$ for all $n \in \mathbb{Z}^+$, $r = 0, 1, 2, ..., n$.

4 Find k such that:

a $4\binom{7}{k} = \binom{9}{k+4}$ **b** $7\binom{k}{3} = 2\binom{k+2}{k-2}$

5 Determine whether the following are examples of combinations or permutations:

a making a 3-digit number using the digits 1, 2, 3, 4, and 5 at most once each

b selecting a committee of 3 people from a list of 5

c selecting the chairperson and treasurer from a committee of 8 people

d selecting 2 different pieces of fruit to take to school from a bowl of 10 pieces.

Example 10 — Self Tutor

How many different teams of 3 can be selected from a squad of 8 if:

a there are no restrictions **b** the team must include the captain?

a There are 8 players for selection and we can choose any 3 of them.
This can be done in $\binom{8}{3} = 56$ different ways.

b The captain must be included *and* we need any 2 of the other 7.
This can be done in $\binom{1}{1} \times \binom{7}{2} = 21$ different ways.

6 How many different teams of 7 can be chosen from a training squad of 15?

7 **a** How many different committees of 5 can be selected from a club of 27 members?
b How many of these committees consist of the president and 4 others?

8 A physics exam contains 7 questions. Students must answer both questions 1 and 2, and any 3 of the remaining 5 questions. How many different selections are possible?

9 A team of 8 is selected from a squad of 14. The captain and vice-captain must both be included. How many different teams can be selected?

10 A team of 11 is to be selected from a squad of 18. How many different teams can be chosen if:
a there are no restrictions
b the team must include the captain and vice-captain
c the team must include exactly one of the captain or vice-captain?

Example 11 — Self Tutor

An equestrian team of 4 is to be chosen from 7 male and 6 female riders. How many different teams can be chosen if:

a there are no restrictions
b there must be two riders of each sex
c at least one rider of each sex is required?

a There are $7 + 6 = 13$ riders up for selection, and we need to choose any 4 of them. So there are $\binom{13}{4} = 715$ different teams.

b The males can be chosen in $\binom{7}{2}$ ways and the females in $\binom{6}{2}$ ways.
$\therefore$ $\binom{7}{2} \times \binom{6}{2} = 315$ different teams can be chosen.

c We could have 3 males and 1 female **or** 2 males and 2 females **or** 1 male and 3 females.
$\therefore$ $\binom{7}{3} \times \binom{6}{1} + \binom{7}{2} \times \binom{6}{2} + \binom{7}{1} \times \binom{6}{3} = 665$ different teams can be chosen.

Note: Since the only other alternatives are 4M *and* 0F or 0M *and* 4F,
we could also have used $\binom{13}{4} - \binom{7}{4} \times \binom{6}{0} - \binom{7}{0} \times \binom{6}{4}$

11 The committee of a table tennis club consists of 3 people chosen from 6 male and 8 female members.

a If there are no restrictions, how many different committees can be chosen?

b How many of the different possible committees consist of:

i 3 males **ii** 2 males and 1 female **iii** 1 male and 2 females **iv** 3 females?

12 A committee of 4 singers is chosen from 7 tenors and 9 sopranos.
Find the number of ways of selecting the committee if:

a there are no restrictions

b it must contain 2 tenors and 2 sopranos

c it must contain at least 2 sopranos.

13 A music class consists of 5 piano players, 7 guitarists, and 4 violinists. A band of 1 piano player, 3 guitarists, and 2 violinists must be chosen to play at a school concert. In how many different ways can the band be chosen?

14 A committee of 5 people is chosen from 10 men and 6 women. Determine the number of ways of selecting the committee if:

a there are no restrictions **b** it must contain 3 men and 2 women

c it must contain all men **d** it must contain at least 3 men.

15 A committee of 8 is chosen from 9 boys and 6 girls. In how many ways can this be done if:

a there are no restrictions **b** there must be 5 boys and 3 girls

c all the girls are selected **d** there are more boys than girls?

16 Answer the **Opening Problem** on page **248**.

17

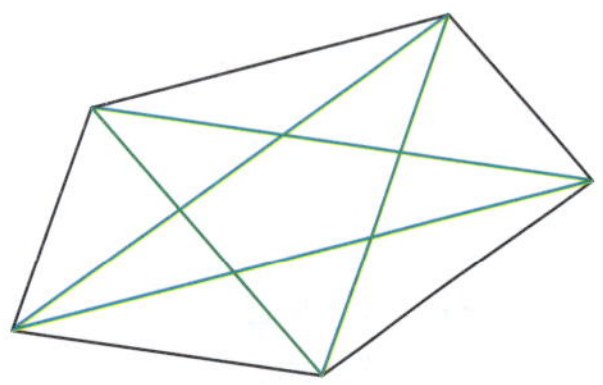

A pentagon has 5 diagonals.
How many diagonals does a 15-sided polygon have?

18 How many 5-digit numbers can be constructed for which the digits are in descending order from left to right?

19 A soccer team consists of 11 players. The team must be sorted into 3 attackers, 3 midfielders, 4 defenders, and 1 goalkeeper for today's match.

a How many possible ways are there to do this?

b Melissa is chosen to play goalkeeper. Linda wants to play attacker, and Robyn does not want to play defender. How many combinations are now possible?

20 In how many ways can 8 people be divided into:

a two equal groups **b** four equal groups?

21 Kristen's school offers 6 Group A subjects, 8 Group B subjects, and 5 Group C subjects. Kristen must study 2 Group A, 2 Group B, and 1 Group C subject. Her sixth subject can come from any group. In how many ways can she make her selection?

F BINOMIAL EXPANSIONS

Consider the cube alongside, which has sides of length $(a+b)$ cm.

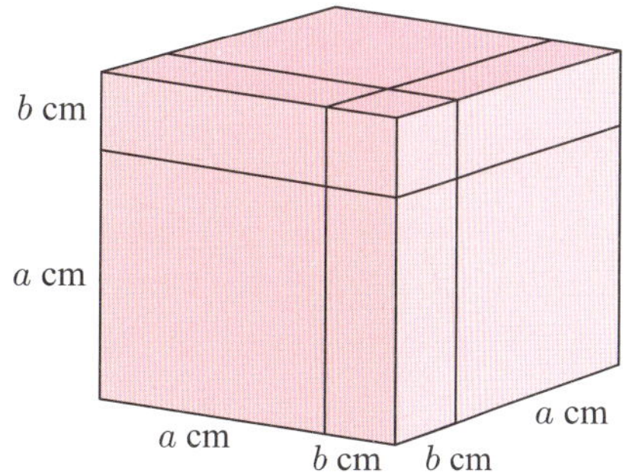

The cube has been subdivided into 8 blocks by making 3 cuts parallel to the cube's surfaces as shown.

We know that the total volume of the cube is $(a+b)^3$ cm^3. However, we can also find an expression for the cube's volume by adding the volumes of the 8 individual blocks.

ANIMATION

We have:

	1 block	$a \times a \times a$
	3 blocks	$a \times a \times b$
	3 blocks	$a \times b \times b$
	1 block	$b \times b \times b$

$\therefore$ the cube's volume $= a^3 + 3a^2b + 3ab^2 + b^3$

$\therefore \quad (a+b)^3 = a^3 + 3a^2b + 3ab^2 + b^3$

> The sum $\boldsymbol{a+b}$ is called a **binomial** as it contains two terms.
>
> Any expression of the form $\boldsymbol{(a+b)^n}$ is called a **power of a binomial**.

All binomials raised to a power can be expanded using the same general principles.

Consider the following algebraic expansions:

$$(a+b)^1 = a+b$$
$$(a+b)^2 = a^2+2ab+b^2$$

$$\begin{aligned}(a+b)^3 &= (a+b)(a+b)^2\\ &= (a+b)(a^2+2ab+b^2)\\ &= a^3+2a^2b+ab^2+a^2b+2ab^2+b^3\\ &= a^3+3a^2b+3ab^2+b^3\end{aligned}$$

> The **binomial expansion** of $\boldsymbol{(a+b)^2}$ is $\boldsymbol{a^2+2ab+b^2}$.
>
> The **binomial expansion** of $\boldsymbol{(a+b)^3}$ is $\boldsymbol{a^3+3a^2b+3ab^2+b^3}$.

Discovery 2 — The binomial expansion

What to do:

1 Expand $(a+b)^4$ in the same way as for $(a+b)^3$ above.
Hence expand $(a+b)^5$ and $(a+b)^6$.

2 The cubic expansion $(a+b)^3 = a^3+3a^2b+3ab^2+b^3$ contains 4 terms. They are written in order so that the powers of a decrease. We observe that their coefficients are: 1 3 3 1

- **a** With the terms written in this order, what happens to the powers of b?
- **b** Does the pattern in **a** continue for the expansions of $(a+b)^4$, $(a+b)^5$, and $(a+b)^6$?

c Use your results to continue this pattern of coefficients up to the case $n = 6$.

$n = 1$			1		1		
$n = 2$		1		2		1	
$n = 3$	1		3		3		1 ← row 3
				⋮			

3 The triangle of coefficients in **c** above is called **Pascal's triangle**.

a How can each row of Pascal's triangle be predicted from the previous one?

b Write a formula for the sum of the numbers in the nth row of Pascal's triangle.

c Hence predict the elements of the 7th row of Pascal's triangle.

d Hence write down the binomial expansion of $(a+b)^7$.

e Check your result algebraically by using $(a+b)^7 = (a+b)(a+b)^6$ and your results from **1**.

You should have observed that in Pascal's triangle, the values on the end of each row are always 1. Each of the remaining values is found by adding the two values diagonally above it.

1 1	row 1
1 2 1	row 2
1 3 3 1	row 3
1 4 6 4 1	row 4
1 5 10 10 5 1	row 5

You should have also found that $(a+b)^4 = a^4 + 4a^3b + 6a^2b^2 + 4ab^3 + b^4$

$$= a^4 + 4a^3b^1 + 6a^2b^2 + 4a^1b^3 + b^4$$

Notice in this expansion that:

- As we look from left to right across the expansion, the powers of a decrease by 1, while the powers of b increase by 1.
- The sum of the powers of a and b in each term of the expansion is 4.
- The number of terms in the expansion is $4 + 1 = 5$.
- The coefficients of the terms are row 4 of Pascal's triangle.

For the expansion of $(a+b)^n$ where $n \in \mathbb{N}$:

- As we look from left to right across the expansion, the powers of a *decrease* by 1, while the powers of b *increase* by 1.
- The sum of the powers of a and b in each term of the expansion is n.
- The number of terms in the expansion is $n + 1$.
- The coefficients of the terms are row n of Pascal's triangle.

In the following Examples we see how the general binomial expansion $(a+b)^n$ may be put to use.

Example 12 — Self Tutor

Use $(a+b)^3 = a^3 + 3a^2b + 3ab^2 + b^3$ to find the binomial expansion of:

a $(x+4)^3$ **b** $(2x-1)^3$

a In the expansion of $(a+b)^3$ we substitute $a = (x)$ and $b = (4)$.

$$\therefore\ (x+4)^3 = (x)^3 + 3(x)^2(4) + 3(x)(4)^2 + (4)^3$$
$$= x^3 + 12x^2 + 48x + 64$$

b We substitute $a = (2x)$ and $b = (-1)$

$$\therefore\ (2x-1)^3 = (2x)^3 + 3(2x)^2(-1) + 3(2x)(-1)^2 + (-1)^3$$
$$= 8x^3 - 12x^2 + 6x - 1$$

Example 13

Self Tutor

Find the:

a 5th row of Pascal's triangle

b binomial expansion of $\left(x - \frac{2}{x}\right)^5$.

a

1	← the 0th row, for $(a+b)^0$
1 1	← the 1st row, for $(a+b)^1$
1 2 1	
1 3 3 1	
1 4 6 4 1	
1 5 10 10 5 1	← the 5th row, for $(a+b)^5$

b Using the coefficients obtained in **a**, $(a+b)^5 = a^5 + 5a^4b + 10a^3b^2 + 10a^2b^3 + 5ab^4 + b^5$

Letting $a = (x)$ and $b = \left(\frac{-2}{x}\right)$, we find

$$\left(x - \frac{2}{x}\right)^5 = (x)^5 + 5(x)^4\left(\frac{-2}{x}\right) + 10(x)^3\left(\frac{-2}{x}\right)^2 + 10(x)^2\left(\frac{-2}{x}\right)^3 + 5(x)\left(\frac{-2}{x}\right)^4 + \left(\frac{-2}{x}\right)^5$$
$$= x^5 - 10x^3 + 40x - \frac{80}{x} + \frac{80}{x^3} - \frac{32}{x^5}$$

EXERCISE 10F

1 Use $(a+b)^3 = a^3 + 3a^2b + 3ab^2 + b^3$ to expand and simplify:

a $(x+2)^3$ **b** $(x-1)^3$ **c** $(1+2x)^3$

d $(3x-2)^3$ **e** $\left(x + \frac{1}{2}\right)^3$ **f** $\left(2x - \frac{1}{x}\right)^3$

2 Use $(a+b)^4 = a^4 + 4a^3b + 6a^2b^2 + 4ab^3 + b^4$ to expand and simplify:

a $(x+3)^4$ **b** $(x-1)^4$ **c** $(2x+1)^4$

d $(2-x)^4$ **e** $(3x+4)^4$ **f** $\left(x - \frac{2}{x}\right)^4$

3 Use the binomial expansion of $(a+b)^5$ to expand and simplify:

a $(x+1)^5$ **b** $(x-y)^5$ **c** $\left(2 + \frac{1}{x}\right)^5$ **d** $\left(x^2 - \frac{2}{x}\right)^5$

4 Expand and simplify $(2+x)^5 + (2-x)^5$.

5 **a** Write down the 6th row of Pascal's triangle.

b Find the binomial expansion of:

i $(x+1)^6$ **ii** $(2-x)^6$ **iii** $\left(x - \frac{1}{x^2}\right)^6$

6 Write down the first three terms of $(1+3x)^6$ in ascending powers of x.

7 Expand and simplify:

a $(1+\sqrt{3})^3$ **b** $(\sqrt{2}+1)^4$ **c** $(1-\sqrt{3})^5$

8 Find $\dfrac{(2+\sqrt{3})^3}{4+\sqrt{3}}$, giving your answer in the form $\dfrac{a+b\sqrt{3}}{c}$, where $a, b, c \in \mathbb{Z}$.

9 **a** Expand $(3-x)^5$. **b** *Hence* find the value of $(2.99)^5$.

10 The first two terms in a binomial expansion are: $(a+b)^3 = 27 - 27\cos x + ...$

a Find a and b. **b** Hence determine the remaining two terms of the expansion.

11 Expand and simplify $(3x+1)(x+2)^4$.

12 Find the coefficient of:

a x^3y^2 in the expansion of $(2x+y)^5$ **b** x^3y^3 in the expansion of $(x+3y)^6$.

G THE BINOMIAL THEOREM

Discovery 3 — The binomial theorem

What to do:

1 Write Pascal's triangle down to row 7.

2 Evaluate:

a $\binom{7}{0}$ **b** $\binom{7}{1}$ **c** $\binom{7}{2}$ **d** $\binom{7}{3}$

e $\binom{7}{4}$ **f** $\binom{7}{5}$ **g** $\binom{7}{6}$ **h** $\binom{7}{7}$

What do you notice?

In **Section F** we saw how the coefficients of the binomial expansion $(a+b)^n$ can be found in the nth row of Pascal's triangle. These coefficients are in fact the values $\binom{n}{r}$ for $r = 0, 1, 2, ..., n$. We call these values **binomial coefficients**.

$$\begin{array}{ccccccccc} & & & 1 & & 1 & & & \\ & & 1 & & 2 & & 1 & & \\ & 1 & & 3 & & 3 & & 1 & \\ 1 & & 4 & & 6 & & 4 & & 1 \end{array} \qquad \begin{array}{ccccccccc} & & & \binom{1}{0} & & \binom{1}{1} & & & \\ & & \binom{2}{0} & & \binom{2}{1} & & \binom{2}{2} & & \\ & \binom{3}{0} & & \binom{3}{1} & & \binom{3}{2} & & \binom{3}{3} & \\ \binom{4}{0} & & \binom{4}{1} & & \binom{4}{2} & & \binom{4}{3} & & \binom{4}{4} \end{array}$$

The **binomial theorem** states that

$$(a+b)^n = a^n + \binom{n}{1}a^{n-1}b + ... + \binom{n}{r}a^{n-r}b^r + ... + b^n$$

where $\binom{n}{r}$ is the **binomial coefficient** of $a^{n-r}b^r$ and $r = 0, 1, 2, 3, ..., n$.

The binomial theorem allows us to perform a binomial expansion or find a particular term in a binomial expansion, without having to draw Pascal's triangle each time.

The **general term** or $(r+1)$th term in the binomial expansion $(a+b)^n$ is $T_{r+1} = \binom{n}{r}a^{n-r}b^r$.

Discussion

The product $(a+b)^n$, $n \in \mathbb{Z}$, consists of n factors $(a+b)$ multiplied together.

Why should the value $\binom{n}{r}$, which is the number of combinations of r objects chosen out of n, be the coefficient of $a^{n-r}b^r$?

Example 14

Self Tutor

Write down the first three and last two terms of the expansion of $(3x-7)^{11}$.
Do not simplify your answer.

$$(3x-7)^{11} = \binom{11}{0}(3x)^{11}(-7)^0 + \binom{11}{1}(3x)^{10}(-7)^1 + \binom{11}{2}(3x)^9(-7)^2 + ...$$
$$.... + \binom{11}{10}(3x)^1(-7)^{10} + \binom{11}{11}(3x)^0(-7)^{11}$$

Example 15

Self Tutor

Find the 6th term of $\left(2x+\frac{1}{x}\right)^8$. Do not simplify your answer.

$a=(2x)$, $b=\left(\frac{1}{x}\right)$, and $n=8$.

Given the general term $T_{r+1} = \binom{n}{r}a^{n-r}b^r$, we let $r=5$.

$$\therefore \ T_6 = \binom{8}{5}(2x)^3\left(\frac{1}{x}\right)^5$$

EXERCISE 10G

1 Write down the first three and last two terms of the following binomial expansions.
Do not simplify your answers.

a $(2x+5)^9$ **b** $(4x-3)^{13}$ **c** $\left(3x-\frac{2}{x}\right)^{16}$

2 Without simplifying, write down:

a the 5th term of $(3x+5)^7$ **b** the 8th term of $(x^2-2)^{11}$

c the 7th term of $\left(x+\frac{5}{x}\right)^{16}$ **d** the 13th term of $\left(\frac{8}{x}-x^2\right)^{17}$.

3 Consider the expression $(2x-1)^7$.

a Without simplifying, write down the:

i 3rd term of the expansion **ii** 5th term of the expansion.

b Write the terms in **a** in simplified form.

Example 16

Self Tutor

In the expansion of $\left(3x - \frac{2}{x}\right)^{10}$, find:

a the coefficient of x^4

b the constant term.

$a = (3x)$, $b = \left(-\frac{2}{x}\right)$, and $n = 10$.

$\therefore$ the general term $T_{r+1} = \binom{10}{r}(3x)^{10-r}\left(-\frac{2}{x}\right)^r$

$= \binom{10}{r}3^{10-r}x^{10-r} \times \frac{(-2)^r}{x^r}$

$= \binom{10}{r}3^{10-r}(-2)^r x^{10-2r}$

a If $10 - 2r = 4$
then $2r = 6$
$\therefore r = 3$

$\therefore T_4 = \binom{10}{3}3^7(-2)^3x^4$

$\therefore$ the coefficient of x^4 is $\binom{10}{3}3^7(-2)^3$ or $-2\,099\,520$.

b If $10 - 2r = 0$
then $2r = 10$
$\therefore r = 5$

$\therefore T_6 = \binom{10}{5}3^5(-2)^5x^0$

$\therefore$ the constant term is $\binom{10}{5}3^5(-2)^5$ or $-1\,959\,552$.

4 Consider the expansion of $(x+3)^7$.

a Write down the general term of the expansion.

b Find the coefficient of x^5.

5 In the expansion of $(2x+3)^{12}$, find:

a the coefficient of x^8 **b** the coefficient of x^5.

6 In the expansion of $(1-3x)^{10}$, find:

a the coefficient of x^3 **b** the coefficient of x^7.

7 In the expansion of $\left(x^2 + \frac{2}{x}\right)^9$, find:

a the coefficient of x^{12} **b** the constant term **c** the coefficient of x^{-6}.

8 Consider the expansion of $(x+a)^8$.

a Write down the general term of the expansion.

b Find a given that the coefficient of x^5 is -448.

9 Find the term independent of x in the expansion of:

a $\left(x + \frac{5}{x}\right)^{10}$ **b** $\left(x - \frac{4}{x^2}\right)^{15}$.

10 Find the coefficient of:

a x^6 in the expansion of $(2-5x^2)^{10}$

b x^{-3} in the expansion of $\left(5x^2+\frac{4}{x}\right)^6$

c x^4y^4 in the expansion of $(x^2+3y)^6$

d x^{15} in the expansion of $\left(x^2-\frac{2}{x}\right)^{15}$.

11 In the expansion of $(k+x)^8$, the coefficient of x^5 is 10 times the coefficient of x^6. Find the value of k.

12 The coefficient of x^5 in the expansion of $(ax-2)^7$ is twice the coefficient of x^5 in the expansion of $(a+x)^9$. Find the value of a.

13 In the expansion of $\left(ax+\frac{b}{x}\right)^6$, the constant term is 20 000, and the coefficient of x^4 is equal to the coefficient of x^2.

a Show that $ab=10$ and $b=\frac{2a}{5}$.

b Find a and b given that they are both positive.

Example 17 — Self Tutor

Find the coefficient of x^4 in the expansion of $(x-5)(3x+1)^5$.

$$\begin{aligned}&(x-5)(3x+1)^5\\&=(x-5)\left[(3x)^5+\binom{5}{1}(3x)^4(1)+\binom{5}{2}(3x)^3(1)^2+....\right]\\&=(x-5)\left(3^5x^5+\binom{5}{1}3^4x^4+\binom{5}{2}3^3x^3+....\right)\end{aligned}$$

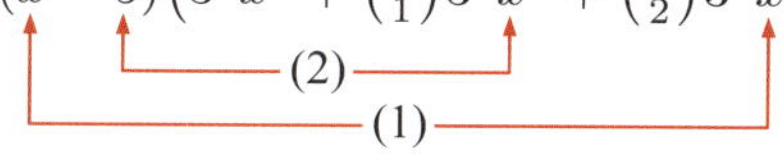

So, the terms containing x^4 are $\binom{5}{2}3^3x^4$ from (1)

and $-5\binom{5}{1}3^4x^4$ from (2)

$\therefore$ the coefficient of x^4 is $\binom{5}{2}3^3-5\binom{5}{1}3^4=-1755$

14 Find:

a the coefficient of x^5 in the expansion of $(x+1)(2x-3)^6$

b the coefficient of x^4 in the expansion of $(3x+4)(x-2)^5$

c the coefficient of x^6 in the expansion of $(x-4)(3x+2)^8$.

15 Find the coefficient of x^4 in the expansion of:

a $(3-2x)^7$

b $(1+3x)(3-2x)^7$

16 Find:

a the coefficient of x^7 in the expansion of $(x^2-3)(2x-5)^8$

b the term independent of x in the expansion of $(1-x^2)\left(x+\frac{2}{x}\right)^6$.

17 When the expansion of $(a+bx)(1-x)^6$ is written in ascending powers of x, the first three terms are $3-20x+cx^2$. Find the values of a, b, and c.

18 In the expansion of $(1+ax)(1+2x)^4$, the coefficients of x^5 and x^2 are equal. Find the value of a.

Example 18 — Self Tutor

Consider the expansion of $(1+3x)^n$, where $n \in \mathbb{Z}^+$.

If the coefficient of x^2 is 90, find the value of n.

$(1+3x)^n$ has general term $T_{r+1} = \binom{n}{r} 1^{n-r}(3x)^r$
$= \binom{n}{r} 3^r x^r$

$\therefore$ $T_3 = \binom{n}{2} 3^2 x^2$ is the x^2 term.

Since the coefficient of x^2 is 90, $\binom{n}{2} \times 9 = 90$

$$\therefore \frac{n(n-1)}{2} = 10$$
$$\therefore n^2 - n = 20$$
$$\therefore n^2 - n - 20 = 0$$
$$\therefore (n-5)(n+4) = 0$$
$$\therefore n = 5 \quad \{n > 0\}$$

$\binom{n}{2} = \dfrac{n(n-1)}{2}$ for all integers $n \geqslant 2$.

19 The coefficient of x^2 in the expansion of $(1+2x)^n$ is 112. Find n.

20 The coefficient of x^2 in the expansion of $\left(1-\dfrac{x}{3}\right)^n$ is $\frac{5}{3}$. Find n.

21 Suppose $(1+kx)^n = 1 - 18x + 135x^2 - \ldots$. Find the values of k and n.

Review set 10A

1 How many different paths lead from A to D?

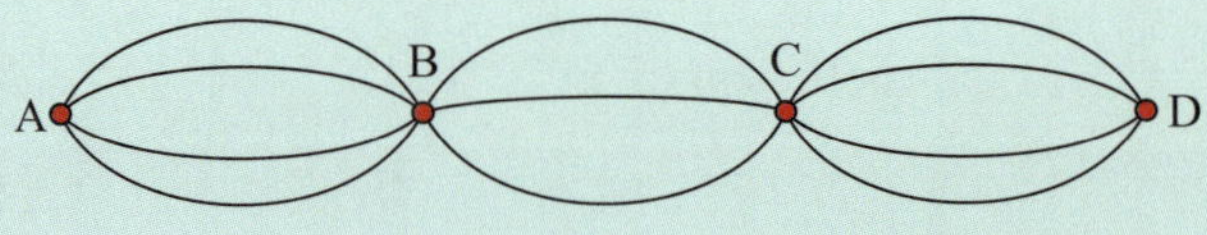

2 Evaluate:

a $4!$ **b** $\dfrac{7!}{5!}$ **c** $\dfrac{9!}{5! \times 4!}$

3 The letters P, Q, R, S, and T are to be arranged in a row. How many of the possible arrangements:

a end with S **b** begin with R and end with Q **c** contain Q and R next to each other?

4 How many 4-digit numbers can be formed using the digits 1, 2, 3, 4, 5, 6, and 7 at most once each if:

a there are no restrictions **b** the number must be even

c the number must be odd and greater than 4000?

5 **a** List the different teams of three people that can be chosen from a squad of six players J, K, L, M, N, and O.

b Show that $\binom{6}{3}$ gives the total number of teams.

6 **a** How many different triangles can be formed by connecting 3 vertices of a regular octagon?

b How many of these triangles are isosceles?

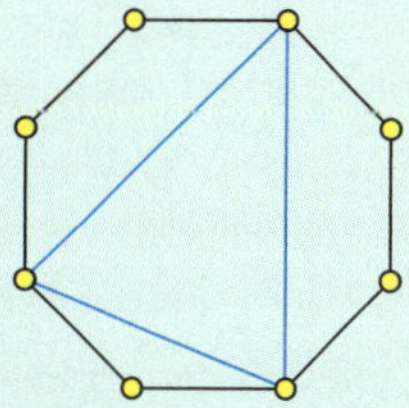

7 A research team of four is chosen from six Englishmen and five Americans.

a If there are no restrictions, how many different teams are possible?

b How many of the teams in **a** have:

i exactly one Englishman **ii** at least two Americans?

8 Use the binomial expansion to expand and simplify: **a** $(x+3y)^3$ **b** $(3-2x)^4$

9 The first two terms in a binomial expansion are: $(a+b)^4 = x^{-3} + 3 + ...$

a Find a and b. **b** Copy and complete the expansion.

10 Expand and simplify $(2\sqrt{3}-1)^5$, giving your answer in the form $a + b\sqrt{3}$, $a, b \in \mathbb{Z}$.

11 Find:

a the coefficient of x^3 in the expansion of $(x+5)^6$

b the coefficient of x^{-2} in the expansion of $\left(x - \frac{3}{x}\right)^8$

c the coefficient of x^4 in the expansion of $(2x+1)(x-3)^5$.

12 In the expansion of $\left(x^2 - \frac{2}{x}\right)^6$, find:

a the coefficient of x^6 **b** the constant term.

13 In the expansion of $(kx-1)^6$, $k \neq 0$, the coefficient of x^4 is equal to four times the coefficient of x^2. Find the possible values of k.

Review set 10B

1 How many different paths lead from A to F?

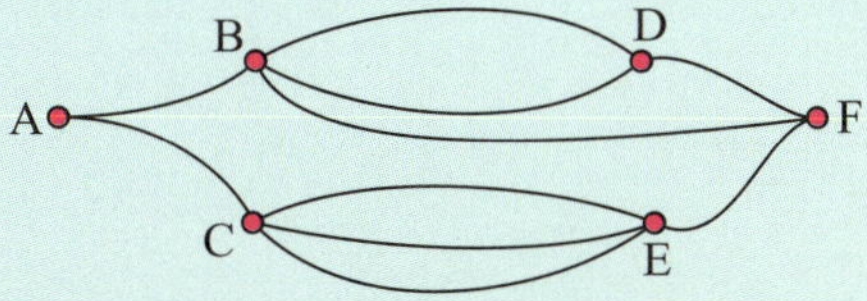

2 Express in factorial form:

a $6 \times 5 \times 4 \times 3 \times 2 \times 1$ **b** $10 \times 9 \times 8$ **c** $\dfrac{12 \times 11 \times 10 \times 9}{4 \times 3 \times 2 \times 1}$

3 How many arrangements containing 4 different letters from the word DRAGONFLY are possible if:

a there are no restrictions

b the letters G and Y must not be included

c the arrangement must start with R and end with N?

4 The Lions and Tigers rugby union teams each have 15 players. At the end of their game, each player shakes hands with every other player, including team mates. How many handshakes take place between:

a players on the same team

b players on opposing teams?

5 Seven people enter a room and sit in a row of seven chairs. In how many ways can the sisters Mimi and Angela sit together in the row?

6 A cricket team of 6 batsmen, 1 wicket keeper, and 4 bowlers is to be chosen from a squad of 9 batsmen, 2 wicket keepers, and 6 bowlers. Calculate the number of different ways in which this can be done.

7 A committee of five is selected from eight engineers, seven scientists, and six mathematicians.

a How many different committees can be selected?

b How many committees consist of:

i all engineers **ii** at least one of each profession

iii the oldest engineer or the oldest scientist, but not both?

8 Write down the first three terms of $(3-2x)^5$ in ascending powers of x.

9 Find $\dfrac{(3-\sqrt{2})^3}{\sqrt{2}+1}$, giving your answer in the form $a+b\sqrt{2}$ where $a, b \in \mathbb{Z}$.

10 In the expansion of $\left(2x+\dfrac{3}{x}\right)^{10}$, find:

a the coefficient of x^4 **b** the constant term **c** the coefficient of x^{-6}.

11 Find the coefficient of x^4 in the expansion of $(x-3)(2x+1)^5$.

12 The coefficient of x^2 in the expansion of $\left(1+\dfrac{x}{2}\right)^n$ is $\frac{21}{4}$. Find n.

13 Suppose $(1+kx)^n = 1-30x+360x^2 - \ldots$. Find k and n.

11

Sequences and series

Contents:

Opening problem

Samantha has built a hollow pyramid using cubes.

There are 4 cubes on the top layer and 8 cubes on the second layer.

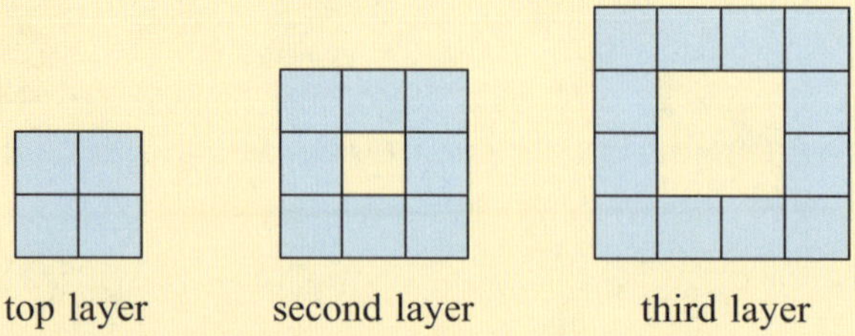

Things to think about:

a How many cubes are in the:

i third layer **ii** fourth layer **iii** nth layer?

b What is the *total* number of cubes in a 10 layer pyramid?

To help understand problems like the **Opening Problem**, we need to study **sequences** and their sums which are called **series**.

A NUMBER SEQUENCES

In mathematics it is important that we can:

- **recognise** a pattern in a set of numbers,
- **describe** the pattern in words, and
- **continue** the pattern.

> A **number sequence** is an ordered list of numbers defined by a rule.
> The numbers in the sequence are said to be its **members** or its **terms**.
> The nth term of a sequence is written as u_n.

We often describe a number sequence in words, giving a rule which connects one term with the next.

For example, the sequence 15, 11, 7, 3, -1, ... can be described by the rule:

($u_1 = 15$, $u_2 = 11$, $u_3 = 7$, $u_4 = 3$, $u_5 = -1$)

"Start with 15, and each term thereafter is 4 less than the previous one."

The next two terms are $u_6 = -1 - 4 = -5$
and $u_7 = -5 - 4 = -9$.

Example 1 — Self Tutor

Write down a rule to describe the sequence, and hence find its next two terms:

a 3, 7, 11, 15, 19, ... **b** 2, 6, 18, 54, ... **c** 0, 1, 1, 2, 3, 5, 8, ...

a Start with 3, and each term thereafter is 4 more than the previous term.
The next two terms are $u_6 = 23$ and $u_7 = 27$.

b Start with 2, and each term thereafter is 3 times the previous term.
The next two terms are $u_5 = 54 \times 3 = 162$ and $u_6 = 162 \times 3 = 486$.

c The first two terms are 0 and 1, and each term thereafter is the sum of the previous two terms.
The next two terms are $u_8 = 5 + 8 = 13$ and $u_9 = 8 + 13 = 21$.

EXERCISE 11A.1

1 Consider the sequence 7, 13, 19, 25, ...

a Write down a rule to describe the sequence.

b State the value of u_1 and u_4.

c Assuming the pattern continues, find u_5 and u_6.

2 Write down a rule to describe the sequence, and hence find its next *two* terms:

a 5, 8, 11, 14, 17, ... **b** 38, 34, 30, 26, 22, ... **c** $\frac{1}{2}$, 2, $3\frac{1}{2}$, 5, $6\frac{1}{2}$, ...

d 3, 6, 12, 24, 48, ... **e** 2, 10, 50, 250, ... **f** 162, 54, 18, 6, ...

3 Find the next *two* terms of:

a 0, 1, 4, 9, 16, ... **b** 1, 4, 9, 16, 25, ... **c** 0, 1, 8, 27, 64, ...

d $\frac{1}{4}$, $\frac{1}{9}$, $\frac{1}{16}$, $\frac{1}{25}$, ... **e** 1, 2, 4, 7, 11, ... **f** 2, 6, 12, 20, 30, ...

4 Write down a rule to describe the sequence, and hence find its next *three* terms:

a 1, 1, 2, 3, 5, 8, ... **b** 1, 3, 4, 7, 11, ... **c** 5, 8, 12, 18, 24, 30, ...

FORMULAE FOR SEQUENCES

An alternative way to describe a sequence is to write a **formula** for the ***n*th term** or **general term** u_n.

For example, the sequence defined by the formula $u_n = n^2 + n$ has $u_1 = 1^2 + 1 = 2$,
$u_2 = 2^2 + 2 = 6$,
$u_3 = 3^3 + 3 = 12$, and so on.

Example 2 Self Tutor

Consider the sequence with general term $u_n = n^2 + 5$.

a Find the first four terms of the sequence.

b Find the 20th term of the sequence.

a $u_1 = 1^2 + 5 = 6$
$u_2 = 2^2 + 5 = 9$
$u_3 = 3^2 + 5 = 14$
$u_4 = 4^2 + 5 = 21$

So, the first four terms are 6, 9, 14, 21.

b The 20th term of the sequence is $u_{20} = 20^2 + 5 = 405$.

EXERCISE 11A.2

1 Consider the sequence with general term $u_n = n^2 + 4n + 1$.

a Find the first four terms of the sequence.
b Find the 10th term of the sequence.
c Which term of the sequence is 61?

2 Find the first *four* terms of the sequence with nth term:

a $u_n = 2n + 5$
b $u_n = 3n + 2$
c $u_n = 6 - 3n$
d $u_n = n^2 + 1$
e $u_n = n^3 + 4$
f $u_n = 3^n$
g $u_n = 5 \times 2^n$
h $u_n = 3^n - n$
i $u_n = \dfrac{n^3}{2n+1}$

3 Consider the sequence with general term $u_n = n^2 + 8n$.

a Find the first four terms of the sequence.
b Explain why the sequence will not contain any prime numbers.
c Find the first term of the sequence which is greater than 1000.

4 **a** Find a general formula for the sequence 1, 4, 9, 16, 25, ...

b *Hence* find a general formula for:

i 4, 9, 16, 25, 36, ...
ii 0, 3, 8, 15, 24, ...
iii $\frac{1}{1}, \frac{1}{4}, \frac{1}{9}, \frac{1}{16}, \frac{1}{25}, ...$
iv $\frac{1}{4}, \frac{2}{9}, \frac{3}{16}, \frac{4}{25}, \frac{5}{36}, ...$

B ARITHMETIC SEQUENCES

In an **arithmetic sequence** or **arithmetic progression**, each term differs from the previous term by the same fixed number.

The difference between successive terms is called the **common difference** d.

For example:

- 3, 8, 13, 18, ... is an arithmetic sequence with common difference 5
  +5 +5 +5
- 20, 17, 14, 11, ... is an arithmetic sequence with common difference -3.
  -3 -3 -3

GENERAL TERM FORMULA

Suppose a sequence is arithmetic with first term u_1 and common difference d.

Then $u_2 = u_1 + d$
$u_3 = u_1 + 2d$
$u_4 = u_1 + 3d$, and so on.

The nth term of an arithmetic sequence with first term u_1 and common difference d is $\boldsymbol{u_n = u_1 + (n-1)d}$.

Example 3 Self Tutor

Consider the sequence 5, 11, 17, 23, ...

a Show that the sequence is arithmetic.

b Find a formula for the general term u_n.

c Find the 50th term of the sequence.

d Find the first term which is greater than 200.

a $11 - 5 = 6$
$17 - 11 = 6$
$23 - 17 = 6$

The difference between successive terms is 6.

$\therefore$ the sequence is arithmetic, with $u_1 = 5$ and $d = 6$.

b $u_n = u_1 + (n-1)d$
$\therefore\ u_n = 5 + 6(n-1)$
$\therefore\ u_n = 6n - 1$

c $u_{50} = 6(50) - 1$
$= 299$

d If $u_n > 200$ then $6n - 1 > 200$
$\therefore\ 6n > 201$
$\therefore\ n > 33\frac{1}{2}$

So, the first term greater than 200 is $u_{34} = 6(34) - 1 = 203$.

EXERCISE 11B

1 Decide whether the following sequences are arithmetic. If the sequence is arithmetic, state the common difference.

a 3, 10, 17, 24, 31, ...

b 4, $6\frac{1}{2}$, 9, $11\frac{1}{2}$, 14, ...

c 26, 22, 18, 14, 10, ...

d $3\frac{1}{3}$, 5, $6\frac{2}{3}$, $7\frac{1}{3}$, 9, ...

e 11.2, 7.3, 3.4, -0.5, -4.4, ...

f -12, -7, -2, 2, 7, 12, ...

g $\sqrt{5}$, $\sqrt{8}$, $\sqrt{11}$, $\sqrt{14}$, $\sqrt{17}$, ...

h $\ln 3$, $\ln 9$, $\ln 27$, $\ln 81$, ...

2 Consider the sequence 3, 7, 11, 15, ...

a Show that the sequence is arithmetic.

b Find a formula for the nth term u_n.

c Find the 60th term of the sequence.

d Find the first term which is greater than 500.

3 Consider the sequence 172, 165, 158, 151, ...

a Show that the sequence is arithmetic.

b Find a formula for the nth term u_n.

c Find the 12th term of the sequence.

d Find the first negative term of the sequence.

4 The general term of a sequence is $u_n = \frac{5n-3}{2}$.

- **a** Find $u_{n+1} - u_n$. Hence show that the sequence is arithmetic.
- **b** Find u_1 and d.
- **c** Find the 20th term of the sequence.
- **d** Which term of the sequence is 101?

Example 4 — Self Tutor

Find k given that $k-5$, $3k$, and $41-k$ are consecutive terms of an arithmetic sequence.

Since the terms are arithmetic, $3k-(k-5) = (41-k)-3k$ {equating differences}

$$\therefore\ 2k+5 = 41-4k$$
$$\therefore\ 6k = 36$$
$$\therefore\ k = 6$$

5 Find k given the consecutive arithmetic terms:

- **a** $8,\ 2k+1,\ 3k-1$
- **b** $k-5,\ 10-k,\ 4k+4$
- **c** $5k+6,\ 25-k,\ k^2$
- **d** $1-k,\ 3k+1,\ 2k^2+4$
- **e** $-\frac{1}{2},\ 2^k,\ 1$
- **f** $4^k+1,\ 3\times 2^k-6,\ -5$
- **g** $\sqrt{k},\ k+1,\ 3k-4$
- **h** $9+\sqrt{k},\ 6,\ 4-\frac{2}{\sqrt{k}}$
- **i** $\log_2 k-3,\ 3\log_8 k+1,\ 7$

6 Suppose k^2-3, k^3+2, and $7+k$ are the first three terms of an arithmetic progression.

- **a** Find the possible values of k.
- **b** For each possible value of k:
  - **i** Find the common difference.
  - **ii** List the first three terms of the progression.
  - **iii** Find the 10th term of the progression.

7 **a** Show that k^2+1, $-2-k^2$, and 3 cannot be consecutive terms of an arithmetic progression.

b Explain this result by considering only the signs of the terms.

Example 5 — Self Tutor

Find the general term u_n for an arithmetic sequence with $u_5 = 18$ and $u_{12} = 81$.

$u_5 = 18 \quad \therefore\ u_1 + 4d = 18 \quad .... (1)$

$u_{12} = 81 \quad \therefore\ u_1 + 11d = 81 \quad .... (2)$

We now solve (1) and (2) simultaneously:

$$\begin{aligned} -u_1 - 4d &= -18 \quad \{-1\times(1)\} \\ u_1 + 11d &= 81 \\ \hline \text{Adding, } \quad 7d &= 63 \\ \therefore\ d &= 9 \end{aligned}$$

In (1): $u_1 + 4(9) = 18$

$\therefore\ u_1 + 36 = 18$

$\therefore\ u_1 = -18$

So, $u_n = u_1 + (n-1)d$

$\therefore\ u_n = -18 + 9(n-1)$

$\therefore\ u_n = 9n - 27$

8 Find the general term u_n for an arithmetic sequence with:

a $u_4 = 32$ and $u_9 = 47$
b $u_6 = 11$ and $u_{10} = -17$
c $u_7 = -6$ and $u_{13} = 23$
d $u_5 = 10\frac{1}{2}$ and $u_{13} = -9\frac{1}{2}$.

9 The 4th term of an arithmetic progression is -2, and the 11th term is $2\frac{1}{2}$. Find the 30th term of the progression.

10 The 6th term of an arithmetic progression is -331, and the 15th term is -217. Find the first positive term of the progression.

11 Suppose $\{u_n\}$ is an arithmetic sequence with common difference d. Decide whether the following sequences are also arithmetic. If the sequence is arithmetic, state the common difference.

a $u_1 + 3,\ u_2 + 3,\ u_3 + 3,\ u_4 + 3,\ ....$
b $5u_1,\ 5u_2,\ 5u_3,\ 5u_4,\ ....$
c $u_1^2,\ u_2^2,\ u_3^2,\ u_4^2,\ ....$
d $u_1 + u_2,\ u_2 + u_3,\ u_3 + u_4,\ ....$
e $\ln(u_1),\ \ln(u_2),\ \ln(u_3),\ \ln(u_4),\ ....$
f $u_{u_1},\ u_{u_2},\ u_{u_3},\ u_{u_4},\ ....$

12 **a** Give an example of an arithmetic progression containing:

i only irrational terms
ii both rational *and* irrational terms.

b Is it possible to construct an arithmetic progression which alternates between rational and irrational terms? Explain your answer.

Activity 1

Consider the family of straight lines with equation $ax + by = c$, where a, b, and c are in arithmetic progression. One such line is graphed alongside.

Graph several other lines of this type. What feature do they all have in common? Can you explain why this occurs?

GRAPHING PACKAGE

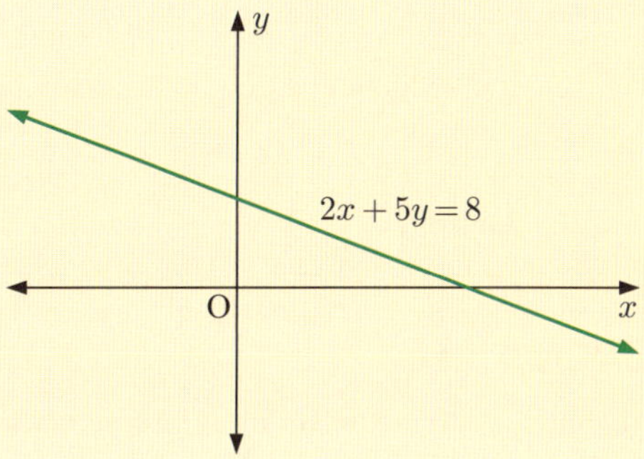

C GEOMETRIC SEQUENCES

In a **geometric sequence** or **geometric progression**, each term can be obtained from the previous term by multiplying by the same non-zero number.

The number that we multiply by to get from one term to the next is called the **common ratio** r.

For example:

- 4, 12, 36, 108, ... is a geometric sequence with common ratio 3
  $\times 3$ $\times 3$ $\times 3$
- 40, -20, 10, -5, ... is a geometric sequence with common ratio $-\frac{1}{2}$.
  $\times -\frac{1}{2}$ $\times -\frac{1}{2}$ $\times -\frac{1}{2}$

GENERAL TERM FORMULA

Suppose a sequence is geometric with first term u_1 and common ratio r.

Then $u_2 = u_1 \times r$
$u_3 = u_1 \times r^2$
$u_4 = u_1 \times r^3$

The nth term of a geometric sequence with first term u_1 and common ratio r is $\boldsymbol{u_n = u_1 r^{n-1}}$.

Example 6

Self Tutor

Consider the sequence 40, 60, 90, 135, ...

a Show that the sequence is geometric.

b Find a formula for the general term u_n.

c Find the 7th term of the sequence.

a $\frac{60}{40} = \frac{3}{2}$, $\frac{90}{60} = \frac{3}{2}$, $\frac{135}{90} = \frac{3}{2}$

The consecutive terms have a common ratio of $\frac{3}{2}$.

$\therefore$ the sequence is geometric with $u_1 = 40$ and $r = \frac{3}{2}$.

b $u_n = u_1 r^{n-1}$

$\therefore u_n = 40\left(\frac{3}{2}\right)^{n-1}$

c $u_7 = 40\left(\frac{3}{2}\right)^6$

$= \frac{3645}{8}$ or $455\frac{5}{8}$

EXERCISE 11C

1 Decide whether the following sequences are geometric. If the sequence is geometric, state the common ratio.

a 3, 15, 75, 375, ...

b 7, -14, 28, -56, ...

c 3, 12, 36, 144, ...

d 46, 4.6, 0.46, 0.046, ...

e 160, -40, 10, -2.5, ...

f $\sqrt{7}$, 7, $7\sqrt{7}$, 49, ...

g 1!, 2!, 3!, 4!, 5!, ...

h $\ln 25$, $\ln 5$, $\ln(\sqrt{5})$, $\ln(\sqrt[4]{5})$, ...

2 Consider the sequence 5, 15, 45, 135, ...

a Show that the sequence is geometric.

b Find a formula for the general term u_n.

c Find the 10th term of the sequence.

3 Consider the sequence 320, -160, 80, -40, ...

a Show that the sequence is geometric.

b Find a formula for the general term u_n.

c Find the 7th term of the sequence.

d Which term of the sequence is $-\frac{5}{8}$?

Example 7

Self Tutor

Find k given that $k-1$, $k+4$, and $3k+2$ are consecutive terms of a geometric sequence.

Since the terms are geometric, $\dfrac{k+4}{k-1} = \dfrac{3k+2}{k+4}$ {equating the common ratio}

$$\therefore\ (k+4)^2 = (3k+2)(k-1)$$
$$\therefore\ k^2+8k+16 = 3k^2-k-2$$
$$\therefore\ 0 = 2k^2-9k-18$$
$$\therefore\ (2k+3)(k-6) = 0$$
$$\therefore\ k = -\tfrac{3}{2} \text{ or } 6$$

4 Find k given the consecutive geometric terms:

a $8,\ k,\ 72$ **b** $7,\ k,\ 35$ **c** $k-3,\ k+1,\ k+7$

d $k-7,\ k+2,\ 5k-2$ **e** $\frac{16}{9},\ 2^k,\ 3^k$ **f** $k+3,\ 6+2k,\ k^2+12$

g $k-3,\ 2\sqrt{k},\ 10$ **h** $2e^{2k}-2,\ 3,\ 3e^{-k}$ **i** $\log_3 k-1,\ 2,\ \log_{\sqrt{3}} k+5$

5 Suppose $k-3$, $k+1$, and $4k-2$ are the first 3 terms of a geometric progression.

a Find the possible values of k.

b For each possible value of k:

i List the first 3 terms of the progression.

ii State the common ratio.

iii Find the 10th term of the progression.

6 **a** Find the positive value of p such that $10-p$, $p-1$, and $p+5$ are the first 3 terms of:

i an arithmetic progression **ii** a geometric progression.

b Explain why we can be sure that the arithmetic and geometric progressions in **a** do not have any terms in common.

Example 8

Self Tutor

Find the general term u_n for a geometric sequence with $u_3 = 3$ and $u_6 = 192$.

$u_3 = 3 \quad \therefore\ u_1r^2 = 3 \quad .... (1)$

$u_6 = 192 \quad \therefore\ u_1r^5 = 192 \quad .... (2)$

Now $\dfrac{u_1r^5}{u_1r^2} = \dfrac{192}{3}$ {(2) ÷ (1)}

$$\therefore\ r^3 = 64$$
$$\therefore\ r = \sqrt[3]{64}$$
$$\therefore\ r = 4$$

Using (1), $u_1(4^2) = 3$

$$\therefore\ u_1 = \tfrac{3}{16}$$

Thus, $u_n = \frac{3}{16} \times 4^{n-1}$

7 Find the general term u_n for a geometric sequence with:

a $u_2 = 5$ and $u_5 = 40$

b $u_3 = 54$ and $u_6 = 2$

c $u_3 = 10$ and $u_5 = 250$

d $u_4 = 6$ and $u_8 = \frac{3}{8}$

e $u_4 = \frac{3}{8}$ and $u_7 = -\frac{81}{64}$

f $u_2 = -\frac{18}{35}$ and $u_4 = -\frac{648}{1715}$

8 A geometric progression has 2nd term 120 and 5th term $-\frac{405}{8}$.

a Find the first term and the common ratio of the progression.

b Find the 7th term of the progression.

9 Give an example of a geometric progression which:

a contains only irrational terms

b alternates between rational and irrational terms.

10 Suppose $\{u_n\}$ is a geometric sequence with common ratio r. Decide whether the following sequences are geometric, arithmetic, or neither. If the sequence is geometric or arithmetic, state the common ratio or common difference.

a $2u_1, 2u_2, 2u_3, 2u_4, ...$

b $u_1 - 3,\ u_2 - 3,\ u_3 - 3,\ u_4 - 3,\ ....$

c $\ln(u_1),\ \ln(u_2),\ \ln(u_3),\ \ln(u_4),\ ....$

d $u_1^3,\ u_2^3,\ u_3^3,\ u_4^3,\ ....$

e $u_1, u_3, u_5, u_7, ...$

f $u_1u_2,\ u_3u_4,\ u_5u_6,\ u_7u_8,\ ....$

D SERIES

A **series** is the sum of the terms of a sequence.

For a finite sequence $u_1, u_2, u_3, ..., u_n$ with n terms, the corresponding series is $u_1 + u_2 + u_3 + ... + u_n$.

The sum of this series is $S_n = u_1 + u_2 + u_3 + ... + u_n$.

For example, consider the sequence 1, 3, 6, 10, 15, which has 5 terms. The corresponding series is $1 + 3 + 6 + 10 + 15$, and the sum of the series is $S_5 = 35$.

EXERCISE 11D

1 Consider the sequence 3, 5, 8, 11, 19. Find:

a u_3 **b** S_3 **c** u_5 **d** S_5.

2 For each of the following sequences, find:

i the first 4 terms of the sequence **ii** S_4.

a $u_n = 2n + 3$

b $u_n = n^2 - 6$

c $u_n = 7 \times 2^{n-1}$

d $u_1 = 3,\ u_n = 3 \times u_{n-1} - 4,\ n \geqslant 2$

3 A sequence has $S_5 = 21$ and $S_6 = 33$. Find the value of u_6.

4 Consider a sequence with general term $u_n = \dfrac{1}{n(n+1)}$. Let S_n be the sum of the first n terms of the sequence.

a Find the values of S_1, S_2, S_3, and S_4 as fractions.

b Hence conjecture the value of $\dfrac{1}{1 \times 2} + \dfrac{1}{2 \times 3} + \dfrac{1}{3 \times 4} + \dfrac{1}{4 \times 5} + ... + \dfrac{1}{100 \times 101}$.

E ARITHMETIC SERIES

An **arithmetic series** is the sum of the terms of an arithmetic sequence.

Consider the arithmetic sequence 5, 10, 15, ..., 90, 95, 100. The first term is $u_1 = 5$ and the last term is $u_{20} = 100$.

The sum of the terms of this sequence is $S_{20} = 5 + 10 + 15 + ... + 90 + 95 + 100$

However, we can also write $S_{20} = 100 + 95 + 90 + ... + 15 + 10 + 5$ {reversing the terms}

Adding these equations gives $2 \times S_{20} = \underbrace{105 + 105 + 105 + ... + 105 + 105 + 105}_{\text{20 of these}}$

$\therefore\ 2 \times S_{20} = 20 \times 105$

$\therefore\ S_{20} = \frac{20}{2} \times 105 = 1050$

The sum of a finite arithmetic series with first term u_1, common difference d, and last term u_n, is

$$S_n = \frac{n}{2}(u_1 + u_n)$$

$$\text{or}\quad S_n = \frac{n}{2}(2u_1 + (n-1)d) \qquad \{\text{using } u_n = u_1 + (n-1)d\}$$

Example 9 Self Tutor

An arithmetic series has 8 terms. The first term is 3 and the last term is 17. Find the sum of the series.

The series is arithmetic with $u_1 = 3$ and $u_8 = 17$.

Now $S_n = \frac{n}{2}(u_1 + u_n)$

$\therefore\ S_8 = \frac{8}{2}(3 + 17)$

$= 80$

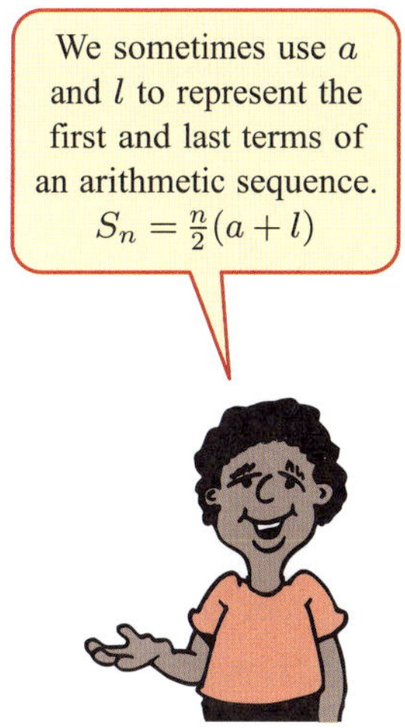

Example 10 Self Tutor

Find the sum of $1 + 5 + 9 + 13 + ...$ to 30 terms.

The series is arithmetic with $u_1 = 1$, $d = 4$, and $n = 30$.

Now $S_n = \frac{n}{2}(2u_1 + (n-1)d)$

$\therefore\ S_{30} = \frac{30}{2}(2 \times 1 + 29 \times 4)$

$= 1770$

Example 11 Self Tutor

Find the sum $6+10+14+18+....+102$.

The series is arithmetic with $u_1=6$, $d=4$, and $u_n=102$.

$$u_n=102$$
$$\therefore\ u_1+(n-1)d=102$$
$$\therefore\ 6+4(n-1)=102$$
$$\therefore\ 4n=100$$
$$\therefore\ n=25$$

Now $S_n=\frac{n}{2}(u_1+u_n)$

$$\therefore\ S_{25}=\tfrac{25}{2}(6+102)$$
$$=1350$$

EXERCISE 11E

1 Find the value of $5+8+11+14+17+20$:

a by adding the terms directly

b using $S_n=\frac{n}{2}(u_1+u_n)$

c using $S_n=\frac{n}{2}(2u_1+(n-1)d)$.

2 An arithmetic series has 12 terms. The first term is 10 and the last term is 65. Find the sum of the series.

3 An arithmetic series has 20 terms. The first term is 30 and the last term is -8. Find the sum of the series.

4 Find the sum of:

a $3+7+11+15+....$ to 10 terms

b $40+35+30+25+....$ to 15 terms

c $8+11+14+17+....$ to 50 terms

d $-6+3+12+21+....$ to 40 terms

e $21+19+17+15+....$ to 60 terms

f $7+1+(-5)+(-11)+....$ to 30 terms

g $5+5\frac{1}{2}+6+6\frac{1}{2}+....$ to 25 terms

h $20+19\frac{1}{2}+19+18\frac{1}{2}+....$ to 50 terms.

5 An arithmetic sequence has $S_1=4$ and $S_2=11$. Find S_{40}.

6 Consider the series $4+9+14+19+....+119$.

a How many terms are in the series?

b Find the sum of the series.

7 Find each sum:

a $7+9+11+13+....+55$

b $10+13+16+19+....+100$

c $87+83+79+75+....+15$

d $-5+1+7+13+....+109$

e $12+7+2+(-3)+....+(-58)$

f $6+7\frac{1}{2}+9+10\frac{1}{2}+....+30$

8 **a** Show that the sum of the first n multiples of 4 is $2n(n+1)$.

b Hence, find $4+8+12+16+....+80$.

9 Jim is saving money to buy a car. He puts \$20 in the bank in the first week, then \$25 in the second week, then \$30 in the third week, and so on.

a How much will Jim put into the bank in the 10th week?

b How much money will Jim have saved in total after 20 weeks?

10 The arrangement of numbers alongside is called a 3×3 **magic square**. The numbers from 1 to 9 are placed so that the numbers in each row, column, and main diagonal all add up to the same number, known as the **magic constant**. For a 3×3 magic square, the magic constant is 15.

2	7	6
9	5	1
4	3	8

An $n \times n$ magic square contains the numbers from 1 to n^2.

a Find, in terms of n, the magic constant for an $n \times n$ magic square.

b Check that your formula is correct for $n = 3$.

c Find the magic constant for an 8×8 magic square.

11 Consider the series $5 + 9 + 13 + 17 + ...$

a Find a formula for S_n.

b Find n such that $S_n = 230$.

12 Suppose e^m, $3e^{-m}$, and 1 are the first 3 terms of an arithmetic progression.

a Find m.

b Find the sum of the first 20 terms of the progression.

13 An arithmetic progression has first term 4 and common difference -3. Find the smallest n such that S_n is less than -100.

Discussion

Is it possible to find the sum of an infinite arithmetic series?

F GEOMETRIC SERIES

A **geometric series** is the sum of the terms of a geometric sequence.

For example: 1, 2, 4, 8, 16, ... , 1024 is a geometric sequence.

$1 + 2 + 4 + 8 + 16 + ... + 1024$ is a geometric series.

If we are adding the first n terms of a geometric sequence, we say we have a **finite geometric series**.

If we are adding all of the terms of a geometric sequence which goes on forever, we say we have an **infinite geometric series**.

SUM OF A FINITE GEOMETRIC SERIES

The sum of the first n terms of a geometric series is

$$S_n = u_1 + u_1 r + u_1 r^2 + u_1 r^3 + ... + u_1 r^{n-1}$$

For a finite geometric series with $r \neq 1$,

$$\boldsymbol{S_n = \frac{u_1(r^n - 1)}{r - 1}} \quad \text{or} \quad \boldsymbol{S_n = \frac{u_1(1 - r^n)}{1 - r}}.$$

Proof:

$$\text{If} \quad S_n = u_1 + u_1 r + u_1 r^2 + u_1 r^3 + ... + u_1 r^{n-2} + u_1 r^{n-1} \qquad (*)$$

$$\text{then} \quad rS_n = (u_1 r + u_1 r^2 + u_1 r^3 + u_1 r^4 + ... + u_1 r^{n-1}) + u_1 r^n$$

$$\therefore \quad rS_n = (S_n - u_1) + u_1 r^n \qquad \{\text{from } (*)\}$$

$$\therefore \quad rS_n - S_n = u_1 r^n - u_1$$

$$\therefore \quad S_n(r-1) = u_1(r^n - 1)$$

$$\therefore \quad S_n = \frac{u_1(r^n - 1)}{r - 1} \quad \text{or} \quad \frac{u_1(1 - r^n)}{1 - r} \quad \text{provided } r \neq 1.$$

Example 12

Self Tutor

Find the sum of:

a $3 + 6 + 12 + 24 + ...$ to 10 terms

b $8 - 4 + 2 - 1 + ...$ to 7 terms.

a The series is geometric with $u_1 = 3$ and $r = 2$.

$$S_n = \frac{u_1(r^n - 1)}{r - 1}$$

$$\therefore \quad S_{10} = \frac{3(2^{10} - 1)}{2 - 1} = 3069$$

b The series is geometric with $u_1 = 8$ and $r = -\frac{1}{2}$.

$$S_n = \frac{u_1(1 - r^n)}{1 - r}$$

$$\therefore \quad S_7 = \frac{8\left(1 - \left(-\frac{1}{2}\right)^7\right)}{1 - \left(-\frac{1}{2}\right)} = \frac{8\left(\frac{129}{128}\right)}{\frac{3}{2}} = \frac{43}{8}$$

EXERCISE 11F.1

1 Find $5 + 10 + 20 + 40 + 80 + 160$:

a by adding the terms directly

b using $S_n = \dfrac{u_1(r^n - 1)}{r - 1}$.

2 Find the sum of:

a $1+3+9+27+....$ to 8 terms

b $2+10+50+250+....$ to 10 terms

c $24+12+6+3+....$ to 9 terms

d $1+\sqrt{2}+2+2\sqrt{2}+....$ to 12 terms

e $4-8+16-32+....$ to 10 terms

f $81-27+9-3+....$ to 8 terms

g $11\sqrt{11}+11+\sqrt{11}+1+....$ to 9 terms

h $16-4+1-\frac{1}{4}+....$ to 10 terms.

3 A geometric series has $S_1=2$ and $S_2=8$.

a Find the first term u_1 and the common ratio r.

b Find the sum of the first 10 terms of the series.

4 Consider the geometric series $7+14+28+....+3584$.

a Find the number of terms in the series.

b Hence find the sum of the series.

5 A geometric series has first term u_1 and common ratio $r=-1$.

Show that $S_n = \begin{cases} u_1 & \text{if } n \text{ is odd} \\ 0 & \text{if } n \text{ is even} \end{cases}$.

6 The 2nd term of a geometric progression is 162, and the 4th term is 72.

a Find the possible values for the common ratio r.

b For each possible value of r, find the sum of the first 8 terms of the progression.

7 A company made \$50 000 profit in their first year of operation. In each subsequent year, their profit increased by 10%. Find, to the nearest \$1000:

a the profit made in the 10th year

b the total profit made during the first 10 years.

8 Doug is marooned on a desert island with only 10 L of fresh water.
He drinks 2.5 L on the first day, but realises he will soon run out of water if he drinks that much each day. He decides that each day he will drink $\frac{3}{4}$ of the amount he drank the previous day.

a How much water will Doug drink on the:

i 5th day **ii** 10th day **iii** 15th day?

b Describe what happens to the amount of water Doug drinks each day.

c How much water will Doug have drunk in total after:

i 10 days **ii** 20 days **iii** 30 days?

d Do you think Doug will ever run out of water? Explain your answer.

SUM OF AN INFINITE GEOMETRIC SERIES

An **infinite geometric series** is the sum of the terms of a geometric sequence which continues indefinitely.

Examples of infinite geometric series include $2+8+32+128+....$ and $10+5+2\frac{1}{2}+1\frac{1}{4}+....$

If $r > 1$ **or** $r < -1$, the terms in the series get larger and larger. The sum of the series becomes infinitely large, and cannot be found. We say that the series **diverges**.

If $-1 < r < 1$, the terms in the series get smaller and smaller, and the sum of the series **converges** to a finite value.

Consider $S_n = \frac{u_1(1-r^n)}{1-r}$ as n gets very large.

For $-1 < r < 1$, r^n approaches zero, so S_n approaches the value $\frac{u_1}{1-r}$.

If $-1 < r < 1$, the sum of an infinite geometric series with first term u_1 and common ratio r is $\boldsymbol{S_\infty = \frac{u_1}{1-r}}$.

Example 13 — Self Tutor

Find the sum $2 + 1 + \frac{1}{2} + \frac{1}{4} + ...$

This is an infinite geometric series with $u_1 = 2$ and $r = \frac{1}{2}$.

Since $-1 < r < 1$, the series converges.

The sum is $S_\infty = \frac{u_1}{1-r} = \frac{2}{1-\frac{1}{2}} = 4$

EXERCISE 11F.2

1 Decide whether the following infinite geometric series will converge or diverge:

a $7 + 14 + 28 + 56 + ...$

b $6 + 3 + 1\frac{1}{2} + \frac{3}{4} + ...$

c $1 - \sqrt{3} + 3 - 3\sqrt{3} + ...$

d $80 - 8 + 0.8 - 0.08 + ...$

2 Consider the infinite geometric series $9 + 6 + 4 + \frac{8}{3} + ...$

a Find: **i** S_5 **ii** S_{10} **iii** S_{20}.

b Predict the sum of the infinite geometric series.

c Check your answer to **b** by finding the sum $S_\infty = \frac{u_1}{1-r}$.

3 Find each sum:

a $16 + 8 + 4 + 2 + ...$

b $1 + \frac{1}{5} + \frac{1}{25} + \frac{1}{125} + ...$

c $36 - 12 + 4 - \frac{4}{3} + ...$

d $32 + 24 + 18 + \frac{27}{2} + ...$

e $72 - 12 + 2 - \frac{1}{3} + ...$

f $0.6 + 0.06 + 0.006 + 0.0006 + ...$

4 **a** Without evaluating either sum, explain why $8 - 4 + 2 - 1 + \frac{1}{2} - \frac{1}{4} + ... = 4 + 1 + \frac{1}{4} + ...$

b Verify this fact by evaluating each sum.

5 Consider a rectangle with area 1 unit^2.
The rectangle is divided into thirds, with one third coloured blue, and another third coloured red. The rectangle is then rotated $90°$, and the process is repeated on the remaining unshaded third.
Suppose this process continues indefinitely.

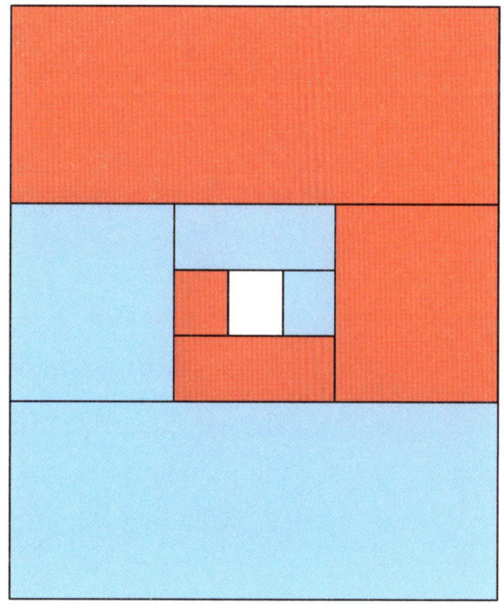

- **a** Show that the total blue shaded area $= \frac{1}{3} + \frac{1}{9} + \frac{1}{27} + ...$
- **b** Explain why the blue shaded area = red shaded area.
- **c** Hence, explain why $\frac{1}{3} + \frac{1}{9} + \frac{1}{27} + ... = \frac{1}{2}$.
- **d** Check this fact using $S_\infty = \dfrac{u_1}{1-r}$.

6 Find the sum of $5 + \dfrac{5}{\sqrt{2}} + \dfrac{5}{2} + \dfrac{5}{2\sqrt{2}} + ...$, giving your answer in the form $a + b\sqrt{2}$, $a, b \in \mathbb{Z}$.

7 The second term of a convergent infinite geometric series is 3. The sum of the series is 16. Show that there are two possible series, and find the first term and the common ratio in each case.

8 The 3rd term of a geometric sequence is e^2, and the 6th term is $-\dfrac{8}{e}$. Find the sum of the infinite geometric series.

9 Suppose $u_1 + u_2 + u_3 + u_4 + ...$ is a convergent infinite geometric series, with $u_1, u_2, u_3, u_4, ... > 0$.
- **a** Explain why $u_1 - u_2 + u_3 - u_4 + ...$ and $\sqrt{u_1} + \sqrt{u_2} + \sqrt{u_3} + \sqrt{u_4} + ...$ must also be convergent infinite geometric series.
- **b** Given $u_1 - u_2 + u_3 - u_4 + ... = \frac{64}{5}$ and $\sqrt{u_1} + \sqrt{u_2} + \sqrt{u_3} + \sqrt{u_4} + ... = 8$, find $u_1 + u_2 + u_3 + u_4 + ...$.

10 Show that $\frac{1}{4} + \frac{2}{8} + \frac{3}{16} + \frac{4}{32} + \frac{5}{64} + ... = 1$.

Activity 2 — *The harmonic series*

The **harmonic series** is the sum of the reciprocals of the positive integers:

$$1 + \tfrac{1}{2} + \tfrac{1}{3} + \tfrac{1}{4} + \tfrac{1}{5} + \tfrac{1}{6} + \tfrac{1}{7} + \tfrac{1}{8} + ...$$

What to do:

1 Explain why the harmonic series is neither arithmetic nor geometric.

2
- **a** For the harmonic series, calculate by direct evaluation:
  - **i** S_3
  - **ii** S_5
  - **iii** S_{10}
  - **iv** S_{15}
- **b** Does the series appear to be converging?

3 By comparing the harmonic series with the series $1 + \underbrace{\tfrac{1}{2}}_{\text{1 term}} + \underbrace{\tfrac{1}{4} + \tfrac{1}{4}}_{\text{2 terms}} + \underbrace{\tfrac{1}{8} + \tfrac{1}{8} + \tfrac{1}{8} + \tfrac{1}{8}}_{\text{4 terms}} + ...$,
show that the harmonic series does not converge to a particular value, but instead diverges.

4 The Swiss mathematician **Leonhard Euler** (1707 - 1783) wrote the harmonic series as a product of geometric series involving prime numbers:

$$1+\tfrac{1}{2}+\tfrac{1}{3}+\tfrac{1}{4}+\tfrac{1}{5}+\tfrac{1}{6}+\tfrac{1}{7}+\tfrac{1}{8}+....$$
$$=\left(1+\frac{1}{2}+\frac{1}{2^2}+....\right)\times\left(1+\frac{1}{3}+\frac{1}{3^2}+....\right)\times\left(1+\frac{1}{5}+\frac{1}{5^2}+....\right)\times ....$$
$$\times\left(1+\frac{1}{p_i}+\frac{1}{p_i^{\,2}}+....\right)\times ....$$

where p_i is the ith prime number.

- **a** Explain why this result is true.
- **b** Are the geometric series in this result convergent or divergent?
- **c** Hence prove that there are infinitely many prime numbers.

Review set 11A

1 Find the first four terms of the sequence with general term:

- **a** $u_n = 8 - 3n$
- **b** $u_n = n^2 - \dfrac{1}{n}$

2 Consider the sequence 5, 13, 21, 29, ...

- **a** Show that the sequence is arithmetic.
- **b** Find a formula for the nth term u_n.
- **c** Find the 15th term of the sequence.
- **d** Find the first term which is greater than 1000.

3 The 5th term of an arithmetic progression is 13, and the 11th term is -14.

- **a** Find the first term and common difference of the progression.
- **b** Find the 20th term of the progression.

4 Decide whether the following sequences are geometric. If the sequence is geometric, state the common ratio.

- **a** $\sqrt{5}, 1, \frac{1}{\sqrt{5}}, \frac{1}{5}, ...$
- **b** $5, -20, 80, -320, ...$
- **c** $\pi, \pi^2, \pi^4, \pi^8, ...$
- **d** $-\dfrac{4}{e}, 2, -e, \dfrac{e^2}{2}, ...$

5 Find k given the consecutive geometric terms:

- **a** $k+5,\ k-7,\ k-1$
- **b** $k-3,\ 2k-2,\ 5k+1$
- **c** $-k,\ 4-k^2,\ -2k$

6 Find the sum of:

- **a** $21+25+29+33+....$ to 20 terms
- **b** $40+34+28+22+....$ to 30 terms.

7 Find the sum of the positive terms of the arithmetic progression with first term 86 and common difference -4.

8 Find the sum of:

- **a** $5+20+80+320+....$ to 10 terms
- **b** $18-12+8-\frac{16}{3}+....$ to 9 terms.

9 The first term of a geometric progression is $4\frac{1}{2}$, and the sum of the first 3 terms is $6\frac{1}{2}$.

- **a** Find the possible values for the common ratio r.
- **b** For each possible value of r:
  - **i** Find the sum of the first 5 terms of the progression.
  - **ii** If possible, find the sum to infinity of the progression.

10 Find each sum:

- **a** $25 + 5 + 1 + \frac{1}{5} + ...$
- **b** $24 - 12 + 6 - 3 + ...$
- **c** $6 + 2\sqrt{3} + 2 + \frac{2}{\sqrt{3}} + ...$
- **d** $\ln\left(\frac{1}{3^8}\right) - \ln\left(\frac{1}{3^4}\right) + \ln\left(\frac{1}{3^2}\right) - \ln\left(\frac{1}{3}\right) + ...$

Review set 11B

1 Consider the sequence with general term $u_n = n^2 + 3n$.

- **a** Find the first four terms of the sequence.
- **b** Explain why all of the terms of the sequence are even.
- **c** Find the first term of the sequence which is greater than 600.

2 Find the general term for an arithmetic sequence with:

- **a** $u_3 = 10$ and $u_8 = 45$
- **b** $u_4 = 9$ and $u_{12} = -20$

3 Suppose $k - 7$, $13 - k$, and $4k - 2$ are the first three terms of an arithmetic progression.

- **a** Find k.
- **b** Find the 25th term of the progression.
- **c** Which term of the progression is 498?

4 Consider the sequence $108, -72, 48, -31, ...$

- **a** Show that the sequence is geometric.
- **b** Find the 7th term of the sequence.
- **c** Find the sum of the first 7 terms of the sequence.

5 Suppose $\{u_n\}$ is a finite geometric sequence. Show that the product of the terms is $\sqrt{u_1^n \times u_n^n}$.

6 Find the sum of:

- **a** $6 + 11 + 16 + 21 + ... + 101$
- **b** $80 + 73 + 66 + 59 + ...$ to 20 terms
- **c** $17 + 14 + 11 + 8 + ... + (-31)$
- **d** $16 + 24 + 36 + 54 + ...$ to 10 terms.

7 An arithmetic series has $S_2 = -7$ and $S_4 = -2$. Find the value of S_6.

8 An arithmetic progression has first term 5 and common difference 6. Find the smallest n such that S_n is greater than 300.

9 The 3rd term of a geometric progression is $\frac{10}{9}$, and the 5th term is $\frac{2}{5}$. Given that the common ratio is positive, find:

- **a** the first term and common ratio
- **b** the sum to infinity of the progression.

10 Consider the sequence m, π, $\frac{\pi^2}{m}$, $\frac{\pi^3}{m^2}$, ... where m is a constant.

a Show that the sequence is geometric.

b For what values of m is the sequence convergent?

c Find the sum of the series when $m = 4$.

d Find m such that the sum of the series is 4π.

12

Vectors

Contents:

Opening problem

A bird starts flying towards its nest, which is 40 km to the east and 30 km to the north. The bird is flying at 20 km h^{-1}.

Things to think about:

- **a** How can we use an array of numbers to represent the speed *and* direction of the bird?
- **b** On what *bearing* is the bird travelling?
- **c** Can you describe the *position* of the bird, relative to its starting position, after t hours of flying?
- **d** How long will the bird take to reach its nest?

A VECTORS AND SCALARS

In the **Opening Problem**, the motion of the bird is determined by both its speed *and* its direction.

> Quantities which have only magnitude are called **scalars**.
>
> Quantities which have both magnitude and direction are called **vectors**.

The *speed* of the bird is a scalar. It describes how fast the bird is flying.

The *velocity* of the bird is a vector. It includes both its speed and also its direction.

Other examples of vector quantities are:

- acceleration
- force
- displacement
- momentum

From previous courses, you should have seen how we can represent a vector quantity using a **directed line segment** or **arrow**. The **length** of the arrow represents the size or magnitude of the quantity, and the **arrowhead** shows its direction.

POSITION VECTORS

Consider the vector from the origin O to the point A. We call this the **position vector** of point A.

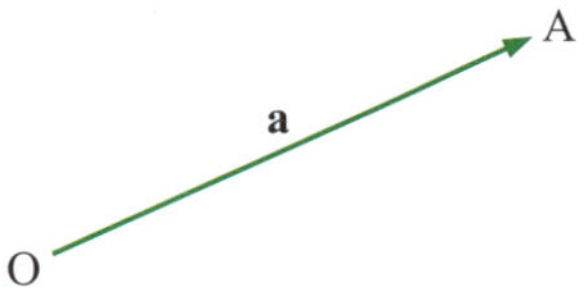

- The **position vector** of A can be represented by $\overrightarrow{OA}$ or $\mathbf{a}$ (bold used in textbooks) or $\vec{a}$ (used by students).
- The **magnitude** or **length** of **a** can be represented by $|\overrightarrow{OA}|$ or OA or $|\mathbf{a}|$ or $|\vec{a}|$.

Now consider the vector from point A to point B. We say that:

- $\overrightarrow{AB}$ is the vector which **originates** at A and **terminates** at B
- $\overrightarrow{AB}$ is the **position vector** of B relative to A.

VECTORS IN THE PLANE

When we plot points in the Cartesian plane, we move first in the x-direction and then in the y-direction.

For example, to plot the point P(4, 2), we start at the origin, move 4 units in the x-direction, and then 2 units in the y-direction.

The vector from O to P is $\overrightarrow{OP} = \begin{pmatrix} 4 \\ 2 \end{pmatrix}$.

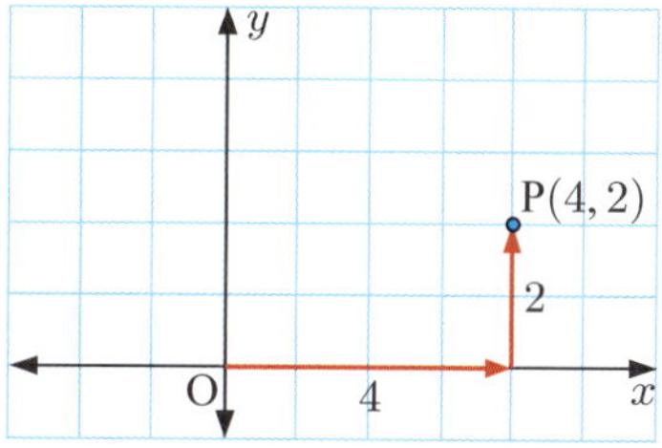

Suppose that $\mathbf{i} = \begin{pmatrix} 1 \\ 0 \end{pmatrix}$ is a vector of length 1 unit in the positive x-direction

and that $\mathbf{j} = \begin{pmatrix} 0 \\ 1 \end{pmatrix}$ is a vector of length 1 unit in the positive y-direction.

i and **j** are called **unit vectors** because they have length 1.

We can see that moving from O to P is equivalent to 4 lots of **i** plus 2 lots of **j**.

$$\overrightarrow{OP} = 4\mathbf{i} + 2\mathbf{j}$$

$$\therefore \begin{pmatrix} 4 \\ 2 \end{pmatrix} = 4\begin{pmatrix} 1 \\ 0 \end{pmatrix} + 2\begin{pmatrix} 0 \\ 1 \end{pmatrix}$$

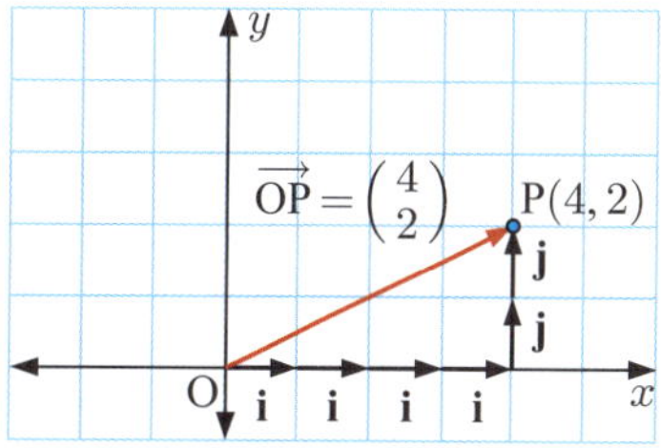

The point P(x, y) has **position vector** $\overrightarrow{OP} = \begin{pmatrix} x \\ y \end{pmatrix} = x\mathbf{i} + y\mathbf{j}$

component form — unit vector form

where $\mathbf{i} = \begin{pmatrix} 1 \\ 0 \end{pmatrix}$ is the **base unit vector** in the x-direction

and $\mathbf{j} = \begin{pmatrix} 0 \\ 1 \end{pmatrix}$ is the **base unit vector** in the y-direction.

All vectors in the plane can be described in terms of the base unit vectors **i** and **j**.

For example: $\mathbf{a} = 5\mathbf{i} - \mathbf{j}$

$\mathbf{b} = -3\mathbf{i} + 2\mathbf{j}$

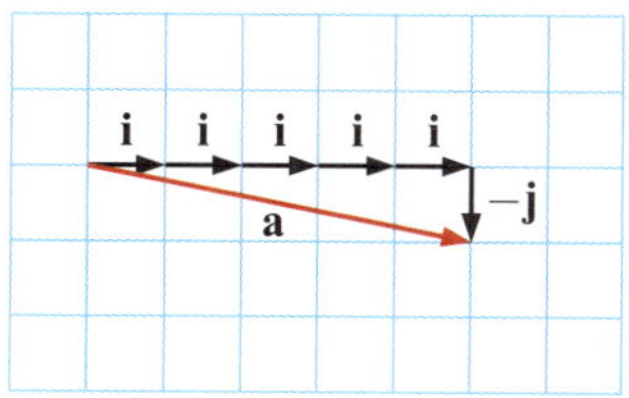

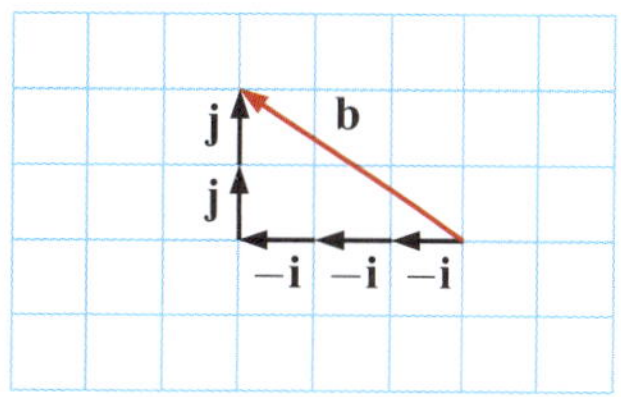

THE ZERO VECTOR

The **zero vector**, $\mathbf{0}$, is a vector of length 0.

It is the only vector with no direction.

In component form, $\mathbf{0} = \begin{pmatrix} 0 \\ 0 \end{pmatrix}$.

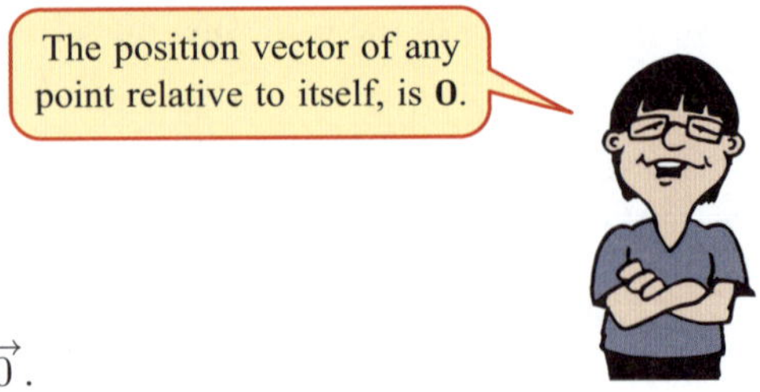

When we write the zero vector by hand, we usually write $\vec{0}$.

VECTOR EQUALITY

Two vectors are **equal** if they have the same magnitude and direction.

In component form, their x-components are equal *and* their y-components are equal.

Equal vectors are **parallel** and in the same direction, and are **equal in length**. The arrows that represent them are translations of one another.

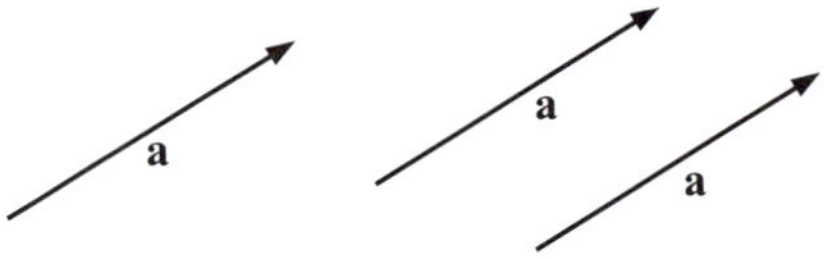

Example 1

Self Tutor

a Write vectors $\overrightarrow{AB}$ and $\overrightarrow{CD}$ in component form and in unit vector form.

b Comment on your answers in **a**.

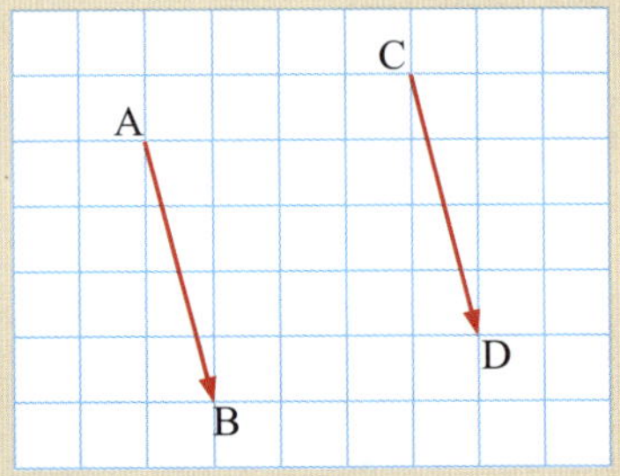

a $\overrightarrow{AB} = \begin{pmatrix} 1 \\ -4 \end{pmatrix} = \mathbf{i} - 4\mathbf{j}$ $\qquad$ $\overrightarrow{CD} = \begin{pmatrix} 1 \\ -4 \end{pmatrix} = \mathbf{i} - 4\mathbf{j}$

b The vectors $\overrightarrow{AB}$ and $\overrightarrow{CD}$ are equal.

EXERCISE 12A

1 Write the illustrated vectors in component form and in unit vector form:

a

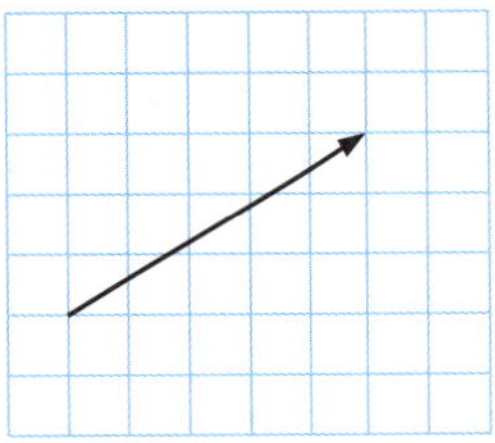

b

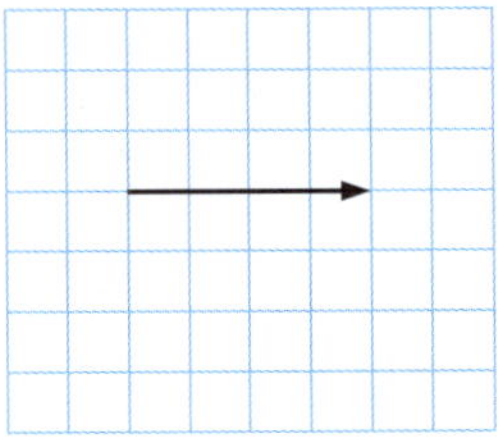

c

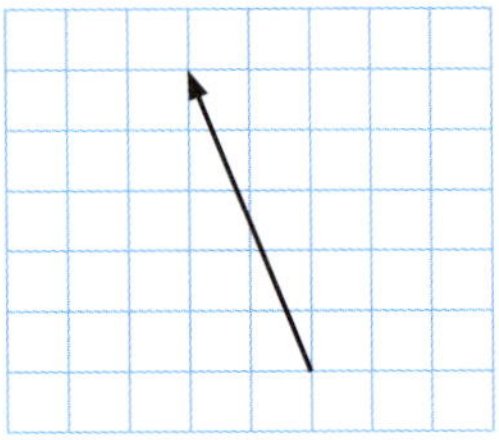

d

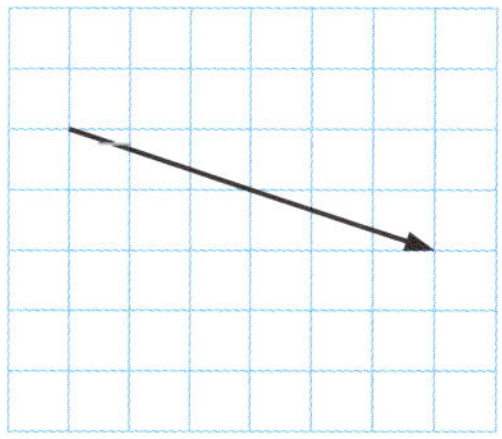

e

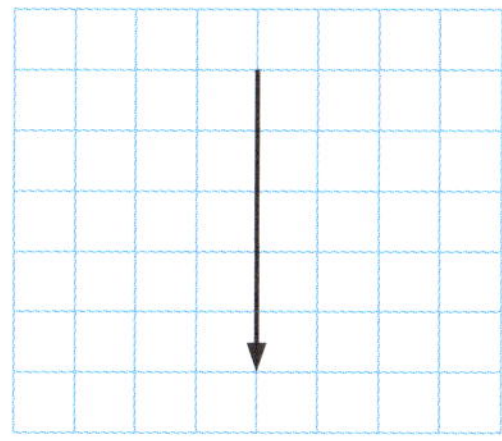

f

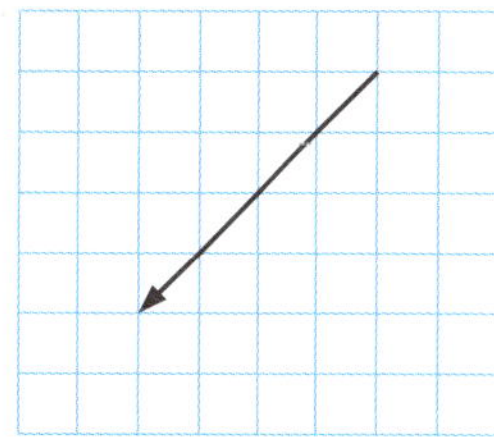

2 Write each vector in unit vector form, and illustrate it using an arrow diagram:

a $\begin{pmatrix} 2 \\ 5 \end{pmatrix}$ **b** $\begin{pmatrix} 1 \\ -3 \end{pmatrix}$ **c** $\begin{pmatrix} -5 \\ 0 \end{pmatrix}$ **d** $\begin{pmatrix} -3 \\ -2 \end{pmatrix}$

3 Write each vector in component form, and illustrate it using an arrow diagram:

a $3\mathbf{i} + 4\mathbf{j}$ **b** $2\mathbf{i} - \mathbf{j}$ **c** $3\mathbf{j}$ **d** $-5\mathbf{i} - 4\mathbf{j}$

4

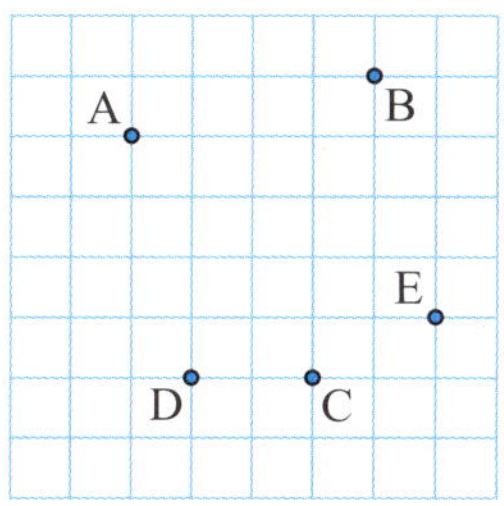

a Find in component form and in unit vector form:

i $\overrightarrow{AB}$ **ii** $\overrightarrow{BA}$ **iii** $\overrightarrow{BC}$

iv $\overrightarrow{DC}$ **v** $\overrightarrow{AC}$ **vi** $\overrightarrow{DE}$

b Which two vectors in **a** are equal? Explain your answer.

5 Consider the vector $\overrightarrow{PQ} = \begin{pmatrix} 2 \\ -3 \end{pmatrix}$.

a Write $\overrightarrow{PQ}$ in unit vector form. **b** Draw an arrow diagram of $\overrightarrow{PQ}$.

c Find $\overrightarrow{QP}$. **d** Does $\overrightarrow{PQ} = \overrightarrow{QP}$? Explain your answer.

6 Find k such that:

a $\begin{pmatrix} k+2 \\ k^2 \end{pmatrix} = \begin{pmatrix} 4 \\ 4 \end{pmatrix}$ **b** $\begin{pmatrix} 2 \\ k+1 \end{pmatrix} = \begin{pmatrix} k^2+1 \\ 0 \end{pmatrix}$ **c** $\begin{pmatrix} 3 \\ k^2+6 \end{pmatrix} = \begin{pmatrix} k^2-2k \\ 5k \end{pmatrix}$

7 Find a and b such that:

a $\begin{pmatrix} a \\ b+2 \end{pmatrix} = \begin{pmatrix} 3a-4 \\ 7 \end{pmatrix}$ **b** $\begin{pmatrix} a \\ b-4 \end{pmatrix} = \begin{pmatrix} b+3 \\ 5a-3 \end{pmatrix}$ **c** $\begin{pmatrix} 2a+3b \\ a \end{pmatrix} = \begin{pmatrix} -4 \\ b^2-3 \end{pmatrix}$

B THE MAGNITUDE OF A VECTOR

Consider the vector $\mathbf{a} = \begin{pmatrix} 3 \\ 5 \end{pmatrix} = 3\mathbf{i} + 5\mathbf{j}$.

The **magnitude** or **length** of **a** is represented by $|\,\mathbf{a}\,|$.

By Pythagoras, $|\,\mathbf{a}\,|^2 = 3^2 + 5^2 = 9 + 25 = 34$

$\therefore\ |\,\mathbf{a}\,| = \sqrt{34}$ units $\{$since $|\,\mathbf{a}\,| > 0\}$

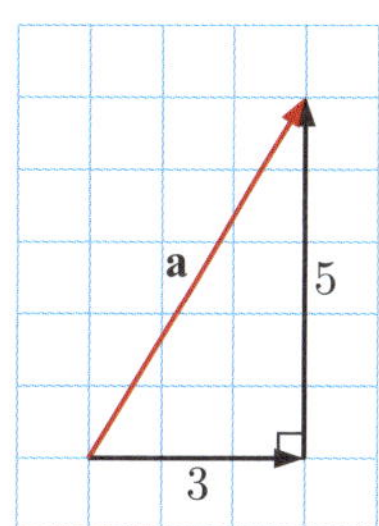

If $\mathbf{a} = \begin{pmatrix} a_1 \\ a_2 \end{pmatrix} = a_1\mathbf{i} + a_2\mathbf{j}$, the **magnitude** or **length** of $\mathbf{a}$ is $|\mathbf{a}| = \sqrt{a_1^2 + a_2^2}$.

Example 2 — Self Tutor

If $\mathbf{p} = \begin{pmatrix} 2 \\ -1 \end{pmatrix}$ and $\mathbf{q} = -3\mathbf{i} + 7\mathbf{j}$, find: **a** $|\mathbf{p}|$ **b** $|\mathbf{q}|$

a $\mathbf{p} = \begin{pmatrix} 2 \\ -1 \end{pmatrix}$

$\therefore |\mathbf{p}| = \sqrt{2^2 + (-1)^2}$

$= \sqrt{5}$ units

b $\mathbf{q} = -3\mathbf{i} + 7\mathbf{j} = \begin{pmatrix} -3 \\ 7 \end{pmatrix}$

$\therefore |\mathbf{q}| = \sqrt{(-3)^2 + 7^2}$

$= \sqrt{58}$ units

UNIT VECTORS

A **unit vector** is any vector which has a length of 1 unit.

For example, we have seen that $\mathbf{i} = \begin{pmatrix} 1 \\ 0 \end{pmatrix}$ and $\mathbf{j} = \begin{pmatrix} 0 \\ 1 \end{pmatrix}$ are unit vectors in the positive x and y-directions respectively.

Example 3 — Self Tutor

Find k given that $\begin{pmatrix} k \\ \frac{1}{4} \end{pmatrix}$ is a unit vector.

Since $\begin{pmatrix} k \\ \frac{1}{4} \end{pmatrix}$ is a unit vector, $\sqrt{k^2 + (\frac{1}{4})^2} = 1$

$\therefore \sqrt{k^2 + \frac{1}{16}} = 1$

$\therefore k^2 + \frac{1}{16} = 1$ {squaring both sides}

$\therefore k^2 = \frac{15}{16}$

$\therefore k = \pm\frac{\sqrt{15}}{4}$

EXERCISE 12B

1 Find the magnitude of:

a $\begin{pmatrix} 2 \\ 1 \end{pmatrix}$ **b** $\begin{pmatrix} 3 \\ -4 \end{pmatrix}$ **c** $\begin{pmatrix} 0 \\ 3 \end{pmatrix}$ **d** $\begin{pmatrix} 1 \\ -1 \end{pmatrix}$ **e** $\begin{pmatrix} -6 \\ -5 \end{pmatrix}$

2 Find the length of:

a $\mathbf{i} - \mathbf{j}$ **b** $2\mathbf{i} + 3\mathbf{j}$ **c** $-3\mathbf{i} + 8\mathbf{j}$ **d** $-5\mathbf{j}$ **e** $-6\mathbf{i} - 8\mathbf{j}$

3 Which of the following are unit vectors?

a $\begin{pmatrix} -1 \\ 0 \end{pmatrix}$ **b** $\begin{pmatrix} \frac{1}{2} \\ \frac{1}{2} \end{pmatrix}$ **c** $\begin{pmatrix} \frac{3}{5} \\ -\frac{4}{5} \end{pmatrix}$ **d** $\begin{pmatrix} \frac{1}{3} \\ -\frac{2}{3} \end{pmatrix}$ **e** $\begin{pmatrix} -\frac{1}{\sqrt{2}} \\ \frac{1}{\sqrt{2}} \end{pmatrix}$

4 Find k for the unit vectors:

a $\begin{pmatrix} 0 \\ k \end{pmatrix}$ **b** $\begin{pmatrix} k \\ 0 \end{pmatrix}$ **c** $\begin{pmatrix} k \\ 1 \end{pmatrix}$ **d** $\begin{pmatrix} k \\ \frac{1}{3} \end{pmatrix}$ **e** $\begin{pmatrix} -\frac{2}{5} \\ k \end{pmatrix}$

5 Suppose $\mathbf{v} = \begin{pmatrix} -7 \\ k \end{pmatrix}$ and $|\mathbf{v}| = \sqrt{65}$ units. Find the possible values of k.

6 Suppose $\mathbf{a} = 5\mathbf{i} - 4\mathbf{j}$, $\mathbf{b} = -2\mathbf{i} + p\mathbf{j}$, and $|\mathbf{a}| = |\mathbf{b}|$. Find the possible values of p.

C OPERATIONS WITH PLANE VECTORS

VECTOR ADDITION

Given vectors **a** and **b**, we can construct the vector $\mathbf{a} + \mathbf{b}$ by following these steps:

Step 1: Draw **a**.

Step 2: At the arrowhead end of **a**, draw **b**.

Step 3: Join the beginning of **a** to the arrowhead end of **b**. This is vector $\mathbf{a} + \mathbf{b}$.

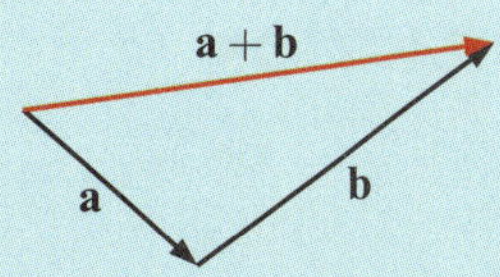

DEMO

Consider adding vectors $\mathbf{a} = \begin{pmatrix} a_1 \\ a_2 \end{pmatrix}$ and $\mathbf{b} = \begin{pmatrix} b_1 \\ b_2 \end{pmatrix}$.

Notice that:

- the horizontal step for $\mathbf{a} + \mathbf{b}$ is $a_1 + b_1$
- the vertical step for $\mathbf{a} + \mathbf{b}$ is $a_2 + b_2$.

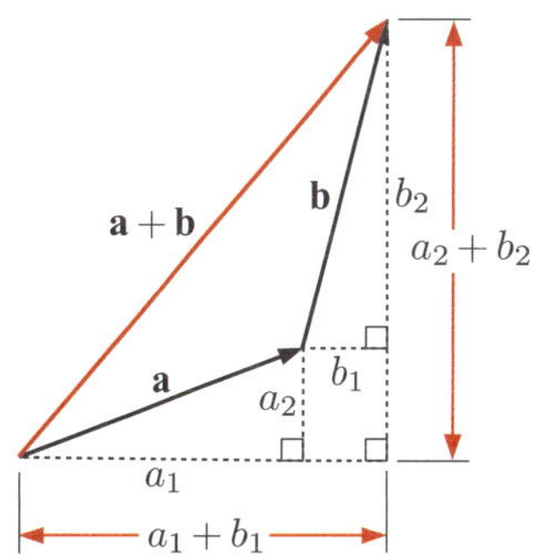

If $\mathbf{a} = \begin{pmatrix} a_1 \\ a_2 \end{pmatrix}$ and $\mathbf{b} = \begin{pmatrix} b_1 \\ b_2 \end{pmatrix}$ then $\mathbf{a} + \mathbf{b} = \begin{pmatrix} a_1 + b_1 \\ a_2 + b_2 \end{pmatrix}$.

Example 4

Self Tutor

If $\mathbf{a} = \begin{pmatrix} 2 \\ 4 \end{pmatrix}$ and $\mathbf{b} = \begin{pmatrix} 5 \\ -1 \end{pmatrix}$, find $\mathbf{a} + \mathbf{b}$. Check your answer graphically.

$$\mathbf{a} + \mathbf{b} = \begin{pmatrix} 2 \\ 4 \end{pmatrix} + \begin{pmatrix} 5 \\ -1 \end{pmatrix}$$
$$= \begin{pmatrix} 2+5 \\ 4+(-1) \end{pmatrix}$$
$$= \begin{pmatrix} 7 \\ 3 \end{pmatrix}$$

Graphical check:

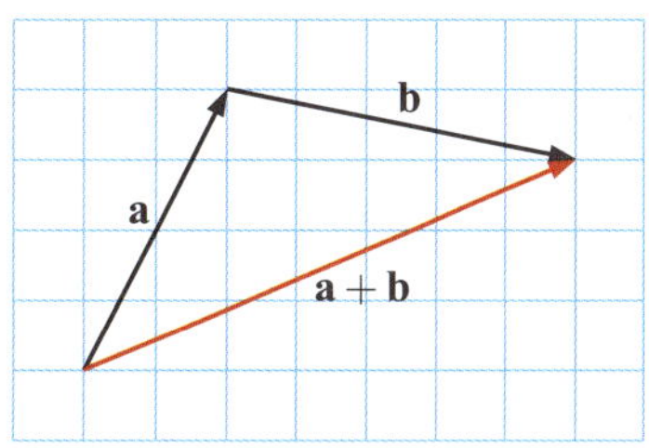

NEGATIVE VECTORS

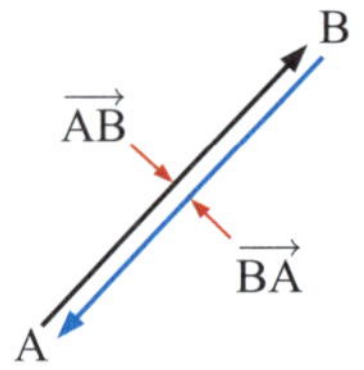

$\overrightarrow{AB}$ and $\overrightarrow{BA}$ have the same length, but they have opposite directions.

We say that $\overrightarrow{BA}$ is the **negative** of $\overrightarrow{AB}$, and write $\overrightarrow{BA} = -\overrightarrow{AB}$.

In the diagram we see the vector $\mathbf{a} = \begin{pmatrix} 3 \\ 4 \end{pmatrix}$ and its negative $-\mathbf{a} = \begin{pmatrix} -3 \\ -4 \end{pmatrix}$.

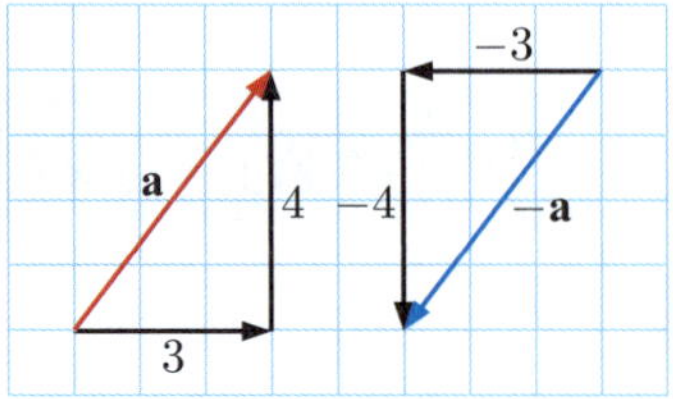

If $\mathbf{a} = \begin{pmatrix} a_1 \\ a_2 \end{pmatrix}$ then $-\mathbf{a} = \begin{pmatrix} -a_1 \\ -a_2 \end{pmatrix}$.

Notice that $\mathbf{a} + (-\mathbf{a}) = \begin{pmatrix} a_1 \\ a_2 \end{pmatrix} + \begin{pmatrix} -a_1 \\ -a_2 \end{pmatrix} = \begin{pmatrix} a_1 - a_1 \\ a_2 - a_2 \end{pmatrix} = \begin{pmatrix} 0 \\ 0 \end{pmatrix} = \mathbf{0}$.

VECTOR SUBTRACTION

To subtract one vector from another, we **add its negative**.

If $\mathbf{a} = \begin{pmatrix} a_1 \\ a_2 \end{pmatrix}$ and $\mathbf{b} = \begin{pmatrix} b_1 \\ b_2 \end{pmatrix}$

then
$$\begin{aligned} \mathbf{a} - \mathbf{b} &= \mathbf{a} + (-\mathbf{b}) \\ &= \begin{pmatrix} a_1 \\ a_2 \end{pmatrix} + \begin{pmatrix} -b_1 \\ -b_2 \end{pmatrix} \\ &= \begin{pmatrix} a_1 - b_1 \\ a_2 - b_2 \end{pmatrix} \end{aligned}$$

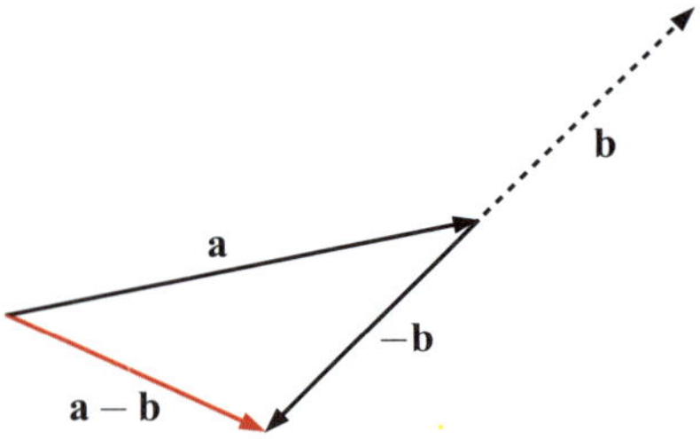

If $\mathbf{a} = \begin{pmatrix} a_1 \\ a_2 \end{pmatrix}$ and $\mathbf{b} = \begin{pmatrix} b_1 \\ b_2 \end{pmatrix}$, then $\mathbf{a} - \mathbf{b} = \begin{pmatrix} a_1 - b_1 \\ a_2 - b_2 \end{pmatrix}$.

Example 5 — Self Tutor

Given $\mathbf{a} = \begin{pmatrix} 4 \\ 1 \end{pmatrix}$, $\mathbf{b} = \begin{pmatrix} 2 \\ -2 \end{pmatrix}$, and $\mathbf{c} = \begin{pmatrix} -3 \\ 5 \end{pmatrix}$, find:

a $\mathbf{a} - \mathbf{b}$

b $\mathbf{b} - \mathbf{c} - \mathbf{a}$

a
$$\begin{aligned} &\mathbf{a} - \mathbf{b} \\ &= \begin{pmatrix} 4 \\ 1 \end{pmatrix} - \begin{pmatrix} 2 \\ -2 \end{pmatrix} \\ &= \begin{pmatrix} 4-2 \\ 1+2 \end{pmatrix} \\ &= \begin{pmatrix} 2 \\ 3 \end{pmatrix} \end{aligned}$$

b
$$\begin{aligned} &\mathbf{b} - \mathbf{c} - \mathbf{a} \\ &= \begin{pmatrix} 2 \\ -2 \end{pmatrix} - \begin{pmatrix} -3 \\ 5 \end{pmatrix} - \begin{pmatrix} 4 \\ 1 \end{pmatrix} \\ &= \begin{pmatrix} 2+3-4 \\ -2-5-1 \end{pmatrix} \\ &= \begin{pmatrix} 1 \\ -8 \end{pmatrix} \end{aligned}$$

SCALAR MULTIPLICATION

We can multiply vectors by scalars such as 2 and -3, or in fact any $k \in \mathbb{R}$.

If $\mathbf{a}$ is a vector, we define

$2\mathbf{a} = \mathbf{a} + \mathbf{a}$ and $3\mathbf{a} = \mathbf{a} + \mathbf{a} + \mathbf{a}$

so $-3\mathbf{a} = 3(-\mathbf{a}) = (-\mathbf{a}) + (-\mathbf{a}) + (-\mathbf{a})$.

A **scalar** is a non-vector quantity. It has size but no direction.

If $\mathbf{a}$ is 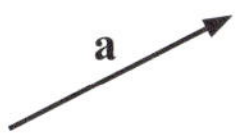 then

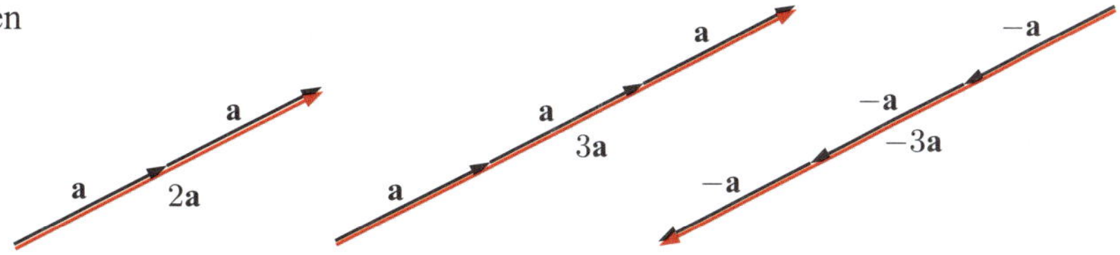

So, $2\mathbf{a}$ is in the same direction as $\mathbf{a}$ but is twice as long as $\mathbf{a}$

$3\mathbf{a}$ is in the same direction as $\mathbf{a}$ but is three times longer than $\mathbf{a}$

$-3\mathbf{a}$ has the opposite direction to $\mathbf{a}$ and is three times longer than $\mathbf{a}$.

If $\mathbf{a}$ is a vector and k is a scalar, then $k\mathbf{a}$ is also a vector and we are performing **scalar multiplication**.

- If $k > 0$, $k\mathbf{a}$ and $\mathbf{a}$ have the same direction.
- If $k < 0$, $k\mathbf{a}$ and $\mathbf{a}$ have opposite directions.
- If $k = 0$, $k\mathbf{a} = \mathbf{0}$, the zero vector.

VECTOR SCALAR MULTIPLICATION

If $\mathbf{a} = \begin{pmatrix} a_1 \\ a_2 \end{pmatrix}$, then

$$\begin{aligned} 2\mathbf{a} &= \mathbf{a} + \mathbf{a} \\ &= \begin{pmatrix} a_1 \\ a_2 \end{pmatrix} + \begin{pmatrix} a_1 \\ a_2 \end{pmatrix} \\ &= \begin{pmatrix} 2a_1 \\ 2a_2 \end{pmatrix} \end{aligned}$$

If k is any scalar and $\mathbf{a} = \begin{pmatrix} a_1 \\ a_2 \end{pmatrix}$, then $k\mathbf{a} = \begin{pmatrix} ka_1 \\ ka_2 \end{pmatrix}$.

Notice that:

- $(-1)\mathbf{a} = \begin{pmatrix} (-1)\,a_1 \\ (-1)\,a_2 \end{pmatrix} = \begin{pmatrix} -a_1 \\ -a_2 \end{pmatrix} = -\mathbf{a}$
- $0\mathbf{a} = \begin{pmatrix} (0)\,a_1 \\ (0)\,a_2 \end{pmatrix} = \begin{pmatrix} 0 \\ 0 \end{pmatrix} = \mathbf{0}$

Example 6 **Self Tutor**

If $\mathbf{p} = \begin{pmatrix} 6 \\ 2 \end{pmatrix}$ and $\mathbf{q} = \begin{pmatrix} 5 \\ -1 \end{pmatrix}$, find:

a $4\mathbf{p}$ **b** $\mathbf{p} + 3\mathbf{q}$ **c** $\frac{1}{2}\mathbf{p} - 2\mathbf{q}$

a $$\begin{aligned} &4\mathbf{p} \\ &= 4\begin{pmatrix} 6 \\ 2 \end{pmatrix} \\ &= \begin{pmatrix} 4 \times 6 \\ 4 \times 2 \end{pmatrix} \\ &= \begin{pmatrix} 24 \\ 8 \end{pmatrix} \end{aligned}$$

b $$\begin{aligned} &\mathbf{p} + 3\mathbf{q} \\ &= \begin{pmatrix} 6 \\ 2 \end{pmatrix} + 3\begin{pmatrix} 5 \\ -1 \end{pmatrix} \\ &= \begin{pmatrix} 6 \\ 2 \end{pmatrix} + \begin{pmatrix} 15 \\ -3 \end{pmatrix} \\ &= \begin{pmatrix} 21 \\ -1 \end{pmatrix} \end{aligned}$$

c $$\begin{aligned} &\tfrac{1}{2}\mathbf{p} - 2\mathbf{q} \\ &= \tfrac{1}{2}\begin{pmatrix} 6 \\ 2 \end{pmatrix} - 2\begin{pmatrix} 5 \\ -1 \end{pmatrix} \\ &= \begin{pmatrix} 3 \\ 1 \end{pmatrix} - \begin{pmatrix} 10 \\ -2 \end{pmatrix} \\ &= \begin{pmatrix} -7 \\ 3 \end{pmatrix} \end{aligned}$$

EXERCISE 12C

1 Consider the vectors **a** and **b** alongside.

  a Write **a** and **b** in component form.

  b Find $\mathbf{a} + \mathbf{b}$ and $\mathbf{a} - \mathbf{b}$, and illustrate your answers.

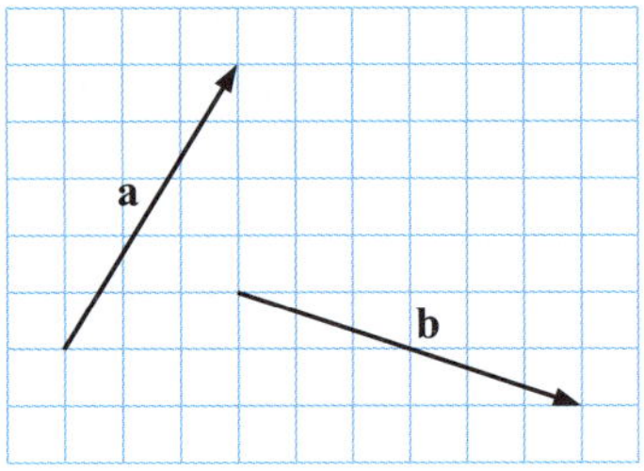

2 If $\mathbf{p} = \begin{pmatrix} 2 \\ 3 \end{pmatrix}$, $\mathbf{q} = \begin{pmatrix} -1 \\ 5 \end{pmatrix}$, and $\mathbf{r} = \begin{pmatrix} 3 \\ -2 \end{pmatrix}$, find:

  a $\mathbf{p} + \mathbf{q}$ **b** $\mathbf{q} + \mathbf{p}$ **c** $\mathbf{p} + \mathbf{r}$ **d** $\mathbf{r} + \mathbf{p}$

  e $\mathbf{q} + \mathbf{r}$ **f** $\mathbf{r} + \mathbf{q}$ **g** $\mathbf{p} + \mathbf{p}$ **h** $\mathbf{p} + \mathbf{q} + \mathbf{r}$

3 Given $\mathbf{a} = \begin{pmatrix} -3 \\ 2 \end{pmatrix}$, $\mathbf{b} = \begin{pmatrix} 1 \\ 6 \end{pmatrix}$, and $\mathbf{c} = \begin{pmatrix} 4 \\ 0 \end{pmatrix}$, find:

  a $\mathbf{a} - \mathbf{b}$ **b** $\mathbf{b} - \mathbf{a}$ **c** $\mathbf{b} - \mathbf{c}$

  d $\mathbf{c} - \mathbf{a}$ **e** $\mathbf{a} - \mathbf{b} - \mathbf{c}$ **f** $\mathbf{a} + \mathbf{c} - \mathbf{b}$

4 Consider the vector **v** alongside.

  a Write **v** in component form.

  b Find $2\mathbf{v}$ and $-3\mathbf{v}$, and illustrate your answers.

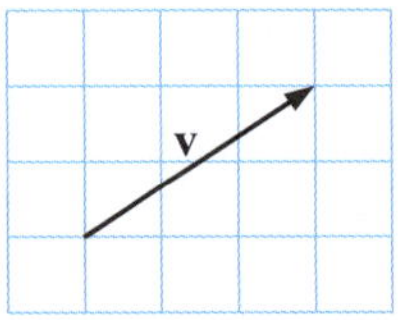

5 For $\mathbf{a} = \begin{pmatrix} 4 \\ 2 \end{pmatrix}$, find:

  a $2\mathbf{a}$ **b** $-\mathbf{a}$ **c** $5\mathbf{a}$ **d** $-3\mathbf{a}$ **e** $\frac{1}{2}\mathbf{a}$ **f** $-\frac{1}{4}\mathbf{a}$

6 For $\mathbf{a} = \begin{pmatrix} 2 \\ 6 \end{pmatrix}$, $\mathbf{b} = \begin{pmatrix} 5 \\ -3 \end{pmatrix}$, and $\mathbf{c} = \begin{pmatrix} -1 \\ 6 \end{pmatrix}$, find:

a $3\mathbf{b}$ **b** $-2\mathbf{c}$ **c** $\frac{1}{2}\mathbf{a}$ **d** $\mathbf{a} + 2\mathbf{b}$

e $\mathbf{c} + 3\mathbf{a}$ **f** $\mathbf{b} - 4\mathbf{a}$ **g** $3\mathbf{a} - 4\mathbf{c}$ **h** $\frac{1}{3}\mathbf{a} - 5\mathbf{b}$

7 Consider $\mathbf{a} = \begin{pmatrix} 1 \\ 2 \end{pmatrix}$ and $\mathbf{b} = \begin{pmatrix} 3 \\ -1 \end{pmatrix}$. Find geometrically and then comment on the results:

a $\mathbf{a} + \mathbf{a} + \mathbf{a} + \mathbf{b} + \mathbf{b}$ **b** $\mathbf{a} + \mathbf{b} + \mathbf{a} + \mathbf{b} + \mathbf{a}$ **c** $\mathbf{b} + \mathbf{a} + \mathbf{a} + \mathbf{b} + \mathbf{a}$

8 Let $\mathbf{p} = \begin{pmatrix} 3 \\ 1 \end{pmatrix}$, $\mathbf{q} = \begin{pmatrix} 1 \\ -2 \end{pmatrix}$, and $\mathbf{r} = \begin{pmatrix} 6 \\ -5 \end{pmatrix}$.

a Show that $\mathbf{p} + 2\mathbf{q} = \mathbf{r} - \mathbf{q}$.

b Illustrate this result graphically.

Example 7 Self Tutor

If $\mathbf{a} = 2\mathbf{i} - 3\mathbf{j}$ and $\mathbf{b} = \mathbf{i} + 4\mathbf{j}$, find $|\mathbf{a} - 3\mathbf{b}|$.

$$\begin{aligned} \mathbf{a} - 3\mathbf{b} &= 2\mathbf{i} - 3\mathbf{j} - 3(\mathbf{i} + 4\mathbf{j}) \\ &= 2\mathbf{i} - 3\mathbf{j} - 3\mathbf{i} - 12\mathbf{j} \\ &= -\mathbf{i} - 15\mathbf{j} \\ \therefore\ |\mathbf{a} - 3\mathbf{b}| &= \sqrt{(-1)^2 + (-15)^2} \\ &= \sqrt{226} \text{ units} \end{aligned}$$

9 For $\mathbf{r} = 3\mathbf{i} + 4\mathbf{j}$ and $\mathbf{s} = -\mathbf{i} + 6\mathbf{j}$, find:

a $|\mathbf{r}|$ **b** $|\mathbf{s}|$ **c** $\mathbf{r} + \mathbf{s}$ **d** $|\mathbf{r} + \mathbf{s}|$

e $\mathbf{r} - \mathbf{s}$ **f** $|\mathbf{r} - \mathbf{s}|$ **g** $|\mathbf{s} - 2\mathbf{r}|$ **h** $|3\mathbf{r} - 4\mathbf{s}|$

10 Suppose $\mathbf{a} = 2\mathbf{i} - \mathbf{j}$, $\mathbf{b} = \mathbf{i} + 3\mathbf{j}$, and $\mathbf{c} = -4\mathbf{i}$. Find:

a $\mathbf{a} + \mathbf{b}$ **b** $3\mathbf{b} + \mathbf{c}$ **c** $\mathbf{a} - \mathbf{c}$

d $2\mathbf{b} - \mathbf{a}$ **e** $|\mathbf{c} + 2\mathbf{a}|$ **f** $|-2\mathbf{b}|$

11 Suppose $\mathbf{a} = \begin{pmatrix} 2 \\ 3 \end{pmatrix}$, $\mathbf{b} = \begin{pmatrix} 3 \\ -4 \end{pmatrix}$, and $\mathbf{c} = \begin{pmatrix} 5 \\ k \end{pmatrix}$. Find k such that:

a $|\mathbf{c}| = |\mathbf{a} - \mathbf{b}|$ **b** $|2\mathbf{b}| = |\mathbf{a} + \mathbf{c}|$

12 Suppose $\mathbf{a} = \mathbf{i} + 3\mathbf{j}$ and $\mathbf{b} = 4\mathbf{i} - \mathbf{j}$.

a Write $\mathbf{a} + k\mathbf{b}$ in terms of k.

b Hence find k such that $\mathbf{a} + k\mathbf{b} = 17\mathbf{i} - \mathbf{j}$.

13 Let $\mathbf{r} = \begin{pmatrix} -2 \\ 1 \end{pmatrix}$ and $\mathbf{s} = \begin{pmatrix} 4 \\ -1 \end{pmatrix}$. Find k_1 and k_2 such that $k_1(\mathbf{r} + \mathbf{s}) + 5(\mathbf{r} - \mathbf{s}) = k_2(4\mathbf{r} - \mathbf{s})$.

D THE VECTOR BETWEEN TWO POINTS

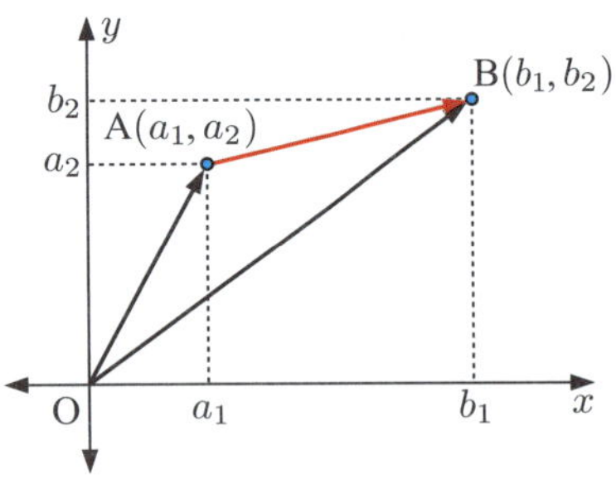

Suppose point A has position vector $\overrightarrow{OA} = \begin{pmatrix} a_1 \\ a_2 \end{pmatrix}$

and point B has position vector $\overrightarrow{OB} = \begin{pmatrix} b_1 \\ b_2 \end{pmatrix}$.

$$\begin{aligned} \therefore \quad \overrightarrow{AB} &= \overrightarrow{AO} + \overrightarrow{OB} \\ &= -\overrightarrow{OA} + \overrightarrow{OB} \\ &= \overrightarrow{OB} - \overrightarrow{OA} \\ &= \begin{pmatrix} b_1 \\ b_2 \end{pmatrix} - \begin{pmatrix} a_1 \\ a_2 \end{pmatrix} \\ &= \begin{pmatrix} b_1 - a_1 \\ b_2 - a_2 \end{pmatrix} \end{aligned}$$

The **position vector of B relative to A** is $\overrightarrow{AB} = \overrightarrow{OB} - \overrightarrow{OA} = \begin{pmatrix} b_1 - a_1 \\ b_2 - a_2 \end{pmatrix}$.

In general, for two points A and B with position vectors **a** and **b** respectively:

$$\begin{aligned} \overrightarrow{AB} &= -\mathbf{a} + \mathbf{b} \\ &= \mathbf{b} - \mathbf{a} \\ &= \begin{pmatrix} b_1 - a_1 \\ b_2 - a_2 \end{pmatrix} \end{aligned} \quad \text{and} \quad \begin{aligned} \overrightarrow{BA} &= -\mathbf{b} + \mathbf{a} \\ &= \mathbf{a} - \mathbf{b} \\ &= \begin{pmatrix} a_1 - b_1 \\ a_2 - b_2 \end{pmatrix} \end{aligned}$$

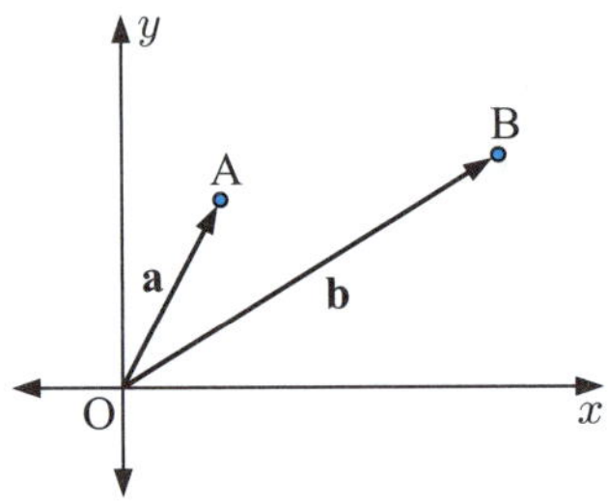

Example 8

Self Tutor

Given points A(4, 1), B(−2, 5), and C(3, −3), find the position vector of:

a A relative to O
b B relative to A
c A relative to C.

a The position vector of A relative to O is $\overrightarrow{OA} = \begin{pmatrix} 4-0 \\ 1-0 \end{pmatrix} = \begin{pmatrix} 4 \\ 1 \end{pmatrix}$.

b The position vector of B relative to A is $\overrightarrow{AB} = \begin{pmatrix} -2-4 \\ 5-1 \end{pmatrix} = \begin{pmatrix} -6 \\ 4 \end{pmatrix}$.

c The position vector of A relative to C is $\overrightarrow{CA} = \begin{pmatrix} 4-3 \\ 1-(-3) \end{pmatrix} = \begin{pmatrix} 1 \\ 4 \end{pmatrix}$.

EXERCISE 12D

1 Find $\overrightarrow{AB}$ given:

a A(1, 2) and B(6, 3)
b A(5, 4) and B(7, −1)
c A(−2, 4) and B(3, −5)
d A(−4, 3) and B(0, −2)
e A(−1, −5) and B(6, −4)
f A(−8, −5) and B(−2, 9)

2 Given points P(2, 4), Q(−1, 3), and R(7, −5), find the position vector of:

a Q relative to O

b R relative to P

c P relative to Q.

3 O, P, and Q are points on the Cartesian plane. Find $\overrightarrow{PQ}$ given:

a $\overrightarrow{OP} = \begin{pmatrix} 2 \\ 3 \end{pmatrix}$ and $\overrightarrow{OQ} = \begin{pmatrix} 4 \\ 7 \end{pmatrix}$

b $\overrightarrow{OP} = \begin{pmatrix} 5 \\ 2 \end{pmatrix}$ and $\overrightarrow{OQ} = \begin{pmatrix} 6 \\ -3 \end{pmatrix}$

c $\overrightarrow{OP} = 2\mathbf{i} - 4\mathbf{j}$ and $\overrightarrow{OQ} = -3\mathbf{i} + \mathbf{j}$

d $\overrightarrow{OP} = \mathbf{i} - 7\mathbf{j}$ and $\overrightarrow{OQ} = 4\mathbf{i} - 10\mathbf{j}$

4 O, A, and B are points such that $\overrightarrow{OA} = -\mathbf{i} + 6\mathbf{j}$ and $\overrightarrow{OB} = 3\mathbf{i} - 4\mathbf{j}$. Find:

a $\overrightarrow{AB}$

b $|\overrightarrow{AB}|$

c $\overrightarrow{BA}$

d $|\overrightarrow{BA}|$

5 Relative to an origin O, the position vector of point P is $3\mathbf{i} - 5\mathbf{j}$, and the position vector of point Q is $-2\mathbf{i} - 2\mathbf{j}$.

a Find $\overrightarrow{PQ}$.

b Find the position vector of M, the midpoint of PQ.

6 A(2, −5) and B(k, −1) are two points which are 7 units apart.

a Find $\overrightarrow{AB}$ and $|\overrightarrow{AB}|$ in terms of k.

b Hence find the two possible values of k.

c Show, by illustration, why k should have two possible values.

7 O, A, B, and C are points such that $\overrightarrow{OA} = \mathbf{a}$, $\overrightarrow{OB} = \mathbf{b}$, and $\overrightarrow{OC} = 2\mathbf{a} + 5\mathbf{b}$.

a Find, in terms of **a** and **b**:

i $\overrightarrow{AB}$

ii $\overrightarrow{BC}$

b Given that $\mathbf{a} = \mathbf{i} + 3\mathbf{j}$ and $\mathbf{b} = 4\mathbf{i} - \mathbf{j}$, find $|\overrightarrow{AC}|$.

8

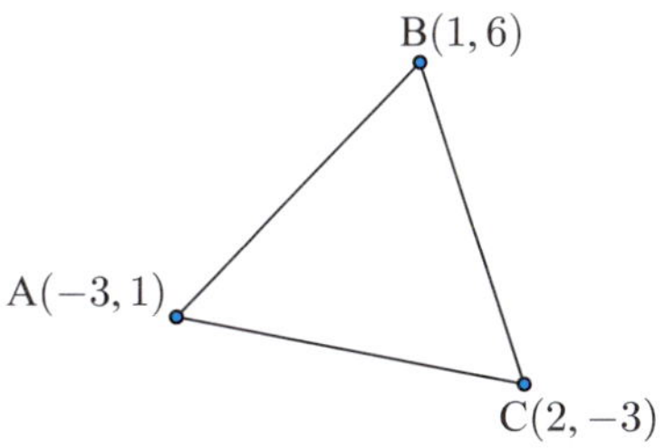

a Find $\overrightarrow{AB}$ and $\overrightarrow{AC}$.

b Explain why $\overrightarrow{BC} = -\overrightarrow{AB} + \overrightarrow{AC}$.

c Hence find $\overrightarrow{BC}$.

d Check your answer to **c** by direct evaluation.

9 **a** Given $\overrightarrow{BA} = \begin{pmatrix} 1 \\ -5 \end{pmatrix}$ and $\overrightarrow{BC} = \begin{pmatrix} -2 \\ 3 \end{pmatrix}$, find $\overrightarrow{AC}$.

b Given $\overrightarrow{AB} = \begin{pmatrix} -5 \\ 2 \end{pmatrix}$ and $\overrightarrow{CA} = \begin{pmatrix} 4 \\ -4 \end{pmatrix}$, find $\overrightarrow{CB}$.

c Given $\overrightarrow{PQ} = \begin{pmatrix} 0 \\ 2 \end{pmatrix}$, $\overrightarrow{RQ} = \begin{pmatrix} -1 \\ 4 \end{pmatrix}$, and $\overrightarrow{RS} = \begin{pmatrix} -2 \\ -6 \end{pmatrix}$, find $\overrightarrow{SP}$.

E PARALLELISM

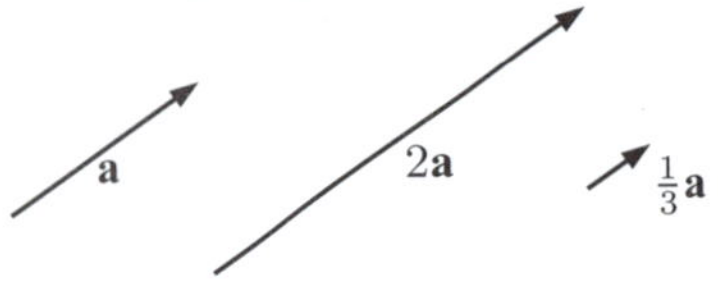

are parallel vectors of different length.

Two non-zero vectors are **parallel** if and only if one is a scalar multiple of the other.

Given any non-zero vector **a** and non-zero scalar k, the vector $k\mathbf{a}$ is parallel to **a**.

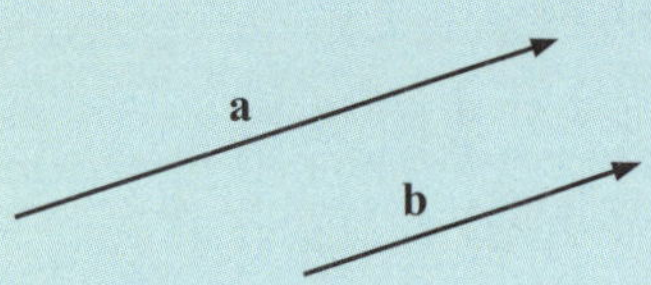

- If **a** is parallel to **b**, then there exists a scalar k such that $\mathbf{a} = k\mathbf{b}$.
- If $\mathbf{a} = k\mathbf{b}$ for some scalar k, then
  - **a** is parallel to **b**, and
  - $|\mathbf{a}| = |k|\,|\mathbf{b}|$.

Example 9

Self Tutor

Find r given that $\mathbf{a} = \begin{pmatrix} -1 \\ r \end{pmatrix}$ is parallel to $\mathbf{b} = \begin{pmatrix} 2 \\ -3 \end{pmatrix}$.

Since **a** and **b** are parallel, $\mathbf{a} = k\mathbf{b}$ for some scalar k.

$$\therefore \begin{pmatrix} -1 \\ r \end{pmatrix} = k\begin{pmatrix} 2 \\ -3 \end{pmatrix}$$

$$\therefore -1 = 2k \text{ and } r = -3k$$

$$\therefore k = -\tfrac{1}{2}$$

$$\therefore r = -3(-\tfrac{1}{2}) = \tfrac{3}{2}$$

UNIT VECTORS

Given a non-zero vector **a**, its magnitude $|\mathbf{a}|$ is a scalar quantity.

If we multiply **a** by the scalar $\dfrac{1}{|\mathbf{a}|}$, we obtain the parallel vector $\dfrac{1}{|\mathbf{a}|}\mathbf{a}$ with length 1.

- A unit vector in the direction of **a** is $\dfrac{1}{|\mathbf{a}|}\mathbf{a}$.
- A vector of length k in the same direction as **a** is $\dfrac{k}{|\mathbf{a}|}\mathbf{a}$.
- A vector of length k which is *parallel to* **a** could be $\pm\dfrac{k}{|\mathbf{a}|}\mathbf{a}$.

Example 10

Self Tutor

If $\mathbf{a} = 2\mathbf{i} + 3\mathbf{j}$, find:

a a unit vector in the direction of $\mathbf{a}$

b a vector of length 6 units in the direction of $\mathbf{a}$

c vectors of length 6 units which are parallel to $\mathbf{a}$.

a $|\mathbf{a}| = \sqrt{2^2 + 3^2}$
$= \sqrt{13}$ units

$\therefore$ the unit vector is $\frac{1}{\sqrt{13}}(2\mathbf{i} + 3\mathbf{j})$
$= \frac{2}{\sqrt{13}}\mathbf{i} + \frac{3}{\sqrt{13}}\mathbf{j}$

b A vector of length 6 units in the direction of $\mathbf{a}$ is $\frac{6}{\sqrt{13}}(2\mathbf{i} + 3\mathbf{j})$
$= \frac{12}{\sqrt{13}}\mathbf{i} + \frac{18}{\sqrt{13}}\mathbf{j}$

c The vectors of length 6 units which are parallel to $\mathbf{a}$ are $\frac{12}{\sqrt{13}}\mathbf{i} + \frac{18}{\sqrt{13}}\mathbf{j}$ and $-\frac{12}{\sqrt{13}}\mathbf{i} - \frac{18}{\sqrt{13}}\mathbf{j}$.

EXERCISE 12E.1

1 Determine whether the following pairs of vectors are parallel:

a $\begin{pmatrix} 5 \\ 3 \end{pmatrix}$ and $\begin{pmatrix} 10 \\ 6 \end{pmatrix}$ **b** $\begin{pmatrix} 1 \\ -3 \end{pmatrix}$ and $\begin{pmatrix} -3 \\ -9 \end{pmatrix}$ **c** $\begin{pmatrix} 2 \\ 12 \end{pmatrix}$ and $\begin{pmatrix} \frac{1}{2} \\ 3 \end{pmatrix}$

d $\begin{pmatrix} -4 \\ 3 \end{pmatrix}$ and $\begin{pmatrix} 8 \\ -6 \end{pmatrix}$ **e** $\begin{pmatrix} 10 \\ -15 \end{pmatrix}$ and $\begin{pmatrix} -3 \\ 2 \end{pmatrix}$ **f** $\begin{pmatrix} 6 \\ 0 \end{pmatrix}$ and $\begin{pmatrix} -5 \\ 0 \end{pmatrix}$

2 Find m given that:

a $\begin{pmatrix} 20 \\ m \end{pmatrix}$ and $\begin{pmatrix} -5 \\ 3 \end{pmatrix}$ are parallel **b** $\begin{pmatrix} 3 \\ 2 \end{pmatrix}$ and $\begin{pmatrix} 9 \\ m \end{pmatrix}$ are parallel

c $\begin{pmatrix} -4 \\ 1 \end{pmatrix}$ and $\begin{pmatrix} m \\ -5 \end{pmatrix}$ are parallel **d** $\begin{pmatrix} 3 \\ m \end{pmatrix}$ and $\begin{pmatrix} m+2 \\ 8 \end{pmatrix}$ are parallel.

3 Consider a vector $\mathbf{v}$. Write an expression for the vector:

a in the same direction as $\mathbf{v}$ and twice its length

b in the opposite direction to $\mathbf{v}$ and a third of its length.

4 **a** Write each of these vectors in component form.

b Which of the vectors $\mathbf{a}$, $\mathbf{b}$, $\mathbf{c}$, $\mathbf{d}$, and $\mathbf{e}$, are:

i parallel to $\mathbf{v}$

ii in the same direction as $\mathbf{v}$

iii the same length as $\mathbf{v}$

iv equal to $\mathbf{v}$?

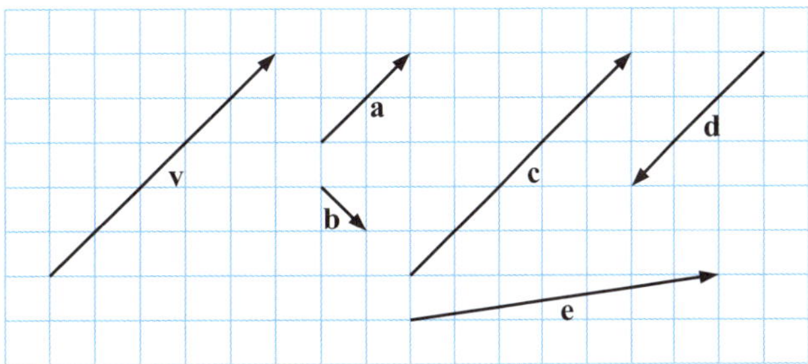

5 What can be deduced from the following?

a $\overrightarrow{PQ} = 4\overrightarrow{RS}$ **b** $\overrightarrow{AB} = -\overrightarrow{CD}$ **c** $\overrightarrow{KL} = \frac{1}{2}\overrightarrow{MN}$

6 Find the unit vector in the direction of:

a $\mathbf{i} + 3\mathbf{j}$ **b** $5\mathbf{i} - 2\mathbf{j}$ **c** $-4\mathbf{i} + \mathbf{j}$

7 **a** Write the vector $\mathbf{v}$ alongside in unit vector form.

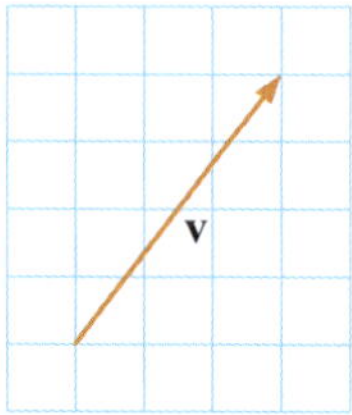

b Find:

i a unit vector in the direction of $\mathbf{v}$

ii a vector of length 3 units in the direction of $\mathbf{v}$

iii a vector of length 10 units in the opposite direction to $\mathbf{v}$.

c Draw each of the vectors in **b**.

8 If $\mathbf{a} = 3\mathbf{i} - 5\mathbf{j}$, find:

a a unit vector in the direction of $\mathbf{a}$

b a vector of length 5 units in the direction of $\mathbf{a}$

c vectors of length 8 units which are parallel to $\mathbf{a}$.

9 Find the vector $\mathbf{v}$ which has:

a the same direction as $\begin{pmatrix} -4 \\ 2 \end{pmatrix}$ and length 3 units

b the opposite direction to $\begin{pmatrix} -6 \\ -1 \end{pmatrix}$ and length 4 units.

10 Relative to an origin O, the position vector of P is $2\mathbf{i} - 3\mathbf{j}$, and the position vector of Q is $-3\mathbf{i} + \mathbf{j}$. Find:

a $\overrightarrow{PQ}$ **b** a unit vector in the direction of $\overrightarrow{PQ}$.

11 Suppose $\overrightarrow{OP} = \mathbf{p}$, $\overrightarrow{OQ} = \mathbf{q}$, $\overrightarrow{OR} = 3\mathbf{p} + \mathbf{q}$, and $\overrightarrow{OS} = -4\mathbf{p} - \mathbf{q}$.

a Find $\overrightarrow{PR}$ and $\overrightarrow{QS}$ in terms of $\mathbf{p}$ and $\mathbf{q}$.

b Hence show that $\overrightarrow{PR}$ and $\overrightarrow{QS}$ are parallel.

c Do $\overrightarrow{PR}$ and $\overrightarrow{QS}$ have the same direction or opposite directions? Explain your answer.

12 A is $(5, 3)$ and point B is 6 units from A in the direction $\begin{pmatrix} 2 \\ -1 \end{pmatrix}$.

a Find $\overrightarrow{AB}$. **b** Find $\overrightarrow{OB}$ using $\overrightarrow{OB} = \overrightarrow{OA} + \overrightarrow{AB}$.

c Hence find the coordinates of B.

COLLINEAR POINTS

Three or more points are said to be **collinear** if they lie on the same straight line.

A, B, and C are **collinear** if $\overrightarrow{AB} = k\overrightarrow{BC}$ for some scalar k.

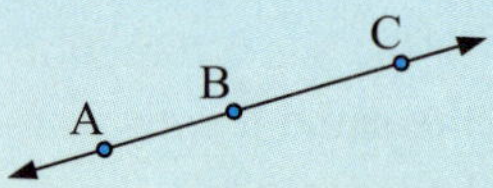

Example 11 Self Tutor

Show that A(−3, −1), B(1, 1), and C(3, 2) are collinear.

$\overrightarrow{AB} = \begin{pmatrix} 1-(-3) \\ 1-(-1) \end{pmatrix} = \begin{pmatrix} 4 \\ 2 \end{pmatrix}$

$\overrightarrow{BC} = \begin{pmatrix} 3-1 \\ 2-1 \end{pmatrix} = \begin{pmatrix} 2 \\ 1 \end{pmatrix}$

Now $\overrightarrow{AB} = 2\overrightarrow{BC}$, so $\overrightarrow{AB}$ is parallel to $\overrightarrow{BC}$.

∴ A, B, and C are collinear.

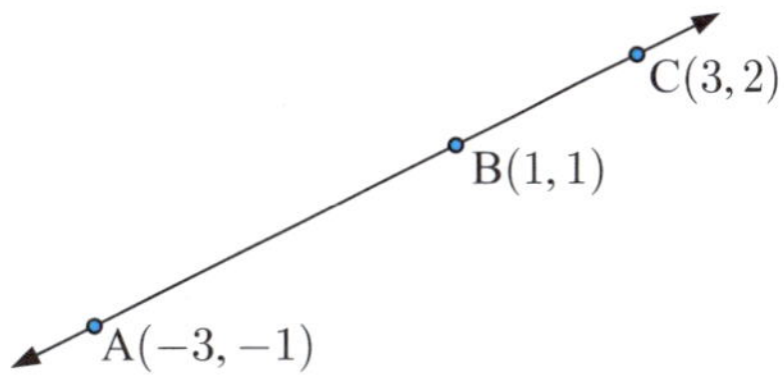

EXERCISE 12E.2

1 Show that the following sets of points are collinear:

a A(−5, 2), B(1, −4), C(3, −6)

b P(8, 3), Q(−4, −6), R(4, 0)

c X(7, −3), Y(−3, 2), Z(−7, 4)

2 Find m such that the points (−8, −7), (−5, 2), and $(m, -1)$ are collinear.

3 O, A, B, and C are four points such that $\overrightarrow{OA} = \mathbf{a}$, $\overrightarrow{OB} = \mathbf{b}$, and $\overrightarrow{OC} = 4\mathbf{a} - 3\mathbf{b}$.

a Find $\overrightarrow{AB}$ and $\overrightarrow{BC}$ in terms of **a** and **b**.

b Hence show that A, B, and C are collinear.

4 The four points O, P, Q, and R are such that $\overrightarrow{OP} = 4\mathbf{a}$, $\overrightarrow{OQ} = 10\mathbf{b}$, and $\overrightarrow{OR} = 25\mathbf{b} - 6\mathbf{a}$. Show that Q lies on the line PR.

F LINES

We have seen in Cartesian geometry that we can determine the **equation of a line** using its **direction** and any **fixed point** on the line. We can do the same using vectors.

Suppose a line passes through a fixed point A with position vector **a**, and that the line is parallel to the vector **b**.

Consider any point R on the line with position vector **r**.

By vector addition, $\mathbf{r} = \mathbf{a} + \overrightarrow{AR}$

∴ $\mathbf{r} = \mathbf{a} + t\mathbf{b}$ for some $t \in \mathbb{R}$

{since $\overrightarrow{AR}$ is parallel to **b**}

DEMO

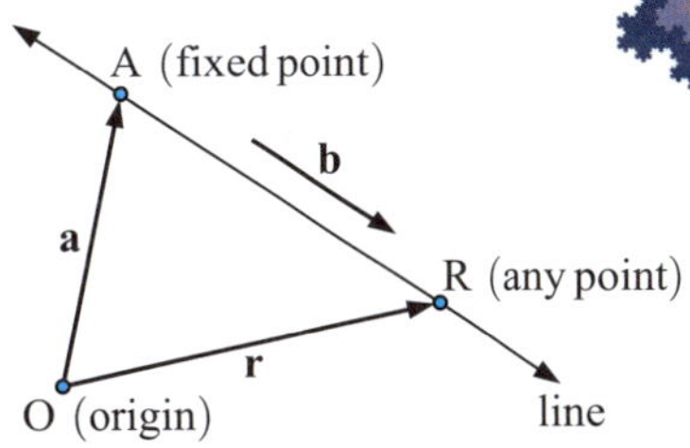

Suppose a line passes through a fixed point $A(a_1, a_2)$ with position vector **a**, and that the line is parallel to the vector $\mathbf{b} = \begin{pmatrix} b_1 \\ b_2 \end{pmatrix}$. If $R(x, y)$ with position vector **r** is any point on the line, then:

- $\mathbf{r} = \mathbf{a} + t\mathbf{b}$, $t \in \mathbb{R}$ or $\begin{pmatrix} x \\ y \end{pmatrix} = \begin{pmatrix} a_1 \\ a_2 \end{pmatrix} + t\begin{pmatrix} b_1 \\ b_2 \end{pmatrix}$ is the **vector equation** of the line.

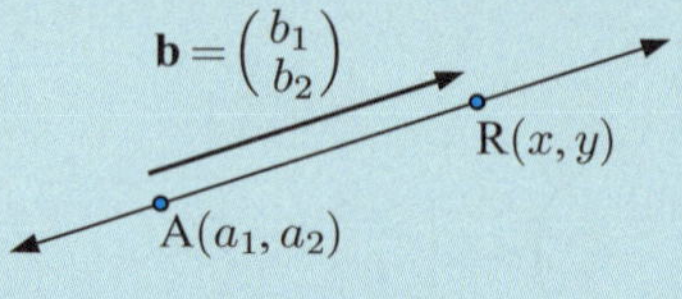

- The gradient of the line is $m = \dfrac{b_2}{b_1}$.
- Since $\begin{pmatrix} x \\ y \end{pmatrix} = \begin{pmatrix} a_1 + b_1t \\ a_2 + b_2t \end{pmatrix}$, the **parametric equations** of the line are $x = a_1 + b_1t$ and $y = a_2 + b_2t$, where $t \in \mathbb{R}$ is the **parameter**.
  Each point on the line corresponds to exactly one value of t.
- We can convert these equations into Cartesian form by equating t values.
  Using $t = \dfrac{x - a_1}{b_1} = \dfrac{y - a_2}{b_2}$ we obtain $b_2x - b_1y = b_2a_1 - b_1a_2$ which is the **Cartesian equation** of the line.

It is possible to convert between vectors and Cartesian equations. However, in 3 and higher dimensions, vectors are much simpler to use.

The equations of lines do not need to be written in parametric form for the syllabus.

Example 12

Self Tutor

A line passes through the point $A(6, 3)$ and has direction vector $\begin{pmatrix} -1 \\ 4 \end{pmatrix}$. Describe the line using:

a a vector equation **b** parametric equations **c** a Cartesian equation.

a The vector equation is $\mathbf{r} = \mathbf{a} + t\mathbf{b}$ where

$\mathbf{a} = \overrightarrow{OA} = \begin{pmatrix} 6 \\ 3 \end{pmatrix}$, $\mathbf{b} = \begin{pmatrix} -1 \\ 4 \end{pmatrix}$, and $t \in \mathbb{R}$.

$\therefore \begin{pmatrix} x \\ y \end{pmatrix} = \begin{pmatrix} 6 \\ 3 \end{pmatrix} + t\begin{pmatrix} -1 \\ 4 \end{pmatrix}$, $t \in \mathbb{R}$

b $x = 6 - t$ and $y = 3 + 4t$, $t \in \mathbb{R}$

c Now $t = \dfrac{x - 6}{-1} = \dfrac{y - 3}{4}$

$\therefore 4x - 24 = -y + 3$

$\therefore 4x + y = 27$

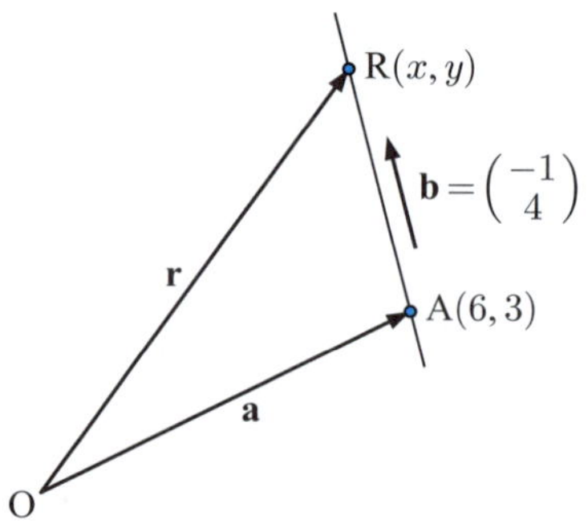

NON-UNIQUENESS OF THE VECTOR EQUATION OF A LINE

In the Example on the previous page, we could have used any non-zero scalar multiple of $\begin{pmatrix} -1 \\ 4 \end{pmatrix}$ to obtain the vector equation of the line.

For example, using the direction vector $(-1)\begin{pmatrix} -1 \\ 4 \end{pmatrix} = \begin{pmatrix} 1 \\ -4 \end{pmatrix}$, we could have equivalently written the vector equation of the line as $\begin{pmatrix} x \\ y \end{pmatrix} = \begin{pmatrix} 6 \\ 3 \end{pmatrix} + s\begin{pmatrix} 1 \\ -4 \end{pmatrix}$, $s \in \mathbb{R}$.

We used a different parameter s since the same values of s and t correspond to different points. For example:

When $t = 1$, we have $\begin{pmatrix} x \\ y \end{pmatrix} = \begin{pmatrix} 6 \\ 3 \end{pmatrix} + \begin{pmatrix} -1 \\ 4 \end{pmatrix} = \begin{pmatrix} 5 \\ 7 \end{pmatrix}$

When $s = 1$, we have $\begin{pmatrix} x \\ y \end{pmatrix} = \begin{pmatrix} 6 \\ 3 \end{pmatrix} + \begin{pmatrix} 1 \\ -4 \end{pmatrix} = \begin{pmatrix} 7 \\ -1 \end{pmatrix}$.

Discussion

1 In the vector equations $\begin{pmatrix} x \\ y \end{pmatrix} = \begin{pmatrix} 6 \\ 3 \end{pmatrix} + t\begin{pmatrix} -1 \\ 4 \end{pmatrix}$ and $\begin{pmatrix} x \\ y \end{pmatrix} = \begin{pmatrix} 6 \\ 3 \end{pmatrix} + s\begin{pmatrix} 1 \\ -4 \end{pmatrix}$, how are t and s related?

2 Are the parametric equations of lines unique?

3 Can we make the same conclusion about the Cartesian equations of lines?

EXERCISE 12F

1 Consider the lines:

L_1: $\mathbf{r} = \begin{pmatrix} 2 \\ 3 \end{pmatrix} + t\begin{pmatrix} -5 \\ 1 \end{pmatrix}$, $t \in \mathbb{R}$ and L_2: $2x + 10y = 7$

a Find the direction vector of each line.

b Hence explain why L_1 and L_2 are parallel.

2 Describe each of the following lines using:

i a vector equation **ii** parametric equations **iii** a Cartesian equation

a a line with direction $\begin{pmatrix} 5 \\ -6 \end{pmatrix}$ which passes through $(8, 2)$

b a line parallel to $-4\mathbf{i} + 9\mathbf{j}$ which cuts the x-axis at 3

c a line passing through $(-7, -1)$ and $(-1, 2)$.

3 A line has vector equation $\mathbf{r} = \begin{pmatrix} 2 \\ -1 \end{pmatrix} + t\begin{pmatrix} 3 \\ 1 \end{pmatrix}$, $t \in \mathbb{R}$.

a Find the point on the line corresponding to:

i $t = 0$ **ii** $t = 2$ **iii** $t = -3$

b Find the gradient of the line.

4 A line passes through $(2, 6)$ with direction vector $\begin{pmatrix} 10 \\ 5 \end{pmatrix}$.

a Write parametric equations for the line using the parameter t.

b Find the points on the line for which $t = 0, 1, 3, -1$, and -4.

c Draw a diagram to illustrate your answers to **b**.

5 Does $(-3, 7)$ lie on the line with vector equation $\mathbf{r} = \begin{pmatrix} 6 \\ -3 \end{pmatrix} + t\begin{pmatrix} 2 \\ 4 \end{pmatrix}$?

6 (k, k) lies on the line with parametric equations $x = 4 + 2t$, $y = 1 - t$. Find k.

7 At what point does the line with vector equation $\mathbf{r} = \begin{pmatrix} 6 \\ 7 \end{pmatrix} + t\begin{pmatrix} -2 \\ 1 \end{pmatrix}$ cut:

a the x-axis **b** the y-axis?

8 Consider the lines L_1: $\mathbf{r} = \begin{pmatrix} -4 \\ 2 \end{pmatrix} + t\begin{pmatrix} 6 \\ -4 \end{pmatrix}$, $t \in \mathbb{R}$

L_2: $\mathbf{r} = \begin{pmatrix} 2 \\ -2 \end{pmatrix} + s\begin{pmatrix} -3 \\ 2 \end{pmatrix}$, $s \in \mathbb{R}$.

a Find the Cartesian equation of each line.

b What can you deduce about L_1 and L_2?

c How are the parameters t and s related?

G CONSTANT VELOCITY PROBLEMS

A cyclist is at A$(1, 15)$. He rides with constant speed along a straight road. One hour later, he is at B$(11, -6)$. The distance units are kilometres.

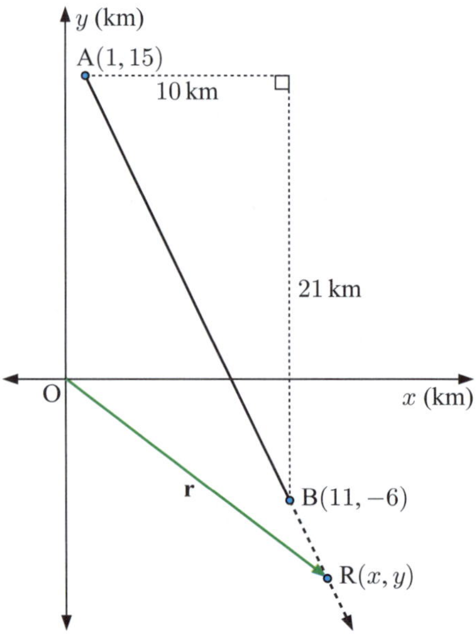

In this case:

- The **initial position** of the cyclist is given by the position vector $\mathbf{a} = \begin{pmatrix} 1 \\ 15 \end{pmatrix}$.
- The **velocity** of the cyclist is given by the vector $\mathbf{b} = \begin{pmatrix} 10 \\ -21 \end{pmatrix}$ since he travels 10 km east and 21 km south in one hour.

Suppose that t hours after leaving A, the cyclist is at R(x, y).

$$\overrightarrow{OR} = \overrightarrow{OA} + \overrightarrow{AR}$$

$$\therefore \mathbf{r} = \begin{pmatrix} 1 \\ 15 \end{pmatrix} + t\begin{pmatrix} 10 \\ -21 \end{pmatrix} \quad \text{for } t \geqslant 0$$

$\therefore \begin{pmatrix} x \\ y \end{pmatrix} = \begin{pmatrix} 1 \\ 15 \end{pmatrix} + t\begin{pmatrix} 10 \\ -21 \end{pmatrix}$ is the **vector equation** of the cyclist's path.

If an object has initial position vector **a** and moves with constant velocity **b**, its position at time t is given by

$$\mathbf{r} = \mathbf{a} + t\mathbf{b} \quad \text{for } t \geqslant 0.$$

The **speed** of the object is $|\mathbf{b}|$.

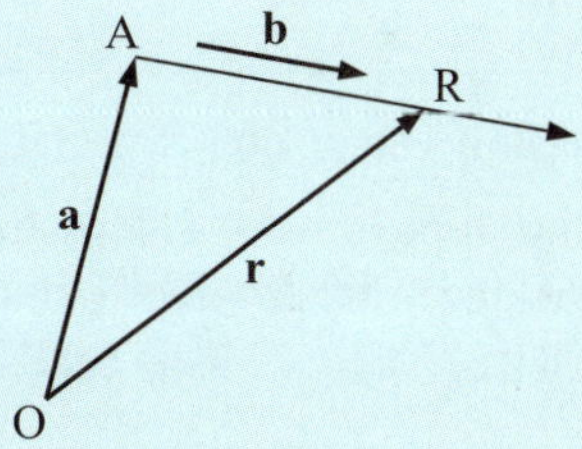

Example 13

Self Tutor

$\begin{pmatrix} x \\ y \end{pmatrix} = \begin{pmatrix} -3 \\ -5 \end{pmatrix} + t\begin{pmatrix} 1 \\ 2 \end{pmatrix}$ is the vector equation of the path of an object.

The time t is in seconds, $t \geqslant 0$. The distance units are metres.

- **a** Find the object's initial position.
- **b** Plot the path of the object for $t = 0, 1, 2, 3$.
- **c** Find the velocity vector of the object.
- **d** Find the object's speed.
- **e** Suppose the object continues in the same direction but increases its speed to 6 m s^{-1}. State its new velocity vector.

a When $t = 0$, $\begin{pmatrix} x \\ y \end{pmatrix} = \begin{pmatrix} -3 \\ -5 \end{pmatrix}$

$\therefore$ the initial position is $(-3, -5)$.

b

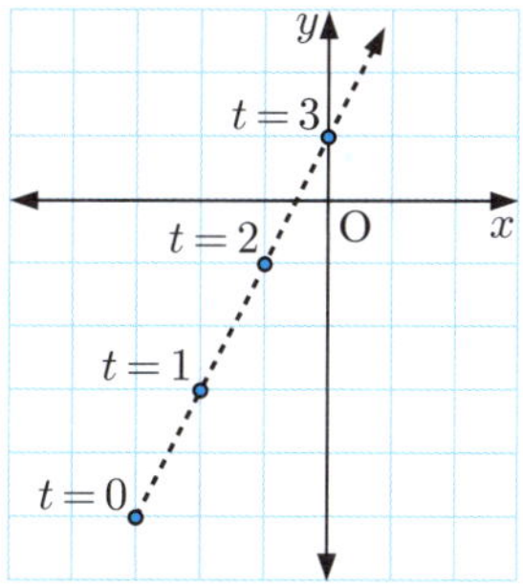

c The velocity vector is $\begin{pmatrix} 1 \\ 2 \end{pmatrix}$.

d The speed is $\left|\begin{pmatrix} 1 \\ 2 \end{pmatrix}\right| = \sqrt{1^2 + 2^2}$

$= \sqrt{5}\text{ m s}^{-1}$.

e Previously, the speed was $\sqrt{5}\text{ m s}^{-1}$ and the velocity vector was $\begin{pmatrix} 1 \\ 2 \end{pmatrix}$.

$\therefore$ the new velocity vector is $\frac{6}{\sqrt{5}}\begin{pmatrix} 1 \\ 2 \end{pmatrix}$.

We can also describe an object's direction by comparing it with the true north direction. We call this a **bearing**.

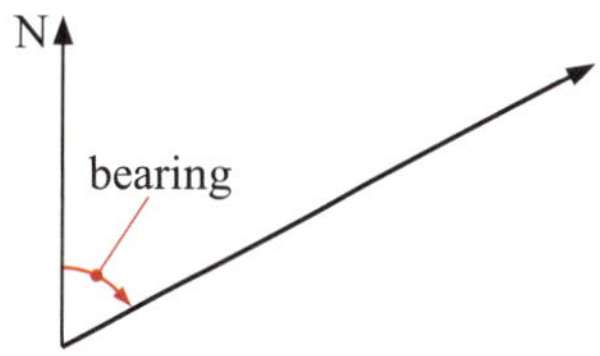

Bearing angles are always measured in the **clockwise** direction and are always written as 3 digits.

For example, we write 062° instead of 62°.

Example 14

Self Tutor

Suppose $\mathbf{i}$ is the unit vector due east and $\mathbf{j}$ is the unit vector due north.

An object has initial position $-\mathbf{i} + 6\mathbf{j}$ relative to an origin O, and moves with velocity vector $3\mathbf{i} - 2\mathbf{j}$. The distances are in metres, and time is measured in minutes.

a Find the position vector of the object after t minutes.

b Find the direction of the object as a bearing to the nearest degree.

c At what time will the object be due east of O?

a

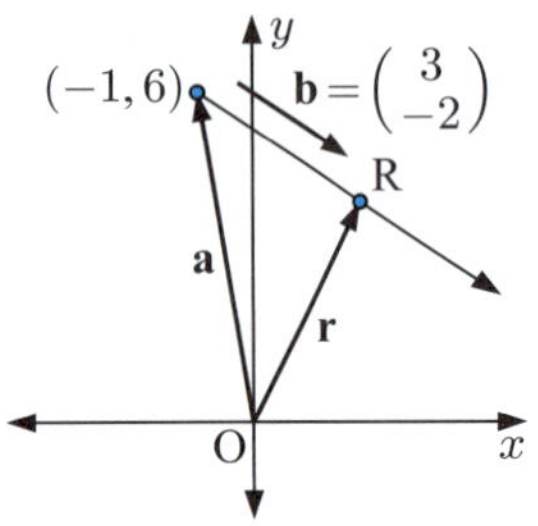

$\mathbf{r} = \mathbf{a} + t\mathbf{b}$

$\therefore \begin{pmatrix} x \\ y \end{pmatrix} = \begin{pmatrix} -1 \\ 6 \end{pmatrix} + t\begin{pmatrix} 3 \\ -2 \end{pmatrix}$

$\therefore$ the position vector of the object after t minutes is $\begin{pmatrix} -1+3t \\ 6-2t \end{pmatrix}$.

b $\tan\theta = \frac{2}{3}$

$\therefore \theta \approx 33.7°$

$\therefore$ the bearing of the object $\approx 90° + 33.7° \approx 124°$

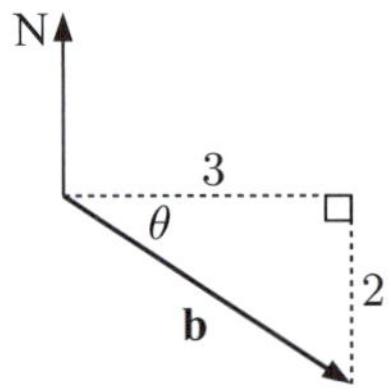

c When the object is due east of O, y must be zero.

$\therefore 6 - 2t = 0$

$\therefore t = 3$

When $t = 3$, $x = -1 + 3(3) = 8$. So, the object is due east of O at this time.

$\therefore$ the object is due east of O after 3 minutes.

EXERCISE 12G

1 Each of the following lines represents the path of a moving object. t is measured in seconds, $t \geqslant 0$, and distances are measured in metres. In each case, find:

i the initial position **ii** the velocity vector **iii** the speed of the object.

a $\begin{pmatrix} x \\ y \end{pmatrix} = \begin{pmatrix} 9 \\ -7 \end{pmatrix} + t\begin{pmatrix} -2 \\ -8 \end{pmatrix}$

b $\begin{pmatrix} x \\ y \end{pmatrix} = \begin{pmatrix} 2 \\ -5 \end{pmatrix} + t\begin{pmatrix} -1 \\ 4 \end{pmatrix}$

c $x = -3 + 10t,\ y = 4 + t$

d $x = -5 + 4t,\ y = 3 - 7t$

2 An object moves with vector equation $\begin{pmatrix} x \\ y \end{pmatrix} = \begin{pmatrix} 4 \\ 1 \end{pmatrix} + t\begin{pmatrix} -3 \\ 1 \end{pmatrix}$, $t \geqslant 0$ seconds.

The distances are in metres.

a Find the object's initial position.

b Plot the path of the object for $t = 0, 1, 2, 3$.

c Find the speed of the object.

d Find the position of the object after 8 seconds.

e If the object continues in the same direction but increases its speed to $7\ \text{m}\,\text{s}^{-1}$, state its new velocity vector.

3 Suppose $\mathbf{i}$ is the unit vector due east and $\mathbf{j}$ is the unit vector due north. Relative to an origin O, a train has initial position vector $-8\mathbf{i}+3\mathbf{j}$ and moves with velocity vector $2\mathbf{i}+\mathbf{j}$.
The distance units are kilometres and time is measured in minutes.

- **a** Find the position vector of the train after t minutes.
- **b** Hence find the position of the train after 2 minutes.
- **c** Find the direction of the train's motion. Write your answer as a bearing to the nearest degree.
- **d** At what time will the train be due north of O?

4 Answer the **Opening Problem** on page **292**.

5 Object A moves with vector equation $\begin{pmatrix} x \\ y \end{pmatrix} = \begin{pmatrix} -7 \\ 3 \end{pmatrix} + t\begin{pmatrix} 4 \\ -3 \end{pmatrix}$.

Object B moves with vector equation $\begin{pmatrix} x \\ y \end{pmatrix} = \begin{pmatrix} -10 \\ -12 \end{pmatrix} + t\begin{pmatrix} 5 \\ 2 \end{pmatrix}$.

The distance units are metres, and the time t is measured in seconds.

- **a** Find the initial position of each object.
- **b** Find the speed of each object.
- **c** At what time will the objects meet?
- **d** Find the point where the objects will meet.

6 Suppose $\mathbf{i}$ represents 1 km due east, and $\mathbf{j}$ represents 1 km due north.
Relative to base camp O, Alistair is camping at $(4, -2)$, and Bianca is camping at $(-4, -4)$. They both leave their campsites at the same time. Alistair walks with velocity vector $-3\mathbf{i}+p\mathbf{j}$ where $p>0$. Bianca walks with velocity vector $\mathbf{i}+q\mathbf{j}$.
The distance units are in kilometres and time is measured in hours.

- **a** Given that Alistair's speed is 5 km h^{-1}, find p.
- **b** At what time will Alistair be due east of the base camp?
- **c** Find Alistair's direction. Give your answer as a bearing to the nearest degree.
- **d** Given that the campers meet, find:
  - **i** the time at which they meet
  - **ii** the position at which they meet
  - **iii** the value of q.

7 A cyclist is initially at $(10, -20)$, and travels on the bearing $050°$ at 8 m s^{-1}. At the same time, a runner leaves the point $(50, p)$, and runs on the bearing $110°$ at 5 m s^{-1}. The distance units are metres, and the time units are seconds.

- **a** Show that the velocity vector of the cyclist is $\begin{pmatrix} 8\cos 40° \\ 8\sin 40° \end{pmatrix}$.
- **b** Find the velocity vector for the runner.
- **c** Find the position of the cyclist after 5 seconds.
- **d** Given that the cyclist and the runner meet, find:
  - **i** the time at which they meet
  - **ii** the position at which they meet
  - **iii** the value of p.

Review set 12A

1 **a** Write the given vectors in component form and in unit vector form.

b Find in unit vector form:

i $\mathbf{a}+\mathbf{b}$ **ii** $2\mathbf{a}-\mathbf{b}$

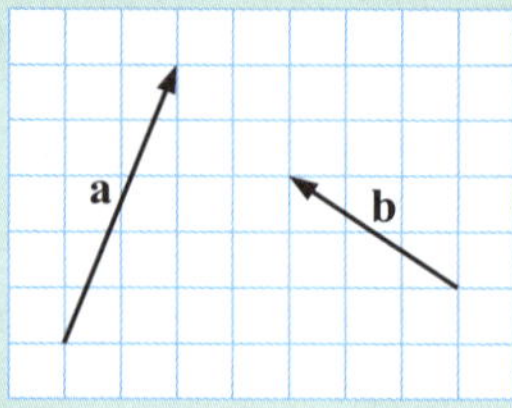

2 Consider the vector $3\mathbf{i}-\mathbf{j}$.

a Write the vector in component form.

b Illustrate the vector using an arrow diagram.

c Write the negative of the vector.

d Find the length of the vector.

3 Find k if the following are unit vectors: **a** $\begin{pmatrix} k \\ \frac{2}{5} \end{pmatrix}$ **b** $\begin{pmatrix} k \\ k+1 \end{pmatrix}$

4 Find the vector which is:

a 3 units long and has the same direction as $\begin{pmatrix} 4 \\ -1 \end{pmatrix}$

b 7 units long and has the opposite direction to $\begin{pmatrix} -5 \\ 3 \end{pmatrix}$.

5 For $\mathbf{a}=\begin{pmatrix} 3 \\ -3 \end{pmatrix}$, $\mathbf{b}=\begin{pmatrix} 1 \\ 5 \end{pmatrix}$, and $\mathbf{c}=\begin{pmatrix} -4 \\ -2 \end{pmatrix}$, find:

a $\mathbf{a}+\mathbf{c}-\mathbf{b}$

b $2\mathbf{b}-\frac{1}{2}\mathbf{c}$

c $|\mathbf{c}-\mathbf{a}|$

6 Given points A(3, 1), B(5, −2), and C(8, 4), find:

a $\overrightarrow{AB}$

b $\overrightarrow{CB}$

c $|\overrightarrow{AC}|$

7 Let $\mathbf{r}=\begin{pmatrix} 1 \\ -2 \end{pmatrix}$ and $\mathbf{s}=\begin{pmatrix} 2 \\ 3 \end{pmatrix}$. Find m and n such that $3(\mathbf{r}+\mathbf{s})+m(\mathbf{r}-\mathbf{s})=n(4\mathbf{s}-\mathbf{r})$.

8 Find k such that the points $(-7, 5)$, $(-3, 3)$, and $(k, 0)$ are collinear.

9 For the line that passes through $(-6, 3)$ with direction $\begin{pmatrix} 4 \\ -3 \end{pmatrix}$, write down the corresponding:

a vector equation

b parametric equations

c Cartesian equation.

10 $(-3, m)$ lies on the line with vector equation $\begin{pmatrix} x \\ y \end{pmatrix}=\begin{pmatrix} 18 \\ -2 \end{pmatrix}+t\begin{pmatrix} -7 \\ 4 \end{pmatrix}$. Find m.

11 Relative to an origin O, a ferry has initial position vector $70\mathbf{i}+80\mathbf{j}$. It moves with velocity vector $-2\mathbf{i}-4\mathbf{j}$ across a river. The distance units are metres, and the time units are seconds.

a Find the position vector of the ferry after t seconds.

b Hence find the position of the ferry after 15 seconds.

c Find the speed of the ferry.

d The ferry's destination has position vector $-26\mathbf{i}-112\mathbf{j}$.

i How long will the ferry take to complete its journey?

ii Find the distance the ferry has travelled in this time.

Review set 12B

1 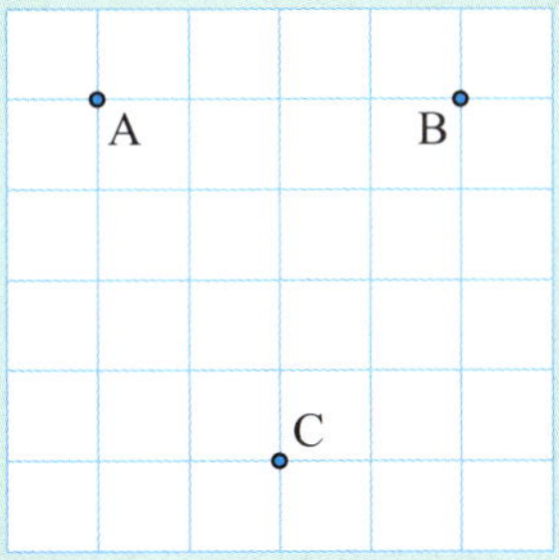

a Find, in component form and in unit vector form:

i $\overrightarrow{AB}$ **ii** $\overrightarrow{BC}$ **iii** $\overrightarrow{CA}$

b Which two vectors in **a** have the same length? Explain your answer.

2 Find a and b such that:

a $\begin{pmatrix} a+1 \\ 2b-5 \end{pmatrix} = \begin{pmatrix} 13 \\ 9 \end{pmatrix}$ **b** $2\begin{pmatrix} a \\ 3a-b \end{pmatrix} = \begin{pmatrix} b^2 \\ 1 \end{pmatrix}$

3 If $\mathbf{r} = \begin{pmatrix} 4 \\ 1 \end{pmatrix}$ and $\mathbf{s} = \begin{pmatrix} -3 \\ 2 \end{pmatrix}$, find:

a $|\mathbf{s}|$ **b** $|\mathbf{r}+\mathbf{s}|$ **c** $|2\mathbf{s}-\mathbf{r}|$

4 Suppose $\mathbf{p} = \begin{pmatrix} 3 \\ -5 \end{pmatrix}$ and $\mathbf{q} = \begin{pmatrix} 4 \\ k \end{pmatrix}$. Find k such that:

a $|\mathbf{p}| = |\mathbf{q}|$ **b** $|2\mathbf{p}| = |\mathbf{p}-\mathbf{q}|$

5 If $\overrightarrow{AB} = \begin{pmatrix} 5 \\ -4 \end{pmatrix}$ and $\overrightarrow{AC} = \begin{pmatrix} -2 \\ 7 \end{pmatrix}$, find $\overrightarrow{BC}$.

6 Suppose $\mathbf{v}$ is a vector of length $\sqrt{11}$ units. Write, in terms of $\mathbf{v}$:

a a unit vector in the direction of $\mathbf{v}$

b a vector of length 2 units in the same direction as $\mathbf{v}$

c a vector of length 4 units in the opposite direction to $\mathbf{v}$.

7 Relative to an origin O, the position vector of A is $-2\mathbf{i} - 5\mathbf{j}$, and the position vector of B is $4\mathbf{i} + \mathbf{j}$. Find:

a $\overrightarrow{AB}$ **b** $|\overrightarrow{AB}|$ **c** a unit vector in the direction of $\overrightarrow{AB}$.

8 Find n if $\begin{pmatrix} -2 \\ n \end{pmatrix}$ and $\begin{pmatrix} 10 \\ 35 \end{pmatrix}$ are parallel vectors.

9 O, P, Q, and R are four points such that $\overrightarrow{OP} = \mathbf{p}$, $\overrightarrow{OQ} = \mathbf{q}$, and $\overrightarrow{OR} = -2\mathbf{p} + 3\mathbf{q}$.

a Write $\overrightarrow{PQ}$ and $\overrightarrow{QR}$ in terms of $\mathbf{p}$ and $\mathbf{q}$.

b What can you conclude about the points P, Q, and R?

10 Find the vector equation of the line which cuts the y-axis at $(0, 8)$ and has direction $5\mathbf{i} + 4\mathbf{j}$.

11 Suppose **i** represents the unit vector due east, and **j** represents the unit vector due north.

Jerome and Molly are playing with their remote controlled boats on a lake. Relative to the centre of the lake O, Jerome launches his boat from $(10, 42)$ with velocity vector $-2\mathbf{i} - 3\mathbf{j}$. At the same time, Molly launches her boat from $(40, -18)$ with velocity vector $k\mathbf{i} + \mathbf{j}$, $k < 0$. The distance units are metres, and the time units are seconds.

a Molly's boat travelled at $\sqrt{17}\ \text{m s}^{-1}$. Find k.

b Find the direction in which Jerome's boat is travelling. Write your answer as a bearing to the nearest degree.

c At what time will Jerome's boat be due north of the centre of the lake?

d At what time will the boats collide?

e How far will the boats be from the centre of the lake when they collide?

13

Introduction to differential calculus

Contents:

Opening problem

A stone is dropped from the top of a cliff. The graph shows the distance travelled by the stone in the 5 seconds until it hits the ground below.

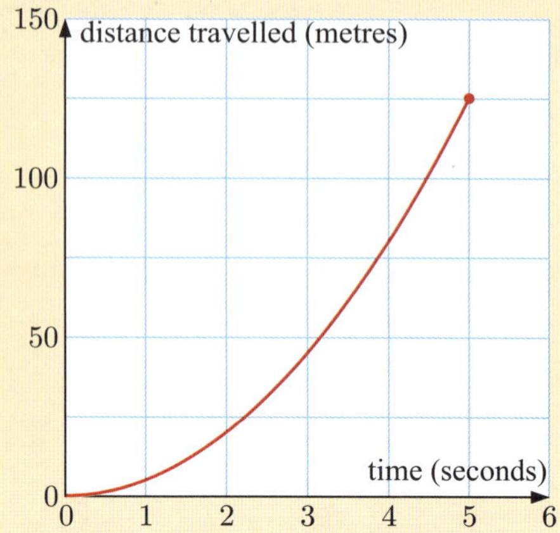

Things to think about:

a What total distance did the stone travel?

b What was the *average* speed of the stone over the 5 second period?

c Did the speed of the stone increase, decrease, or remain constant over time? What feature of the graph leads you to your answer?

d How can we measure the *instantaneous* speed of the stone at any given time?

Calculus is a major branch of mathematics which builds on algebra, trigonometry, and analytic geometry. It has widespread applications in science, engineering, and financial mathematics.

The study of calculus is divided into two fields, **differential calculus** and **integral calculus**. These fields are linked by the **Fundamental Theorem of Calculus** which we will study later in the course.

A LIMITS

The concept of a **limit** is essential in differential calculus. We will see that calculating limits is necessary for finding the gradient of a tangent to a curve at any point on the curve.

The table alongside shows values for $f(x) = x^2$ where x is less than 2, but increasing and getting closer and closer to 2.

x	1	1.9	1.99	1.999	1.9999
$f(x)$	1	3.61	3.9601	3.996 00	3.999 60

We say that "as x approaches 2 from the left, $f(x)$ approaches 4 from below".

We can construct a similar table of values where x is greater than 2, but decreasing and getting closer and closer to 2.

x	3	2.1	2.01	2.001	2.0001
$f(x)$	9	4.41	4.0401	4.004 00	4.000 40

We say that "as x approaches 2 from the right, $f(x)$ approaches 4 from above".

So, as x approaches 2 from either direction, $f(x)$ approaches a *limit* of 4. We write this as $\lim_{x \to 2} x^2 = 4$.

INFORMAL DEFINITION OF A LIMIT

The following definition of a limit is informal but adequate for the purposes of this course:

> If $f(x)$ can be made as close as we like to some real number A by making x sufficiently close to (but not equal to) a, then we say that $f(x)$ has a **limit** of A as x approaches a, and we write
>
> $$\lim_{x \to a} f(x) = A.$$
>
> In this case, $f(x)$ is said to **converge** to A as x approaches a.

Even if the function $f(x)$ is undefined at $x = a$, we may still be able to find the limit as $x \to a$. We are only interested in the behaviour of $f(x)$ as x *approaches* a.

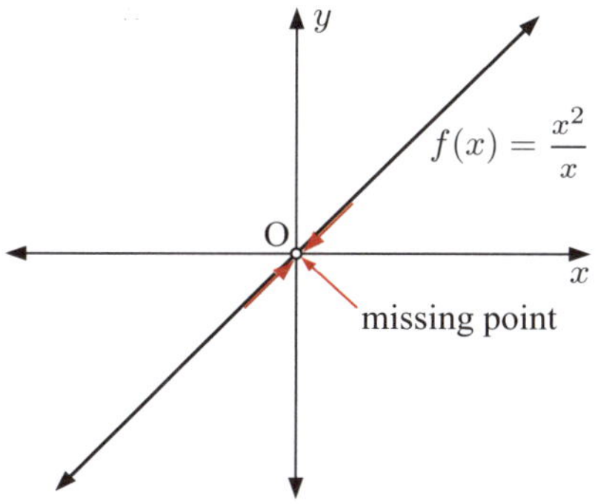

Example 1 **Self Tutor**

Evaluate:

a $\lim\limits_{x \to 2} (x^3 - 1)$

b $\lim\limits_{x \to 0} \dfrac{x^2 - x}{x}$

a As $x \to 2$,
x^3 approaches 8
$\therefore$ $x^3 - 1$ approaches 7
$\therefore$ $\lim\limits_{x \to 2} (x^3 - 1) = 7$

b $\lim\limits_{x \to 0} \dfrac{x^2 - x}{x}$
$= \lim\limits_{x \to 0} \dfrac{\cancel{x}(x-1)}{\cancel{x}}$
$= \lim\limits_{x \to 0} (x - 1)$ {as $x \neq 0$}
$= -1$

In **b**, we can divide the numerator and denominator by x, since we are only considering $x \to 0$, not $x = 0$.

EXERCISE 13A

1 Evaluate:

a $\lim\limits_{x \to 3} (x^2 + 2)$
b $\lim\limits_{x \to 5} (2x - 4)$
c $\lim\limits_{x \to 0} (3x + 5)$
d $\lim\limits_{x \to 0} (x - 3)$
e $\lim\limits_{x \to 0} \dfrac{x + 4}{x + 1}$
f $\lim\limits_{x \to 0} (x^2 + 4x + 2)$

The evaluation of limits is not required for this syllabus.

2 Evaluate:

a $\lim\limits_{x \to 0} \dfrac{x(x + 2)}{x}$
b $\lim\limits_{x \to 0} \dfrac{x^2 - 5x}{x}$
c $\lim\limits_{h \to 0} \dfrac{h^2 + 3h}{h}$
d $\lim\limits_{x \to 0} \dfrac{4x}{x}$
e $\lim\limits_{h \to 0} \left(-\dfrac{h}{h}\right)$
f $\lim\limits_{x \to 0} \dfrac{x^2 + 6x}{2x}$
g $\lim\limits_{h \to 0} \dfrac{3h^2 - 4h}{h}$
h $\lim\limits_{x \to 0} \dfrac{x^3 - 2x}{x}$
i $\lim\limits_{h \to 0} \dfrac{h^3 - 5h^2 + 8h}{h}$

B RATES OF CHANGE

A **rate** is a comparison between two quantities with different units.

For example, we can say that:

- a jogger's heart rate is 90 *beats* per *minute*
- a cricket team's run rate is 5.7 *runs* per *over*
- a tree grows at a rate of 20 *centimetres* per *year*.

Speed is a commonly used rate. It is the rate of change in distance per unit of time.

You should be familiar with the formula:

$$\textbf{average speed} = \frac{\textbf{distance travelled}}{\textbf{time taken}}$$

However, if a car has an average speed of 60 km h^{-1} for a journey, it does not mean that the car travels at exactly 60 km h^{-1} the entire time.

In fact, its speed will probably vary continuously throughout the journey.

So, how can we calculate the car's speed at any particular time?

Suppose we are given a graph of the car's distance travelled against time taken. If this graph is a straight line, then we know the speed is constant and is given by the *gradient* of the line.

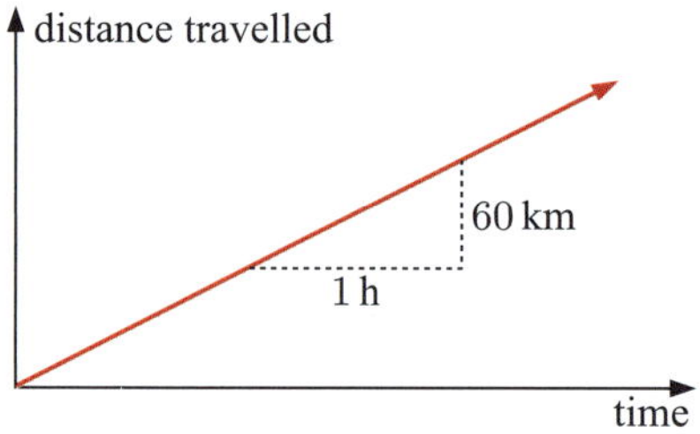

If the graph is a curve, then the car's instantaneous speed is given by the *gradient of the tangent* to the curve at that time.

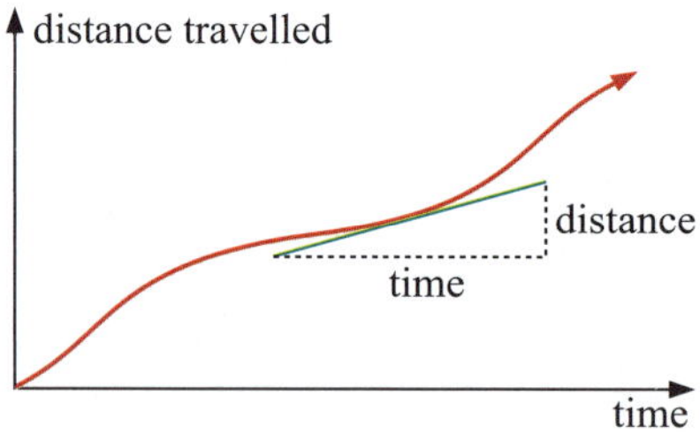

In the context of functions, we say that:

The **instantaneous rate of change** in $f(x)$ at any point A on the curve is the **gradient of the tangent** at A.

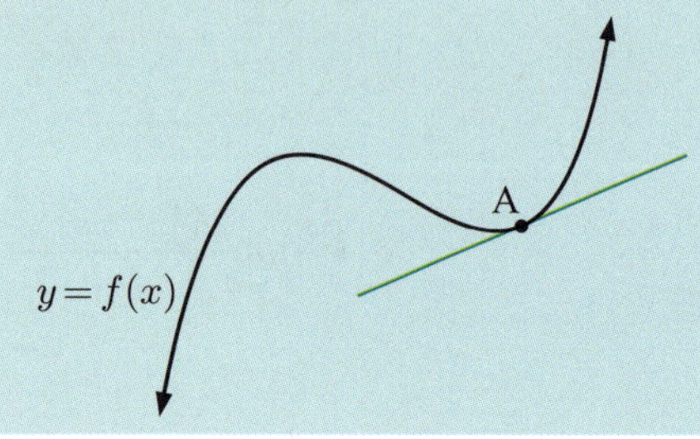

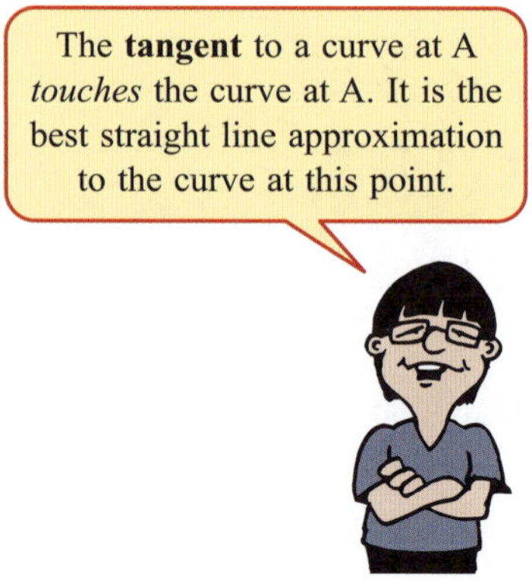

Example 2

Self Tutor

Use the tangents drawn to find the instantaneous rate of change in $y = f(x)$ at:

a A **b** B

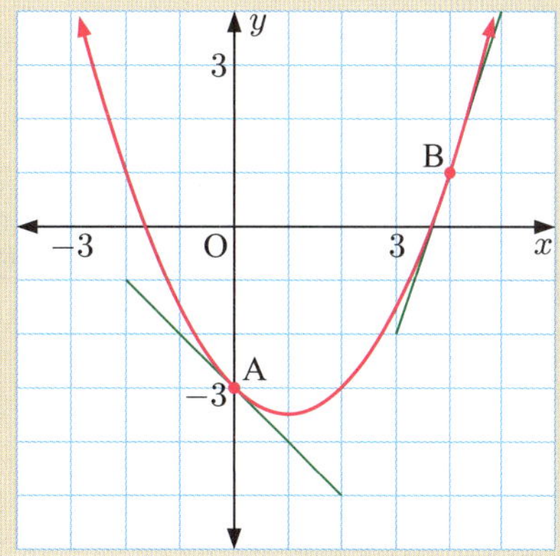

a The tangent at A has gradient -1.

$\therefore$ the instantaneous rate of change at A is -1.

b The tangent at B has gradient 3.

$\therefore$ the instantaneous rate of change at B is 3.

EXERCISE 13B

1 The graph of $f(x) = \frac{8}{x}$ is shown alongside.

Use the tangents drawn to find the instantaneous rate of change in $f(x)$ at:

a $x = -2$ **b** $x = 4$.

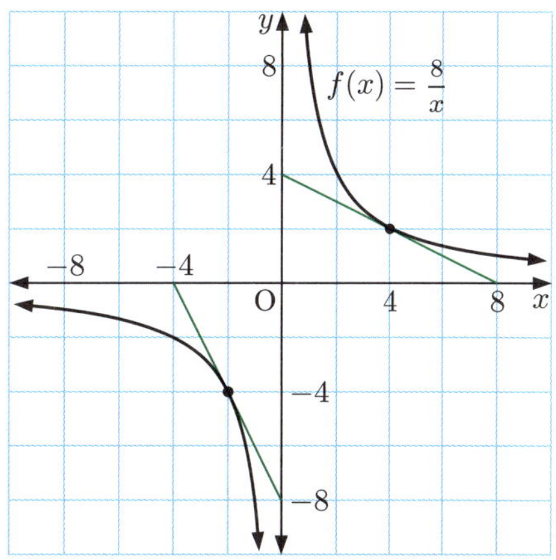

2

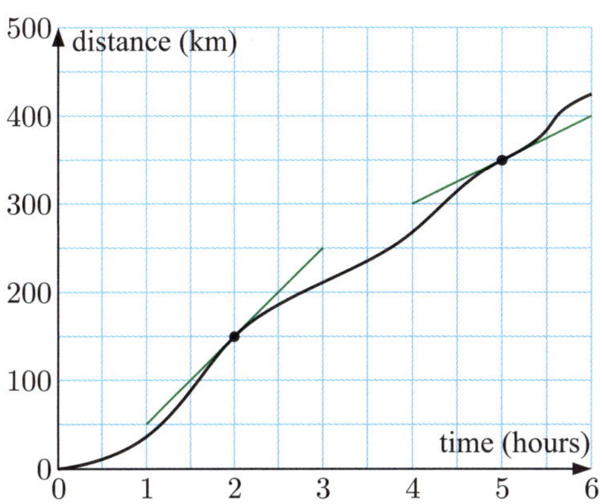

This graph shows the distance travelled by a car. Use the tangents drawn to find the car's instantaneous speed after:

a 2 hours **b** 5 hours.

3 **a** Draw an accurate sketch of $y = -x^2$ on fine grid paper for $-2 \leqslant x \leqslant 2$.
As accurately as possible, draw the tangent to $y = -x^2$ at $x = -1$.

b Hence find the instantaneous rate of change in $y = -x^2$ when $x = -1$.

PRINTABLE GRAPH

C FINDING THE GRADIENT OF THE TANGENT

Drawing a tangent on a graph and measuring its gradient can be time-consuming and inaccurate. We therefore seek a more efficient and accurate method for finding the gradient of a tangent.

Suppose we wish to find the gradient of the tangent to a function $y = f(x)$ at a fixed point $\mathrm{F}(a,\, f(a))$.

In the diagram, the tangent is shown in green. We want to know the gradient of this line.

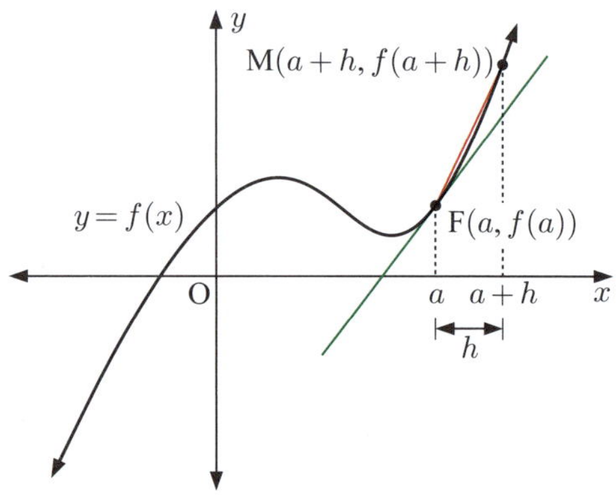

Let M be a point on the curve close to F. We let M have x-coordinate $a + h$, so M is $(a + h,\, f(a + h))$.

$$\text{The gradient of FM} = \frac{f(a+h) - f(a)}{(a+h) - a} = \frac{f(a+h) - f(a)}{h}$$

Suppose we move M along the curve so it gets closer and closer to F. This means that h gets smaller and smaller, and $h \to 0$ as M approaches F. At the same time, the line segment FM becomes more and more like the tangent at F.

$\therefore$ the gradient of FM $\to$ the gradient of the tangent at F.

DEMO

$$\text{In fact, the gradient of the tangent at F} = \lim_{h \to 0} (\text{gradient of FM}) = \lim_{h \to 0} \frac{f(a+h) - f(a)}{h}$$

We can summarise this result as follows:

The gradient of the tangent to the curve $y = f(x)$ at the point where $x = a$ is

$$\lim_{h \to 0} \frac{f(a+h) - f(a)}{h}.$$

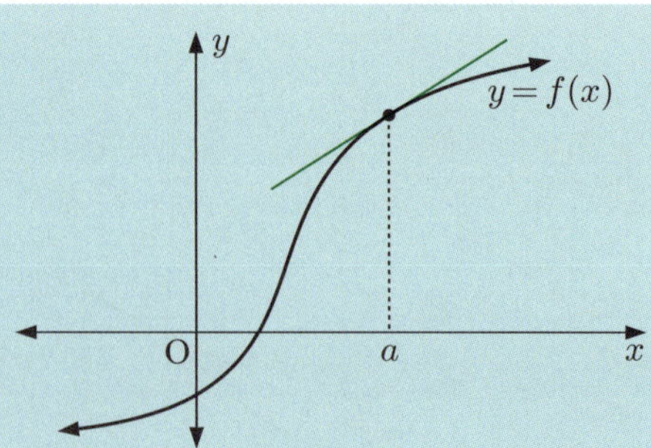

Example 3

Self Tutor

Find the gradient of the tangent to $f(x) = x^2 + 1$ at the point $(1, 2)$.

Let F be the point $(1, 2)$ and M have the x-coordinate $1 + h$, so M is $(1 + h, (1 + h)^2 + 1)$.

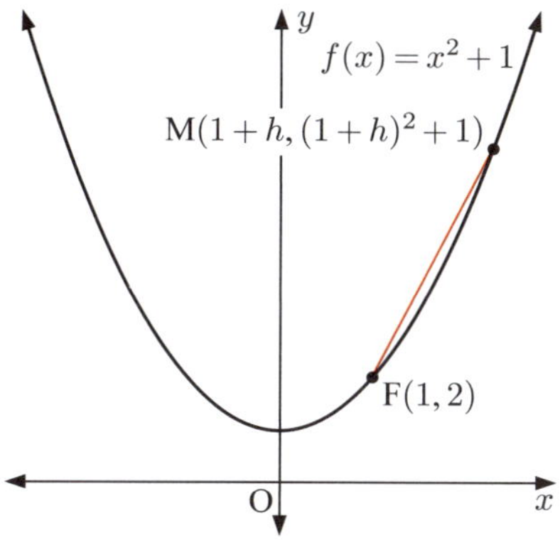

The gradient of the tangent at F

$$= \lim_{h \to 0} \frac{f(1+h) - f(1)}{h}$$

$$= \lim_{h \to 0} \frac{[(1+h)^2 + 1] - 2}{h}$$

$$= \lim_{h \to 0} \frac{1 + 2h + h^2 + 1 - 2}{h}$$

$$= \lim_{h \to 0} \frac{h(2+h)}{h}$$

$$= \lim_{h \to 0} (2 + h) \quad \{\text{as } h \neq 0\}$$

$$= 2$$

EXERCISE 13C

1 F(2, 4) lies on the graph of $f(x) = x^2$. M also lies on the graph, and has x-coordinate $2 + h$.

a State the y-coordinate of M.

b Show that the gradient of the line segment FM is $4 + h$.

c Hence find the gradient of FM where M has coordinates:

i $(3, 9)$ **ii** $(2.5, 6.25)$ **iii** $(2.1, 4.41)$ **iv** $(2.01, 4.0401)$

d What do you suspect is the gradient of the tangent to $f(x) = x^2$ at the point $(2, 4)$?

e Use limits to find the gradient of the tangent to $f(x) = x^2$ at the point $(2, 4)$.

2

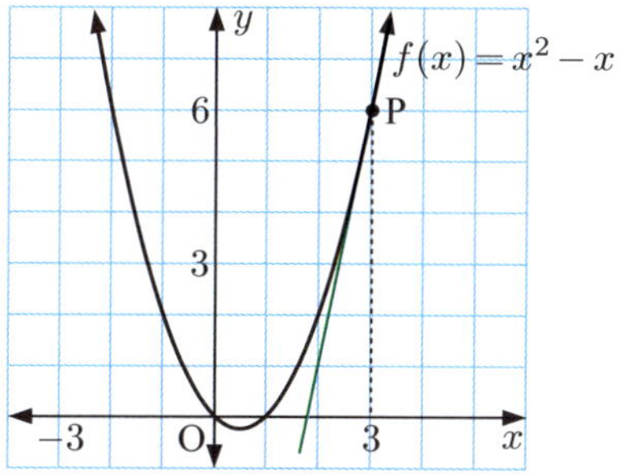

The graph of $f(x) = x^2 - x$ is shown alongside.

The point P lies on the graph of $y = f(x)$, and has x-coordinate 3.

a State the y-coordinate of P.

b From the graph, what do you suspect is the gradient of the tangent to $f(x) = x^2 - x$ when $x = 3$?

c Use limits to find the gradient of the tangent to $f(x) = x^2 - x$ when $x = 3$.

3 Use limits to find the gradient of the tangent to:

a $f(x) = x^2 + 2x$ at the point $(1, 3)$

b $f(x) = x^3$ at the point $(2, 8)$

c $f(x) = \frac{3}{x}$ at the point where $x = 3$

d $f(x) = x^2 - 2x + 3$ at the point where $x = 1$.

4 **a** Use limits to find the gradient of the tangent to $f(x) = \frac{1}{2}x^2$ at the point where:

i $x = 1$ **ii** $x = 2$ **iii** $x = 3$.

b Predict the gradient of the tangent to $f(x) = \frac{1}{2}x^2$ at the point where $x = a$.

D THE DERIVATIVE FUNCTION

For a non-linear function with equation $y = f(x)$, the gradients of the tangents at various points are different.

We can therefore write a **gradient function** which gives the gradient of the tangent to $y = f(x)$ at $x = a$, for any point a in the domain of f.

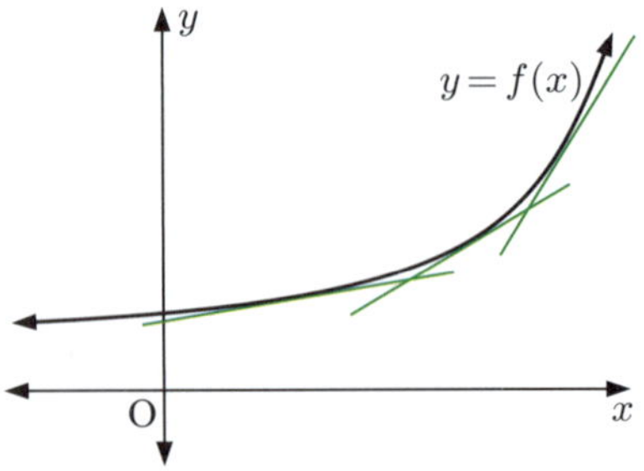

The gradient function of $y = f(x)$ is called its **derivative function** or simply **derivative** and is labelled $f'(x)$.

We read the derivative function as "f dashed x".

The value $f'(a)$ is the gradient of the tangent to $y = f(x)$ at the point where $x = a$.

Example 4

Self Tutor

For the given graph, find $f'(2)$.

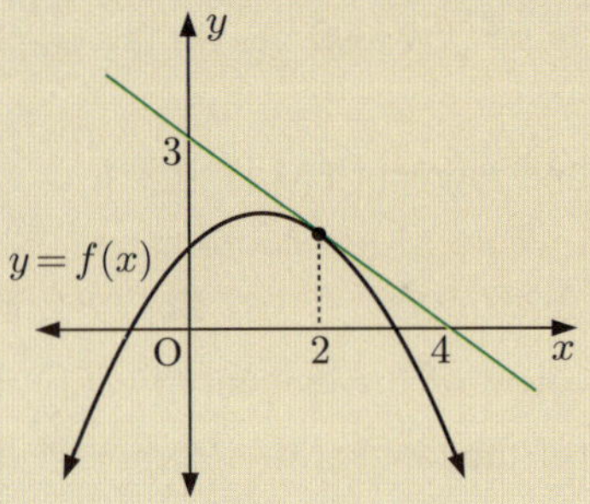

$f'(2)$ is the gradient of the tangent to the curve $y = f(x)$ at the point where $x = 2$.

The tangent passes through $(4, 0)$ and $(0, 3)$.

$\therefore \quad f'(2) =$ gradient of the tangent

$$= \frac{3-0}{0-4}$$

$$= -\tfrac{3}{4}$$

EXERCISE 13D.1

1 For the given graph, find $f'(1)$.

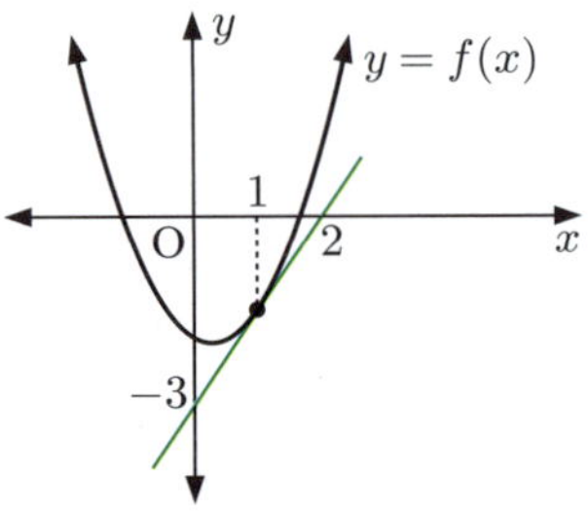

2 A curve $y = f(x)$ has gradient function $f'(x) = 2x - 5$. Find $f'(3)$ and $f'(-1)$. Interpret your answers.

3 For the graph of $y = f(x)$ alongside, the derivative function is $f'(x) = x^2 + 2x - 1$.
Find the gradient of each illustrated tangent.

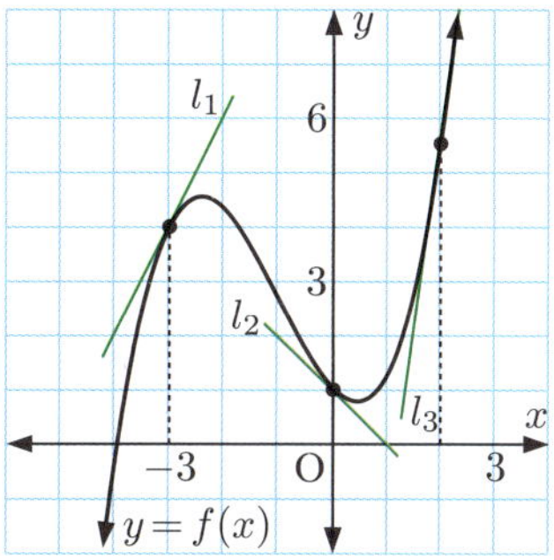

4

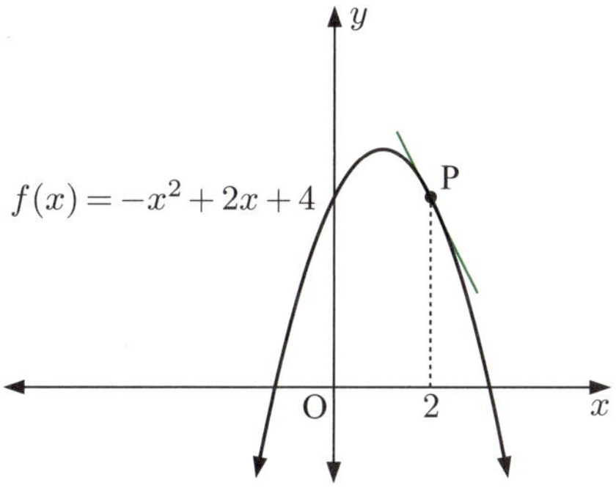

The function $f(x) = -x^2 + 2x + 4$ has derivative function $f'(x) = -2x + 2$.

a State the coordinates of P.

b Find the gradient of the illustrated tangent.

c Find the point on $y = f(x)$ at which the tangent:

  i has gradient 4

  ii is horizontal.

FINDING THE DERIVATIVE FUNCTION FROM FIRST PRINCIPLES

To find the derivative function $f'(x)$ of the function $f(x)$, we use limits to find the gradient of the tangent to the curve at a general point $(x, f(x))$.

Suppose F is the point $(x, f(x))$ and M is the point $(x + h, f(x + h))$.

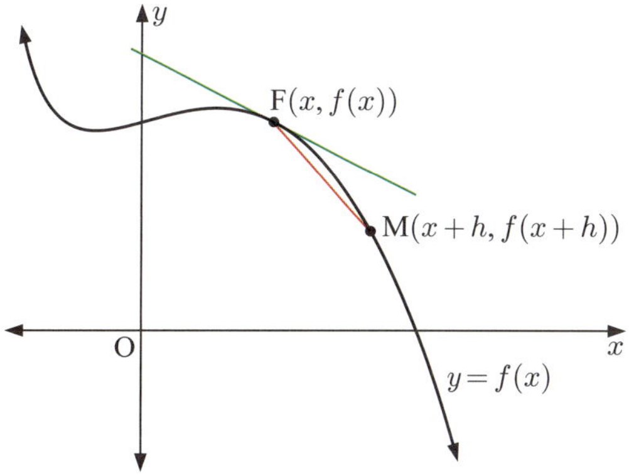

$$\text{The gradient of FM} = \frac{f(x+h) - f(x)}{(x+h) - x} = \frac{f(x+h) - f(x)}{h}$$

If we let M approach F, then $h \to 0$ and the gradient of FM approaches the gradient of the tangent at F.

$\therefore$ the gradient of the tangent at $(x, f(x))$

$$= \lim_{h \to 0} \frac{f(x+h) - f(x)}{h}$$

The **derivative function** of $y = f(x)$ is defined as $\boldsymbol{f'(x) = \lim_{h \to 0} \frac{f(x+h) - f(x)}{h}}$.

The process of finding the derivative function is called **differentiation**.

When we find the derivative using limits, we say we are performing **differentiation from first principles**.

Example 5

Self Tutor

Use first principles to find the gradient function $f'(x)$ of $f(x) = x^2 - 3x$.

$$\begin{aligned} f'(x) &= \lim_{h\to 0} \frac{f(x+h)-f(x)}{h} \\ &= \lim_{h\to 0} \frac{\left[(x+h)^2 - 3(x+h)\right] - \left[x^2 - 3x\right]}{h} \\ &= \lim_{h\to 0} \frac{x^2 + 2xh + h^2 - 3x - 3h - x^2 + 3x}{h} \\ &= \lim_{h\to 0} \frac{h(2x+h-3)}{h} \\ &= \lim_{h\to 0} (2x+h-3) \qquad \{\text{as } h \neq 0\} \\ &= 2x - 3 \end{aligned}$$

ALTERNATIVE NOTATION

If we are given a function $f(x)$ then $f'(x)$ represents the derivative function.

If we are given y in terms of x then y' or $\dfrac{dy}{dx}$ are commonly used to represent the derivative.

$\dfrac{dy}{dx}$ reads "dee y by dee x" or "the derivative of y with respect to x".

$\dfrac{dy}{dx}$ is **not a fraction**. However, the notation $\dfrac{dy}{dx}$ is a result of taking the limit of a fraction. If we replace h by δx and $f(x+h) - f(x)$ by δy, then

$$f'(x) = \lim_{h\to 0} \frac{f(x+h)-f(x)}{h} \quad \text{becomes}$$

$$\begin{aligned} f'(x) &= \lim_{\delta x\to 0} \frac{\delta y}{\delta x} \\ &= \frac{dy}{dx}. \end{aligned}$$

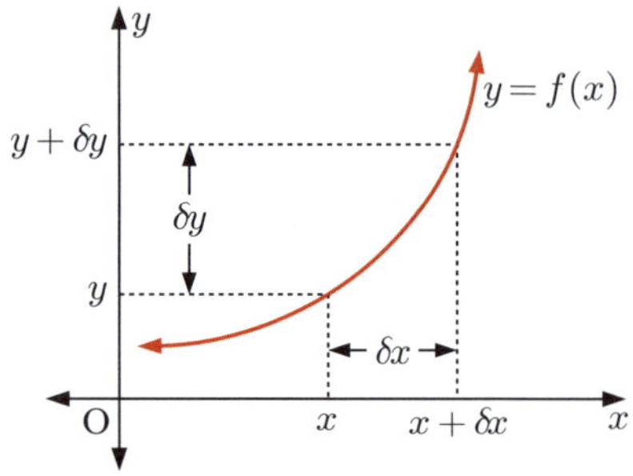

THE DERIVATIVE WHEN $x = a$

Substituting a real number a into $f'(x)$ gives us $f'(a)$, which is the gradient of the tangent to $y = f(x)$ at the point where $x = a$.

Example 6

Self Tutor

a Given $f(x) = 2x^3$, find $f'(x)$.

b Find $f'(1)$, and interpret your answer.

a
$$\begin{aligned} f'(x) = \lim_{h\to 0} \frac{f(x+h)-f(x)}{h} &= \lim_{h\to 0} \frac{2(x+h)^3 - 2x^3}{h} \\ &= \lim_{h\to 0} \frac{2(x^3 + 3x^2h + 3xh^2 + h^3) - 2x^3}{h} \qquad \{\text{binomial expansion}\} \\ &= \lim_{h\to 0} \frac{2x^3 + 6x^2h + 6xh^2 + 2h^3 - 2x^3}{h} \\ &= \lim_{h\to 0} \frac{h(6x^2 + 6xh + 2h^2)}{h} \\ &= \lim_{h\to 0} (6x^2 + 6xh + 2h^2) \qquad \{\text{since } h \neq 0\} \\ &= 6x^2 \end{aligned}$$

b $f'(1) = 6$

The gradient of the tangent to $f(x) = 2x^3$ at the point where $x = 1$, is 6.

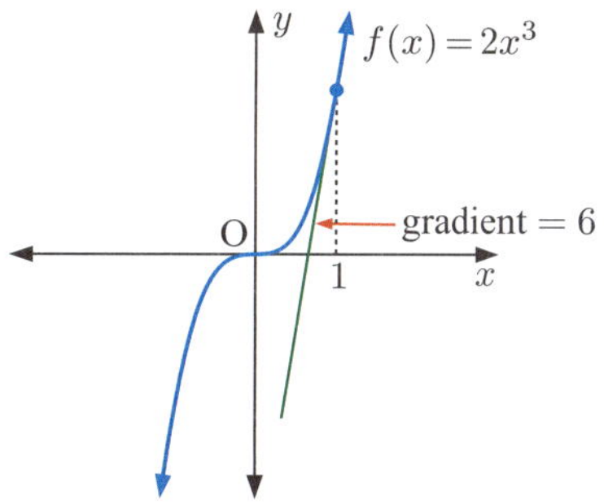

EXERCISE 13D.2

1 Find, from first principles, the derivative function of:

a $f(x) = 3x$ **b** $f(x) = x^2$ **c** $f(x) = -x$

2 Find $\frac{dy}{dx}$ from first principles, given that:

a $y = 5 - 2x$ **b** $y = x^2 + 4x$ **c** $y = x^3$

3 **a** Find $f'(x)$ for $f(x) = 2x^2 - x$.

b Find $f'(-1)$ and $f'(2)$. Interpret your answers.

4 **a** Find $f'(x)$ for:

i $f(x) = 7$ **ii** $f(x) = -4$

b What do you suspect is the derivative function of $f(x) = c$, where c is a constant? Explain your answer geometrically.

5 The graph of $f(x) = x^3 - x$ is shown alongside.

a Find $f'(x)$.

b Hence find the gradient of the illustrated tangent.

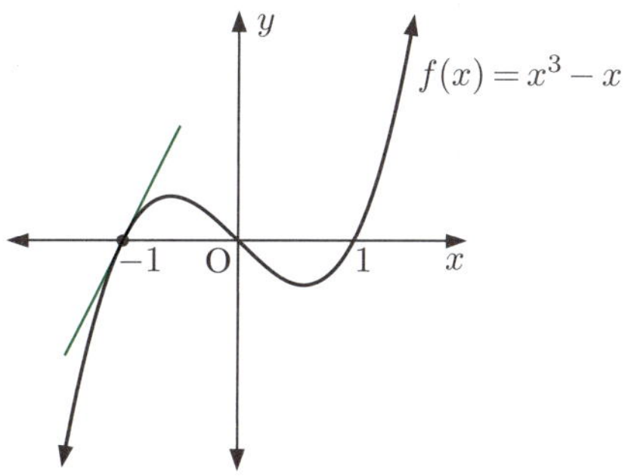

6 Suppose $f(x) = 5x^2$ and $g(x) = x^2$.

a Find $g'(x)$ and $f'(x)$.

b Show that in general, if $f(x) = cg(x)$ where c is a constant, then $f'(x) = cg'(x)$.

Hint: $\lim\limits_{h \to 0} cg(x) = c \lim\limits_{h \to 0} g(x)$.

7 **a** Find $f'(x)$ for:

i $f(x) = 2x^2$ **ii** $f(x) = 9x$ **iii** $f(x) = 2x^2 + 9x$

b Show that in general, if $f(x) = g(x) + h(x)$, then $f'(x) = g'(x) + h'(x)$.

Hint: $\lim\limits_{h \to 0} (f(x) + g(x)) = \lim\limits_{h \to 0} f(x) + \lim\limits_{h \to 0} g(x)$.

Historical note — Gottfried Leibniz

Writing the derivative function as $\frac{dy}{dx}$ is sometimes referred to as **Leibniz notation**, named after the German mathematician **Gottfried Leibniz** (1646 - 1716).

Leibniz studied Philosophy and Law at the University of Leipzig. He visited Paris in 1672, where he met Dutch mathematician **Christiaan Huygens**. Under Huygens' mentorship, Leibniz began to make his own contributions to mathematics.

In 1674, Leibniz began working on calculus, and he published his first work on the subject in 1684.

Gottfried Leibniz

From 1699 until his death in 1716, Leibniz was involved in a dispute with British mathematician **Sir Isaac Newton** regarding the discovery of calculus. Although Newton did not publish his work on calculus until 1687, he claimed to have first worked on a form of calculus, which he called the *method of fluxions*, in 1666. Newton accused Leibniz of plagiarising his work. Today, it is generally accepted that Leibniz and Newton each discovered calculus independently.

In addition to his work on calculus, Leibniz developed a counting wheel used in the first mechanical calculators. He also made significant contributions to science, philosophy, and politics.

Review set 13A

1 Evaluate:

a $\lim\limits_{x \to 2} (x^2 + 5)$ **b** $\lim\limits_{x \to -1} \frac{2x+3}{x-1}$ **c** $\lim\limits_{x \to 0} \frac{x^3 - x}{x}$

2 The graph of $f(x) = \frac{1}{3}x^3 - x^2 - \frac{2}{3}$ is shown alongside. Use the tangents drawn to find the instantaneous rate of change in $f(x)$ at:

a $x = -1$ **b** $x = 2$

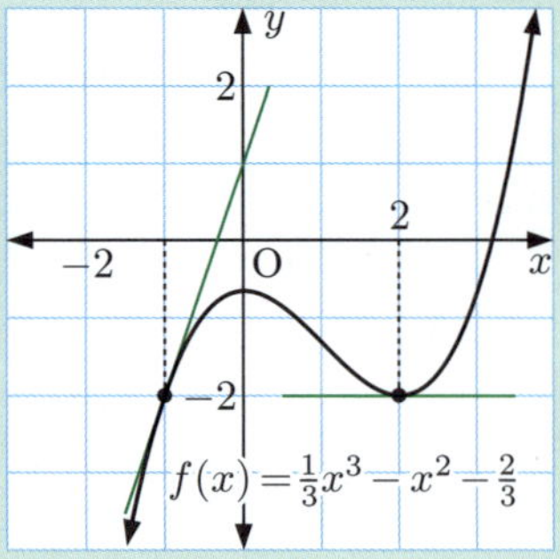

3

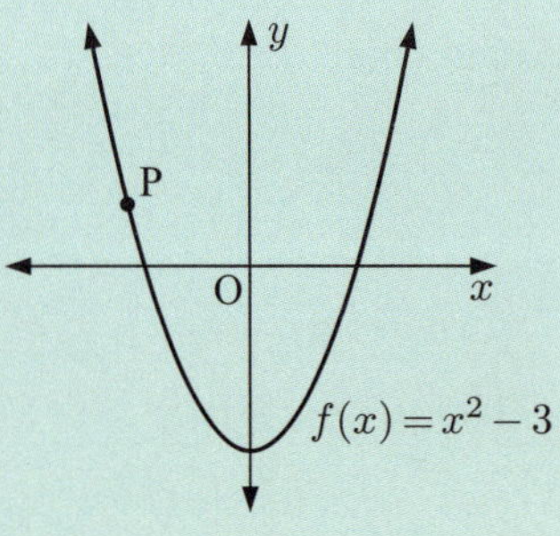

The point P on the graph of $f(x) = x^2 - 3$ has x-coordinate -2.

a Find the y-coordinate of P.

b Use limits to find the gradient of the tangent to $f(x) = x^2 - 3$ at point P.

4 Find the gradient of the tangent to:

a $f(x) = x^2 + 3x$ at the point $(1, 4)$ **b** $f(x) = x^3 - x$ at the point $(2, 6)$.

5 For the given graph, find $f'(1)$.

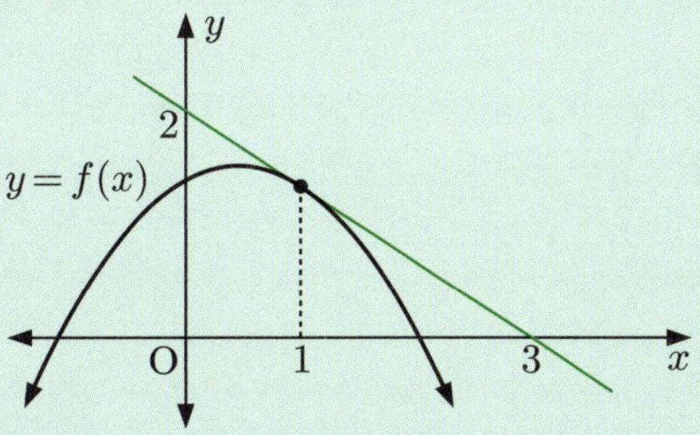

6 A curve $y = f(x)$ has gradient function $f'(x) = 4x + 7$. Find $f'(-3)$, and interpret your answer.

7 Find, from first principles, the derivative function of:

a $f(x) = 5x$ **b** $f(x) = x^2 - 2x$

8 The graph of $f(x) = 3x^2 - 4x$ is shown alongside.

a Find $f'(x)$ from first principles.

b Hence find the gradient of the illustrated tangent.

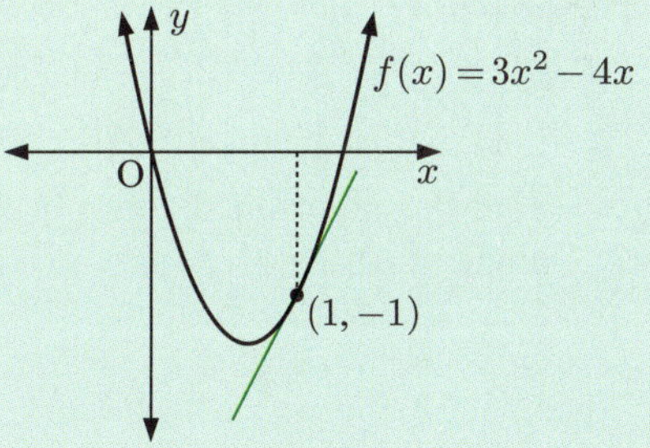

Review set 13B

1 Evaluate:

a $\lim\limits_{x \to 3} (2x^2 - 5x)$ **b** $\lim\limits_{h \to 0} \dfrac{h^2 - 4h}{h}$ **c** $\lim\limits_{x \to 0} \dfrac{3x^3 - x^2 + 2x}{x}$

2 **a** Draw an accurate sketch of $y = x^2 + 2$ on fine grid paper for $-2 \leqslant x \leqslant 2$. As accurately as possible, draw the tangent to $y = x^2 + 2$ at $x = 1$.

b Hence find the instantaneous rate of change in $y = x^2 + 2$ when $x = 1$.

PRINTABLE GRAPH

3 Find the gradient of the tangent to:

a $f(x) = \dfrac{2}{x}$ at the point where $x = -2$ **b** $f(x) = 2x^3$ at the point $(-1, -2)$.

4 For the given graph, find $f'(2)$.

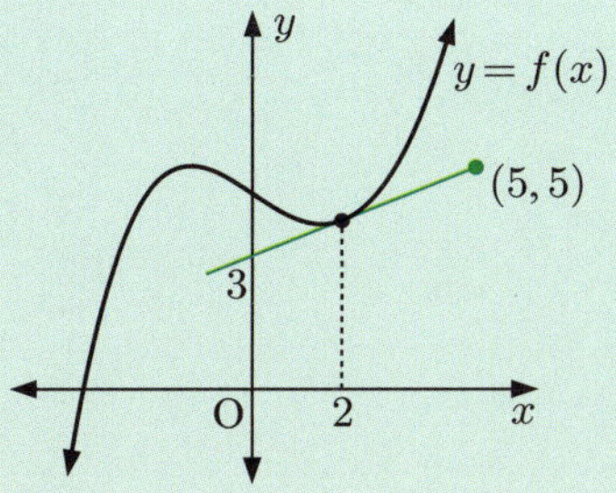

5 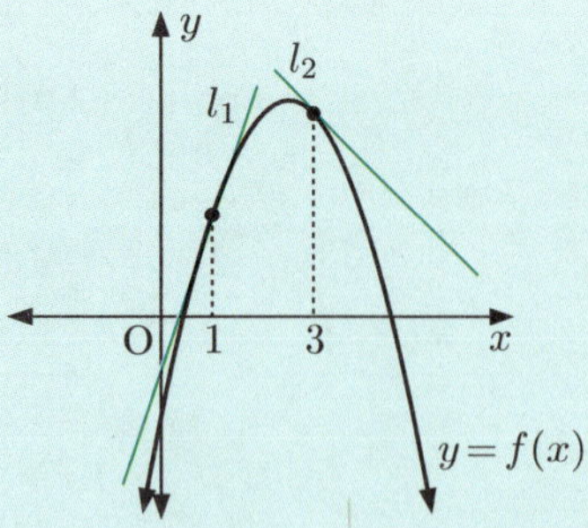

For the graph of $y = f(x)$ alongside, the derivative function is $f'(x) = 5 - 2x$.

Find the gradient of each of the illustrated tangents.

6 Find $\frac{dy}{dx}$ from first principles, given that:

a $y = 4 - 7x$

b $y = 2x^2 - 5$

7 **a** Find $f'(x)$ for $f(x) = x^3 + 2x$.

b Find $f'(0)$ and $f'(2)$. Interpret your answers.

8 The graph of $f(x) = -2x^2 - 4x$ is shown alongside.

a Find $f'(x)$ from first principles.

b Find the gradient of the illustrated tangent.

c Find the point on the graph where the tangent has gradient -12.

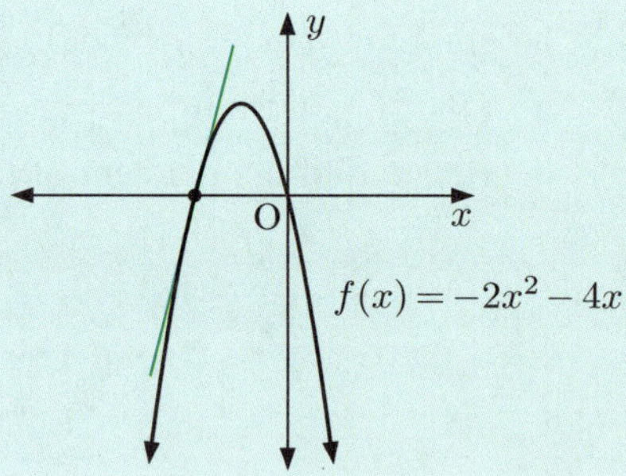

14

Rules of differentiation

Contents:

Opening problem

In the previous Chapter, we discovered how to find derivative functions from first principles.

Things to think about:

a Using first principles, what is the derivative function of:

i $f(x) = x$ **ii** $f(x) = x^2$ **iii** $f(x) = x^3$ **iv** $f(x) = x^4$?

b Can you predict a formula for the derivative function of $f(x) = x^n$ where $n \in \mathbb{Z}^+$?

c Will your formula be valid for other $n \in \mathbb{R}$?

The process of finding derivative functions using first principles can be quite time-consuming. Fortunately, there are rules we can use to differentiate functions without having to use first principles. We will explore some of these rules in this Chapter.

A RULES FOR DIFFERENTIATION

Differentiation is the process of finding a derivative or gradient function.

Given a function $f(x)$, we obtain $f'(x)$ by **differentiating with respect to** x.

In the **Opening Problem**, you should have obtained the derivatives in the table alongside for some functions of the form $f(x) = x^n$.

$f(x)$	$f'(x)$
x^1	$1 = 1x^0$
x^2	$2x = 2x^1$
x^3	$3x^2$
x^4	$4x^3$

These results suggest that:

If $f(x) = x^n$, $n \in \mathbb{Z}^+$, then $f'(x) = nx^{n-1}$.

Proof:

Let $f(x) = x^n$, $n \in \mathbb{Z}^+$.

$$\begin{aligned}
\therefore\ f'(x) &= \lim_{h \to 0} \frac{f(x+h) - f(x)}{h} \\
&= \lim_{h \to 0} \frac{(x+h)^n - x^n}{h} \\
&= \lim_{h \to 0} \frac{\left[\binom{n}{0}x^n + \binom{n}{1}x^{n-1}h + \binom{n}{2}x^{n-2}h^2 + ... + \binom{n}{n}x^0h^n\right] - x^n}{h} && \{\text{binomial expansion}\} \\
&= \lim_{h \to 0} \frac{h\left[nx^{n-1} + \binom{n}{2}x^{n-2}h + ... + \binom{n}{n}x^0h^{n-1}\right]}{h} \\
&= \lim_{h \to 0} \left(nx^{n-1} + \binom{n}{2}x^{n-2}h + ... + \binom{n}{n}x^0h^{n-1}\right) && \{\text{since } h \neq 0\} \\
&= nx^{n-1}
\end{aligned}$$

In fact, this rule is true for all real $n \neq 0$.

For example:

- if $f(x) = x^8$ then $f'(x) = 8x^7$
- if $f(x) = x^{\frac{5}{2}}$ then $f'(x) = \frac{5}{2}x^{\frac{3}{2}}$
- if $f(x) = x^{-2}$ then $f'(x) = -2x^{-3}$.

The table below summarises the most basic rules of differentiation.

$f(x)$	$f'(x)$	*Name of rule*
c (a constant)	0	**differentiating a constant**
x^n	nx^{n-1}	**differentiating x^n**
$cu(x)$	$cu'(x)$	**constant times a function**
$u(x)+v(x)$	$u'(x)+v'(x)$	**addition rule**

Proof of the last two rules:

- Let $f(x) = cu(x)$ where $c \in \mathbb{R}$

$$\begin{aligned} \therefore\ f'(x) &= \lim_{h\to 0} \frac{f(x+h)-f(x)}{h} \\ &= \lim_{h\to 0} \frac{cu(x+h)-cu(x)}{h} \\ &= \lim_{h\to 0} c\left[\frac{u(x+h)-u(x)}{h}\right] \\ &= c\lim_{h\to 0} \frac{u(x+h)-u(x)}{h} \\ &= cu'(x) \end{aligned}$$

- Let $f(x) = u(x)+v(x)$

$$\begin{aligned} \therefore\ f'(x) &= \lim_{h\to 0} \frac{f(x+h)-f(x)}{h} \\ &= \lim_{h\to 0} \left(\frac{u(x+h)+v(x+h)-[u(x)+v(x)]}{h}\right) \\ &= \lim_{h\to 0} \left(\frac{u(x+h)-u(x)+v(x+h)-v(x)}{h}\right) \\ &= \lim_{h\to 0} \frac{u(x+h)-u(x)}{h} + \lim_{h\to 0} \frac{v(x+h)-v(x)}{h} \\ &= u'(x)+v'(x) \end{aligned}$$

Using the rules we have now developed we can differentiate sums of powers of x.

For example, if $f(x) = 2x^3+5x^2-7x+4$ then

$$\begin{aligned} f'(x) &= 2(3x^2)+5(2x)-7(1)+0 \\ &= 6x^2+10x-7 \end{aligned}$$

Example 1

Self Tutor

Use the rules of differentiation to find $f'(x)$ if:

a $f(x) = x^3+2x^2-3x+4$

b $f(x) = 3x - \dfrac{5}{x^2} - \dfrac{1}{x^3}$

a

$$\begin{aligned} f(x) &= x^3+2x^2-3x+4 \\ \therefore\ f'(x) &= 3x^2+2(2x)-3(1)+0 \\ &= 3x^2+4x-3 \end{aligned}$$

b

$$\begin{aligned} f(x) &= 3x - \frac{5}{x^2} - \frac{1}{x^3} \\ &= 3x-5x^{-2}-x^{-3} \\ \therefore\ f'(x) &= 3(1)-5(-2x^{-3})-(-3x^{-4}) \\ &= 3+10x^{-3}+3x^{-4} \\ &= 3+\frac{10}{x^3}+\frac{3}{x^4} \end{aligned}$$

EXERCISE 14A

1 Find $f'(x)$ if:

a $f(x)=x^4$ **b** $f(x)=x^7$ **c** $f(x)=x^{10}$

d $f(x)=5x$ **e** $f(x)=3$ **f** $f(x)=4x^2$

g $f(x)=6x^3$ **h** $f(x)=-2x$ **i** $f(x)=-\frac{1}{2}x^{10}$

j $f(x)=x^2-3x$ **k** $f(x)=2x+5$ **l** $f(x)=3x^2-x-4$

m $f(x)=x^3+2x^2-6x+1$ **n** $f(x)=2x^3-4x^2-5x$ **o** $f(x)=\frac{1}{2}x^3-\frac{1}{3}x^2+2$

2 Differentiate with respect to x:

a $\dfrac{1}{x}$ **b** $\dfrac{1}{x^3}$ **c** $\dfrac{1}{x^7}$ **d** $\dfrac{2}{x^2}$

e $-\dfrac{5}{x}$ **f** $\dfrac{3}{x^4}$ **g** $x+\dfrac{6}{x}$ **h** $\dfrac{1}{x^2}-\dfrac{3}{x^5}$

i $x^2-\dfrac{4}{x^3}+\dfrac{1}{x^4}$ **j** $\dfrac{6}{x}-\dfrac{1}{x^2}+\dfrac{2}{x^5}$ **k** $\dfrac{1}{7x}$ **l** $x-\dfrac{2}{3x^2}$

Example 2 — Self Tutor

Find $f'(x)$ if:

a $f(x)=5x+2\sqrt{x}$ **b** $f(x)=\dfrac{x+4}{\sqrt{x}}$

a
$$\begin{aligned} f(x)&=5x+2\sqrt{x}\\ &=5x+2x^{\frac{1}{2}}\\ \therefore\ f'(x)&=5(1)+2(\tfrac{1}{2}x^{-\frac{1}{2}})\\ &=5+x^{-\frac{1}{2}}\\ &=5+\frac{1}{\sqrt{x}}\end{aligned}$$

b
$$\begin{aligned} f(x)&=\frac{x+4}{\sqrt{x}}\\ &=\frac{x+4}{x^{\frac{1}{2}}}\\ &=x^{\frac{1}{2}}+4x^{-\frac{1}{2}}\\ \therefore\ f'(x)&=\tfrac{1}{2}x^{-\frac{1}{2}}+4(-\tfrac{1}{2}x^{-\frac{3}{2}})\\ &=\tfrac{1}{2}x^{-\frac{1}{2}}-2x^{-\frac{3}{2}}\\ &=\frac{1}{2\sqrt{x}}-\frac{2}{x\sqrt{x}}\end{aligned}$$

3 Find $f'(x)$ if:

a $f(x)=6\sqrt{x}$ **b** $f(x)=\sqrt[4]{x}$ **c** $f(x)=\dfrac{2}{\sqrt{x}}$

d $f(x)=\dfrac{1}{\sqrt[3]{x}}$ **e** $f(x)=x^2+8\sqrt{x}$ **f** $f(x)=7x-\frac{1}{2}\sqrt{x}$

g $f(x)=2x\sqrt{x}$ **h** $f(x)=\dfrac{2}{x}-\dfrac{1}{x\sqrt{x}}$ **i** $f(x)=\sqrt{x}+4x^2\sqrt{x}$

4 Find $\frac{dy}{dx}$ if:

a $y = x(x+5)$ **b** $y = (x-3)^2$ **c** $y = \frac{x^2+4}{x}$

d $y = \frac{3x-2}{\sqrt{x}}$ **e** $y = (2x+5)^2$ **f** $y = \frac{(\sqrt{x}-1)^2}{2\sqrt{x}}$

5 Find:

a $\frac{d}{dx}(5x^2 - 4x + 2)$

b $\frac{d}{dx}\left(8\sqrt{x} + \frac{3}{x}\right)$

c $\frac{d}{dx}\left(\frac{x^3-6}{x^2}\right)$

$\frac{d}{dx}$(....) reads "the derivative of (....) with respect to x".

6 Find:

a $\frac{dy}{dx}$ if $y = 3x^3 - 5x$ **b** $\frac{dy}{dt}$ if $y = 7t - \frac{2}{t^2}$

c $\frac{du}{dx}$ if $u = 4\sqrt{x} - \frac{1}{x}$ **d** $\frac{dP}{dt}$ if $P = (t+4)^2$

e $\frac{dT}{dx}$ if $T = \frac{(2x+1)(x-2)}{x}$ **f** $\frac{dy}{du}$ if $y = \frac{(u-2)^2}{\sqrt{u}}$

Example 3 Self Tutor

Consider the curve $f(x) = \sqrt{x}$ whose graph is drawn alongside.

a Find $f'(x)$.

b Find the gradient of the tangent at the point where $x = 1$.

c Find the point on the curve $f(x) = \sqrt{x}$ at which the tangent has gradient $\frac{1}{4}$.

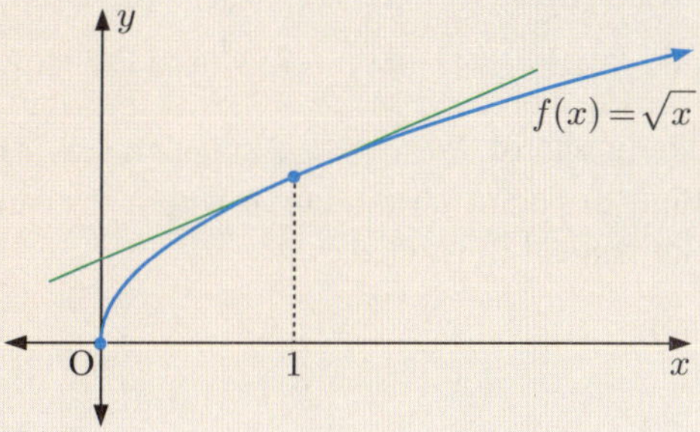

a $f(x) = \sqrt{x} = x^{\frac{1}{2}}$

$\therefore f'(x) = \frac{1}{2}x^{-\frac{1}{2}} = \frac{1}{2\sqrt{x}}$

b $f'(1) = \frac{1}{2\sqrt{1}} = \frac{1}{2}$

$\therefore$ the tangent at the point where $x = 1$ has gradient $\frac{1}{2}$.

c If the gradient of the tangent is $\frac{1}{4}$, then

$f'(x) = \frac{1}{4}$

$\therefore \frac{1}{2\sqrt{x}} = \frac{1}{4}$

$\therefore \sqrt{x} = 2$

$\therefore x = 4$

$f(4) = \sqrt{4} = 2$

So, the tangent at $(4, 2)$ has gradient $\frac{1}{4}$.

7 Consider $f(x) = x^3 + 2x^2 - 3x + 1$.

a Find $f'(x)$. **b** Find $f(2)$ and $f'(2)$.

c Copy and complete:

For $f(x) = x^3 + 2x^2 - 3x + 1$, the gradient of the tangent at $(2, ...)$ is ....

8 Find the gradient of the tangent to:

a $f(x) = 3x^2$ at $x = -1$

b $f(x) = \frac{6}{x}$ at $x = 2$

c $f(x) = x^3 + 2x + 1$ at $x = 1$

d $f(x) = x^2 + 7x$ at $x = -2$

e $f(x) = \frac{x^2+1}{x}$ at $x = 2$

f $f(x) = \sqrt{x} + \frac{8}{x}$ at $x = 4$.

9 In the graph alongside, find the gradient of:

a line 1

b line 2

c line 3.

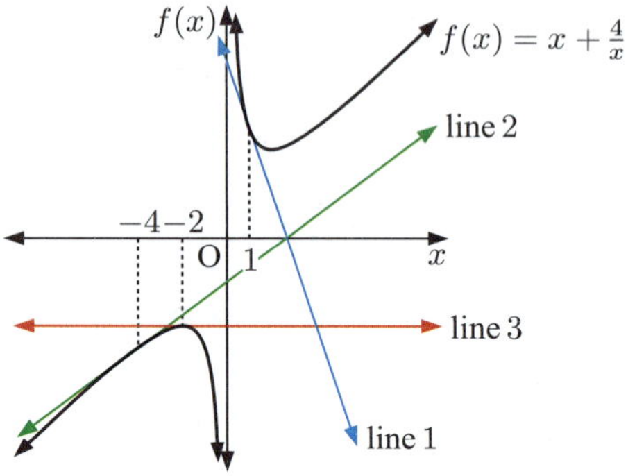

10 Find the coordinates of the point(s) on:

a $f(x) = x^2 + 4x + 5$ where the tangent has gradient 0

b $f(x) = \sqrt{x}$ where the tangent has gradient $\frac{1}{2}$

c $f(x) = x^3 + x^2 - 1$ where the tangent has gradient 1

d $f(x) = x^3 - 3x + 1$ where the tangent has gradient 9

e $f(x) = x^3 - 6x^2 + 7$ where the tangent is horizontal.

11 The graph of $y = x^3 + \frac{3}{2}x^2 - 2x + 1$ is shown alongside. Find the point on the curve where the tangent is parallel to the tangent illustrated.

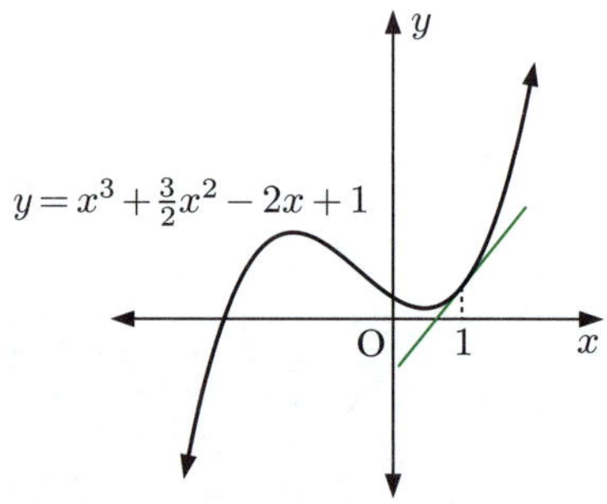

Example 4 — Self Tutor

$f(x) = 4x^3 + ax^2 - 9x + b$ is divisible by $x - 1$, and $f'(x)$ is divisible by $2x - 1$. Find a and b.

$f(x) = 4x^3 + ax^2 - 9x + b$

$\therefore\ f'(x) = 12x^2 + 2ax - 9$

Now $f(1) = 0$ and $f'(\frac{1}{2}) = 0$ {Factor theorem}

$\therefore\ 4 + a - 9 + b = 0$ and $12(\frac{1}{4}) + 2a(\frac{1}{2}) - 9 = 0$

$\therefore\ a + b = 5$ and $3 + a - 9 = 0$

$\therefore\ a = 6$ and $b = -1$.

12 $f(x) = x^3 + ax^2 - 3x + b$ is divisible by $x + 1$, and $f'(x)$ is divisible by $x + 3$. Find a and b.

13 Suppose $p(x) = 4x^3 + mx^2 + nx - 3$ where $p(x)$ and $p'(x)$ are both divisible by $2x + 1$.

a Find m and n.

b Fully factorise $p(x)$.

B THE CHAIN RULE

In **Chapter 2** we defined the **composite** of two functions g and f as $(g \circ f)(x)$ or $gf(x)$.

We can often write complicated functions as the composite of two or more simpler functions.

For example $y = (x^2 + 3x)^4$ could be rewritten as $y = u^4$ where $u = x^2 + 3x$, or as $y = gf(x)$ where $g(x) = x^4$ and $f(x) = x^2 + 3x$.

Example 5 — Self Tutor

Find:

a $gf(x)$ if $g(x) = \dfrac{1}{x}$ and $f(x) = x^2 - 2$

b $g(x)$ and $f(x)$ such that $gf(x) = \sqrt{3x + 5}$.

a $gf(x) = g(x^2 - 2)$
$$= \frac{1}{x^2 - 2}$$

b If we let $f(x) = 3x + 5$,

then $gf(x) = \sqrt{3x+5} = \sqrt{f(x)}$

$\therefore\ g(x) = \sqrt{x}$ and $f(x) = 3x + 5$

EXERCISE 14B.1

1 Find $gf(x)$ if:

a $g(x) = x^3$ and $f(x) = 4x + 1$

b $g(x) = 4x + 1$ and $f(x) = x^3$

c $g(x) = \dfrac{1}{x}$ and $f(x) = 2 - x^2$

d $g(x) = 2 - x^2$ and $f(x) = \dfrac{1}{x}$

e $g(x) = \sqrt{x}$ and $f(x) = 5x - 3$

f $g(x) = 5x - 3$ and $f(x) = \sqrt{x}$.

2 Find $g(x)$ and $f(x)$ such that $gf(x)$ is:

a $(6x - 1)^2$

b $\dfrac{2}{x^2 - 3}$

c $\sqrt{4x - 7}$

d $\dfrac{1}{\sqrt{8 - x^2}}$

DERIVATIVES OF COMPOSITE FUNCTIONS

The reason we are interested in writing complicated functions as composite functions is to make finding derivatives easier.

Discovery 1 — Differentiating composite functions

The purpose of this Discovery is to learn how to differentiate composite functions.

Based on the rule "if $y = x^n$ then $\frac{dy}{dx} = nx^{n-1}$", we might suspect that if $y = (1-x)^2$ then $\frac{dy}{dx} = 2(1-x)^1$.

What to do:

1 Expand $y = (1-x)^2$ and hence find $\frac{dy}{dx}$. Compare your result with $2(1-x)^1$.

2 Expand $y = (1-2x)^2$ and hence find $\frac{dy}{dx}$. Compare your result with $2(1-2x)^1$.

3 Expand $y = (1-ax)^2$ where a is a constant. Hence find $\frac{dy}{dx}$. Compare your result with $2(1-ax)^1$.

4 Suppose $y = u^2$.

a Find $\frac{dy}{du}$.

b Now suppose $u = 1 - ax$, so $y = (1-ax)^2$.

i Find $\frac{du}{dx}$.

ii Write $\frac{dy}{du}$ from **a** in terms of x.

iii Hence find $\frac{dy}{du} \times \frac{du}{dx}$.

iv Compare your answer to the result in **3**.

c If $y = u^2$ where u is a function of x, what do you suspect $\frac{dy}{dx}$ will be equal to?

5 Expand $y = (2+x^2)^2$ and hence find $\frac{dy}{dx}$.

Does your answer agree with the rule you suggested in **4 c**?

6 Consider $y = (3x-1)^3$.

a Expand the brackets and hence find $\frac{dy}{dx}$.

b If we let $u = 3x - 1$, then $y = u^3$.

i Find $\frac{du}{dx}$.

ii Find $\frac{dy}{du}$, and write it in terms of x.

iii Hence find $\frac{dy}{du} \times \frac{du}{dx}$.

iv Compare your answer to the result in **a**.

7 Copy and complete: "If y is a function of u, and u is a function of x, then $\frac{dy}{dx} = ...$"

THE CHAIN RULE

If $y = g(u)$ where $u = f(x)$ then $\dfrac{dy}{dx} = \dfrac{dy}{du}\dfrac{du}{dx}$.

This rule allows us to differentiate complicated functions much faster.

For example, for any function $f(x)$:

If $y = [f(x)]^n$ then $\dfrac{dy}{dx} = n[f(x)]^{n-1} \times f'(x)$.

Example 6 — Self Tutor

Find $\dfrac{dy}{dx}$ if:

a $y = (x^3 + 2x)^5$

b $y = \dfrac{2}{4x-3}$

a
$$y = (x^3+2x)^5$$
$$\therefore\ y = u^5 \text{ where } u = x^3 + 2x$$
$$\text{Now } \frac{dy}{dx} = \frac{dy}{du}\frac{du}{dx} \quad \{\text{chain rule}\}$$
$$= 5u^4(3x^2+2)$$
$$= 5(x^3+2x)^4(3x^2+2)$$

b
$$y = \frac{2}{4x-3}$$
$$\therefore\ y = 2u^{-1} \text{ where } u = 4x-3$$
$$\text{Now } \frac{dy}{dx} = \frac{dy}{du}\frac{du}{dx} \quad \{\text{chain rule}\}$$
$$= -2u^{-2}(4)$$
$$= -\frac{8}{u^2}$$
$$= -\frac{8}{(4x-3)^2}$$

EXERCISE 14B.2

1 Write in the form au^n, clearly stating what u is:

a $3(2x-5)^4$ **b** $\dfrac{5}{x^2-2}$ **c** $\sqrt{7x-1}$

d $\dfrac{2}{(8-x)^3}$ **e** $-\dfrac{4}{\sqrt{x+1}}$ **f** $\sqrt[3]{x^2+5x-2}$

2 Find $\dfrac{dy}{dx}$ for:

a $y = (3x-2)^2$ **b** $y = \dfrac{1}{2x-4}$ **c** $y = (x^2-3)^5$

d $y = \dfrac{4}{(5x+1)^2}$ **e** $y = \sqrt{4x-5}$ **f** $y = \sqrt{x^3-7x}$

g $y = \dfrac{2}{x^2+x-1}$ **h** $y = \dfrac{1}{\sqrt{6-x}}$ **i** $y = \sqrt[3]{x^2+x}$

3 Consider $y = \sqrt{2x+5}$.

a Find $\dfrac{dy}{dx}$.

b Hence find the gradient of the tangent to $y = \sqrt{2x+5}$ when $x = 2$.

4 Find the gradient of the tangent to:

a $y = (2x-3)^3$ at $x = -1$

b $y = \sqrt{5-x}$ at $x = 1$

c $y = \dfrac{1}{3x+1}$ at $x = 0$

d $y = \sqrt[3]{x^2-1}$ at $x = 3$

e $y = \dfrac{2}{\sqrt{1-2x}}$ at $x = 0$

f $y = \left(x - \dfrac{2}{x}\right)^4$ at $x = 4$.

5 The tangent to $y = \sqrt{ax-3}$ when $x = 6$ has gradient $\frac{1}{3}$. Find the possible values of a.

6 The graph of $y = \dfrac{1}{x^2-k}$ is shown alongside.

a Find k.

b Find $\dfrac{dy}{dx}$.

c Find the gradient of the tangent at:

i P

ii Q

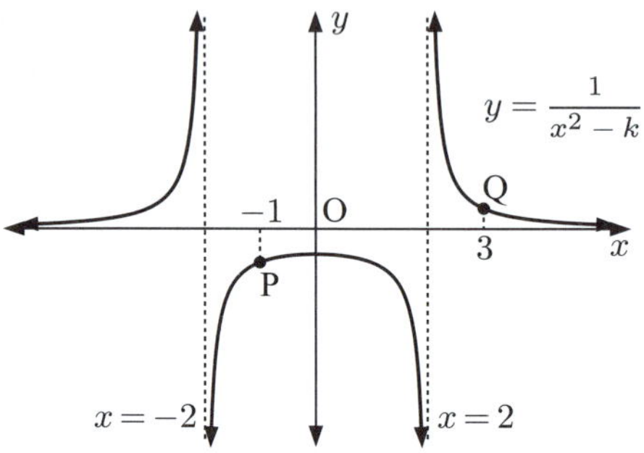

7 Suppose $f(x) = \dfrac{a}{bx+1}$ where $f(1) = 2$ and $f'(1) = -\frac{3}{2}$. Find a and b.

C THE PRODUCT RULE

We have seen the addition rule:

$$\text{If } f(x) = u(x) + v(x) \quad \text{then} \quad f'(x) = u'(x) + v'(x).$$

We now consider the case $f(x) = u(x)\,v(x)$. Is $f'(x) = u'(x)\,v'(x)$?

In other words, does the derivative of a product of two functions equal the product of the derivatives of the two functions?

Discovery 2 — The product rule

Suppose $u(x)$ and $v(x)$ are two functions of x, and that $f(x) = u(x)\,v(x)$ is the product of these functions.

The purpose of this Discovery is to find a rule for determining $f'(x)$.

What to do:

1 Suppose $u(x) = 2$ and $v(x) = x$, so $f(x) = 2x$.

a Find $f'(x)$ by direct differentiation.

b Find $u'(x)$ and $v'(x)$.

c Does $f'(x) = u'(x)\,v'(x)$?

2 Suppose $u(x) = x$ and $v(x) = x^2$, so $f(x) = x^3$.

a Find $f'(x)$ by direct differentiation.

b Find $u'(x)$ and $v'(x)$.

c Does $f'(x) = u'(x)\,v'(x)$?

3 Copy and complete the following table, finding $f'(x)$ by direct differentiation.

$f(x)$	$f'(x)$	$u(x)$	$v(x)$	$u'(x)$	$v'(x)$	$u'(x)\,v(x)+u(x)\,v'(x)$
$2x$		2	x			
x^3		x	x^2			
$2x(x-3)$		$2x$	$x-3$			
$(x^2-1)(3x+1)$		x^2-1	$3x+1$			

4 Copy and complete: "If $f(x)=u(x)\,v(x)$ then $f'(x)=$ ......"

THE PRODUCT RULE

If $\boldsymbol{f(x)=u(x)\,v(x)}$ then $\boldsymbol{f'(x)=u'(x)\,v(x)+u(x)\,v'(x)}$.

Alternatively, if $\boldsymbol{y=uv}$ where u and v are functions of x, then $\boldsymbol{\dfrac{dy}{dx}=\dfrac{du}{dx}\,v+u\,\dfrac{dv}{dx}}$.

Example 7 — Self Tutor

Find $\dfrac{dy}{dx}$ if: **a** $y=x\sqrt{2x-1}$ **b** $y=x^4(3x+1)^3$

a $y=x\sqrt{2x-1}$ is the product of $u=x$ and $v=(2x-1)^{\frac{1}{2}}$

$\therefore$ $u'=1$ and $v'=\frac{1}{2}(2x-1)^{-\frac{1}{2}}(2)$ {chain rule}

$=(2x-1)^{-\frac{1}{2}}$

Now $\dfrac{dy}{dx}=u'v+uv'$ {product rule}

$=(1)(2x-1)^{\frac{1}{2}}+x(2x-1)^{-\frac{1}{2}}$

$=\sqrt{2x-1}+\dfrac{x}{\sqrt{2x-1}}$

b $y=x^4(3x+1)^3$ is the product of $u=x^4$ and $v=(3x+1)^3$

$\therefore$ $u'=4x^3$ and $v'=3(3x+1)^2(3)$ {chain rule}

$=9(3x+1)^2$

Now $\dfrac{dy}{dx}=u'v+uv'$ {product rule}

$=4x^3(3x+1)^3+x^4\times 9(3x+1)^2$

$=4x^3(3x+1)^3+9x^4(3x+1)^2$

EXERCISE 14C

1 Use the product rule to differentiate:

a $f(x)=x(x+3)$ **b** $f(x)=4x(x-1)$ **c** $f(x)=x^2(x-2)$

d $f(x)=x\sqrt{x+1}$ **e** $f(x)=(x+3)(x^2-1)$ **f** $f(x)=(2x+3)(x-4)$

g $f(x)=x^2\sqrt{x-2}$ **h** $f(x)=x^3(2x+5)^2$ **i** $f(x)=5x\sqrt{x^2-6}$

2 Find $\frac{dy}{dx}$ using the product rule:

a $y = x(2x^2 - 3)$
b $y = 3x(x-4)^3$
c $y = x^2\sqrt{1-2x}$
d $y = (3-x)(4x+1)$
e $y = 2x\sqrt{x+5}$
f $y = \sqrt{x}(x-1)^3$
g $y = x^2(2-x)^3$
h $y = 3x^3\sqrt{x+2}$
i $y = \sqrt{x+1}(7-x^2)^3$

3 Find the gradient of the tangent to:

a $y = x^3(2x+1)^2$ at $x = 1$
b $y = 3x\sqrt{x-2}$ at $x = 3$
c $y = \sqrt{x}(x^2-1)^2$ at $x = 2$
d $y = -x^3\sqrt{2-x}$ at $x = -2$.

4 Suppose $f(x) = \sqrt{x}(2x-7)^3$.

a Write $f'(x)$ in the form $f'(x) = \frac{(2x-7)^2(ax+b)}{2\sqrt{x}}$, where $a, b \in \mathbb{Z}$.
b Find the gradient of the tangent to $y = f(x)$ when $x = 4$.
c Find the x-coordinates of all points on $y = f(x)$ where the tangent is horizontal.

5 The graph of $y = x\sqrt{x+1}$ is shown alongside.
Find exactly the x-coordinate of the point where the tangent is perpendicular to the illustrated tangent.

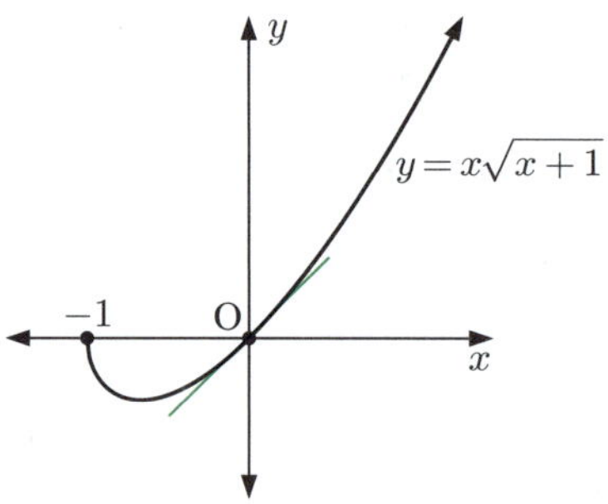

D THE QUOTIENT RULE

Expressions like $\frac{2x}{3x-1}$, $\frac{x^2}{4-x}$, and $\frac{\sqrt{x}+5}{(x-1)^2}$ are called **quotients** because they represent the division of one function by another.

Quotient functions have the form $Q(x) = \frac{u(x)}{v(x)}$, which can be written as

$$Q(x) = u(x) \times \left(\frac{1}{v(x)}\right)$$

$$\therefore \quad Q'(x) = u'(x) \times \left(\frac{1}{v(x)}\right) + u(x) \times \left(\frac{1}{v(x)}\right)' \qquad \{\text{product rule}\}$$

$$= \frac{u'(x)}{v(x)} + u(x) \times \left(-[v(x)]^{-2} \times v'(x)\right) \qquad \{\text{chain rule}\}$$

$$= \frac{u'(x)}{v(x)} - \frac{u(x)\,v'(x)}{[v(x)]^2}$$

$$= \frac{u'(x)\,v(x) - u(x)\,v'(x)}{[v(x)]^2}$$

THE QUOTIENT RULE

If $Q(x) = \dfrac{u(x)}{v(x)}$ then $Q'(x) = \dfrac{u'(x)\,v(x) - u(x)\,v'(x)}{[v(x)]^2}$.

Alternatively, if $y = \dfrac{u}{v}$ where u and v are functions of x, then $\dfrac{dy}{dx} = \dfrac{u'v - uv'}{v^2}$.

Example 8

Self Tutor

Use the quotient rule to find $\dfrac{dy}{dx}$ if:

a $y = \dfrac{x^2 - 2}{3x - 1}$

b $y = \dfrac{\sqrt{x}}{4 - x}$

a $y = \dfrac{x^2 - 2}{3x - 1}$ is a quotient with $u = x^2 - 2$ and $v = 3x - 1$

$\therefore\ u' = 2x$ and $v' = 3$

Now $\dfrac{dy}{dx} = \dfrac{u'v - uv'}{v^2}$ {quotient rule}

$$= \frac{2x(3x-1) - (x^2-2)\times 3}{(3x-1)^2}$$

$$= \frac{6x^2 - 2x - 3x^2 + 6}{(3x-1)^2}$$

$$= \frac{3x^2 - 2x + 6}{(3x-1)^2}$$

b $y = \dfrac{\sqrt{x}}{4 - x}$ is a quotient with $u = x^{\frac{1}{2}}$ and $v = 4 - x$

$\therefore\ u' = \frac{1}{2}x^{-\frac{1}{2}}$ and $v' = -1$

Now $\dfrac{dy}{dx} = \dfrac{u'v - uv'}{v^2}$ {quotient rule}

$$= \frac{\frac{1}{2}x^{-\frac{1}{2}}(4-x) - x^{\frac{1}{2}}\times(-1)}{(4-x)^2}$$

$$= \frac{\frac{4-x}{2\sqrt{x}} + \sqrt{x}\left(\frac{2\sqrt{x}}{2\sqrt{x}}\right)}{(4-x)^2}$$

$$= \frac{4 - x + 2x}{2\sqrt{x}(4-x)^2}$$

$$= \frac{x + 4}{2\sqrt{x}(4-x)^2}$$

EXERCISE 14D

1 Suppose $y = \dfrac{2x^2 - 3}{x}$. Find $\dfrac{dy}{dx}$ using the quotient rule.

Check your answer by writing $y = 2x - \dfrac{3}{x}$ and then differentiating term by term.

2 Use the quotient rule to find $\frac{dy}{dx}$ if:

a $y = \frac{4x+3}{x+1}$ **b** $y = \frac{2x}{3-x}$ **c** $y = \frac{x^2}{2x-3}$

d $y = \frac{x+5}{x^2-1}$ **e** $y = \frac{x^2-x}{2x^2+5}$ **f** $y = \frac{\sqrt{x}}{x+2}$

g $y = \frac{3x}{\sqrt{x}-1}$ **h** $y = \frac{\sqrt{x}}{(3x+1)^2}$ **i** $y = \frac{4x-1}{\sqrt{x}+2}$

3 The graph of $y = \frac{x+2}{2x-3}$ is shown alongside.

a Find $\frac{dy}{dx}$.

b Find the gradient of the tangent at:

i P **ii** Q

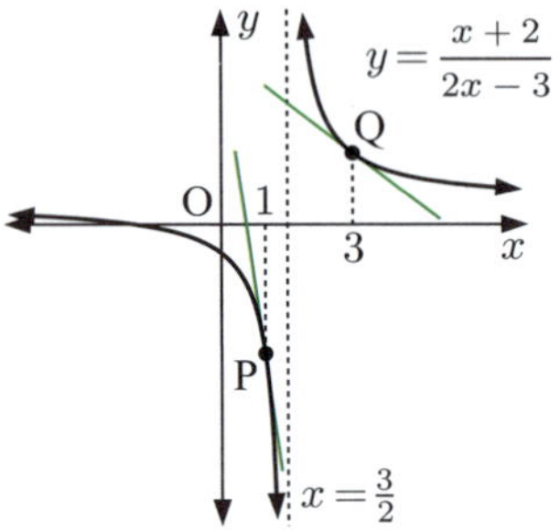

4 Find the gradient of the tangent to:

a $y = \frac{x}{x-3}$ at $x = 4$ **b** $y = \frac{x^2}{2x-5}$ at $x = 2$

c $y = \frac{2\sqrt{x}}{x^2+1}$ at $x = 1$ **d** $y = \frac{3x+1}{\sqrt{x}-4}$ at $x = 8$.

5 $f(x) = \frac{x^2+kx-1}{x+3}$ has gradient function $f'(x) = \frac{x^2+6x+13}{(x+3)^2}$.

a Find k.

b Find the gradient of the tangent to $y = f(x)$ when $x = -2$.

6 **a** If $y = \frac{4x+3}{x^2+1}$, show that $\frac{dy}{dx} = \frac{-4x^2-6x+4}{(x^2+1)^2}$.

b Find the coordinates of the points on $y = \frac{4x+3}{x^2+1}$ at which the tangent is horizontal.

E DERIVATIVES OF EXPONENTIAL FUNCTIONS

In **Chapter 4** we saw that the simplest **exponential functions** have the form $f(x) = b^x$ where b is any positive constant, $b \neq 1$.

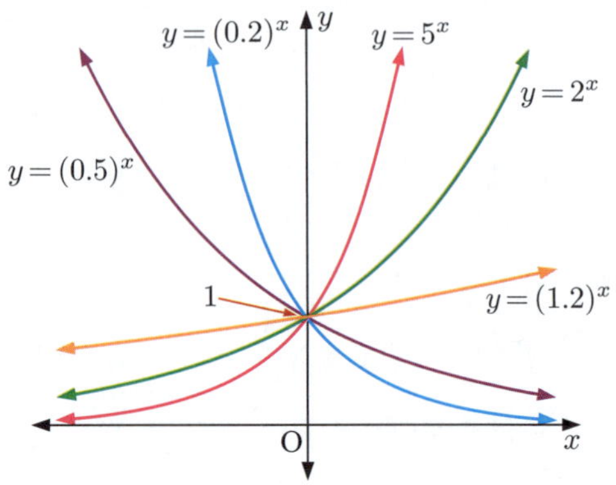

Discovery 3 — The derivative of $y = b^x$

The purpose of this Discovery is to observe the nature of the derivatives of $f(x) = b^x$ for various values of b.

What to do:

1 Use the software provided to help fill in the table for $y = 2^x$:

x	y	$\frac{dy}{dx}$	$\frac{dy}{dx} \div y$
0			
0.5			
1			
1.5			
2			

CALCULUS DEMO

2 Repeat **1** for the following functions:

a $y = 3^x$ **b** $y = 5^x$ **c** $y = (0.5)^x$

3 Use your observations from **1** and **2** to write a statement about the derivative of the general exponential $y = b^x$ for $b > 0$, $b \neq 1$.

From the **Discovery** you should have found that:

If $f(x) = b^x$ then $f'(x) = f'(0) \times b^x$.

When we studied exponential functions in **Chapter 4**, we found a special value of b for which $f'(0) = 1$.

This value was the **natural exponential** $e \approx 2.718\,28....$

This value is special because the function e^x is its own derivative.

If $f(x) = e^x$, then $f'(x) = e^x$.

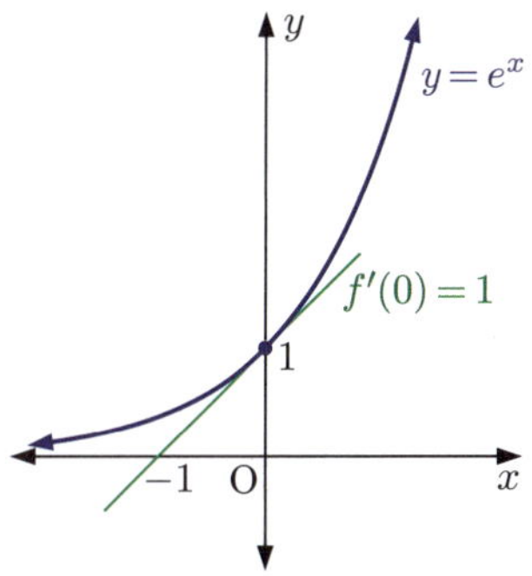

THE DERIVATIVE OF $e^{f(x)}$

The functions e^{-x}, e^{2x+3}, and e^{-x^2} all have the form $e^{f(x)}$.

Suppose $y = e^{f(x)} = e^u$ where $u = f(x)$.

$$\begin{aligned} \text{Now } \frac{dy}{dx} &= \frac{dy}{du}\frac{du}{dx} \quad \{\text{chain rule}\} \\ &= e^u \frac{du}{dx} \\ &= e^{f(x)} \times f'(x) \end{aligned}$$

Function	*Derivative*
e^x	e^x
$e^{f(x)}$	$e^{f(x)} \times f'(x)$

Example 9 — Self Tutor

Find $f'(x)$ for:

a $f(x) = 5e^x + e^{4x}$ **b** $f(x) = x^3 e^{2x}$ **c** $f(x) = \dfrac{x}{e^x}$

a
$$\begin{aligned} f(x) &= 5e^x + e^{4x} \\ \therefore\ f'(x) &= 5e^x + e^{4x}(4) \\ &= 5e^x + 4e^{4x} \end{aligned}$$

b
$$\begin{aligned} f(x) &= x^3 e^{2x} \\ \therefore\ f'(x) &= 3x^2 e^{2x} + x^3 e^{2x}(2) \qquad \{\text{product rule}\} \\ &= 3x^2 e^{2x} + 2x^3 e^{2x} \end{aligned}$$

c
$$\begin{aligned} f(x) &= \frac{x}{e^x} \\ \therefore\ f'(x) &= \frac{(1)e^x - xe^x}{(e^x)^2} \qquad \{\text{quotient rule}\} \\ &= \frac{\cancel{e^x}(1-x)}{(e^x)^{\cancel{2}}} \\ &= \frac{1-x}{e^x} \end{aligned}$$

EXERCISE 14E

1 Find $f'(x)$ for:

a $f(x) = 3e^x$ **b** $f(x) = -e^x$ **c** $f(x) = e^{4x}$

d $f(x) = \exp(2x)$ **e** $f(x) = e^{-\frac{x}{5}}$ **f** $f(x) = e^{x^2}$

g $f(x) = 2e^x + 5e^{-x}$ **h** $f(x) = e^{2x+1} - e^{\frac{1}{x}}$ **i** $f(x) = 7e^{3-4x^2}$

2 Find the derivative of:

a $x^2 e^x$ **b** $2xe^{4x}$ **c** $(3x-1)e^{-2x}$ **d** $e^{5x-1}\sqrt{x}$

e $\dfrac{e^x}{x}$ **f** $\dfrac{x^3}{e^{2x}}$ **g** $\dfrac{2e^x - 1}{\sqrt{x}}$ **h** $\dfrac{3e^x + 1}{e^{-x} - 1}$

3 Suppose $y = e^{2x}\sqrt{x+1}$.

a Find $\dfrac{dy}{dx}$, giving your answer in the form $\dfrac{dy}{dx} = \dfrac{e^{2x}(ax+b)}{2\sqrt{x+1}}$ where a and b are integers.

b Hence find the gradient of the tangent to $y = e^{2x}\sqrt{x+1}$ at:

i $x = 0$ **ii** $x = 3$

4 Find the gradient of the tangent to:

a $y = xe^{2x}$ at $x = 0$

b $y = \dfrac{4x-1}{e^x}$ at $x = -1$

c $y = \dfrac{e^{3x}}{3} - xe^{3x}$ at $x = \ln 2$

d $y = (e^x - 2x)^5$ at $x = 0$.

5 Given $f(x) = ke^{3x} + 5x$ and $f'(0) = -1$, find k.

6 The graph of $y = \dfrac{e^{\frac{x}{2}}}{\sqrt{x}}$ is shown alongside.

a Find the gradient of the illustrated tangent.

b Find the coordinates of P, the point at which the tangent to the curve is horizontal.

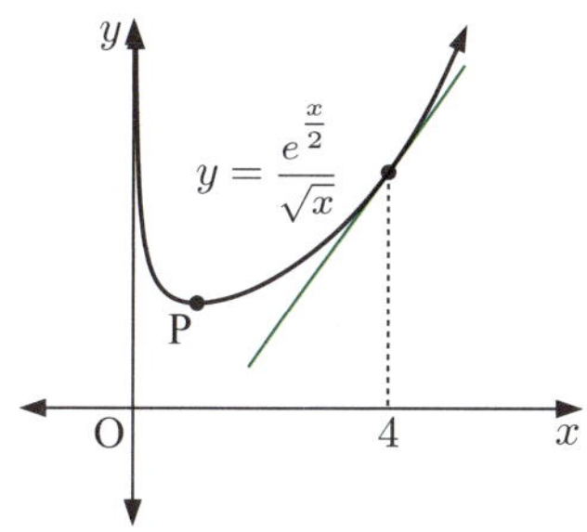

7 Suppose the graph of $f(x) = e^{kx}$ *touches* the line $y = x$. Find k and the coordinates of the point of contact.

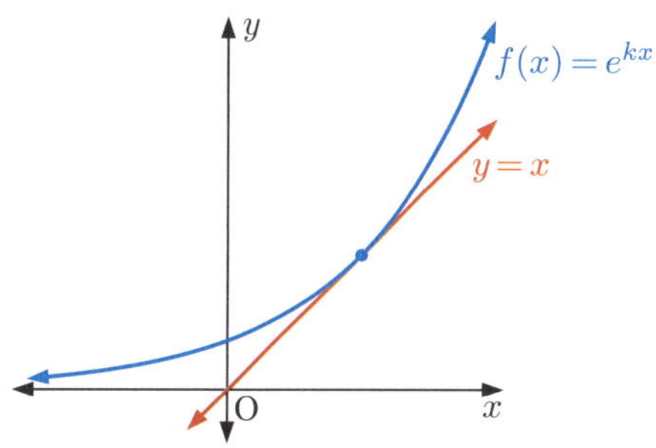

F DERIVATIVES OF LOGARITHMIC FUNCTIONS

In **Chapter 5**, we saw that the logarithmic function $y = \ln x$ is the inverse function of the exponential function $y = e^x$. The graph of $y = \ln x$ is the reflection of $y = e^x$ in the line $y = x$.

The graphs of $f(x) = \ln x$ and $g(x) = e^x$ are shown alongside.

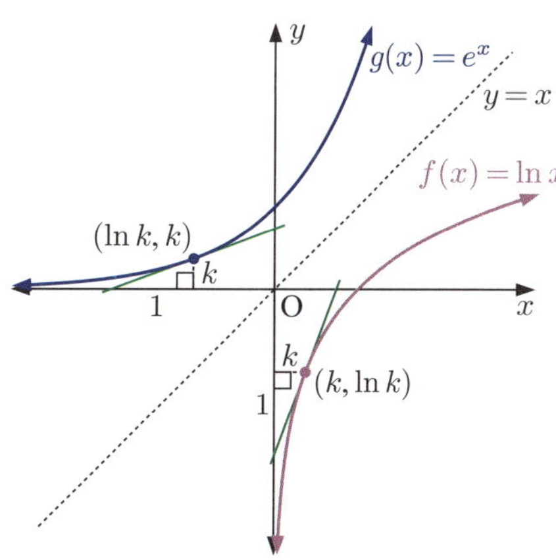

To find the derivative of $f(x) = \ln x$, consider a general point $(k, \ln k)$ on the curve. Since $(k, \ln k)$ lies on $f(x) = \ln x$, the point $(\ln k, k)$ lies on its inverse function $g(x) = e^x$.

Now $g'(x) = e^x$, so $g'(\ln k) = e^{\ln k} = k$.

$\therefore$ the tangent to $g(x) = e^x$ at $(\ln k, k)$ has gradient k.

Using the symmetry of the graphs about the line $y = x$, the tangent to $f(x) = \ln x$ at $(k, \ln k)$ must have gradient $\dfrac{1}{k}$.

We hence deduce that $f'(k) = \dfrac{1}{k}$.

Since this argument is valid at any point on the curve, we conclude:

If $f(x) = \ln x$, then $f'(x) = \dfrac{1}{x}$.

THE DERIVATIVE OF $\ln f(x)$

Suppose $y = \ln f(x)$

$\therefore\ y = \ln u$ where $u = f(x)$.

Now $\dfrac{dy}{dx} = \dfrac{dy}{du}\dfrac{du}{dx}$ {chain rule}

$\therefore\ \dfrac{dy}{dx} = \dfrac{1}{u}\dfrac{du}{dx}$

$= \dfrac{f'(x)}{f(x)}$

Function	*Derivative*
$\ln x$	$\dfrac{1}{x}$
$\ln f(x)$	$\dfrac{f'(x)}{f(x)}$

Example 10 Self Tutor

Find $f'(x)$ for:

a $f(x) = 5\ln x$ **b** $f(x) = \ln(x^2 - 4)$ **c** $f(x) = \dfrac{\ln x}{x}$

a $f(x) = 5\ln x$

$\therefore\ f'(x) = 5 \times \dfrac{1}{x}$

$= \dfrac{5}{x}$

b $f(x) = \ln(x^2 - 4)$

$\therefore\ f'(x) = \dfrac{2x}{x^2 - 4}$

c $f(x) = \dfrac{\ln x}{x}$

$\therefore\ f'(x) = \dfrac{\frac{1}{x}(x) - \ln x(1)}{x^2}$ {quotient rule}

$= \dfrac{1 - \ln x}{x^2}$

The laws of logarithms can help us to differentiate some logarithmic functions more easily.

For $a > 0,\ b > 0,\ n \in \mathbb{R}$: $\quad \ln(ab) = \ln a + \ln b$

$\ln\left(\dfrac{a}{b}\right) = \ln a - \ln b$

$\ln(a^n) = n\ln a$

Example 11 Self Tutor

Differentiate with respect to x:

a $y = \ln(x^3e^{2x})$ **b** $\ln\left(\dfrac{5x - 4}{2x + 7}\right)$

a $y = \ln(x^3e^{2x})$

$= \ln x^3 + \ln e^{2x}$ $\{\ln(ab) = \ln a + \ln b\}$

$= 3\ln x + 2x$

$\therefore\ \dfrac{dy}{dx} = \dfrac{3}{x} + 2$

b $y = \ln\left(\dfrac{5x-4}{2x+7}\right)$

$= \ln(5x-4) - \ln(2x+7) \qquad \{\ln\left(\frac{a}{b}\right) = \ln a - \ln b\}$

$\therefore \dfrac{dy}{dx} = \dfrac{5}{5x-4} - \dfrac{2}{2x+7}$

EXERCISE 14F

1 Find $\dfrac{dy}{dx}$ for:

a $y = 3\ln x$ **b** $y = -\frac{1}{2}\ln x$ **c** $y = 2x - 7\ln x$

d $y = \ln(5x-1)$ **e** $y = \ln(2-7x)$ **f** $y = \ln(x^2+3)$

g $y = \ln(4x^2 - x)$ **h** $y = \ln(x^2 - 5x + 2)$ **i** $y = \ln(e^x)$

j $y = \ln(\sqrt{x})$ **k** $y = x^5 + \ln(4\sqrt{x})$ **l** $y = e^{5x} - \ln\left(\dfrac{1}{x}\right)$

2 Find $f'(x)$ for:

a $f(x) = x\ln x$ **b** $f(x) = 3x^2\ln x$ **c** $f(x) = e^x\ln x$

d $f(x) = \dfrac{\ln x}{x^2}$ **e** $f(x) = 7x\ln(3x+1)$ **f** $f(x) = \dfrac{\ln 5x}{x}$

g $f(x) = \dfrac{\ln(x^3)}{\sqrt{x}}$ **h** $f(x) = 6\sqrt{x}\ln(5x^2-1)$ **i** $f(x) = \dfrac{2}{\ln x}$

3 Use the laws of logarithms to help differentiate with respect to x:

a $y = \ln(x^4e^{-x})$ **b** $y = \ln\left(\dfrac{x+3}{2x-7}\right)$

c $y = \ln(e^{2x}\sqrt{x})$ **d** $y = \ln\sqrt{3x-5}$

e $f(x) = \ln\left(\dfrac{x^2+1}{4x-3}\right)$ **f** $f(x) = \ln\left(\dfrac{2x^2}{1-x}\right)$

g $f(x) = \ln\left((2x+5)^6\right)$ **h** $y = \ln(x^2\sqrt{x-7})$

i $y = \ln\left(\dfrac{5e^{3x}}{x^4}\right)$ **j** $f(x) = \ln\left(\dfrac{(3x-2)^5}{x+2}\right)$

4 Find the gradient of each illustrated tangent.

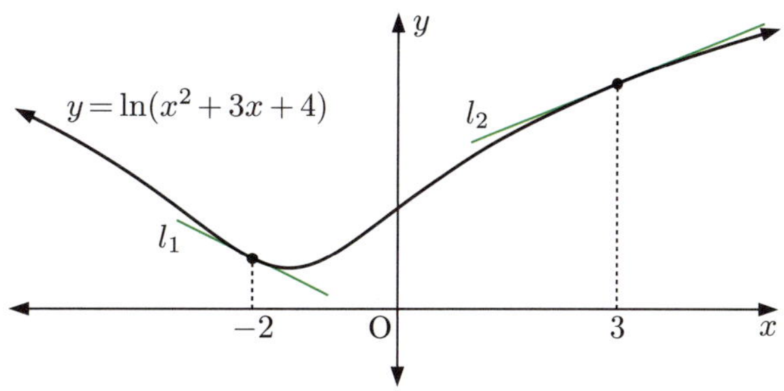

5 Find the gradient of the tangent to:

a $y = \ln(2x-1)$ at $x = 5$ **b** $y = \ln(3x + e^x)$ at $x = 0$

c $y = 2x\ln x$ at $x = 3$ **d** $y = \dfrac{\ln(5x^2)}{x}$ at $x = e$.

6 Suppose $f(x) = (5x + b)\ln x$, and $f'(1) = 7$. Find:

a the value of b

b the gradient of the tangent to $y = f(x)$ at $x = e^2$.

7 The graph of $f(x) = \dfrac{x^2}{\ln x} - 2$ is shown alongside.

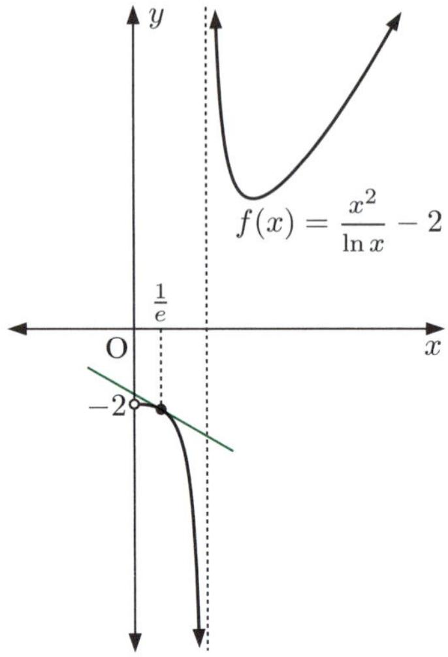

Lee differentiated the function as follows:

$$\begin{aligned} f'(x) &= \frac{2x(\ln x) - x^2\left(\frac{1}{x}\right)}{(\ln x)^2} \qquad \{\text{quotient rule}\} \\ &= \frac{2x\ln x - x}{2\ln x} \\ &= \frac{x(2\ln x - 1)}{2\ln x} \end{aligned}$$

a Use Lee's derivative function to find $f'(\frac{1}{e})$.

b Use the graph to explain why Lee's derivative function must be incorrect.

c Identify the error Lee has made.

d Calculate the correct value of $f'(\frac{1}{e})$.

8 Consider the function $f(x) = \ln\left(\dfrac{x+3}{2x-1}\right)$.

a Find the domain of the function.

b Find the gradient of the tangent to $y = f(x)$ at $x = 1$.

c Find the point(s) at which the tangent has gradient $-\frac{1}{7}$.

G DERIVATIVES OF TRIGONOMETRIC FUNCTIONS

In **Chapter 9** we saw that sine and cosine curves arise naturally from motion in a circle.

Click on the icon to observe the motion of point P around the unit circle. Observe the graphs of P's height relative to the x-axis, and then P's horizontal displacement from the y-axis. The resulting graphs are those of $y = \sin t$ and $y = \cos t$.

DEMO

Discovery 4 — *Derivatives of* $\sin x$ *and* $\cos x$

Our aim is to use a computer demonstration to investigate the derivatives of $\sin x$ and $\cos x$.

What to do:

1 Click on the icon to observe the graph of $y = \sin x$. A tangent with x-step of length 1 unit moves across the curve, and its y-step is translated onto the gradient graph. Predict the derivative of the function $y = \sin x$.

DERIVATIVES DEMO

2 Repeat the process in **1** for the graph of $y = \cos x$. Hence predict the derivative of the function $y = \cos x$.

From the **Discovery** you should have deduced that:

For x in radians: If $f(x) = \sin x$ then $f'(x) = \cos x$.
If $f(x) = \cos x$ then $f'(x) = -\sin x$.

THE DERIVATIVE OF $\tan x$

DERIVATIVE DEMO

Consider $y = \tan x = \dfrac{\sin x}{\cos x}$

Let $u = \sin x$ and $v = \cos x$

$\therefore$ $u' = \cos x$ and $v' = -\sin x$

$$\therefore \quad \frac{dy}{dx} = \frac{u'v - uv'}{v^2} \qquad \{\text{quotient rule}\}$$

$$= \frac{\cos x \cos x - \sin x(-\sin x)}{[\cos x]^2}$$

$$= \frac{\cos^2 x + \sin^2 x}{\cos^2 x}$$

$$= \frac{1}{\cos^2 x} \qquad \{\sin^2 x + \cos^2 x = 1\}$$

$$= \sec^2 x$$

Function	*Derivative*
$\sin x$	$\cos x$
$\cos x$	$-\sin x$
$\tan x$	$\sec^2 x$

THE DERIVATIVES OF $\sin[f(x)]$, $\cos[f(x)]$, AND $\tan[f(x)]$

Suppose $y = \sin[f(x)]$

Let $u = f(x)$, so $y = \sin u$.

$$\text{Now} \quad \frac{dy}{dx} = \frac{dy}{du}\frac{du}{dx} \qquad \{\text{chain rule}\}$$

$$\therefore \quad \frac{dy}{dx} = \cos u \times f'(x)$$

$$= \cos[f(x)] \times f'(x)$$

We can perform the same procedure for $\cos[f(x)]$ and $\tan[f(x)]$, giving the following results:

Function	*Derivative*
$\sin[f(x)]$	$\cos[f(x)]\,f'(x)$
$\cos[f(x)]$	$-\sin[f(x)]\,f'(x)$
$\tan[f(x)]$	$\sec^2[f(x)]\,f'(x)$

Example 12 — Self Tutor

Differentiate with respect to x: **a** $5\cos x$ **b** $4\tan 2x$ **c** $x\sin(x^2)$

a $\dfrac{d}{dx}(5\cos x) = -5\sin x$

b $\dfrac{d}{dx}(4\tan 2x) = 4\sec^2 2x \times 2$
$= 8\sec^2 2x$

c $\dfrac{d}{dx}(x\sin(x^2)) = (1)\sin(x^2) + x(\cos(x^2) \times 2x)$ {product rule}
$= \sin(x^2) + 2x^2\cos(x^2)$

EXERCISE 14G

1 Find $\frac{dy}{dx}$ for:

a $y = 2\sin x$ **b** $y = -\tan x$ **c** $y = \frac{1}{3}\cos x$

d $y = \tan 3x$ **e** $y = \sin 5x$ **f** $y = 2\cos 4x$

g $y = \sin(2x - 1)$ **h** $y = \cos(3 - x)$ **i** $y = 3\tan\frac{x}{2}$

j $y = \sin x - \cos x$ **k** $y = 3\cos x - \tan 2x$ **l** $y = \cos \pi x + 2\sin(4 - x)$

2 Find $f'(x)$ for:

a $f(x) = \cos(x^2)$ **b** $f(x) = \sin\sqrt{x}$ **c** $f(x) = \tan\frac{1}{x}$

d $f(x) = \sin(x^2 - x)$ **e** $f(x) = 4\cos(\ln x)$ **f** $f(x) = 2\tan(x^3) + \sin(x + 1)$

g $f(x) = \tan(e^x) - 3\cos 2x$ **h** $f(x) = \cos^2 x$ **i** $f(x) = \sqrt{\sin x}$

3 Differentiate with respect to x:

a $x\sin x$ **b** $e^x\cos x$ **c** $x^2\tan x$ **d** $\sin x\cos x$

e $\frac{\cos x}{x}$ **f** $\frac{x^2}{\sin x}$ **g** $\frac{\sin x}{1 + \cos x}$ **h** $\frac{\tan x}{\sqrt{x}}$

i $\cos x + \ln x \sin x$ **j** $e^{-2x}\cos 3x$ **k** $2\tan x - \frac{\sin 2x}{x}$ **l** $\frac{2 + \cos x}{5\sin x}$

4 The graph of $y = \cos 2x$ is shown alongside.

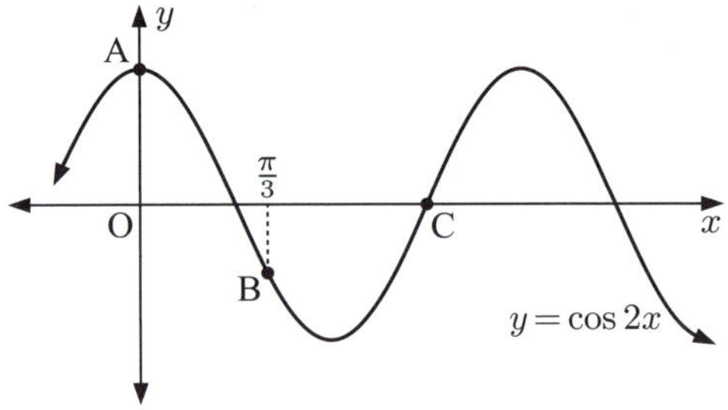

a Find the coordinates of:

i A **ii** B **iii** C

b Find $\frac{dy}{dx}$.

c Find the gradient of the tangent to $y = \cos 2x$ at:

i A **ii** B **iii** C

5 Find the gradient of the tangent to:

a $y = \sin x + 3\cos x$ at $x = 0$ **b** $y = \tan\left(x - \frac{\pi}{4}\right)$ at $x = \frac{\pi}{2}$

c $y = 2x\cos x$ at $x = \frac{\pi}{3}$ **d** $y = \frac{1}{\sin x}$ at $x = -\frac{\pi}{6}$.

H SMALL INCREMENTS AND APPROXIMATIONS

In the previous Chapter we considered a small change δy in y which occurred due to a small change δx in x.

By taking the limit of $\frac{\delta y}{\delta x}$ as $\delta x \to 0$, we defined the derivative function $\frac{dy}{dx}$.

For small increments δx and δy, we know that

$$\frac{\delta y}{\delta x} \approx \frac{dy}{dx}$$

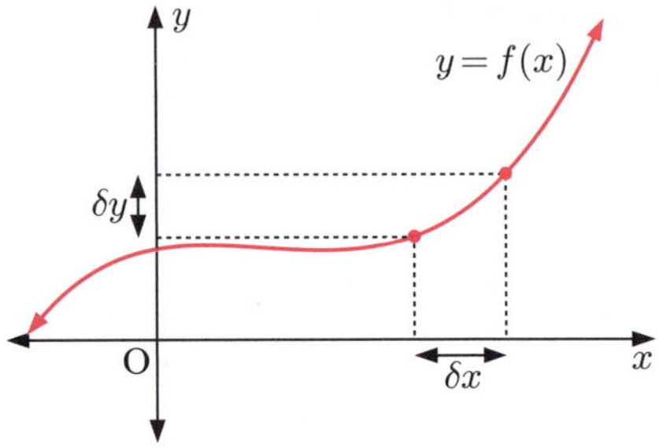

So, if we are given a small change δx or δy in one variable, we can use this small increments approximation to approximate the change in the other variable.

Example 13

Self Tutor

Suppose $y = 2x^3$.

a Find $\frac{dy}{dx}$.

b Hence approximate:

  i the change in y when x increases from 3 to 3.02

  ii the change in x when y decreases from 16 to 15.95.

a $y = 2x^3$

$\therefore \frac{dy}{dx} = 6x^2$

b **i** $\frac{\delta y}{\delta x} \approx \frac{dy}{dx}$

$\therefore \delta y \approx \delta x \times \frac{dy}{dx}$

When $x = 3$, $\frac{dy}{dx} = 6 \times 3^2 = 54$

$\therefore$ for a change $\delta x = 0.02$, $\delta y \approx 0.02 \times 54 \approx 1.08$

$\therefore$ y increases by approximately 1.08.

ii When $y = 16$, $2x^3 = 16$

$\therefore x^3 = 8$

$\therefore x = 2$

$\therefore \frac{dy}{dx} = 6 \times 2^2 = 24$

Now $\frac{\delta y}{\delta x} \approx \frac{dy}{dx}$

$\therefore$ for a change $\delta y \approx -0.05$,

$\frac{-0.05}{\delta x} \approx 24$

$\therefore \delta x \approx \frac{-0.05}{24} \approx -0.002\,08$

$\therefore$ x decreases by approximately 0.002 08.

EXERCISE 14H

1 Suppose $y = 4x^2$.

  a Find $\frac{dy}{dx}$.

  b Hence approximate the change in y when x increases from 2 to 2.01.

2 Suppose $y = \frac{12}{x}$.

a Find $\frac{dy}{dx}$.

b Hence approximate:

i the change in y when x decreases from 3 to 2.97

ii the change in x when y increases from 6 to 6.02.

3 The variables x and y are related by the equation $y = \frac{4+x}{2-x}$.

a Find $\frac{dy}{dx}$.

b Hence approximate:

i the change in y when x increases from 2.99 to 3

ii the change in x when y decreases from 1.06 to 1.05.

4 Suppose $y = e^{3x}$.

a Find $\frac{dy}{dx}$.

b Approximate the change in y when x increases from 1 to 1.01.

c Write $\frac{dy}{dx}$ in terms of y.

d Hence approximate the change in x when y decreases from 5 to 4.97.

5 Suppose $y = x^5$.

a Approximate the change in y when x increases from 2 to 2.02.

b Hence estimate the value of 2.02^5. Use your calculator to check the accuracy of your estimate.

6 The graph of $y = \sin x$ is shown alongside.

a If x decreases from $\frac{\pi}{6}$, would you expect y to increase or decrease? Explain your answer.

b Find the change in y as x decreases from $\frac{\pi}{6}$ by $\frac{\pi}{120}$. Write your answer correct to 5 significant figures.

c Hence estimate the value of $\sin \frac{19\pi}{120}$.

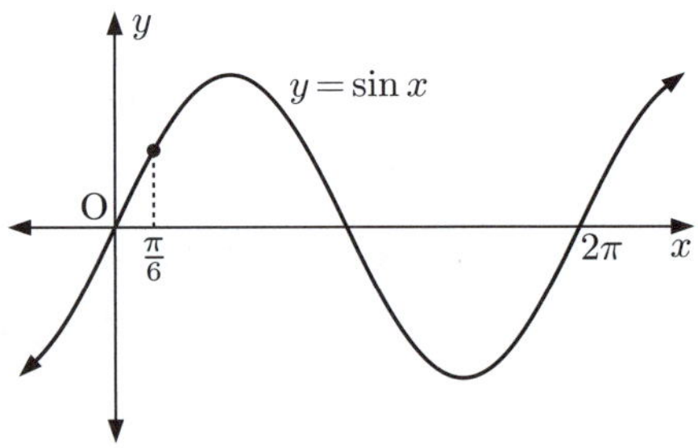

Discussion

- What is actually calculated when we use the small increments formula to estimate δy given δx?
- Why does the formula generally only provide accurate estimates of δy for small values of δx?
- Why does the formula give the *exact* values of y if the function $f(x)$ is linear?

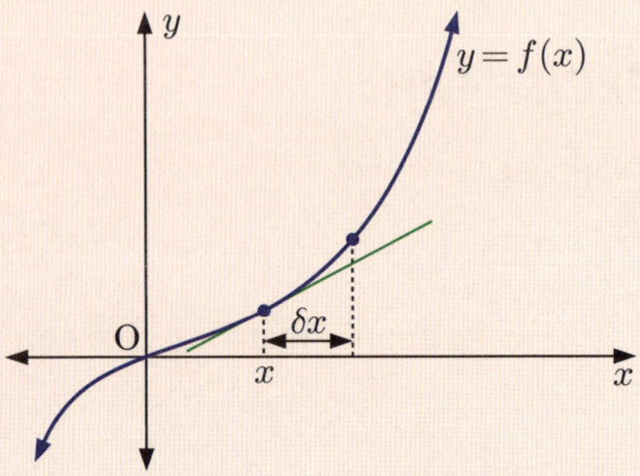

Activity — Euler's method

Euler's method, named after the Swiss mathematician **Leonhard Euler** (1707 - 1783), is a method for approximating solutions to *differential equations*.

Leonhard Euler

Given a derivative function $f'(x)$ and an initial point (x_0, y_0), Euler's method is used to approximate the value of y at other values of x.

The method works by estimating y in a series of small x-steps of size h from the initial point. We label these x values $x_1 = x_0 + h$, $x_2 = x_0 + 2h$, and so on, and the corresponding estimates of y as y_1, y_2, and so on.

Given the initial point $P_0(x_0, y_0)$, we calculate $y_1 = y_0 + hf'(x_0)$.

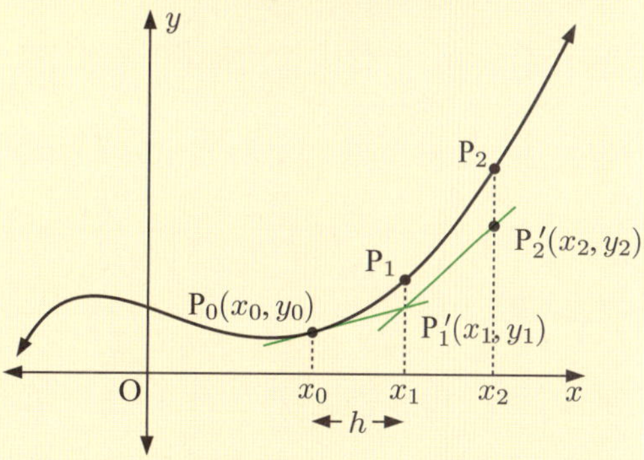

This corresponds to the point P_1' on the graph alongside. Provided h is sufficiently small, this will be a good approximation to the point P_1 on the curve.

We then use (x_1, y_1) as our starting point, and repeat the process to calculate $y_2 = y_1 + hf'(x_1)$.

More generally, we calculate $y_{n+1} = y_n + hf'(x_n)$.

What to do:

1 If $f(x) = \frac{1}{3}x^3 - \frac{1}{3}$ then $f'(x) = x^2$ and $f(1) = 0$.

a Calculate $f(1.2)$ directly.

b Use Euler's method for two steps with $h = 0.1$ to estimate $f(1.2)$.

n	x_n	y_n	$f'(x_n)$
0	1	0	
1			
2			

2 If $f(x) = -\sin x$ then $f'(x) = -\cos x$ and $f(0) = 0$.

a Calculate $f(0.5)$ directly.

b Use Euler's method for five steps with $h = 0.1$ to estimate $f(0.5)$.

3 Suppose $f'(x) = e^x + 1$ and $f(0) = 1$.

a Estimate $f(0.5)$ using Euler's method for:

i two steps with $h = 0.25$ **ii** five steps with $h = 0.1$.

b Which of your estimates do you expect to be more accurate? Explain your answer.

c Estimate $f(0.5)$ using a spreadsheet which applies Euler's method for:

i 50 steps with $h = 0.01$ **ii** 500 steps with $h = 0.001$.

I SECOND DERIVATIVES

Given a function $f(x)$, the derivative $f'(x)$ is known as the **first derivative**.

The **second derivative** of $f(x)$ is the derivative of $f'(x)$, or **the derivative of the first derivative**.

We use $f''(x)$ or y'' or $\frac{d^2y}{dx^2}$ to represent the second derivative.

$f''(x)$ reads "*f double dashed x*".

$\frac{d^2y}{dx^2} = \frac{d}{dx}\left(\frac{dy}{dx}\right)$ reads "*dee two y by dee x squared*".

Example 14 Self Tutor

Find $f''(x)$ given $f(x) = x^2 + \ln x$.

$$\begin{aligned} f(x) &= x^2 + \ln x \\ \therefore\ f'(x) &= 2x + \frac{1}{x} \\ &= 2x + x^{-1} \\ \therefore\ f''(x) &= 2 - x^{-2} \\ &= 2 - \frac{1}{x^2} \end{aligned}$$

EXERCISE 14I

1 Suppose $f(x) = 2x^4 - 5x^2$. Find:

a $f'(x)$ **b** $f''(x)$

2 A curve has equation $y = 3\ln x + x^3$. Find:

a $\frac{dy}{dx}$ **b** $\frac{d^2y}{dx^2}$

3 Find $f''(x)$ given that:

a $f(x) = 4x^2 - x$ **b** $f(x) = x^5 - \frac{2}{x}$ **c** $f(x) = 4x^3 - 3x^2 + 6x - 1$

d $f(x) = 2\sqrt{x}$ **e** $f(x) = 4\ln x - \frac{1}{\sqrt{x}}$ **f** $f(x) = (2x-3)^7$

g $f(x) = e^x$ **h** $f(x) = 2\sin x$ **i** $f(x) = e^{3x} - \cos 2x$

4 Find $\frac{d^2y}{dx^2}$ given that:

a $y = e^{2x} - x$ **b** $y = 2x^2 + \frac{1}{1-x}$ **c** $y = \ln(x^2+1)$

d $y = 3xe^x$ **e** $y = \frac{x+4}{2-x}$ **f** $y = \frac{3x-1}{x+2} - x^2$

g $y = \sin^2 x$ **h** $y = (x^2-1)^5$ **i** $y = \tan x$

5 Given $f(x) = x^3 - 5x^2 + 6x - 2$, find:

a $f(2)$ **b** $f'(2)$ **c** $f''(2)$

6 Given $f(x) = x \sin 2x$, find:

a $f(\frac{\pi}{6})$ **b** $f'(\frac{\pi}{6})$ **c** $f''(\frac{\pi}{6})$

7 Suppose $f(x) = \dfrac{x}{x^2+2}$.

a Find $f''(x)$.

b Hence find x such that $f''(x) = 0$.

8 Find $\dfrac{d^2y}{dx^2}$ given:

a $y = -\ln x$ **b** $y = x \ln x$ **c** $y = (\ln x)^2$ **d** $y = \ln[(x+2)(x-5)]$

9 Suppose $f(x) = e^x \sin x$. Find x on $0 \leqslant x \leqslant 2\pi$ such that:

a $f(x) = 0$ **b** $f'(x) = 0$ **c** $f''(x) = 0$

10 Suppose a quadratic function $f(x)$ has $f(1) = 5$, $f'(1) = 3$, and $f''(1) = 4$. Find $f(x)$.

11 If $y = 4e^{3x} + 3e^{2x}$, show that $\dfrac{d^2y}{dx^2} - 5\dfrac{dy}{dx} + 6y = 0$.

12 If $y = \cos(3x - 1)$, show that $\dfrac{d^2y}{dx^2} + 9y = 0$.

Discussion

- If $f'(x) = f(x)$, then does $f''(x) = f(x)$ also?
- Can you find a function $f(x)$ such that $f''(x) = f(x)$ but $f'(x) \neq f(x)$?

Review set 14A

1 Find $f'(x)$ if:

a $f(x) = 3x^4 - 2x^2$ **b** $f(x) = \dfrac{2}{x} - \dfrac{5}{x^3}$ **c** $f(x) = 8\sqrt{x} - \dfrac{3}{\sqrt{x}}$

2 Find $\dfrac{dy}{dx}$ if:

a $y = (5x-2)^6$ **b** $y = \dfrac{4}{(6-x)^2}$ **c** $y = \sqrt{x^2 - 3x}$

3 Find the gradient of the tangent to:

a $f(x) = x\sqrt{x+4}$ at $x = 5$ **b** $f(x) = \dfrac{2x-1}{x+1}$ at $x = -3$.

4 Find $f'(x)$ for:

a $f(x) = e^{3x-1}$ **b** $f(x) = \ln(4 - 5x)$ **c** $f(x) = \sin(x^2 - x)$

5 Find the gradient of the illustrated tangent:

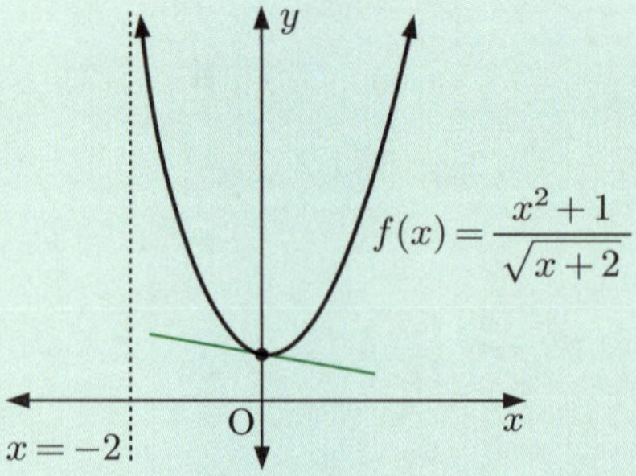

6 Find:

a $\dfrac{d}{dx}(\sqrt{x}\sin x)$ **b** $\dfrac{d}{dx}\left(\dfrac{\ln x}{e^x}\right)$

7 The function $f(x)=\dfrac{x+a}{2x+b}$ has $f(-2)=-2$ and $f'(-2)=-5$. Find a and b.

8 In the graph of $y = x^3+kx^2-5x$ alongside, the illustrated tangents are parallel. Find:

a the value of k

b the gradients of the tangents.

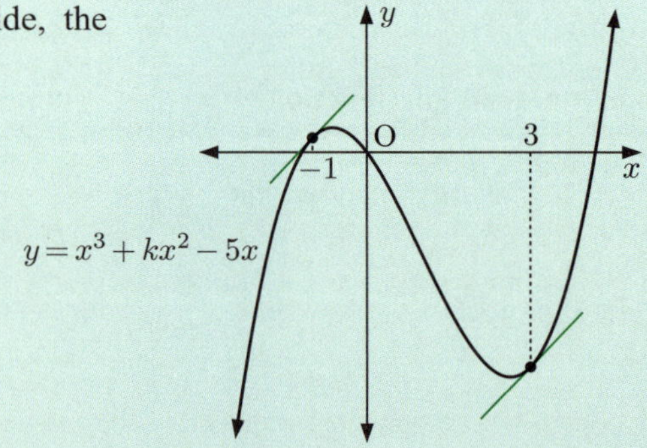

9 Use the laws of logarithms to help differentiate with respect to x:

a $y=\ln\left(\dfrac{1-2x}{x^2+3}\right)$ **b** $y=\ln(x^3\cos x)$

10 Consider the function $f(x)=\dfrac{e^x}{\cos x}$.

a Find $f'(x)$.

b Find $f'(0)$ and interpret your answer.

c Find the points on the interval $0\leqslant x\leqslant 2\pi$ at which the tangent to $y=f(x)$ is horizontal.

11 Suppose $y=x^4$.

a Find $\dfrac{dy}{dx}$.

b Approximate the change in y when x increases from 3 to 3.02.

c Hence estimate the value of 3.02^4.

12 Find $f''(x)$ given that:

a $f(x)=7x^2-\dfrac{2}{x}$ **b** $f(x)=e^{\sin x}$

Review set 14B

1 Find $\frac{dy}{dx}$ if:

a $y = 3x^5 - 4x^2 - 7$ **b** $y = (x-4)^2$ **c** $y = \frac{5x^2+1}{x}$

2 Find:

a $\frac{dy}{dt}$ if $y = \sqrt[3]{t}$ **b** $\frac{dM}{dx}$ if $M = (7-3x)^5$

3 $f(x) = x^3 + ax^2 + 9x + b$ is divisible by $x - 2$, and $f'(x)$ is divisible by $x - 3$. Find a and b.

4 Find the gradient of the tangent to:

a $f(x) = x^2 - \frac{1}{x^2}$ at $x = -1$ **b** $f(x) = x\ln(x+1)$ at $x = 3$.

5 Find the gradients of the illustrated tangents.

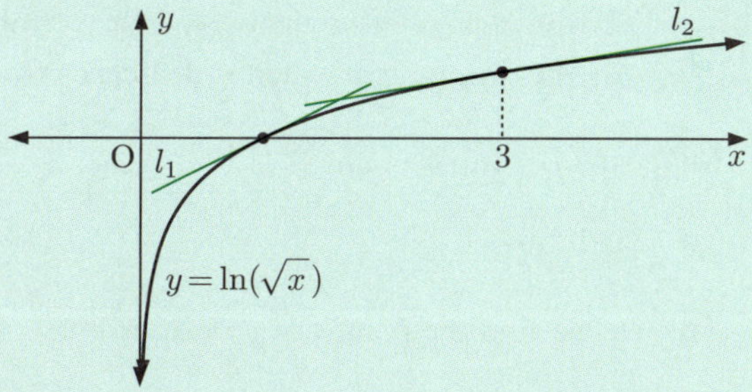

6 Suppose $f(x) = e^{2x}\sqrt{6x+1}$.

a Write $f'(x)$ in the form $f'(x) = \frac{e^{2x}(ax+b)}{\sqrt{6x+1}}$ where $a, b \in \mathbb{Z}$.

b Find the gradient of the tangent to $y = f(x)$ at $x = \frac{1}{2}$.

7 Differentiate with respect to x:

a $\frac{2x^2+1}{x^2-1}$ **b** $\sin 4x - x\cos x$ **c** $5\ln(6x^2-1)$ **d** $\tan^2 x$

8 The graph of $y = xe^{-x}$ is shown alongside.
The illustrated tangents are perpendicular. Find the value of k.

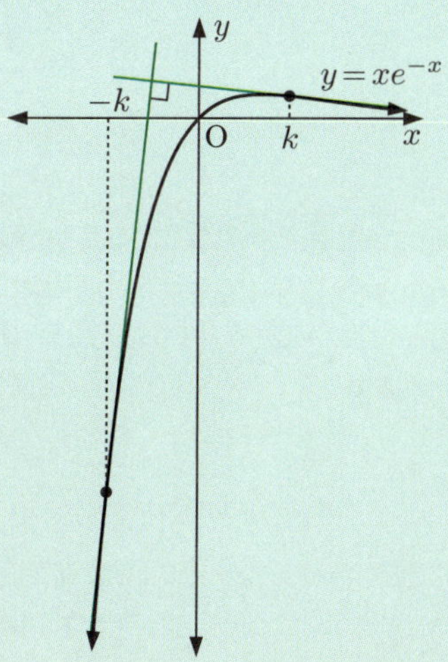

9 Suppose $f(x) = \ln\left(\frac{x}{x-2}\right)$.

a Find $f'(x)$.

b Find the gradient of the tangent to $y = f(x)$ at $x = 6$.

c Find the points at which the tangent has gradient $-\frac{1}{4}$.

10 Consider the function $f(x) = \tan x - \ln(\cos x)$.

a Show that $f'(x) = \tan^2 x + \tan x + 1$.

b Hence find the point on $y = f(x)$ on the interval $0 \leqslant x \leqslant \frac{\pi}{2}$ for which $f'(x) = 3$.

11 Suppose $y = \frac{3+2x}{1-x}$.

a Find $\frac{dy}{dx}$.

b Hence approximate:

i the change in y when x increases from 2 to 2.03

ii the change in x when y decreases from -3 to -3.01.

12 Suppose $y = \ln(e^x - 2)$.

a Find $\frac{d^2y}{dx^2}$.

b Hence find the point on $y = \ln(e^x - 2)$ such that $\frac{d^2y}{dx^2} = -2$.

15

Applications of differential calculus

Contents:

Opening problem

The effectiveness of an injection t minutes after it is administered is given by $E(t) = 5te^{-\frac{t}{4}}$ units. The graph of $E(t)$ is shown alongside.

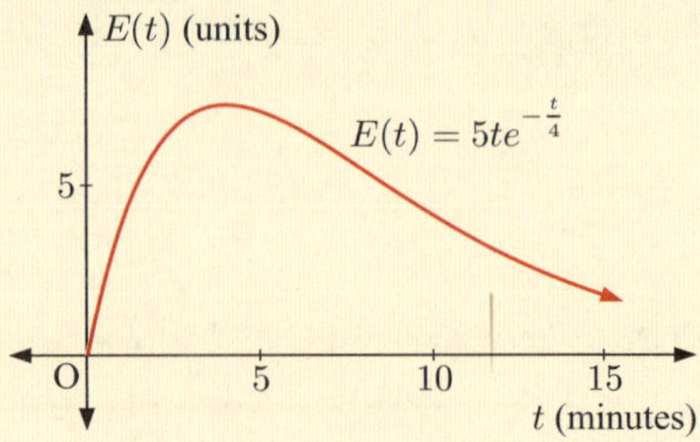

Things to think about:

a What does the sign of $E'(t)$ tell us about how the injection's effectiveness is changing?

b How can we identify the time when the injection is most effective?

We have seen how to differentiate many types of functions, so we now consider their applications.

In this Chapter we will use derivatives to find:

- tangents and normals to curves
- turning points and stationary inflections.

We will then look at applying these techniques to real world problems including:

- kinematics or motion problems
- rates of change
- optimisation of different quantities.

A TANGENTS AND NORMALS

TANGENTS

We have seen that a **tangent** to a curve is a straight line which *touches* the curve.

Suppose $A(a, f(a))$ lies on the curve $y = f(x)$.

The tangent to $y = f(x)$ at A has gradient $f'(a)$.

$\therefore$ the equation of this tangent is
$y - f(a) = f'(a)(x - a)$ {gradient-point form}

or $y = f(a) + f'(a)(x - a)$.

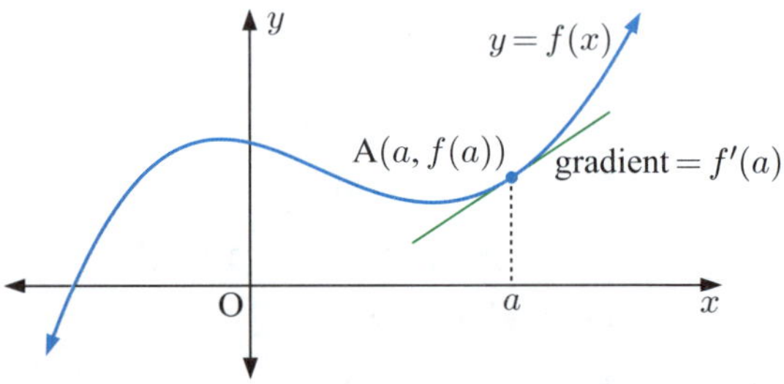

Example 1

Self Tutor

Find the equation of the tangent to $f(x) = 4x - x^2$ at the point where $x = 1$.

$f(1) = 4(1) - 1^2 = 3$

$\therefore$ the point of contact is $(1, 3)$.

Now $f'(x) = 4 - 2x$

$\therefore$ the gradient of the tangent at $x = 1$ is $f'(1) = 4 - 2(1) = 2$

$\therefore$ the tangent has equation $y = 3 + 2(x - 1)$
which is $y = 2x + 1$.

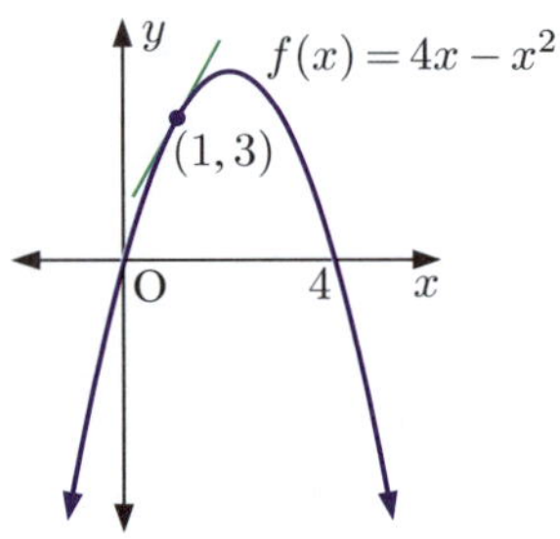

NORMALS

A **normal** to a curve is a line which is perpendicular to the tangent at the point of contact.

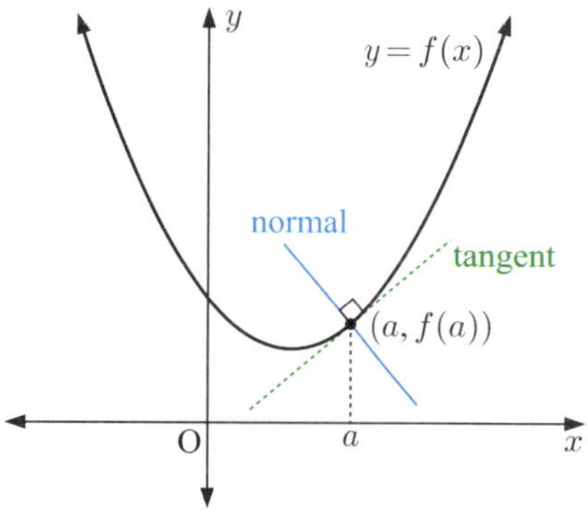

The gradients of perpendicular lines are negative reciprocals of each other, so:

The gradient of the normal to the curve at $x = a$ is $-\dfrac{1}{f'(a)}$.

We can use this gradient and the point of contact to find the equation of the normal.

Example 2

Self Tutor

Find the equation of the normal to $y = \dfrac{10}{x}$ at the point where $x = 2$.

When $x = 2$, $y = \frac{10}{2} = 5$. So, the point of contact is $(2, 5)$.

Now $y = 10x^{-1}$, $\dfrac{dy}{dx} = -10x^{-2} = -\dfrac{10}{x^2}$

When $x = 2$, $\dfrac{dy}{dx} = -\dfrac{10}{2^2} = -\frac{5}{2}$

$\therefore$ the normal at $(2, 5)$ has gradient $\frac{2}{5}$.

$\therefore$ the equation of the normal is

$2x - 5y = 2(2) - 5(5)$

or $2x - 5y = -21$

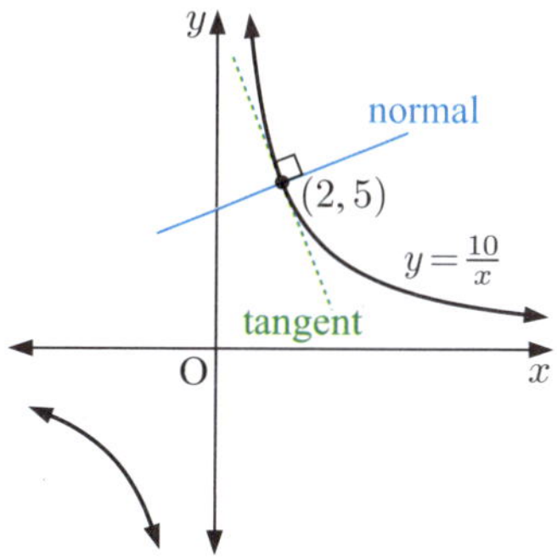

A line with gradient $\dfrac{A}{B}$ passing through (x_0, y_0) has equation $Ax - By = Ax_0 - By_0$.

EXERCISE 15A

1 Find the equation of the tangent to:

a $y = x^2$ at $x = 3$

b $y = 2x^2 - x$ at $x = 1$

c $y = x^3 - 2x + 5$ at $x = -2$

d $y = \dfrac{6}{x}$ at $x = 3$

e $y = 4\sqrt{x}$ at $x = 9$

f $y = 5x - \dfrac{3}{x^2}$ at $x = -1$.

2 The graph of $f(x) = x^3 - 2\sqrt{x} - 3$ is shown alongside.

a Find the equation of the illustrated tangent.

b Find the coordinates of P and Q.

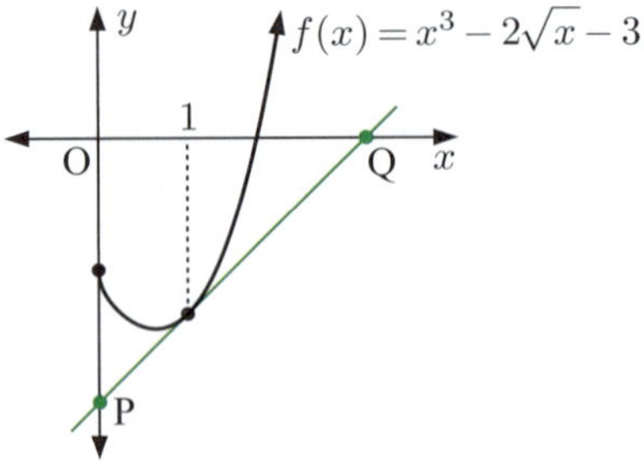

3 The tangent to $y = \frac{4}{x} - \frac{2}{x^2}$ at the point where $x = 2$ cuts the x-axis at A and the y-axis at B. Find the area of triangle OAB, where O is the origin.

4 Find the equation of the normal to:

a $y = x^2 + x$ at the point $(2, 6)$

b $y = 4x^2 + 5x + 2$ at $x = -1$

c $y = \frac{4}{x}$ at the point $(4, 1)$

d $y = \sqrt{x} - 3x$ at $x = 1$

e $y = x^3 - 5x^2 + 6x - 2$ at the point $(0, -2)$

f $y = 5x\sqrt{x}$ at $x = 4$.

5

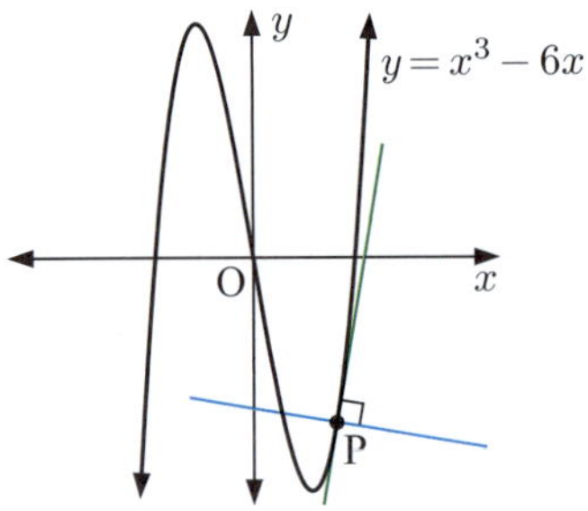

In the graph of $y = x^3 - 6x$ alongside, the point P has x-coordinate 2.

Find:

a the y-coordinate of P

b the equation of the tangent to the curve at P

c the equation of the normal to the curve at P.

6 **a** Find the equation of the normal to $f(x) = -\frac{1}{\sqrt{x}} + \frac{x}{8} + 2$ at the point $(4, 2)$.

b Find the points at which the normal in **a** cuts the axes.

7 **a** Find the point at which the tangent to $y = 3x - \frac{2}{x^2}$ has gradient 4.

b Find the equation of the tangent at this point.

8 The normal to the curve $y = 2x^2 + bx + c$ at the point where $x = 2$ has equation $x + 5y = 17$. Find the values of b and c.

9 Find the equation of the tangent to:

a $y = \sqrt{3x + 1}$ at $(1, 2)$

b $y = \frac{1}{(5 - 2x)^2}$ at $x = 3$

c $y = \frac{x + 1}{2 - x}$ at $x = 0$

d $y = \frac{\sqrt{x}}{x + 4}$ at $(1, \frac{1}{5})$.

10 Find the equation of the normal to:

a $y = \sqrt{6x + 4}$ at $x = 0$

b $y = x^2\sqrt{x + 1}$ at $x = 3$

c $f(x) = \frac{x}{x^2 - 1}$ at $(2, \frac{2}{3})$

d $f(x) = \frac{2x + 1}{x - 1}$ at $(4, 3)$.

11 The tangent to the curve $y = \sqrt{ax + b}$ at the point where $x = -1$ has equation $y = 3x + 4$. Find the values of a and b.

12 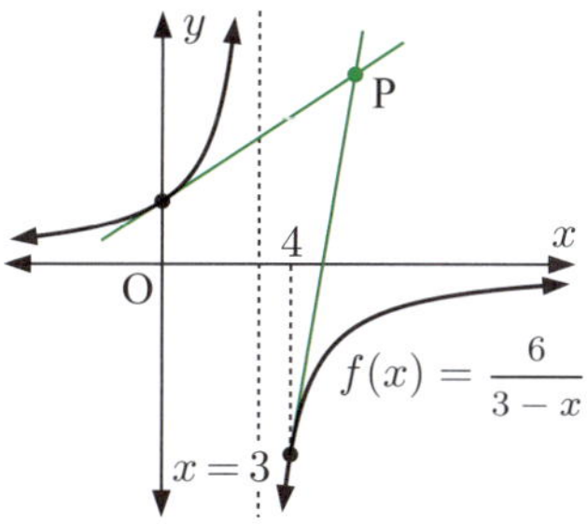

The graph of $f(x) = \frac{6}{3-x}$ is shown alongside.

Find the coordinates of P, the intersection of the tangents to $f(x)$ at $x = 0$ and $x = 4$.

13 Consider the function $y = \sqrt{9 - x^2}$.

a Find the domain of the function.

b Show that *any* normal to this function passes through the origin.

c Explain the result in **b** geometrically.

Hint: If $y = \sqrt{9 - x^2}$ then $x^2 + y^2 = 9$.

Example 3 — Self Tutor

Find the equation of:

a the tangent to $f(x) = \ln(2x^2 + 1)$ at the point where $x = 1$

b the normal to $f(x) = \sin 2x$ at the point where $x = \frac{\pi}{6}$.

a $f(1) = \ln(2(1)^2 + 1) = \ln 3$

$\therefore$ the point of contact is $(1, \ln 3)$.

Now if $f(x) = \ln(2x^2 + 1)$ then $f'(x) = \frac{4x}{2x^2 + 1}$

$\therefore$ the tangent at $(1, \ln 3)$ has gradient $f'(1) = \frac{4}{3}$.

$\therefore$ the tangent has equation $4x - 3y = 4(1) - 3(\ln 3)$

which is $4x - 3y = 4 - 3\ln 3$.

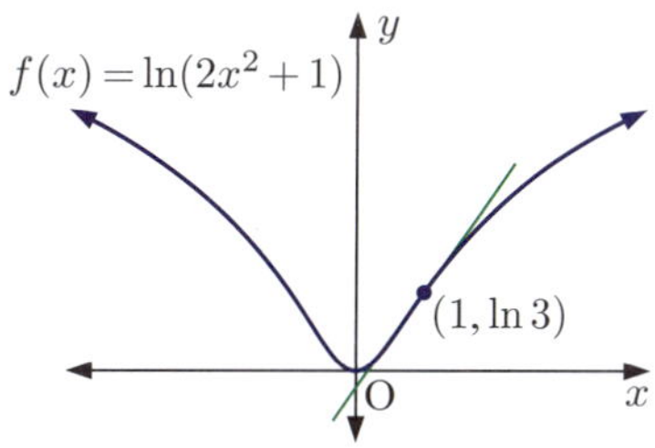

b $f\left(\frac{\pi}{6}\right) = \sin \frac{\pi}{3} = \frac{\sqrt{3}}{2}$

$\therefore$ the point of contact is $\left(\frac{\pi}{6}, \frac{\sqrt{3}}{2}\right)$.

Now if $f(x) = \sin 2x$ then $f'(x) = 2\cos 2x$

$\therefore f'\left(\frac{\pi}{6}\right) = 2\cos \frac{\pi}{3} = 1$

$\therefore$ the normal at $\left(\frac{\pi}{6}, \frac{\sqrt{3}}{2}\right)$ has gradient -1.

$\therefore$ the normal has equation $x + y = \frac{\pi}{6} + \frac{\sqrt{3}}{2}$.

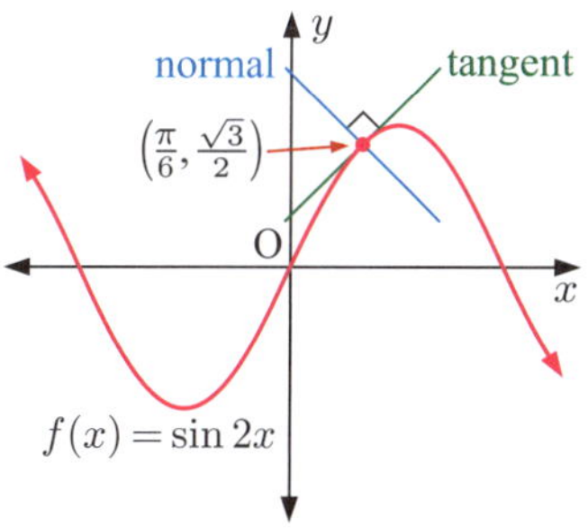

14 Find the equation of the tangent to:

a $y = \ln x$ at $(1, 0)$

b $y = e^{2x}$ at $x = \ln 2$

c $y = \ln(x^2 - 3x)$ at $x = 4$

d $y = xe^{-x}$ at $x = 2$

e $y = \sin\left(x + \frac{\pi}{2}\right)$ at $\left(\frac{\pi}{3}, \frac{1}{2}\right)$

f $y = 2\cos 3x$ at $x = \frac{\pi}{6}$

g $y = \tan x - 2$ at $x = \frac{\pi}{4}$

h $y = \frac{x + 1}{\cos x + 1}$ at $x = 0$.

15 Find the equation of the normal to:

a $f(x) = e^{x-3}$ at $x = 4$

b $f(x) = \ln(4 - x)$ at $(3, 0)$

c $f(x) = \sin x - \cos 2x$ at $x = \frac{\pi}{6}$

d $f(x) = \dfrac{\ln(x^2 + 2)}{x + 1}$ at $x = 0$.

16 The graph of $y = \ln(5 - 2x)$ is shown alongside. Find the area of triangle PAB.

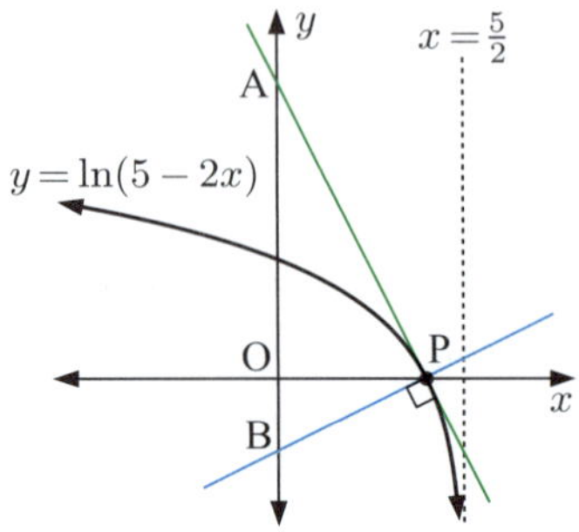

17 The tangent to $f(x) = ae^{bx} \cos x$ at the point where $x = 0$ has equation $6x - y = -2$. Find the values of a and b.

Example 4

Self Tutor

Find where the tangent to $y = x^3 - x^2 - 8x - 4$ at $(-1, 2)$ meets the curve again.

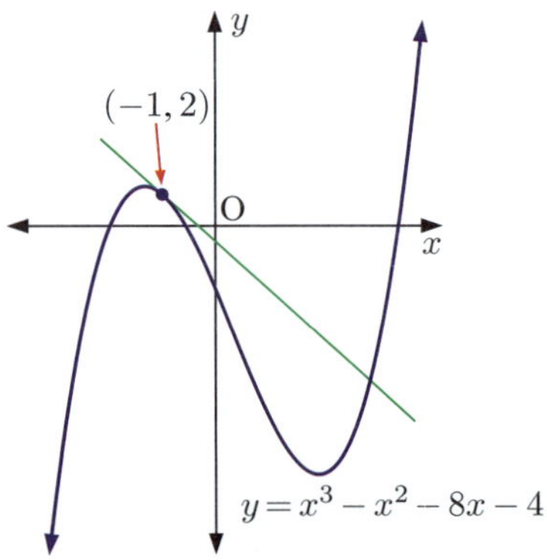

Let $f(x) = x^3 - x^2 - 8x - 4$

$\therefore\ f'(x) = 3x^2 - 2x - 8$

$\therefore\ f'(-1) = 3(-1)^2 - 2(-1) - 8 = -3$

$\therefore$ the equation of the tangent at $(-1, 2)$ is

$3x + y = 3(-1) + 2$

or $y = -3x - 1$

The curve meets the tangent when

$x^3 - x^2 - 8x - 4 = -3x - 1$

$\therefore\ x^3 - x^2 - 5x - 3 = 0$

$\therefore\ (x + 1)^2(x - 3) = 0$

{the tangent *touches* the curve at $x = -1$, so $(x + 1)^2$ must be a factor}

$\therefore$ the tangent meets the curve again when $x = 3$.

Now $f(3) = (3)^3 - (3)^2 - 8(3) - 4 = -10$

$\therefore$ the tangent meets the curve again at $(3, -10)$.

18 **a** Find the equation of the tangent to $y = x^3 - 5x + 1$ at $(1, -3)$.

b Find where this tangent meets the curve again.

19 Find where the tangent to $y = x^3 + 2x^2 - x$ at $(-2, 2)$ meets the curve again.

20 **a** Find the equation of the normal to $y = x^2 - 5x + 1$ at $(3, -5)$.

b Find where this normal meets the curve again.

21 The graph of $f(x) = \frac{1}{x^2}$ is shown alongside. The tangent at P meets the curve again at Q.

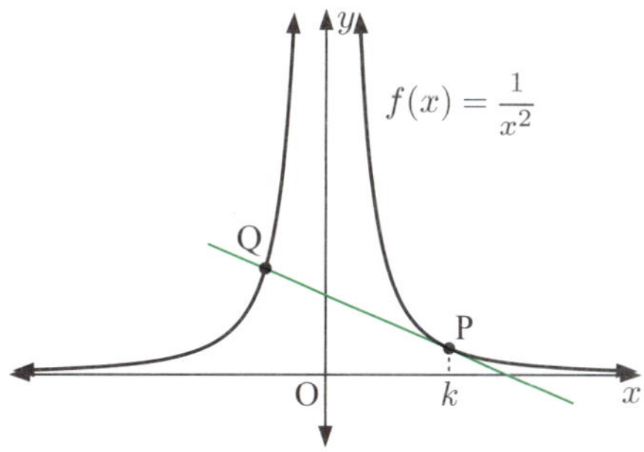

a Find, in terms of k:

i the equation of the illustrated tangent

ii the x-coordinate of Q.

b Find the exact value of k such that the illustrated tangent is the *normal* to the curve at Q.

22 Find, correct to 2 decimal places, the angle between the tangents to $y = 3e^{-x}$ and $y = 2 + e^x$ at their point of intersection.

23 A cubic has three real roots. Prove that the tangent line at the average of any two roots of the cubic, passes through the third root.

Hint: Let $f(x) = a(x - \alpha)(x - \beta)(x - \gamma)$.

Discussion

- Does a tangent to a quadratic ever meet the curve again?
- Does a normal to a quadratic always meet the curve again?
- Does a tangent to a cubic always meet the curve again?
- Does a normal to a cubic always meet the curve again?

B INCREASING AND DECREASING FUNCTIONS

The concepts of increasing and decreasing functions are closely linked to **intervals** or subsets of a function's domain.

Suppose S is an interval in the domain of $f(x)$, so $f(x)$ is defined for all x in S.

- $f(x)$ is **increasing** on S $\Leftrightarrow$ $f(a) \leqslant f(b)$ for all $a, b \in S$ such that $a < b$.
- $f(x)$ is **decreasing** on S $\Leftrightarrow$ $f(a) \geqslant f(b)$ for all $a, b \in S$ such that $a < b$.

We can determine intervals where a curve $y = f(x)$ is increasing or decreasing by considering the derivative function $f'(x)$.

For most functions that we deal with in this course:

- $f(x)$ is **increasing** on S $\Leftrightarrow$ $f'(x) \geqslant 0$ for all x in S
- $f(x)$ is **decreasing** on S $\Leftrightarrow$ $f'(x) \leqslant 0$ for all x in S.

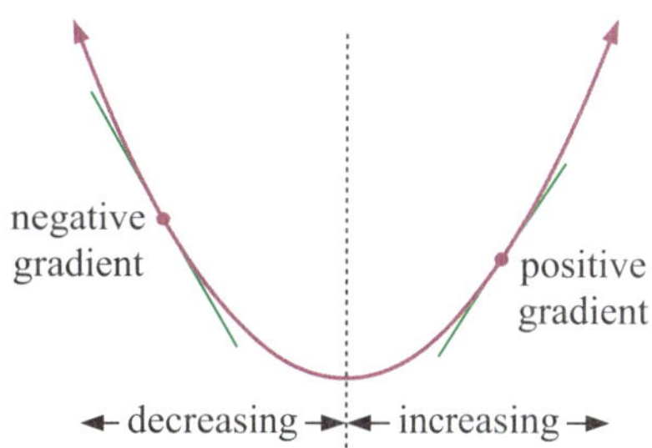

SIGN DIAGRAMS

In **Chapter 2** we drew sign diagrams to show intervals where a function $f(x)$ lies above or below the x-axis.

Drawing a sign diagram of the *derivative function* $f'(x)$ allows us to show intervals where $f(x)$ is increasing or decreasing.

Example 5

Self Tutor

Find the intervals where the following functions are increasing or decreasing:

a $f(x) = x^3 - 3x^2 - 9x$ **b** $f(x) = \dfrac{1}{x} + \dfrac{2}{x^2}$

a
$$\begin{aligned} f(x) &= x^3 - 3x^2 - 9x \\ \therefore\ f'(x) &= 3x^2 - 6x - 9 \\ &= 3(x^2 - 2x - 3) \\ &= 3(x+1)(x-3) \end{aligned}$$

which has sign diagram:

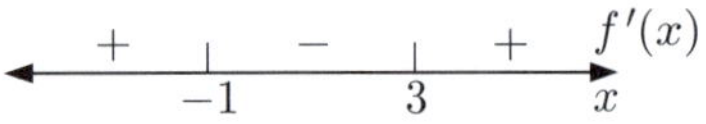

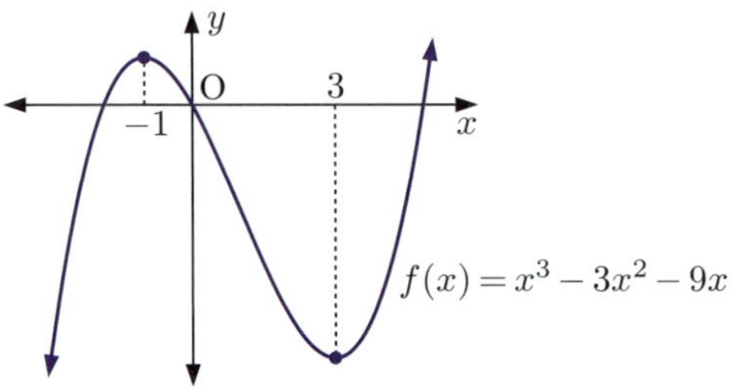

So, $f(x)$ is increasing for $x \leqslant -1$ and $x \geqslant 3$, and decreasing for $-1 \leqslant x \leqslant 3$.

b
$$\begin{aligned} f(x) &= \frac{1}{x} + \frac{2}{x^2} = x^{-1} + 2x^{-2} \\ \therefore\ f'(x) &= -x^{-2} - 4x^{-3} \\ &= -\frac{1}{x^2} - \frac{4}{x^3} \\ &= -\frac{x+4}{x^3} \end{aligned}$$

which has sign diagram:

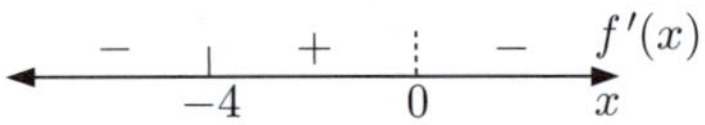

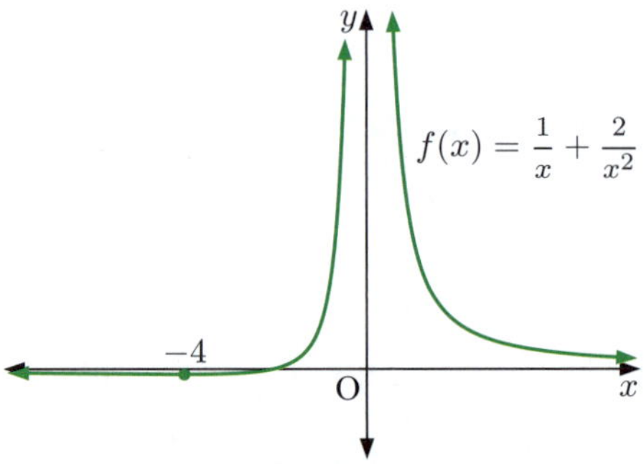

So, $f(x)$ is increasing for $-4 \leqslant x < 0$, and decreasing for $x \leqslant -4$ and $x > 0$.

Remember that $f(x)$ must be defined for all x on an interval before we can classify the function as increasing or decreasing on that interval. We need to take care with vertical asymptotes and other values for x where the function is not defined.

EXERCISE 15B

1 The graph of $f(x) = -x^3 + 3x^2 + 24x - 10$ is shown alongside.

  a Find $f'(x)$, and draw its sign diagram.

  b Find the intervals where $f(x)$ is increasing or decreasing.

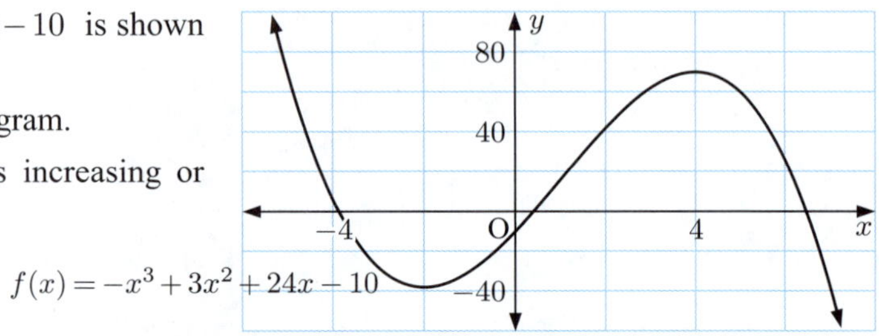

2 Find intervals where $f(x)$ is increasing or decreasing:

a $f(x) = x^2 + 2x + 5$ **b** $f(x) = x^3 - 3x$ **c** $f(x) = -x^2 + 5x - 4$

d $f(x) = \frac{6}{x}$ **e** $f(x) = \frac{1}{x} - \frac{3}{x^2}$ **f** $f(x) = 4\sqrt{x}$

g $f(x) = x + \frac{5}{x}$ **h** $f(x) = \ln(x^2 + 3)$ **i** $f(x) = xe^x$ **j** $f(x) = \frac{x+2}{x-3}$

3 Consider the function $f(x) = -x^3 + 2x^2 - 3x - 7$.

a Find $f'(x)$.

b Show that $f'(x) < 0$ for all x, and explain the significance of this result.

C STATIONARY POINTS

A **stationary point** of a function is a point where $f'(x) = 0$.
The tangent to the function at any stationary point is horizontal.

A stationary point could be a **maximum turning point**, a **minimum turning point**, or a **stationary inflection point**.

For example, the graph of $y = f(x)$ alongside has three stationary points where the tangent to the graph is horizontal.

The graph is:

- increasing before the maximum turning point and decreasing after it
- decreasing before the minimum turning point and increasing after it
- decreasing either side of the stationary inflection point.

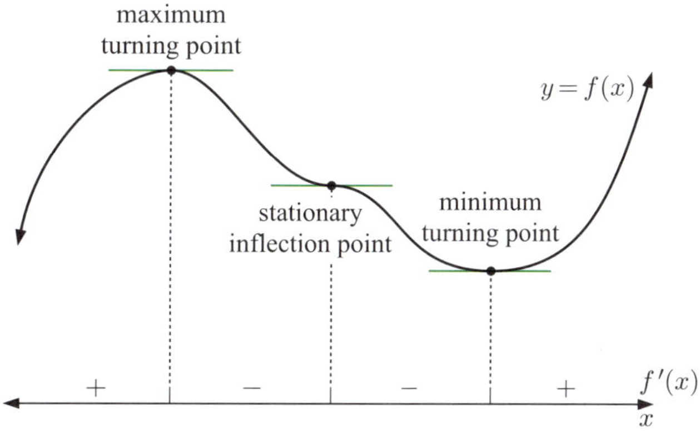

In the sign diagram of $f'(x)$, we see that the sign changes at a turning point, and it is unchanged at a stationary inflection.

Stationary point where $f'(a) = 0$	*Sign diagram of* $f'(x)$ *near* $x = a$	*Shape of curve near* $x = a$
maximum turning point	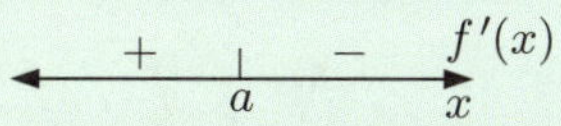	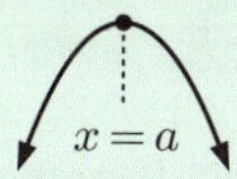
minimum turning point	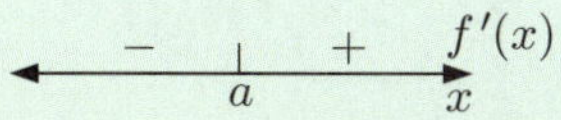	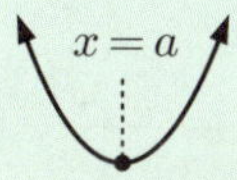
stationary inflection	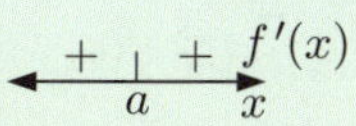 or 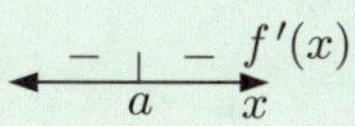	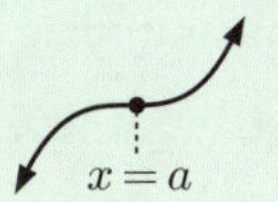 or 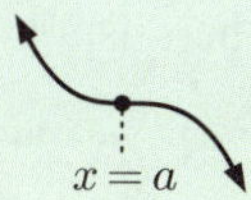

Example 6 — Self Tutor

Consider the function $f(x) = x^3 + 6x^2 + 12x + 5$.

a Find the y-intercept.

b Find and classify any stationary points.

c Hence sketch the graph of $y = f(x)$.

a $f(0) = 5$, so the y-intercept is 5.

b
$$f(x) = x^3 + 6x^2 + 12x + 5$$
$$\therefore\ f'(x) = 3x^2 + 12x + 12$$
$$= 3(x^2 + 4x + 4)$$
$$= 3(x+2)^2 \quad \text{which has sign diagram:}$$

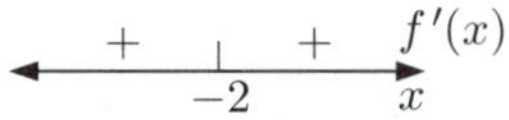

$f(-2) = (-2)^3 + 6(-2)^2 + 12(-2) + 5 = -3$

$\therefore$ there is a stationary inflection point at $(-2, -3)$.

c

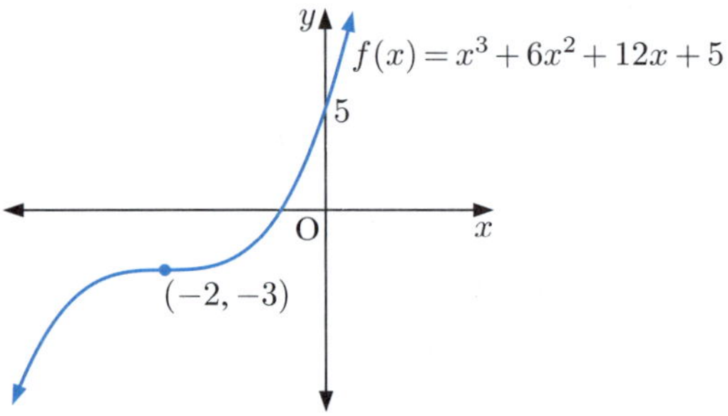

Example 7 — Self Tutor

Find and classify all stationary points of $f(x) = \dfrac{x^2+1}{x}$.

$$f(x) = \frac{x^2+1}{x}$$
$$\therefore\ f'(x) = \frac{2x(x) - (x^2+1)}{x^2} \quad \{\text{quotient rule}\}$$
$$= \frac{x^2-1}{x^2}$$
$$= \frac{(x+1)(x-1)}{x^2}$$

$f'(x)$ has sign diagram:

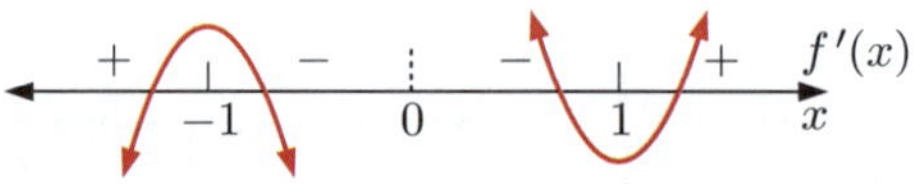

Now $f(-1) = \dfrac{(-1)^2+1}{(-1)} = -2$ and $f(1) = \dfrac{1^2+1}{1} = 2$

$\therefore$ there is a maximum turning point at $(-1, -2)$ and a minimum turning point at $(1, 2)$.

THE SECOND DERIVATIVE TEST

The second derivative of a function can be used to determine the nature of its stationary points.

Suppose a function $f(x)$ has a stationary point at $x = a$.

- If $f''(a) > 0$, then it is a **minimum turning point**.
- If $f''(a) < 0$, then it is a **maximum turning point**.
- If $f''(a) = 0$, then it could be a **maximum turning point**, a **minimum turning point**, or a **stationary inflection point**.

Example 8

Self Tutor

Find and classify all stationary points of $f(x) = 2x^3 + 3x^2 - 12$.

$$f(x) = 2x^3 + 3x^2 - 12$$
$$\therefore\ f'(x) = 6x^2 + 6x$$
$$= 6x(x+1)$$
$$\therefore\ f'(x) = 0 \text{ when } x = 0 \text{ or } x = -1$$

Now $f(0) = 2(0)^3 + 3(0)^2 - 12 = -12$
$f(-1) = 2(-1)^3 + 3(-1)^2 - 12 = -11$

Also, $f''(x) = 12x + 6$
$\therefore\ f''(0) = 12(0) + 6 = 6$ which is > 0
and $f''(-1) = 12(-1) + 6 = -6$ which is < 0

So, there is a minimum turning point at $(0, -12)$ and a maximum turning point at $(-1, -11)$.

EXERCISE 15C

1 Consider the graph of $y = f(x)$ alongside.

a Classify the stationary points P, Q, and R.

b Draw a sign diagram for:

i $f(x)$ **ii** $f'(x)$.

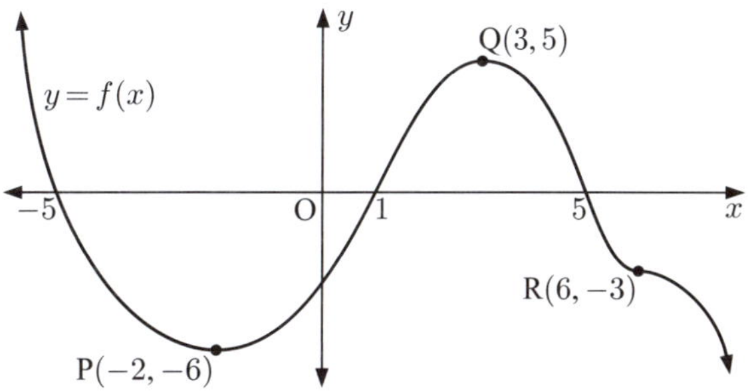

2

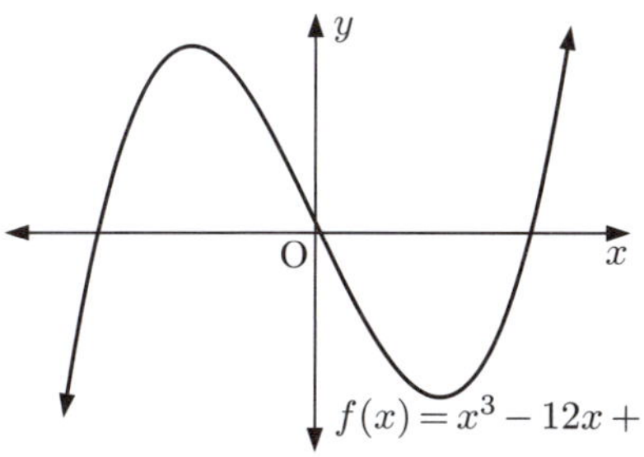

The graph of $f(x) = x^3 - 12x + 1$ is shown alongside.

a Find $f'(x)$, and draw its sign diagram.

b Find the intervals where $f(x)$ is increasing or decreasing.

c Find and classify the stationary points.

3 For each of the following functions, find and classify any stationary points. Sketch the function, showing all important features.

a $f(x) = x^2 + 3x$
b $f(x) = -x^3 - 2$
c $f(x) = x^3 - 9x^2 + 24x - 11$
d $f(x) = x^4 - 8x^2 + 3$
e $f(x) = x - 2\sqrt{x}$
f $f(x) = -\frac{2}{x} + \frac{1}{x^2}$
g $f(x) = x^4 + 8x^3 + 18x^2 - 15$
h $f(x) = x^2 + \frac{16}{x}$

GRAPHING PACKAGE

4 $f(x) = x^3 + ax^2 - 3x + 2$ has a turning point at $x = -3$.

a Find a.
b Find $f''(x)$.
c Use the second derivative test to determine whether the turning point at $x = -3$ is a maximum or minimum turning point.

5 $f(x) = x^4 + ax + b$ has a stationary point at $(1, -2)$.

a Find a and b.
b Show that the function has no other stationary points.
c Determine the nature of the stationary point.

Example 9

Self Tutor

Find the exact position and nature of the stationary points of $y = x^2e^x$.

$y = x^2e^x$

$\therefore \frac{dy}{dx} = 2xe^x + x^2e^x$ {product rule}

$= xe^x(2 + x)$

$\frac{dy}{dx}$ has sign diagram:

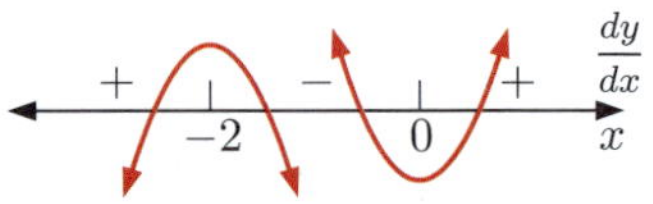

When $x = -2$, $y = (-2)^2e^{-2} = \frac{4}{e^2}$

When $x = 0$, $y = 0$

$\therefore$ there is a maximum turning point at $(-2, \frac{4}{e^2})$, and a minimum turning point at $(0, 0)$.

6 The graph of $y = \frac{x+2}{e^x}$ is shown alongside.

C is a stationary point.

Find the exact coordinates of:

a A **b** B **c** C

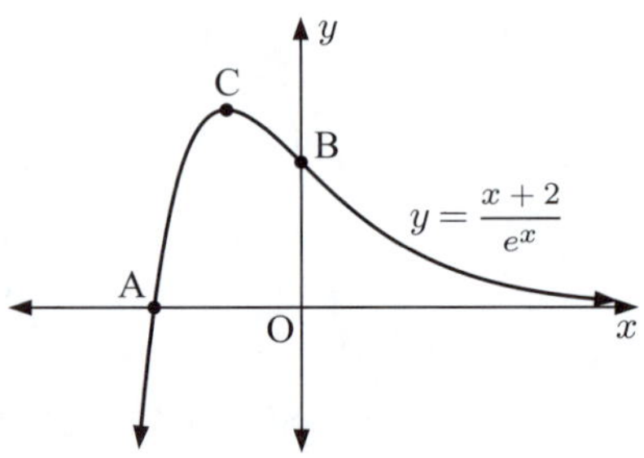

7 Find the exact position and nature of the stationary point(s) of:

a $y = 3xe^{2x}$ **b** $y = \dfrac{x}{x^2+1}$ **c** $y = \ln(x^2 + 4x + 6)$

d $y = x^2 \ln x$ **e** $y = \dfrac{e^x}{x^2 + 6x + 10}$ **f** $y = (x-1)\sqrt{15 - x^2}$

8 Consider the function $y = \dfrac{3x-1}{x+2} - 7x$.

a Find $\dfrac{dy}{dx}$ and $\dfrac{d^2y}{dx^2}$.

b Locate and describe the stationary points of the curve.

9 The function $f(x) = e^x(x^2 + ax + b)$ has a stationary point at $(-3, 0)$.

a Find a and b.

b Find the coordinates of the other stationary point.

10 The graph of $f(x) = 2\sin\left(x + \frac{\pi}{6}\right)$ on $0 \leqslant x \leqslant 2\pi$ is shown alongside.

a Find $f'(x)$.

b Find the coordinates of the stationary points A and B.

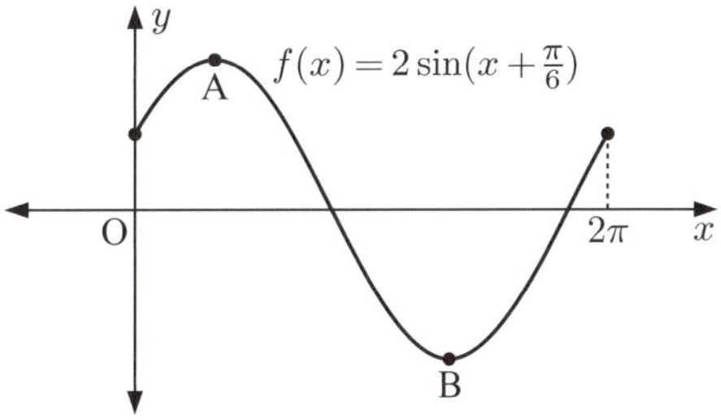

11 For each of the following, determine the position and nature of the stationary points on the interval $0 \leqslant x \leqslant 2\pi$, then display them on a graph of the function.

GRAPHING PACKAGE

a $f(x) = 3\cos x$ **b** $f(x) = \sin 2x$ **c** $f(x) = \cos\left(x - \frac{\pi}{4}\right)$

d $f(x) = \sin x + \cos x$ **e** $f(x) = \cos^2 x$ **f** $f(x) = \sin x + \sqrt{3}\cos x$

12 Show that the graph of $y = \tan x$ does not have any stationary points.

13 Find the position and nature of the stationary points of each curve on the interval $-\pi \leqslant x \leqslant \pi$:

a $y = x + 2\sin x$ **b** $y = \dfrac{1}{2 - \sin x}$ **c** $y = \dfrac{e^x}{\cos x}$

14 Consider the function $f(x) = \cos^2 x + \sin x - 2$.

a Find $f'(x)$.

b Identify the stationary points of $f(x)$ on the interval $0 \leqslant x \leqslant 2\pi$.

c Find $f''(x)$.

d Use the second derivative test to determine the nature of the stationary points in **b**.

15 The function $f(x) = e^{3x} + ae^{2x} + be^x + c$ has a stationary inflection point at $(\ln 2, 3)$. Find a, b, and c.

D KINEMATICS

Kinematics is the study of motion.

In this Section we will consider the motion of objects moving in a straight line. Some examples include the motion of a toy train moving back and forth along a straight piece of track, or an elevator moving up and down.

DISPLACEMENT

The **displacement** of an object is its *position* relative to an origin O. It is a vector quantity, whose sign indicates the direction of the object from O.

In the diagram alongside, P has displacement 3 units, and Q has displacement -2 units.

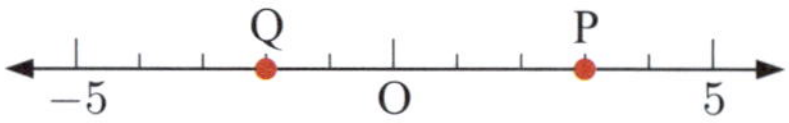

We can use a **displacement function** $s(t)$ to describe the displacement s of an object at any time $t \geqslant 0$.

For example, suppose an object P moves with displacement function $s(t) = t^3 - 5t - 5$ cm, where t is in seconds. We can find the position of P after 0, 1, 2, and 3 seconds:

$$s(0) = -5, \quad s(1) = -9, \quad s(2) = -7, \quad s(3) = 7$$

To appreciate the motion of P we draw a **motion graph**. You can also view the motion by clicking on the icon.

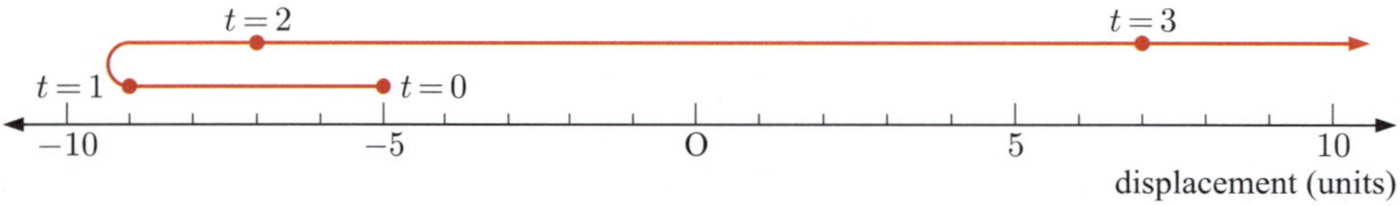

DEMO

VELOCITY

The **velocity** of an object is its rate of change of displacement. It is a vector quantity whose sign indicates the direction of motion.

The **average velocity** of an object moving in a straight line in the time interval from $t = t_1$ to $t = t_2$ is the ratio of the change in displacement to the time taken.

If $s(t)$ is the displacement function then **average velocity** $= \dfrac{s(t_2) - s(t_1)}{t_2 - t_1}$.

In **Chapter 13** we established that the instantaneous rate of change of a quantity is given by its derivative.

If $s(t)$ is the displacement function of an object moving in a straight line, then the **instantaneous velocity** or **velocity function** of the object at time t is $v(t) = s'(t)$.

ACCELERATION

The **acceleration** of an object is its rate of change of velocity.

If an object moves in a straight line with displacement function $s(t)$ and velocity function $v(t)$, then:

- the **average acceleration** for the time interval from $t = t_1$ to $t = t_2$ is given by

$$\textbf{average acceleration} = \frac{v(t_2) - v(t_1)}{t_2 - t_1}$$

- The **instantaneous acceleration** or **acceleration function** of the object is $a(t) = v'(t) = s''(t)$.

For example, if an object moves with displacement function $s(t) = t^3 - 5t - 5$ then its velocity function $v(t) = s'(t) = 3t^2 - 5$, and its acceleration function $a(t) = v'(t) = 6t$.

UNITS

Each time we differentiate with respect to time t, we calculate a rate per unit of time. So, if displacement is measured in metres and time in seconds, then:

- the units of velocity are m s^{-1}
- the units of acceleration are m s^{-2}.

Discussion

- What is the relationship between the displacement function $s(t)$ and the acceleration function $a(t)$?
- How are the units of velocity and acceleration related to their formulae? You may wish to research "dimensional analysis".

Example 10

Self Tutor

A particle moves in a straight line with displacement function $s(t) = -2t^3 + 3t^2 + 4$ metres, where $t \geqslant 0$, t in seconds.

a Find the displacement of the particle after 1 second.

b Find the particle's velocity and acceleration functions.

c Find the velocity and acceleration of the particle after 2 seconds.

a $s(1) = -2(1)^3 + 3(1)^2 + 4 = 5$ m

After 1 second, the particle has displacement 5 m.

b $v(t) = s'(t) = -6t^2 + 6t \text{ m s}^{-1}$

$a(t) = v'(t) = -12t + 6 \text{ m s}^{-2}$

c $v(2) = -6(2)^2 + 6(2) = -12 \text{ m s}^{-1}$

$a(2) = -12(2) + 6 = -18 \text{ m s}^{-2}$

After 2 seconds, the particle has velocity -12 m s^{-1} and acceleration -18 m s^{-2}.

SPEED

We have seen that velocities have size (magnitude) and sign (direction). In contrast, *speed* simply measures *how fast* something is travelling, regardless of the direction of travel. Speed is a *scalar* quantity which has size but no sign. Speed cannot be negative.

> The **speed** at any instant is the magnitude of the object's velocity.
> If an object has velocity v then its speed is $|v|$.

In the previous Example, after 2 seconds the particle has velocity $-12\ \text{m s}^{-1}$, so its speed is $12\ \text{m s}^{-1}$.

EXERCISE 15D.1

1 An object moves in a straight line with displacement $s(t) = t^2 - 3t + 5$ metres at time $t \geqslant 0$ seconds.

- **a** Find the displacement of the object after:
  - **i** 1 second
  - **ii** 3 seconds.
- **b** Find the particle's velocity and acceleration functions.
- **c** Find the velocity and acceleration of the particle after 4 seconds.

2 A particle moves in a straight line with displacement $s(t) = t^3 - 2t^2 - 10t$ centimetres at time $t \geqslant 0$ seconds.

- **a** Find $v(t)$ and $a(t)$.
- **b** Find the acceleration of the particle after 2 seconds.
- **c** At what time does the particle have velocity $5\ \text{cm s}^{-1}$?
- **d** Find the speed of the particle after 1 second.

3 As a bus leaves the station, its displacement after t seconds is $s(t) = \frac{1}{2}t^2 + \sqrt{t}$ metres.

- **a** How far has the bus travelled after 4 seconds?
- **b** Find the velocity of the bus after 5 seconds.
- **c** At what time does the bus have acceleration $0.75\ \text{m s}^{-2}$?

4 A particle P travels in a straight line such that its velocity after t seconds is $v = (e^{\frac{t}{3}} - 2)^3\ \text{m s}^{-1}$, $t \geqslant 0$.

- **a** Find the initial speed of P.
- **b** Find the exact time at which P has velocity $8\ \text{m s}^{-1}$.
- **c** Find the acceleration of P after 3 seconds.

5 A bus must perform a series of braking and accelerating manoeuvres as part of a safety test. The velocity of the bus t seconds after starting the test is $v(t) = 10 - 6\cos\left(\frac{\pi}{6}t\right)\ \text{m s}^{-1}$, $t \geqslant 0$.

- **a** Find the initial speed of the bus.
- **b** At what time does the speed of the bus first reach $13\ \text{m s}^{-1}$?
- **c** Find the acceleration of the bus after:
  - **i** 2 seconds
  - **ii** 7 seconds.

Activity 1 Displacement, velocity, and acceleration graphs

In this Activity we examine the motion of a projectile which is fired in a vertical direction. The projectile is affected by gravity, which is responsible for the projectile's constant acceleration.

MOTION DEMO

We then extend the Activity to consider other cases of motion in a straight line.

What to do:

1. Click on the icon to examine vertical projectile motion.
Observe first the displacement along the line, then look at the velocity which is the rate of change in displacement. When is the velocity positive and when is it negative?
2. Examine the following graphs and comment on their shapes:
   - *displacement* v *time*
   - *velocity* v *time*
   - *acceleration* v *time*
3. Pick from the menu or construct functions of your own choosing to investigate the relationship between displacement, velocity, and acceleration.

SIGN INTERPRETATION

By constructing **sign diagrams**, we can determine the times at which an object's displacement, velocity, and acceleration are positive, negative, or zero.

The tables below describe how we interpret the signs of $s(t)$, $v(t)$, and $a(t)$:

SIGNS OF $s(t)$:

$s(t)$	*Interpretation*
$=0$	P is at O
>0	P is located to the right of O
<0	P is located to the left of O

SIGNS OF $v(t)$:

$v(t)$	*Interpretation*
$=0$	P is instantaneously at rest
>0	P is moving to the right
<0	P is moving to the left

SIGNS OF $a(t)$:

$a(t)$	*Interpretation*
>0	velocity is increasing
<0	velocity is decreasing
$=0$	velocity may be a maximum or minimum or possibly constant

Example 11

Self Tutor

A particle moves in a straight line with position relative to O given by $s(t) = t^3 - t^2 - 8t + 3$ m, where t is in seconds, $t \geqslant 0$.

a Find expressions for the particle's velocity and acceleration, and draw a sign diagram for each of them.

b Find the initial conditions and hence describe the motion at this instant.

c Find the time at which the particle changes direction, and the position of the particle at this time.

d Draw a motion diagram for the particle.

e Find the total distance travelled in the first 3 seconds.

a $s(t) = t^3 - t^2 - 8t + 3$ m

$\therefore\ v(t) = 3t^2 - 2t - 8$

$= (3t+4)(t-2)\ \text{m s}^{-1}$

which has sign diagram:

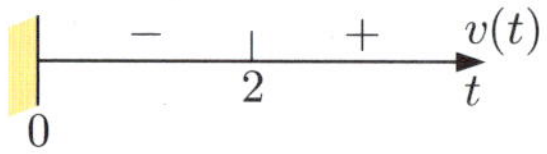

Since $t \geqslant 0$, the stationary point at $t = -\frac{4}{3}$ is not required.

and $a(t) = 6t - 2$

$= 2(3t-1)\ \text{m s}^{-2}$

which has sign diagram:

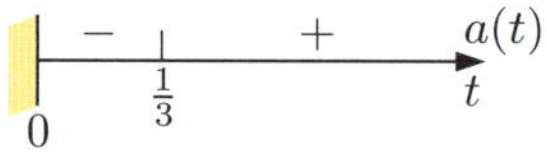

b $s(0) = 3$ m

$v(0) = -8\ \text{m s}^{-1}$

$a(0) = -2\ \text{m s}^{-2}$

$\therefore$ the particle is initially 3 m to the right of O, moving to the left with speed 8 m s^{-1}. It is accelerating at 2 m s^{-2} to the left, so its speed is increasing.

c Since $v(t)$ changes sign when $t = 2$, the particle changes direction at this instant.

$s(2) = (2)^3 - (2)^2 - 8(2) + 3 = -9$, so the particle changes direction when it is 9 m to the left of O.

The particle changes direction when its velocity changes sign.

d

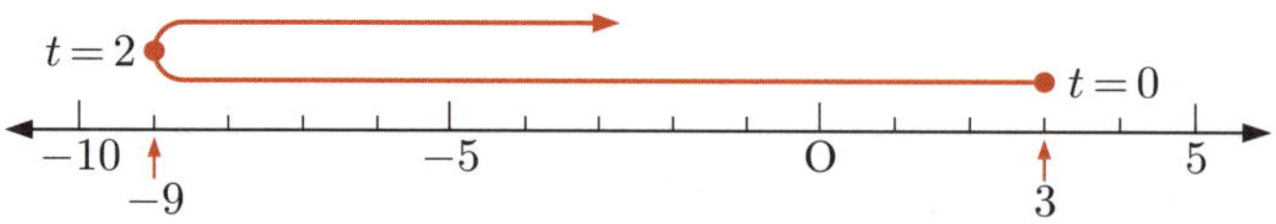

e $s(3) = (3)^3 - (3)^2 - 8(3) + 3 = -3$

In the first 3 seconds, the particle travels 12 m to the left, then 6 m to the right.

$\therefore$ total distance travelled $= 12 + 6 = 18$ m.

EXERCISE 15D.2

1 An object moves with displacement $s(t) = t^2 - 6t + 4$ cm from O, where t is in seconds, $t \geqslant 0$.

a Find expressions for the object's velocity and acceleration, and draw a sign diagram for each function.

b Find the initial conditions and hence describe the motion at this instant.

c Find the time at which the object changes direction, and the position of the object at this time.

d Draw a motion diagram for the object.

e Find the displacement of the object after 5 seconds.

f Hence find the total distance travelled in the first 5 seconds.

2 A particle moves with displacement $s(t) = t^3 - 5t^2 + 7t + 2$ m from O, where t is in seconds, $t \geqslant 0$.

a Find the velocity and acceleration functions for the particle, and draw a sign diagram for each function.

b Describe the initial motion of the particle.

c Find the time intervals in which:

i the particle is moving to the left
ii the particle is moving to the right
iii the particle's velocity is increasing
iv the particle's velocity is decreasing.

d Find the acceleration of the particle after 3 seconds.

e Find the particle's position when it first changes direction.

f Find the total distance travelled in the first 2 seconds.

3 A cricket ball is thrown vertically into the air. Its height above the ground after t seconds is $s(t) = -4.9t^2 + 19.6t + 1.5$ metres.

a Find the velocity and acceleration functions for the ball, and draw their sign diagrams.

b Find the initial position and velocity of the ball.

c Find the time at which the ball is momentarily at rest.

d Find the maximum height reached by the ball.

e Find the total distance travelled by the ball in the first 3 seconds.

4 In an experiment, an object is fired vertically from the Earth's surface. From the results, a two-dimensional graph of the position $s(t)$ metres above the Earth's surface is plotted, where t is the time in seconds. It is noted that the graph is *parabolic*, so $s(t) = at^2 + bt + c$ for some constants a, b, c.
Assuming that gravitational acceleration is the constant g and that the initial velocity is $v(0)$, show that:

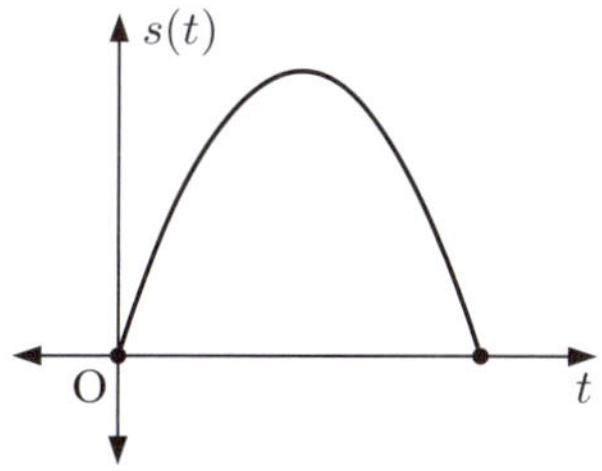

a $s(t) = \frac{1}{2}gt^2 + v(0) \times t$
b $v(t) = gt + v(0)$

5 A particle moves with velocity $v(t) = \ln(t^2 - 5t + 7)$ m s^{-1}, where t is in seconds, $t \geqslant 0$.

a Find the initial velocity of the particle.

b Find the acceleration of the particle after 1 second.

c Find the times at which the particle changes direction.

d When is the particle moving to the left?

e Find the times when the particle has acceleration 1 m s^{-2}.

6 A top fuel racing car sprints along a straight section of track.

Its speed after t seconds is $v(t) = 70te^{-\frac{t}{5}}$ m s^{-1}.

a Find the velocity of the car after:

i 1 second **ii** 2 seconds.

b Find the acceleration function $a(t)$.

c How long do you think the race lasted? Explain your answer.

7 An object moves with displacement $s(t) = 5\sin 3t$ cm from O, where t is in seconds, $t \geqslant 0$.

a Find $v(t)$.

b Draw a sign diagram for $s(t)$ and $v(t)$ on the interval $0 \leqslant t \leqslant \pi$.

c Find the time intervals on $0 \leqslant t \leqslant \pi$ in which:

i the object is to the right of O **ii** the object is to the left of O

iii the object is moving to the right **iv** the object is moving to the left.

d Find the speed of the object when $t = \frac{2\pi}{9}$.

e Find the acceleration of the object when it first changes direction.

8 Kerry is flying her new drone. The height of the drone above ground level t seconds after taking off, is $h(t) = 0.5t + \sin t$ m.

a Find the height of the drone above ground level after 5 seconds.

b Find the speed at which the drone leaves the ground.

c At what times in the first 10 seconds does the drone change direction?

d Find the first time that the drone has acceleration $0.5\ \text{m s}^{-2}$.

e Find, correct to 3 significant figures, the distance travelled by the drone in the first 5 seconds.

E RATES OF CHANGE

There are countless examples in the real world where quantities vary with time, or with respect to some other variable.

For example:
- the volume of water in a lake varies continuously
- the height of a person varies as the person grows up
- the weight of a spherical orange increases as its radius increases.

We have already seen that if $y = f(x)$ then $f'(x)$ or $\frac{dy}{dx}$ is the gradient of the tangent to $y = f(x)$ at the given point.

$\frac{dy}{dx}$ gives the **rate of change in y with respect to x.**

We can therefore use the derivative of a function to tell us the **rate** at which something is happening.

For example:

- $\frac{dT}{dt}$ or $T'(t)$ could be the instantaneous rate of change in temperature over time, with units °C per minute.
- $\frac{dV}{dt}$ or $V'(t)$ could be a car's instantaneous rate of change in value, with units dollars per year.

Example 12

Self Tutor

The temperature of a bowl of soup t minutes after it is served, is given by $T(t) = 20 + 50e^{-\frac{t}{10}}$ °C.

a Find the rate at which the temperature of the soup is changing after:

i 5 minutes **ii** 10 minutes.

b Show that $T'(t) < 0$ for all $t \geqslant 0$. Comment on the significance of this result.

a $T(t) = 20 + 50e^{-\frac{t}{10}}$ $\quad \therefore \quad T'(t) = 50\left(-\frac{1}{10}\right) e^{-\frac{t}{10}}$
$= -5e^{-\frac{t}{10}}$

i $T'(5) = -5e^{-0.5}$
≈ -3.03

After 5 minutes, the temperature is decreasing at ≈ 3.03°C per minute.

ii $T'(10) = -5e^{-1}$
≈ -1.84

After 10 minutes, the temperature is decreasing at ≈ 1.84°C per minute.

b Since $e^{-\frac{t}{10}} > 0$ for all t, $T'(t) = -5e^{-\frac{t}{10}} < 0$ for all $t \geqslant 0$.
This means that the temperature of the soup is always decreasing.

EXERCISE 15E

1 The height of a giraffe t years after birth is $H(t) = 6 - \frac{4}{t+1}$ metres.

a Find the height of the giraffe:

i at birth **ii** after 1 year.

b Find $H'(t)$.

c Find the rate at which the giraffe is growing after:

i 1 year **ii** 3 years.

d Show that $H'(t) > 0$ for all $t \geqslant 0$. Comment on the significance of your answer.

2 The value of a house t years after it is built, is $V = 500\,000e^{\frac{t}{20}}$ dollars.

a Find the value of the house after:

i 2 years **ii** 5 years.

b Find $\frac{dV}{dt}$.

c Find the rate at which the value of the house is changing after:

i 5 years **ii** 10 years.

d Show that:

i $\frac{dV}{dt} > 0$ for all $t \geqslant 0$ **ii** $\frac{d^2V}{dt^2} > 0$ for all $t \geqslant 0$.

Comment on the significance of each result.

3 Adam is camping with his friends. At a distance x metres from the campfire, the temperature is $T = \dfrac{1000}{x^2+2}$ °C.

a Find the temperature 4 metres from the campfire.

b Find $\dfrac{dT}{dx}$.

c Find the rate at which the temperature changes at a distance of:

i 2 metres **ii** 5 metres from the campfire.

d Show that $\dfrac{dT}{dx} < 0$ for all $x \geqslant 0$. Comment on the significance of this result.

4 If a car is travelling at v m s^{-1} before the brakes are applied, it will come to rest in a distance $D = \frac{1}{8}v^2$ metres.

a Find the stopping distance for the car if it is travelling at:

i 10 m s^{-1} **ii** 16 m s^{-1}

b Find the rate at which the stopping distance is changing at a speed of:

i 8 m s^{-1} **ii** 20 m s^{-1}

c Show that:

i $\dfrac{dD}{dv} > 0$ for all $v > 0$ **ii** $\dfrac{d^2D}{dv^2} > 0$ for all $v > 0$.

Explain the significance of each result.

5 Studies have shown that the average head circumference of a baby t days after birth can be modelled by $C = 10\ln(0.2t + 20)$ cm.

a Find the average head circumference of a baby:

i 5 days after birth **ii** 8 weeks after birth.

b Find the rate at which a baby's head circumference is changing after:

i 20 days **ii** 100 days.

c Show that:

i $\dfrac{dC}{dt} > 0$ for all $t \geqslant 0$ **ii** $\dfrac{d^2C}{dt^2} < 0$ for all $t \geqslant 0$.

Explain the significance of each result.

6 Shelley records the price of petrol at a service station each day. The price t days after she starts recording is $120 + 15\sin\left(\frac{\pi}{3.5}t\right)$ cents per litre.

a Find the price of petrol after:

i 3 days **ii** 15 days.

b Find the rate at which the petrol price is changing after:

i 6 days **ii** 10 days.

7 The population of a parrot species in a rainforest t years after it is introduced, is given by $P = \frac{7500}{1+10e^{-1.3t}}$, $t \geqslant 0$.

a Find $\frac{dP}{dt}$.

b Find the rate at which the population is growing after: **i** 6 months **ii** 3 years.

c Explain why $\frac{dP}{dt} > 0$ for all t.

d Describe what happens to $\frac{dP}{dt}$ as $t \to \infty$. Discuss the significance of your answer.

Example 13 — Self Tutor

Find the rate of change in the area of triangle ABC as θ changes, at the time when $\theta = 30°$.

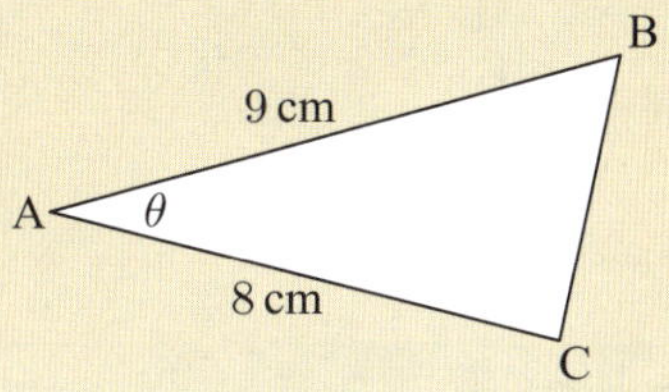

Area $A = \frac{1}{2} \times 8 \times 9 \times \sin\theta$ $\{\text{Area} = \frac{1}{2}bc\sin A\}$

$\therefore A = 36\sin\theta$ cm^2

$\therefore \frac{dA}{d\theta} = 36\cos\theta$

When $\theta = \frac{\pi}{6}$, $\cos\theta = \frac{\sqrt{3}}{2}$

$\therefore \frac{dA}{d\theta} = 18\sqrt{3}$ cm^2 per radian

8 A circle of radius r m has area A m^2.

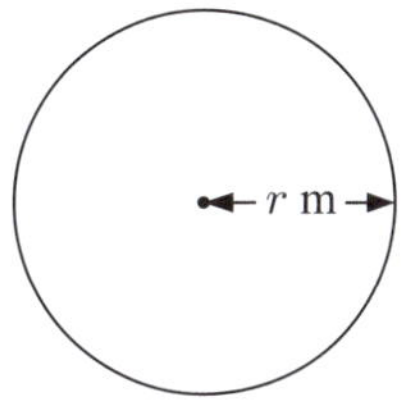

a Write A in terms of r.

b Find $\frac{dA}{dr}$.

c Find the rate of change in the area of the circle with respect to the radius, when $r = 5$.

9

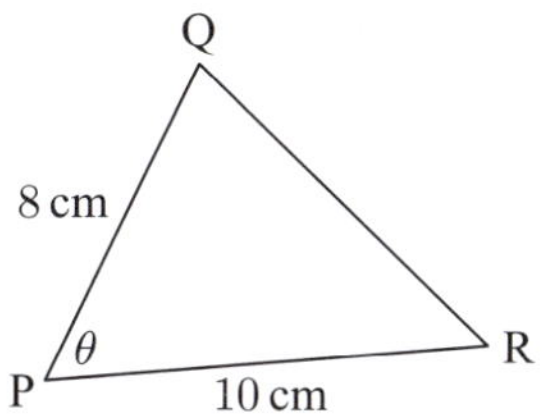

Find the rate of change in the area of triangle PQR as θ changes, at the time when $\theta = 60°$.

10 A regular hexagon has sides of length x cm. Find the rate at which the area of the hexagon is changing as x changes, when $x = 12$.

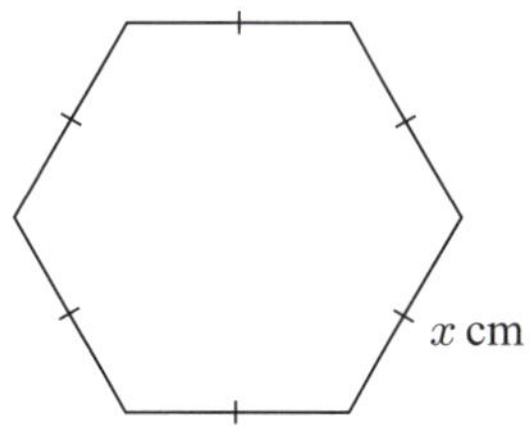

11

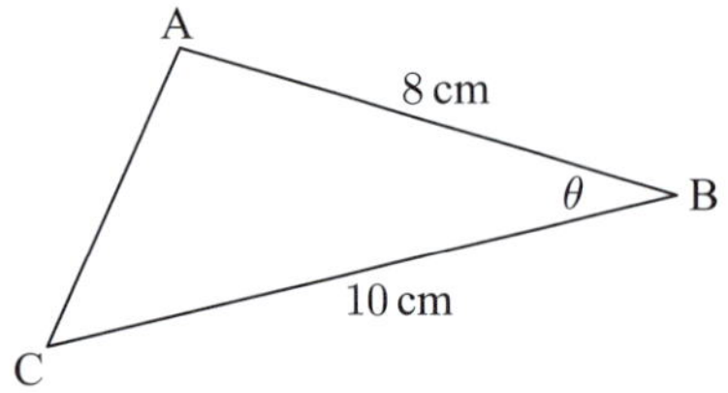

Find the rate of change in the length of AC as θ changes, at the time when $\theta = 30°$.

12 The circle alongside has radius 7 cm. The shaded area is A cm^2.

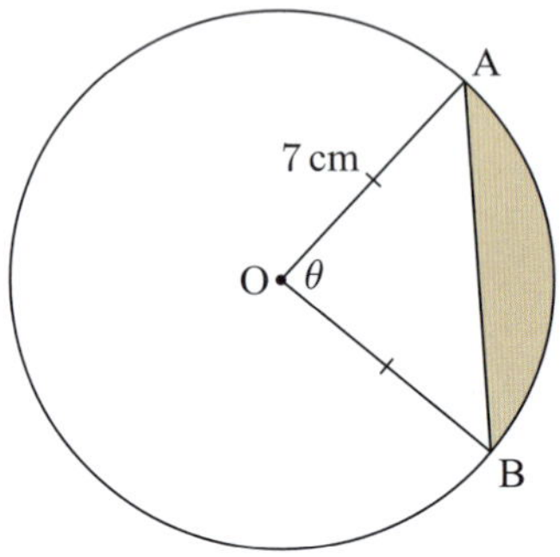

a Write A in terms of θ.

b Find the rate of change in the shaded area as θ changes, when:

i $\theta = 45°$ **ii** $\theta = \frac{5\pi}{6}$

iii triangle OAB is equilateral.

F OPTIMISATION

There are many problems for which we need to find the **maximum** or **minimum** value of a function. The solution is often referred to as the **optimal** solution and the process is called **optimisation**.

We have seen previously that we can locate the maximum and minimum turning points of $y = f(x)$ by finding where $f'(x) = 0$. However, to make sure we find the optimal solution on a fixed domain, we must also examine the value of the function at the end points of the interval under consideration.

For example, over the interval $a \leqslant x \leqslant b$, $y = f(x)$ is maximised at the maximum turning point at $x = p$, and minimised at the end point $x = b$.

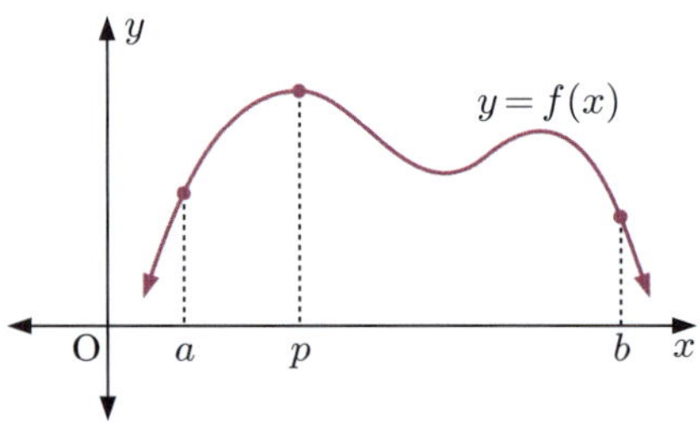

OPTIMISATION PROBLEM SOLVING METHOD

Step 1: Draw a large, clear diagram of the situation.

Step 2: Construct a **formula** with the variable to be optimised as the subject. It should be written in terms of one convenient variable, for example x. Write down what domain restrictions there are on x.

Step 3: Find the **first derivative** and find the values of x which make the first derivative **zero**.

Step 4: For each stationary point, use a sign diagram or the second derivative test to determine if you have a maximum or minimum.

Step 5: Identify the optimal solution, also considering end points where appropriate.

Step 6: Write your answer in a sentence, making sure you specifically answer the question.

Example 14

Self Tutor

An isosceles triangle has perimeter 12 cm. Find the length of the equal sides which maximises the area of the triangle.

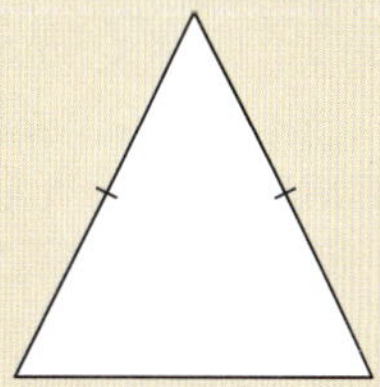

Step 1: Let x cm be the length of the equal sides.

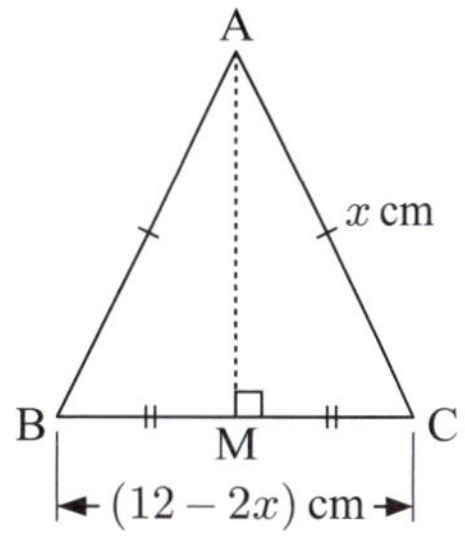

Step 2: Since the perimeter is 12 cm, the length of the third side is $(12-2x)$ cm.

$\therefore$ MC $= (6-x)$ cm

Now $\text{AM}^2 = x^2-(6-x)^2$ {Pythagoras}

$\therefore$ $\text{AM}^2 = 12x-36$

$\therefore$ $\text{AM} = \sqrt{12x-36}$ cm {AM > 0}

$\therefore$ the area of the triangle is $A = \frac{1}{2}(12-2x)\sqrt{12x-36}$

$= (6-x)\sqrt{12x-36}$ cm^2

Since AM > 0 and MC > 0, $\sqrt{12x-36}>0$ and $6-x>0$

$\therefore$ $12(x-3)>0$ and $x<6$

$\therefore$ $x>3$ and $x<6$

$\therefore$ $3<x<6$

Step 3: $\dfrac{dA}{dx} = (-1)\sqrt{12x-36}+(6-x)\frac{1}{2}(12x-36)^{-\frac{1}{2}}(12)$ {product rule}

$= -\sqrt{12x-36}+\dfrac{6(6-x)}{\sqrt{12x-36}}$

$= \dfrac{-(12x-36)+36-6x}{\sqrt{12x-36}}$

$= \dfrac{72-18x}{\sqrt{12x-36}}$

$\therefore$ $\dfrac{dA}{dx}=0$ when $72-18x=0$

$\therefore$ $x=4$

Step 4: $\dfrac{dA}{dx}$ has sign diagram:

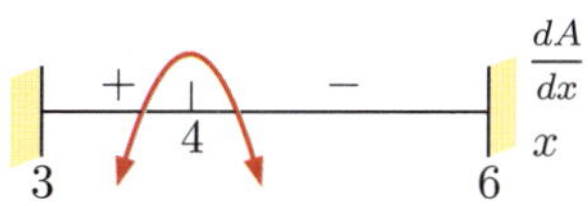

Step 5: There is a maximum turning point when $x=4$.

Step 6: The maximum area is obtained when the equal sides have length 4 cm. In this case the triangle is equilateral.

EXERCISE 15F

1 40 m of fence is used to construct three sides of a rectangular enclosure. The fourth side is an existing wall.

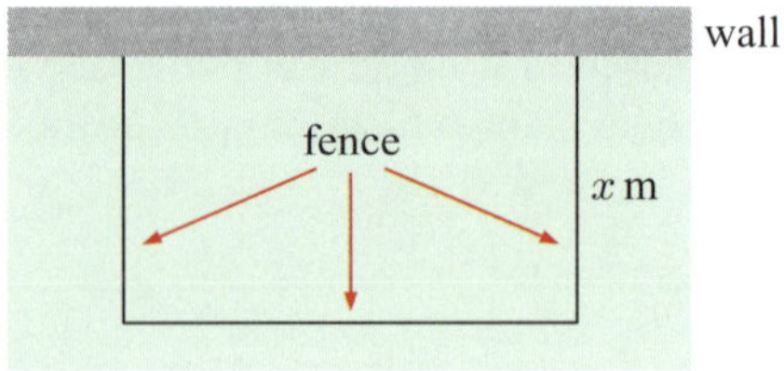

- **a** If the sides of the enclosure adjacent to the wall are x m long, show that the area of the enclosure is $A = x(40 - 2x)$ m^2.
- **b** Find the value of x which maximises A.
- **c** State the dimensions of the enclosure which maximise the area.

2 Square corners are cut from a piece of 18 cm by 18 cm tinplate, which is then folded to form an open dish. Let the side lengths of the cut out squares be x cm.

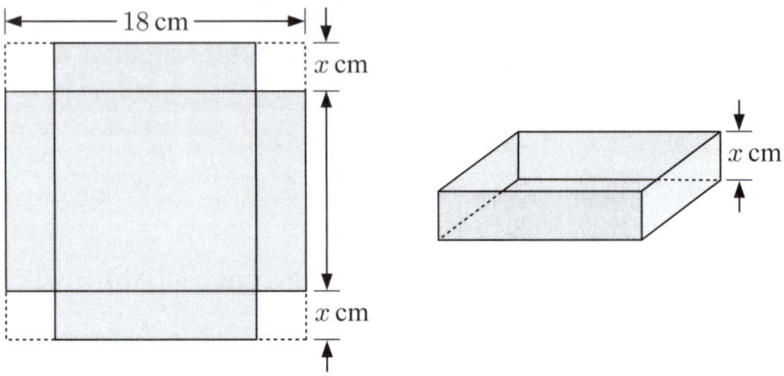

- **a** What possible values can x take?
- **b** Show that the volume of the dish is given by $V = x(18 - 2x)^2$ cm^3.
- **c** What size squares should be cut out to produce the dish of greatest volume?

3 Find the angle θ which maximises the area of this triangle.

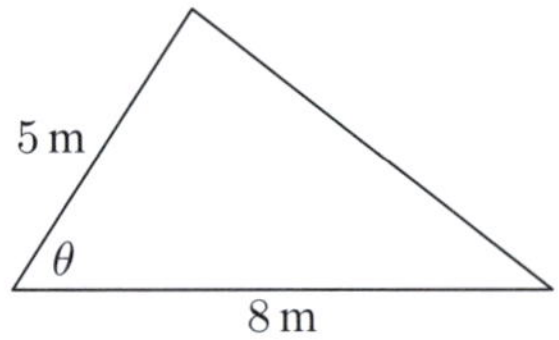

4 A psychologist claims that the ability A to memorise simple facts during infancy years can be calculated using the formula $A(t) = t \ln t + 1$ where $0 < t \leqslant 5$, t being the age of the child in years. At what age is the child's memorising ability a minimum?

5
- **a** Show that the trapezium alongside has area $A = (4 + x)\sqrt{16 - x^2}$ m^2.
- **b** Find the value of x which maximises the area of the trapezium.

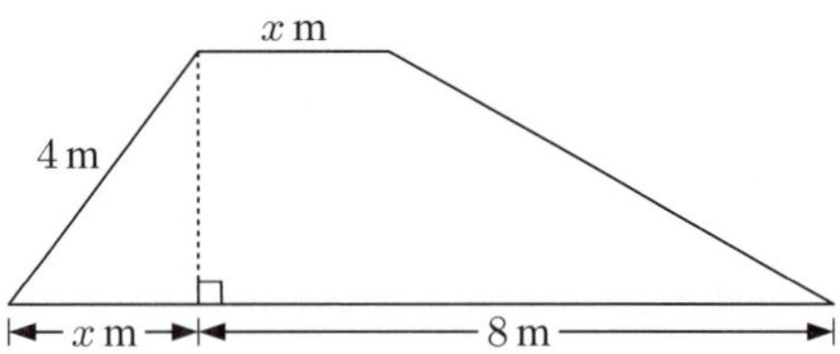

6 A rectangle is positioned on the x-axis under the graph of $y = 9 - x^2$, as shown.

- **a** Find the coordinates of point P.
- **b** What possible values can a have?
- **c** Find, in terms of a, the area of the shaded rectangle.
- **d** Find the maximum possible area of the rectangle.

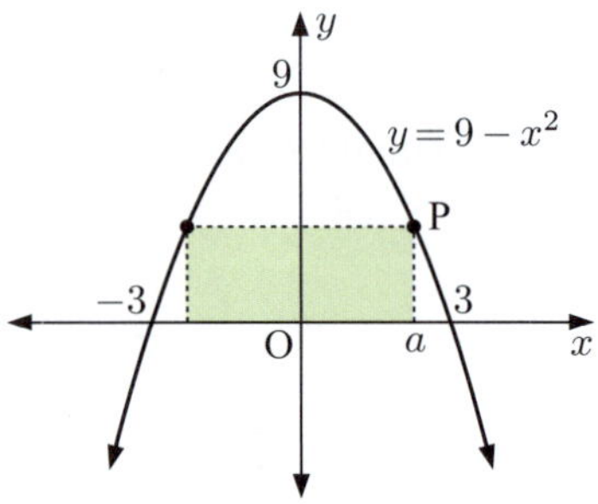

7 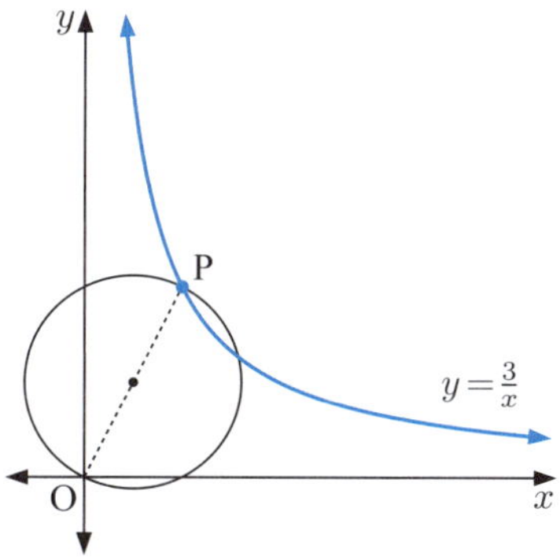

P is a point on the graph of $y = \dfrac{3}{x}$. A circle with diameter OP is drawn as shown. Find the coordinates of P that minimise the area of the circle.

8 Consider the **Opening Problem** on page **362**.

a Find $E'(t)$.

b Find the rate at which the effectiveness of the injection is changing after:

i 1 minute **ii** 4 minutes.

c Find the time at which the injection is most effective.

9 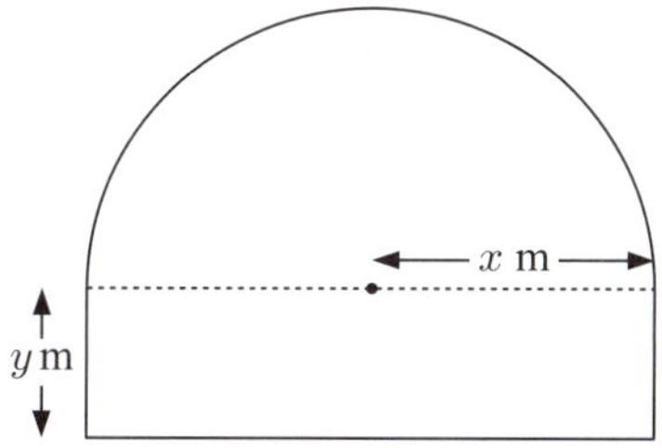

a The figure alongside has area 30 m^2. Show that the perimeter of the figure is $P = \left(2 + \frac{\pi}{2}\right)x + \dfrac{30}{x}$ m.

b Find the value of x which minimises the perimeter of the figure.

10 The isosceles trapezium alongside has three sides of length 20 cm.

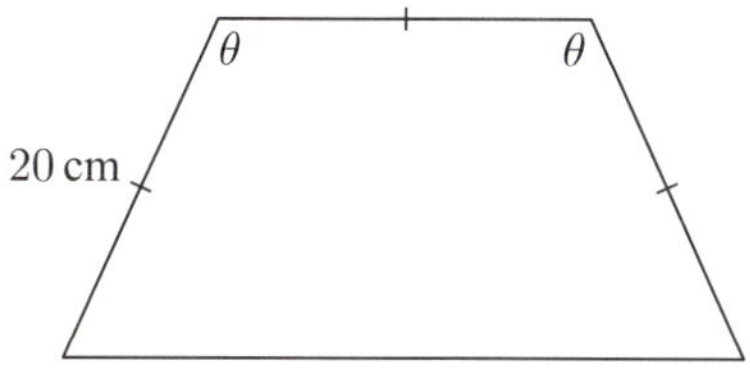

a Show that the area of the trapezium is
$A = 400\cos\left(\theta - \frac{\pi}{2}\right)\left[1 + \sin\left(\theta - \frac{\pi}{2}\right)\right]$ cm^2.

b Find $\dfrac{dA}{d\theta}$.

c Find the angle θ which maximises the area of the trapezium.

11 A swimming pool is to be lifted over a house and placed in a backyard with a crane as shown.

a Find y in terms of x.

b Write an expression for the crane length in terms of x.

c Find the distance the crane should be placed from the house in order to minimise the crane length needed.

d Hence find the minimum crane length.

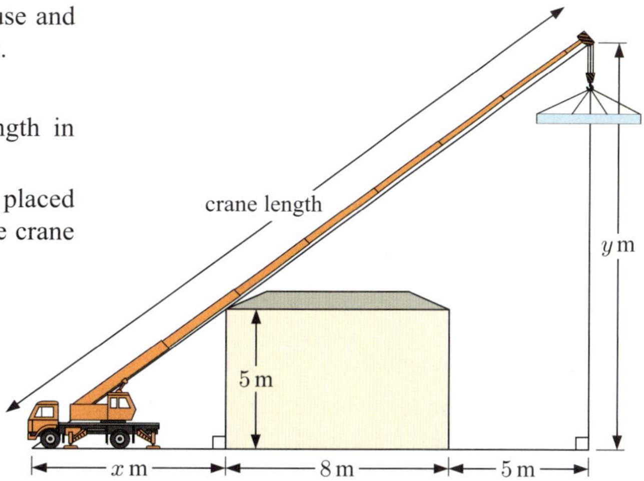

Activity 2 — A square and a circle

Suppose a 40 cm long wire AB is cut at a point P. The section AP is formed into a square, and the section BP is formed into a circle.

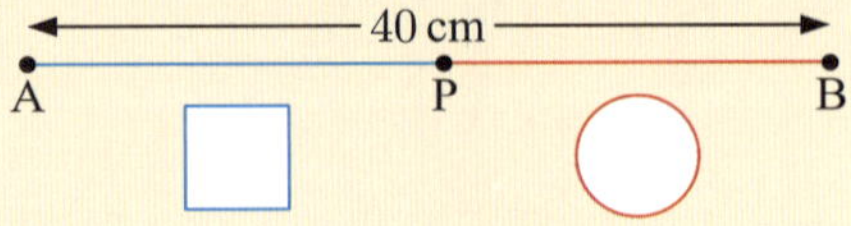

What to do:

1 Find the total area of the square and circle if AP equals:

a 10 cm **b** 20 cm **c** 30 cm

2 In small groups, discuss where you think the cut P should be made to minimise the total area of the square and circle. Should the cut be made at the midpoint of AB, or somewhere else?

3 Use calculus to determine where the cut should be made to minimise the total area of the square and circle.

G CONNECTED RATES OF CHANGE

Discussion

Suppose a balloon is filled with air from a pump. Air flows into the balloon at a constant rate.

- How do we know that the *volume* of the balloon increases at a constant rate?
- As the volume of the balloon increases, what happens to the *radius* of the balloon? Does the radius also increase at a constant rate, or does its rate of increase change over time?
- Suppose the pump is adjusted, so that the rate of air flow increases over time. How would this affect the rate at which the radius increases?

Assume the balloon in the **Discussion** above is spherical. Its volume V and radius r are connected by the equation $V = \frac{4}{3}\pi r^3$. We can use this information to explore the relationship between their *rates of change* $\frac{dV}{dt}$ and $\frac{dr}{dt}$.

Problems involving related variables which change over time are called **connected rates of change** problems.

To solve connected rates of change problems, we follow these steps:

Step 1: Write an **equation** connecting the variables. You may need to use:
- perimeter, surface area, or volume formulae
- Pythagoras' theorem
- trigonometry.

Step 2: Use the **chain rule** to differentiate the equation with respect to time t.

Step 3: Substitute the values for the **particular case** corresponding to some instant in time, and solve to find the required unknown.

Example 15

Self Tutor

Variables x and y are connected by the equation $y = x^4 - 3x^2$.

Given that x is increasing at the rate of 2 units per second, find the rate of change in y when $x = 1$.

$y = x^4 - 3x^2$

Differentiating with respect to t gives $\dfrac{dy}{dt} = \dfrac{dy}{dx}\dfrac{dx}{dt}$ {chain rule}

$$\therefore \quad \frac{dy}{dt} = (4x^3 - 6x)\frac{dx}{dt}$$

Particular case:

When $x = 1$ and $\dfrac{dx}{dt} = 2$, $\dfrac{dy}{dt} = \big(4(1)^3 - 6(1)\big) \times 2$

$= -4$ units per second

At the instant when $x = 1$, y is decreasing at the rate of 4 units per second.

Example 16

Self Tutor

Air is pumped into a spherical balloon at the constant rate of 50 cm^3 per second.

Find the rate at which the radius of the balloon is increasing when the radius is:

a 5 cm **b** 10 cm.

Suppose the balloon has volume V cm^3 and radius r cm.

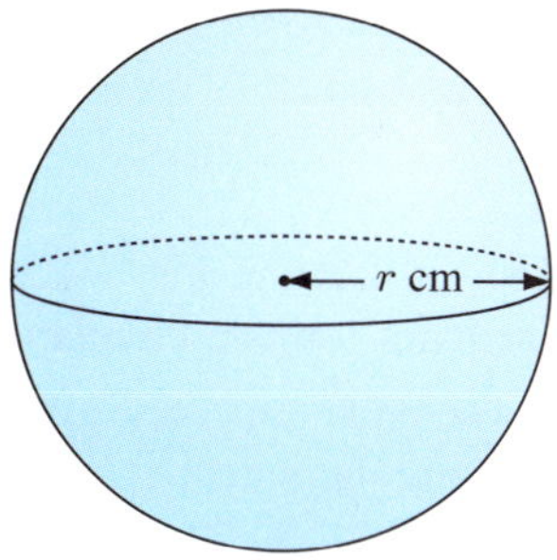

$\therefore \quad V = \frac{4}{3}\pi r^3$

Differentiating with respect to t gives

$$\frac{dV}{dt} = \frac{dV}{dr}\frac{dr}{dt} \quad \text{\{chain rule\}}$$

$$\therefore \quad \frac{dV}{dt} = (4\pi r^2)\frac{dr}{dt}$$

a When $r = 5$ and $\dfrac{dV}{dt} = 50$, $50 = 4\pi(5)^2\dfrac{dr}{dt}$

$$\therefore \quad 50 = 100\pi\frac{dr}{dt}$$

$$\therefore \quad \frac{dr}{dt} = \frac{1}{2\pi} \approx 0.159$$

The radius is increasing at ≈ 0.159 cm per second at this instant.

b When $r = 10$ and $\dfrac{dV}{dt} = 50$, $50 = 4\pi(10)^2\dfrac{dr}{dt}$

$$\therefore \quad 50 = 400\pi\frac{dr}{dt}$$

$$\therefore \quad \frac{dr}{dt} = \frac{1}{8\pi} \approx 0.0398$$

The radius is increasing at ≈ 0.0398 cm per second at this instant.

Warning:

You **must not** substitute values for the particular case too early. Otherwise you will incorrectly treat variables as constants. The differentiated equation in fully generalised form must be established first.

EXERCISE 15G

1 The variables x and y are connected by the equation $y = 2x^3 - 4x$.

- **a** Differentiate this equation with respect to t.
- **b** Given that x is increasing at the rate of 3 units per second, find the rate of change of y when $x = 1$.

2 The variables x and y are connected by the equation $y = (3x - 1)^4$.

- **a** If x is increasing at 2 units per second, find the rate of change of y when $x = 0$.
- **b** If y is increasing at 50 units per second, find the rate of change of x when $x = 2$.

3 The variables x and y are connected by the equation $y = \ln x$.
Given that x is increasing at 4 units per second, find the rate of change of y when:

- **a** $x = 5$
- **b** $y = 2$.

4 A cube of side length x m has volume V m^3.

- **a** Write down the equation connecting V and x.
- **b** The cube is expanding, its side lengths increasing at 2 m s^{-1}. Find the rate of change in the cube's volume, at the instant when its sides are:
  - **i** 1 m long
  - **ii** 4 m long.

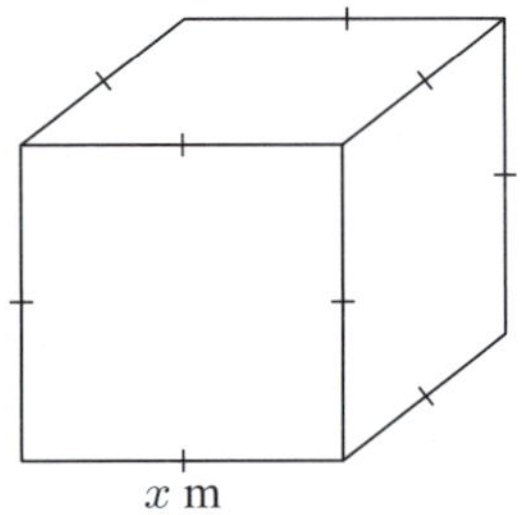

5 While Tom is painting his house, he knocks over a can of paint. The paint forms a circular patch on the ground.

- **a** Write an equation connecting the area A and the radius r of the circular patch.
- **b** The area of the patch increases at 20 cm^2 per second. At what rate does the radius of the patch change, when the radius is:
  - **i** 5 cm
  - **ii** 12 cm?

6

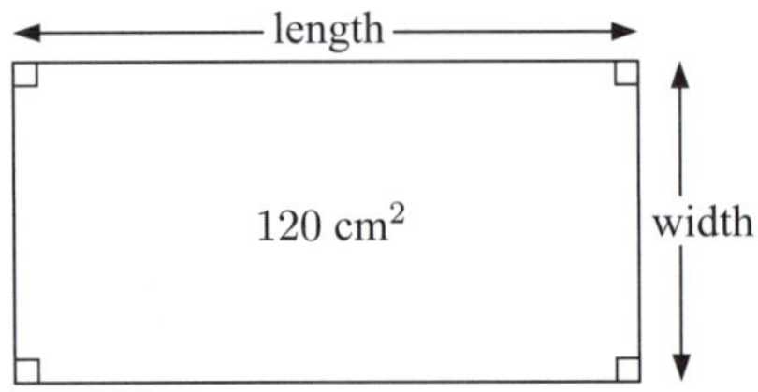

The length of a rectangle is increasing at 3 cm s^{-1}, but the area of the rectangle remains constant at 120 cm^2.
Find the rate at which the width of the rectangle is changing, when the length is 20 cm.

7 The radius of a sphere is decreasing at 5 cm per minute. Find the rate at which the sphere's surface area is changing at the instant when the radius is 6 cm.

8 In the right angled triangle ABC, AB is 4 cm long, and the length of AC increases at 0.5 cm s^{-1}.
Find the rate at which the length of the hypotenuse BC is changing when:

a AC = 3 cm **b** AC = 5 cm.

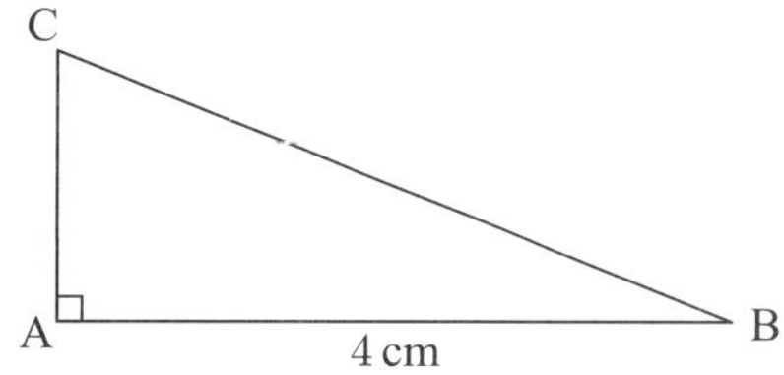

Example 17

Self Tutor

Triangle PQR is right angled at P, and PQ = 10 cm. $\widehat{PQR}$ increases at a constant rate of 2° per minute. At what rate is PR changing at the instant when $\widehat{PQR}$ measures 60°?

Let $\widehat{PQR} = \theta$ and $PR = x$ cm

Now $\tan\theta = \frac{x}{10}$

$\therefore\ x = 10\tan\theta$

Differentiating with respect to t gives

$$\frac{dx}{dt} = 10\sec^2\theta\,\frac{d\theta}{dt}$$

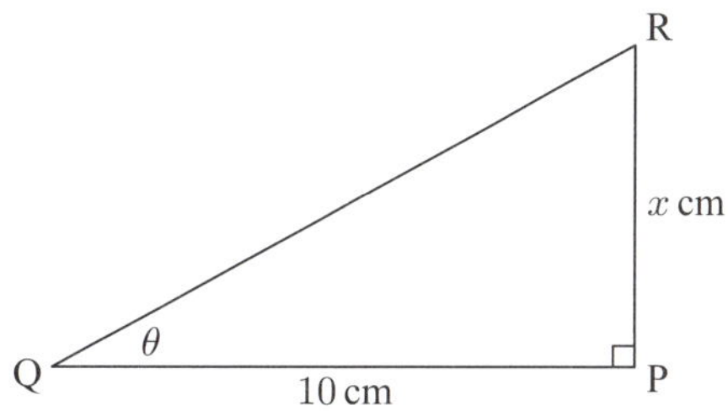

Particular case:

When $\theta = 60° = \frac{\pi}{3}$ and $\frac{d\theta}{dt} = 2°$ per minute $= \frac{\pi}{90}$ radians per minute,

$$\frac{dx}{dt} = \frac{10}{\cos^2\left(\frac{\pi}{3}\right)} \times \frac{\pi}{90}$$

$$= \frac{10}{\frac{1}{4}} \times \frac{\pi}{90}$$

$= \frac{4\pi}{9} \approx 1.40$ $\therefore$ PR is increasing at ≈ 1.40 cm per minute.

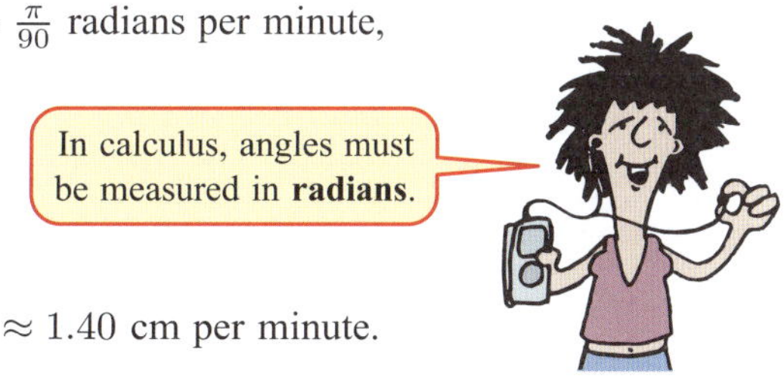

9 Consider a right angled triangle ABC in which BC = 8 cm. $\widehat{BCA}$ increases at 1° per minute. Let $\widehat{BCA} = \theta$ and $AC = x$ cm.

a Write x in terms of θ.

b At what rate is AC changing when $\theta = 30°$?

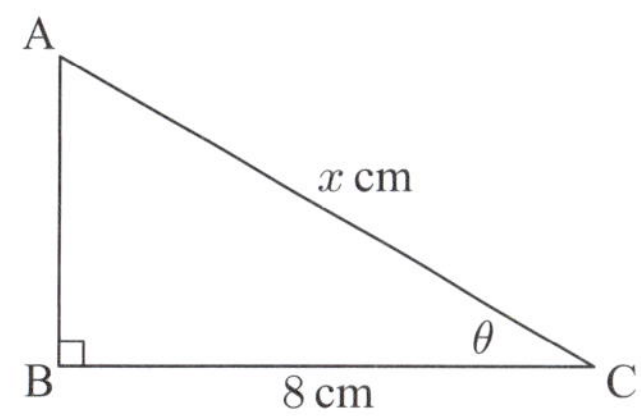

10

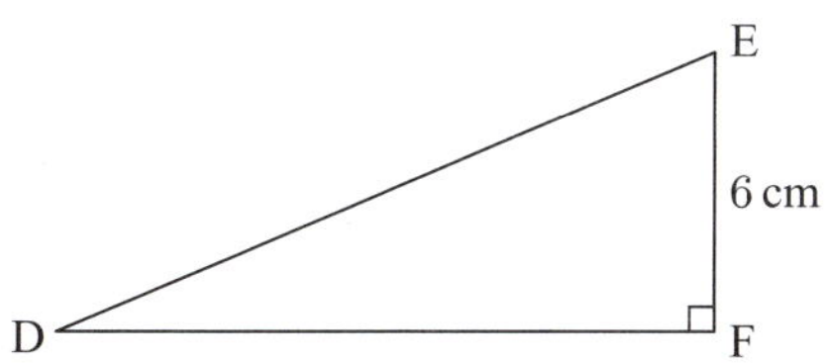

In the right angled triangle DEF, $\widehat{EDF}$ increases at 3° per minute. At what rate is DF changing at the instant when $\widehat{EDF} = 60°$?

11 Triangle PQR is right angled at Q, and QR = 12 cm. The hypotenuse PR increases in length at 0.5 cm per second. At what rate is $\widehat{QPR}$ changing at the instant when $\widehat{QPR} = 45°$?

12 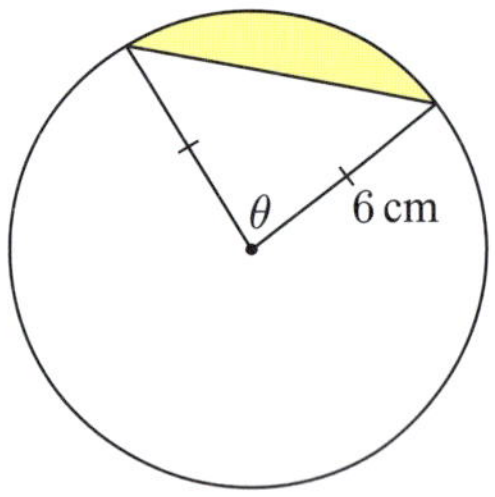

a Show that the shaded area is $A = 18(\theta - \sin\theta)$ cm^2.

b Given that θ increases at $\frac{\pi}{36}$ radians per minute, find the rate at which A changes when:

i $\theta = \frac{\pi}{6}$ **ii** $\theta = \frac{2\pi}{3}$

13 Given that ϕ increases at $3°$ per second, find the rate at which BC changes when:

a $\phi = \frac{\pi}{4}$ **b** $\phi = \frac{\pi}{2}$

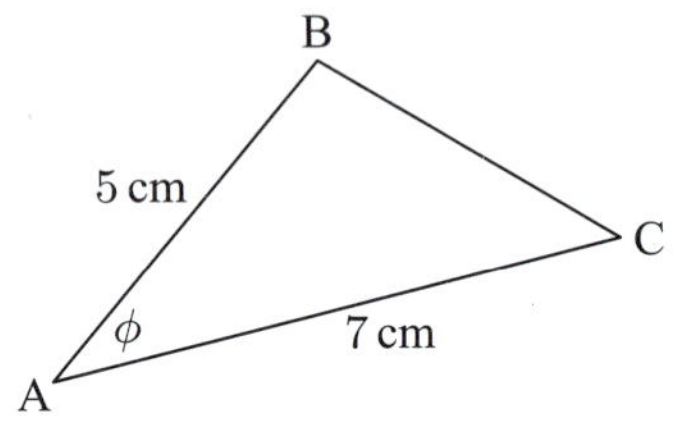

14 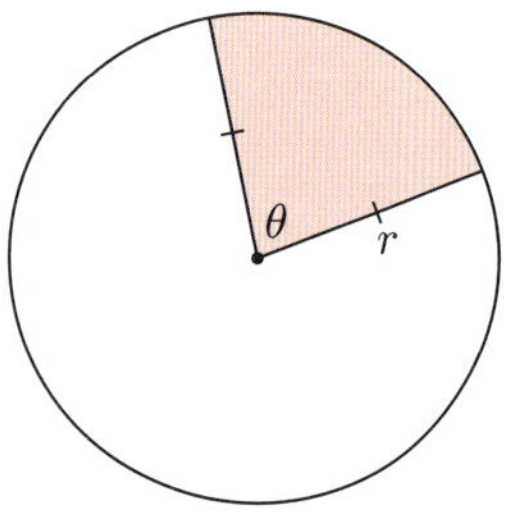

The radius r of this circle increases at 2 cm per second. However, the perimeter of the shaded region remains constant at 30 cm.

a Write r in terms of θ.

b Find the rate at which θ changes at the instant when:

i $\theta = \frac{\pi}{6}$ **ii** $r = 5$

Review set 15A

1 Find the equation of the tangent to:

a $y = x^2 - x + 5$ at $x = 2$

b $y = x - \frac{4}{x}$ at $x = -4$

c $y = \sqrt{x+2}$ at $x = 7$

d $y = \frac{x}{3x+2}$ at $x = -1$.

2 The graph of $f(x) = e^{x^2}$ is shown alongside.

a Find the equation of the illustrated tangent.

b Find the equation of the normal at P.

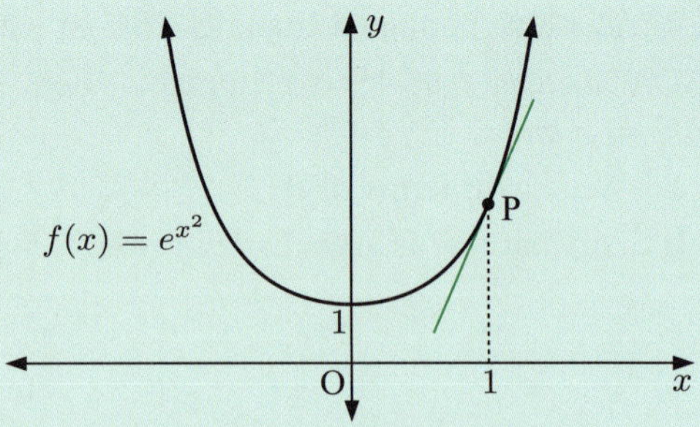

3 Consider the curve $y = ax + \frac{b}{x^2}$ where a and b are constants. The tangent to this curve at the point where $x = 1$, has equation $y = -8x + 15$. Find the values of a and b.

4 Consider the function $f(x) = x^3 - 6x$.

a Find $f'(x)$, and draw its sign diagram.

b Find intervals where $f(x)$ is decreasing.

5 Find intervals where $f(x)$ is increasing or decreasing:

a $f(x) = -x^2 + 6x - 2$ **b** $f(x) = x^2 - \frac{2}{x}$

6 The graph of $y = \ln(\sin x + 3)$ is shown alongside. Find the exact coordinates of:

a A **b** B **c** C

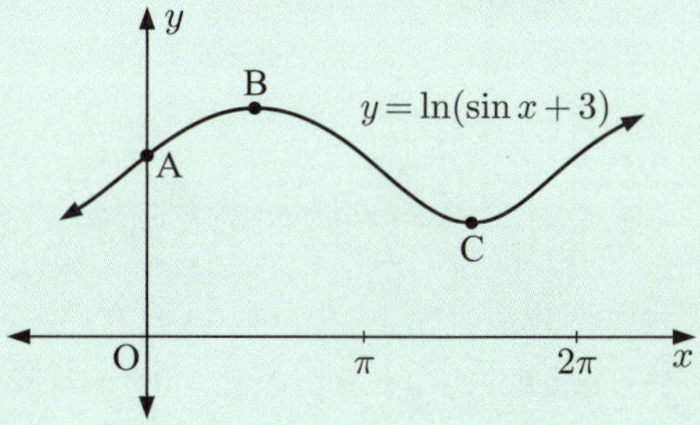

7 Consider the function $f(x) = -3x^4 + 4x^3 + 12x^2$.

a Find and classify the stationary points.

b Sketch the function, showing all important features.

8 A particle moves in a straight line with displacement function $s(t) = t^3 - 6t + 3$ cm, where t is in seconds, $t \geqslant 0$.

a Find the velocity function $v(t)$.

b Find the speed of the particle after 1 second.

c At what time is the particle momentarily at rest?

d Find the acceleration of the particle after 2 seconds.

9 A pebble is thrown vertically into the air. Its position above ground level after t seconds is given by $s(t) = -4.9t^2 + 24.5t + 1$ metres.

a Find the velocity and acceleration functions for the pebble, and draw their sign diagrams.

b Find the initial position and velocity of the pebble.

c Find the maximum height reached by the pebble.

d Find the total distance travelled by the pebble in the first 4 seconds.

10 Henry's bucket holds 10 litres of water. He accidentally makes a hole in his bucket, and water starts to leak out. The volume of water left in the bucket t seconds after the hole is made, is given by $V = 10\,000\left(1 - \frac{t}{40}\right)^2$ millilitres where $0 \leqslant t \leqslant 40$.

a Find the amount of water in the bucket after:

i 10 seconds **ii** 20 seconds.

b At what rate is the volume of water changing after:

i 10 seconds **ii** 20 seconds?

11 Find the rate of change in the area of triangle ABC as θ changes, at the time when $\theta = 45°$.

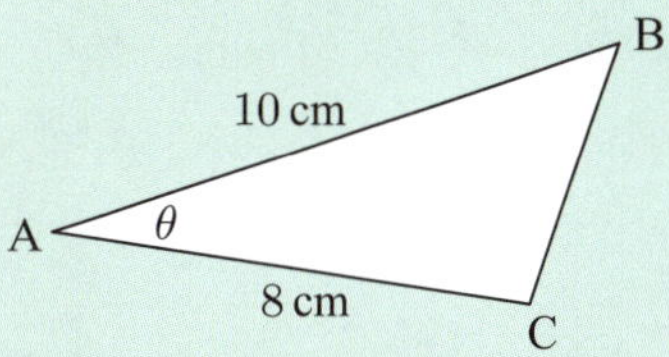

12 Infinitely many rectangles which sit on the x-axis can be inscribed under the curve $y = \cos 4x$.
Determine the coordinates of C such that rectangle ABCD has maximum perimeter.

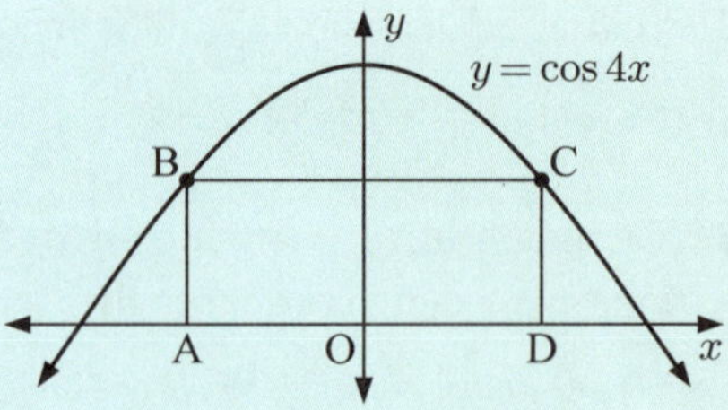

13 The variables x and y are connected by the equation $y = \dfrac{1}{2x+3}$.
Given that x is increasing by 5 units per second, find the rate of change of y when $x = 1$.

14 A cone is 12 cm high, and its base radius r cm is increasing at 0.5 cm per second. Find the rate at which the volume is changing, when the base radius is 4 cm.

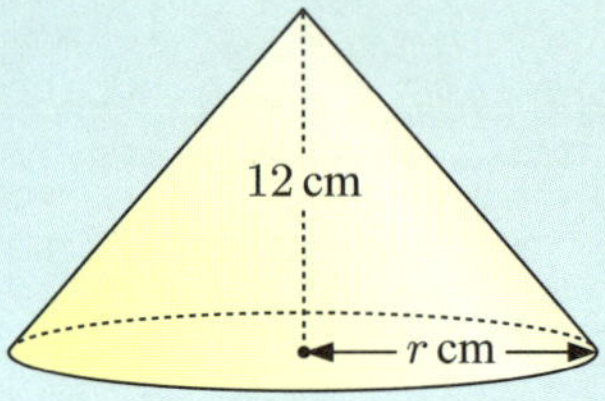

Review set 15B

1 Find the equation of the normal to:

a $y = -x^2 - x + 3$ at $x = -1$

b $y = x \ln x$ at $x = e$.

2 The tangent to $y = \dfrac{15}{x}$ at the point where $x = 5$ cuts the x-axis at A and the y-axis at B. Find the area of triangle OAB, where O is the origin.

3 The graph of $f(x) = x^3 - 4x + 2$ is shown alongside. Find:

a the equation of the illustrated tangent

b the coordinates of Q.

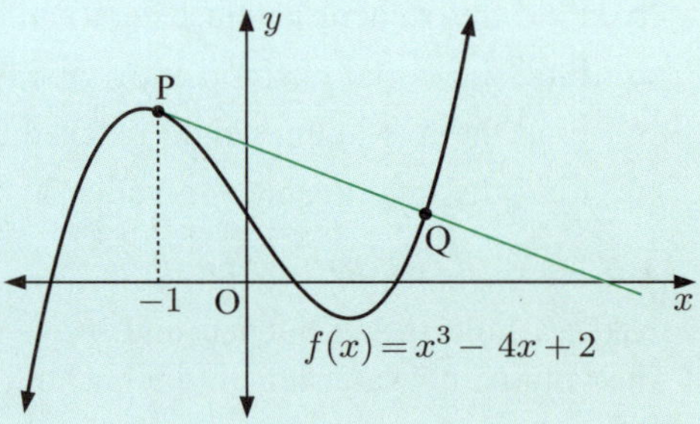

4 Consider the function $f(x) = 2x^3 - 5x^2 + 6x - 3$.

a Find $f'(x)$.

b Show that $f'(x) > 0$ for all x, and explain the significance of this result.

5 The graph of $f(x) = 4x^2 + \dfrac{1}{x}$ is shown alongside.

a Find intervals where $f(x)$ is decreasing.

b Find the coordinates of P.

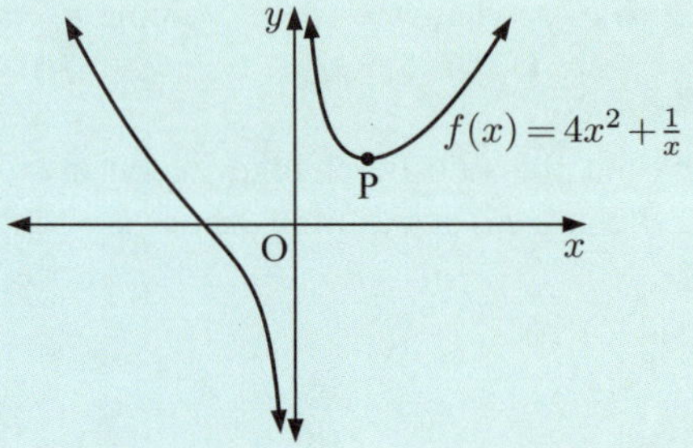

6 Find the exact position and nature of the stationary point(s) of:

a $y = 9x - x^3$ **b** $y = x^3 e^x$ **c** $y = \dfrac{x}{\ln x}$

7 The function $f(x) = \dfrac{x+a}{x^2-5}$ has a stationary point at $x = -1$.

a Find a.

b Determine the position and nature of the other stationary point.

8 The graph of $y = \sin x + \tan x + 2$ is shown alongside.
Find the coordinates of the stationary inflection points P and Q.

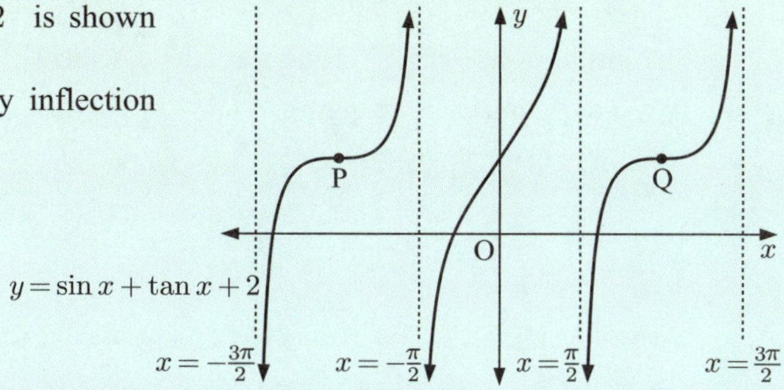

9 An object moves with displacement function $s(t) = e^{5t} + 3t$ cm, where t is in seconds, $t \geqslant 0$.

a Find the velocity function $v(t)$.

b Show that the object is never at rest.

c Find the acceleration of the object after 1 second.

d Find the total distance travelled by the object in the first 2 seconds.

10 AB is a diameter of a circle with radius 4 cm.

a Show that the shaded region has area $A = 8(\sin 2\theta + 2\theta)$ cm^2.

b Suppose θ increases at $\frac{\pi}{60}$ radians per minute. Find the rate at which the area is changing at the instant when:

i $\theta = \frac{\pi}{6}$ **ii** $\theta = \frac{\pi}{4}$.

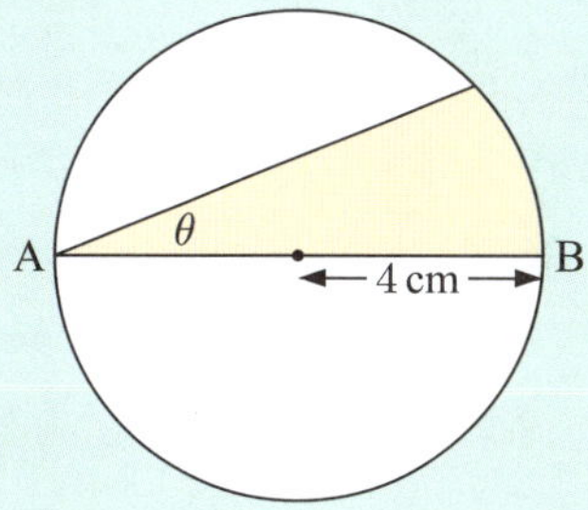

11

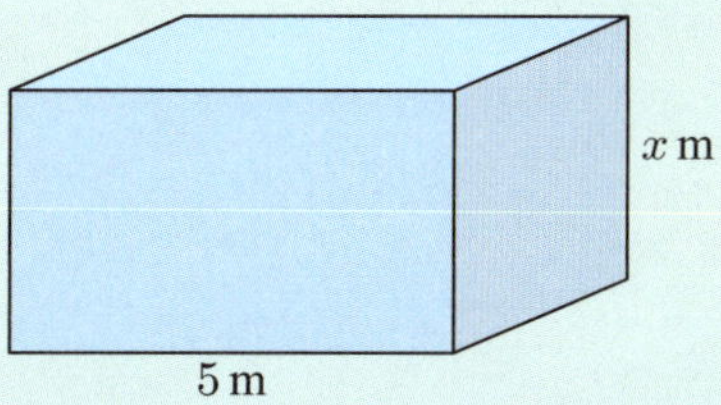

A rectangular prism is 5 m long and x m high. Its surface area is 200 m^2.

a Show that the prism has volume $V = \dfrac{500x - 25x^2}{x+5}$ m^3.

b Find, correct to 3 significant figures, the value of x which maximises the volume of the prism.

12 Line l is drawn through the origin, at an angle θ to the x-axis.
AP and BQ are drawn perpendicular to l.
Find the value of θ which maximises the distance AP + BQ.

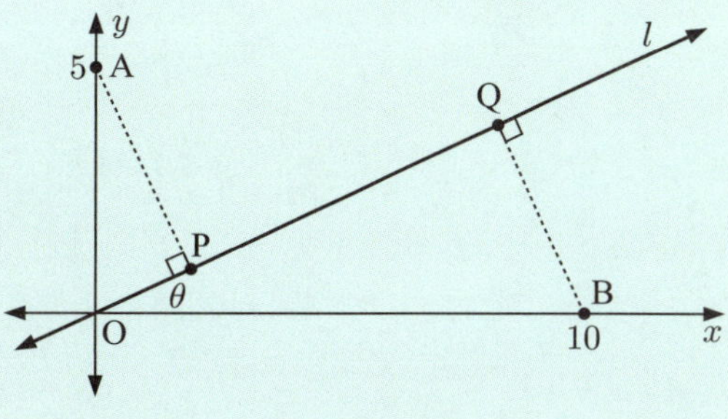

13 The variables x and y are connected by the equation $y = \sqrt{2x+4}$.
Given that y is increasing at 3 units per second, find the rate of change in x when $x = 5$.

14 An isosceles triangle has base 10 cm and base angles θ.

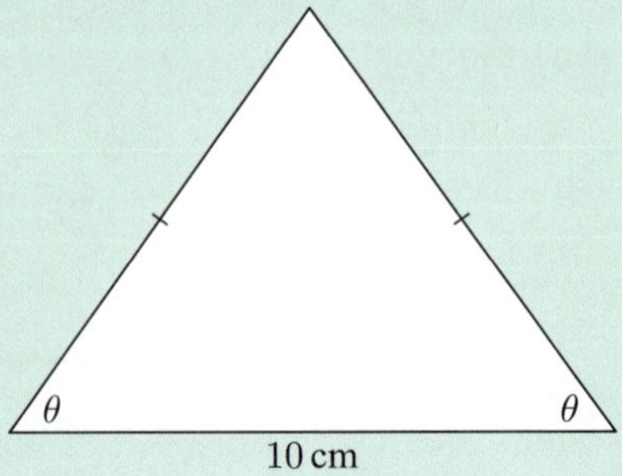

a Write an expression for:

- **i** the perimeter of the triangle
- **ii** the area of the triangle.

b Suppose θ increases at a rate of $5°$ per minute. At the instant when $\theta = \frac{\pi}{6}$, find the rate at which:

- **i** the perimeter is changing
- **ii** the area is changing.

16

Integration

Contents:

Opening problem

The graphs of $y = 3x$ and $y = x^2$ are shown alongside. The area between each graph and the x-axis from $x = 0$ to $x = 1$ has been shaded.

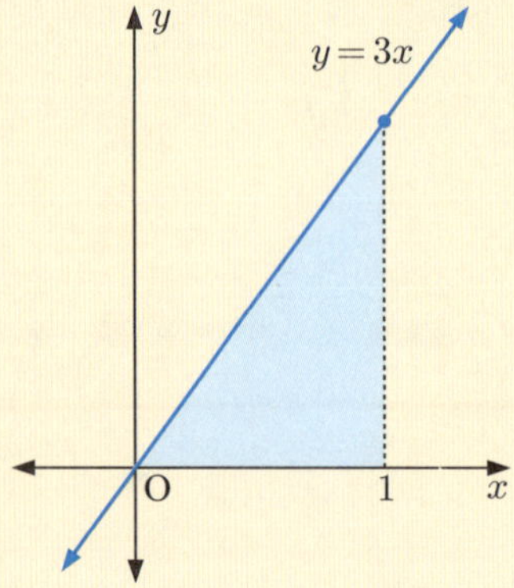

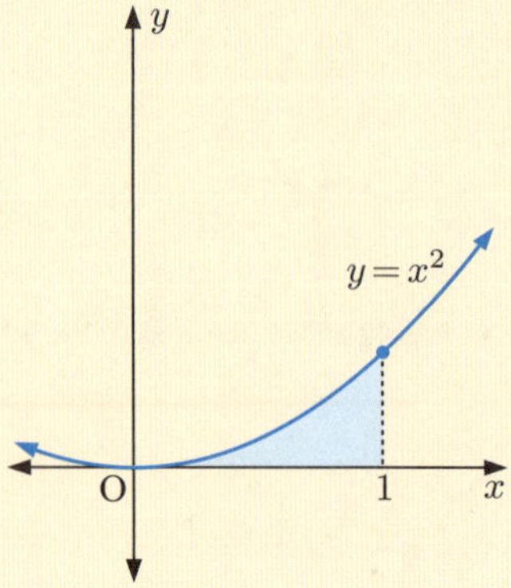

Things to think about:

a If you were asked to calculate the shaded area under $y = 3x$, what method would you use? Could you use the same method to calculate the shaded area under $y = x^2$?

b Can you name a function whose derivative is x^2? How can this function be used to find the shaded area under $y = x^2$?

In this Chapter we consider **integral calculus**. This involves **antidifferentiation**, which is the reverse process of differentiation. Integral calculus also has many useful applications, including:

- finding areas of shapes with curved boundaries
- finding volumes of revolution
- finding distances travelled from velocity functions
- solving problems in economics, biology, and statistics
- solving differential equations.

A THE AREA UNDER A CURVE

The task of finding the area under a curve has been important to mathematicians for thousands of years.

For a positive continuous function $f(x)$, the shaded area A between $y = f(x)$ and the x-axis for $a \leqslant x \leqslant b$ is called the **definite integral** of $f(x)$ from a to b.

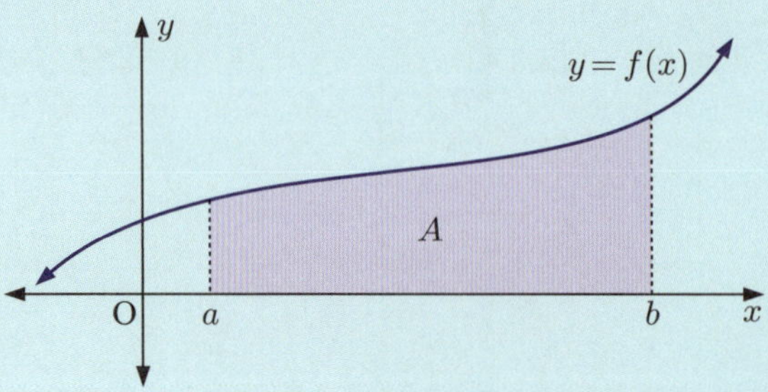

We write $A = \int_a^b f(x)\,dx$.

For example, the area between $y = 2x$ and the x-axis for $0 \leqslant x \leqslant 3$, is $\frac{1}{2} \times 3 \times 6 = 9$ units^2.

We write $\int_0^3 2x\,dx = 9$.

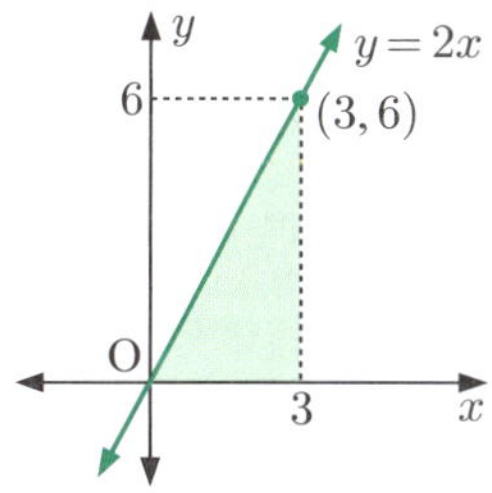

Example 1

Self Tutor

Use a known area formula to find $\int_2^6 (x+1)\,dx$.

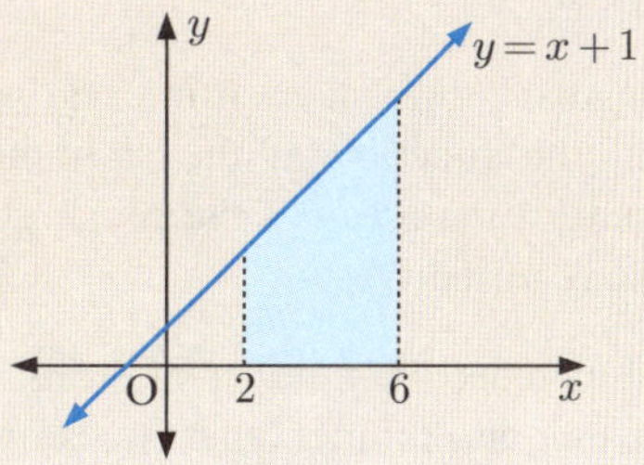

The shaded region is a trapezium.

$$\therefore \int_2^6 (x+1)\,dx = \left(\frac{3+7}{2}\right) \times 4$$
$$= 20$$

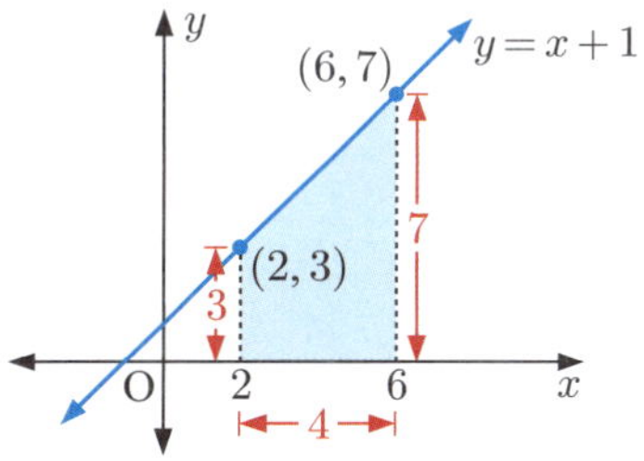

Discovery — The area under $y = x^2$ from $x = 0$ to $x = 1$

In the **Opening Problem**, we considered the area under $y = x^2$ from $x = 0$ to $x = 1$. This region contains a curved edge, so its area cannot be found using standard area formulae.

However, in this Discovery we will *estimate* the area by approximating the region using shapes with straight sides.

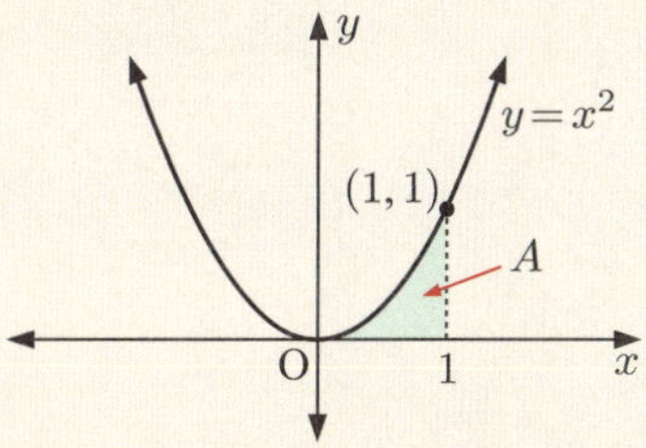

What to do:

1. Explain why $A < \frac{1}{2}$ units^2.
2. Use the graph alongside to explain why $A < \frac{3}{8}$ units^2.
3. Subdivide the interval $0 \leqslant x \leqslant 1$ into 3, 4, and then 5 equal parts, finding an upper bound for A in each case.
4. Consider the upper bounds for A found in **1**, **2**, and **3**. These numbers are approaching the actual value of A. Can you predict what this value is?

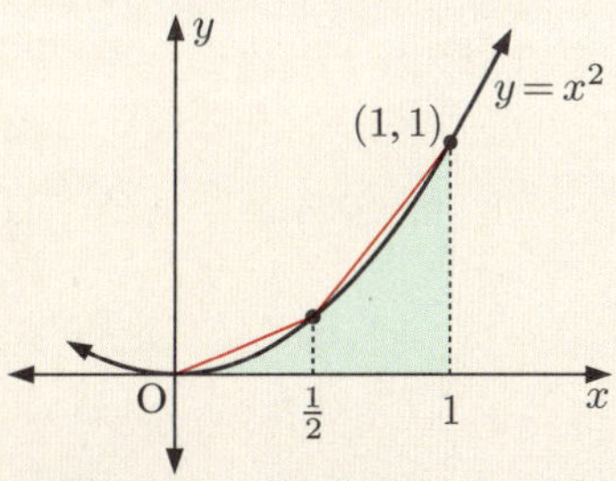

From the **Discovery** on the previous page, you may have guessed that the area between the curve $y = x^2$ and the x-axis from $x = 0$ to $x = 1$ is $\frac{1}{3}$ units^2.

To prove that this is the case, we use our knowledge of limits, and also the formula $1^2 + 2^2 + 3^2 + ... + n^2 = \dfrac{n(n+1)(2n+1)}{6}$.

Consider a more general case where we want to find the area between $y = x^2$ and the x-axis from $x = 0$ to $x = a$.

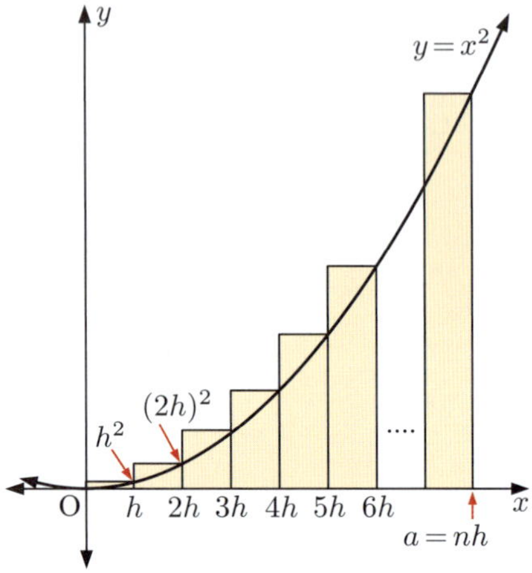

We subdivide the interval $0 \leqslant x \leqslant a$ into n equal intervals of width h units, so $a = nh$.

We draw rectangular strips on each interval with height equal to the value of the function at the right hand side of the interval. The sum of the areas of the rectangles will therefore overestimate the required area.

However, if we increase the number of strips n, the rectangles will become thinner and thinner, and the estimate will become more accurate. In fact, in the limit as $n \to \infty$ and $h \to 0$, the rectangles will give the *exact* area.

DEMO

The sum of the areas of the rectangles is

$$\begin{aligned} S &= h(h^2) + h(2h)^2 + h(3h)^2 + ... + h(nh)^2 \\ &= h^3(1^2 + 2^2 + 3^2 + ... + n^2) \\ &= \left(\frac{a}{n}\right)^3 \frac{n(n+1)(2n+1)}{6} \qquad \{\text{using formula}\} \\ &= \frac{a^3}{6}\left(\frac{n+1}{n}\right)\left(\frac{2n+1}{n}\right) \\ &= \frac{a^3}{6}\left(1 + \frac{1}{n}\right)\left(2 + \frac{1}{n}\right) \end{aligned}$$

Now $h = \dfrac{a}{n}$.

$\therefore$ as $h \to 0$, $\dfrac{1}{n} \to 0$.

$\therefore$ $\displaystyle\lim_{h \to 0} S = \frac{a^3}{6}(1)(2) = \frac{a^3}{3}$.

So, the area between $y = x^2$ and the x-axis from $x = 0$ to $x = a$, is $\dfrac{a^3}{3}$ units^2.

When $a = 1$, the area is $\frac{1}{3}$ units^2, as seen in the **Discovery**. Hence $\displaystyle\int_0^1 x^2\,dx = \frac{1}{3}$.

EXERCISE 16A

1 Use known area formulae to find:

a $\int_0^4 3\,dx$

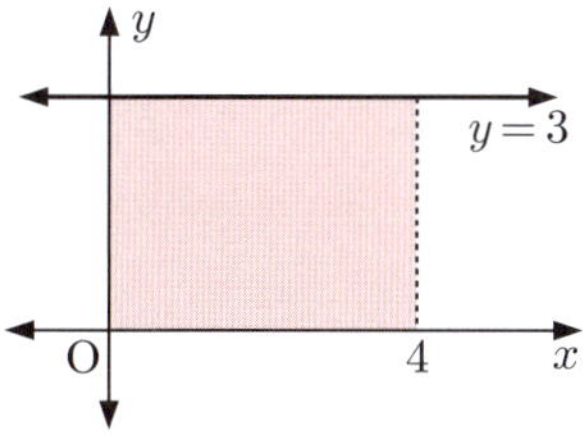

b $\int_0^2 5x\,dx$

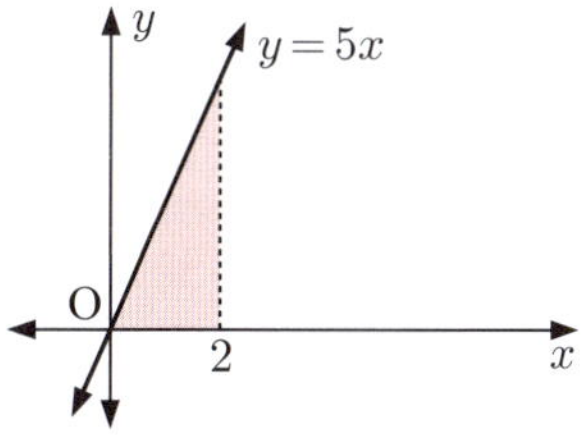

c $\int_1^5 (2x+1)\,dx$

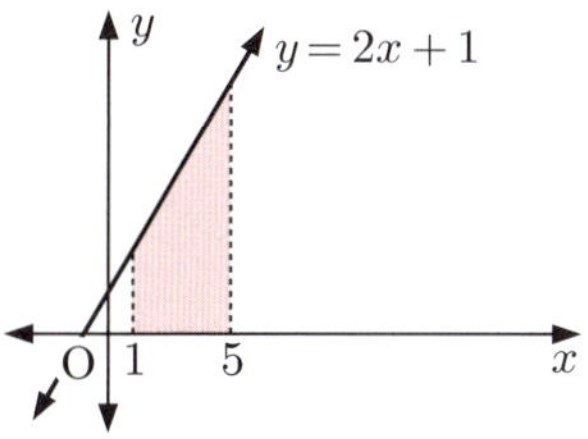

d $\int_{-2}^3 (4-x)\,dx$

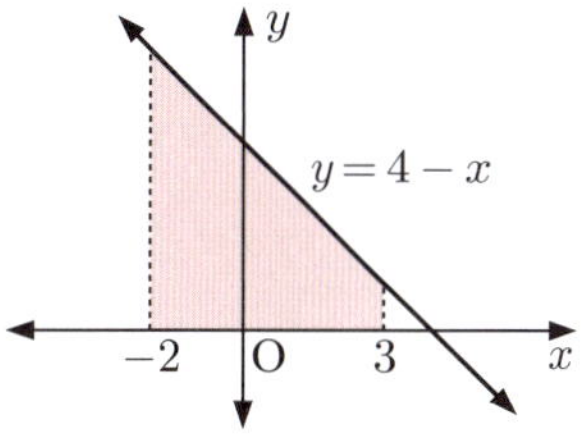

2 Use the known area formulae to find:

a $\int_0^a 1\,dx$

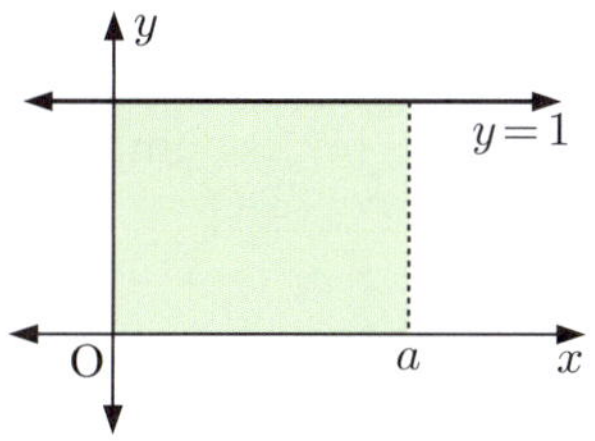

b $\int_0^a x\,dx$

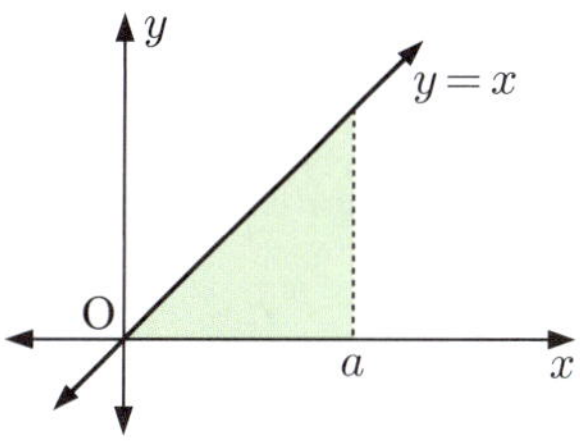

3 Consider the area between $y = x^3$ and the x-axis from $x = 0$ to $x = a$.

Suppose the interval $0 \leqslant x \leqslant a$ is divided into n equal intervals. Rectangular strips are drawn on each interval with height equal to the value of the function at the right hand side of the interval.

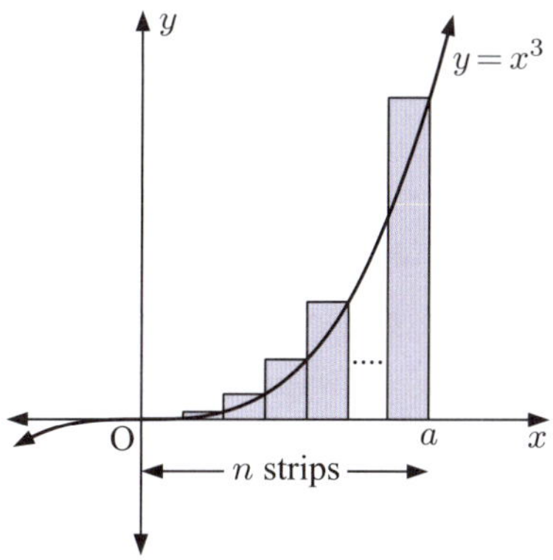

a Explain why, for finite n, the area between $y = x^3$ and the x-axis from $x = 0$ to $x = a$ must be less than the sum of the areas of the strips.

b Show that the sum of the areas of the strips

$S = \left(\frac{a}{n}\right)^4 (1^3 + 2^3 + ... + n^3)$.

c Use the formula $1^3 + 2^3 + 3^3 + ... + n^3 = \frac{n^2(n+1)^2}{4}$ to show that $S = \frac{a^4}{4}\left(1 + \frac{1}{n}\right)^2$.

d Hence find the exact value of $\int_0^a x^3\,dx$.

4 Use the previous results to predict the value of:

a $\int_0^a x^4\,dx$

b $\int_0^a x^5\,dx$

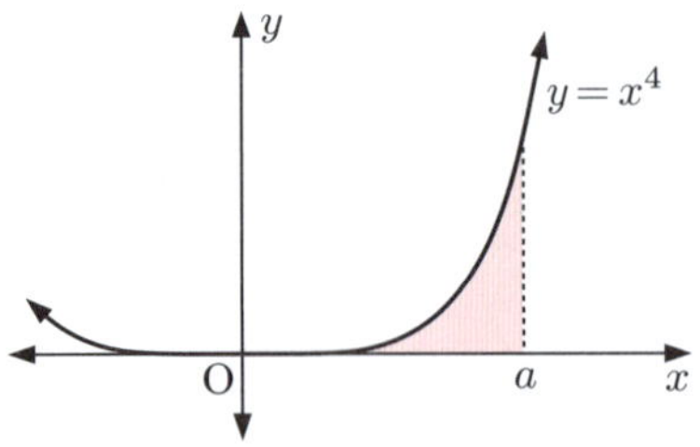

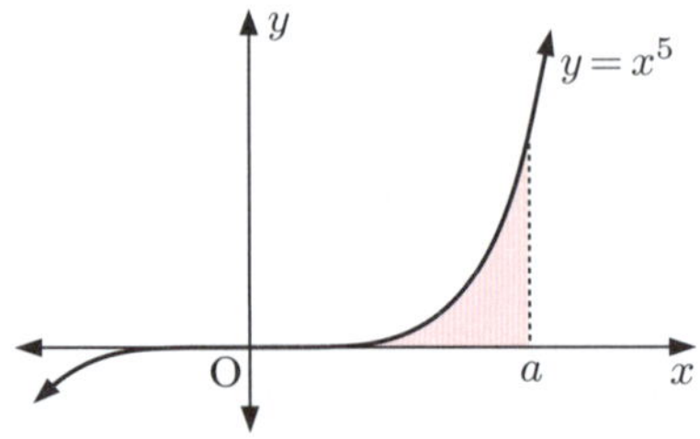

B INTEGRATION

The table alongside summarises the results obtained in the previous Exercise.

Looking carefully at the results, you may be able to see a pattern.

For example, for the case $y = x^2$, the area is of the form $\frac{x^3}{3}$.

The derivative of $\frac{x^3}{3}$ is $\frac{3x^2}{3} = x^2$, which was the original function.

This pattern also holds for the other functions in the table.

Function	*Area*
$y = 1$	a
$y = x$	$\frac{a^2}{2}$
$y = x^2$	$\frac{a^3}{3}$
$y = x^3$	$\frac{a^4}{4}$

This observation suggests that, to find the area under $y = f(x)$, we need to find a function whose derivative is $f(x)$. This is the reverse process of differentiation, and we call such a process **antidifferentiation** or **integration**.

Integration is the reverse process of differentiation.

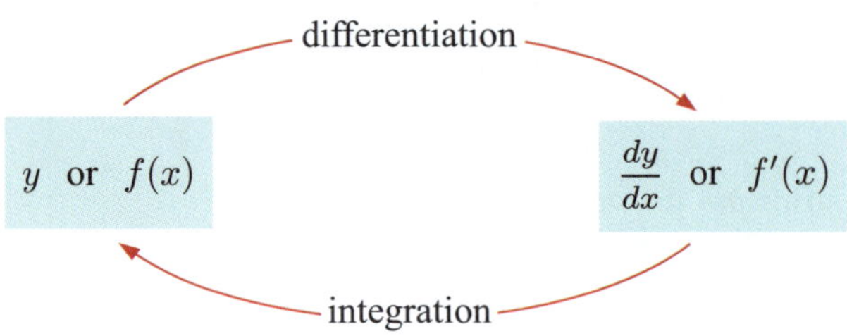

To integrate x^2, we must find a function for which x^2 is the derivative. We know the derivative of $\frac{1}{3}x^3$ is x^2. However, $\frac{1}{3}x^3 - 1$, $\frac{1}{3}x^3 + 10$, and $\frac{1}{3}x^3 - 5$ also have the derivative x^2.

In fact, all functions of the form $\frac{1}{3}x^3 + c$ where c is any real constant, have the derivative x^2.

We say that $\frac{1}{3}x^3 + c$ is the **integral** of x^2 with respect to x, and write $\int x^2\,dx = \frac{1}{3}x^3 + c$.

If $F(x)$ and $f(x)$ are functions such that $F'(x) = f(x)$ then:

- $f(x)$ is the **derivative** of $F(x)$ and
- $F(x)$ is the **integral** or **antiderivative** of $f(x)$.

$\int f(x)\,dx$ reads "the integral of $f(x)$ with respect to x".

If $\boldsymbol{F'(x) = f(x)}$ then $\boldsymbol{\int f(x)\,dx = F(x) + c}$.

Example 2 — Self Tutor

a Find $F'(x)$ for $F(x) = \frac{x^4}{4}$, and hence find $\int x^3\,dx$.

b Find $F'(x)$ for $F(x) = x^{\frac{3}{2}}$, and hence find $\int x^{\frac{1}{2}}\,dx$.

a If $F(x) = \frac{x^4}{4}$, then $F'(x) = \frac{4x^3}{4} = x^3$

$\therefore \int x^3\,dx = \frac{x^4}{4} + c$

b If $F(x) = x^{\frac{3}{2}}$, then $F'(x) = \frac{3}{2}x^{\frac{1}{2}}$

$\therefore$ if $F(x) = \frac{2}{3}x^{\frac{3}{2}}$, then $F'(x) = \frac{2}{3} \times \frac{3}{2}x^{\frac{1}{2}} = x^{\frac{1}{2}}$

$\therefore \int x^{\frac{1}{2}}\,dx = \frac{2}{3}x^{\frac{3}{2}} + c$

EXERCISE 16B

1 **a** Find $F'(x)$ for $F(x) = \frac{x^2}{2}$, and hence find $\int x\,dx$.

b Find $F'(x)$ for $F(x) = \frac{x^5}{5}$, and hence find $\int x^4\,dx$.

c Find $F'(x)$ for $F(x) = x^8$, and hence find $\int x^7\,dx$.

d Find $F'(x)$ for $F(x) = x^{-2}$, and hence find $\int x^{-3}\,dx$.

e Find $F'(x)$ for $F(x) = x^{\frac{1}{2}}$, and hence find $\int x^{-\frac{1}{2}}\,dx$.

f Use your results to predict a formula for $\int x^n\,dx$.

g Does your formula work for $n = -1$? Explain your answer.

2 **a** Find $F'(x)$ for $F(x) = e^x$, and hence find $\int e^x\,dx$.

b Find $F'(x)$ for $F(x) = \frac{1}{2}e^{2x}$, and hence find $\int e^{2x}\,dx$.

c Find $F'(x)$ for $F(x) = e^{\frac{x}{3}}$, and hence find $\int e^{\frac{x}{3}}\,dx$.

d Use your results to predict a formula for $\int e^{kx}\,dx$.

3 **a** Find $F'(x)$ for $F(x) = 5x$, and hence find $\int 5\,dx$.

b Predict a formula for $\int k\,dx$, where k is a constant.

4 **a** Find $F'(x)$ for $F(x) = \frac{1}{7}x^7$.

b Hence find:

i $\int x^6\,dx$ **ii** $\int 2x^6\,dx$ **iii** $\int 5x^6\,dx$ **iv** $\int -x^6\,dx$

c Copy and complete: If k is a constant, then $\int k\,f(x)\,dx = ...$

5 **a** Find $F'(x)$ for $F(x) = 3x^2 + 8x$, and hence find $\int (6x+8)\,dx$.

b Use your previous results to find:

i $\int 6x\,dx$ **ii** $\int 8\,dx$

c Copy and complete: $\int (f(x) + g(x))\,dx = ...$

Example 3

Self Tutor

Suppose $y = \sqrt{3x+1}$.

a Find $\frac{dy}{dx}$.

b Hence find $\int \frac{1}{\sqrt{3x+1}}\,dx$.

a

$$y = \sqrt{3x+1} = (3x+1)^{\frac{1}{2}}$$

$$\therefore \frac{dy}{dx} = \tfrac{1}{2}(3x+1)^{-\frac{1}{2}}(3) \quad \{\text{chain rule}\}$$

$$= \frac{3}{2\sqrt{3x+1}}$$

b Using **a**,

$$\int \frac{3}{2\sqrt{3x+1}}\,dx = \sqrt{3x+1} + c$$

$$\therefore \tfrac{3}{2}\int \frac{1}{\sqrt{3x+1}}\,dx = \sqrt{3x+1} + c$$

$$\therefore \int \frac{1}{\sqrt{3x+1}}\,dx = \tfrac{2}{3}\sqrt{3x+1} + c$$

6 Suppose $y = \sqrt{1-x}$.

a Find $\frac{dy}{dx}$.

b Hence find $\int \frac{1}{\sqrt{1-x}}\,dx$.

7 **a** Find $\frac{d}{dx}(\sqrt{2x+5})$. **b** Hence find $\int \frac{1}{\sqrt{2x+5}}\,dx$.

8 **a** **i** Find $\frac{d}{dx}(\ln x)$. **ii** Hence find $\int \frac{1}{x}\,dx$ for $x > 0$.

b **i** Find $\frac{d}{dx}(\ln(-x))$. **ii** Hence find $\int \frac{1}{x}\,dx$ for $x < 0$.

9 Suppose $y = \sin x$.

a Find $\frac{dy}{dx}$. **b** Hence find $\int \cos x\,dx$.

10 **a** Find $\frac{d}{dx}(\cos x)$. **b** Hence find $\int \sin x\,dx$.

11 Suppose $y = \tan x$.

a Find $\frac{dy}{dx}$. **b** Hence find $\int \sec^2 x\,dx$.

12 **a** Find $\frac{d}{dx}(\sin 3x)$. **b** Hence find $\int \cos 3x\,dx$.

C RULES FOR INTEGRATION

In the previous Exercise, you should have discovered some general rules to help integrate functions more efficiently. These rules can be deduced by reviewing the rules for differentiation.

For k a constant, $\frac{d}{dx}(kx+c) = k$ $\quad\therefore\quad$ $\boldsymbol{\int k\,dx = kx + c}$

If $n \neq -1$, $\frac{d}{dx}\left(\frac{x^{n+1}}{n+1} + c\right) = \frac{(n+1)x^n}{n+1} = x^n$ $\quad\therefore\quad$ $\boldsymbol{\int x^n\,dx = \frac{x^{n+1}}{n+1} + c,\ n \neq -1}$

$\frac{d}{dx}(e^x + c) = e^x$ $\quad\therefore\quad$ $\boldsymbol{\int e^x\,dx = e^x + c}$

For $x > 0$, $\frac{d}{dx}(\ln x + c) = \frac{1}{x}$

For $x < 0$, $\frac{d}{dx}(\ln(-x) + c) = \frac{-1}{-x} = \frac{1}{x}$

$\therefore \int \frac{1}{x}\,dx = \begin{cases} \ln x & \text{if } x > 0 \\ \ln(-x) & \text{if } x < 0 \end{cases}$ $\quad\therefore\quad$ $\boldsymbol{\int \frac{1}{x}\,dx = \ln|x| + c,\ x \neq 0}$

$\frac{d}{dx}(\sin x + c) = \cos x$ $\quad\therefore\quad$ $\boldsymbol{\int \cos x\,dx = \sin x + c}$

$\frac{d}{dx}(-\cos x + c) = \sin x$ $\quad\therefore\quad$ $\boldsymbol{\int \sin x\,dx = -\cos x + c}$

Function	*Integral*
k, a constant	$kx + c$
$x^n,\ n \neq -1$	$\dfrac{x^{n+1}}{n+1} + c$
e^x	$e^x + c$
$\dfrac{1}{x}$	$\ln\lvert x\rvert + c$
$\cos x$	$\sin x + c$
$\sin x$	$-\cos x + c$

You should have also discovered the following useful rules:

- $\displaystyle\int k\,f(x)\,dx = k\int f(x)\,dx$ $\{k$ is a constant$\}$
- $\displaystyle\int (f(x) \pm g(x))\,dx = \int f(x)\,dx \pm \int g(x)\,dx$

Example 4

Self Tutor

Integrate with respect to x:

a $x^3 + 3x + 4$ **b** $(x^2 - 4)^2$ **c** $\dfrac{\sqrt{x} + 5}{x}$

a
$$\int (x^3 + 3x + 4)\,dx$$
$$= \int x^3\,dx + \int 3x\,dx + \int 4\,dx$$
$$= \frac{x^4}{4} + \frac{3x^2}{2} + 4x + c$$

b
$$\int (x^2 - 4)^2\,dx$$
$$= \int (x^4 - 8x^2 + 16)\,dx$$
$$= \frac{x^5}{5} - \frac{8x^3}{3} + 16x + c$$

c
$$\int \frac{\sqrt{x} + 5}{x}\,dx$$
$$= \int \left(x^{-\frac{1}{2}} + \frac{5}{x}\right)dx$$
$$= \frac{x^{\frac{1}{2}}}{\frac{1}{2}} + 5\ln\lvert x\rvert + c$$
$$= 2\sqrt{x} + 5\ln\lvert x\rvert + c$$

EXERCISE 16C.1

1 Integrate with respect to x:

a x^{10} **b** $8x^3$ **c** $-4x^8$

d $\dfrac{1}{x^4}$ **e** $6x - 2$ **f** $3x^2$

g $\dfrac{6}{x}$ **h** $2x^2 + 1$ **i** $x^2 + 3x - 2$

j $x^3 - x$ **k** $5x^2 - \dfrac{2}{x}$ **l** $10x + \dfrac{5}{x^2}$

m $5\sqrt{x}$ **n** $-\dfrac{3}{\sqrt{x}}$ **o** $2\sqrt{x} + \dfrac{7}{x}$

p $2x^3 + 3x - \dfrac{4}{x^2}$ **q** $x^2 + \dfrac{1}{\sqrt{x}}$ **r** $3\sqrt{x} - \dfrac{2}{x^3}$

2 Find:

a $\int (x-2)^2\,dx$ **b** $\int x(x-6)\,dx$ **c** $\int (x^2-3)^2\,dx$

d $\int \left(\frac{3}{x}+x\right)^2 dx$ **e** $\int \left(\frac{4}{x}+1\right)^2 dx$ **f** $\int \frac{4x^2+3x}{x}\,dx$

g $\int \frac{2x^2+5}{x^2}\,dx$ **h** $\int \frac{x^2+2x-3}{x}\,dx$ **i** $\int \frac{x^3+4x^2-1}{x}\,dx$

j $\int \frac{6x+4}{\sqrt{x}}\,dx$ **k** $\int \frac{x^2-2}{\sqrt{x}}\,dx$ **l** $\int \frac{\left(3-\frac{1}{\sqrt{x}}\right)^2}{\sqrt{x}}\,dx$

3 Use the binomial expansion to help find:

a $\int (x+2)^3\,dx$ **b** $\int (3x+1)^3\,dx$ **c** $\int (2-x)^3\,dx$ **d** $\int (x-1)^4\,dx$

Example 5 — Self Tutor

Find: **a** $\int 4\sin x\,dx$ **b** $\int \left(5e^x - \frac{1}{2}\cos x\right) dx$

a $\int 4\sin x\,dx$

$= 4(-\cos x)+c$

$= -4\cos x + c$

b $\int \left(5e^x - \frac{1}{2}\cos x\right) dx$

$= 5e^x - \frac{1}{2}\sin x + c$

4 Find:

a $\int 2\cos x\,dx$ **b** $\int 7e^x\,dx$ **c** $\int (5\sin x - e^x)\,dx$

d $\int (\sin x + \cos x)\,dx$ **e** $\int \left(6\cos x - \frac{e^x}{4}\right) dx$ **f** $\int \left(2e^x - \frac{1}{3}\sin x\right) dx$

5 Find $f(x)$ if:

a $f'(x) = 5x^3 - 2x + 6$ **b** $f'(x) = x^2\sqrt{x} - \frac{2}{x^4}$

c $f'(x) = x^3 - \cos x$ **d** $f'(x) = 6x - 4e^x$

e $f'(x) = \sin x - \frac{4}{x}$ **f** $f'(x) = 2e^x - \frac{7-x^6}{x}$

g $f'(x) = \frac{1}{\sqrt{x}} - 3\sin x$ **h** $f'(x) = 3\cos x - \sin x + \frac{1}{5}e^x$

6 Find y if:

a $\frac{dy}{dx} = \frac{4}{x^3} - 10x$ **b** $\frac{dy}{dx} = \frac{x^3+5x-1}{x^2}$ **c** $\frac{dy}{dx} = \frac{2e^x - \sqrt{x}}{3}$

d $\frac{dy}{dx} = 6\cos x - 2\sin x$ **e** $\frac{dy}{dx} = \frac{x^2+3}{\sqrt{x}} - 8e^x$ **f** $\frac{dy}{dx} = (2x + x^{-1})^2 - \frac{1}{2}\cos x$

PARTICULAR VALUES

We can find the constant of integration c if we are given a particular value of the function.

Example 6

Self Tutor

Find $f(x)$ given that:

a $f'(x) = x^2 - 5x + 2$ and $f(0) = 3$

b $f'(x) = 3e^x + \cos x$ and $f(0) = 1$.

a Since $f'(x) = x^2 - 5x + 2$,

$$f(x) = \int (x^2 - 5x + 2)\, dx$$

$$\therefore\ f(x) = \frac{x^3}{3} - \frac{5x^2}{2} + 2x + c$$

But $f(0) = 3$

$\therefore\ c = 3$

Thus $f(x) = \dfrac{x^3}{3} - \dfrac{5x^2}{2} + 2x + 3$

b Since $f'(x) = 3e^x + \cos x$,

$$f(x) = \int (3e^x + \cos x)\, dx$$

$$\therefore\ f(x) = 3e^x + \sin x + c$$

But $f(0) = 1$

$\therefore\ 3(1) + 0 + c = 1$

$\therefore\ c = -2$

Thus $f(x) = 3e^x + \sin x - 2$

If we are given a second derivative, we need to integrate twice to find the function. This creates two integrating constants, so we need two other facts about the curve in order to determine these constants.

Example 7

Self Tutor

Find $f(x)$ given that $f''(x) = 12x^2 - 4$, $f'(0) = -1$, and $f(1) = 4$.

If $f''(x) = 12x^2 - 4$

then $f'(x) = \dfrac{12x^3}{3} - 4x + c$ {integrating with respect to x}

$\therefore\ f'(x) = 4x^3 - 4x + c$

But $f'(0) = -1$, so $c = -1$

Thus $f'(x) = 4x^3 - 4x - 1$

$\therefore\ f(x) = \dfrac{4x^4}{4} - \dfrac{4x^2}{2} - x + d$ {integrating again}

$\therefore\ f(x) = x^4 - 2x^2 - x + d$

But $f(1) = 4$, so $1 - 2 - 1 + d = 4$ and hence $d = 6$

Thus $f(x) = x^4 - 2x^2 - x + 6$

EXERCISE 16C.2

1 Find $f(x)$ given that:

a $f'(x) = 4x - 1$ and $f(0) = 5$

b $f'(x) = x + 2$ and $f(0) = -4$

c $f'(x) = \dfrac{6}{x^2}$ and $f(1) = 0$

d $f'(x) = 3\sqrt{x}$ and $f(4) = 5$.

2 Find y given that:

a $\dfrac{dy}{dx} = x^2 - 2x$ and $y = 0$ when $x = 1$

b $\dfrac{dy}{dx} = e^x + 3$ and $y = 4$ when $x = 0$

c $\dfrac{dy}{dx} = \sqrt{x} + \dfrac{3}{x}$ and $y = 1$ when $x = 1$

d $\dfrac{dy}{dx} = 2\sin x - 6\cos x$ and $y = 7$ when $x = \frac{\pi}{6}$.

3 A curve passes through the point $(1, 4)$ and has gradient function $\dfrac{dy}{dx} = 5x^{-\frac{1}{2}}$.
Find the equation of the curve.

4 The function $f(x)$ has derivative $f'(x) = 3x - 4e^x$. Given that the curve $y = f(x)$ passes through $(0, -2)$, find the equation of the curve.

5 Find $f(x)$ given that:

a $f''(x) = 6x - 2$, $f'(0) = 4$, and $f(1) = -1$

b $f''(x) = \dfrac{1}{\sqrt{x}} - 3$, $f'(1) = 3$, and $f(4) = 3$

c $f''(x) = \dfrac{2}{x^2} - \dfrac{6}{x^3}$, $f'(1) = 1$, and $f(e^2) = -4$

d $f''(x) = \sin x - 3\cos x$, $f(0) = 2$, and $f(-\frac{\pi}{2}) = \pi$.

6 The second derivative of a function is $\dfrac{d^2y}{dx^2} = \cos x - 5e^x$. When $x = 0$, $\dfrac{dy}{dx} = 1$ and $y = -10$.
Find the equation of the curve.

D INTEGRATING $f(ax+b)$

In **Chapter 14** we used the chain rule to find derivatives such as:

- $\dfrac{d}{dx}\left((2x+5)^3\right) = 3(2x+5)^2 \times 2$
- $\dfrac{d}{dx}\left(e^{3x+1}\right) = 3e^{3x+1}$
- $\dfrac{d}{dx}(\sin 5x) = 5\cos 5x$

By applying the chain rule in reverse, we can integrate more complicated functions.

Notice that $\dfrac{d}{dx}\left(\dfrac{1}{a}e^{ax+b}\right) = \dfrac{1}{a}e^{ax+b} \times a = e^{ax+b}$

$$\therefore \quad \int e^{ax+b}\,dx = \frac{1}{a}e^{ax+b} + c \quad \text{for } a \neq 0$$

Likewise if $n \neq -1$,

$$\begin{aligned}\frac{d}{dx}\left(\frac{1}{a(n+1)}(ax+b)^{n+1}\right) &= \frac{1}{a(n+1)}(n+1)(ax+b)^n \times a \\ &= (ax+b)^n\end{aligned}$$

$$\therefore \quad \int (ax+b)^n\,dx = \frac{1}{a}\,\frac{(ax+b)^{n+1}}{(n+1)} + c \quad \text{for } n \neq -1$$

We can perform the same process for $\cos(ax+b)$ and $\sin(ax+b)$:

$$\frac{d}{dx}\left(\frac{1}{a}\sin(ax+b)\right) = \frac{1}{a}\cos(ax+b)\times a$$
$$= \cos(ax+b)$$

So, $$\int \cos(ax+b)\,dx = \frac{1}{a}\sin(ax+b)+c \quad \text{for } a \neq 0.$$

Likewise we can show $$\int \sin(ax+b)\,dx = -\frac{1}{a}\cos(ax+b)+c \quad \text{for } a \neq 0.$$

Finally, $\frac{d}{dx}\left(\frac{1}{a}\ln(ax+b)\right) = \frac{1}{a}\left(\frac{a}{ax+b}\right) = \frac{1}{ax+b}$ for $ax+b>0$, $a \neq 0$

$$\therefore \int \frac{1}{ax+b}\,dx = \frac{1}{a}\ln(ax+b)+c \quad \text{for } ax+b>0,\ a\neq 0$$

We can similarly show that $\int \frac{1}{ax+b}\,dx = \frac{1}{a}\ln(-(ax+b))+c$ for $ax+b<0$, $a \neq 0$

$$\therefore \int \frac{1}{ax+b}\,dx = \frac{1}{a}\ln|ax+b|+c \quad \text{for } a\neq 0.$$

For a, b constants with $a \neq 0$, we have:

Function	*Integral*
e^{ax+b}	$\frac{1}{a}e^{ax+b}+c$
$(ax+b)^n$, $n \neq -1$	$\frac{1}{a}\frac{(ax+b)^{n+1}}{n+1}+c$
$\cos(ax+b)$	$\frac{1}{a}\sin(ax+b)+c$
$\sin(ax+b)$	$-\frac{1}{a}\cos(ax+b)+c$
$\frac{1}{ax+b}$	$\frac{1}{a}\ln\|ax+b\|+c$

Example 8

Self Tutor

Find: **a** $\int (3x-2)^5\,dx$ **b** $\int \left(\frac{1}{\sqrt{4x+7}}+\frac{1}{5x-1}\right)dx$

a $\int (3x-2)^5\,dx$

$= \frac{1}{3}\times\frac{(3x-2)^6}{6}+c$

$= \frac{1}{18}(3x-2)^6+c$

b $\int \left(\frac{1}{\sqrt{4x+7}}+\frac{1}{5x-1}\right)dx$

$= \int \left((4x+7)^{-\frac{1}{2}}+\frac{1}{5x-1}\right)dx$

$= \frac{1}{4}\times\frac{(4x+7)^{\frac{1}{2}}}{\frac{1}{2}}+\frac{1}{5}\ln|5x-1|+c$

$= \frac{1}{2}\sqrt{4x+7}+\frac{1}{5}\ln|5x-1|+c$

EXERCISE 16D

1 Find:

a $\int (2x-1)^4\,dx$ **b** $\int (5x+2)^3\,dx$ **c** $\int (3-x)^5\,dx$

d $\int (4+3x)^4\,dx$ **e** $\int 2(4x-3)^5\,dx$ **f** $\int (5-2x)^6\,dx$

2 Integrate with respect to x:

a $\frac{1}{(3x+1)^2}$ **b** $\sqrt{6x-1}$ **c** $\frac{4}{(5x-3)^3}$

d $\sqrt{2-4x}$ **e** $\frac{1}{\sqrt{2x+3}}$ **f** $\frac{6}{\sqrt{1-8x}}$

3 Find:

a $\int \frac{1}{3x+5}\,dx$ **b** $\int \frac{1}{4x-1}\,dx$ **c** $\int \frac{1}{7-2x}\,dx$

d $\int \frac{3}{8x-7}\,dx$ **e** $\int \left(\frac{3}{\sqrt{x}} - \frac{1}{x+5}\right) dx$ **f** $\int \left(\frac{2}{1-6x} + \sqrt{x+1}\right) dx$

Example 9 — Self Tutor

Integrate with respect to x:

a $e^{5x+1} - 6e^{2x}$ **b** $\cos 4x + 3\sin(2x+\pi)$

a $\int (e^{5x+1} - 6e^{2x})\,dx$

$= \frac{1}{5}e^{5x+1} - 6(\frac{1}{2})e^{2x} + c$

$= \frac{1}{5}e^{5x+1} - 3e^{2x} + c$

b $\int (\cos 4x + 3\sin(2x+\pi))\,dx$

$= \frac{1}{4}\sin 4x + 3(-\frac{1}{2})\cos(2x+\pi) + c$

$= \frac{1}{4}\sin 4x - \frac{3}{2}\cos(2x+\pi) + c$

4 Find:

a $\int e^{3x}\,dx$ **b** $\int e^{4x-1}\,dx$ **c** $\int e^{9-2x}\,dx$

d $\int (e^{2x} + e^{-4x})\,dx$ **e** $\int (4e^{2x+1} - e^{-x})\,dx$ **f** $\int (e^{x+1} - 5)^2\,dx$

g $\int (e^{6x-2} + (3x-2)^3)\,dx$ **h** $\int (e^{2x} + e^{-2x})^2\,dx$ **i** $\int \left(5e^{1-3x} - \frac{1}{\sqrt{4x+1}}\right) dx$

5 Integrate with respect to x:

a $\cos 4x$ **b** $\sin 3x$ **c** $\cos\left(2x + \frac{\pi}{2}\right)$

d $2\cos 3x + \sin 6x$ **e** $5\cos\left(\frac{\pi}{2} - x\right)$ **f** $\frac{1}{6}\sin 2x + \cos\left(x - \frac{\pi}{6}\right)$

g $\sin 5x - x^2$ **h** $e^{7x} - 3\cos 4x$ **i** $2\sin(\pi - 3x) + \sqrt{9-x}$

6 Find $f(x)$ given $f'(x) = (3x-1)^4$ and $f(0) = 0$.

7 A curve has gradient function $\frac{dy}{dx} = (4x-3)^{-\frac{1}{2}}$. Given that it passes through $(3, 5)$, find the equation of the curve.

8 A curve has gradient function $\frac{dy}{dx} = \frac{3}{2x-6}$. Given that it passes through $(5, \ln 40)$, find the equation of the curve.

9 The function $f(x)$ has derivative $f'(x) = 6x - 4e^{2x}$.

a Given that the curve $y = f(x)$ passes through $\mathrm{P}(0, 3)$, find the equation of the curve.

b Find the equation of the tangent to the curve at P.

10 A curve has gradient function $\frac{dy}{dx} = 6\sin 3x$.

a Given that the curve passes through $\mathrm{P}\left(\frac{\pi}{2}, -1\right)$, find the equation of the curve.

b Find the equation of the normal to the curve at:

i P **ii** the point where $x = \frac{\pi}{9}$.

11 A function has second derivative $\frac{d^2y}{dx^2} = 8e^{-4x}$. Given that $\frac{dy}{dx} = -6$ when $x = 0$, and that the curve passes through $\left(1, \frac{1}{2e^4}\right)$, find the equation of the curve.

12 Suppose $f''(x) = \cos ax + \sin(x - a)$, where $0 \leqslant a \leqslant 2\pi$ is a constant. The curve $y = f(x)$ has a stationary inflection point at $(0, 1)$.

a Find a. **b** Find the equation of the curve.

13 Consider a function f such that $f''(x) = \sin ax$, $a \in \mathbb{R}$. For what values of a will the tangents to $y = f(x)$ at $x = \frac{\pi}{2}$ and $x = \pi$ have the same gradient?

14 A function has derivative $f'(x) = \cos(2x + a)$ where $0 \leqslant a \leqslant 2\pi$. Given that $f(0) = 2$ and $f(\frac{\pi}{2}) = 1$, find the function $f(x)$.

E THE DEFINITE INTEGRAL

We have seen that the area between the curve $y = x^2$ and the x-axis from $x = 0$ to $x = a$, is $\frac{a^3}{3}$.

Similarly, the area between the curve and the x-axis from $x = 0$ to $x = b$ is $\frac{b^3}{3}$.

So, the shaded area $\int_a^b x^2\,dx = \frac{b^3}{3} - \frac{a^3}{3}$.

This can also be written as $F(b) - F(a)$, where $F(x) = \frac{x^3}{3}$ is the antiderivative of $f(x) = x^2$.

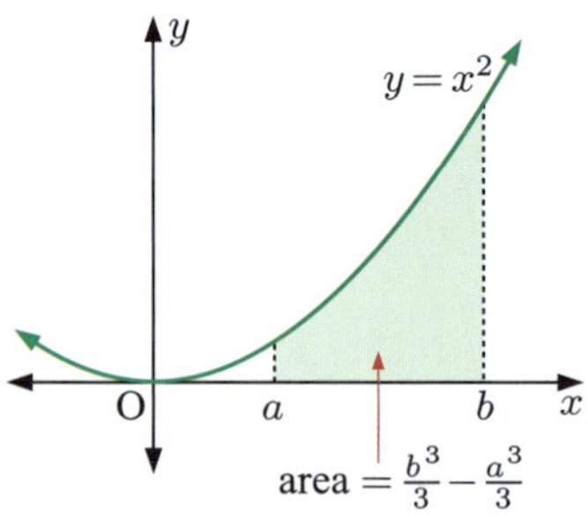

THE FUNDAMENTAL THEOREM OF CALCULUS

If $F(x)$ is the antiderivative or integral of $f(x)$ then the **definite integral** of $f(x)$ on the interval $a \leqslant x \leqslant b$ is

$$\int_a^b f(x)\,dx = F(b) - F(a).$$

Example 10 — Self Tutor

Evaluate $\int_1^2 x^2\,dx$.

If $f(x)=x^2$, then $F(x)=\frac{x^3}{3}$.

$$\therefore \int_1^2 x^2\,dx = F(2)-F(1)$$
$$= \left(\frac{2^3}{3}\right)-\left(\frac{1^3}{3}\right)$$
$$= \frac{7}{3}$$
$$= 2\tfrac{1}{3}$$

We often write $F(b)-F(a)$ as $\left[F(x)\right]_a^b$, so $\int_a^b f(x)\,dx = \left[F(x)\right]_a^b = F(b)-F(a)$.

Example 11 — Self Tutor

Evaluate:

a $\int_0^1 (e^{3x}-1)\,dx$

b $\int_{\frac{\pi}{3}}^{\frac{\pi}{2}} (\cos x+\sin 2x)\,dx$

a $\int_0^1 (e^{3x}-1)\,dx$
$$= \left[\frac{e^{3x}}{3}-x\right]_0^1$$
$$= \left(\frac{e^3}{3}-1\right)-\left(\frac{e^0}{3}-0\right)$$
$$= \frac{e^3}{3}-1-\frac{1}{3}$$
$$= \frac{e^3-4}{3}$$

b $\int_{\frac{\pi}{3}}^{\frac{\pi}{2}} (\cos x+\sin 2x)\,dx$
$$= \left[\sin x - \tfrac{1}{2}\cos 2x\right]_{\frac{\pi}{3}}^{\frac{\pi}{2}}$$
$$= \left(\sin\tfrac{\pi}{2}-\tfrac{1}{2}\cos\pi\right)-\left(\sin\tfrac{\pi}{3}-\tfrac{1}{2}\cos\tfrac{2\pi}{3}\right)$$
$$= 1-\tfrac{1}{2}(-1)-\tfrac{\sqrt{3}}{2}+\tfrac{1}{2}\left(-\tfrac{1}{2}\right)$$
$$= \tfrac{5}{4}-\tfrac{\sqrt{3}}{2}$$
$$= \frac{5-2\sqrt{3}}{4}$$

EXERCISE 16E

1 **a** Find $\int 6x\,dx$. **b** Hence find $\int_1^2 6x\,dx$.

2 **a** Find $\int (4x+5)\,dx$. **b** Hence find $\int_{-1}^1 (4x+5)\,dx$.

3 Evaluate:

a $\int_0^2 3x^2\,dx$ **b** $\int_2^5 x\,dx$ **c** $\int_1^3 (2x+4)\,dx$

d $\int_{-1}^2 (x^2+3)\,dx$ **e** $\int_1^6 (2-x)\,dx$ **f** $\int_{0.5}^1 \frac{1}{x^2}\,dx$

g $\int_1^3 \left(4-\frac{2}{x^3}\right)dx$ **h** $\int_1^4 \sqrt{x}\,dx$ **i** $\int_4^9 \frac{4}{\sqrt{x}}\,dx$

4 Evaluate:

a $\int_1^4 \left(\sqrt{x}-\frac{1}{\sqrt{x}}\right)dx$ **b** $\int_0^2 (2x+3)^2\,dx$ **c** $\int_{-2}^{-1} \frac{x^2+4}{x^2}\,dx$

d $\int_{-1}^1 (x^3-x)\,dx$ **e** $\int_4^9 \frac{\sqrt{x}-x^2}{x}\,dx$ **f** $\int_1^4 \sqrt{x}(x+1)\,dx$

5 Find:

a $\int_2^5 \frac{1}{x}\,dx$ **b** $\int_{-6}^{-4} \frac{3}{x}\,dx$ **c** $\int_e^{e^3} \frac{6}{x}\,dx$

d $\int_3^7 \frac{4}{2x+5}\,dx$ **e** $\int_{-2}^0 \frac{5}{3x-1}\,dx$ **f** $\int_{-2}^3 \frac{7}{11-3x}\,dx$

6 Find:

a $\int_0^1 e^x\,dx$ **b** $\int_1^{\ln 3} 4e^x\,dx$ **c** $\int_0^2 (e^x+2)\,dx$

d $\int_0^3 e^{2x}\,dx$ **e** $\int_1^3 e^{\frac{x}{3}}\,dx$ **f** $\int_{-2}^1 e^{5x-2}\,dx$

g $\int_{1.5}^2 e^{4x-1}\,dx$ **h** $\int_0^1 (e^{-x+1}-2x)\,dx$ **i** $\int_0^{\ln 2} (e^{2x}+5e^x)\,dx$

7 Evaluate:

a $\int_0^{\frac{\pi}{6}} \cos x\,dx$ **b** $\int_0^{\frac{\pi}{2}} 2\sin x\,dx$ **c** $\int_{-\frac{\pi}{2}}^{\pi} 5\cos x\,dx$

d $\int_0^{\frac{\pi}{2}} (x-\sin x)\,dx$ **e** $\int_{\frac{\pi}{6}}^{\pi} \cos 3x\,dx$ **f** $\int_0^{\frac{\pi}{2}} (2-\sin 2x)\,dx$

g $\int_0^{\frac{\pi}{2}} \cos\frac{x}{3}\,dx$ **h** $\int_{-\frac{\pi}{6}}^{\frac{\pi}{3}} (\sin 2x - 2\cos 3x)\,dx$ **i** $\int_0^{\pi} \cos\left(x+\frac{\pi}{2}\right)dx$

j $\int_{-\frac{\pi}{2}}^0 \sin(5x-\pi)\,dx$ **k** $\int_{\frac{\pi}{4}}^{\frac{\pi}{3}} \cos(\pi-2x)\,dx$ **l** $\int_{\frac{\pi}{4}}^{\pi} \left(\sin\frac{2x}{3}-\sqrt{x}\right)dx$

8 Find m given that:

a $\int_1^m \frac{1}{x^2}\,dx = \frac{1}{2}$ **b** $\int_{\ln 2}^m e^{2x}\,dx = 1$ **c** $\int_0^m \sin\frac{x}{2}\,dx = 1,\ 0 \leqslant m \leqslant \pi$

9 Use the Fundamental Theorem of Calculus to prove that:

a $\displaystyle\int_a^a f(x)\,dx = 0$

b $\displaystyle\int_a^b f(x)\,dx = -\int_b^a f(x)\,dx$

c $\displaystyle\int_a^b f(x)\,dx + \int_b^c f(x)\,dx = \int_a^c f(x)\,dx$

d $\displaystyle\int_a^b k\,f(x)\,dx = k\int_a^b f(x)\,dx$ where k is a constant

e $\displaystyle\int_a^b (f(x)+g(x))\,dx = \int_a^b f(x)\,dx + \int_a^b g(x)\,dx$

10 Given that $\displaystyle\int_0^2 f(x)\,dx = 5$ and $\displaystyle\int_2^5 f(x)\,dx = 9$, find:

a $\displaystyle\int_0^5 f(x)\,dx$ **b** $\displaystyle\int_5^2 f(x)\,dx$ **c** $\displaystyle\int_3^3 f(x)\,dx$ **d** $\displaystyle\int_0^2 4f(x)\,dx$

11 Suppose $\displaystyle\int_1^3 f(x)\,dx = 6$ and $\displaystyle\int_1^3 g(x)\,dx = 2$. Find:

a $\displaystyle\int_1^3 3f(x)\,dx$ **b** $\displaystyle\int_1^3 (f(x)+g(x))\,dx$ **c** $\displaystyle\int_1^3 (f(x)-2g(x))\,dx$

12

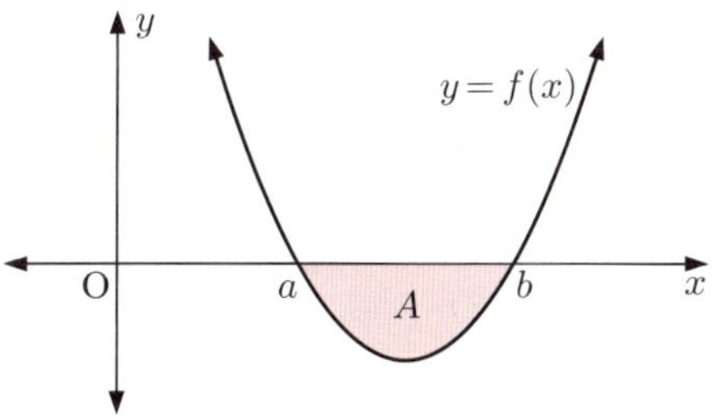

Suppose $f(x) \leqslant 0$ on an interval $a \leqslant x \leqslant b$.
By reflecting $y = f(x)$ in the x-axis, explain why the shaded area $A = -\displaystyle\int_a^b f(x)\,dx$.

13 Evaluate the following integrals using areas:

a $\displaystyle\int_0^3 f(x)\,dx$ **b** $\displaystyle\int_3^5 f(x)\,dx$

c $\displaystyle\int_3^9 f(x)\,dx$ **d** $\displaystyle\int_0^7 f(x)\,dx$

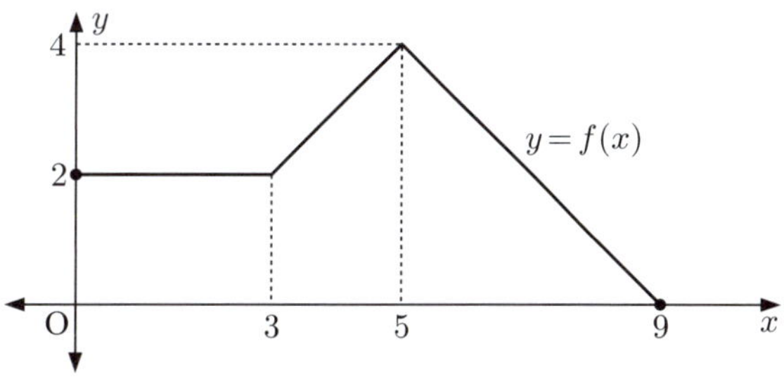

14 Evaluate the following integrals using areas:

a $\displaystyle\int_3^6 f(x)\,dx$ **b** $\displaystyle\int_6^9 f(x)\,dx$

c $\displaystyle\int_0^6 f(x)\,dx$ **d** $\displaystyle\int_3^9 f(x)\,dx$

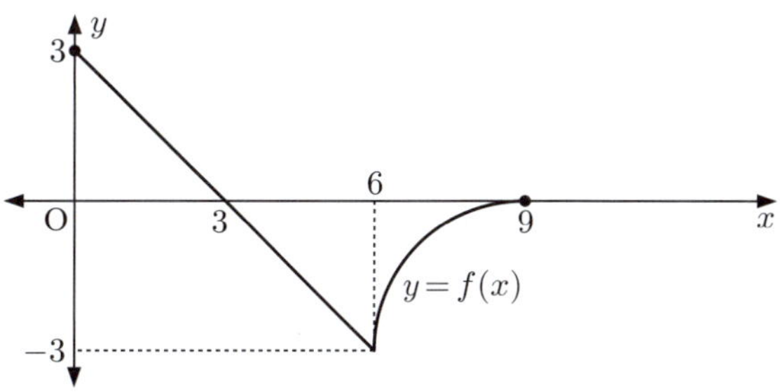

Example 12

Self Tutor

a Show that $\dfrac{d}{dx}\left(xe^{2x}-\dfrac{e^{2x}}{2}\right)=2xe^{2x}$.

b Hence find $\displaystyle\int_0^1 2xe^{2x}\,dx$.

a $\dfrac{d}{dx}\left(xe^{2x}-\dfrac{e^{2x}}{2}\right)=(1)e^{2x}+x(2e^{2x})-e^{2x}$ {product rule}

$=2xe^{2x}$

b $\displaystyle\int_0^1 2xe^{2x}\,dx=\left[xe^{2x}-\frac{e^{2x}}{2}\right]_0^1$ {using **a**}

$=(1)e^2-\dfrac{e^2}{2}-\left(0-\dfrac{e^0}{2}\right)$

$=\frac{1}{2}e^2+\frac{1}{2}$

$=\dfrac{e^2+1}{2}$

15 **a** Show that $\dfrac{d}{dx}\left(\dfrac{2x^2-1}{x^2+1}\right)=\dfrac{6x}{(x^2+1)^2}$.

b Hence find $\displaystyle\int_0^2 \frac{6x}{(x^2+1)^2}\,dx$.

16 **a** Show that $\dfrac{d}{dx}(x\sqrt{x+1})=\dfrac{3x+2}{2\sqrt{x+1}}$.

b Hence find $\displaystyle\int_0^3 \frac{3x+2}{\sqrt{x+1}}\,dx$.

17 **a** Find $\dfrac{d}{dx}\left(e^{x^3}\right)$.

b Hence find $\displaystyle\int x^2e^{x^3}\,dx$.

c Find the exact value of a such that $\displaystyle\int_0^a x^2e^{x^3}\,dx=3$.

Review set 16A

1 Use an area formula to find $\displaystyle\int_0^4 2x\,dx$.

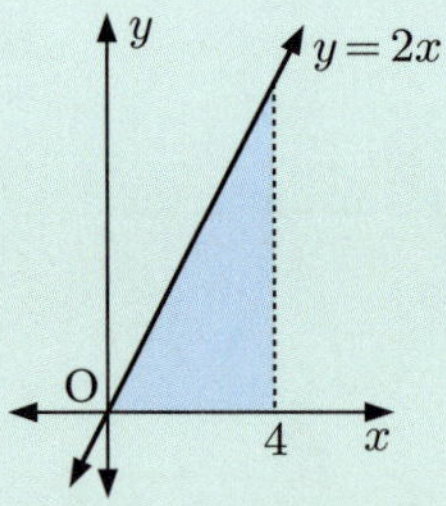

2 Suppose $y=\sqrt{2x-3}$.

a Find $\dfrac{dy}{dx}$.

b Hence find $\displaystyle\int \frac{1}{\sqrt{2x-3}}\,dx$.

3 Integrate with respect to x:

a x^9

b $6x^2+x-1$

c $3\sqrt{x}+\dfrac{4}{x}$

d $\dfrac{8}{x^3}-7$

e $\left(3+\dfrac{2}{x}\right)^2$

f $\dfrac{5x-4}{\sqrt{x}}$

4 Find $f(x)$ given that:

a $f'(x)=1-3e^x$ and $f(0)=1$

b $f'(x)=2\sin x-x$ and $f(0)=-6$.

5 Find:

a $\int (2x+4)^3\,dx$ **b** $\int (7-x)^5\,dx$ **c** $\int \sqrt{5x-2}\,dx$

d $\int \frac{1}{4x-3}\,dx$ **e** $\int e^{6-3x}\,dx$ **f** $\int \left(\sqrt{2-x}+e^{\frac{x}{5}}\right)dx$

6 The curve $y=f(x)$ has the gradient function $f'(x)=1-\cos 3x$.

a Given that the curve passes through $P\left(\frac{\pi}{2}, \frac{\pi}{2}\right)$, find the equation of the curve.

b Find the equation of the tangent to the curve at P.

7 **a** Find $\int x(x+4)\,dx$.

b Hence find $\int_0^2 x(x+4)\,dx$.

8 Evaluate:

a $\int_2^4 (3x-1)\,dx$ **b** $\int_4^9 \sqrt{x}\,dx$ **c** $\int_3^4 \frac{5x^2-6}{x^2}\,dx$

d $\int_{0.5}^1 e^{2x+1}\,dx$ **e** $\int_0^{\ln 3} (e^x-3)^2\,dx$ **f** $\int_0^{\frac{\pi}{2}} \sin\left(2x-\frac{\pi}{2}\right)dx$

9 Find m given that: **a** $\int_m^5 \frac{1}{\sqrt{x+4}}\,dx=4$ **b** $\int_0^m \frac{4}{3x+2}\,dx=\ln 16$.

10 Evaluate the following integrals using areas:

a $\int_0^4 f(x)\,dx$ **b** $\int_4^8 f(x)\,dx$

c $\int_2^7 f(x)\,dx$

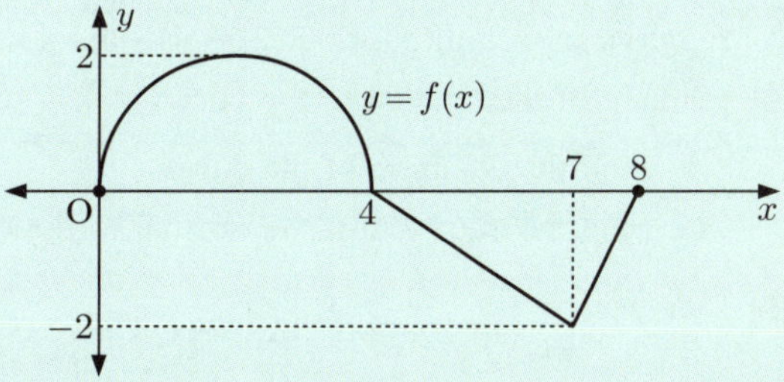

11 **a** Find $\frac{d}{dx}\left(\ln(2x^2+5)\right)$ and hence find $\int \frac{x}{2x^2+5}\,dx$.

b Evaluate $\int_3^4 \frac{x}{2x^2+5}\,dx$, giving your answer in the form $a\ln b$ where $a, b \in \mathbb{Q}$.

Review set 16B

1 Use an area formula to find $\int_{-1}^2 (x+3)\,dx$.

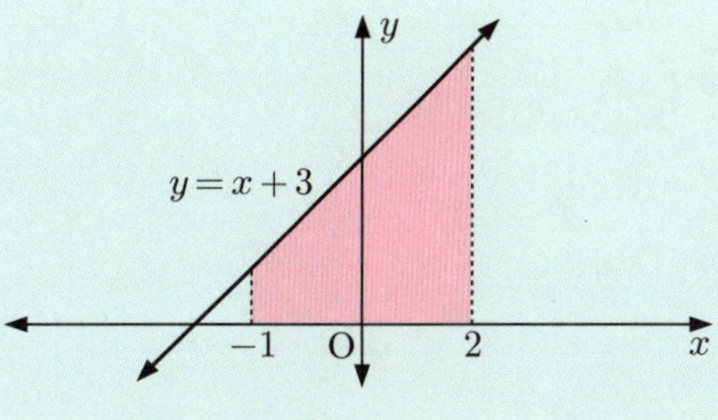

2 Suppose $f(x) = x\ln x - x$.

a Find $f'(x)$. **b** Hence find $\int \ln x \, dx$.

3 Find:

a $\int (3 - x^5)\, dx$ **b** $\int \sqrt{x}(x-1)\, dx$ **c** $\int 5\cos x \, dx$

d $\int (-2e^x - \sin x)\, dx$ **e** $\int \left(\frac{1}{4}\cos x - \frac{3}{x^4} + \frac{2}{x}\right) dx$ **f** $\int (e^x + 1)^2 \, dx$

4 Find y if:

a $\frac{dy}{dx} = \frac{5e^x - 2\sqrt{x}}{3}$ **b** $\frac{dy}{dx} = 5\cos x - 4\sin 3x$

5 Find $f(x)$ given that:

a $f'(x) = \frac{2}{\sqrt{x}} + 6$ and $f(1) = 7$

b $f''(x) = \sin x - e^x$, $f'(0) = 2$, and $f(0) = 5$.

6 Integrate with respect to x:

a $\frac{1}{(2-x)^4}$ **b** $-\frac{6}{\sqrt{2x-1}}$ **c** $e^{2x} - e^{-3x}$

d $\cos\frac{x}{2} - 7x^2$ **e** $\sin 4x - \frac{5}{e^x} + \frac{6}{1-4x}$ **f** $\frac{1}{10}\cos\left(2x - \frac{\pi}{4}\right) + \frac{8}{x^2}$

7 A curve passes through P(2, −4). Its second derivative is $\frac{d^2y}{dx^2} = 3e^{\frac{x}{2}}$. When $x = 0$, the gradient of the tangent to the curve is 4.

a Find the equation of the curve.

b Find the equation of the normal to the curve at P.

8 Evaluate:

a $\int_1^4 (5x - 3)\, dx$ **b** $\int_3^4 \left(x^2 - \frac{1}{x} + \frac{2}{x^2}\right) dx$ **c** $\int_0^{\ln 5} e^{2x}(e - 1)\, dx$

d $\int_0^{\pi} \sin\frac{x}{2}\, dx$ **e** $\int_{-2}^{5} (\sqrt{x+11} - 4)\, dx$ **f** $\int_{-\frac{\pi}{3}}^{\frac{\pi}{3}} \sin\left(3x - \frac{\pi}{6}\right) dx$

9 Given that $\int_{-2}^{3} f(x)\, dx = 7$ and $\int_{3}^{5} f(x)\, dx = 4$, find:

a $\int_5^3 f(x)\, dx$ **b** $\int_{-2}^{5} f(x)\, dx$ **c** $\int_{-2}^{3} 4f(x)\, dx$

10 **a** Show that $\frac{d}{dx}(\cot x) = -\operatorname{cosec}^2 x \, dx$.

b Evaluate $\int_{\frac{\pi}{4}}^{\frac{\pi}{3}} \operatorname{cosec}^2 x \, dx$.

17

Applications of integration

Contents:

Opening problem

This graph shows the velocity of a Formula One race car as it enters the pit lane.

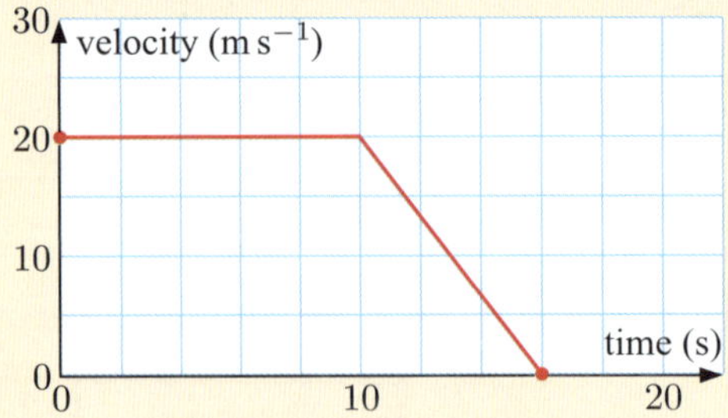

Things to think about:

a For the first 10 seconds:
- **i** how *fast* does the car move
- **ii** how *far* does the car travel
- **iii** what is the area under the velocity graph?

b In total, how far did the car travel in the pit lane?

c Suppose we are given the velocity function $v(t)$ of a moving object. How can we determine the distance the object travels in a particular time interval?

In this Chapter we further explore the relationship between integration and area, and consider other applications of integral calculus including kinematics.

A THE AREA UNDER A CURVE

We have already established in **Chapter 16** that:

If $f(x)$ is positive and continuous on the interval $a \leqslant x \leqslant b$, then the area A bounded by $y = f(x)$, the x-axis, and the vertical lines $x = a$ and $x = b$ is given by

$$\int_a^b f(x)\,dx = F(b) - F(a), \quad \text{where } F(x) = \int f(x)\,dx.$$

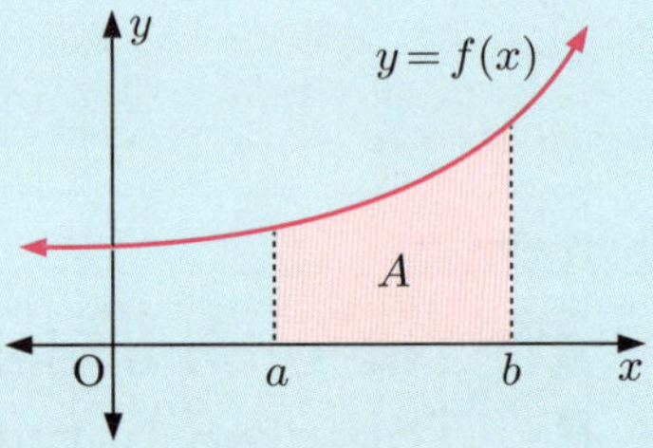

Example 1

Self Tutor

Find the area of the region enclosed by $y = 3x$, the x-axis, $x = 0$, and $x = 2$ by using:

a a geometric argument **b** integration.

a

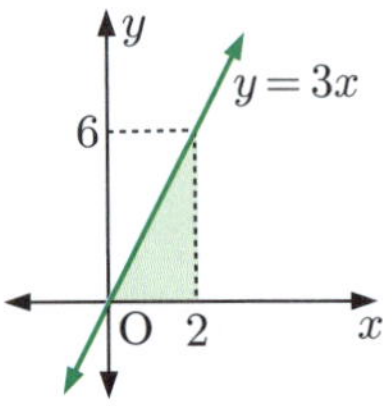

$$\text{Area} = \tfrac{1}{2} \times 2 \times 6$$
$$= 6 \text{ units}^2$$

b

$$\text{Area} = \int_0^2 3x\,dx$$
$$= \left[\tfrac{3}{2}x^2\right]_0^2$$
$$= \tfrac{3}{2}(2)^2 - \tfrac{3}{2}(0)^2$$
$$= 6 \text{ units}^2$$

Example 2

Self Tutor

Find the shaded area.

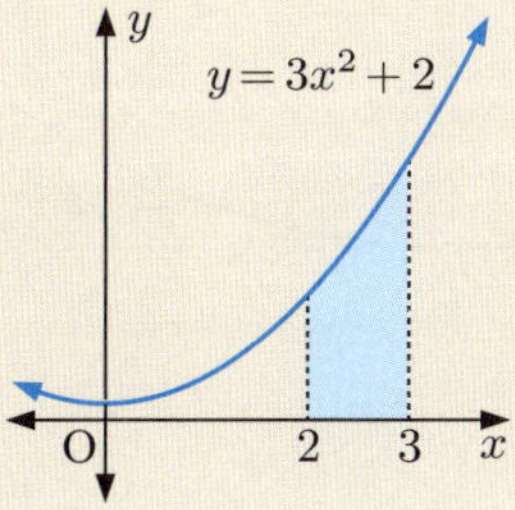

$$\text{Area} = \int_2^3 (3x^2 + 2)\, dx$$
$$= \left[x^3 + 2x\right]_2^3$$
$$= (27 + 6) - (8 + 4)$$
$$= 21 \text{ units}^2$$

EXERCISE 17A.1

1 Find each shaded area using:

i a geometric argument **ii** integration.

a

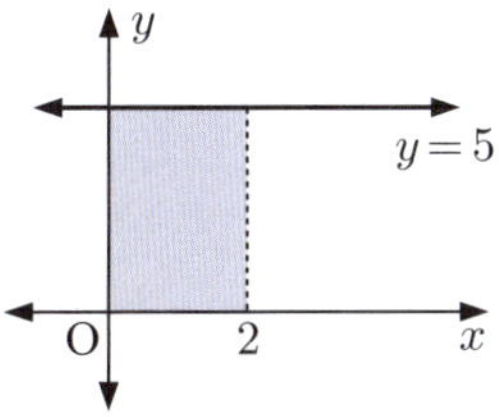

b

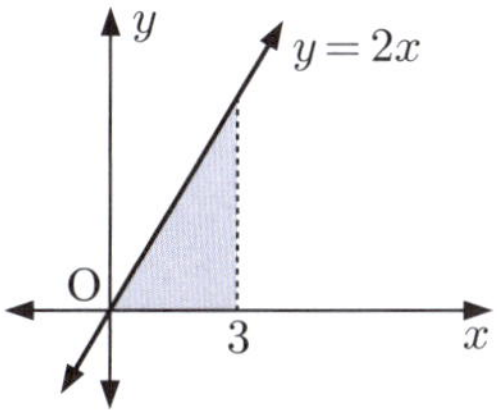

c

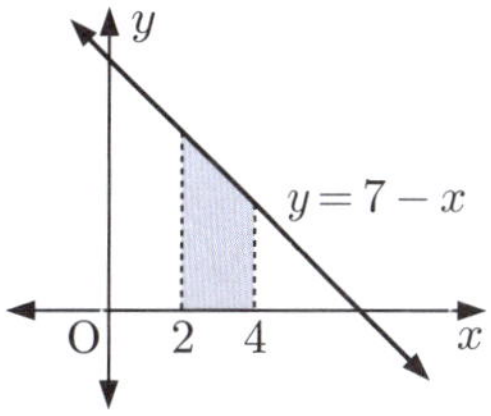

2 Find the shaded area:

a

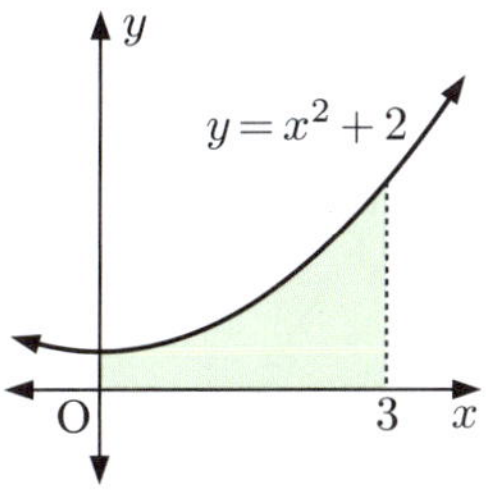

b

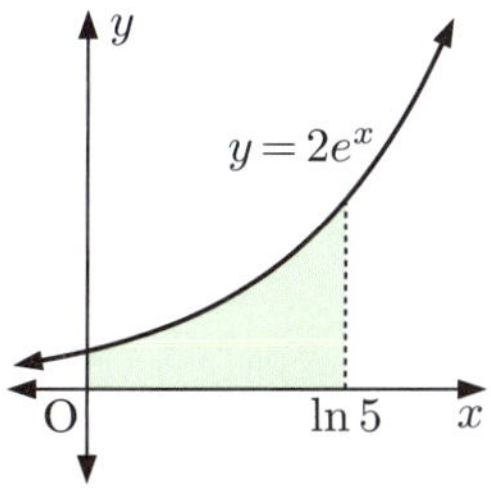

c

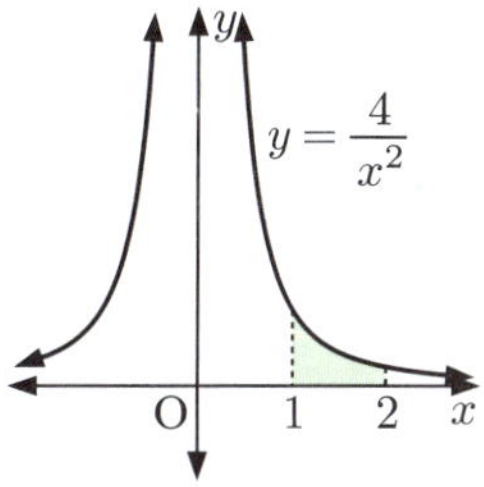

d

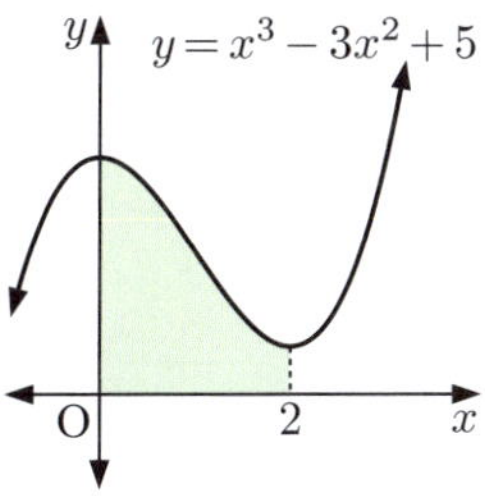

e

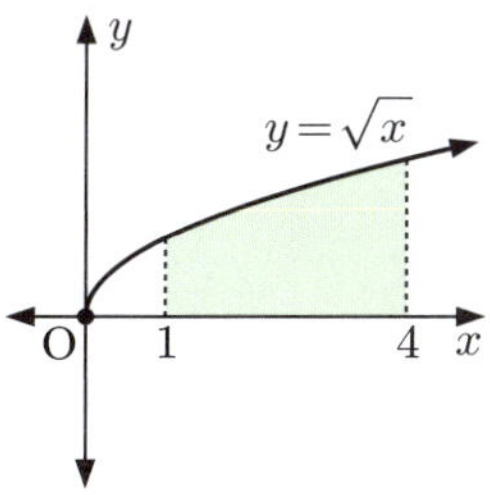

f

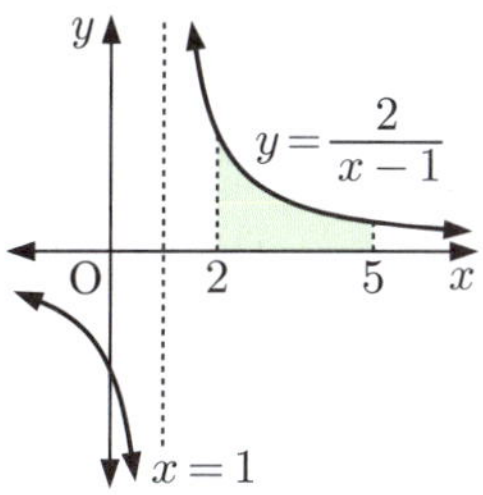

g

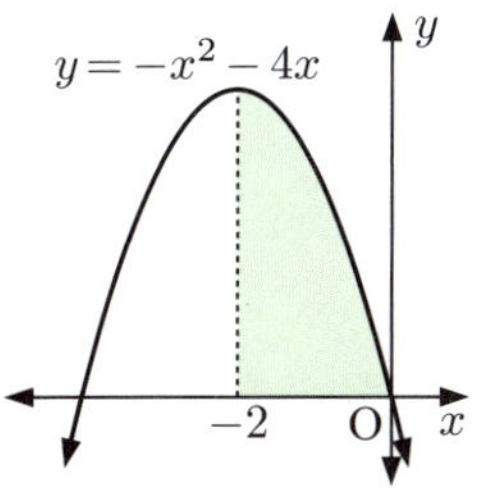

h

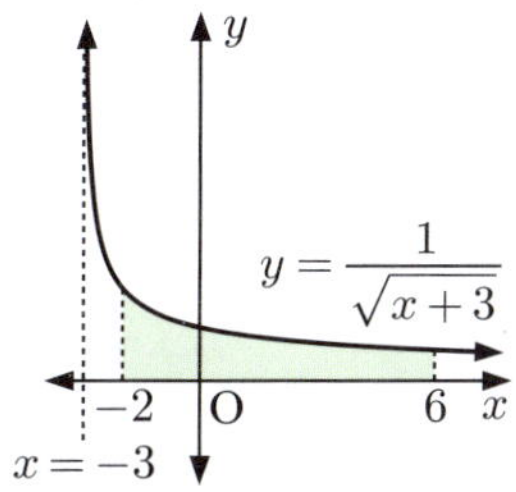

i

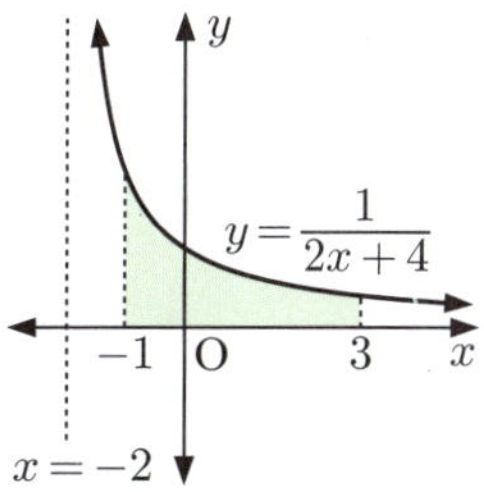

3 Which of the shaded regions is larger?

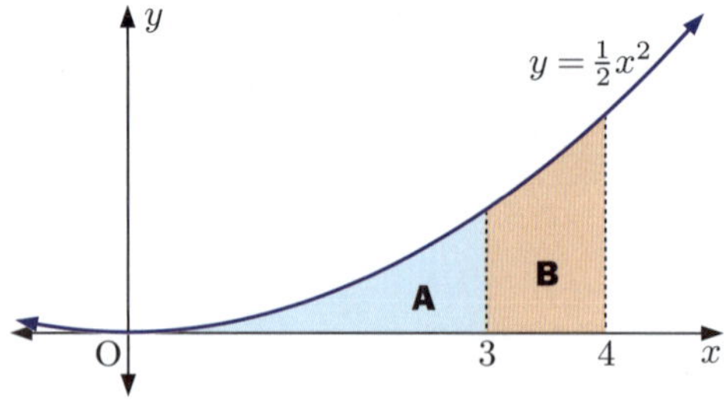

4 Find the area of the region bounded by:

a $y = e^{2x}$, the x-axis, $x = 1$, and $x = 4$

b $y = \cos x$, the x-axis, $x = 0$, and $x = \frac{\pi}{2}$

c $y = \dfrac{2}{5-x}$, the x-axis, $x = 2$, and $x = 4$

d $y = \sqrt{20-x}$, the x-axis, $x = 4$, and $x = 11$.

5 Find the shaded area:

a

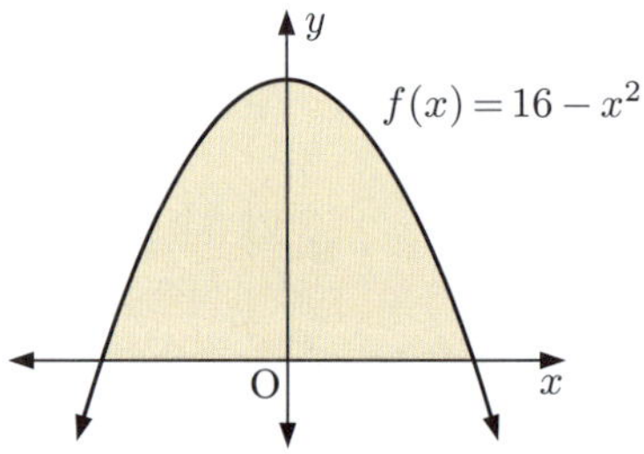

b

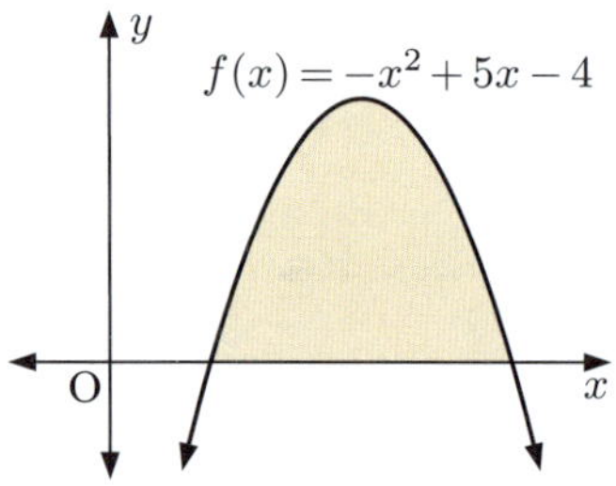

c

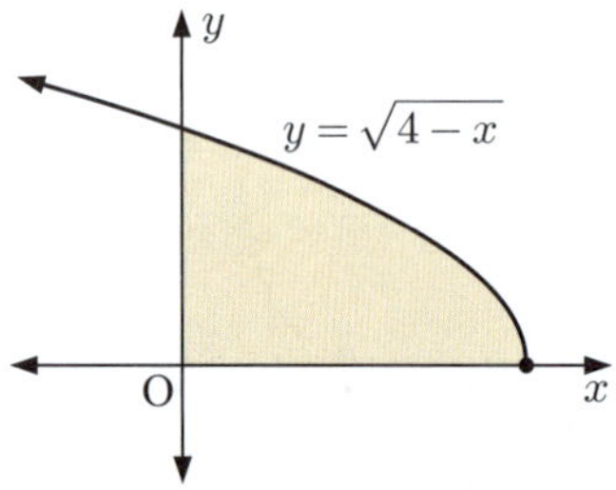

d

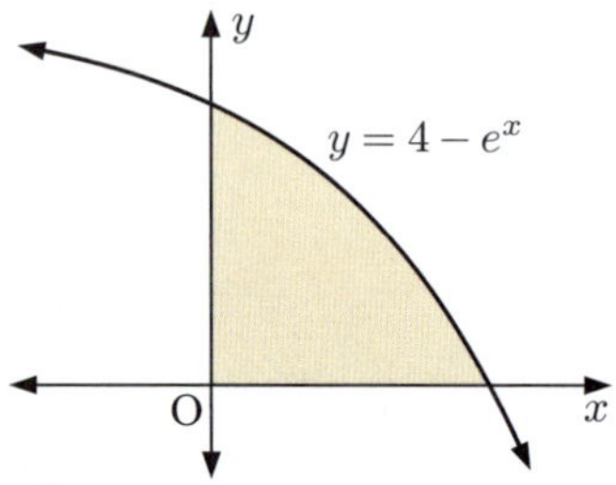

6 The graph of $y = \sin \frac{x}{2}$ shown has a maximum turning point at P, and cuts the x-axis at Q.

a Find the coordinates of P and Q.

b Find the shaded area.

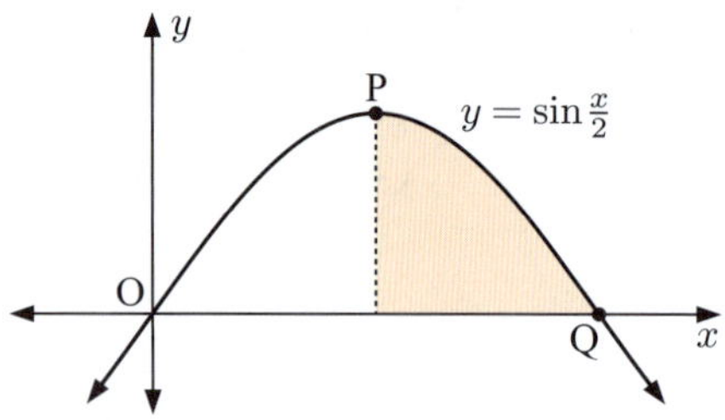

7

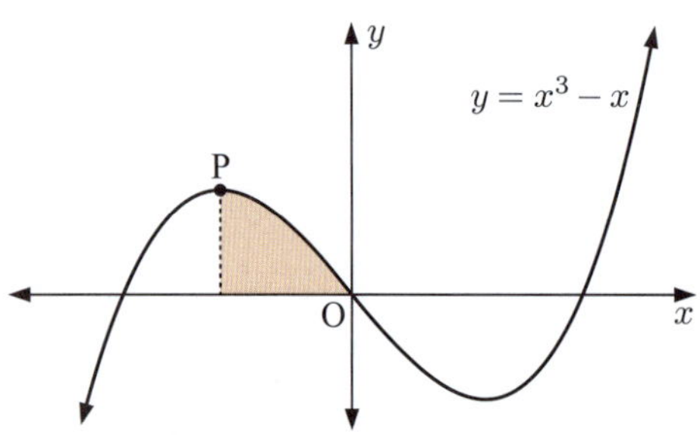

The graph of $y = x^3 - x$ shown has a maximum turning point at P.

a Find the coordinates of P.

b Find the shaded area.

8 In the graph of $f(x) = 2\cos\left(x - \frac{\pi}{3}\right) + 3$ shown, A is a maximum turning point and B is a minimum turning point. Find the area of:

a the blue region **b** the red region.

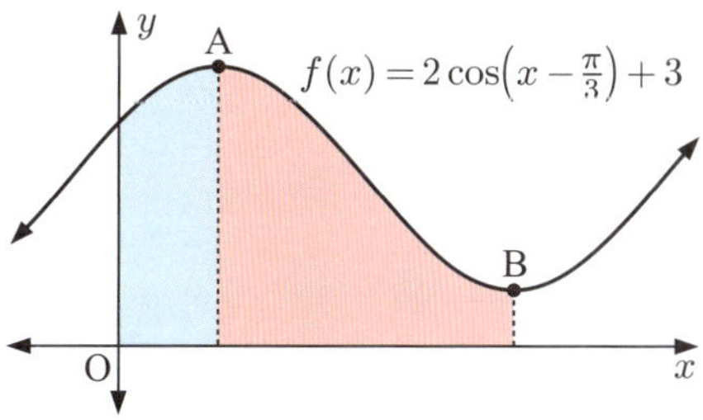

9 **a** Show that $\dfrac{d}{dx}\left(\dfrac{\cos x}{1 - \sin x}\right) = \dfrac{1}{1 - \sin x}$.

b Hence find the area bounded by the graph of $y = \dfrac{1}{1 - \sin x}$ and the x-axis, between $x = 0$ and $x = \frac{\pi}{4}$.

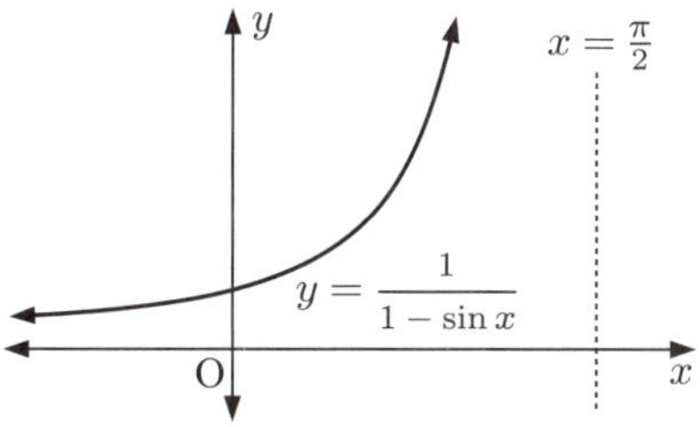

10

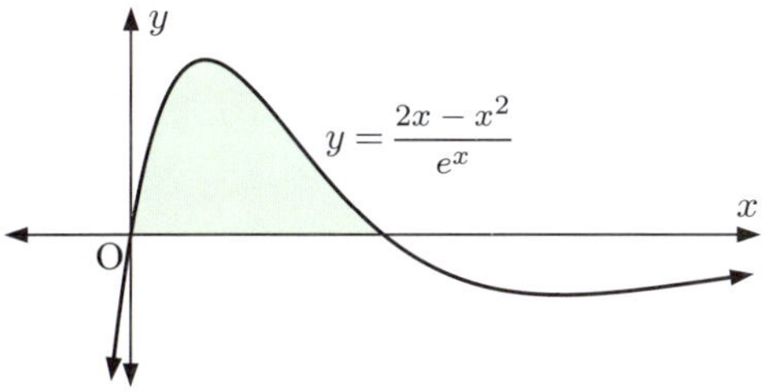

a Show that $\dfrac{d}{dx}\left(\dfrac{x^2}{e^x}\right) = \dfrac{2x - x^2}{e^x}$.

b Hence find the shaded area.

11 Find k such that:

a the shaded area is 4 units^2

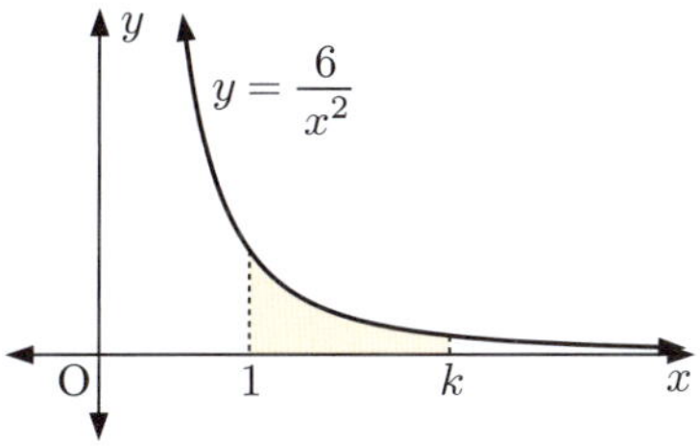

b the shaded area is 7 units^2.

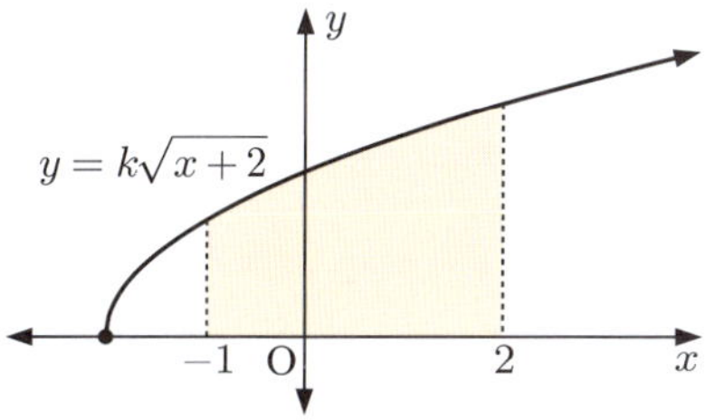

12 **a** Find the coordinates of:

i A **ii** B

b Find the shaded area.

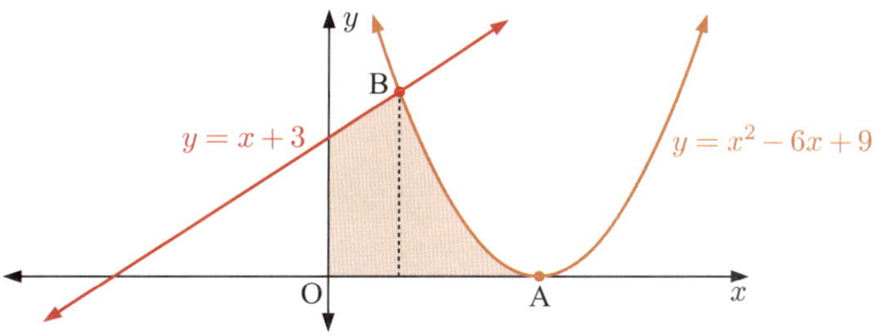

13

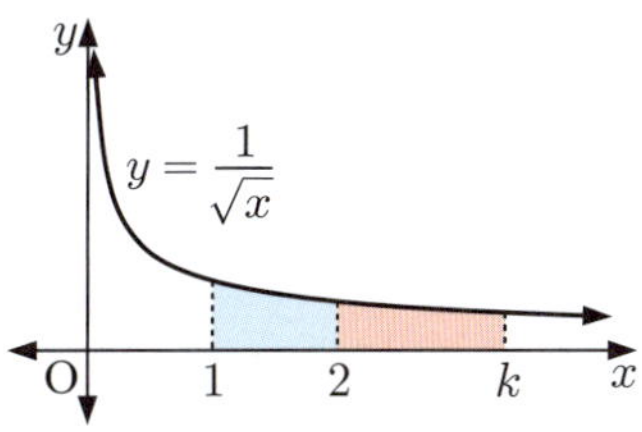

The blue and red shaded areas are equal. Find k, giving your answer in the form $a + b\sqrt{2}$, where $a, b \in \mathbb{Q}$.

14 Show that, for any point P on the curve $y = x^2$, the tangent at P divides the shaded area in the ratio $1 : 3$.

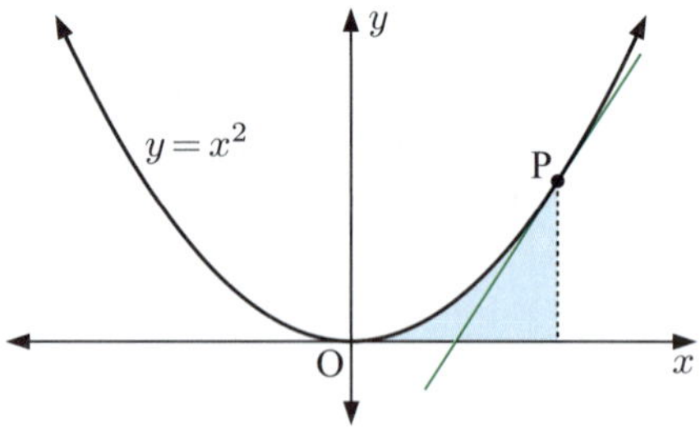

THE AREA BETWEEN $y = f(x)$ AND THE x-AXIS WHEN $f(x) \leqslant 0$

In the graph alongside, the function $y = f(x)$ lies *below* the x-axis for $a \leqslant x \leqslant b$. The shaded area A_1 does *not* equal $\int_a^b f(x)\,dx$, as this is only true if $f(x) \geqslant 0$ on $a \leqslant x \leqslant b$.

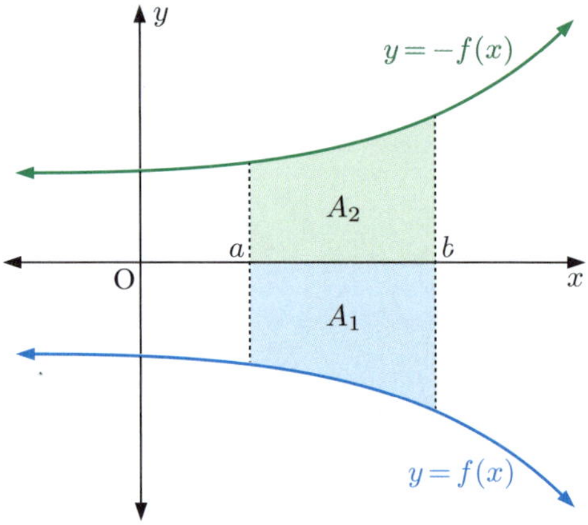

By reflecting $y = f(x)$ in the x-axis, we generate the graph of $y = -f(x)$. This graph is positive for $a \leqslant x \leqslant b$, and the area A_2 must equal A_1.

Now $A_2 = \int_a^b (-f(x))\,dx$.

$\therefore \; A_1 = \int_a^b (-f(x))\,dx = -\int_a^b f(x)\,dx$.

If a function $y = f(x)$ lies below the x-axis for $a \leqslant x \leqslant b$, then the area bounded by $y = f(x)$, the x-axis, and the lines $x = a$ and $x = b$ is given by

$$A = -\int_a^b f(x)\,dx.$$

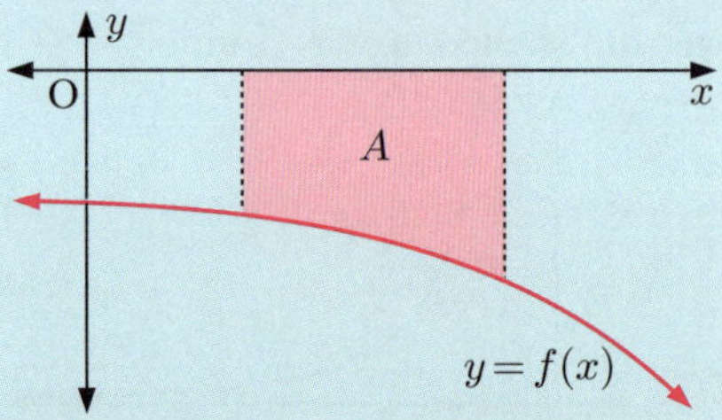

Example 3

Self Tutor

Calculate the shaded area.

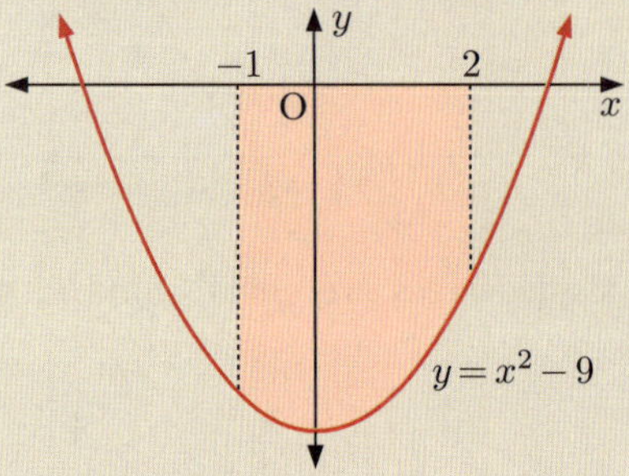

$$\begin{aligned}\text{Area} &= -\int_{-1}^{2} (x^2 - 9)\,dx \\ &= -\left[\frac{x^3}{3} - 9x\right]_{-1}^{2} \\ &= -\left(\left(\tfrac{8}{3} - 18\right) - \left(-\tfrac{1}{3} + 9\right)\right) \\ &= -(-24) \\ &= 24 \text{ units}^2\end{aligned}$$

EXERCISE 17A.2

1 Write an integral which could be used to calculate the shaded area.

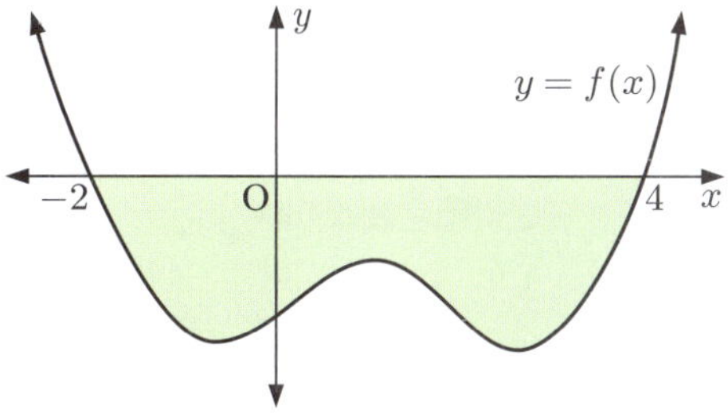

2 Find the shaded area:

a

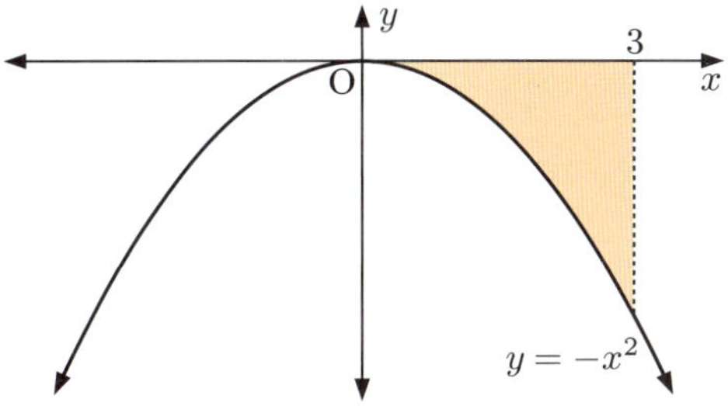

b

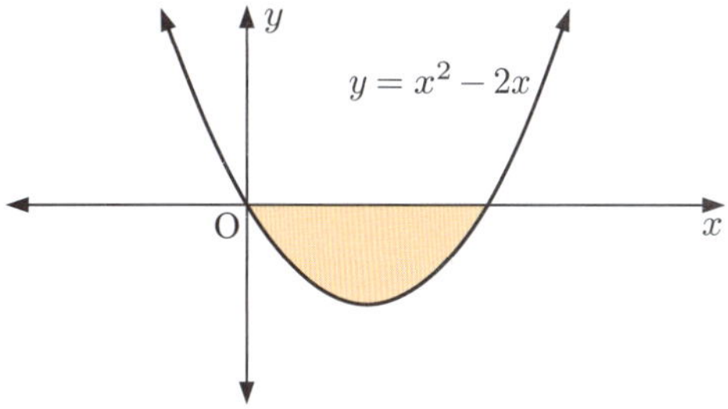

c

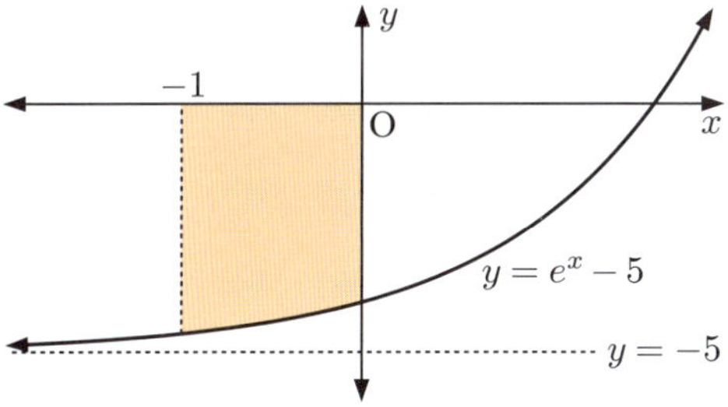

d

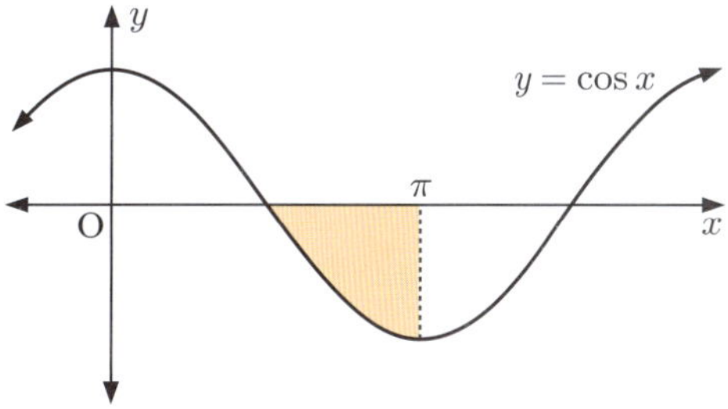

e

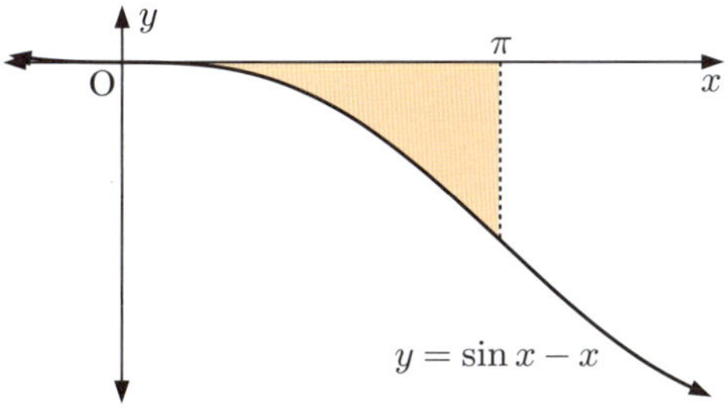

f

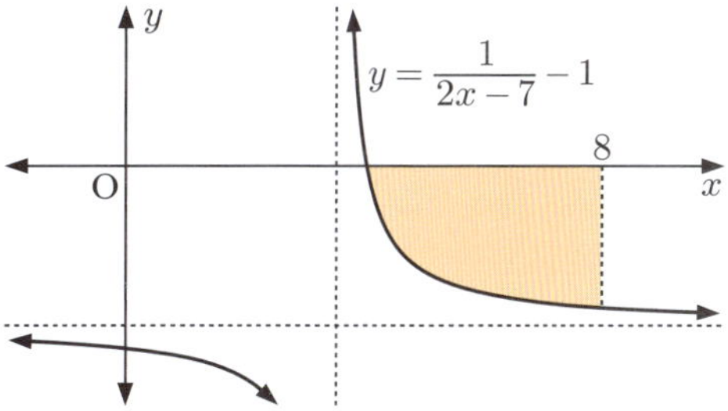

3 In the graph alongside, P is a minimum turning point.

a Find the coordinates of P.

b Find the shaded area.

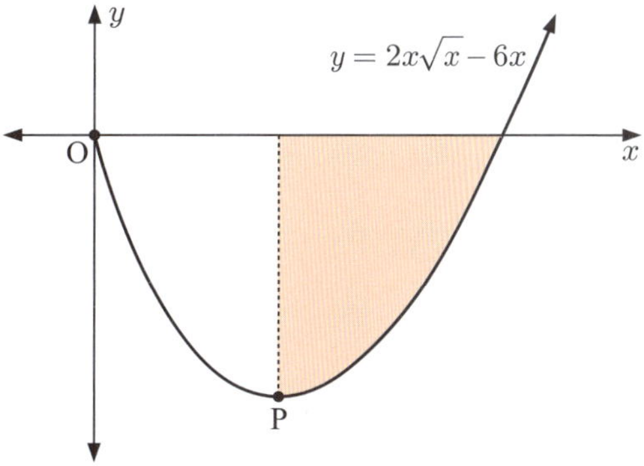

Example 4

Self Tutor

Find the total area of the regions bounded by $f(x) = 4x^3 - 12x^2 - 16x$ and the x-axis.

$$\begin{aligned} f(x) &= 4x^3 - 12x^2 - 16x \\ &= 4x(x^2 - 3x - 4) \\ &= 4x(x+1)(x-4) \end{aligned}$$

$\therefore$ $y = f(x)$ cuts the x-axis at -1, 0, and 4.

Total area

$$\begin{aligned} &= \int_{-1}^{0} (4x^3 - 12x^2 - 16x)\,dx - \int_{0}^{4} (4x^3 - 12x^2 - 16x)\,dx \\ &= \left[x^4 - 4x^3 - 8x^2\right]_{-1}^{0} - \left[x^4 - 4x^3 - 8x^2\right]_{0}^{4} \\ &= (0 - (-3)) - (-128 - 0) \\ &= 3 + 128 \\ &= 131 \text{ units}^2 \end{aligned}$$

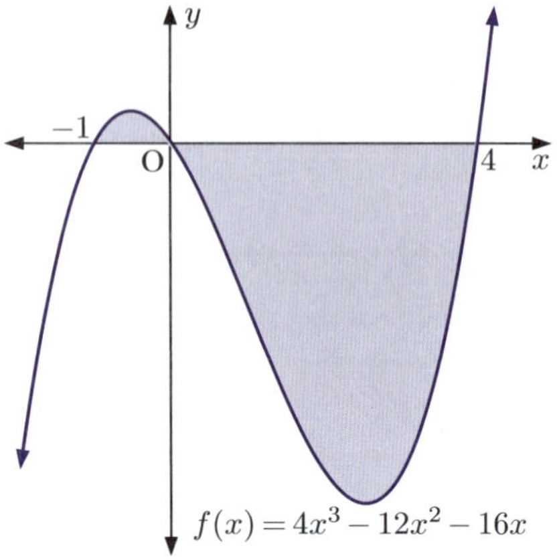

4 Consider the graph of $y = 2\sin 2x$ shown.

a Find the coordinates of A and B.

b Find the total shaded area.

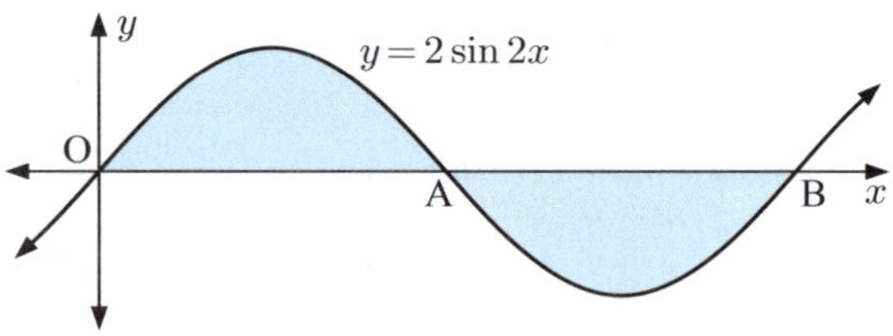

5 The graph of $y = x^3 - 5x^2 + 2x + 8$ is shown alongside.

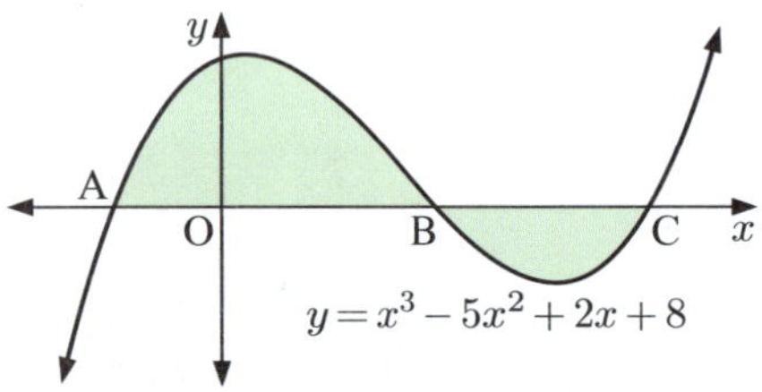

a Given that A has coordinates $(-1, 0)$, find the coordinates of B and C.

b Find the total shaded area.

6 State, in terms of the areas A_1 and A_2, what is represented by:

a $\displaystyle\int_{-3}^{2} f(x)\,dx$

b $\displaystyle\int_{-3}^{1} f(x)\,dx - \int_{1}^{2} f(x)\,dx$

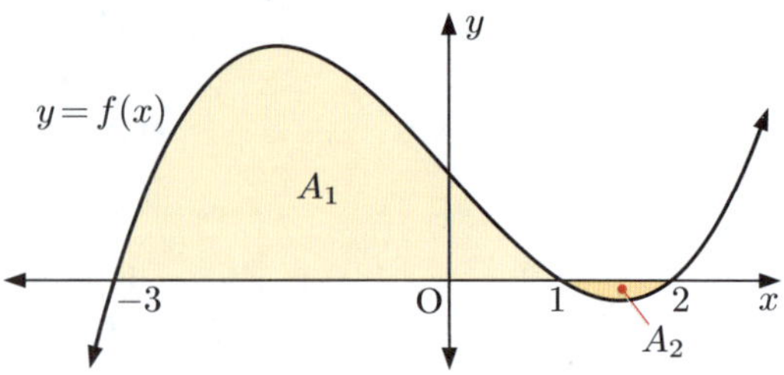

7 In the graph alongside, $\displaystyle\int_{-4}^{5} f(x)\,dx = -3$.

a Which of A_1 or A_2 is larger? Explain your answer.

b Given that $\displaystyle\int_{2}^{5} f(x)\,dx = 8$, find the total area of the shaded region.

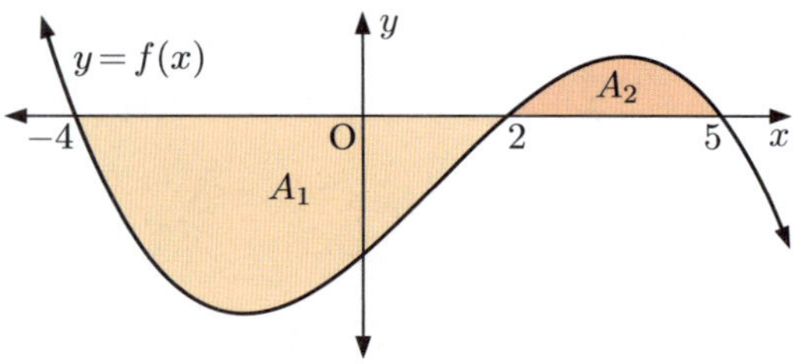

Discussion

Suppose $\int_a^b f(x)\,dx = 5$. Discuss whether each of the following statements is definitely true, possibly true, or definitely false.

- $f(x)$ lies entirely above the x-axis over $a \leqslant x \leqslant b$.
- $f(x)$ lies entirely below the x-axis over $a \leqslant x \leqslant b$.
- The total area between $y = f(x)$ and the x-axis over $a \leqslant x \leqslant b$ is 5 units^2.
- The total area between $y = f(x)$ and the x-axis over $a \leqslant x \leqslant b$ is greater than 5 units^2.
- The total area between $y = f(x)$ and the x-axis over $a \leqslant x \leqslant b$ is less than 5 units^2.
- The area enclosed by $y = f(x)$ *above* the x-axis is 5 units^2 greater than the area *below* the x-axis over $a \leqslant x \leqslant b$.

B THE AREA BETWEEN TWO FUNCTIONS

Consider two functions $f(x)$ and $g(x)$ where $f(x) \geqslant g(x)$ for all $a \leqslant x \leqslant b$.

The area between the two functions on the interval $a \leqslant x \leqslant b$ is given by

$$A = \int_a^b [f(x) - g(x)]\,dx.$$

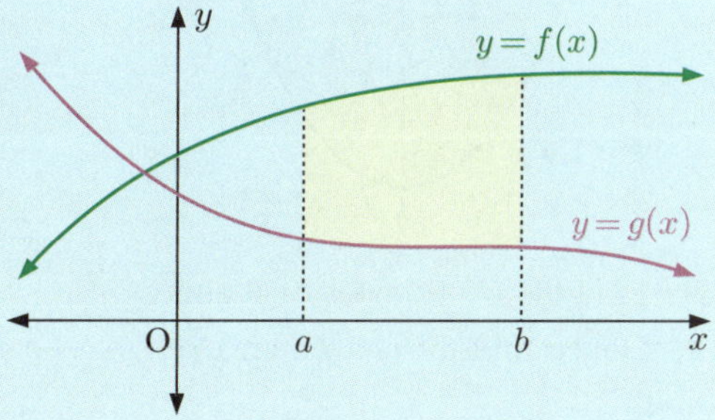

Proof:

If necessary, we translate each curve k units upwards until they are both above the x-axis on the interval $a \leqslant x \leqslant b$. This has no effect on the area between the functions.

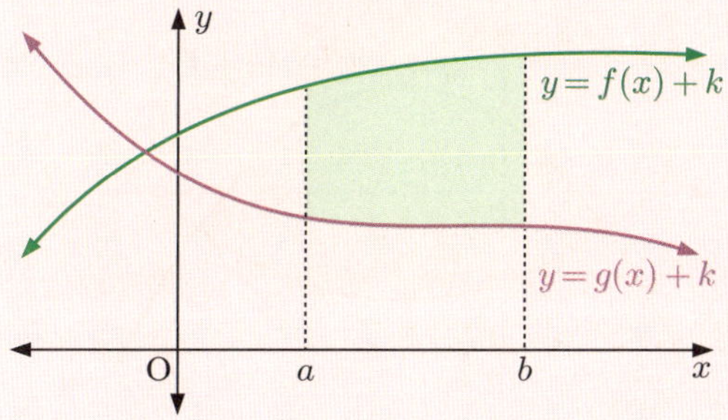

The area of the shaded region

$$= \int_a^b [f(x) + k]\,dx - \int_a^b [g(x) + k]\,dx$$

$$= \int_a^b [f(x) - g(x)]\,dx$$

Notice that if $y = g(x)$ is negative on the interval between two x-intercepts a and b, we can let $f(x) = 0$ and hence derive the formula we saw in the last Section:

$$\text{Area} = -\int_a^b g(x)\,dx.$$

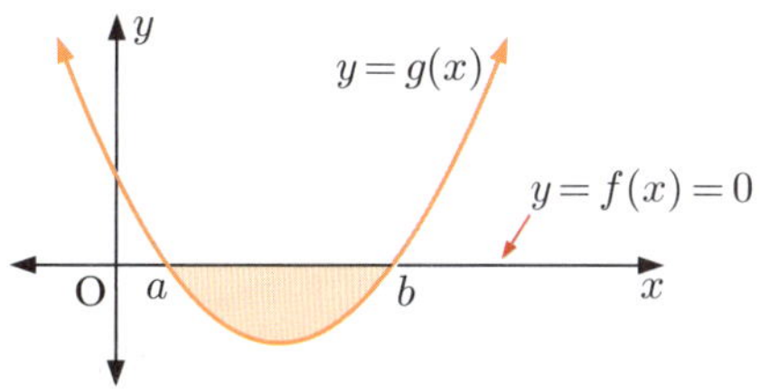

Example 5

Self Tutor

Find the shaded area:

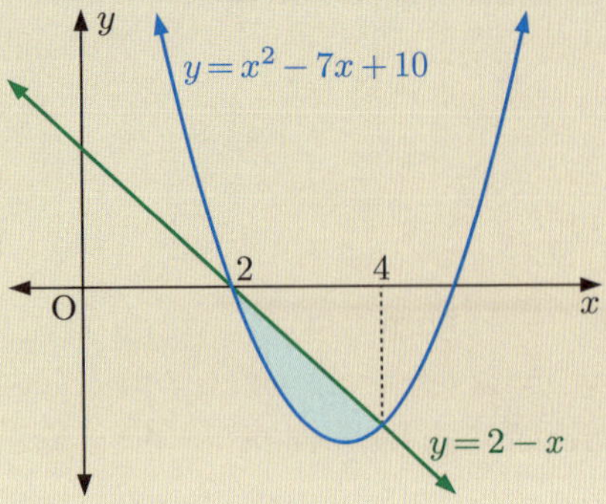

Since $2 - x \geqslant x^2 - 7x + 10$ on the interval $2 \leqslant x \leqslant 4$,

$$\text{area} = \int_2^4 \big((2-x) - (x^2 - 7x + 10)\big)\,dx$$
$$= \int_2^4 (-x^2 + 6x - 8)\,dx$$
$$= \left[-\frac{x^3}{3} + 3x^2 - 8x\right]_2^4$$
$$= \left(-\tfrac{64}{3} + 48 - 32\right) - \left(-\tfrac{8}{3} + 12 - 16\right)$$
$$= 1\tfrac{1}{3}\text{ units}^2$$

EXERCISE 17B

1 Find the shaded area:

a

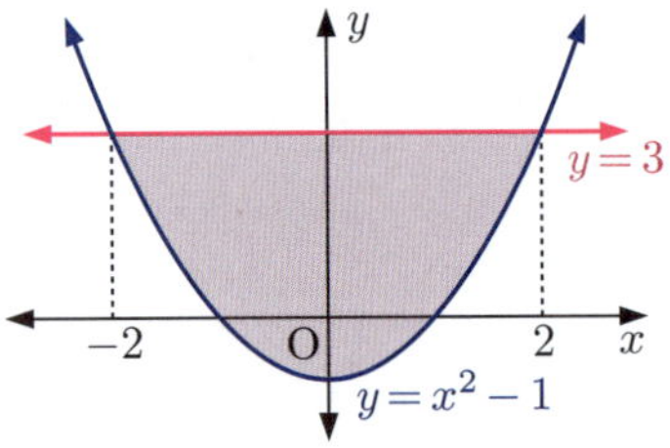

b

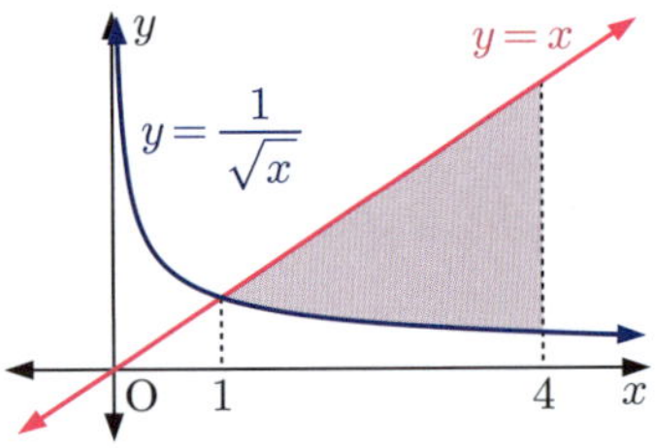

2 **a** Find the coordinates of P and Q.

b Hence find the shaded area.

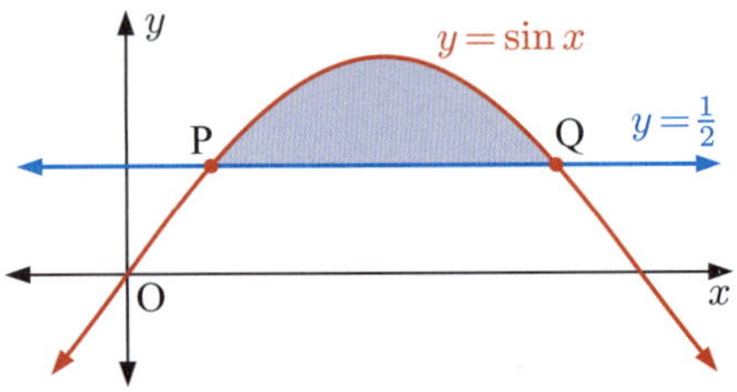

3 Find the shaded area:

a

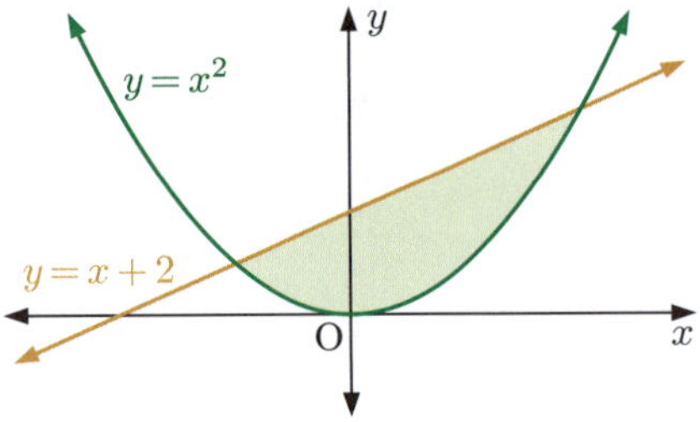

b

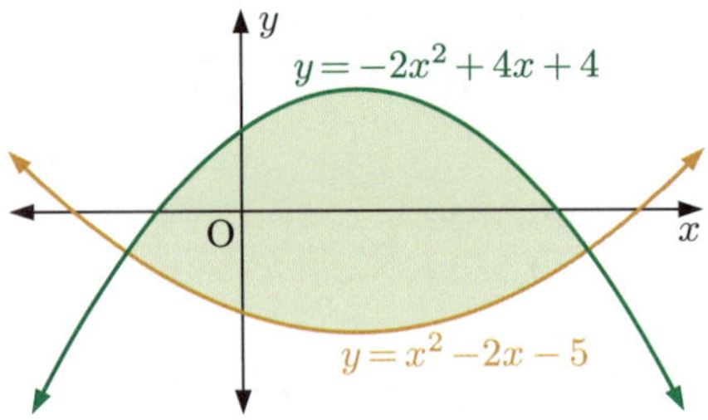

c

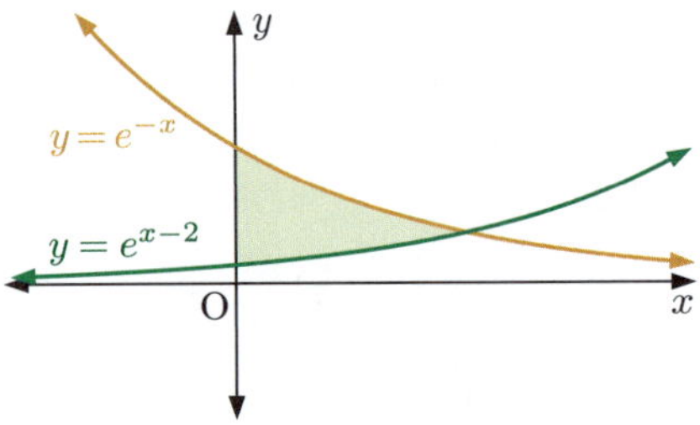

d

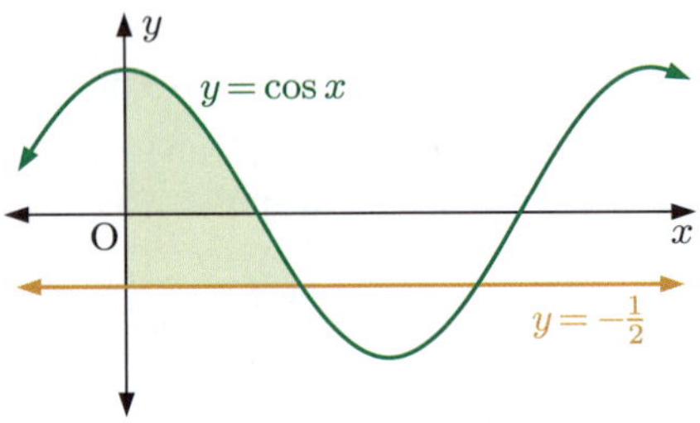

4 **a** Sketch the graphs of $y = x^2$ and $y = 3x$ on the same set of axes.
b Find the coordinates of the points of intersection.
c Find the area of the region enclosed by the functions.

5 **a** Sketch the graphs of $y = \sqrt{x}$ and $y = \frac{x}{2}$ on the same set of axes.
b Find the coordinates of the points of intersection.
c Find the area of the region enclosed by the curves.

Example 6

Self Tutor

Find the total area of the regions enclosed by $f(x) = x^3 - 4x$ and $g(x) = x^2 + 2x$.

The graphs meet when $x^3 - 4x = x^2 + 2x$

$\therefore\ x^3 - x^2 - 6x = 0$

$\therefore\ x(x^2 - x - 6) = 0$

$\therefore\ x(x+2)(x-3) = 0$

$\therefore$ the graphs meet when $x = -2$, 0, and 3.

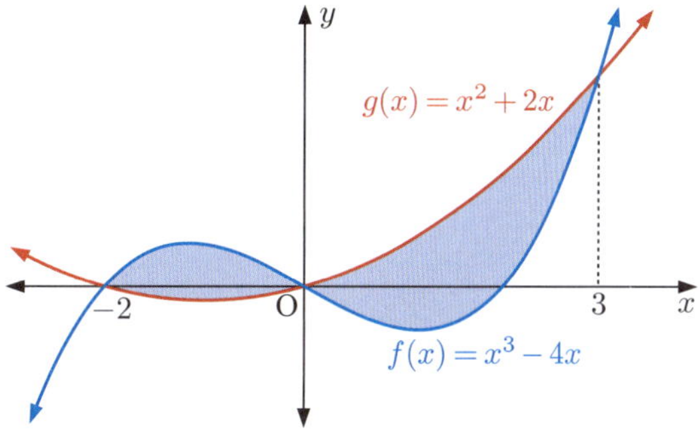

Total area

$$= \int_{-2}^{0} \left((x^3 - 4x) - (x^2 + 2x)\right) dx + \int_{0}^{3} \left((x^2 + 2x) - (x^3 - 4x)\right) dx$$

$$= \int_{-2}^{0} (x^3 - x^2 - 6x)\, dx + \int_{0}^{3} (-x^3 + x^2 + 6x)\, dx$$

$$= \left[\frac{x^4}{4} - \frac{x^3}{3} - 3x^2\right]_{-2}^{0} + \left[-\frac{x^4}{4} + \frac{x^3}{3} + 3x^2\right]_{0}^{3}$$

$$= \left(0 - \left(4 + \tfrac{8}{3} - 12\right)\right) + \left(-\tfrac{81}{4} + 9 + 27 - 0\right)$$

$$= \tfrac{16}{3} + \tfrac{63}{4}$$

$$= 21\tfrac{1}{12} \text{ units}^2$$

6 The graphs of $f(x) = x^3 - 5x^2 + 4$ and $g(x) = 2x^2 - 10x + 4$ are shown alongside.
a Find the coordinates of A, B, and C.
b Find the total area enclosed by the curves.

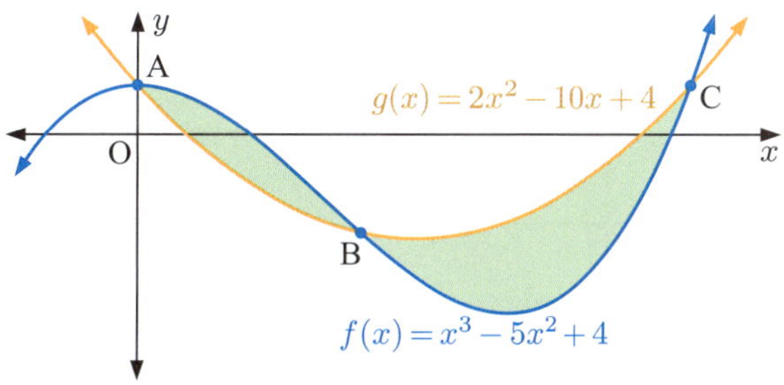

7 Find the total area enclosed by the graphs of $f(x) = x^3 - 3x + 2$ and $g(x) = 2x^2 + 2x - 4$.

8 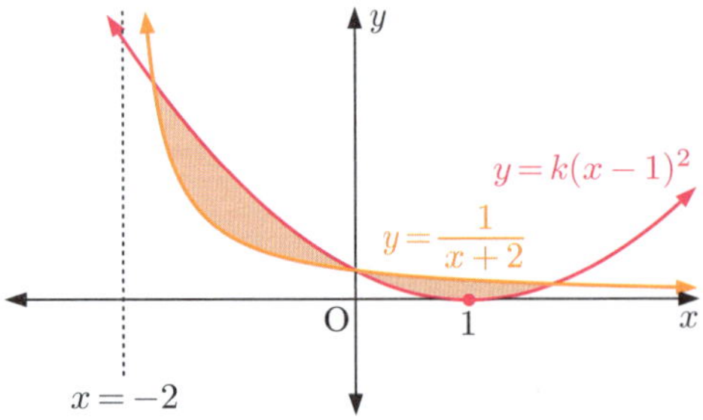

The graphs of $y = \dfrac{1}{x+2}$ and $y = k(x-1)^2$ meet at the y-axis as shown.

a Find the value of k.

b Find the area of the shaded region.

9 The graphs of $y = 4x+1$ and $y = x^3 - 4x^2 + 4x + 1$ are shown alongside. The graphs meet at A. B is a minimum turning point, and the vertical line through B meets $y = 4x+1$ at C.

a Show that the line AC is a tangent to the curve $y = x^3 - 4x^2 + 4x + 1$ at A.

b Find the coordinates of B.

c Find the area of the shaded region.

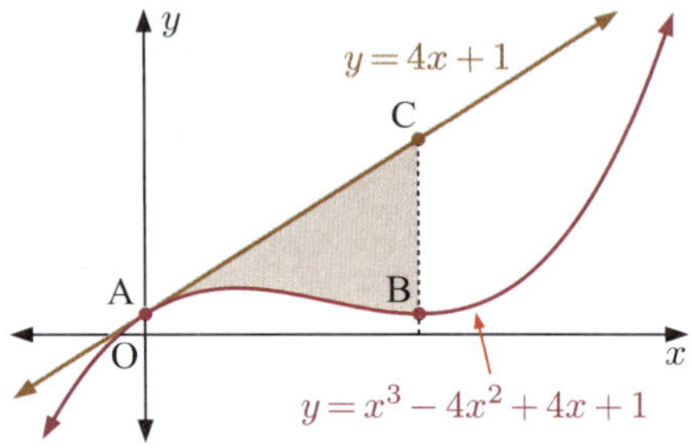

10 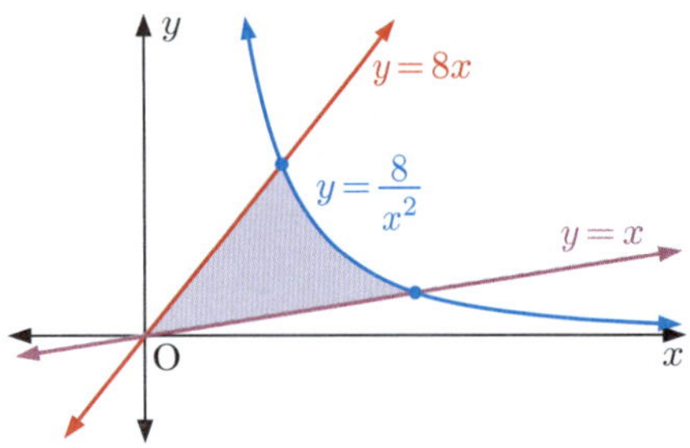

Find the shaded area.

11 Consider the graphs of $f(x) = x^2 + 3x - 4$ and $g(x) = -x^2 - x + 2$.

a Find the coordinates of P and Q.

b Find the area of the shaded region.

c Show that the straight line PQ divides the shaded region in half.

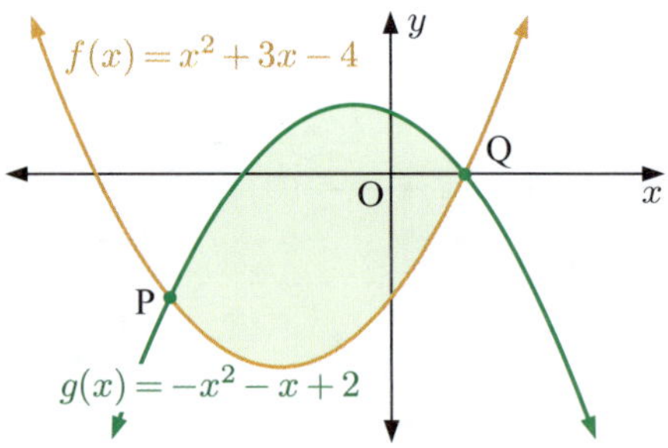

12 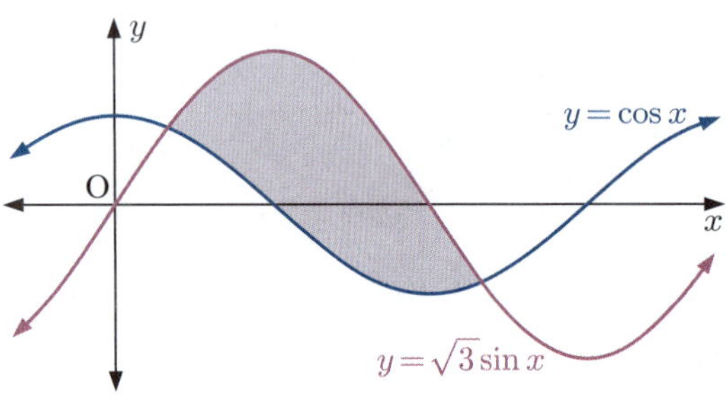

a Show that the shaded region has area 4 units^2.

b What percentage of the shaded region lies above the x-axis?

13 The graph alongside shows the tangents to the curve $y = 5x - x^2$ at A(0, 0) and B(4, 4).
Find the area of the shaded region.

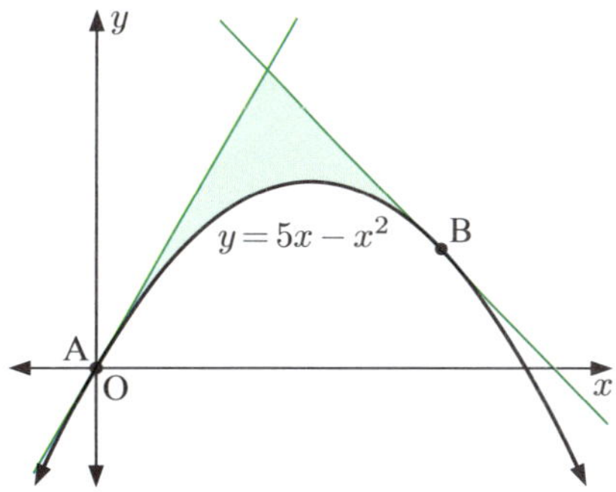

14

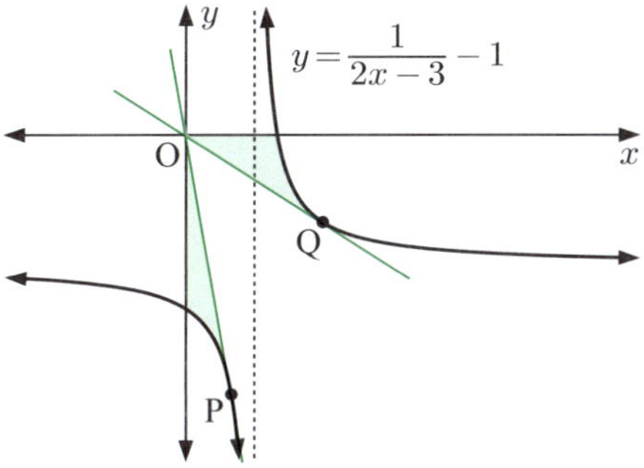

The tangents to $y = \dfrac{1}{2x-3} - 1$ at P and Q pass through the origin. Find the total area of the shaded regions.

C KINEMATICS

DISPLACEMENT AND DISTANCE

In **Chapter 15**, we saw that the velocity $v(t)$ of an object is the derivative of its displacement $s(t)$, so $v(t) = s'(t)$.

The change in displacement of an object from time $t = t_1$ to $t = t_2$ is therefore

$$s(t_2) - s(t_1) = \int_{t_1}^{t_2} s'(t)\, dt \qquad \{\text{Fundamental Theorem of Calculus}\}$$

$$= \int_{t_1}^{t_2} v(t)\, dt$$

If $v(t)$ maintains the same sign in the interval $t_1 \leqslant t \leqslant t_2$, then the object does not change direction.

Suppose an object travels with velocity function $v(t)$.

- If $v(t) \geqslant 0$ on the interval $t_1 \leqslant t \leqslant t_2$,

  distance travelled $= \displaystyle\int_{t_1}^{t_2} v(t)\, dt$

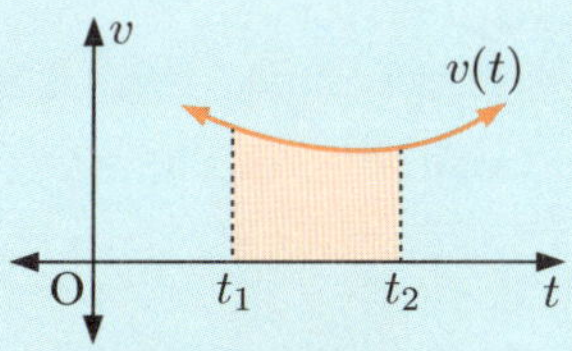

- If $v(t) \leqslant 0$ on the interval $t_1 \leqslant t \leqslant t_2$,

  distance travelled $= -\displaystyle\int_{t_1}^{t_2} v(t)\, dt$

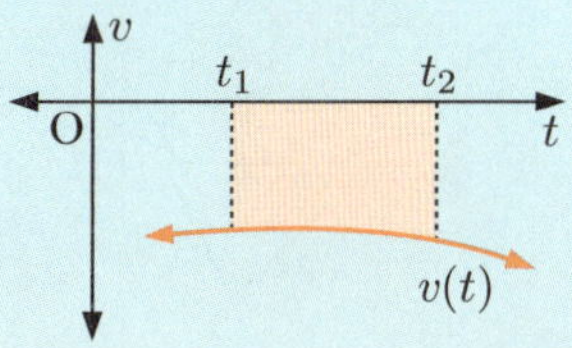

If the velocity changes sign within the time interval, then the object changes direction. We therefore need to add the components of area above and below the t-axis to find the total distance travelled.

Example 7

Self Tutor

An object's velocity-time graph is shown alongside. Find the total distance travelled by the object.

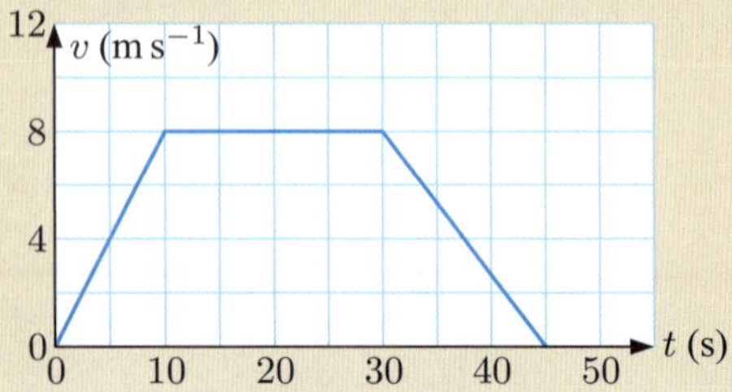

Total distance travelled
$= $ total area under graph
$= $ area $A +$ area $B +$ area C
$= \frac{1}{2} \times 10 \times 8 + 20 \times 8 + \frac{1}{2} \times 15 \times 8$
$= 40 + 160 + 60$
$= 260$ m

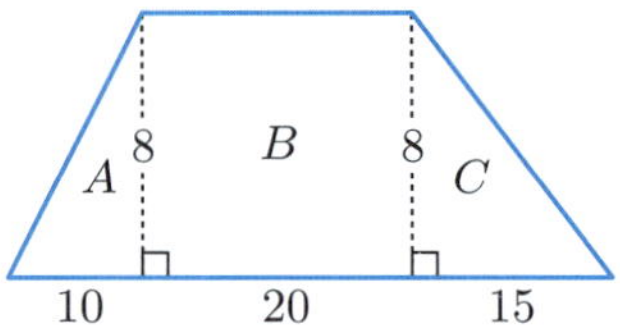

EXERCISE 17C.1

1 For each velocity-time graph, find the total distance travelled by the object.

a

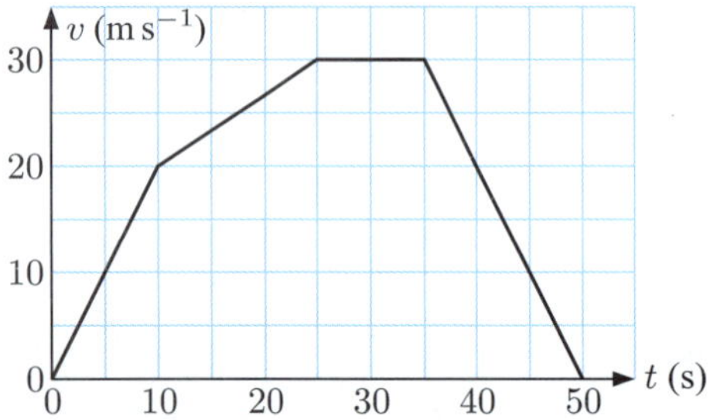

b

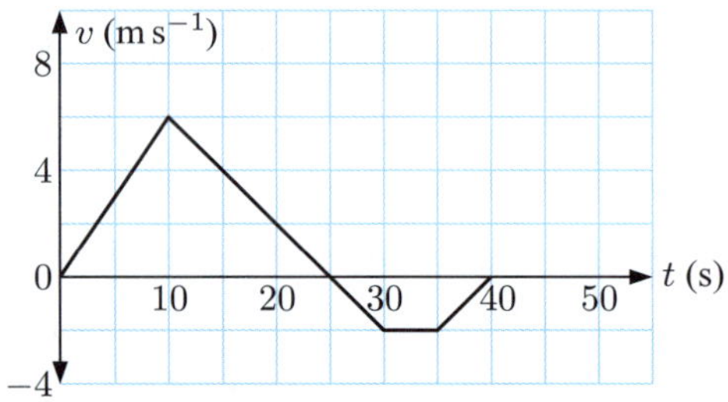

2 A remote-controlled car has the velocity-time graph shown.

a Find the speed of the car after 10 seconds.

b Find the distance travelled by the car in the first:

i 20 seconds **ii** 40 seconds.

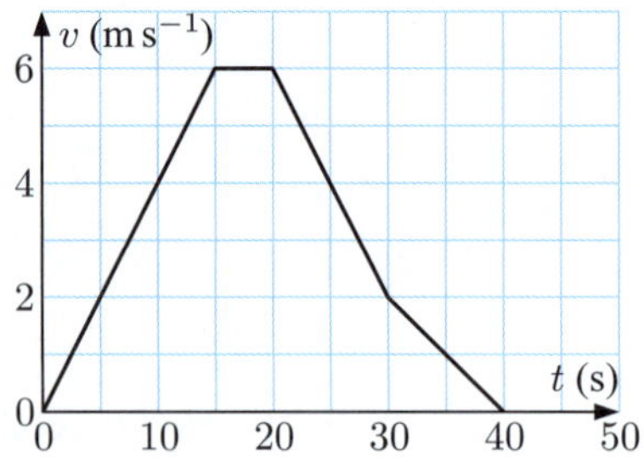

3

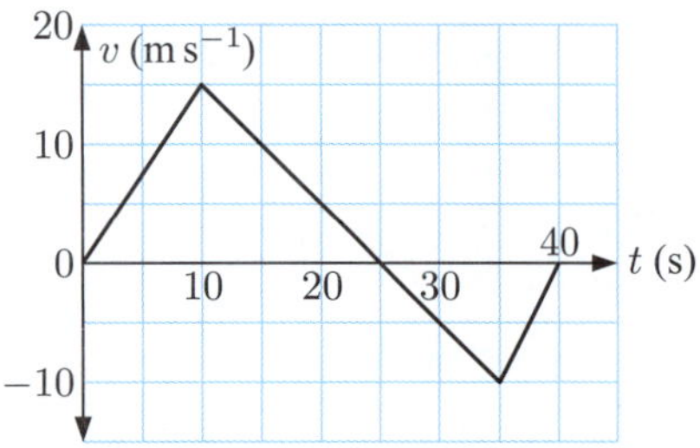

An object has the velocity-time graph shown.

a At what time(s) does the object change direction?

b Find the total distance travelled by the object.

c Find the final *displacement* of the object from its starting point.

4 A jet ski leaves a river bank from rest, and accelerates at a constant rate for 10 seconds until it reaches 10 m s^{-1}. The jet ski maintains this speed for 60 seconds. It then decelerates at a constant rate, coming to rest after 5 seconds.

a Draw a velocity-time graph to show the motion of the jet ski.

b Find the total distance travelled by the jet ski.

5 For the following velocity-time graphs:

i Draw the corresponding displacement-time graph.

ii Find the total distance travelled.

iii Find the final displacement from the starting point.

a

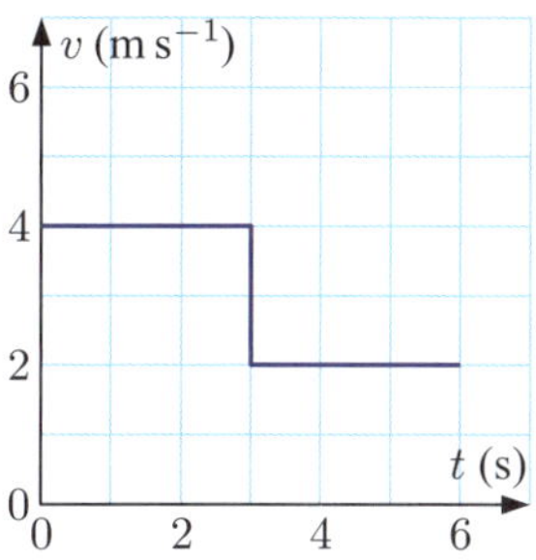

b

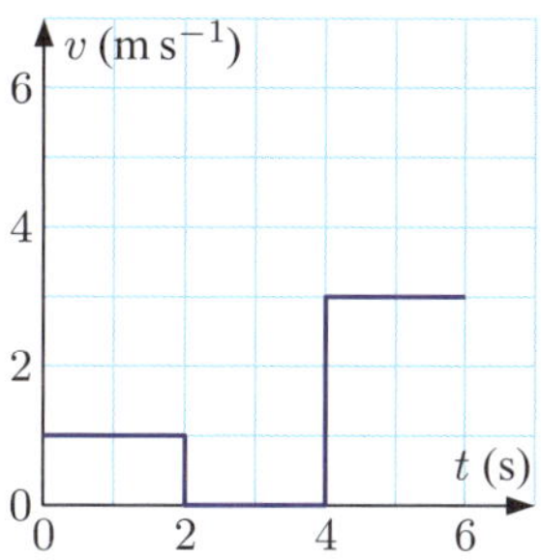

c

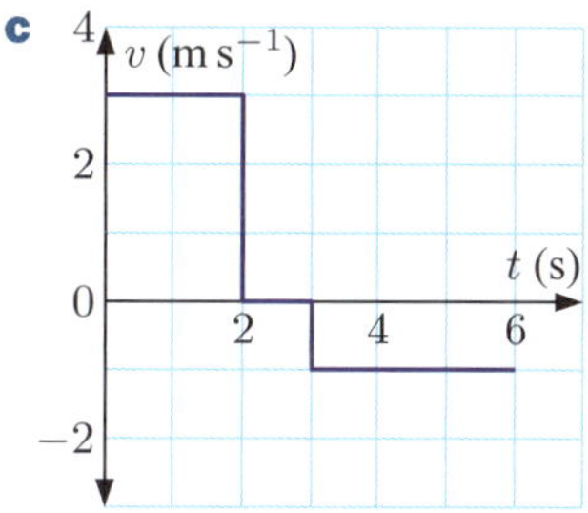

6 A particle travelled at u m s^{-1} for 10 seconds, and then decelerated uniformly for 4 seconds to the velocity $\frac{u}{3}$ m s^{-1}. Over the 14 seconds, the particle travelled a total of 128 m.

a Find the value of u.

b Find the gradient of the line segment PQ. Interpret your answer.

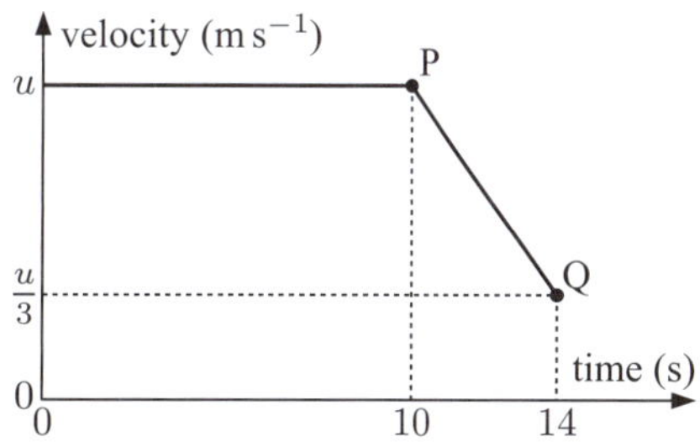

7 When a train leaves a station, it accelerates uniformly for 20 seconds until it reaches a speed of 20 m s^{-1}. It maintains this speed for s seconds. The train then takes 30 seconds to decelerate uniformly until it stops at the next station.

a Draw a velocity-time graph to show the motion of the train.

b Given that the stations are 2 km apart, find the value of s.

DISPLACEMENT AND VELOCITY FUNCTIONS

We have seen previously that the velocity function is the derivative of the displacement function, so $v(t) = s'(t)$.

Given a velocity function, we can determine the displacement function by the integral

$$s(t) = \int v(t)\, dt$$

The constant of integration can be found if we know the position of the object at a particular time.

Example 8

Self Tutor

A particle moves in a straight line with velocity $v(t) = 2t - 4 \text{ m s}^{-1}$. It is initially 2 m to the left of the origin O.

a Find the displacement function $s(t)$.

b How far does the particle travel in the first 5 seconds of motion?

c Find the particle's change in displacement during the first 5 seconds.

a $v(t) = 2t - 4 \text{ m s}^{-1}$

$\therefore\ s(t) = \int (2t - 4)\, dt$

$= t^2 - 4t + c$

Now $s(0) = -2$, so $c = -2$

$\therefore\ s(t) = t^2 - 4t - 2$ m

b $v(t) = 2(t - 2)$ which has sign diagram:

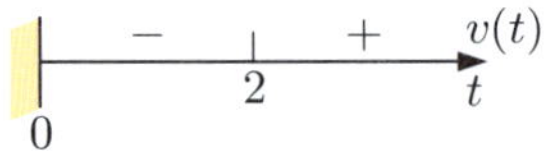

$\therefore$ the particle changes direction when $t = 2$.

Now $s(0) = -2$, $s(2) = -6$, $s(5) = 3$

$\therefore$ total distance travelled $= 4 + 9$

$= 13$ m

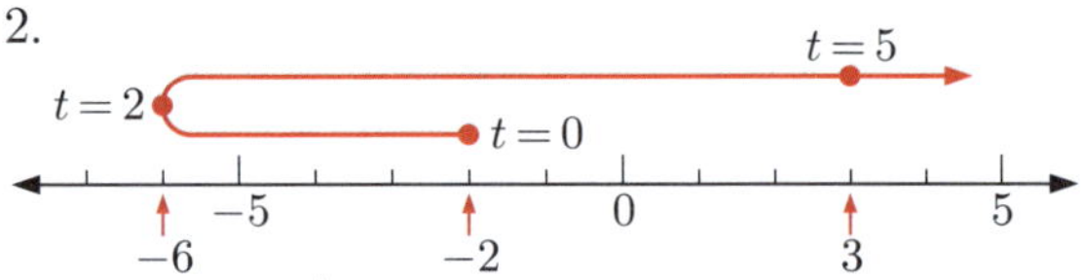

c Change in displacement $= s(5) - s(0)$

$= 3 - (-2)$

$= 5$ m

VELOCITY AND ACCELERATION FUNCTIONS

We have seen previously that the acceleration function is the derivative of the velocity function, so $a(t) = v'(t)$.

Given an acceleration function, we can determine the velocity function by the integral:

$$v(t) = \int a(t)\, dt$$

Summary

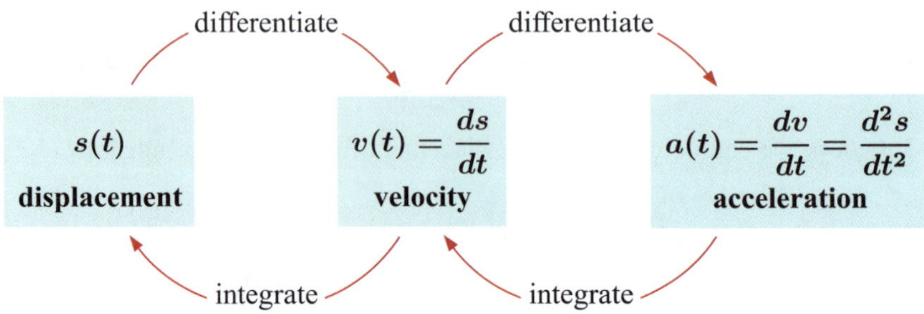

Example 9

Self Tutor

An object is initially at the origin, and moving to the left at 8 m s^{-1}. The object accelerates according to $a(t) = 4t - 6 \text{ m s}^{-2}$.

a Find the:

i velocity function $v(t)$ **ii** displacement function $s(t)$.

b Find the total distance travelled in the first 6 seconds.

a **i** $a(t) = 4t - 6 \text{ m s}^{-2}$

$\therefore v(t) = \int (4t - 6)\, dt$

$= 2t^2 - 6t + c$

Now $v(0) = -8$, so $c = -8$

$\therefore v(t) = 2t^2 - 6t - 8 \text{ m s}^{-1}$

ii $s(t) = \int (2t^2 - 6t - 8)\, dt$

$= \frac{2}{3}t^3 - 3t^2 - 8t + c$

Now $s(0) = 0$, so $c = 0$

$\therefore s(t) = \frac{2}{3}t^3 - 3t^2 - 8t$ m

b $v(t) = 2(t^2 - 3t - 4)$

$= 2(t - 4)(t + 1)$

which has sign diagram:

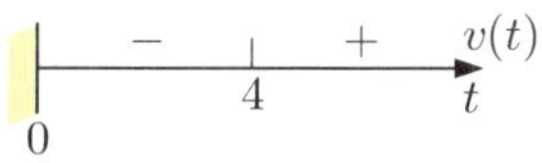

$\therefore$ the object changes direction when $t = 4$.

Now $s(0) = 0$, $s(4) = -37\frac{1}{3}$, $s(6) = -12$

$\therefore$ total distance travelled $= 37\frac{1}{3} + 25\frac{1}{3}$

$= 62\frac{2}{3}$ m

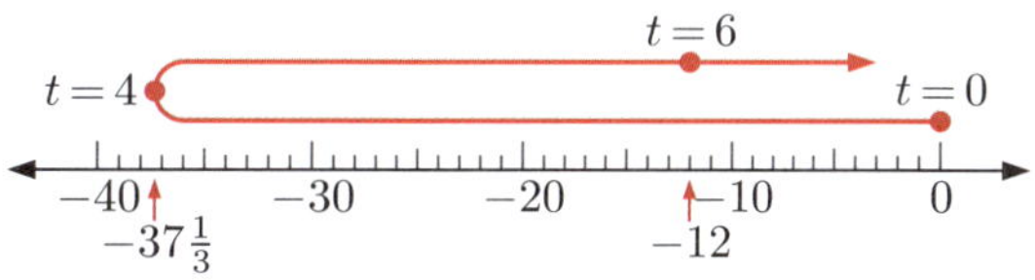

EXERCISE 17C.2

1 A particle moves in a straight line with velocity $v(t) = 2t + 3 \text{ m s}^{-1}$. The particle is initially 4 m to the right of the origin O.

  a Explain why the particle does not change direction.

  b Find the displacement function $s(t)$.

  c Find the distance travelled by the particle in the first 3 seconds.

2 The velocity of an object, t seconds after leaving the origin O, is $v(t) = 6 - 2t \text{ cm s}^{-1}$.

  a Write a formula for the displacement function $s(t)$.

  b Find the distance the object travels in the first 4 seconds.

  c Find the object's change in displacement during the first 4 seconds.

3 A baseball is hit vertically into the air. The velocity of the ball after t seconds is given by $v(t) = 29.4 - 9.8t \text{ m s}^{-1}$.

  a Given that the ball was 1.4 m above ground level when hit, find the displacement function $s(t)$ for the ball.

  b At what time does the ball change direction?

  c Find the total distance travelled by the ball in the first 5 seconds.

4 A particle moves in a straight line with velocity $v(t) = -2t^2 - t + 10 \text{ cm s}^{-1}$.

a Find the speed of the particle after 4 seconds.

b Show that the particle changes direction after 2 seconds.

c Find $\int_0^5 v(t)\, dt$, and interpret your answer.

d Find $\int_0^2 v(t)\, dt - \int_2^5 v(t)\, dt$, and interpret your answer.

5 An object moves with velocity function $v(t) = \frac{12}{3t+1} \text{ m s}^{-1}$. The object is initially 2 m to the right of the origin.

a Find the speed of the object after 1 second.

b Explain why the object is never at rest.

c Find the:

i displacement function $s(t)$ **ii** acceleration function $a(t)$.

d Find the distance travelled by the object in the first 3 seconds.

e Find the acceleration of the object after 1 second.

6 Jarrod is riding a water slide. His velocity t seconds after starting the ride is $v(t) = 2\sqrt{4t+1} \text{ m s}^{-1}$.

a Find Jarrod's:

i initial velocity **ii** velocity after 2 seconds.

b Write an expression for the distance $s(t)$ that Jarrod has travelled after t seconds.

c The water slide is 114 m long.

i How long does Jarrod take to complete the ride?

ii Find Jarrod's speed at the end of the ride.

7 The velocity of a competitive kayaker t seconds after starting a race is $4 - 4e^{-\frac{t}{4}} \text{ m s}^{-1}$.

a Find the speed of the kayaker after 4 seconds.

b Describe what happens to the speed of the kayaker over time.

c How far does the kayaker travel in the first minute?

d Find the acceleration of the kayaker after 10 seconds.

8 The acceleration of a particle is $a(t) = 3 - t \text{ cm s}^{-2}$.

a Find the particle's acceleration after 2 seconds.

b Given that the initial velocity of the particle is 5 cm s^{-1}, find the velocity function $v(t)$.

c Find the speed of the particle after:

i 4 seconds **ii** 8 seconds.

9 An object is initially at the origin, and moving to the left at 2 m s^{-1}. The object accelerates according to the function $a(t) = 4e^{2t} \text{ m s}^{-2}$.

a Find the:

i velocity function $v(t)$ **ii** displacement function $s(t)$.

b Find the time at which $v(t) = 0$.

c Find the distance travelled by the object in the first second.

10 The pendulum of a clock swings back and forth about its equilibrium point. The pendulum is initially at the extreme positive position, 5 cm to the right of equilibrium. The acceleration of the pendulum is given by $a(t) = -80\cos 4t \text{ cm s}^{-2}$.

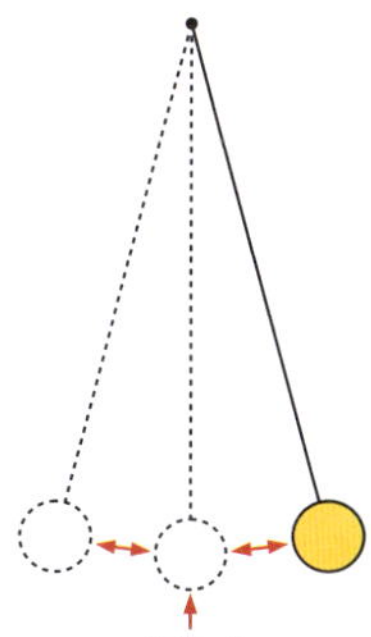

a Explain why $v = 0$ when $t = 0$.

b Find the velocity function $v(t)$.

c Find the displacement function $s(t)$ relative to the equilibrium point.

d Find the speed of the pendulum as it passes through the equilibrium point.

e How long does it take for the pendulum to first return to its starting point?

Review set 17A

1 Find the shaded area:

a

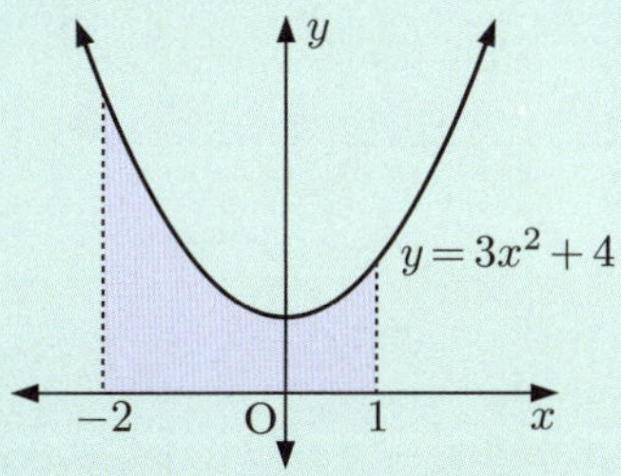

b

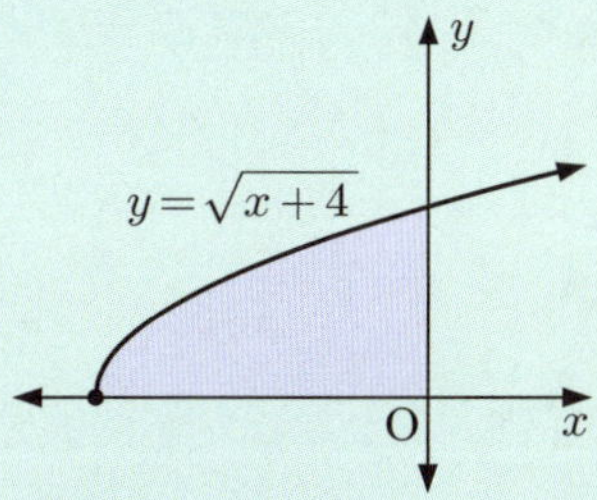

c

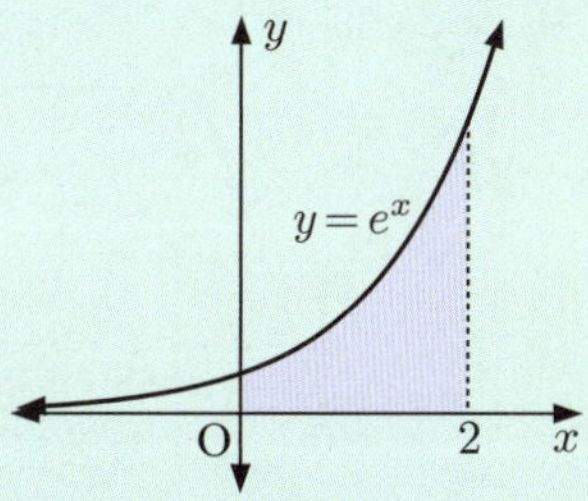

2 The graph of $y = \sin x + \cos x$ is shown alongside. The graph has a maximum turning point at A, and cuts the x-axis at B.

a Find the coordinates of A and B.

b Find the shaded area.

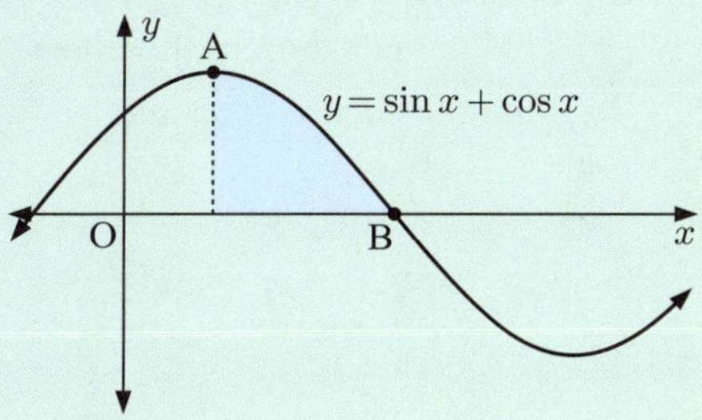

3

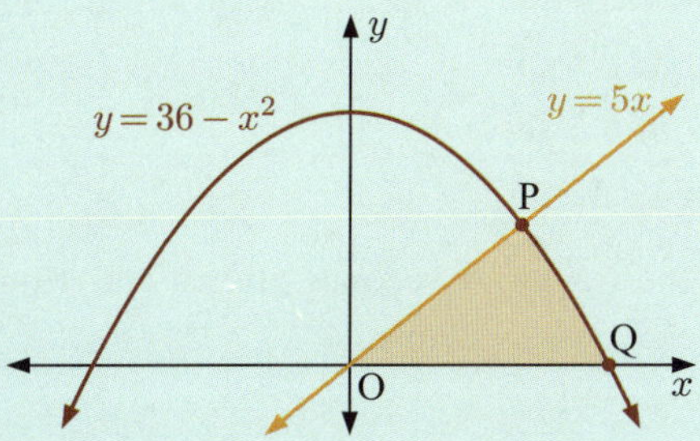

a Find the coordinates of:

i P **ii** Q

b Hence find the shaded area.

4 **a** Find the x-intercepts at A and B.

b Hence find the shaded area.

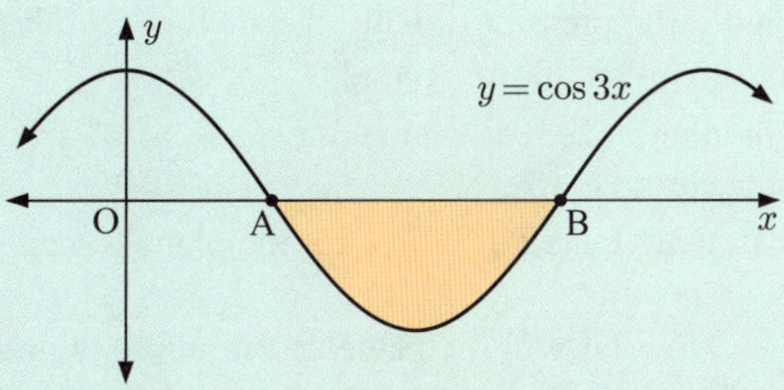

5 **a** On the same set of axes, sketch the graphs of $y = \dfrac{3}{x}$ and $y = 4 - x$.

b Find the coordinates of the points of intersection.

c Find the area of the region enclosed by the functions.

6 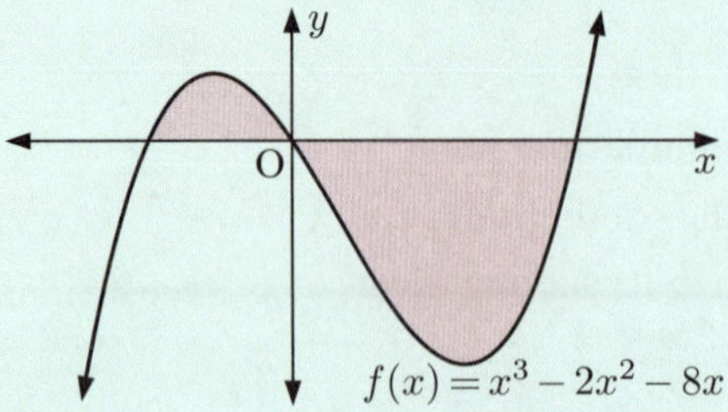

Find the total area of the region enclosed by $f(x) = x^3 - 2x^2 - 8x$ and the x-axis.

7 For each velocity-time graph, find the total distance travelled by the object:

a 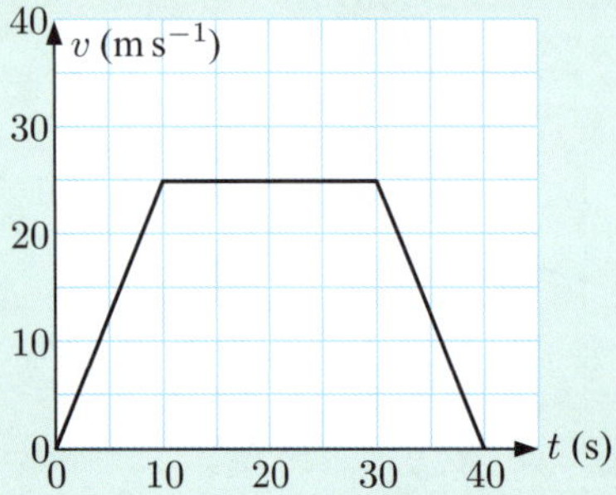

b 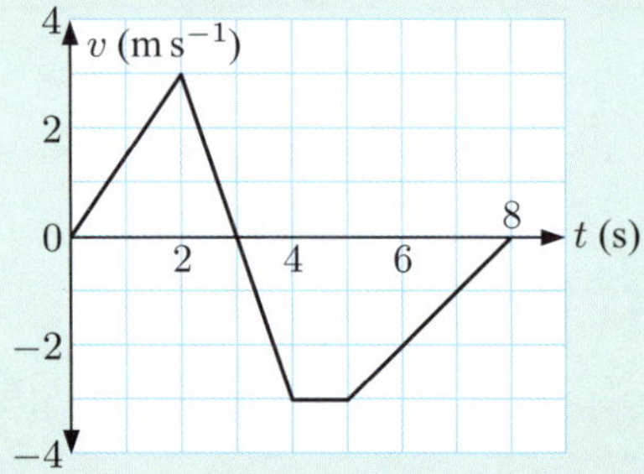

8 The graph of $y = x^2 - 3x + 5$ is shown alongside.

a Find the equation of the illustrated tangent.

b Find the shaded area.

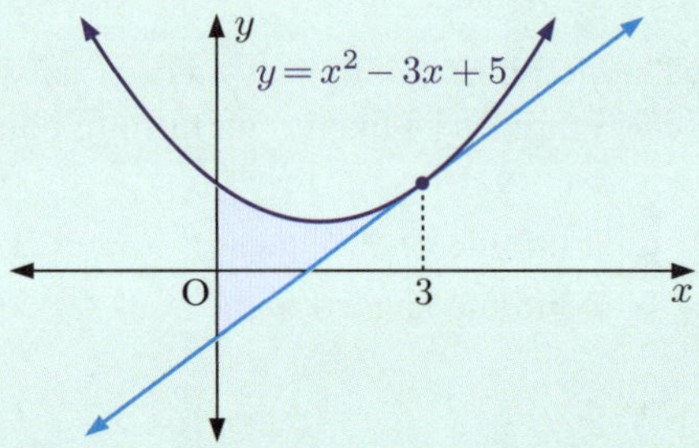

9 The velocity of an object, t seconds after leaving the origin O, is $v(t) = t^2 + 4t - 32$ m s^{-1}.

a Write a formula for the displacement function $s(t)$.

b Find the distance travelled by the object in the first 6 seconds.

c At what time does the object have acceleration 10 m s^{-2}?

10 An aeroplane is about to take off. Its velocity on the runway t seconds after it starts moving, is given by $v(t) = 9t^{\frac{2}{3}}$ m s^{-1}.

a Find the speed of the plane after 8 seconds.

b Find an expression for the distance $s(t)$ the plane has travelled after t seconds.

c The plane needs to achieve a speed of 81 m s^{-1} in order to take off.

i How long will it take for the plane to reach this speed?

ii How far will the plane travel before taking off?

Review set 17B

1

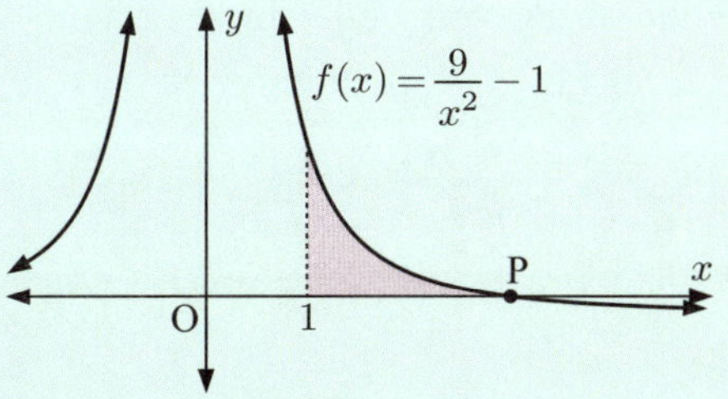

a Find the coordinates of P.

b Hence find the shaded area.

2 **a** Show that $\frac{d}{dx}\left(6xe^{\frac{x}{2}}\right) = 3e^{\frac{x}{2}}(2+x)$.

b Hence find the shaded area in the graph alongside.

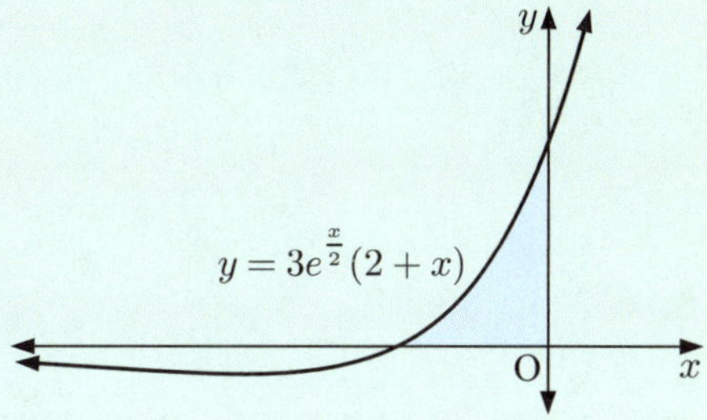

3 Find the shaded area:

a

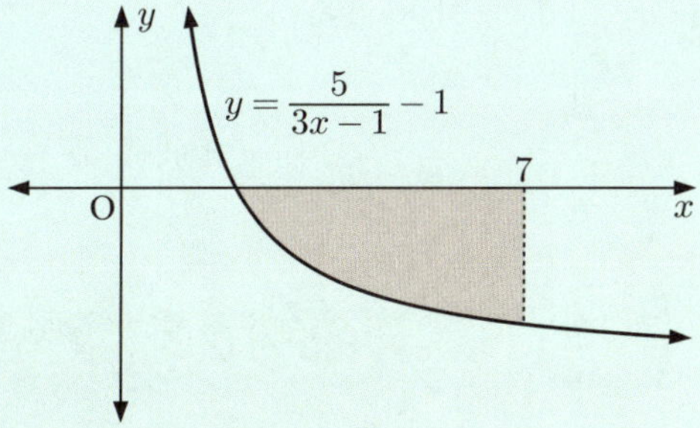

b

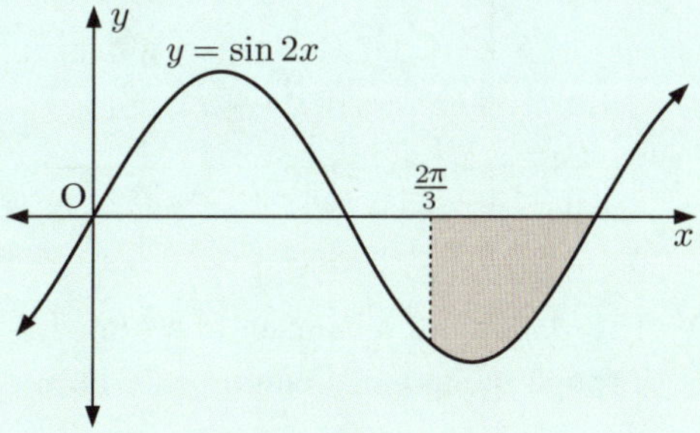

4 The graphs of $y = x^3 - 4x^2 - 4x$ and $y = x^2 - 20$ are shown alongside.

a Given that A has coordinates $(-2, -16)$, find the coordinates of B and C.

b Find the total area of the regions enclosed by the curves.

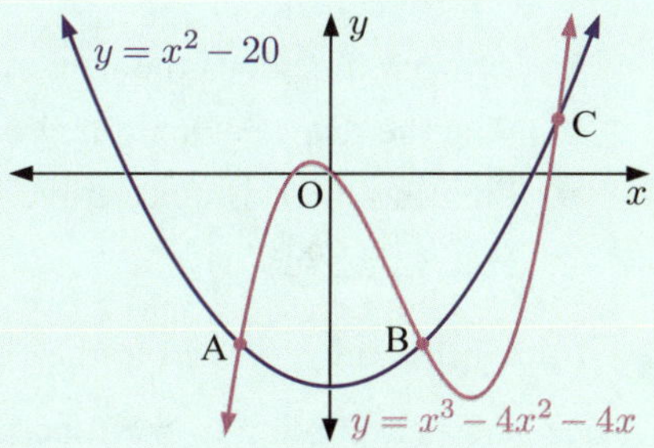

5 Consider the velocity-time graph alongside.

a Draw the corresponding displacement-time graph.

b Find the total distance travelled.

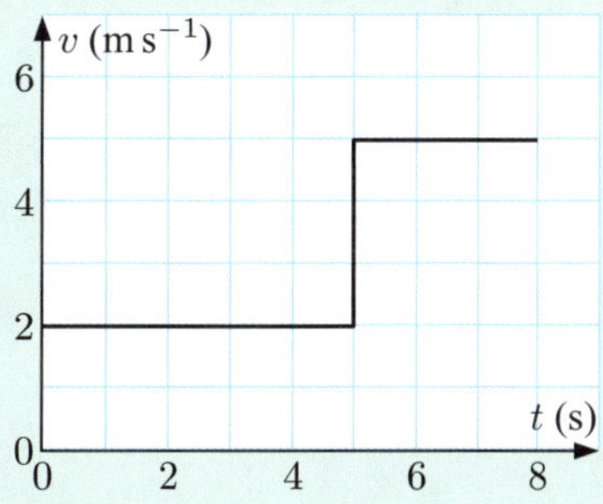

6 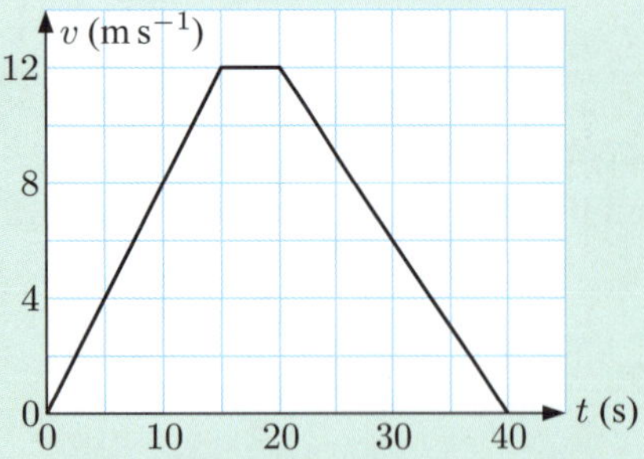

A tram has the velocity-time graph shown alongside.

a Find the speed of the tram after 30 seconds.

b Find the total distance travelled by the tram.

c Find the acceleration of the tram when $t = 10$.

7 **a** Find the coordinates of P and Q.

b Hence find the shaded area.

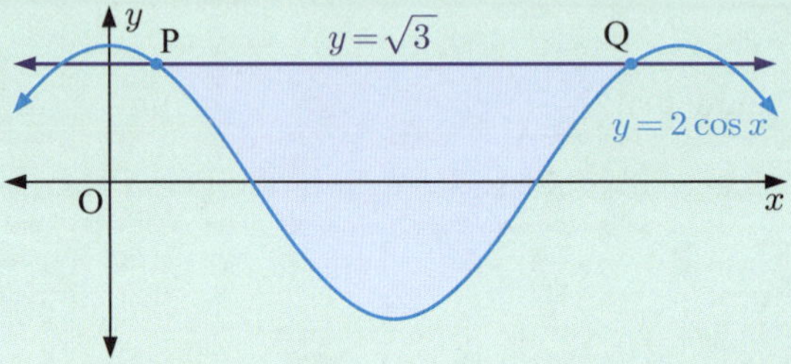

8 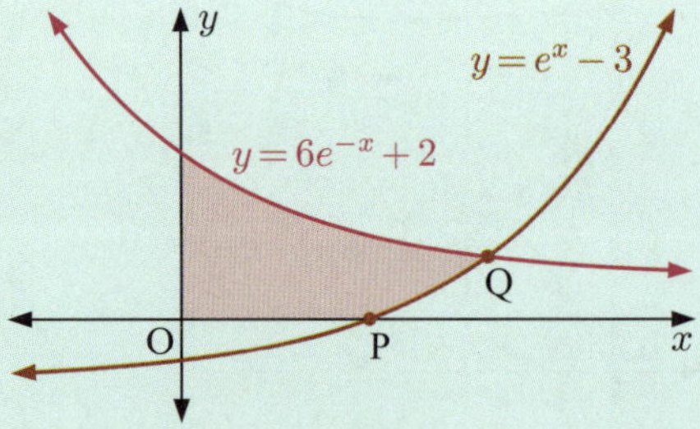

Consider the graphs of $y = e^x - 3$ and $y = 6e^{-x} + 2$.

a Find the coordinates of P and Q.

b Hence find the shaded area.

9 The velocity of a hamster in a tunnel, t seconds after passing through the tunnel's entrance O, is $v(t) = 12 - 3t$ cm s^{-1}.

a Find the speed of the hamster after:

i 1 second **ii** 5 seconds.

b At what time does the hamster change direction?

c Find the displacement function $s(t)$.

d Find the total distance travelled by the hamster in the first 6 seconds.

10 An object is initially 2 m to the right of the origin, and moving to the left at 6 m s^{-1}. The object accelerates according to the function $a(t) = \dfrac{3}{\sqrt{t+1}}$ m s^{-2}.

a Find the:

i velocity function $v(t)$ **ii** displacement function $s(t)$.

b Find the time at which the object changes direction, and the position of the object at that time.

c Find the total distance travelled by the object in the first 8 seconds.

ANSWERS

EXERCISE 1A

1 **a** gradient $= 3$, y-intercept is 5
b gradient $= 4$, y-intercept is -2
c gradient $= \frac{1}{5}$, y-intercept is $\frac{3}{5}$
d gradient $= -7$, y-intercept is -3
e gradient $= \frac{1}{6}$, y-intercept is $\frac{1}{3}$
f gradient $= -\frac{5}{3}$, y-intercept is $\frac{8}{3}$

2 **a** $y = x - 2$ **b** $y = -x + 4$ **c** $y = 2x$
d $y = -\frac{1}{2}x + 3$

3 **a** $y = 4x - 13$ **b** $y = -3x - 5$ **c** $y = -5x + 32$
d $y = \frac{1}{2}x + \frac{7}{2}$ **e** $y = -\frac{1}{3}x + \frac{8}{3}$ **f** $y = 6$

4 **a** $2x - 3y = -11$ **b** $3x - 5y = -23$ **c** $x + 3y = 5$
d $2x + 7y = -2$ **e** $4x - y = -11$ **f** $2x + y = 7$
g $7x + 2y = 18$ **h** $6x - y = -40$

5 **a** $y = \frac{5}{2}x - 2$ **b** $y = -2x + 3$ **c** $y = -2$
d $y = -\frac{1}{5}x + \frac{2}{5}$ **e** $y = \frac{1}{6}x - \frac{11}{6}$ **f** $y = -\frac{2}{3}x - \frac{11}{3}$

6 **a** $x - 3y = -3$ **b** $5x - y = 1$ **c** $x - y = 3$
d $4x - 5y = 10$ **e** $x - 2y = -1$ **f** $2x + 3y = -5$

7 **a** $\sqrt{45}$ units **b** $(-1, \frac{7}{2})$ **c** $\frac{1}{2}$ **d** $y = \frac{1}{2}x + 4$

8 **a** $y = \frac{4}{3}x - 1$ **b** $2x - 3y = -13$ **c** $y = x + 1$
d $2x + y = -2$ **e** $y = -\frac{2}{3}x + 2$ **f** $3x + 7y = -9$

9 **a** $M = \frac{1}{3}p + 2$ **b** $R = -\frac{5}{4}n + 2$ **c** $T = \frac{1}{2}x - 1$
d $F = \frac{1}{10}x + 1$ **e** $H = -\frac{1}{2}z + 2$ **f** $W = -\frac{1}{6}t - 2$

10 **a**
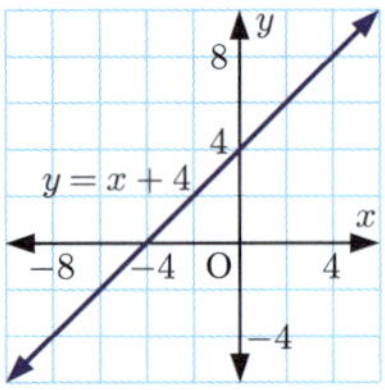

b
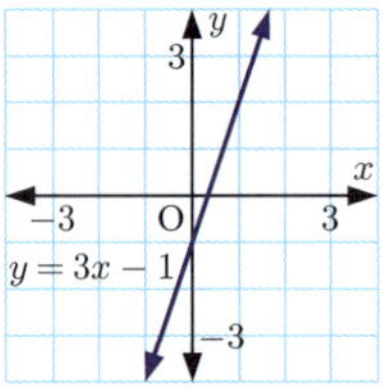

c
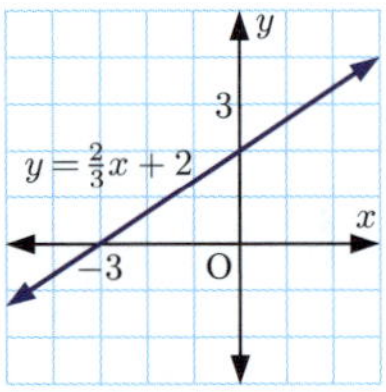

d
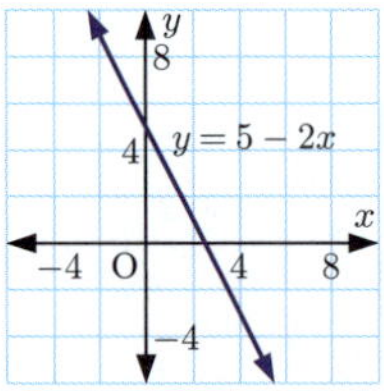

e
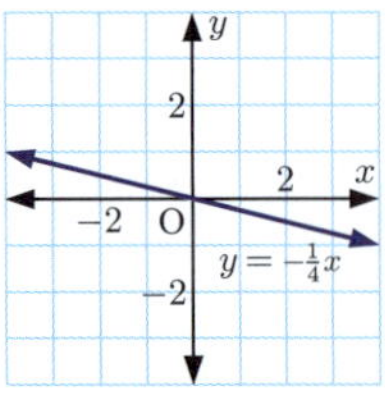

f
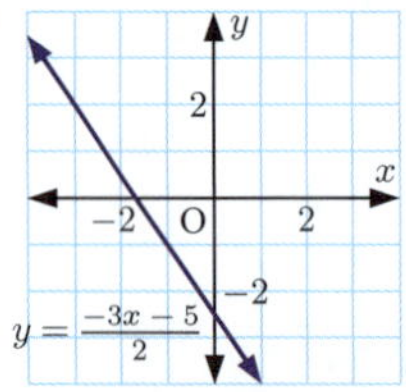

11 **a**
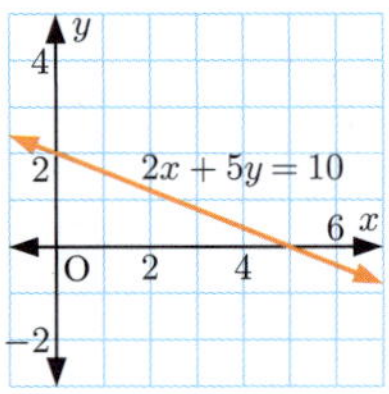

b
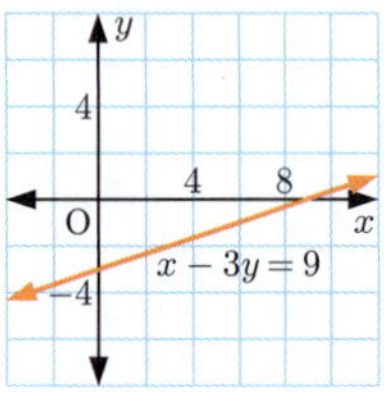

c
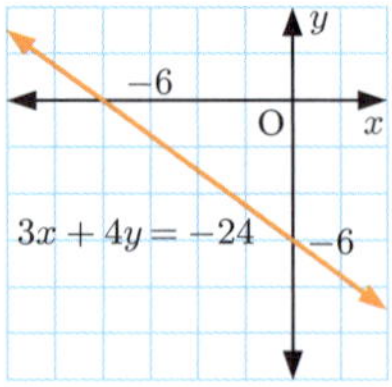

d
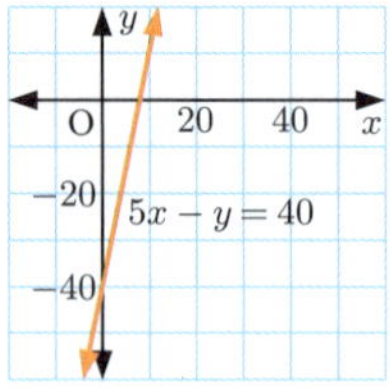

e
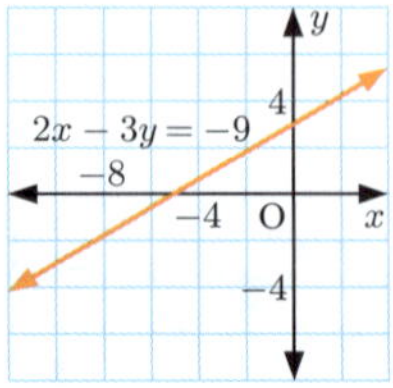

f
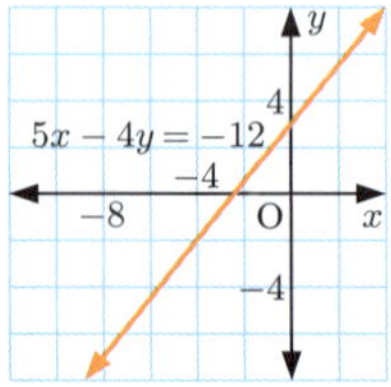

12 **a** $x + 2y = 13$ **b** $(13, 0)$

13 **a** $3x + 5y = 10$ **b** **i** $(\frac{10}{3}, 0)$ **ii** $(0, 2)$

14 54 units^2 **15** **a** $q = 6$ **b** $25\frac{1}{2}$ units^2 **16** $21\frac{3}{5}$ units^2

EXERCISE 1B

1 **a** $\sqrt{160}$ units **b** $(-1, 1)$ **c** -3 **d** $y = \frac{1}{3}x + \frac{4}{3}$

2 **a** $y = x - 4$ **b** $y = 2x + 6$ **c** $y = \frac{6}{5}x + \frac{7}{2}$
d $y = \frac{3}{4}x + \frac{13}{8}$ **e** $x = -\frac{1}{2}$ **f** $y = 1$

3 **a** $C(0, \frac{1}{3})$ **b** $AC = BC = \frac{2\sqrt{85}}{3}$ units **4** 15 units^2

EXERCISE 1C

1 **a** $(-5, -11)$ **b** $(1, 2)$ **c** $(\frac{4}{3}, \frac{7}{3})$
d $(-8, -5)$ **e** $(\frac{6}{7}, -\frac{17}{7})$ **f** $(-9, -\frac{1}{2})$

2 **a** $(1, 3)$ **b** $(2, 3)$ **c** $(6, -3)$
d $(\frac{1}{5}, \frac{9}{5})$ **e** $(-52, -29)$ **f** $(\frac{36}{5}, -\frac{49}{5})$

3 **a**
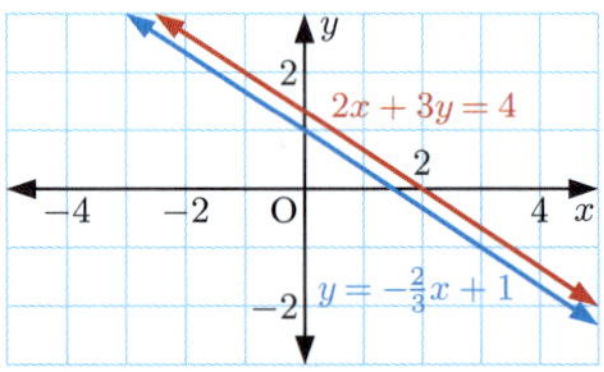

The lines are parallel.
$\therefore$ there is no intersection point.

b
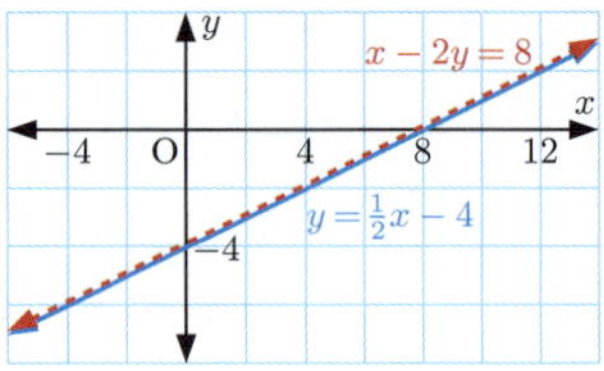

The lines are coincident.
$\therefore$ there are infinitely many intersection points.

4 **a** $(3, -1)$ **b** $(-5, 2)$ **c** $(\frac{3}{11}, -\frac{37}{44})$ **d** $(4, -3)$
e $(-4, -1)$ **f** $(2, -\frac{2}{3})$ **g** $(1, -2)$ **h** $(-3, 7)$
i $(\frac{7}{5}, \frac{2}{5})$

5 **a**
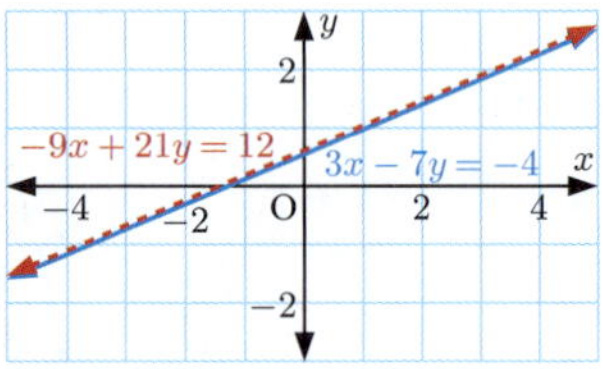

The lines are coincident.
$\therefore$ there are infinitely many intersection points.

b 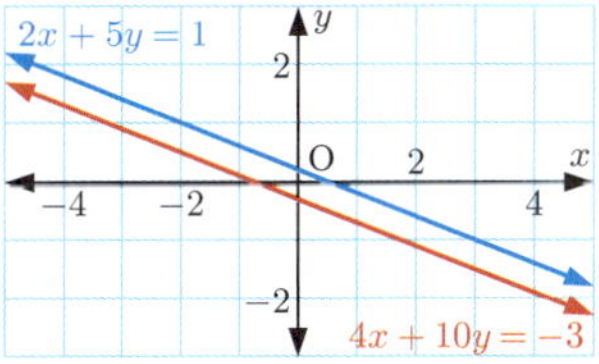

The lines are parallel. $\therefore$ there is no intersection point.

6 **a** l_1: $y = -2x + 9$, l_2: $y = \frac{1}{2}x + 1$ **b** $(\frac{16}{5}, \frac{13}{5})$

7 **a** $3x + 5y = 9$ **b** $(-2, 3)$

8 **a** $(1, 2)$ **b** $(-3, 4)$ **9** $X(4, 2)$

10 **a** $x - 3y = -8$ **b** $y = -3x - 4$ **c** $X(-2, 2)$

11 **a** $S(0, -1)$ **b** 25 units^2

12 **a** $C(-1, 0)$ **b** 26 units^2

13 **a** $y = -\frac{3}{2}x - \frac{5}{2}$

b **i** $P(-1, -1)$ **ii** $(-1, -1) = \left(\frac{-3+1}{2}, \frac{2-4}{2}\right)$

iii isosceles **iv** 26 units^2

14 30 units^2

15 **a** **i** $B(5, 0)$ **ii** $C(7, -4)$ **iii** $N(6, -2)$

b **Hint:** Find the gradients of MN and AC.

c **i** 15 units^2 **ii** 20 units^2

REVIEW SET 1A

1 **a** $\sqrt{40}$ units **b** $(2, 5)$ **c** $x + 3y = 17$

2 $y = -2x + 6$

3 The gradient of a vertical line is undefined.

4 **a** 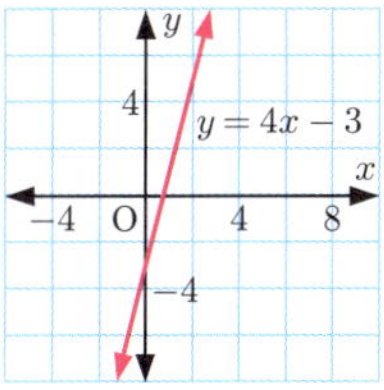

b 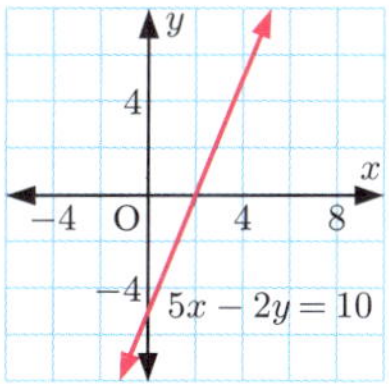

5 **a** $2x - 5y = -7$ **b** $5x + 3y = -5$

6 **a** $x + 2y = 7$ **b** $(7, 0)$

7 **a** $(-3, -7)$ **b** $(\frac{1}{2}, -\frac{13}{8})$

8 **a** $4x + 3y = 9$ **b** $(3, -1)$

9 **a** $R(-3, 0)$ **b** 50 units^2

10 **a** $D(-1, 4)$ **b** $32\frac{1}{2}$ units^2

11 **a** $T(3, 5)$ **b** $y = \frac{1}{2}x + \frac{7}{2}$ **c** $m = 1$, $n = 8$

d 20 units^2

REVIEW SET 1B

1 $5x - 8y = 31$ **2** $3x + y = 7$

3 **a** 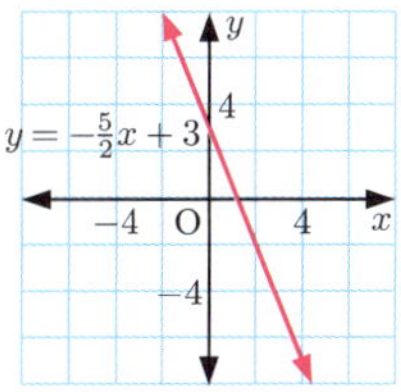

b 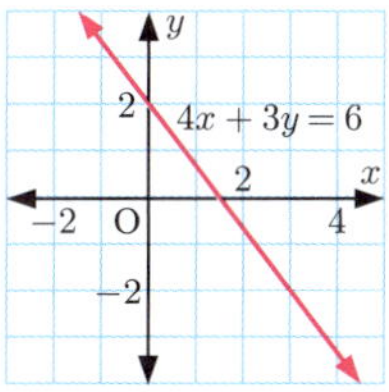

4 **a** $\sqrt{90}$ units **b** $y = \frac{1}{3}x + 3$ **c** $k = -1$ **d** 30 units^2

5 **a** $r = \frac{5}{7}a + 2$ **b** $K = \frac{3}{5}s + 3$ **6** $5x - 6y = -24$

7 **a** $(4, 5)$ **b** $(1, -5)$ **8** $X(3, 2)$

9 **a** **i** $B(2, 12)$ **ii** $C(11, 0)$ **b** 75 units^2

10 **a** $(-2, 5)$ **b** $(32\frac{1}{2}, 20\frac{1}{2})$

11 **b** **i** $y = -2x + 12$ **ii** $X(4, 4)$ **iii** 20 units^2

EXERCISE 2A.1

1 **a** 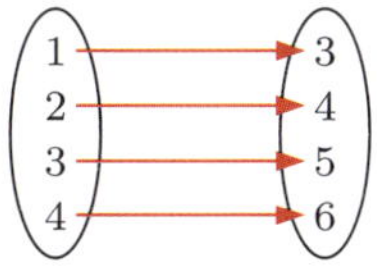

Domain Range

is a function

b 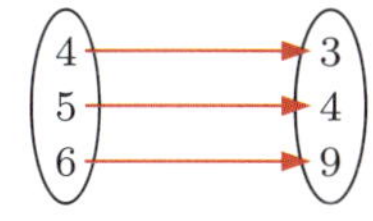

Domain Range

is a function

c 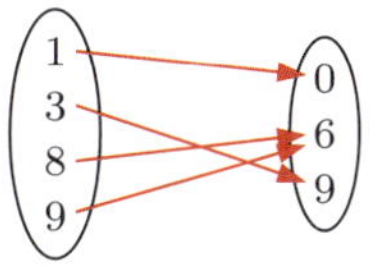

Domain Range

is a function

d 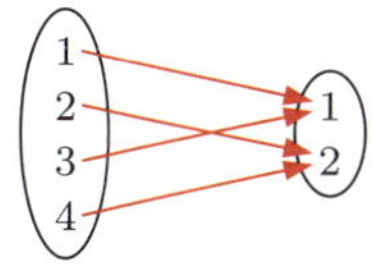

Domain Range

is a function

e 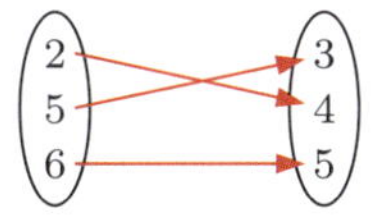

Domain Range

is a function

f 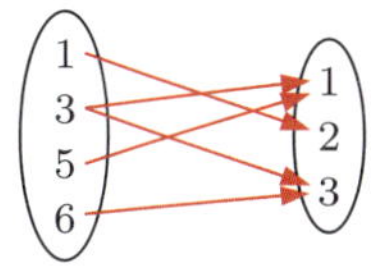

Domain Range

is not a function

2 **a** is a function **b** is a function **c** not a function

d not a function **e** is a function **f** is a function

g is a function **h** not a function **i** not a function

3 If $y^2x = 5$, then $y = \pm\sqrt{\frac{5}{x}}$. For all $x \neq 0$, there are two corresponding values of y.

$\therefore$ $y^2x = 5$ is not a function.

EXERCISE 2A.2

1 **a** one-one **b** not one-one **c** one-one

d not one-one **e** not one-one **f** one-one

2 **a** not a function **b** is a function, is one-one

c is a function, is not one-one

3 If each value in the domain maps to exactly one value in the range, then the relation is a function. However, there are fewer elements in the range than the domain. At least two values in the domain must map to the same value in the range. The function cannot be one-one.

EXERCISE 2B

1 **a** 7 **b** 7 **c** 15 **d** 22 **e** $6\frac{1}{4}$

2 **a** 2 **b** -2 **c** $2\frac{1}{2}$ **d** $-2\frac{1}{2}$ **e** $-2\frac{1}{2}$

3 **a** **i** -2 **ii** 4 **b** **i** $x = 5$ **ii** $x = 0$ or 8

4 **a** 7 **b** $2a + 1$ **c** $-2a + 1$ **d** $2a - 1$

e $2x + 5$ **f** $4x + 1$

5 **a** -1 **b** $-a^2$ **c** $-a^2 - 4a - 4$ **d** $-4x^2$

e $-x^2 + 2x - 1$ **f** $-x$

6 **a** $-9x^2 + 6x + 1$ **b** $-x^2 - 2x + 1$ **c** $-x^2 + 2$

d $-x^2 + 4x - 2$ **e** $-\frac{1}{x^2} + \frac{2}{x} + 1$

f $-x^2 - 2xh - h^2 + 2x + 2h + 1$

7 **a** **i** $\frac{2}{3}$ **ii** $-\frac{3}{8}$ **iii** $-\frac{2}{3}$

b $x = \pm 1$ **c** $\dfrac{x+1}{x^2+2x}$ **d** $x = -\frac{1}{2}$ or 2

8 No. If a relation is a circle, there are values in the domain which correspond to more than one value in the range and hence it is not a function.

9 **a** $H(3) = 288$, the altitude of the balloon 3 minutes before landing is 288 m.

b 6 minutes

10 **a** $a = -\frac{4}{5}$, $b = \frac{22}{5}$ **b** $a = \frac{7}{2}$, $b = -\frac{9}{2}$

c $a = 1$, $b = 2$

EXERCISE 2C.1

1 **a** Domain is $\{x : x \geqslant -1\}$
Range is $\{y : y \leqslant 2\}$

b Domain is $\{x : 1 < x < 4\}$
Range is $\{y : -1 < y < 3\}$

c Domain is $\{x : x \in \mathbb{R}\}$
Range is $\{y : y \in \mathbb{R}\}$

d Domain is $\{x : x \leqslant 2\}$
Range is $\{y : y \in \mathbb{R}\}$

e Domain is $\{x : x \in \mathbb{R}\}$
Range is $\{y : y \leqslant 3\}$

f Domain is $\{x : -2 \leqslant x \leqslant 4\}$
Range is $\{y : -3 \leqslant y \leqslant 3\}$

2 **a** Domain is $\{x : x \neq 3\}$
Range is $\{y : y \neq 2\}$

b Domain is $\{x : -2 < x \leqslant 2\}$
Range is $\{y : -2 \leqslant y \leqslant 2\}$

c Domain is $\{x : 0 \leqslant x \leqslant 1\}$
Range is $\{y : 0 \leqslant y \leqslant 1\}$

3 No. For example, $f(x) = x$ and $f(x) = -x$ both have domain $\{x : x \in \mathbb{R}\}$ and range $\{y : y \in \mathbb{R}\}$.

EXERCISE 2C.2

1 **a** $f(x)$ is defined for $x \neq \frac{5}{3}$, domain is $\{x : x \neq \frac{5}{3}\}$

b $f(x)$ is defined for $x \leqslant 0$, domain is $\{x : x \leqslant 0\}$

c $f(x)$ is defined for $x > -1$, domain is $\{x : x > -1\}$

2 **a** **i**

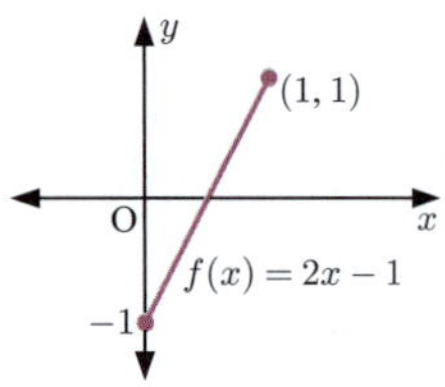

ii

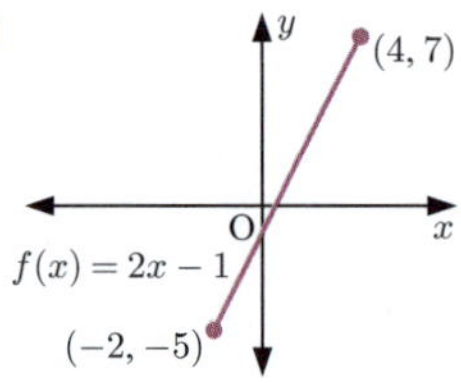

iii

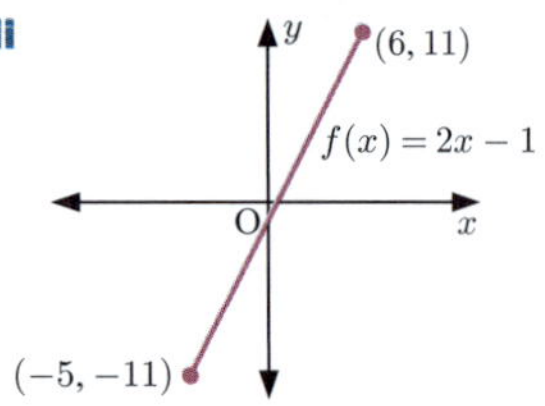

b **i** $\{y : -1 \leqslant y \leqslant 1\}$
ii $\{y : -5 \leqslant y \leqslant 7\}$
iii $\{y : -11 \leqslant y \leqslant 11\}$

c

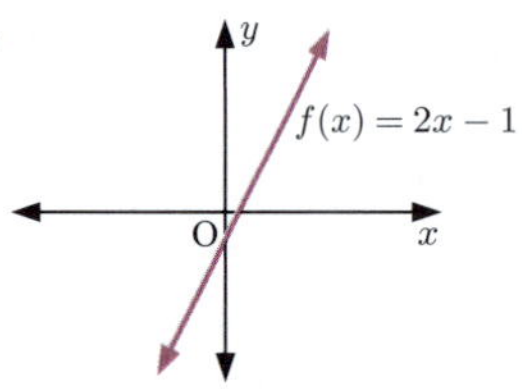

Range is $\{y : y \in \mathbb{R}\}$

3 **a** Domain is $\{x : x \in \mathbb{R}\}$
Range is $\{y : y = 1\}$

b Domain is $\{x : x \neq 0\}$
Range is $\{y : y \neq 0\}$

c Domain is $\{x : x \geqslant -1\}$
Range is $\{y : y \geqslant 0\}$

d Domain is $\{x : x \neq \frac{1}{2}\}$
Range is $\{y : y \neq 0\}$

e Domain is $\{x : x \in \mathbb{R}\}$
Range is $\{y : y \in \mathbb{R}\}$

f Domain is $\{x : x > a\}$
Range is $\{y : y > 0\}$

4 **a**

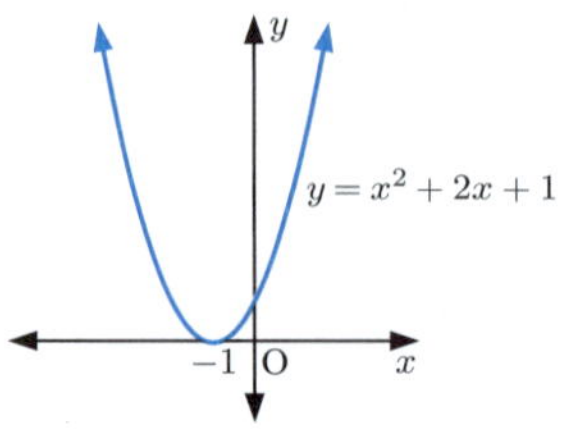

Domain is $\{x : x \in \mathbb{R}\}$
Range is $\{y : y \geqslant 0\}$

b

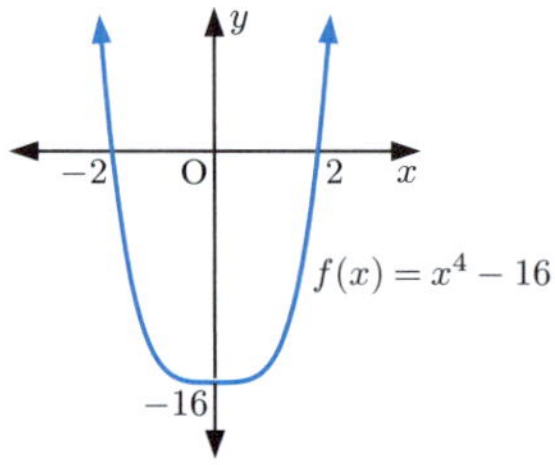

Domain is $\{x : x \in \mathbb{R}\}$
Range is $\{y : y \geqslant -16\}$

c

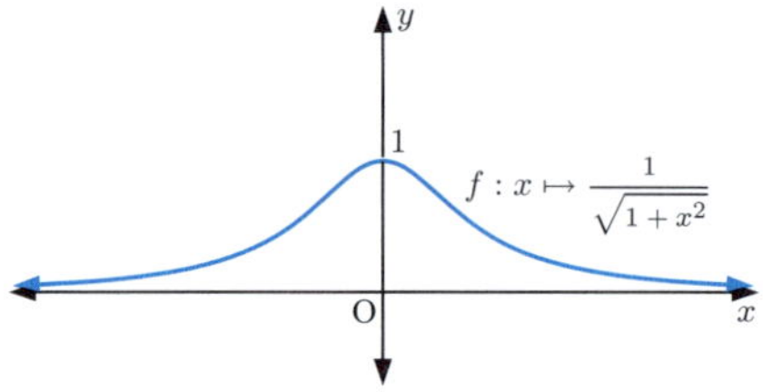

Domain is $\{x : x \in \mathbb{R}\}$
Range is $\{y : 0 < y \leqslant 1\}$

d

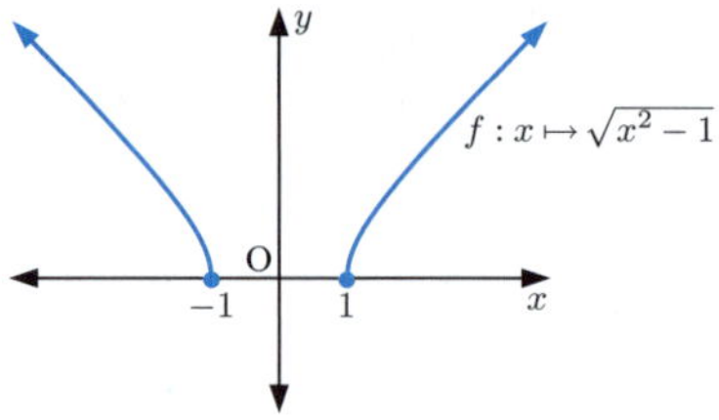

Domain is $\{x : x \leqslant -1 \text{ or } x \geqslant 1\}$
Range is $\{y : y \geqslant 0\}$

e

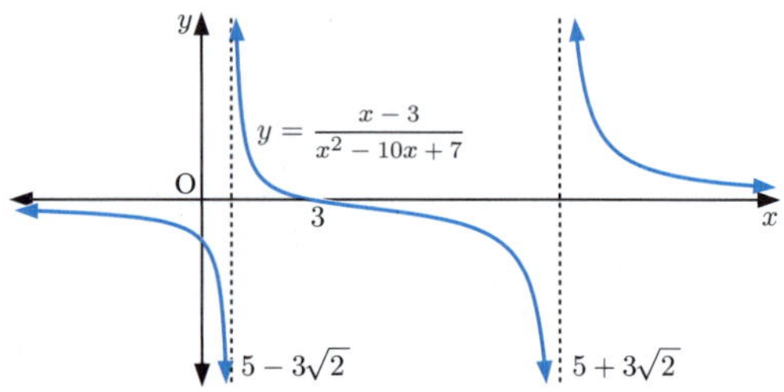

Domain is $\{x : x \neq 5 \pm 3\sqrt{2}\}$
Range is $\{y : y \in \mathbb{R}\}$

f

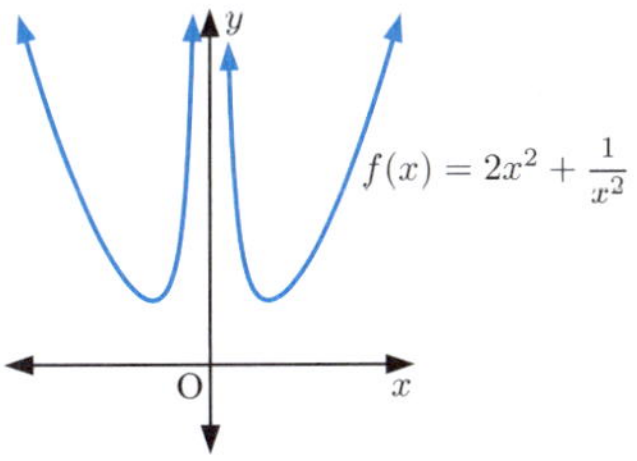

Domain is $\{x : x \neq 0\}$
Range is $\{y : y \geqslant 2.83\}$

g

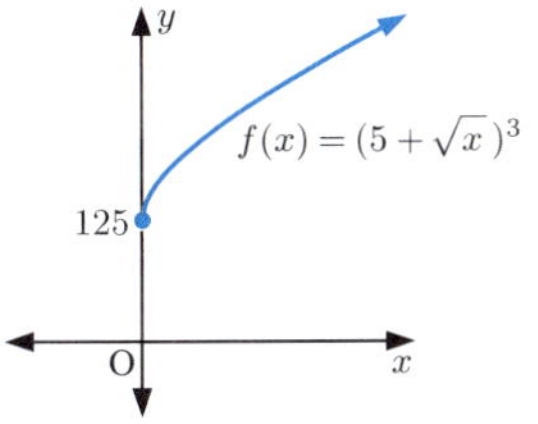

Domain is $\{x : x \geqslant 0\}$
Range is $\{y : y \geqslant 125\}$

h

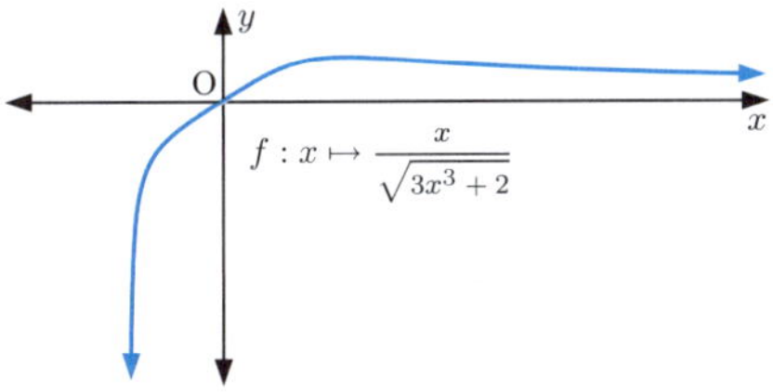

Domain is $\{x : x > -0.87\}$
Range is $\{y : y < 0.45\}$

i

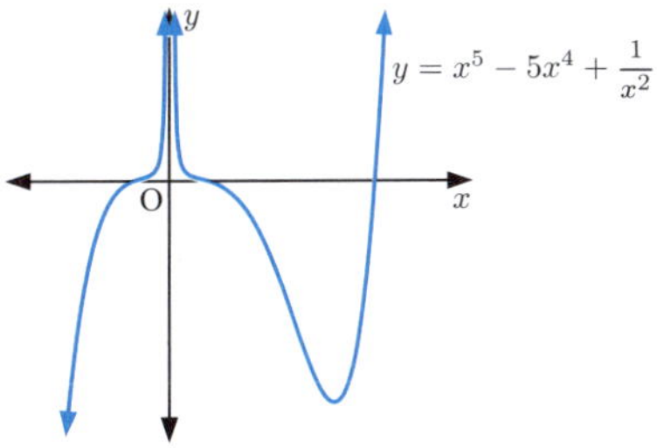

Domain is $\{x : x \neq 0\}$
Range is $\{y : y \in \mathbb{R}\}$

EXERCISE 2D.1

1 **a** 5 **b** 5 **c** 11 **d** 11 **e** 4 **f** 4 **g** 6 **h** 6

2 Yes, the modulus of a number is its size, ignoring its sign. a and $-a$ have the same *size*, so $|a| = |-a|$.

3 **a** 6 **b** 0 **c** $\frac{2}{7}$ **d** $\frac{8}{27}$

4 **a** 1 **b** 6 **c** 4 **d** 3

5 **a** 2 **b** 4 **c** -2 **d** $\frac{7}{3}$

6 **a** 6 **b** 4 **c** 8 **d** 2

EXERCISE 2D.2

1 **a**

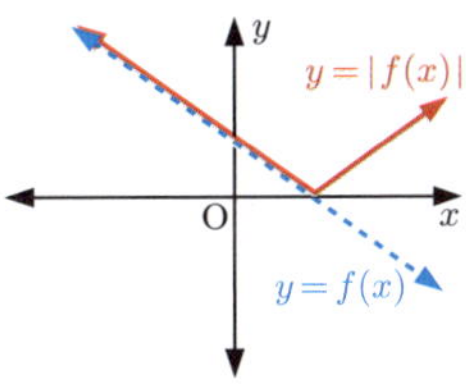

b

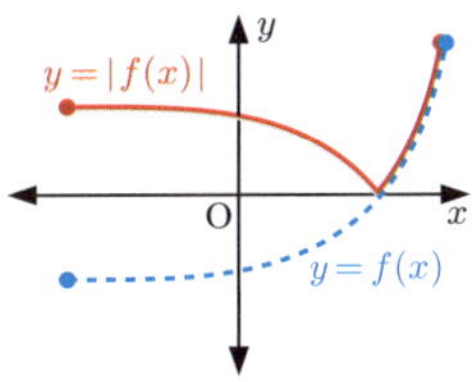

c

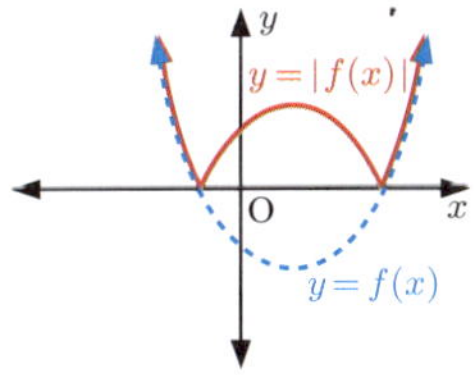

d

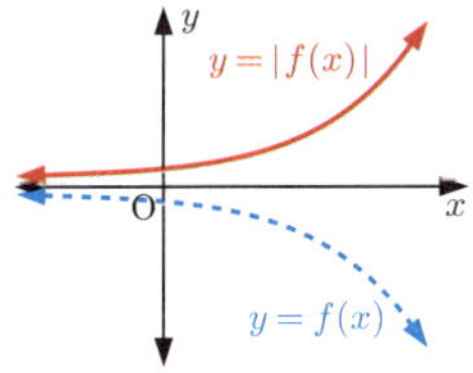

e

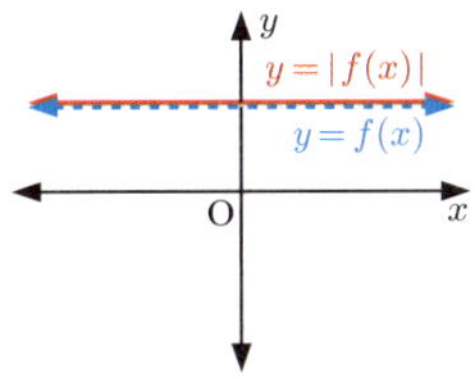

f

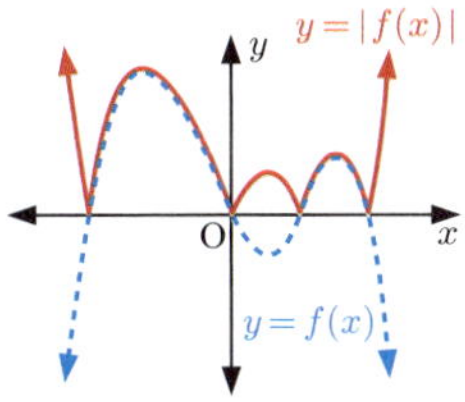

2 function **d** **3** $\{y : 0 \leqslant y \leqslant 6\}$

4 **a** false **b** true **c** true **d** false **e** true

5 **a**

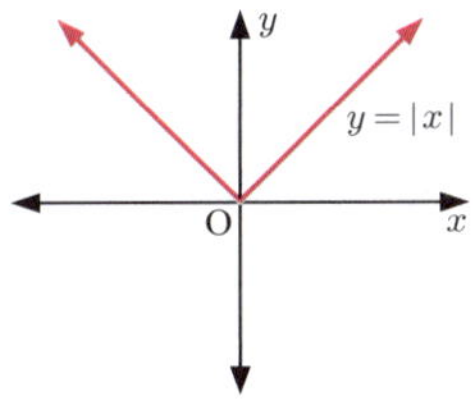

b

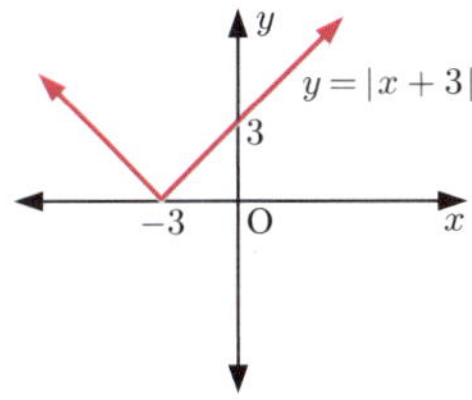

c

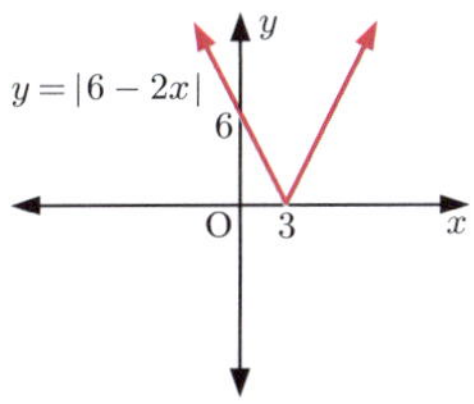

d

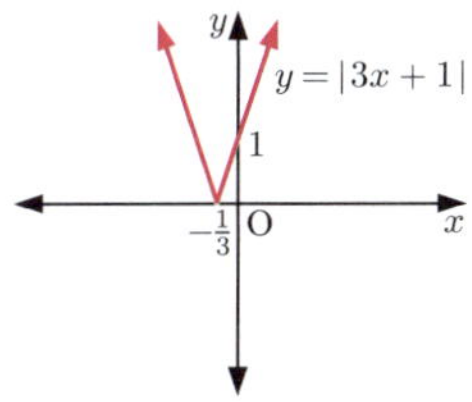

e

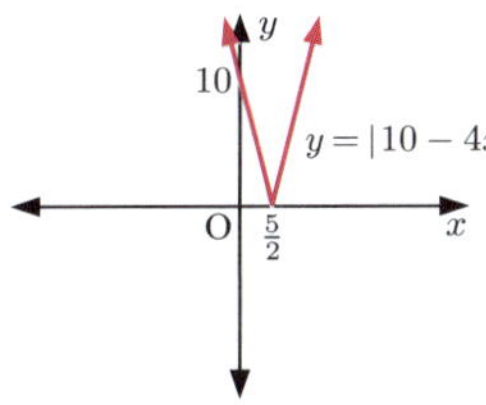

f

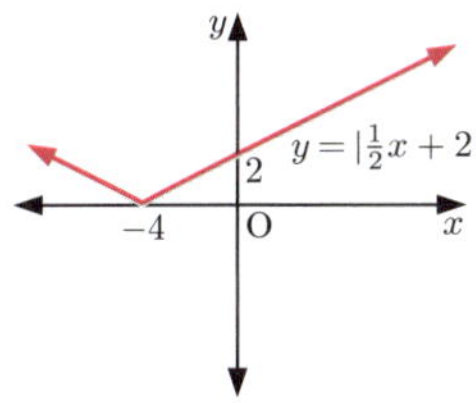

EXERCISE 2E.1

1 **a** $x = \pm 5$ **b** $x = \pm 9$ **c** no solution
d $x = 0$ **e** $x = 2$ or -6 **f** no solution
g $x = 6$ **h** $x = 9$ or -4 **i** $x = \frac{14}{5}$ or -4
j $x = \frac{2}{7}$ or $\frac{8}{7}$ **k** $x = -\frac{1}{2}$ or $-\frac{1}{4}$ **l** $x = 1$ or $-\frac{17}{5}$

2 **a** $x = \frac{1}{3}$ or -13 **b** $x = 4$ or $-\frac{2}{5}$ **c** $x = \frac{5}{7}$ or $-\frac{7}{3}$
d $x = \frac{8}{3}$ or $-\frac{2}{5}$ **e** $x = \frac{3}{5}$ or -11 **f** $x = \frac{8}{3}$ or $-\frac{4}{7}$
g $x = \frac{1}{7}$ or 1 **h** $x = -\frac{1}{3}$ or 1 **i** $x = 0$
j $x = -\frac{1}{2}$ or $-\frac{3}{2}$ **k** $x = \pm 1$ **l** $x = -\frac{5}{2}$ or $\frac{3}{4}$

3 $x - 5 \neq x - 7$ for any value of x, so $x - 5 = -(x - 7)$
$\therefore x = 6$

EXERCISE 2E.2

1 **a** $x = -2$ or 0 **b** $x = -4$ or 2

2 **a** 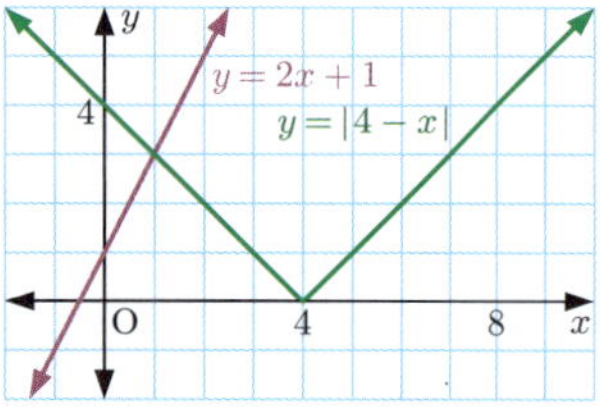

The graphs intersect at $(1, 3)$.

b $x = 1$

3 **a** $x = -3$ or 5 **b** $x = -5$ or -1 **c** $x = -1$ or 4
d no solution **e** $x = -2$ or 10 **f** $x = 1$ or 2

4 **a** $x = 3$ **b** $x = -3$ or 1 **c** no solution
d $x = 1$ **e** $x = 1$ or 4 **f** $x = -4$ or 0

5 **a** 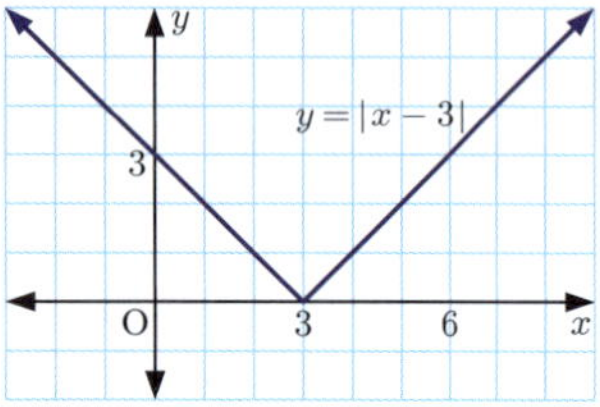

b **i** $0 < m < 1$
ii $m \geqslant 1$ or $m < -1$ or $m = 0$
iii $-1 \leqslant m < 0$

c **i** $x = 2$ or 6
ii $x = 1$

EXERCISE 2F.1

1 **a** $(-7, 4)$, $(1, 4)$
b **i** $-7 < x < 1$ **ii** $x \leqslant -7$ or $x \geqslant 1$

2 **a** 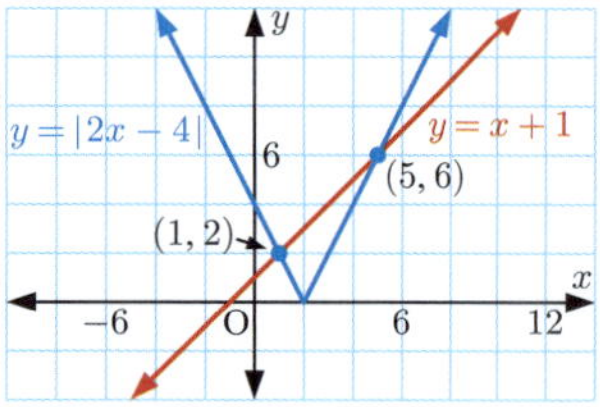

The graphs intersect at points $(1, 2)$ and $(5, 6)$.

b $1 \leqslant x \leqslant 5$

3 **a** 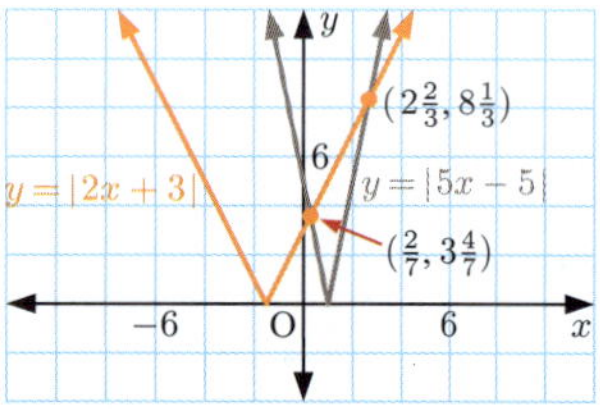

b $(\frac{2}{7}, 3\frac{4}{7})$ and $(2\frac{2}{3}, 8\frac{1}{3})$

c $x < \frac{2}{7}$ or $x > 2\frac{2}{3}$

4 **a** $x < -1$ or $x > 5$ **b** $-3 \leqslant x \leqslant -2$
c $x \leqslant -1$ or $x \geqslant 7$ **d** $x > 4$
e $-6 < x < 0$ **f** $x \geqslant 1$ **g** $-\frac{14}{3} \leqslant x \leqslant -\frac{2}{5}$
h $x < -\frac{7}{2}$ or $x > \frac{9}{4}$ **i** $x < -\frac{15}{14}$ or $x > -\frac{3}{16}$

EXERCISE 2F.2

1 **a** $-5 < x < 5$ **b** $-2 \leqslant x \leqslant 2$ **c** $-1 \leqslant x \leqslant 5$
d $-9 < x < -1$ **e** $-5 \leqslant x \leqslant 2$ **f** $-\frac{7}{3} < x < 3$
g $3 < x < 9$ **h** $-2 \leqslant x \leqslant 3$ **i** $\frac{1}{3} < x < 5$

2 **a** **i** no solution **ii** $x \neq 0$, $x \in \mathbb{R}$
iii no solution **iv** all $x \in \mathbb{R}$
b **i** no solution **ii** all $x \in \mathbb{R}$

3 The statement says that $-7 > 7$, which is false.
The statement should read "If $|x| > 7$ then $x < -7$ or $x > 7$".

4 **a** $x < -3$ or $x > 3$ **b** $x \leqslant -6$ or $x \geqslant 6$
c all $x \in \mathbb{R}$ **d** $x < -12$ or $x > 2$
e $x < -1$ or $x > 7$ **f** $x \leqslant 1$ or $x \geqslant 5$
g $x < -\frac{6}{5}$ or $x > 2$ **h** $x \leqslant -\frac{8}{3}$ or $x \geqslant -\frac{2}{3}$
i all $x \in \mathbb{R}$ **j** $x < 2$ or $x > \frac{14}{3}$
k $x \leqslant -\frac{1}{6}$ or $x \geqslant \frac{1}{2}$ **l** $x \neq 4$

5 **a** $-2 \leqslant x \leqslant 9$ **b** $x < -7$ or $x > -1$
c $-\frac{4}{5} < x < 2$ **d** no solution
e $x \leqslant -\frac{9}{7}$ or $x \geqslant \frac{15}{7}$ **f** $3 < x < \frac{13}{3}$
g $x < -2$ or $x > \frac{5}{2}$ **h** $4 < x < 16$
i $x \leqslant -10$ or $x \geqslant 14$

EXERCISE 2F.3

1 **a** $-3 < x < 3$ **b** $-2 \leqslant x \leqslant 10$ **c** $-\frac{1}{5} < x < 1$
d $x \geqslant 0$ **e** $\frac{12}{7} < x < 4$ **f** $x \geqslant \frac{3}{2}$

2 **a** $x < \frac{4}{3}$ or $x > 4$ **b** $x \leqslant \frac{2}{7}$ or $x \geqslant 4$
c $x < -2$ or $x > \frac{2}{3}$ **d** $x < \frac{1}{2}$
e $x \leqslant \frac{3}{8}$ or $x \geqslant \frac{7}{4}$ **f** $x \geqslant -\frac{4}{7}$

3 **a** $x < -\frac{3}{2}$ or $x > -1$ **b** $x \geqslant 0$
c $x \leqslant -\frac{9}{2}$ or $x \geqslant \frac{1}{6}$ **d** $x \geqslant \frac{3}{4}$
e $-\frac{10}{7} < x < -\frac{2}{3}$ **f** $x < \frac{16}{11}$

4 **b** 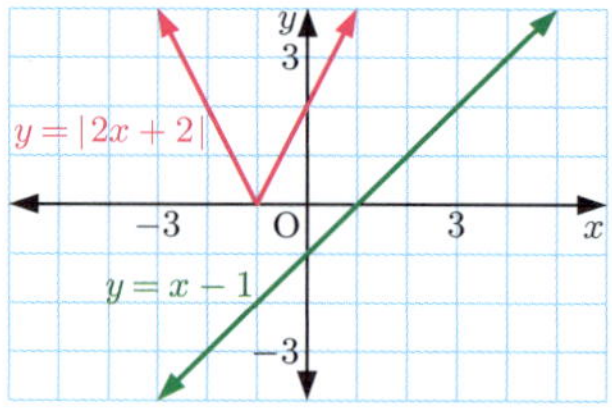

$|2x + 2| > x - 1$ for all $x \in \mathbb{R}$

EXERCISE 2G.1

1 **a** 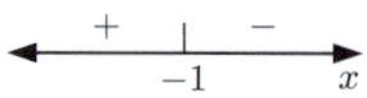

b 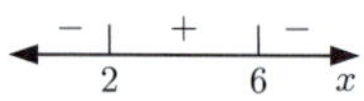

c 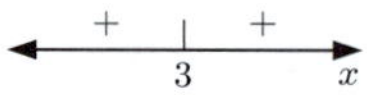

d 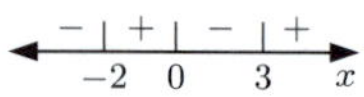

e 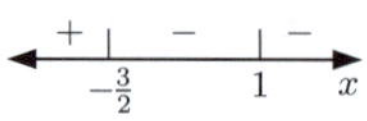

f 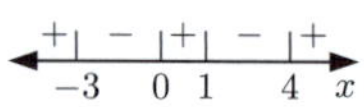

g 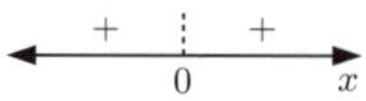

h 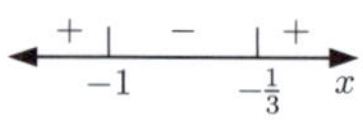

i 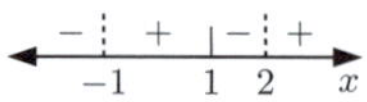

j 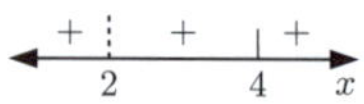

2 **a** 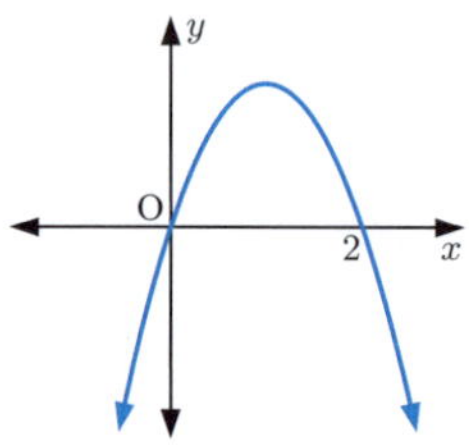

b 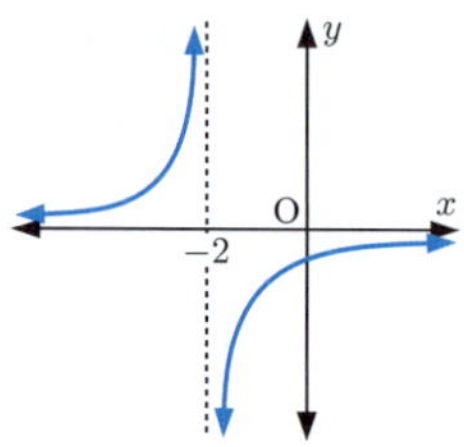

c

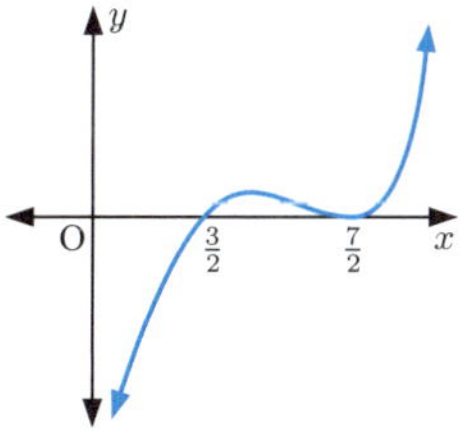

EXERCISE 2G.2

1 **a**

b

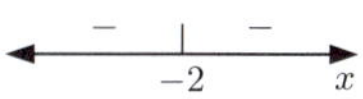

c

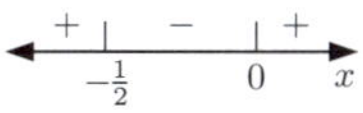

d

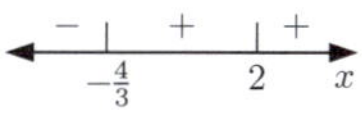

e

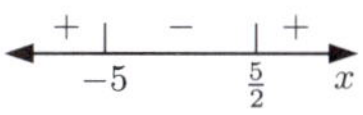

f

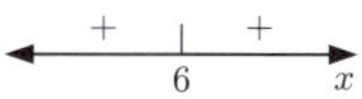

2 **a** $(x+2)(x-2)$

b $(x+3)^2$

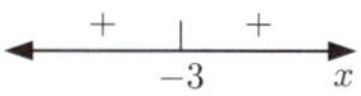

c $(x-3)(x-4)$

3 **a**

b

c

d

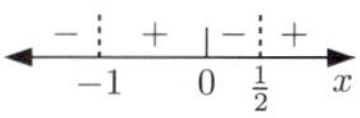

e

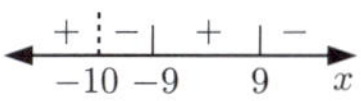

f

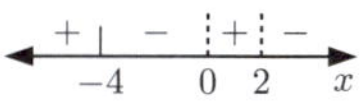

4 **a**

b

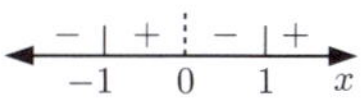

c

EXERCISE 2H

1 **a** $-2-2x^2$ **b** $1+4x^2$ **c** -10 **d** -4

2 **a** $-4x^2-16x-13$ **b** $10-2x^2$ **c** 14 **d** $-\frac{73}{16}$

3 **a** $fg(x)=9-\sqrt{x^2+4}$
Domain is $\{x : x \in \mathbb{R}\}$, Range is $\{y : y \leqslant 7\}$
b 53
c $f^2(x)=9-\sqrt{9-\sqrt{x}}$
Domain is $\{x : 0 \leqslant x \leqslant 81\}$, Range is $\{y : 6 \leqslant y \leqslant 9\}$

4 **a** $-6x-9$ **b** $x=-1$

5 **a** **i** $1-9x^2$ **ii** $1+6x-3x^2$ **b** $x=-\frac{1}{9}$

6 **a** $fg(x)=\dfrac{1}{x-3}$
Domain is $\{x : x \neq 3\}$, Range is $\{y : y \neq 0\}$
b $fg(x)=\sqrt{|x|-1}$
Domain is $\{x : x \leqslant -1 \text{ or } x \geqslant 1\}$
Range is $\{y : y \geqslant 0\}$
c $fg(x)=-\dfrac{1}{x^2+3x+2}$
Domain is $\{x : x \neq -1,\ x \neq -2\}$
Range is $\{y : y \geqslant 4,\ y < 0\}$

7 **a** **i** D_f is $\{x : x \geqslant 0\}$, R_f is $\{y : y \geqslant 0\}$
D_g is $\{x : x \in \mathbb{R}\}$, R_g is $\{y : y \leqslant 1\}$
ii $(f \circ g)(x)=\sqrt{-(x-1)(x-3)}$
Domain is $\{x : 1 \leqslant x \leqslant 3\}$
Range is $\{y : 0 \leqslant y \leqslant 1\}$
b $R_g \cap D_f \neq \varnothing$

8 $g(x)=\frac{1}{2}x+1$ or $g(x)=-\frac{1}{2}x-1$

EXERCISE 2I

1 **a**

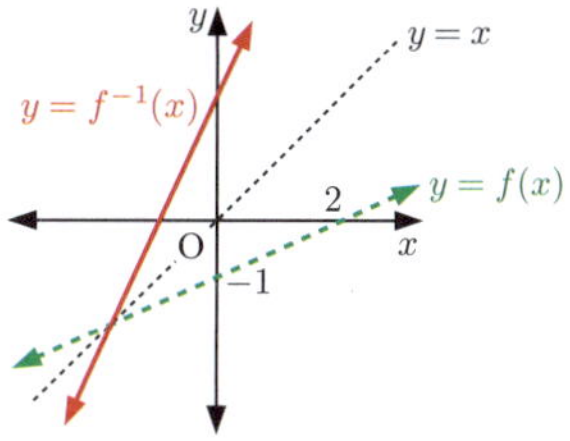

$f(x)$:
Domain is $\{x : x \in \mathbb{R}\}$
Range is $\{y : y \in \mathbb{R}\}$
$f^{-1}(x)$:
Domain is $\{x : x \in \mathbb{R}\}$
Range is $\{y : y \in \mathbb{R}\}$

b

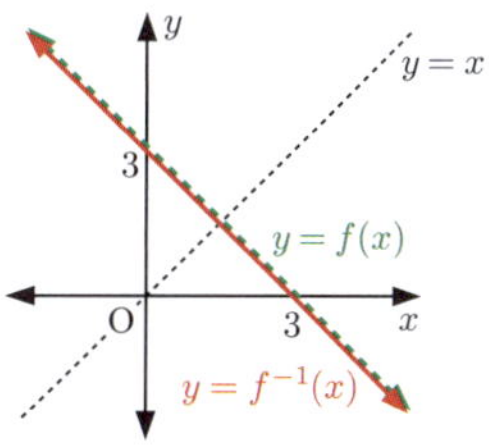

$f(x)$:
Domain is $\{x : x \in \mathbb{R}\}$
Range is $\{y : y \in \mathbb{R}\}$
$f^{-1}(x)$:
Domain is $\{x : x \in \mathbb{R}\}$
Range is $\{y : y \in \mathbb{R}\}$

c $y=f(x)$ does not have an inverse function.

d

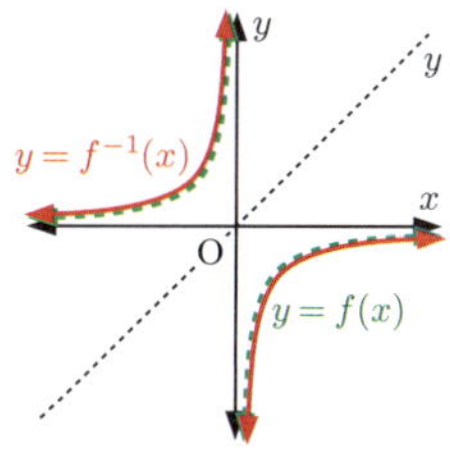

$f(x)$:
Domain is $\{x : x \neq 0\}$
Range is $\{y : y \neq 0\}$
$f^{-1}(x)$:
Domain is $\{x : x \neq 0\}$
Range is $\{y : y \neq 0\}$

e $y=f(x)$ does not have an inverse function.

f

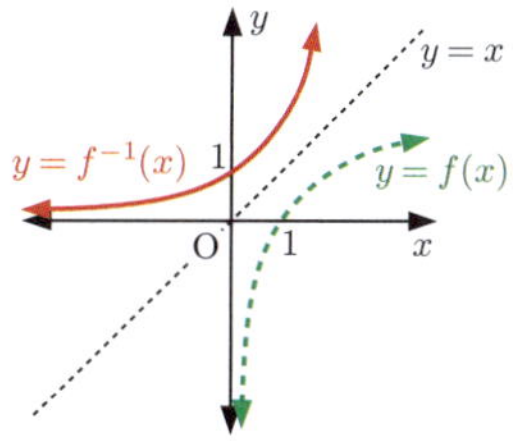

$f(x)$:
Domain is $\{x : x > 0\}$
Range is $\{y : y \in \mathbb{R}\}$
$f^{-1}(x)$:
Domain is $\{x : x \in \mathbb{R}\}$
Range is $\{y : y > 0\}$

g

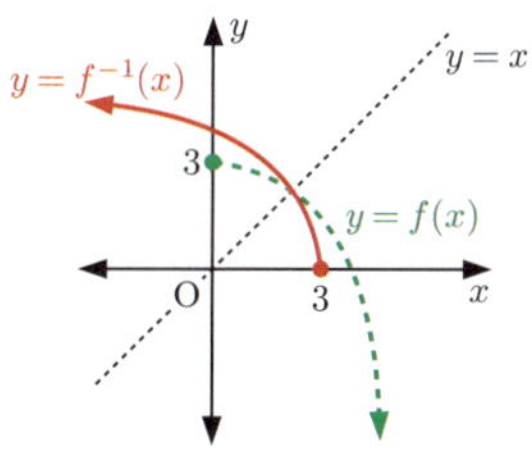

$f(x)$:
Domain is $\{x : x \geqslant 0\}$
Range is $\{y : y \leqslant 3\}$
$f^{-1}(x)$:
Domain is $\{x : x \leqslant 3\}$
Range is $\{y : y \geqslant 0\}$

h

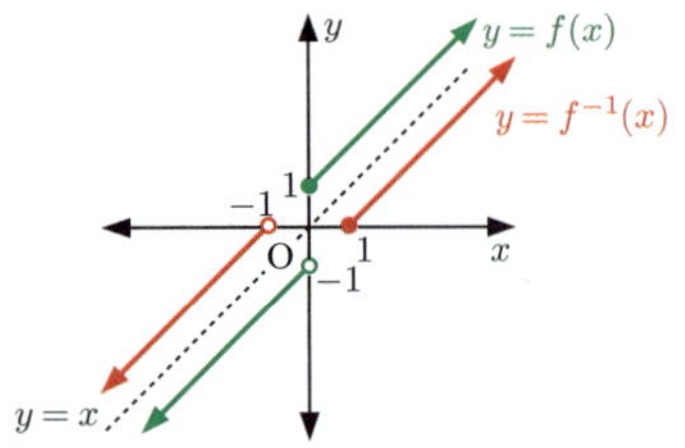

$f(x)$:	Domain is	$\{x : x \in \mathbb{R}\}$
	Range is	$\{y : y < -1 \text{ or } y \geqslant 1\}$
$f^{-1}(x)$:	Domain is	$\{x : x < -1 \text{ or } x \geqslant 1\}$
	Range is	$\{y : y \in \mathbb{R}\}$

i $y = f(x)$ does not have an inverse function.

2 **b** and **d** **3** $\{y : -1 \leqslant y \leqslant 4\}$

4 **a** inverse function is $\{(3, 5), (-1, -4), (1, 1)\}$

b has no inverse function **c** has no inverse function

d inverse function is $\{(-1, -4), (0, 0), (1, 3), (2, 5)\}$

5 **a** **i**

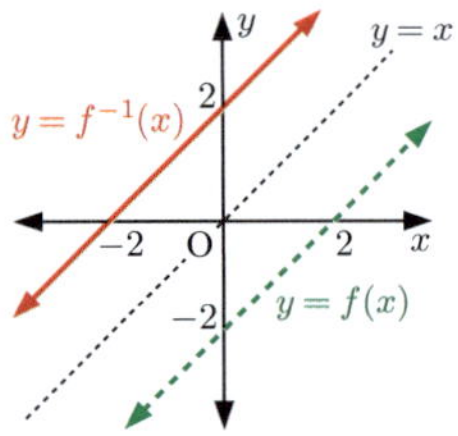

ii $f^{-1}(x) = x + 2$

b **i**

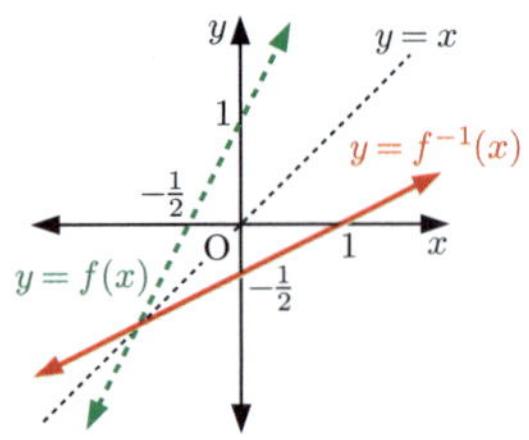

ii $f^{-1}(x) = \dfrac{x-1}{2}$

c **i**

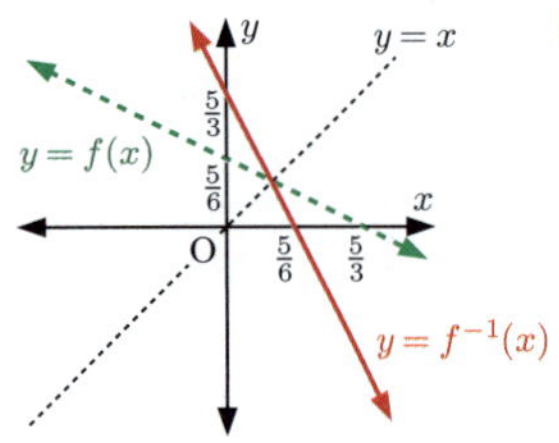

ii $f^{-1}(x) = \dfrac{5-6x}{3}$

d **i**

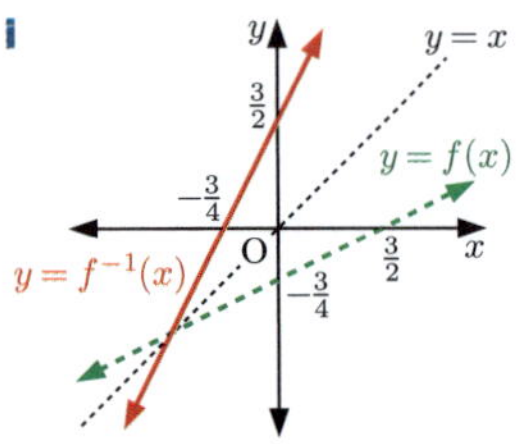

ii $f^{-1}(x) = 2x + \frac{3}{2}$

6 **a** For a self-inverse function, $f = f^{-1}$.
$\therefore$ $f(a) = f^{-1}(a)$ for each a in the domain of f.

b If (a, a) is a point on $y = f(x)$ with inverse $f^{-1}(x)$, then it is invariant.

7 **a** $(f^{-1})^{-1}(x) = f(x) = 4 - 3x$

b

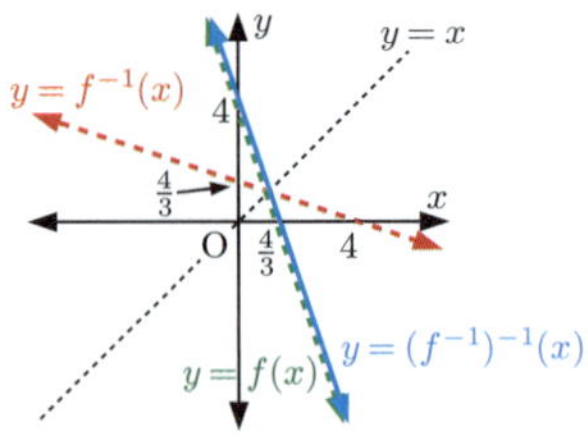

Reflecting $y = f(x)$ in the line $y = x$ produces $y = f^{-1}(x)$.
Reflecting $y = f^{-1}(x)$ in the line $y = x$ again returns us to $y = f(x)$.
$\therefore$ $(f^{-1})^{-1}(x) = f(x)$

8 **a** If a function is one-one, it passes both the vertical line test and the horizontal line test. The inverse is a reflection about $y = x$. This also passes both the vertical and horizontal line tests, so it is also a one-one function.

b $f^{-1}(x) = \dfrac{x-2}{3}$ which is a one-one function.

9 $m \neq 0$ **10** **a** **i** 5 **ii** -1 **iii** -1 **b** $x = 1$

11 $(f^{-1} \circ g^{-1})(x) = (g \circ f)^{-1}(x) = \dfrac{2-x}{3}$

12 **a** **i**

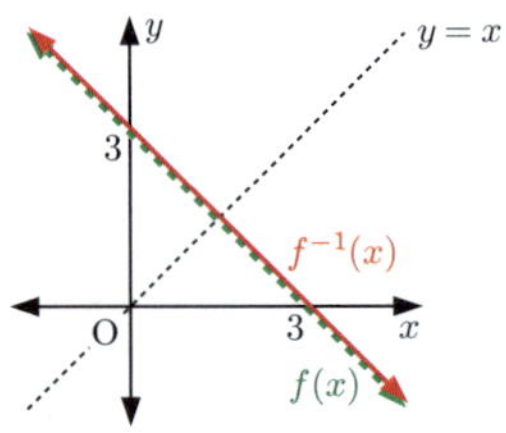

ii $f^{-1}(x) = f(x) = 3 - x$

b **i**

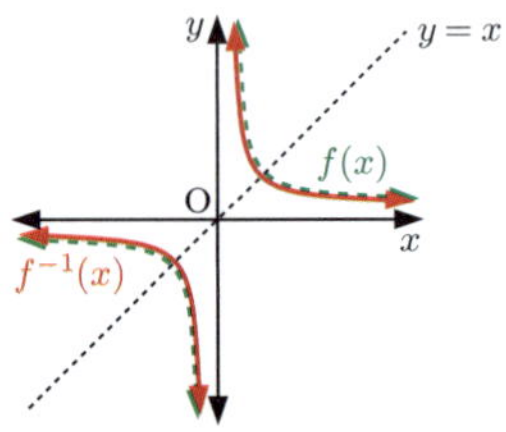

ii $f^{-1}(x) = f(x) = \dfrac{1}{2x}$

c **i**

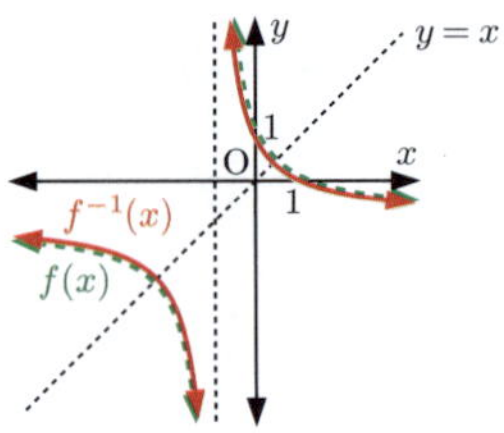

ii $f^{-1}(x) = f(x) = \dfrac{1-x}{1+x}$

REVIEW SET 2A

1 **a**

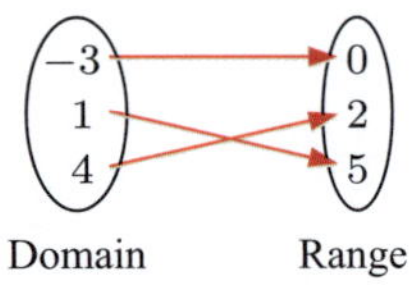

is a function

b

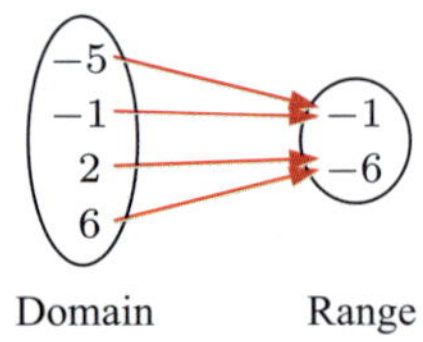

is a function

2 **a** function **b** function **c** function **d** not a function

3 **a** 3 **b** 19 **c** $2a^2+1$ **4** **a** 1 **b** $x=1$

5 **a** **i** Domain is $\{x : x \geqslant -3\}$
Range is $\{y : y \geqslant 2\}$ **ii** one-one

b **i** Domain is $\{x : x \in \mathbb{R}\}$
Range is $\{y : 0 < y \leqslant 3\}$ **ii** not one-one

c **i** Domain is $\{x : x \in \mathbb{R}\}$
Range is $\{y : y = -1,\ 1, \text{ or } 2\}$ **ii** not one-one

6

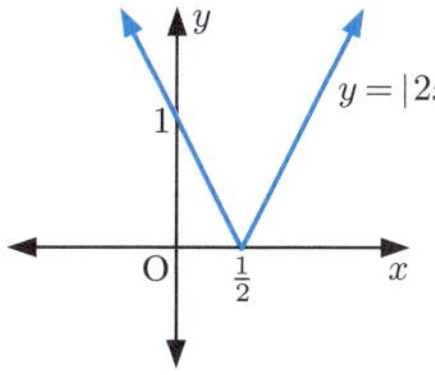

7 **a** $x=-5$ or 6 **b** $x=1$ or 3

8 **a** $-1 < x < \frac{7}{3}$ **b** $x \leqslant 1$ or $x \geqslant 7$

9 **b**

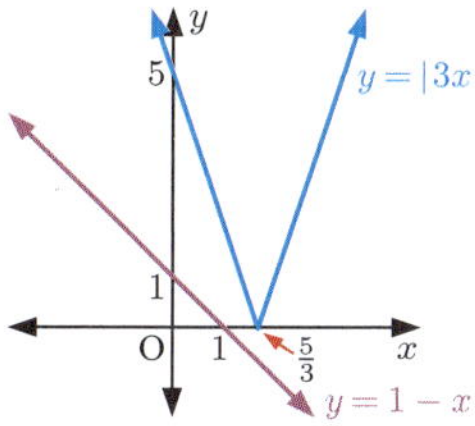

10 **a**

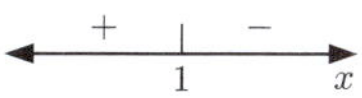

b

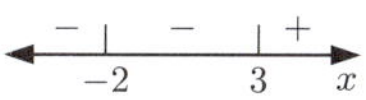

11 **a**

b

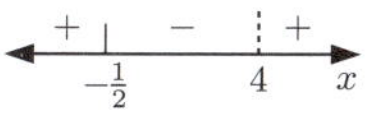

c

12 **a** **i** x^2-4x+2 **ii** x^2+2x-4 **b** $x=-1$

13 $g(x) = \dfrac{8-x^2}{4}$

14 **a** **i**

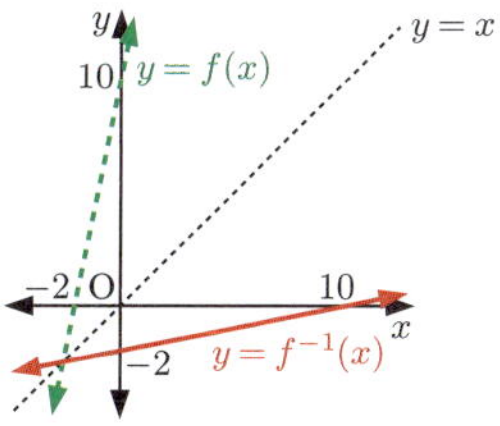

ii $f^{-1}(x) = \dfrac{x-10}{5}$

b **i**

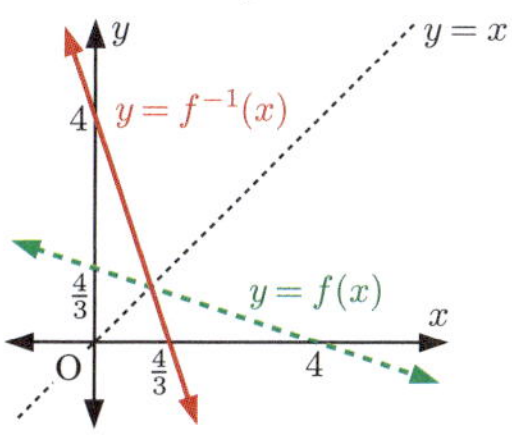

ii $f^{-1}(x) = 4-3x$

15 $a=1$

REVIEW SET 2B

1 **a** is a function, is not one-one **b** is not a function
c is a function, is one-one

2 If $x^2-y^2=1$ then $y=\pm\sqrt{x^2-1}$. For all $x>1$ and $x<-1$, there are two corresponding values of y.
$\therefore$ $x^2-y^2=1$ is not a function.

3 **a** x^2+2x+3 **b** $x-1$ **c** $x^2+2hx+h^2+2$

4 **a** Domain is $\{x : x \neq -4\}$, Range is $\{y : y \neq 0\}$
b Domain is $\{x : x > \frac{1}{3}\}$, Range is $\{y : y < 0\}$

5 **a** $D(2) \approx 1.30$, after 2 hours, the depth of the water in the tank is approximately 1.30 m.
b $t \approx 1.33$, the depth of the water in the tank is 1.2 m after approximately 1.33 hours.
c 1 m

6 $\{y : 0 \leqslant y \leqslant 4\}$

7 **a**

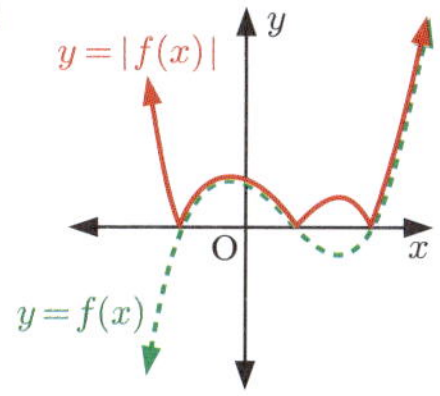

b

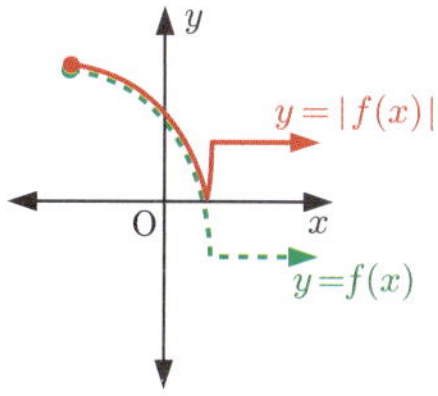

8 **a** $P(-\frac{3}{4}, \frac{13}{4})$, $Q(\frac{5}{2}, \frac{13}{2})$
b **i** $x = -\frac{3}{4}$ or $\frac{5}{2}$ **ii** $x < -\frac{3}{4}$ or $x > \frac{5}{2}$

9 **a** $x=8$ or -20 **b** $x=\frac{7}{5}$ **c** $x=-\frac{7}{3}$ or 1

10 **a** $x < \frac{2}{3}$ or $x > \frac{8}{3}$ **b** $\frac{3}{4} \leqslant x \leqslant \frac{17}{4}$
c $x < -\frac{1}{5}$ or $x > \frac{7}{3}$

11 **a**

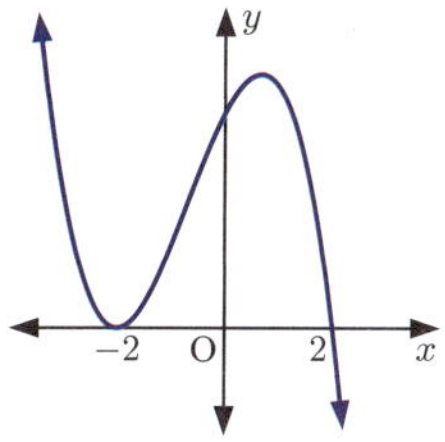

b

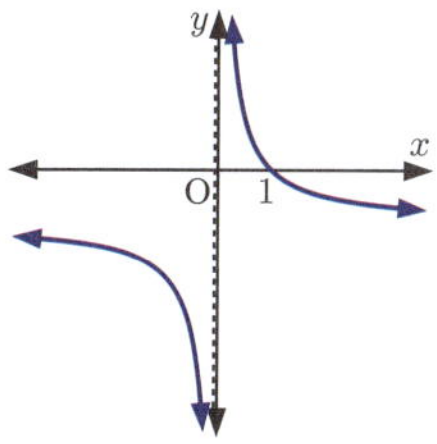

12 **a** $(x+1)(x-3)$

b $(x+6)^2$

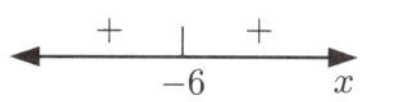

c $x^2(x+1)(x-1)$

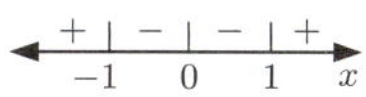

13 **a** $fg(x) = \dfrac{3}{|x+1|}$
Domain is $\{x : x \neq -1\}$, Range is $\{y : y > 0\}$
b $fg(x) = \sqrt{x^2-4x}$
Domain is $\{x : x \leqslant 0 \text{ or } x \geqslant 4\}$, Range is $\{y : y \geqslant 0\}$

14 **a**

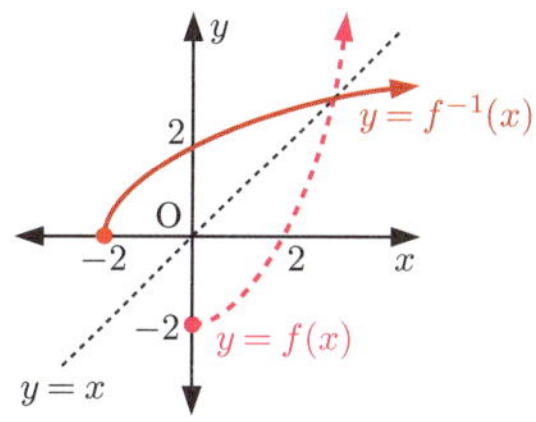

b does not have an inverse, is not one-one

c 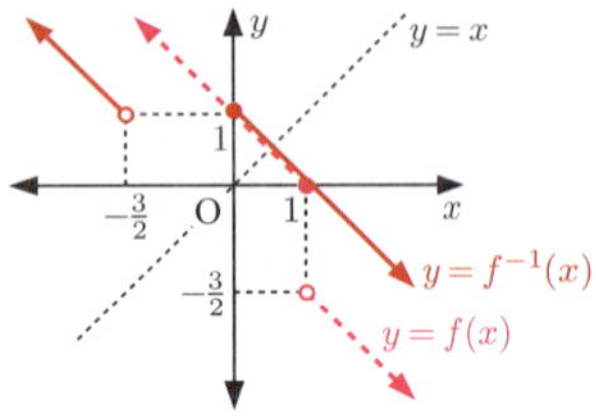

EXERCISE 3A.1

1 **a** $x = 0$ or $-\frac{2}{3}$ **b** $x = 0$ or $-\frac{1}{4}$ **c** $x = 0$ or $\frac{1}{3}$
d $x = 0$ or $\frac{1}{2}$ **e** $x = 0$ or $\frac{1}{5}$ **f** $x = 0$ or $\frac{3}{2}$

2 **a** $x = 3$ or -5 **b** $x = 1$ or 3 **c** $x = 2$ or -4
d $x = 4$ or -3 **e** $x = 6$ or 7 **f** $x = 9$ or -7

3 **a** $x = 4$ or $-\frac{1}{3}$ **b** $x = -\frac{1}{2}$ **c** $x = 1$ or $\frac{3}{2}$
d $x = -1$ or $\frac{1}{4}$ **e** $x = \frac{3}{2}$ or $-\frac{1}{3}$ **f** $x = 4$ or $-\frac{2}{3}$
g $x = -\frac{1}{2}$ or $-\frac{5}{4}$ **h** $x = -\frac{1}{2}$ or $-\frac{5}{2}$ **i** $x = 1$ or $\frac{7}{9}$
j $x = \frac{9}{2}$ or $-\frac{1}{3}$ **k** $x = 1$ or $-\frac{7}{5}$ **l** $x = \frac{3}{5}$ or $-\frac{1}{3}$

4 **a** $x = 2$ or 5 **b** $x = 0$ or -5 **c** $x = \frac{1}{2}$ or $\frac{8}{3}$
d $x = -1$ or $\frac{1}{7}$ **e** $x = 7$ or $-\frac{1}{2}$ **f** $x = 1$

5 **a** $x = \pm 2$ or $\pm\sqrt{7}$ **b** $x = \pm\sqrt{2}$ or $\pm\sqrt{10}$
c $x = \pm\sqrt{6}$ **d** $x = \pm\frac{1}{\sqrt{2}}$ or $\pm\sqrt{3}$
e $x = \pm\frac{\sqrt{3}}{2}$ **f** no real solutions
g $x = -1$ or 2 **h** $x = 9$ or 25 **i** $x = \frac{16}{9}$

6 $x = \frac{1}{2}$ or 3

7 $x = 9$ is a valid solution, but $x = 4$ is not.
Dwayne introduced an invalid solution by squaring both sides.

EXERCISE 3A.2

1 **a** $x = 1$ or 3 **b** $x = -1 \pm \sqrt{5}$ **c** no real solutions
d $x = -6 \pm \sqrt{2}$ **e** $x = 4 \pm \sqrt{6}$ **f** $x = 1 \pm \sqrt{7}$
g no real solutions **h** $x = -\frac{1}{3} \pm \frac{2\sqrt{2}}{3}$ **i** $x = \frac{1}{2} \pm \frac{\sqrt{10}}{2}$

2 **a** $x = -1 \pm \sqrt{5}$ **b** $x = 2 \pm \sqrt{2}$ **c** $x = 3 \pm \sqrt{10}$
d $x = -\frac{5}{2} \pm \frac{\sqrt{17}}{2}$ **e** no real solutions
f $x = -4 \pm \sqrt{19}$ **g** $x = -\frac{3}{2} \pm \frac{3\sqrt{5}}{2}$
h no real solutions **i** no real solutions

3 **a** $x = -\frac{3}{2} \pm \frac{\sqrt{7}}{2}$ **b** $x = \frac{1}{2} \pm \frac{\sqrt{6}}{2}$
c $x = 3 \pm \sqrt{\frac{15}{2}}$ **d** no real solutions
e $x = -\frac{3}{2} \pm \sqrt{\frac{19}{12}}$ **f** $x = \frac{9}{5} \pm \frac{\sqrt{66}}{5}$

4 **a** $x = \frac{1}{2}$ or 1 **b** $x = 2 \pm \frac{3}{\sqrt{2}}$ **c** $x = 1$ or $-\frac{1}{3}$

5 $x = \dfrac{-b \pm \sqrt{b^2 - 4ac}}{2a}$

EXERCISE 3A.3

1 **a** $x = 1 \pm 2\sqrt{6}$ **b** $x = 2 \pm 2\sqrt{7}$ **c** $x = -2 \pm \sqrt{11}$
d $x = \frac{1 \pm 3\sqrt{5}}{2}$ **e** $x = \frac{-5 \pm \sqrt{53}}{2}$ **f** no real solutions
g $x = \frac{3 \pm \sqrt{41}}{8}$ **h** $x = \frac{5 \pm \sqrt{13}}{6}$ **i** $x = \frac{-3 \pm 3\sqrt{17}}{8}$

2 **a** $x = 1 \pm 2\sqrt{3}$ **b** $x = \frac{3 \pm \sqrt{5}}{2}$ **c** $x = \frac{-7 \pm \sqrt{33}}{4}$
d $x = \frac{2 \pm \sqrt{19}}{2}$ **e** no real solutions **f** no real solutions
g $x = \frac{3 \pm \sqrt{5}}{2}$ **h** $x = -1 \pm \sqrt{3}$ **i** $x = 1 \pm \sqrt{2}$

EXERCISE 3B

1 **a** $\{x : -1 \leqslant x \leqslant 1\}$ **b** $\{x : x < -4 \text{ or } x > 2\}$
c $\{x : x \in \mathbb{R}\}$ **d** $\{x : x < -1 \text{ or } x > 0\}$
e $\{x : 0 \leqslant x \leqslant 2\}$ **f** $\{x : x < 0 \text{ or } x > \frac{1}{3}\}$

2 **a** $\{x : -3 < x < 3\}$ **b** $\{x : x \leqslant -2 \text{ or } x \geqslant 2\}$
c $\{x : x = 3\}$ **d** $\{x : x < -2 \text{ or } x > 5\}$
e $\{x : x \leqslant -3 \text{ or } x \geqslant 7\}$ **f** $\{x : x \leqslant -2 \text{ or } x \geqslant \frac{3}{2}\}$
g $\{x : -4 < x < 2\}$ **h** $\{x : x < -\frac{1}{3} \text{ or } x > 1\}$
i $\{x : -7 < x < 1\}$ **j** $\{x : x < -\frac{5}{2} \text{ or } x > \frac{3}{2}\}$
k $\{x : x < -2 \text{ or } x > \frac{2}{5}\}$
l $\{x : \frac{7}{2} - \frac{\sqrt{37}}{2} < x < \frac{7}{2} + \frac{\sqrt{37}}{2}\}$

3 **a** $\square = <$ **b** $\square = \leqslant$ **c** $\square = \geqslant$ or $>$

EXERCISE 3C

1 **a** $\Delta = 64$ **b** 2 distinct rational roots **c** $x = 5$ or -3

2 **a** $\Delta = 0$ **b** a repeated root **c** $x = \frac{5}{2}$

3 **a** 2 distinct rational roots **b** 2 distinct irrational roots
c a repeated root **d** 2 distinct irrational roots
e 2 distinct irrational roots **f** no real roots
g no real roots **h** no real roots
i 2 distinct rational roots

4 **a** $\Delta = 4(4 - k)$ 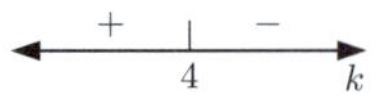

i $k = 4$ **ii** $k < 4$ **iii** $k > 4$

b $\Delta = 4(1 - 3k)$ 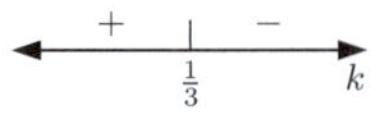

i $k = \frac{1}{3}$ **ii** $k < \frac{1}{3}$ **iii** $k > \frac{1}{3}$

c $\Delta = (5k + 2)(5k - 2)$ 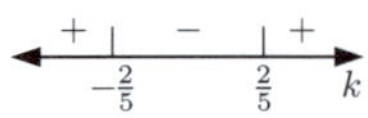

i $k = \pm\frac{2}{5}$ **ii** $k < -\frac{2}{5}$ or $k > \frac{2}{5}$
iii $-\frac{2}{5} < k < \frac{2}{5}$

d $\Delta = (k - 1)^2$ 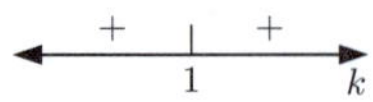

i $k = 1$ **ii** $k \neq 1$ **iii** no values of k

e $\Delta = (k - 7)(k - 15)$ 

i $k = 7$ or 15 **ii** $k < 7$ or $k > 15$
iii $7 < k < 15$

f $\Delta = (2 + \sqrt{3}k)(2 - \sqrt{3}k)$ 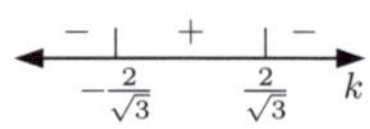

i $k = \pm\frac{2}{\sqrt{3}}$ **ii** $-\frac{2}{\sqrt{3}} < k < \frac{2}{\sqrt{3}}$
iii $k < -\frac{2}{\sqrt{3}}$ or $k > \frac{2}{\sqrt{3}}$

5 If $a > 0$ and $c < 0$, then $-4ac > 0$.
Since $b^2 \geqslant 0$, $\Delta = b^2 - 4ac > 0$.

EXERCISE 3D.1

1 **a**

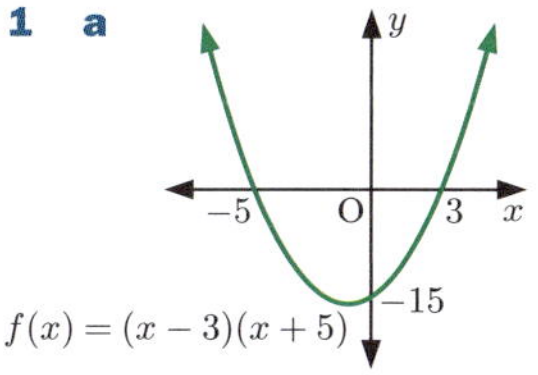

b

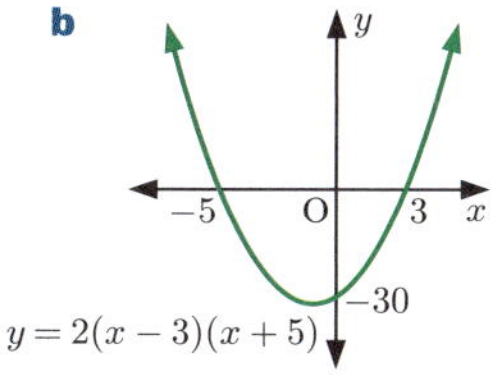

c

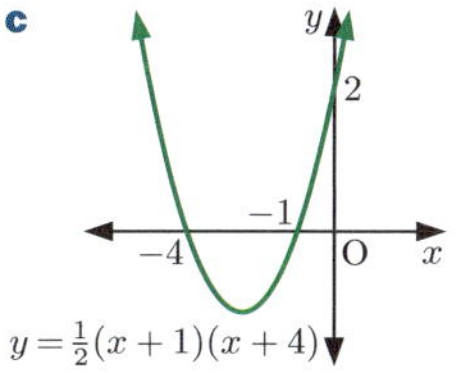

d

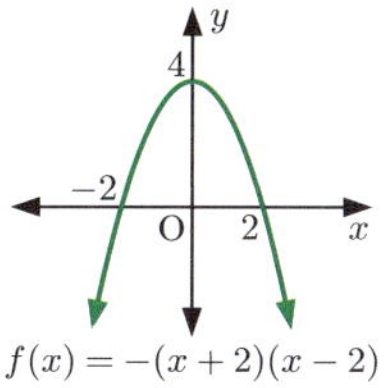

e

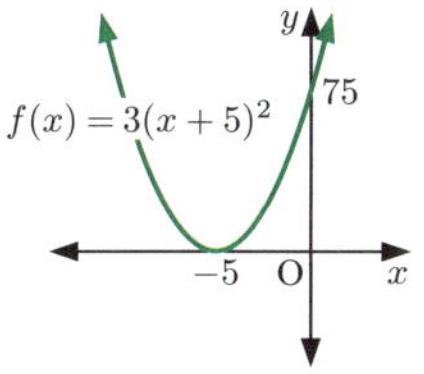

f

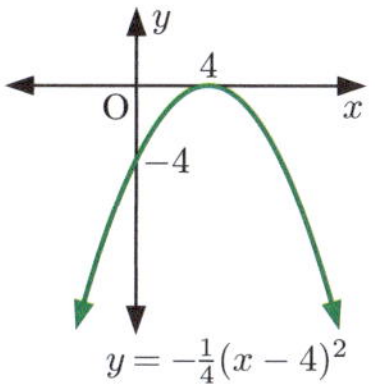

2 **a** $x = -1$ **b** $x = -1$ **c** $x = -\frac{5}{2}$ **d** $x = 0$
e $x = -5$ **f** $x = 4$

3 **a** E **b** C **c** G **d** H **e** A **f** I
g F **h** D **i** B

4 **a**

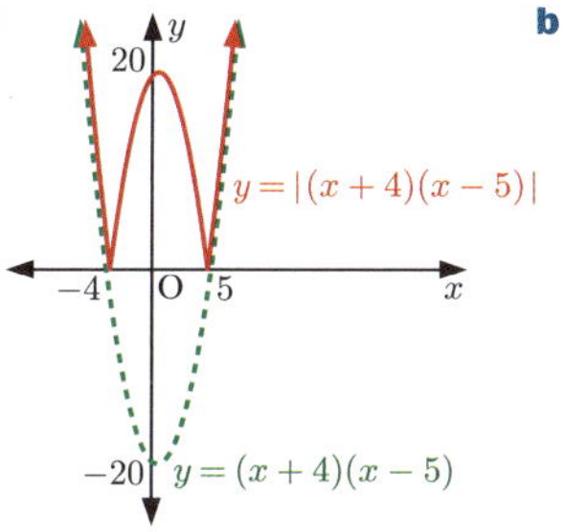

b

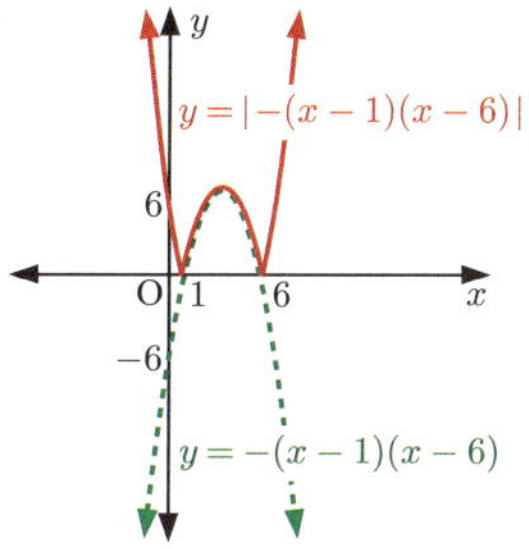

c

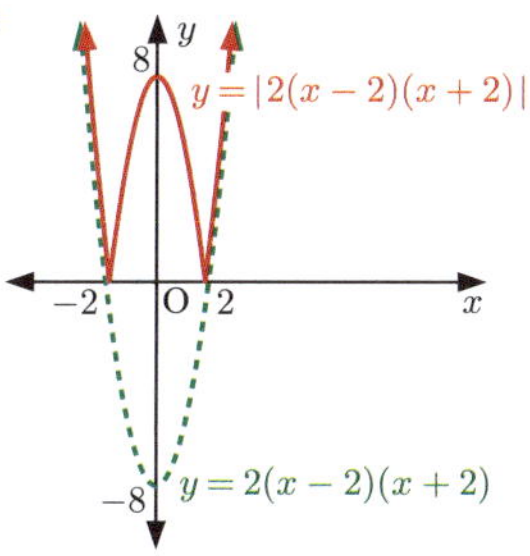

d

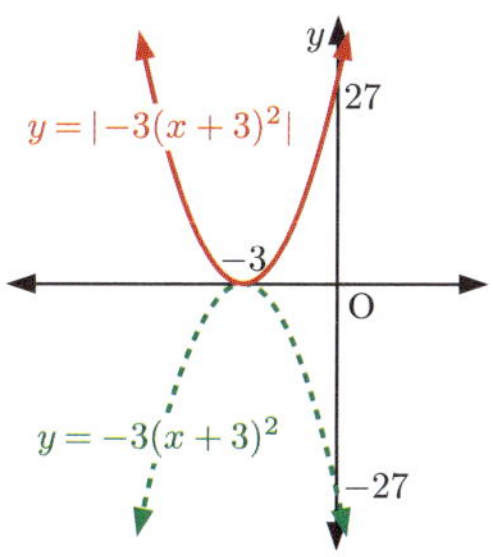

5 **a**

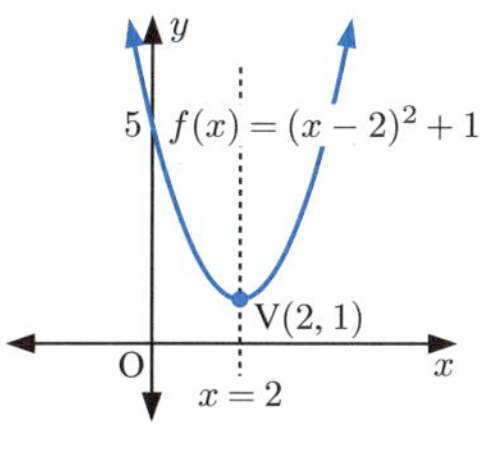

b

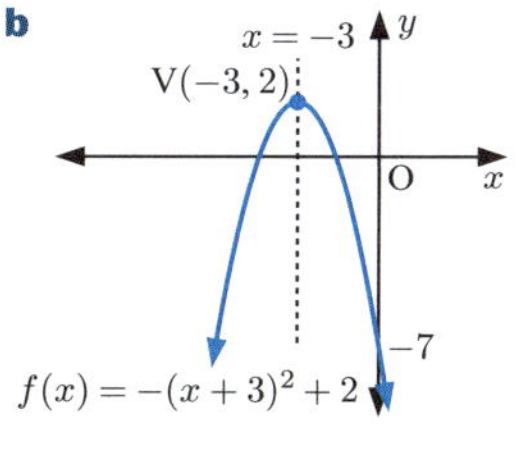

c

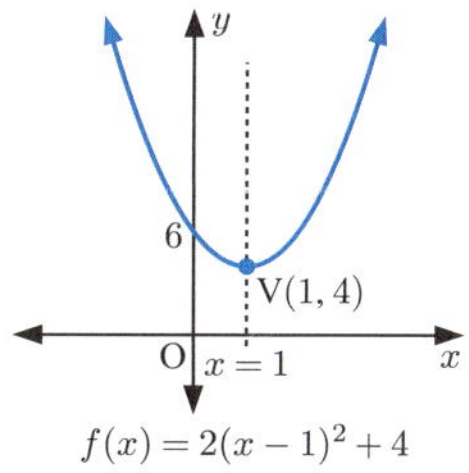

d

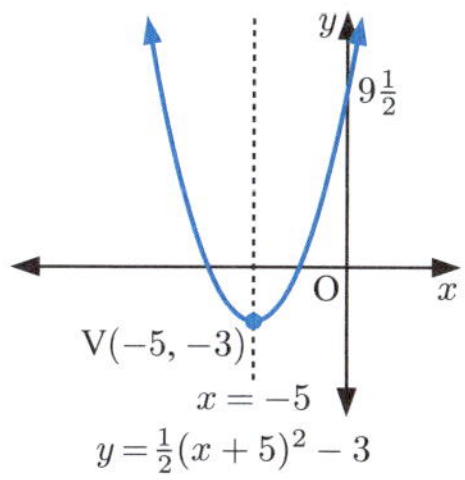

e

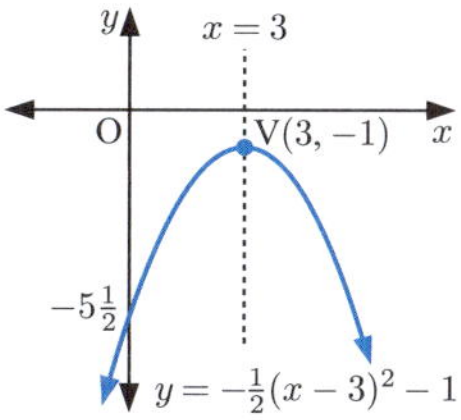

f

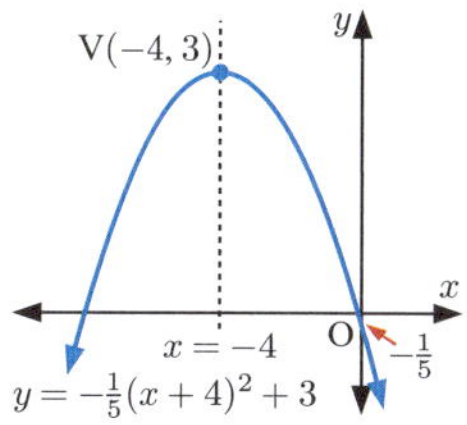

6 **a** E **b** G **c** I **d** F **e** H **f** D
g C **h** B **i** A

EXERCISE 3D.2

1 **a** $f(x) = (x-2)^2 + 3$
b axis of symmetry $x = 2$, vertex $(2, 3)$ **c** 7
d

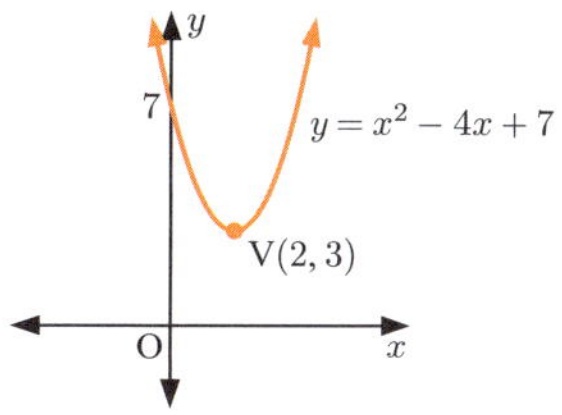

2 **a** $f(x) = (x-1)^2 + 2$

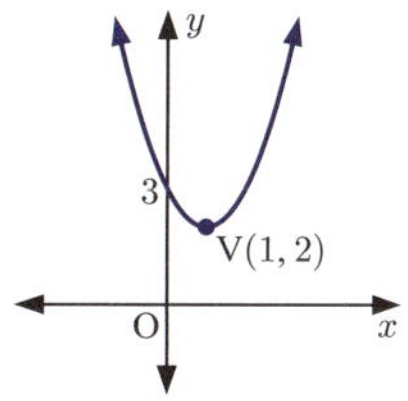

b $f(x) = (x-3)^2 - 8$

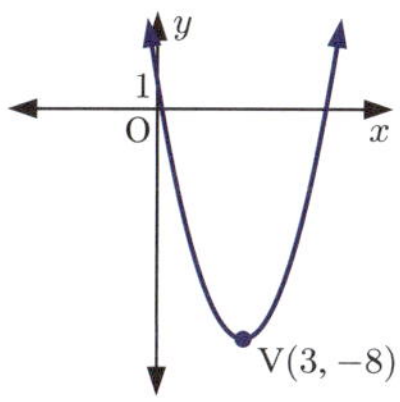

c $f(x) = (x-1)^2 - 1$

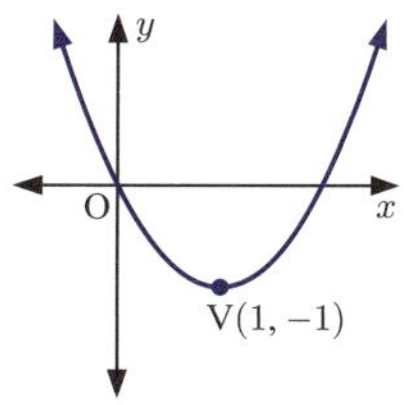

d $f(x) = (x-\frac{3}{2})^2 - \frac{9}{4}$

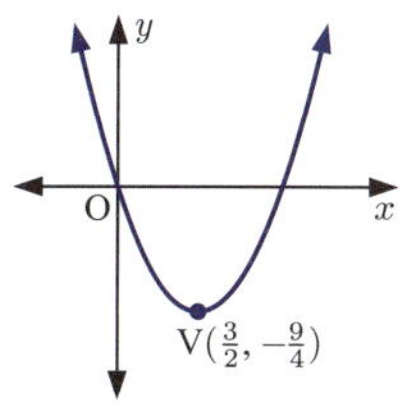

e $f(x) = (x+\frac{5}{2})^2 - \frac{41}{4}$

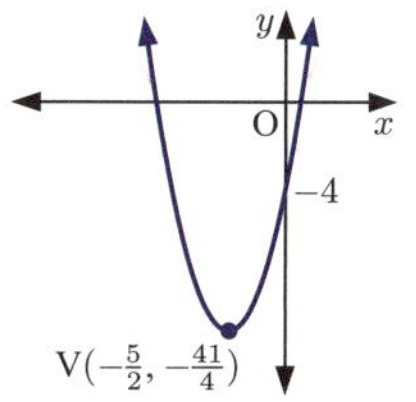

f $f(x) = (x-\frac{3}{2})^2 - \frac{5}{4}$

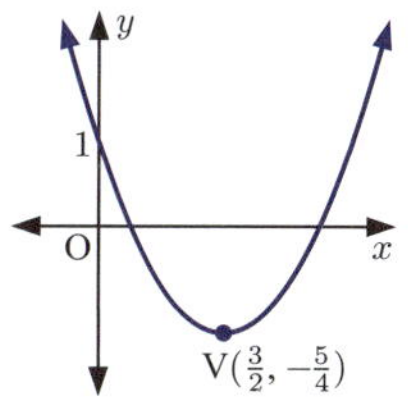

g $f(x) = (x-4)^2 - 14$

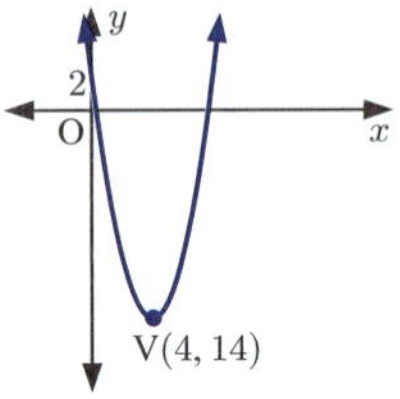

h $f(x) = (x+2)^2 - 7$

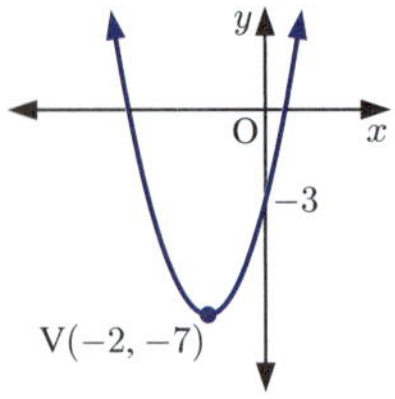

i $f(x) = (x - \frac{7}{2})^2 - \frac{57}{4}$

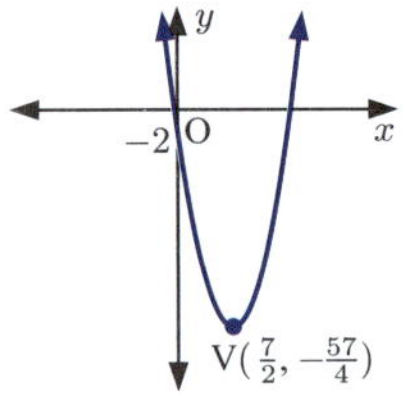

3 **a** **i** $f(x) = 2(x-1)^2 + 4$ **ii** $(1, 4)$ **iii** 6

iv

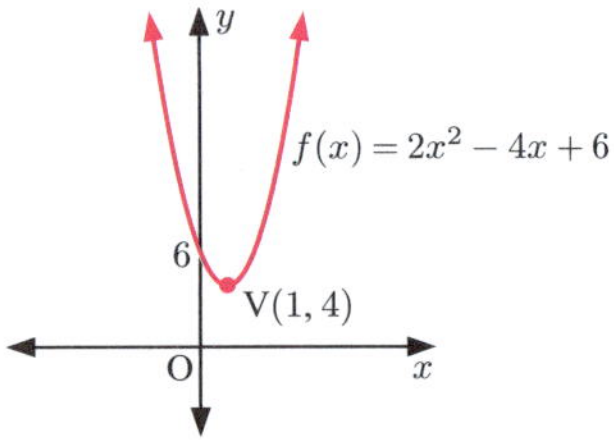

b **i** $f(x) = 2(x+2)^2 - 13$ **ii** $(-2, -13)$ **iii** -5

iv

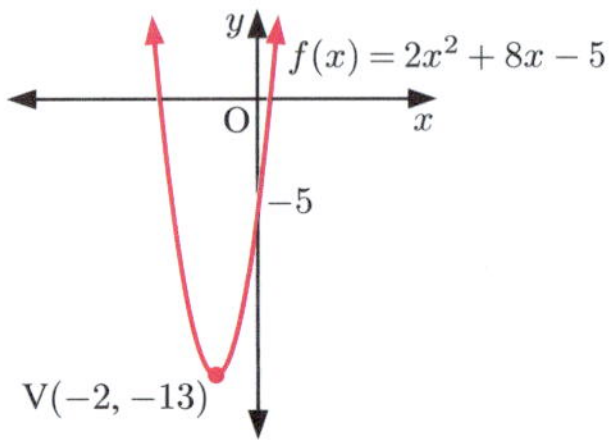

c **i** $f(x) = 3(x+1)^2 + 4$ **ii** $(-1, 4)$ **iii** 7

iv

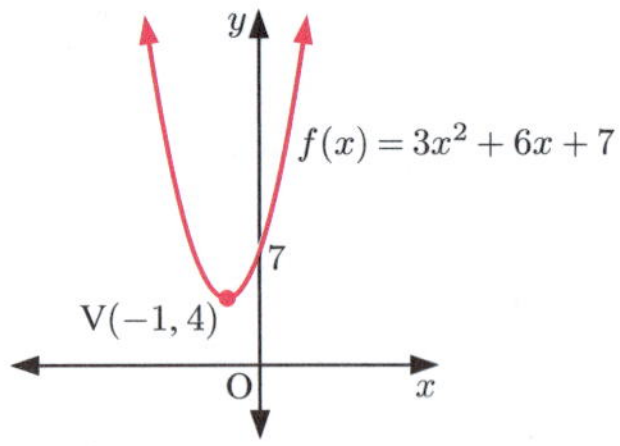

d **i** $f(x) = 4(x-2)^2 - 15$ **ii** $(2, -15)$ **iii** 1

iv

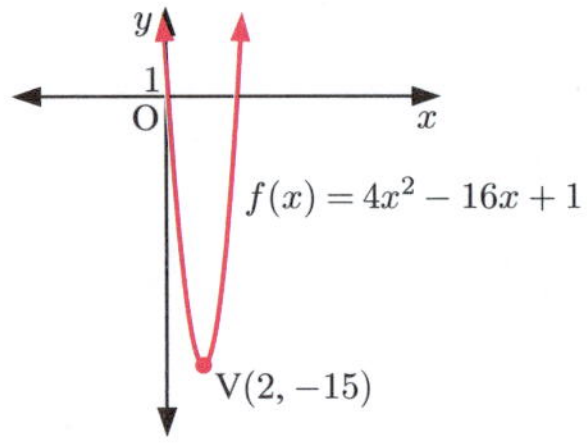

e **i** $f(x) = -(x - \frac{5}{2})^2 + \frac{17}{4}$ **ii** $(\frac{5}{2}, \frac{17}{4})$ **iii** -2

iv

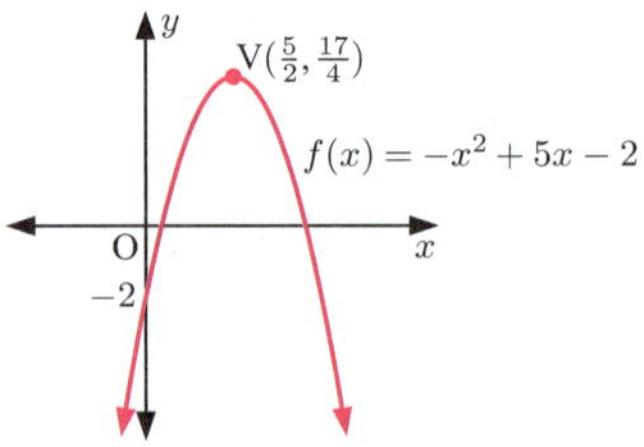

f **i** $f(x) = -2(x - \frac{7}{4})^2 + \frac{25}{8}$ **ii** $(\frac{7}{4}, \frac{25}{8})$ **iii** -3

iv

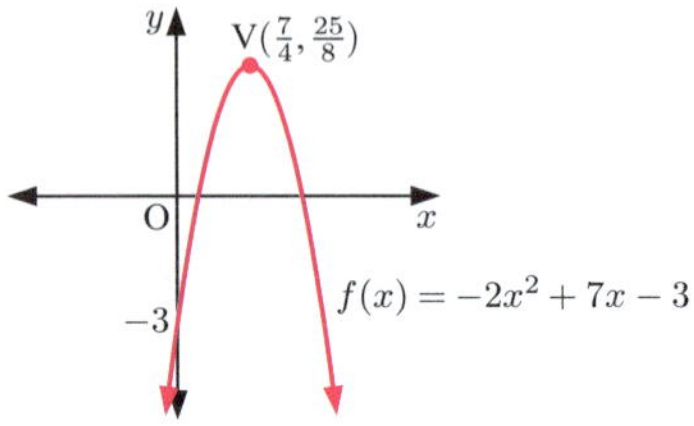

4 $f(x) = a\left(x + \dfrac{b}{2a}\right)^2 + \left(c - \dfrac{b^2}{4a}\right)$

EXERCISE 3D.3

1 **a** $(-2, -3)$ **b** $(2, -2)$ **c** $(0, 7)$ **d** $(\frac{1}{3}, \frac{8}{3})$
e $(0, 6)$ **f** $(-\frac{3}{2}, \frac{29}{4})$ **g** $(-\frac{5}{4}, \frac{57}{8})$ **h** $(-2, -5)$
i $(\frac{1}{4}, -\frac{47}{48})$

2 **a** **i** $x = 1$ **ii** $(1, 0)$ **iii** x-intercept 1, y-intercept 1

iv

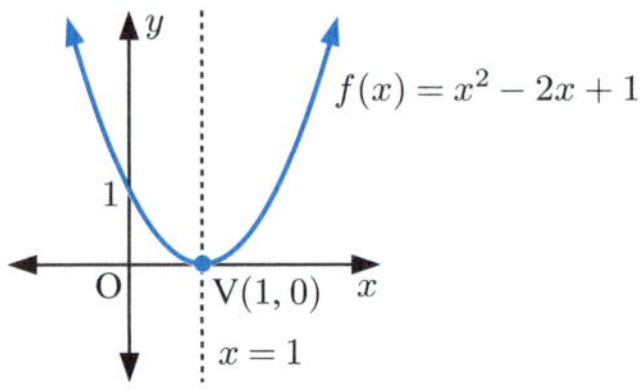

v $\{y : y \geqslant 0\}$

b **i** $x = 2$ **ii** $(2, -1)$
iii x-intercepts 1, 3, y-intercept 3

iv

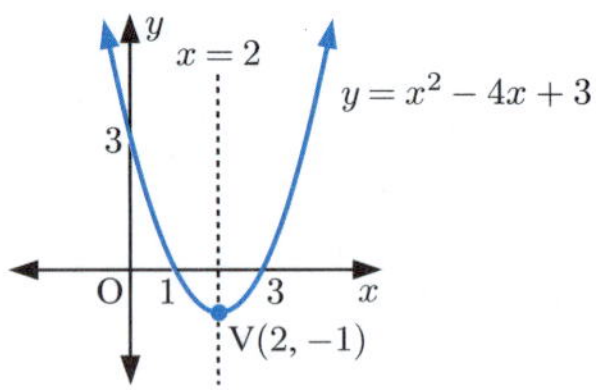

v $\{y : y \geqslant -1\}$

c **i** $x = -\frac{1}{4}$ **ii** $(-\frac{1}{4}, -\frac{65}{8})$
iii x-intercepts $-\frac{1}{4} \pm \frac{\sqrt{65}}{4}$, y-intercept -8

iv

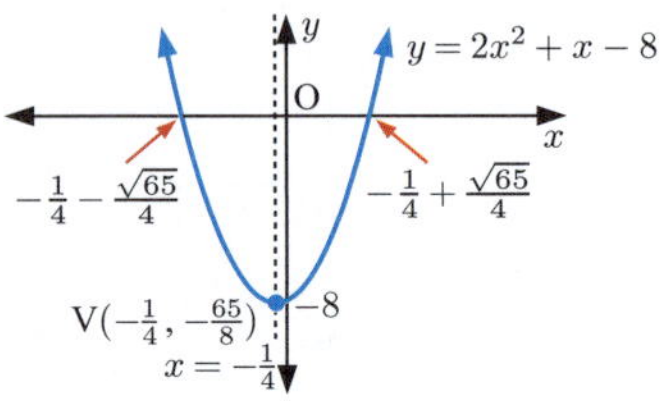

v $\{y : y \geqslant -\frac{65}{8}\}$

d **i** $x = -\frac{1}{2}$ **ii** $(-\frac{1}{2}, \frac{33}{4})$

iii x-intercepts $-\frac{1}{2} \pm \frac{\sqrt{33}}{2}$, y-intercept 8

iv

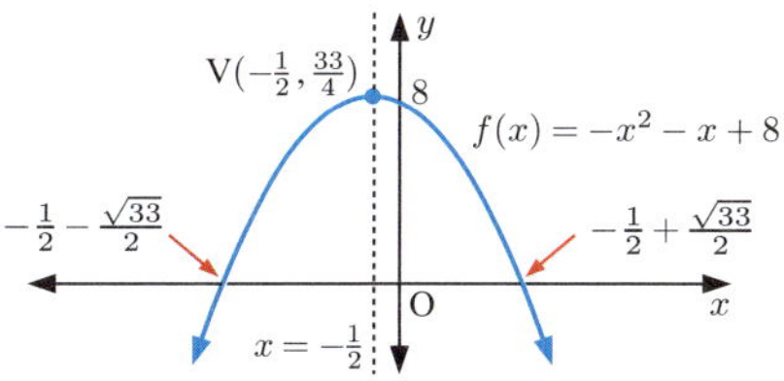

v $\{y : y \leqslant \frac{33}{4}\}$

e **i** $x = -\frac{2}{3}$ **ii** $(-\frac{2}{3}, -\frac{28}{3})$

iii x-intercepts $-\frac{2}{3} \pm \frac{2\sqrt{7}}{3}$, y-intercept -8

iv

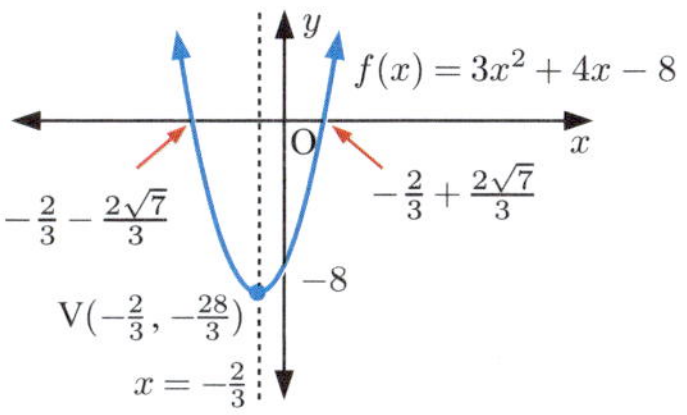

v $\{y : y \geqslant -\frac{28}{3}\}$

f **i** $x = -\frac{7}{4}$ **ii** $(-\frac{7}{4}, \frac{73}{8})$

iii x-intercepts $-\frac{7}{4} \pm \frac{\sqrt{73}}{4}$, y-intercept 3

iv

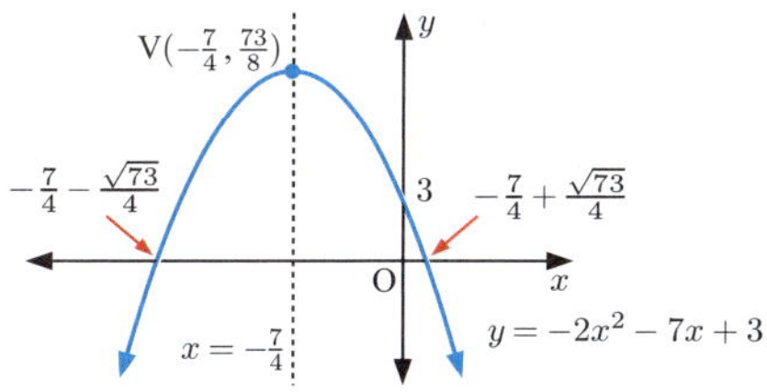

v $\{y : y \leqslant \frac{73}{8}\}$

g **i** $x = 3$ **ii** $(3, \frac{13}{2})$

iii x-intercepts $3 \pm \sqrt{13}$, y-intercept 2

iv

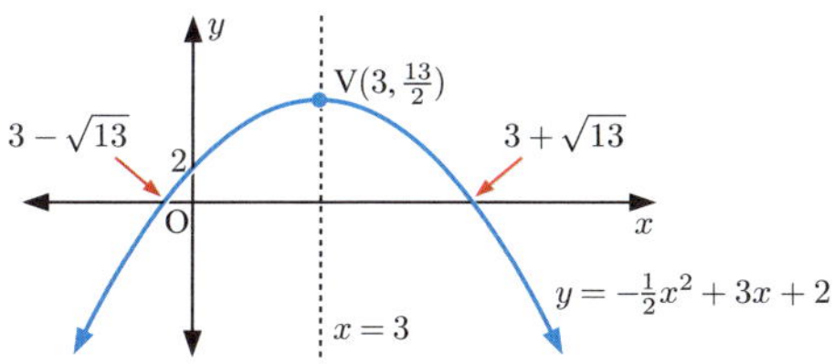

v $\{y : y \leqslant \frac{13}{2}\}$

h **i** $x = -4$ **ii** $(-4, -7)$

iii x-intercepts $-4 \pm 2\sqrt{7}$, y-intercept -3

iv

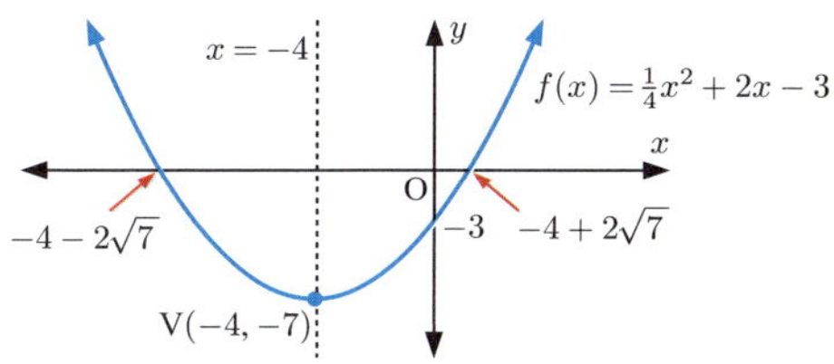

v $\{y : y \geqslant -7\}$

3 **a** **i** $f(x) = (x-2)(x-8)$, roots are 2 and 8

ii $f(x) = (x-5)^2 - 9$, vertex is $(5, -9)$

iii

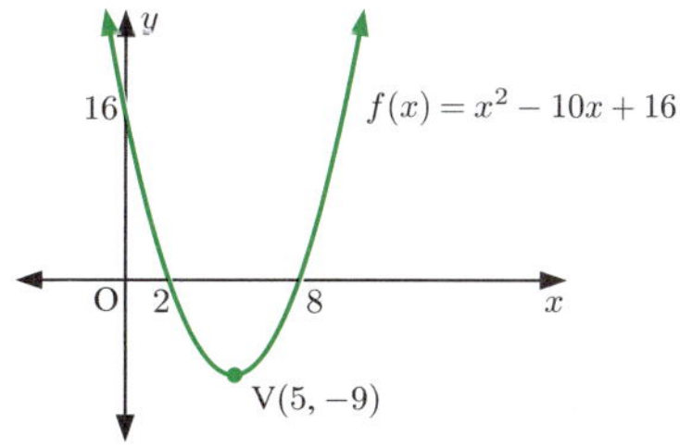

b **i** $y = (x+1)(x+9)$, roots are -1 and -9

ii $y = (x+5)^2 - 16$, vertex is $(-5, -16)$

iii

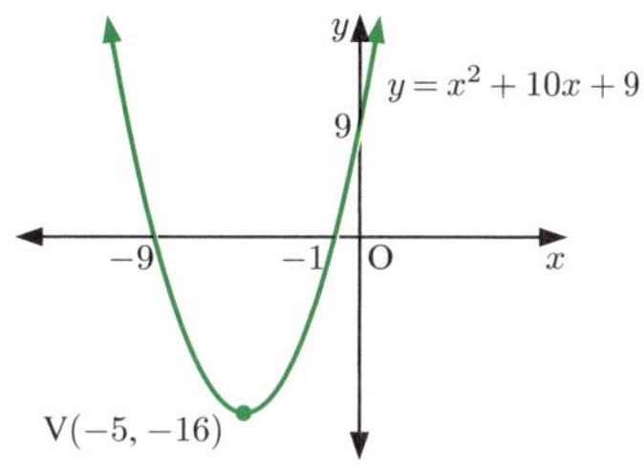

c **i** $f(x) = (x-5)(x-9)$, roots are 5 and 9

ii $f(x) = (x-7)^2 - 4$, vertex is $(7, -4)$

iii

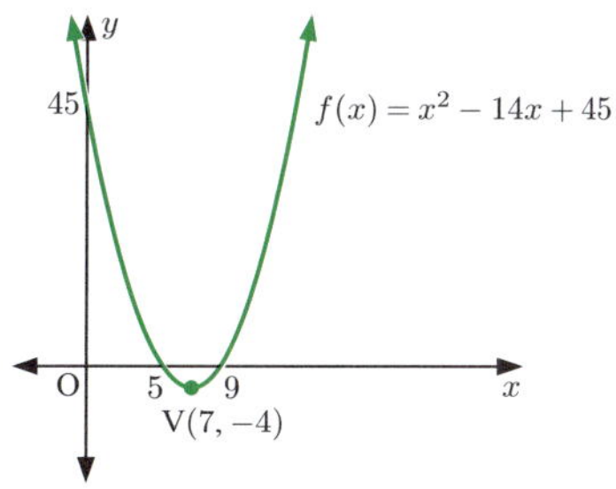

4 **a**

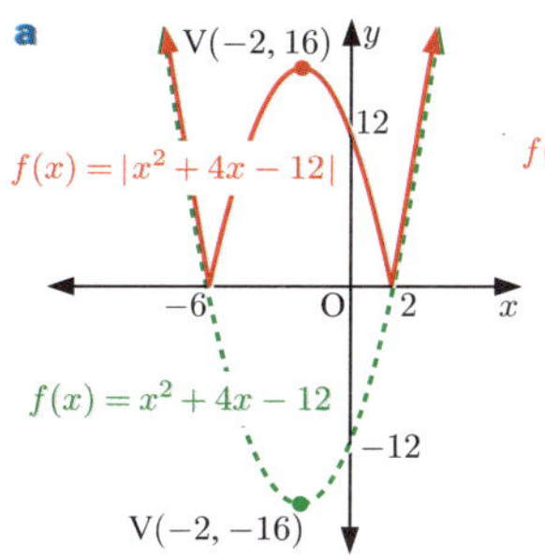

b

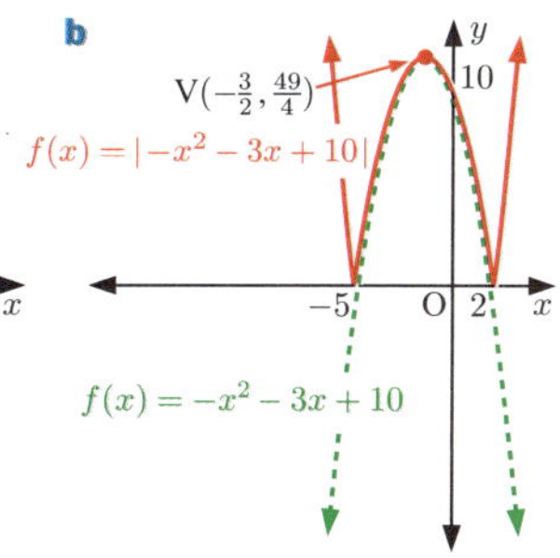

c

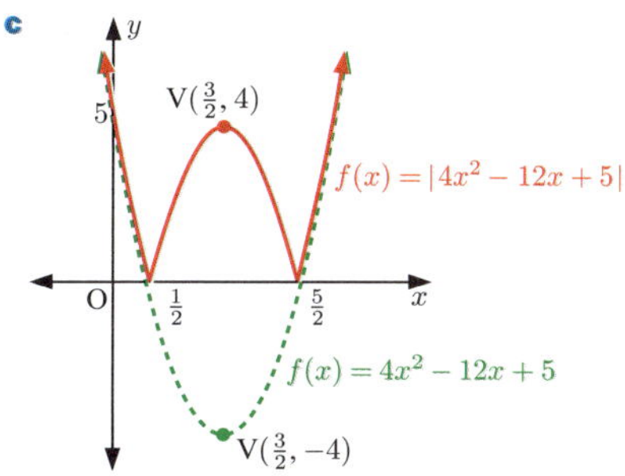

5 **a** $\{y : -10 \leqslant y \leqslant 15\}$ **b** $\{y : 3 \leqslant y \leqslant 19\}$
c $\{y : -13 \leqslant y \leqslant 37\}$ **d** $\{y : -8 \leqslant y \leqslant \frac{49}{4}\}$

EXERCISE 3D.4

1 **a** cuts x-axis twice, concave up
b lies entirely above the x-axis, concave up
c touches x-axis, concave up
d lies entirely below the x-axis, concave down
e cuts x-axis twice, concave down
f lies entirely above the x-axis, concave up
g cuts x-axis twice, concave up
h cuts x-axis twice, concave down
i cuts x-axis twice, concave down

2 **a** concave up
b $\Delta = 5$ which is > 0
c x-intercepts ≈ 0.38 and ≈ 2.62
d 1
e

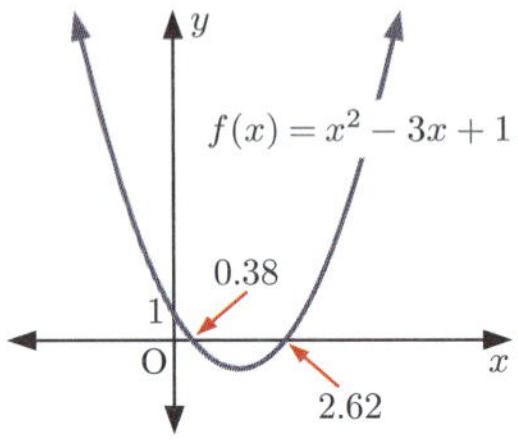

3 **a** $\Delta = -7$ which is < 0
b below
c vertex is $(\frac{5}{4}, -\frac{7}{8})$, y-intercept is -4
d

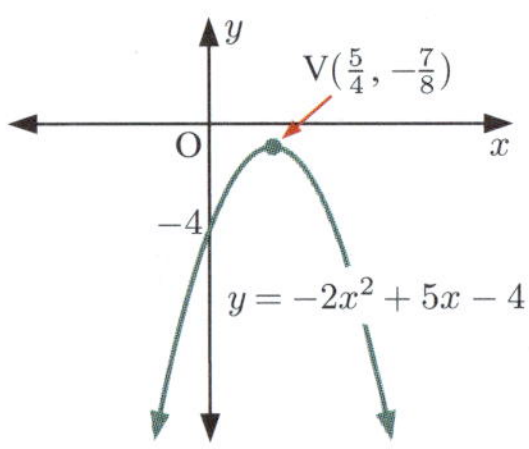

4 **a** $a = 4$ which is > 0 and $\Delta = -12$ which is < 0 so the graph lies entirely above the x-axis.
b $a = -5$ which is < 0 and $\Delta = -91$ which is < 0 so the graph lies entirely below the x-axis.
c $a = 1$ which is > 0 and $\Delta = -4$ which is < 0 so the graph lies entirely above the x-axis.
d Rearranging, $3x^2 - 3x + 1 > 0$
$a = 3$ which is > 0 and $\Delta = -3$ which is < 0 so the graph lies entirely above the x-axis.

5 $-4 < k < 4$

6 $\Delta = k^2 + 4k + 8$ which is always > 0 ($\Delta = -16$, $a = 1$ for this quadratic).
$\therefore$ there are always two distinct solutions.

EXERCISE 3E

1 **a** $y = \frac{1}{2}(x-1)(x-4)$ **b** $y = (x+2)^2$
c $y = \frac{1}{2}(x+5)(x-2)$ **d** $y = -\frac{5}{4}(x-2)^2$
e $y = -\frac{4}{3}(x+3)(x-3)$ **f** $y = -3(x+2)(x-3)$

2 **a** $y = \frac{1}{2}(x+4)(x+2)$ **b** $y = -3(x+1)(x-3)$
c $y = \frac{1}{4}(x-2)^2$

3 **a** $f(x) = -x^2 + 5x - 6$ **b** $f(x) = \frac{1}{2}x^2 + \frac{1}{2}x - 10$
c $f(x) = \frac{3}{2}x^2 - 3x + \frac{3}{2}$

4 **a** $f(x) = 3x^2 - 6x - 9$ **b** $f(x) = \frac{1}{2}x^2 - 2x + \frac{3}{2}$
c $f(x) = \frac{29}{21}x^2 + \frac{232}{21}x + \frac{116}{7}$

5 **a** $y = \frac{1}{2}(x-2)^2 - 1$ **b** $y = -2(x+1)^2 + 7$
c $y = (x+2)^2 + 2$ **d** $y = -\frac{1}{5}(x+3)^2 + 5$
e $y = \frac{3}{2}(x-3)^2 - 2$ **f** $y = -\frac{1}{16}(x-5)^2 - \frac{1}{2}$

EXERCISE 3F.1

1 **a** $(-1, 4)$ and $(3, 8)$ **b** $(-3, 5)$ and $(-2, 6)$
c $(1, 8)$ and $(3, 34)$ **d** $(3, -6)$

2 $c = -6$ **3** $m = 1$ or 5 **4** -1 or 7

5 **a** $c < 1$ **b** For example, $c = 0$

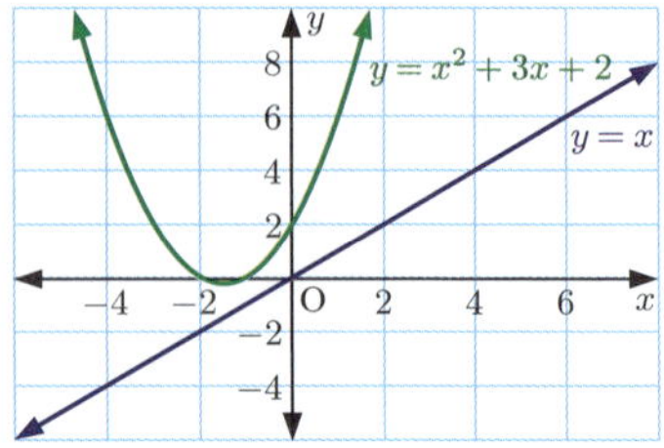

6 **a** $m < -10$ or $m > 2$ **b** $m = -10$ or 2
c $-10 < m < 2$

7 $-1 < k < 8$ **8** $y = x$ and $y = -3x$

EXERCISE 3F.2

1 $(-1, -2)$ and $(\frac{11}{5}, -\frac{2}{5})$ **2** P$(-5, -2)$, Q$(7, -8)$

3 $3\sqrt{2}$ units **4** $y = \frac{1}{2}x$ **5** $(2, -1)$, $(-\frac{4}{3}, -\frac{8}{3})$

6 $5\sqrt{5}$ units **7** $y = \frac{1}{3}x + \frac{13}{3}$

8 **a** $(3, -\frac{3}{2})$, $(4, -1)$ **b** $(1, -3)$, $(\frac{7}{2}, 7)$

9 $(\frac{7}{3}, \frac{5}{2})$ **10** **a** $k = -1$ or 19 **b** $-1 < k < 19$

11 **a** A$(2, 2)$, B$(6, -2)$ **b** $(4, -4)$ **c** no

EXERCISE 3G

1 $x = -3$ or 8 **2** $3 \pm \sqrt{5}$

3 -2 and 5 or 2 and -5 **4** 10 m

5

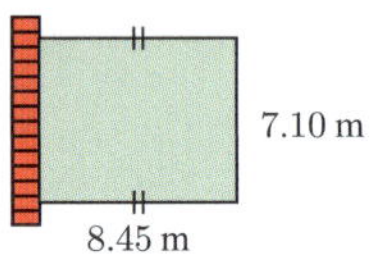

or

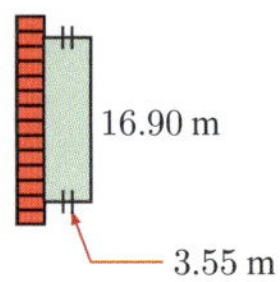

6 17.9 cm **7** base 8 cm, altitude 7 cm **8** 7 cm

9 **a** $n = 24$ **b** $n = 17$

10 $n = 6$ **11** $\frac{2}{5}$ or $\frac{-9}{-6}$ **12** $x = 5$

EXERCISE 3H

1 **a** **i** $x = -1$ **ii** minimum -2
b **i** $x = 2$ **ii** minimum -10
c **i** $x = -1$ **ii** maximum 4
d **i** $x = \frac{1}{4}$ **ii** minimum $4\frac{7}{8}$
e **i** $x = -\frac{2}{5}$ **ii** minimum $-3\frac{4}{5}$
f **i** $x = \frac{3}{4}$ **ii** maximum $1\frac{1}{8}$

2 **a** 7:00 am the next day **b** 9°C **4** 175 m × 350 m

5 **c** 150 m × 100 m **6** $d = 3$

7 **a** $P = \$(-\frac{1}{10}x^2 + 25x - 25)$ **b** 125 toasters

REVIEW SET 3A

1 **a** $x = 1$ or -4 **b** $x = 1$ or $-\frac{3}{2}$ **c** $x = 3$ or $-\frac{9}{4}$
d $x = \pm\sqrt{5}$ **e** $x = \frac{1}{3}$ or $\frac{2}{3}$ **f** $x = \frac{9}{4}$

2 **a** $x = \frac{5 \pm \sqrt{61}}{6}$ **b** $x = 1 \pm \frac{\sqrt{10}}{2}$

3 **a** 3, 5
b $x = 4$
c 5
d $(4, -\frac{1}{3})$
e

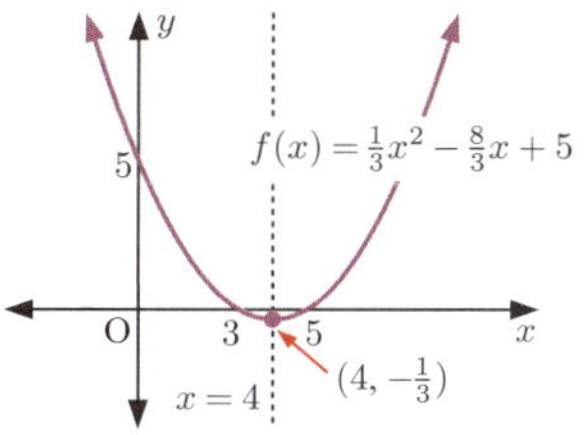

f $\{y : y \geqslant -\frac{1}{3}\}$

4 **a** $-3 < x < 7$ **b** $x \leqslant -\frac{1}{3}$ or $x \geqslant 2$

5 $\{y : -13 \leqslant y \leqslant 12\}$

6 **a** $y = (x+3)(x-2)$ **b** $y = -\frac{4}{3}(x-3)^2$
c $y = \frac{3}{4}(x+2)^2 + 1$

7 **a** $f(x) = \frac{4}{9}x^2 + \frac{16}{9}x + \frac{16}{9}$ **b** $f(x) = \frac{2}{9}x^2 - \frac{4}{3}x - 3$

8 **a** $k = -8$ **b** $k < -8$ or $k > 0$ **c** $-8 < k < 0$

9 **a** $f(x) = 2(x+1)^2 - 5$ **b**

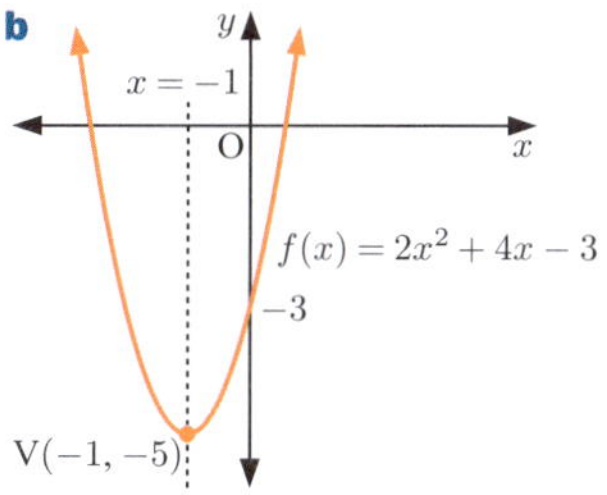

10 **a** $(-2, 7)$ and $(2, 15)$ **b** $(-2, -17)$ and $(1, -5)$

11 $k < 1$ **12** $(-\frac{7}{5}, \frac{26}{5})$ and $(2, -5)$ **13** $y = x - 5$

14 **a** **i** $x = 1$ **ii** minimum -7
b **i** $x = \frac{7}{6}$ **ii** maximum $3\frac{1}{12}$

15 $2 \pm \sqrt{\frac{7}{2}}$ **16** $\frac{3+\sqrt{5}}{2} : 1$

REVIEW SET 3B

1 **a** $x = 3 \pm \sqrt{14}$ **b** $x = 1 \pm \sqrt{3}$ **c** $x = \frac{1}{2} \pm \frac{\sqrt{57}}{6}$

2 **a** $-7 \leqslant x \leqslant 2$ **b** $x < -4$ or $x > \frac{3}{2}$

3 $\{y : -55 \leqslant y \leqslant \frac{11}{2}\}$

4 **a** **i** 1, -3
ii 3
iii

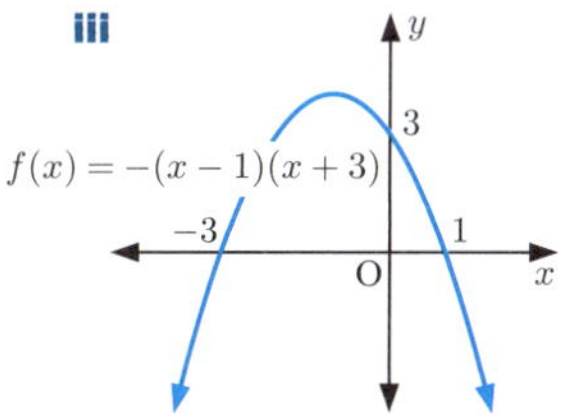

b **i** $-\frac{1}{2}$, 4
ii -2
iii

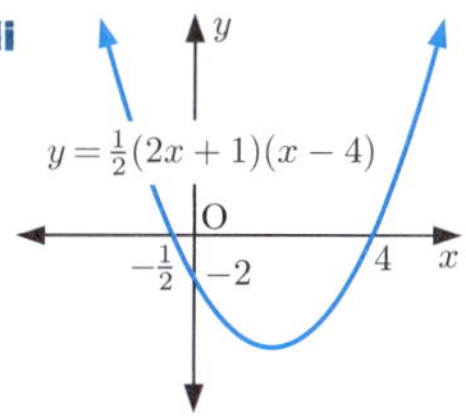

5 **a** **i** $(-3, 2)$
ii 29
iii

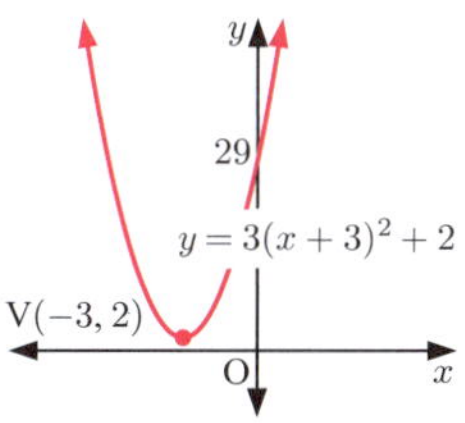

b **i** $(4, 1)$
ii -31
iii

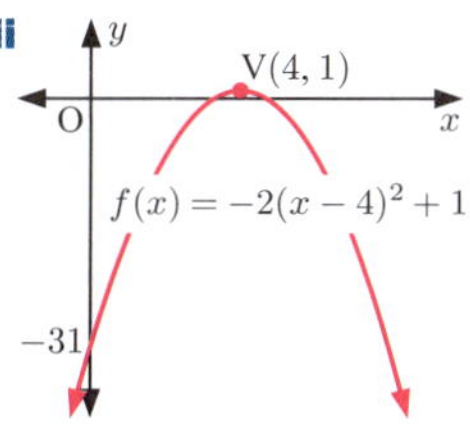

6 **a** $y = -(x+1)^2 + 4$ **b** $y = \frac{1}{5}(x+1)(x-5)$
c $y = \frac{7}{5}(x+1)(x-3)$

7 $m = 1 \pm 2\sqrt{6}$

8

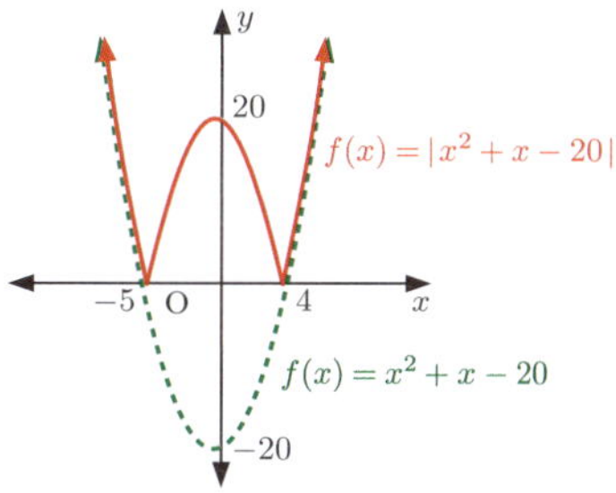

9 **a** cuts x-axis twice, concave up
b lies entirely below the x-axis, concave down
c touches x-axis, concave up

10 $a = \frac{1}{16}$ **11** $\left(\frac{\sqrt{191}-3}{2}\right)$ cm **12** $4\sqrt{5}$ units

13 $(\frac{7}{8}, \frac{1}{2})$ **14** $\frac{16}{5}$ cm

15 **a** $y = 10 - \frac{10}{7}x$ **b** $3\frac{1}{2}$ cm $\times$ 5 cm **16** no

EXERCISE 4A.1

1 **a** 5 **b** $\sqrt{6}$ **c** 7 **d** $\sqrt{15}$
e 12 **f** 30 **g** 45 **h** $\sqrt{11}$
i $\sqrt{5}$ **j** 2 **k** $\sqrt{2}$ **l** 1

2 **a** $2\sqrt{3}$ **b** $3\sqrt{2}$ **c** $3\sqrt{3}$ **d** $2\sqrt{7}$
e $3\sqrt{5}$ **f** $4\sqrt{2}$ **g** $3\sqrt{7}$ **h** $5\sqrt{3}$
i $5\sqrt{5}$ **j** $4\sqrt{5}$ **k** $2\sqrt{33}$ **l** $8\sqrt{2}$

EXERCISE 4A.2

1 **a** $6\sqrt{2}$ **b** $\sqrt{3}$ **c** $-2\sqrt{5}$ **d** $4\sqrt{7}$
e $11\sqrt{6}$ **f** $-2\sqrt{5}$ **g** $4\sqrt{10}$ **h** $\sqrt{3}$

2 **a** $2\sqrt{5} - 5$ **b** $3 + 2\sqrt{3}$ **c** $2\sqrt{7} + 21$
d $12 - \sqrt{6}$ **e** $-3\sqrt{2} - 2$ **f** $24 - 15\sqrt{8}$
g $-5 + 4\sqrt{5}$ **h** $-10\sqrt{3} + 42$

3 **a** $11 + 6\sqrt{3}$ **b** $11 + 16\sqrt{2}$ **c** $-3 + 16\sqrt{5}$
d $-7\sqrt{2}$ **e** $33 - 10\sqrt{10}$ **f** $19\sqrt{7} - 52$

4 **a** $7 + 4\sqrt{3}$ **b** $3 + 2\sqrt{2}$ **c** $21 - 8\sqrt{5}$
d $15 - 6\sqrt{6}$ **e** $21 + 4\sqrt{5}$ **f** $19 + 6\sqrt{2}$
g $76 - 42\sqrt{3}$ **h** $10 + 4\sqrt{6}$

5 **a** 1 **b** 4 **c** 2 **d** 2 **e** 1 **f** -1

EXERCISE 4A.3

1 **a** $\frac{\sqrt{2}}{2}$ **b** $\frac{3\sqrt{5}}{5}$ **c** $\frac{7\sqrt{3}}{3}$ **d** $\frac{11\sqrt{6}}{6}$
e $\frac{9\sqrt{10}}{10}$ **f** $\sqrt{3}$ **g** $5\sqrt{2}$ **h** $6\sqrt{5}$
i $\frac{\sqrt{6}}{3}$ **j** $\frac{3\sqrt{8}}{4}$ **k** $\frac{\sqrt{10}}{5}$ **l** $\frac{\sqrt{21}}{3}$
m $\frac{\sqrt{10}}{6}$ **n** $\frac{\sqrt{6}}{24}$ **o** $\frac{\sqrt{2}}{4}$

2 **a** $\frac{4-\sqrt{3}}{13}$ **b** $\frac{3+\sqrt{7}}{2}$ **c** $\frac{2\sqrt{11}-4}{7}$
d $-1 - \sqrt{6}$ **e** $14 + 7\sqrt{3}$ **f** $\frac{\sqrt{15}-\sqrt{3}}{4}$
g $\frac{7+3\sqrt{3}}{22}$ **h** $\frac{-2\sqrt{7}-7}{3}$ **i** $\frac{15-12\sqrt{5}}{11}$
j $-12 - 4\sqrt{10}$ **k** $\frac{10-2\sqrt{7}}{3}$ **l** $\frac{9+2\sqrt{8}}{7}$

3 **a** $-\frac{9}{7}-\frac{3}{7}\sqrt{2}$ **b** $4-2\sqrt{2}$ **c** $-\frac{2}{23}-\frac{5}{23}\sqrt{2}$
d $-4+2\sqrt{2}$

4 **a** $-2-2\sqrt{3}$ **b** $12-6\sqrt{3}$ **c** $3+2\sqrt{3}$ **d** $-\frac{1}{2}+\frac{5}{6}\sqrt{3}$

5 **a** $(a+b\sqrt{c})(a-b\sqrt{c})=a^2-b^2c$
which is an integer as a, b, and c are integers.
b **i** $\dfrac{-1+2\sqrt{3}}{11}$ **ii** $\dfrac{-6-5\sqrt{2}}{7}$ **iii** $1+\sqrt{2}$

6 **a** $(\sqrt{a}+\sqrt{b})(\sqrt{a}-\sqrt{b})=a-b$
which is an integer as a and b are integers.
b **i** $\sqrt{3}-\sqrt{2}$ **ii** $\dfrac{-3-\sqrt{15}}{2}$ **iii** $\dfrac{2\sqrt{154}-25}{3}$

7 $x=-7+5\sqrt{3}$ **8** $x=\frac{10}{19}+\frac{1}{19}\sqrt{5}$

EXERCISE 4B

1 **a** $2^1=2,\ 2^2=4,\ 2^3=8,\ 2^4=16,\ 2^5=32$
b $(-2)^1=-2,\ (-2)^2=4,\ (-2)^3=-8,\ (-2)^4=16,\ (-2)^5=-32$
c $3^1=3,\ 3^2=9,\ 3^3=27,\ 3^4=81,\ 3^5=243$
d $(-3)^1=-3,\ (-3)^2=9,\ (-3)^3=-27,\ (-3)^4=81,\ (-3)^5=-243$
e $4^1=4,\ 4^2=16,\ 4^3=64,\ 4^4=256,\ 4^5=1024$
f $(-4)^1=-4,\ (-4)^2=16,\ (-4)^3=-64,\ (-4)^4=256,\ (-4)^5=-1024$

2

n	5^n	6^n	7^n
1	5	6	7
2	25	36	49
3	125	216	343
4	625	1296	2401

3 **a** -1 **b** 1 **c** -1 **d** -1
e 1 **f** -1 **g** -1 **h** 64
i -64 **j** -256 **k** -256 **l** -64

4 **a** 256 **b** 6561 **c** $-117\,649$ **d** $117\,649$
e 7776 **f** 7776 **g** $610.351\,562\,5$
h $-610.351\,562\,5$ **i** $-610.351\,562\,5$
j 0.0625 **k** -0.0625 **l** -0.0625

5 **a** 1024 **b** 1024 **c** $50\,625$ **d** $50\,625$
e $7.593\,75$ **f** $7.593\,75$ **g** 4096 **h** 4096
i 125 **j** 125 **k** 1 **l** 1

We see the *index laws*:

- $a^m\times a^n=a^{m+n}$
- $(ab)^n=a^nb^n$
- $\left(\dfrac{a}{b}\right)^n=\dfrac{a^n}{b^n},\quad b\neq 0$
- $(a^m)^n=a^{m\times n}$
- $\dfrac{a^m}{a^n}=a^{m-n},\quad a\neq 0$
- $a^0=1,\quad a\neq 0$

6 8

EXERCISE 4C

1 **a** 3^6 **b** k^5 **c** 5^4 **d** $\dfrac{1}{b^2}$ **e** 6^{15} **f** m^8
g x^7 **h** $\dfrac{1}{q^2}$ **i** 2^{n+3} **j** 11^{2k} **k** 4^{x-6} **l** y^{3z}

2 **a** 2^1 **b** 2^{-1} **c** 2^4 **d** 2^{-4} **e** 2^5 **f** 2^{-5}
g 2^7 **h** 2^{-7}

3 **a** 5^1 **b** 5^{-1} **c** 5^2 **d** 5^{-2} **e** 5^3 **f** 5^{-3}
g 5^0

4 **a** 2^{n+1} **b** 2^{m+2} **c** 2^{3x+4} **d** 2^{y+2} **e** 2^{2k+3}
f 2^{1-a} **g** 2^{a-1} **h** 2^{t-2} **i** 2^{2p-10} **j** 2^{4c-6}

5 **a** 5^{q+1} **b** 5^{2r} **c** 5^{3-k} **d** 5^{2z} **e** 5^{2-y}

6 **a** 3^{4x} **b** 3^{s+4} **c** 3^{4n-4} **d** 3^{w+1} **e** 3^{5-2p}

7 **a** $8x^3$ **b** $16k^2$ **c** x^3y^3 **d** y^4z^4 **e** $\dfrac{p^3}{q^3}$
f $\dfrac{m^2}{4}$ **g** $\dfrac{125}{n^3}$ **h** $\dfrac{9a^2}{c^2}$ **i** $\dfrac{w^4}{16x^4}$ **j** 1

8 **a** $-27x^3$ **b** $16y^8$ **c** $36z^4$ **d** $-64k^6$
e $\dfrac{n^5}{m^{10}}$ **f** $\dfrac{u^6}{9v^2}$ **g** $\dfrac{625w^4}{h^4}$ **h** $\dfrac{49d^6}{b^4}$
i $4x^3y^2$ **j** $32a^5b$ **k** $\dfrac{5a^{12}}{b^2}$ **l** $-\dfrac{2x^{18}}{y^3}$

9 **a** $\dfrac{a}{b^3}$ **b** $\dfrac{1}{p^2q^2}$ **c** $\dfrac{4d^2}{c^2}$ **d** $\dfrac{25x^2}{y^2}$ **e** $\dfrac{a}{3}$
f $\dfrac{p^3}{qr^2}$ **g** $2a^3$ **h** $\dfrac{n^2}{m^5}$ **i** $\dfrac{zu}{w^2}$ **j** $\dfrac{2pr^2}{q}$

10 **a** x^{-n} **b** 2^{-x} **c** 2^{n-1} **d** p^2q^{-x} **e** x^{m+1}

11 **a** $\frac{1}{8}$ **b** $\frac{4}{3}$ **c** $\frac{6}{5}$ **d** $\frac{9}{4}$ **e** 1
f $\frac{82}{9}$ **g** $\frac{4}{25}$ **h** $\frac{81}{4}$

12 **a** $2^{-2}\times 3^{-1}$ **b** 5^{-2} **c** $2^1\times 3^{-2}$
d $2^{-2}\times 3^{-1}\times 5^{-1}$ **e** $2^1\times 5^1\times 3^{-3}$
f $3^{2n-1}\times 2^{-3}$ **g** $2^m\times 3^{m-2}\times 5^{-1}$
h $2^{1-p}\times 3^2\times 5^{-p}$

13 **a** $x=2,\ y=\frac{5}{2}$ **b** $x=6,\ y=\frac{1}{2}$
c $x=2,\ y=-\frac{1}{4}$ or $x=-2,\ y=-\frac{1}{4}$

EXERCISE 4D

1 **a** $2^{\frac{1}{3}}$ **b** $2^{-\frac{1}{2}}$ **c** $2^{\frac{5}{2}}$ **d** 2^{-2} **e** $2^{-\frac{1}{5}}$
f $2^{-\frac{3}{2}}$ **g** $2^{\frac{6}{5}}$ **h** $2^{\frac{1}{3}}$ **i** $2^{-\frac{5}{3}}$ **j** $2^{\frac{8}{3}}$

2 **a** $3^{\frac{1}{5}}$ **b** $3^{-\frac{1}{5}}$ **c** $3^{\frac{5}{2}}$ **d** $3^{\frac{1}{2}}$ **e** $3^{\frac{7}{2}}$
f $3^{\frac{4}{3}}$ **g** $3^{-\frac{3}{2}}$ **h** $3^{-\frac{2}{3}}$ **i** $3^{\frac{17}{5}}$ **j** $3^{-\frac{1}{2}}$

3 **a** $5^{\frac{2}{3}}$ **b** $7^{\frac{1}{2}}$ **c** $11^{\frac{2}{3}}$ **d** $3^{\frac{4}{5}}$ **e** $5^{\frac{3}{4}}$
f $5^{-\frac{2}{3}}$ **g** $7^{-\frac{1}{2}}$ **h** $11^{-\frac{2}{3}}$ **i** $3^{-\frac{4}{5}}$ **j** $5^{-\frac{3}{4}}$

4 **a** 32 **b** 9 **c** 32 **d** 32 **e** 27
f $\frac{1}{3}$ **g** $\frac{1}{125}$ **h** $\frac{1}{8}$ **i** $\frac{1}{16}$ **j** $\frac{1}{81}$

5 **a** ≈ 1.26 **b** ≈ 2.50 **c** ≈ 0.725 **d** ≈ 0.192
e ≈ 1.58

EXERCISE 4E

1 **a** x^3+2x^2 **b** 3^x-9^x **c** x^2+x
d 25^x+1 **e** $3x^5-x^4+x^2$ **f** x^2-x^4+x
g 4^x+2^{-x} **h** $x^{\frac{3}{2}}+4x^{\frac{1}{2}}+1$ **i** $7^{3x}+7^{2x}$

2 **a** $4^x+2^x-2^{x+1}-2$ **b** $9^x-6(3^x)+5$
c 4^x-9 **d** $49^x+7^{x+1}+12$ **e** 4^x-16
f $16^x+2(4^x)+1$ **g** $64^x-6(8^x)+9$
h $49-36^x$ **i** $25^x+10(5^x)+25$

3 **a** $x+8x^{\frac{1}{2}}+15$ **b** $x-16$ **c** $x-2x^{\frac{1}{2}}+1$
d $x^4-2x+\dfrac{1}{x^2}$ **e** $4-x^{\frac{3}{2}}$ **f** $3x+20x^{\frac{1}{2}}-7$
g $x^{\frac{4}{3}}+2x+x^{\frac{2}{3}}$ **h** x^3-2x^2+x **i** $4x-4+\dfrac{1}{x}$

4 **a** $28(3^n)$ **b** $5^n(5^n+1)$ **c** $2^{2n}(2^n-1)$
d $5(4^{n+1})$ **e** $7(7^{n+1}-1)$ **f** $125(5^n+1)$
g $2^n(2^{2n}+1)$ **h** $5^n(1-5^n)$ **i** $27(3^{2n}-1)$

5 **a** $(2^x+3)(2^x-3)$ **b** $(5^x+3)(5^x-3)$
c $(6+2^x)(6-2^x)$ **d** $(7+2^{2x})(7-2^{2x})$
e $(5^x+4^x)(5^x-4^x)$ **f** $(3^x+2)^2$ **g** $(2^x-6)^2$
h $(3^x-7)^2$ **i** $(2^x+8)^2$

6 **a** $(3^x+5)(3^x+3)$ **b** $(2^x+7)(2^x-2)$
c $(3^x-7)(3^x-5)$ **d** $(4^x-9)(4^x+2)$
e $(7^x-4)(7^x-5)$ **f** $(5^x-8)(5^x+3)$

7 **a** 2^n **b** 10^a **c** 3^b **d** $\dfrac{1}{5^n}$ **e** 5^x
f $\left(\frac{3}{4}\right)^a$ **g** 5 **h** 5^n

8 **a** 3^m+1 **b** 6^n+1 **c** 4^n+2^n **d** 4^x-1
e 6^n **f** 5^n **g** 4 **h** 2^n-1
i $\frac{1}{2}$

9 **a** $2^{n+1}n$ **b** -3^{n-1}

EXERCISE 4F

1 **a** $x=4$ **b** $x=2$ **c** $x=2$ **d** $x=0$
e $x=-3$ **f** $x=\frac{1}{2}$ **g** $x=\frac{5}{2}$ **h** $x=1$
i $x=7$ **j** $x=1$ **k** $x=3$ **l** $x=-2$

2 **a** $x=\frac{5}{2}$ **b** $x=\frac{2}{3}$ **c** $x=-\frac{1}{2}$ **d** $x=-\frac{4}{3}$
e $x=\frac{3}{4}$ **f** $x=-1$ **g** $x=\frac{7}{3}$ **h** $x=-\frac{1}{3}$
i $x=0$ **j** $x=3$ **k** $x=-5$ **l** $x=\frac{3}{2}$

3 **a** $x=\frac{7}{5}$ **b** $x=-\frac{7}{4}$ **c** no solution

4 **a** $x=1$ **b** $x=2$ **c** $x=1$ **d** $x=\frac{5}{4}$
e $x=2$ **f** $x=-\frac{9}{7}$

5 **a** $x=2$ **b** $x=2$ **c** $x=3$ **d** $x=2$
e $x=-4$ **f** $x=-1$

6 **a** $x=\pm 2$ **b** $x=-1$ or 3 **c** $x=-1$ or $-\frac{1}{2}$

7 **a** $x=2$ **b** $x=1$ **c** $x=1$ or 4
d $x=0$ or 1 **e** $x=\pm 1$ **f** no solution

8 **a** $x=3,\ y=1$ **b** $x=-2,\ y=-4$
c $x=-3,\ y=2$ or $x=\frac{4}{3},\ y=-\frac{9}{2}$

EXERCISE 4G.1

1 **a**

x	-3	-2	-1	0	1	2	3
y	$\frac{1}{27}$	$\frac{1}{9}$	$\frac{1}{3}$	1	3	9	27

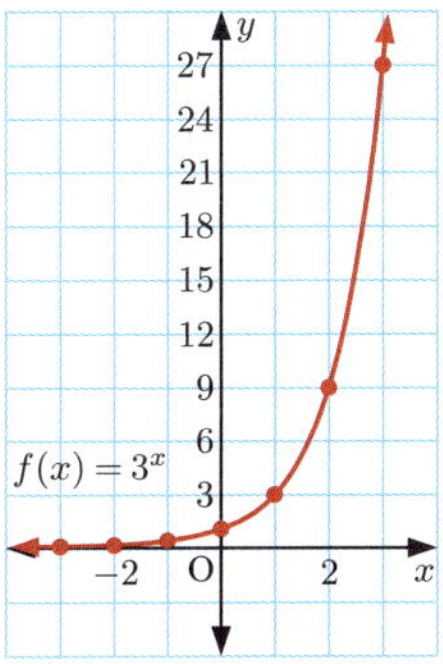

b

x	-3	-2	-1	0	1	2	3
y	$\frac{1}{64}$	$\frac{1}{16}$	$\frac{1}{4}$	1	4	16	64

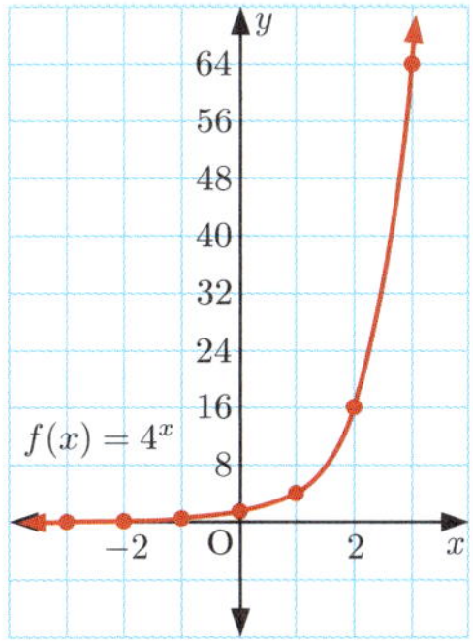

c

x	-3	-2	-1	0	1	2	3
y	8	4	2	1	$\frac{1}{2}$	$\frac{1}{4}$	$\frac{1}{8}$

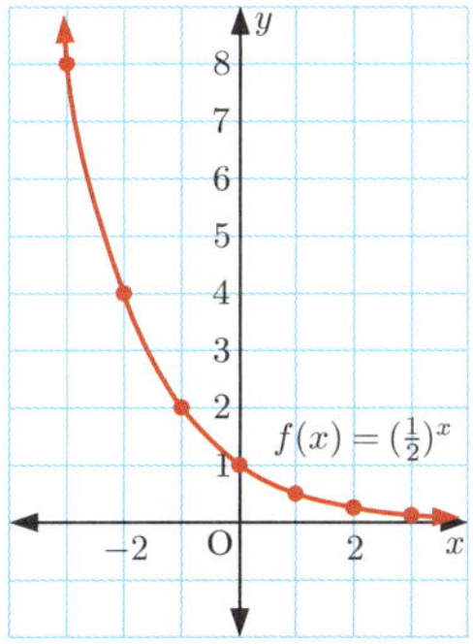

d

x	-3	-2	-1	0	1	2	3
y	27	9	3	1	$\frac{1}{3}$	$\frac{1}{9}$	$\frac{1}{27}$

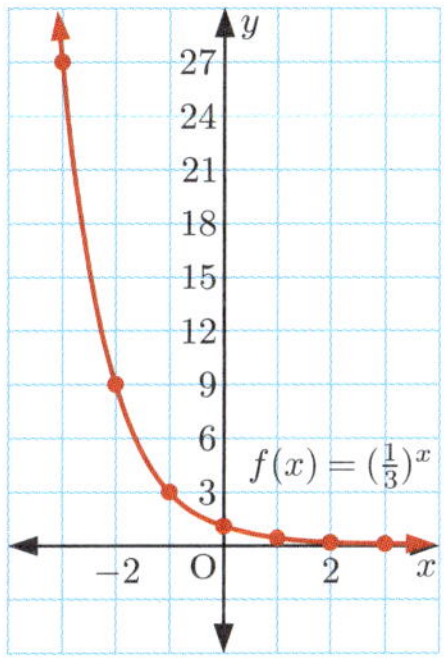

2 **a, b, c, d**

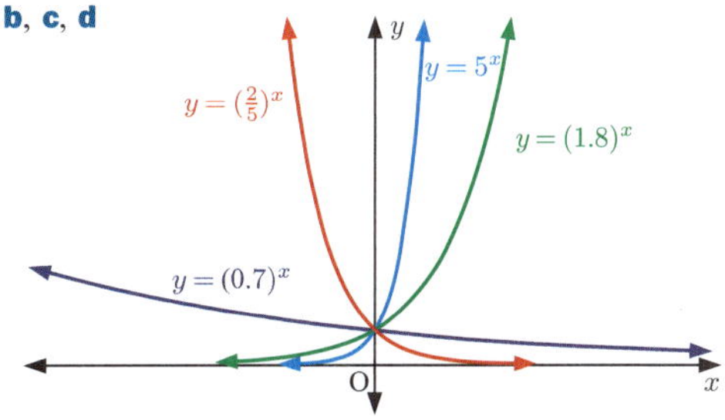

The graph of $y=a^x$ is increasing if $a>1$ and decreasing if $0<a<1$.

3 **a** **i** ≈ 5.2 **ii** ≈ 1.6 **iii** ≈ 0.1 **iv** ≈ 0.6
b **i** $x \approx 1.5$ **ii** $x \approx -0.2$ **iii** $x \approx 1.3$ **iv** $x \approx -1.1$

EXERCISE 4G.2

1 **a** $g(0) = 3$, $g(-1) = 2\frac{1}{5}$ **b** $a = 2$

2 **a**

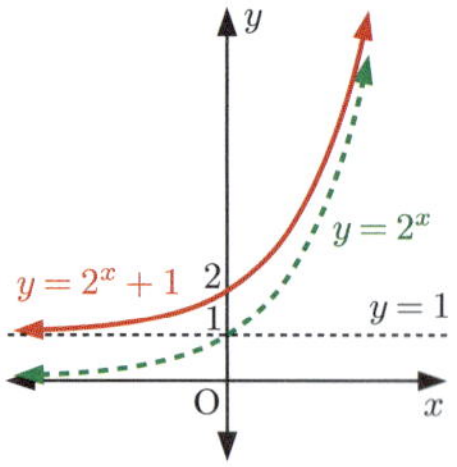

b

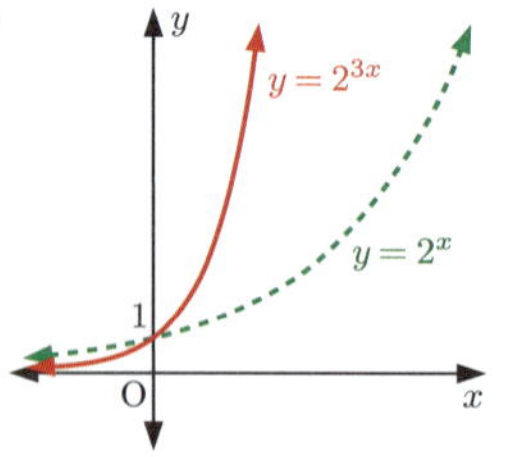

c

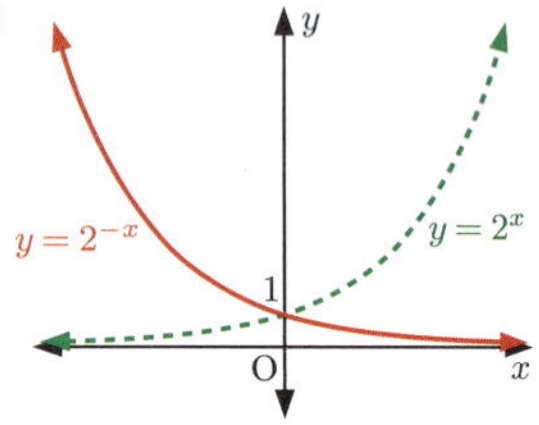

d

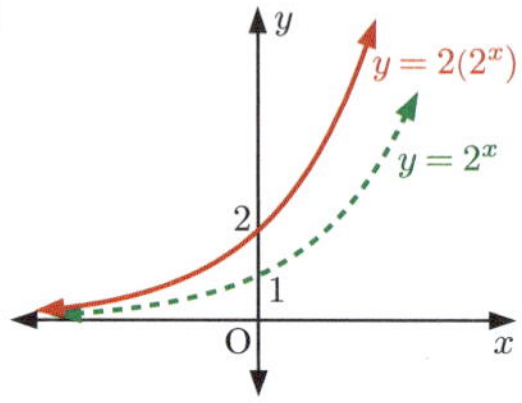

3 **a**

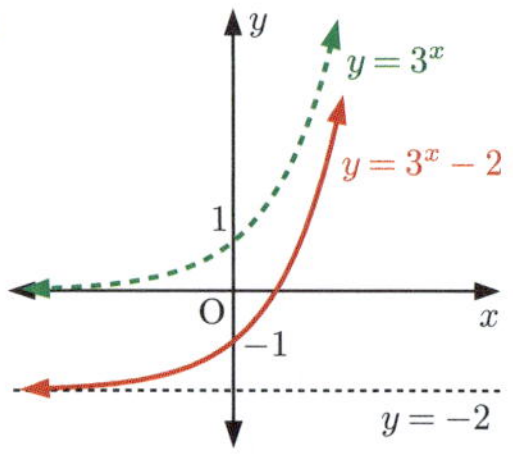

b

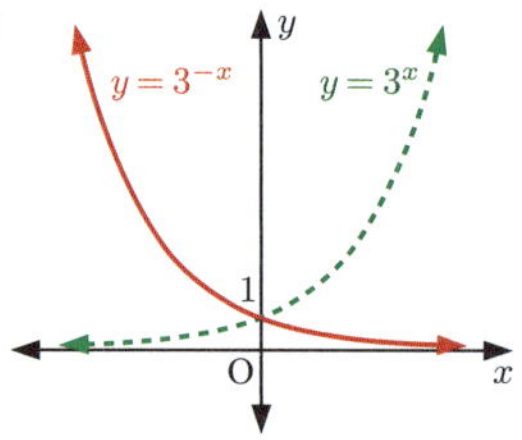

c

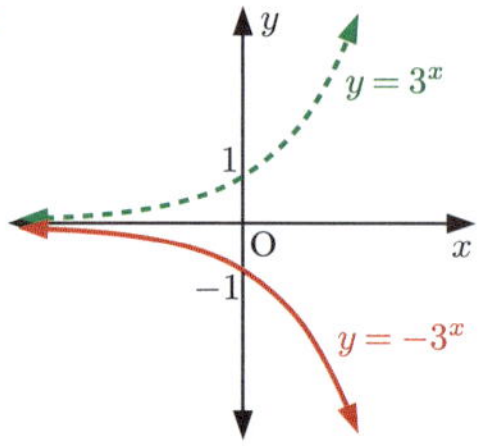

d

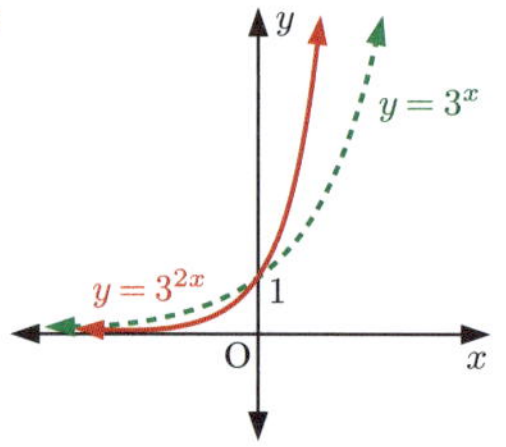

4 **a** **i** 2 **ii** 54
iii $\frac{2}{9}$
b $y = 0$
c

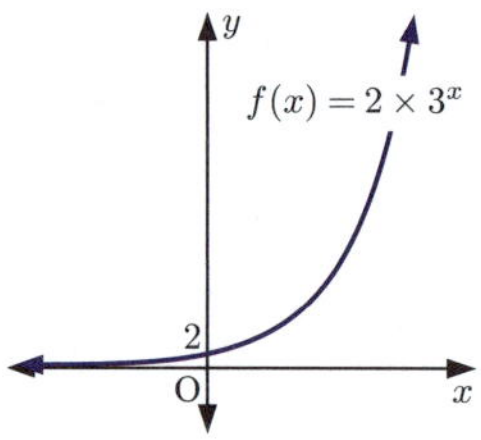

d Domain is $\{x : x \in \mathbb{R}\}$
Range is $\{y : y > 0\}$

5 **a** **i** $y = -3$ **ii** as $x \to \infty$, $y \to \infty$
as $x \to -\infty$, $y \to -3$ from above

iii

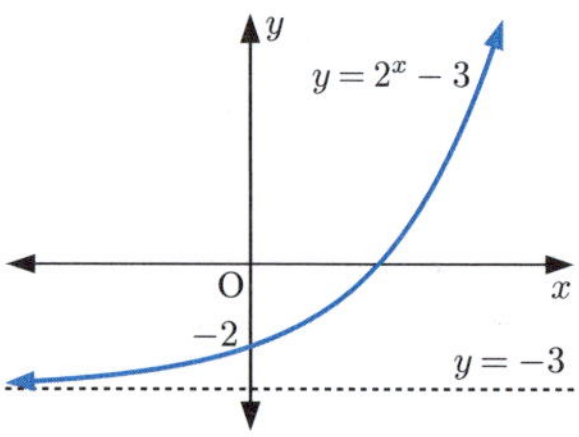

iv Domain is $\{x : x \in \mathbb{R}\}$, Range is $\{y : y > -3\}$

b **i** $y = 5$ **ii** as $x \to \infty$, $y \to -\infty$
as $x \to -\infty$, $y \to 5$ from below

iii

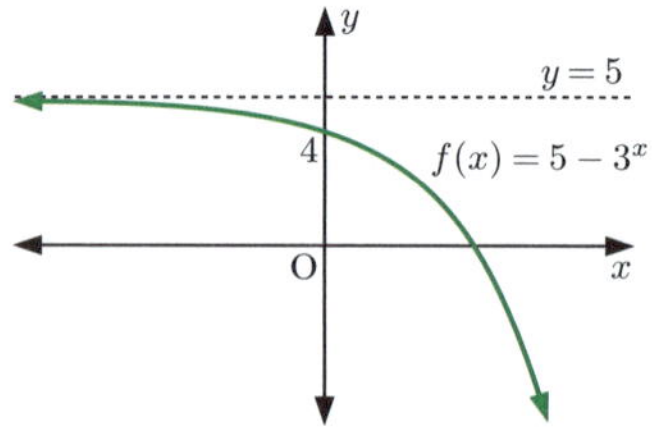

iv Domain is $\{x : x \in \mathbb{R}\}$, Range is $\{y : y < 5\}$

c **i** $y = 1$ **ii** as $x \to \infty$, $y \to \infty$
as $x \to -\infty$, $y \to 1$ from above

iii

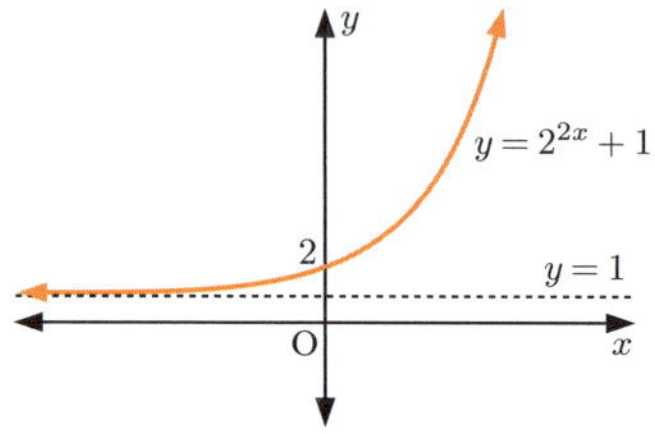

iv Domain is $\{x : x \in \mathbb{R}\}$, Range is $\{y : y > 1\}$

d **i** $y = -1$ **ii** as $x \to \infty$, $y \to \infty$
as $x \to -\infty$, $y \to -1$ from above

iii

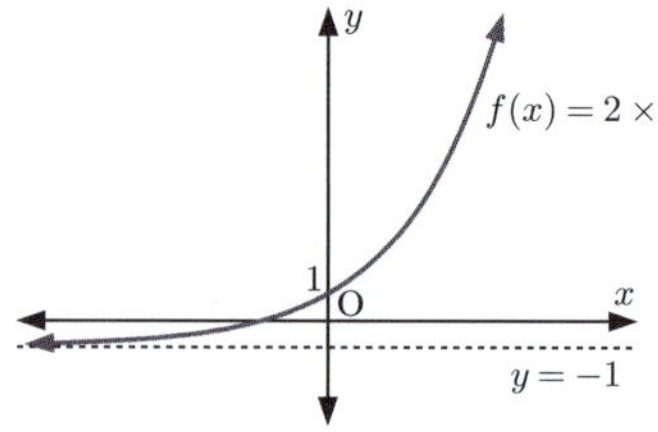

iv Domain is $\{x : x \in \mathbb{R}\}$, Range is $\{y : y > -1\}$

6 **a** **i** \$2000 **ii** \$2120 **iii** \$2382.03 **b** 6%

c

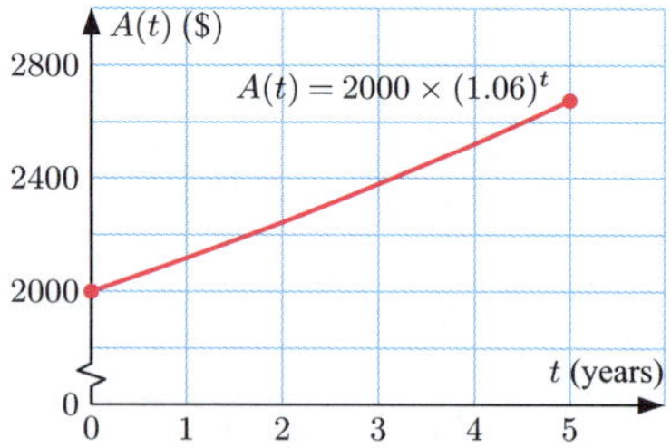

7 **a** 100 000 bacteria **b** 28 minutes
c **i** 400 000 bacteria **ii** $\approx 742\,000\,000$ bacteria

d

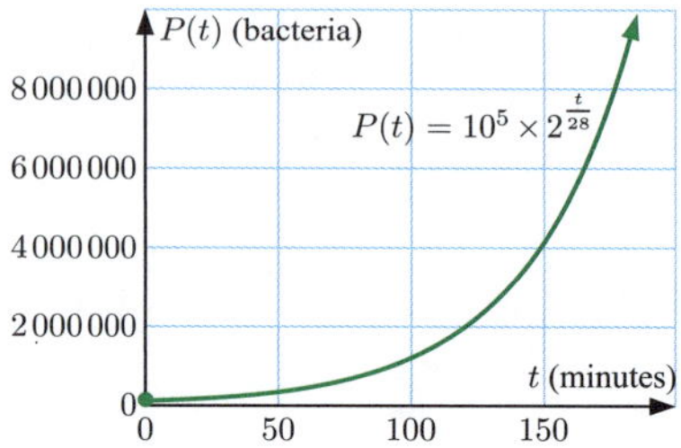

e 168 minutes

8 **a** no

b The area covered each week is 1.2 times greater than the area covered in the previous week, and the initial area covered is 40 cm^2.

c

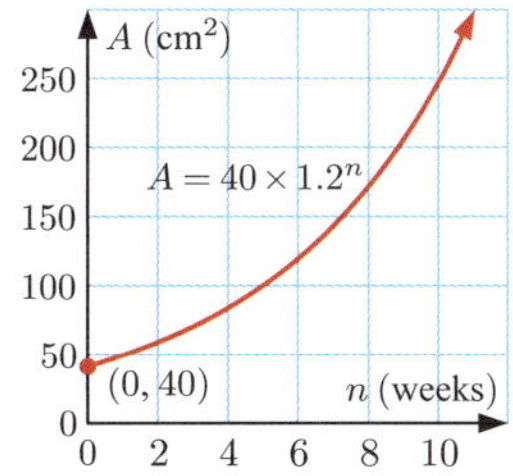

d 69.12 cm^2

e no

9 **a** 4.8 grams

b **i** ≈ 1.52 grams **ii** 0.000 48 grams

c

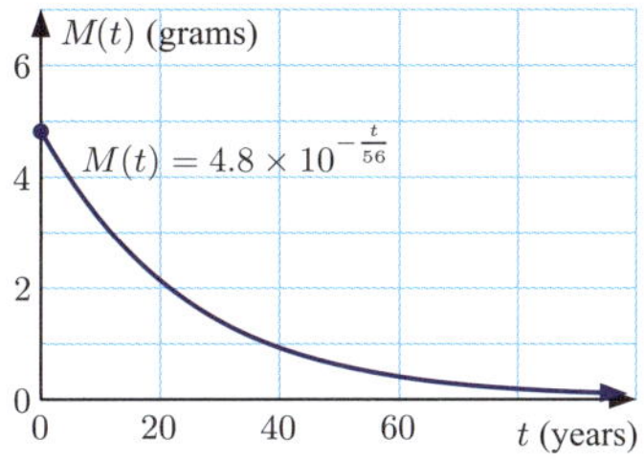

d 448 years

EXERCISE 4H

1 **a** ≈ 7.39 **b** ≈ 53.6 **c** ≈ 74.2 **d** ≈ 0.0498

2 **a** $e^{\frac{1}{2}}$ **b** e^{-5} **c** $e^{-\frac{1}{3}}$ **d** $e^{\frac{5}{2}}$

3 **a** $e^{x+1} + 3e$ **b** $e^{2x} - 2e^x$ **c** $e^{2x} + 2e^x + 1$

4 **a** $x = \frac{1}{3}$ **b** $x = -\frac{1}{2}$ **c** $x = \frac{9}{2}$

5 **a** $fg(x) = e^{3x+2}$, $gf(x) = 3e^x + 2$ **b** $x = -1$

6 The graph of $y = e^x$ lies between $y = 2^x$ and $y = 3^x$.

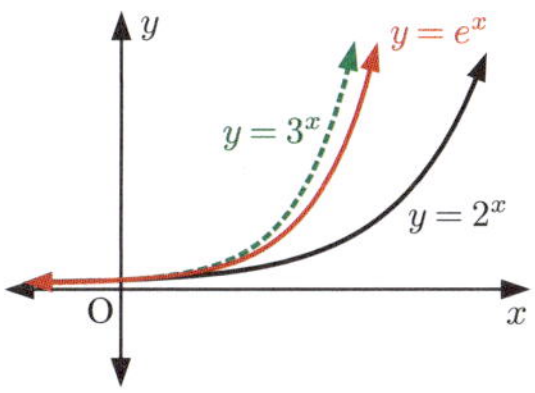

7 **a** $y = a$ **b** $k + a$

8 **a**

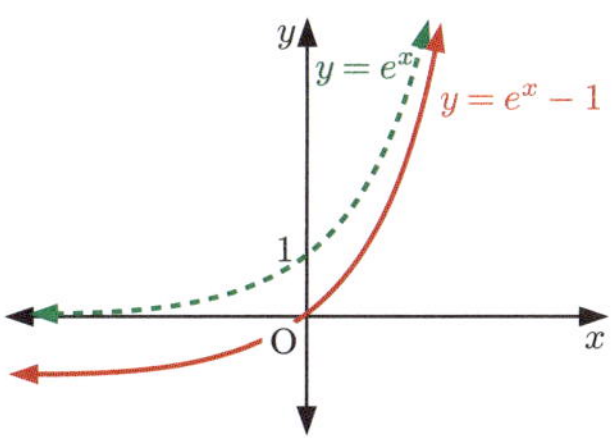

b

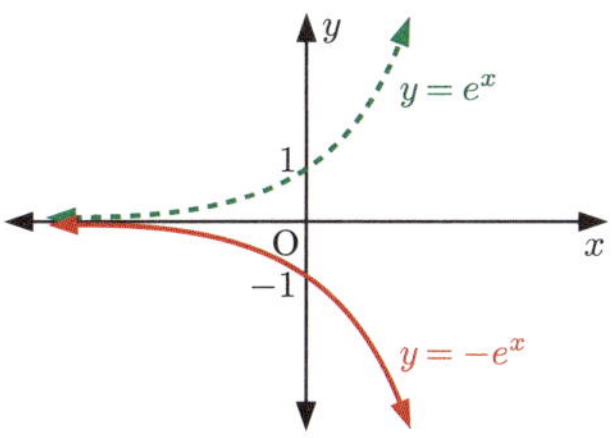

c

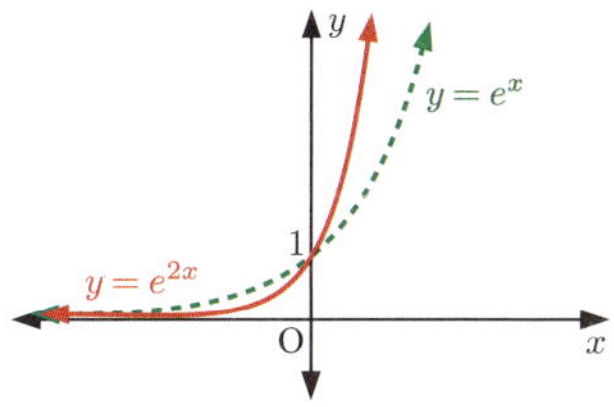

d

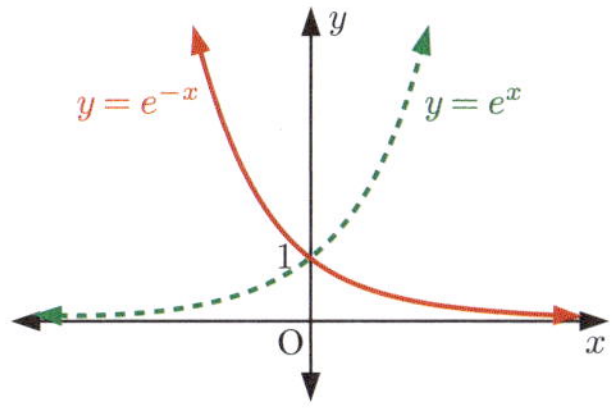

9 **a** $y = 1$ **b** P(0, 3) **c** **i** $k = 2e + 1$ **ii** $k \approx 6.44$

d Domain is $\{x : x \in \mathbb{R}\}$, Range is $\{y : y > 1\}$

10 **a** $y = -5$

b -2

c $y \approx 17.167$

d

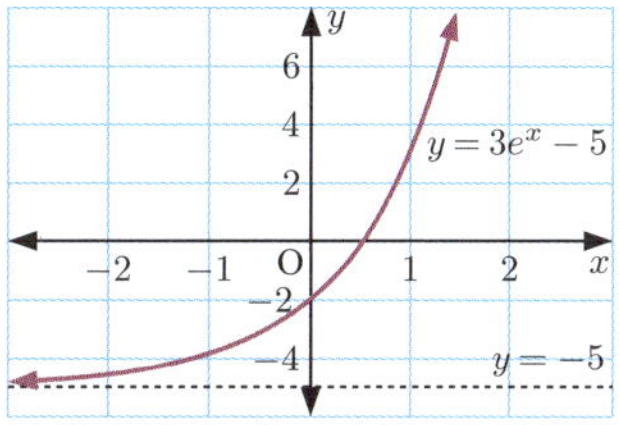

e Domain is $\{x : x \in \mathbb{R}\}$, Range is $\{y : y > -5\}$

11 **a** **i** horizontal asymptote $y = -3$, y-intercept -2

ii

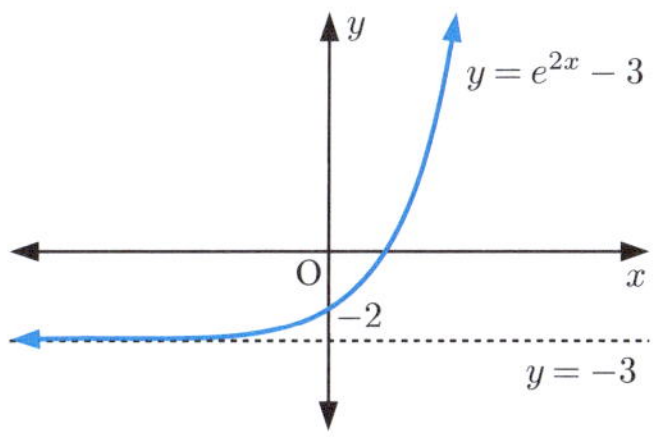

iii Domain is $\{x : x \in \mathbb{R}\}$, Range is $\{y : y > -3\}$

b **i** horizontal asymptote $y = 2$, y-intercept 4

ii

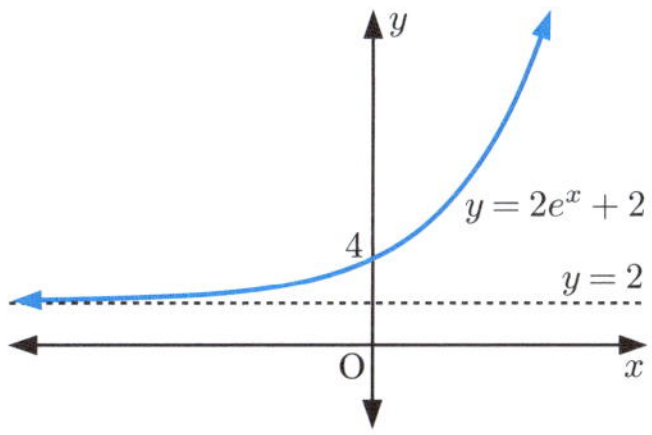

iii Domain is $\{x : x \in \mathbb{R}\}$, Range is $\{y : y > 2\}$

c **i** horizontal asymptote $y = 0$, y-intercept 3

ii

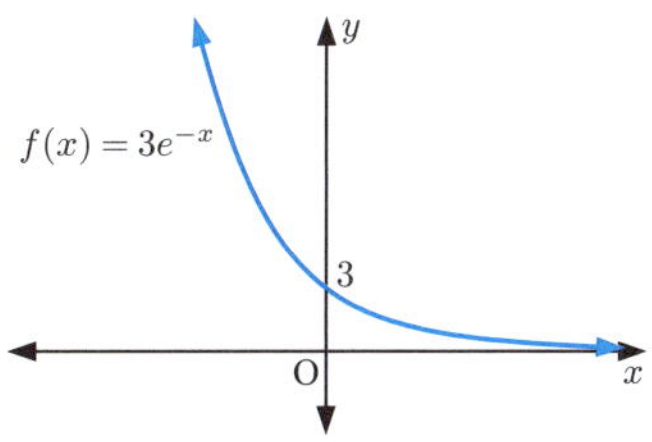

iii Domain is $\{x : x \in \mathbb{R}\}$, Range is $\{y : y > 0\}$

12 **a** C **b** B **c** E **d** A **e** D

13 **a**

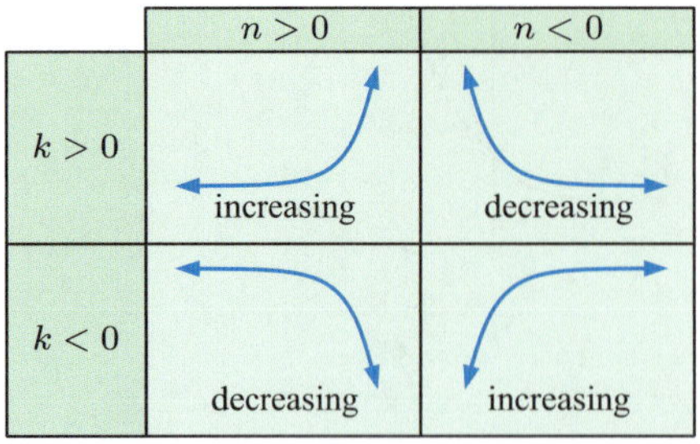

b The graph has an x-intercept if a and k have opposite signs.

14 **a**

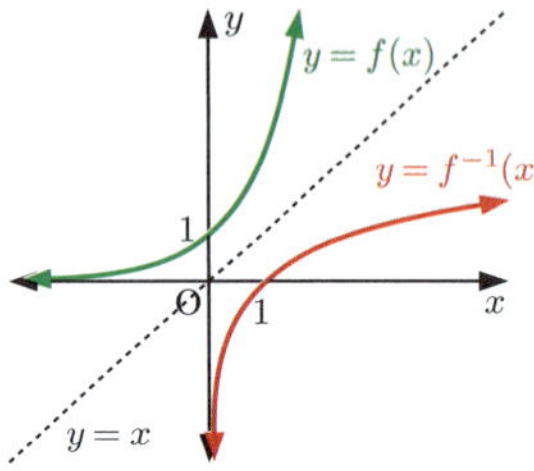

b Domain is $\{x : x > 0\}$, Range is $\{y : y \in \mathbb{R}\}$

REVIEW SET 4A

1 **a** $4\sqrt{7} - 7$ **b** $6 + 2\sqrt{5}$ **c** 1

2 **a** $\frac{2\sqrt{3}}{3}$ **b** $\frac{\sqrt{35}}{5}$ **c** $\frac{\sqrt{7}}{28}$

3 **a** $\dfrac{5+\sqrt{3}}{22}$ **b** $\dfrac{\sqrt{77}+2\sqrt{11}}{3}$ **c** $\dfrac{26+11\sqrt{2}}{7}$

d $\dfrac{-33-14\sqrt{5}}{3}$

4 **a** x^8y^3 **b** $\dfrac{4a}{3b}$ **c** $\dfrac{5c}{32d^7}$

5 **a** **i** 81 **ii** $\frac{1}{3}$ **b** $k = 9$

6 **a** 2^{4-3k} **b** $2^{3+\frac{3}{2}x}$ **7** **a** $\frac{1}{3}$ **b** 128

8 **a** $\dfrac{m}{n^2}$ **b** $\dfrac{1}{m^3n^3}$ **c** $\dfrac{m^2p^2}{n}$ **d** $\dfrac{16n^2}{m^2}$

9 **a** $9 - 6e^x + e^{2x}$ **b** $x - 4$ **c** $2^x + 1$

10 $x = \frac{9}{34} + \frac{1}{34}\sqrt{13}$

11 **a** $x = 5$ **b** $x = -\frac{1}{6}$ **c** $x = \frac{1}{3}$

12 $x = \frac{3}{2}$, $y = 4$

13 **a** $f(1) = -\frac{1}{2}$, $f(-2) = 3$ **b** $a = -4$

14 **a** 27°C **b** 24 minutes

c

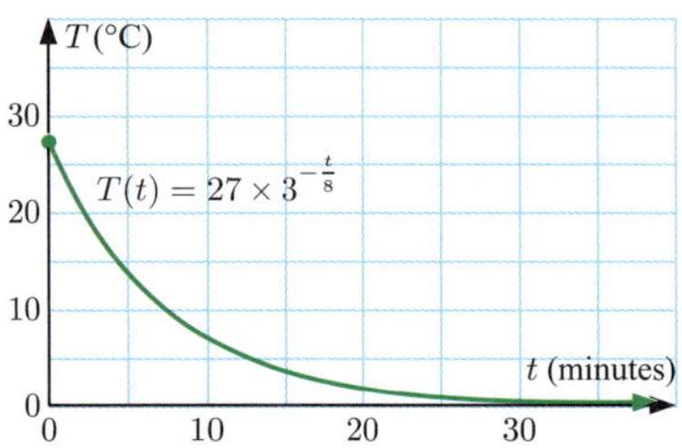

15 **a** $y = -3$ **b** -1 **c** **i** $2e^{-2} - 3$ **ii** ≈ -2.729

16 **a** $\{y : y > -3\}$ **b** $f(0) = -2$ **c** $x = \frac{1}{2}$

REVIEW SET 4B

1 **a** $-2 + 5\sqrt{2}$ **b** $19 - 8\sqrt{3}$ **c** -27

2 **a** $3 - 2\sqrt{2}$ **b** $3 - 2\sqrt{2}$ **c** $3 - 2\sqrt{2}$ **d** $3 - 2\sqrt{2}$

3 **a** a^{21} **b** p^4q^6 **c** $\dfrac{4b}{a^3}$

4 **a** 2^{-3} **b** 2^7 **c** 2^{12}

5 **a** $4m^6$ **b** $-\dfrac{a^9}{b^3}$ **c** $3x^3y^2$ **d** $16ab^{\frac{4}{5}}$

6 2^{2x} **7** **a** 5^0 **b** $5^{\frac{3}{2}}$ **c** $5^{-\frac{1}{4}}$ **d** 5^{2a+6}

8 **a** $x^2 - 1$ **b** $9^x - 5(3^x) - 14$ **c** $x^{\frac{4}{3}} - 2x + x^{\frac{2}{3}}$

9 **a** $24(5^n)$ **b** $7^x(7^x - 1)$ **c** $(2^x - 3)(2^x + 2)$

10 **a** 3^x **b** 28 **c** $2^{n+2}n$

11 **a** $x = -\frac{3}{5}$ **b** $x = 1$ **c** $x = \frac{7}{11}$

12 $x = 4$, $y = -2$ or $x = -3$, $y = \frac{8}{3}$

13

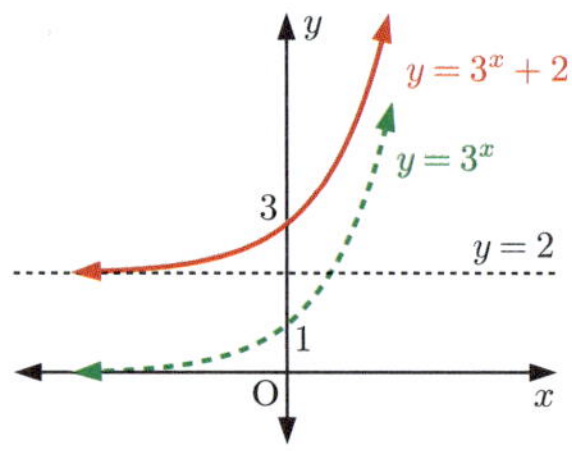

14 **a** $P_0 = 17$

b **i** ≈ 30 birds

ii 272 birds

c

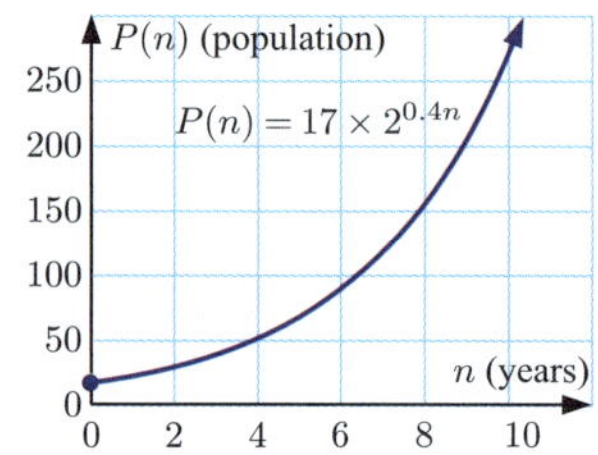

15 **a** horizontal asymptote $y = 1$, y-intercept 2

b

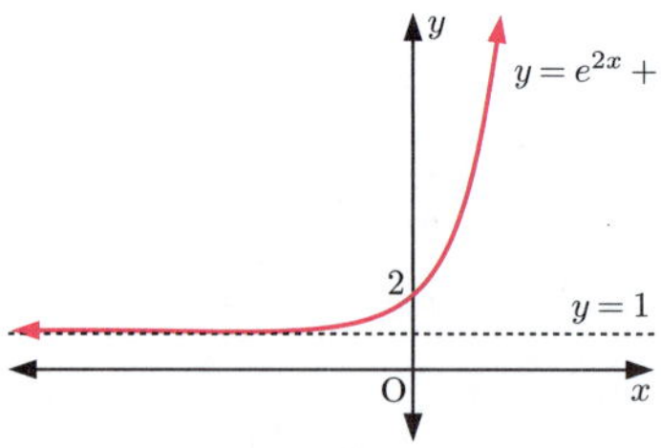

c Domain is $\{x : x \in \mathbb{R}\}$, Range is $\{y : y > 1\}$

16 **a** $\{y : y > 0\}$ **b** $\dfrac{2}{e\sqrt{e}}$ **c** $x = -1 \pm \frac{1}{3}\sqrt{3}$

EXERCISE 5A

1 **a** 2 **b** -2 **c** 4 **d** 0 **e** $\frac{1}{3}$ **f** $-\frac{1}{2}$

g $\frac{5}{2}$ **h** $\frac{5}{4}$ **i** $\frac{3}{2}$ **j** $\frac{3}{2}$ **k** $\frac{1}{2}$ **l** $\frac{3}{4}$

2 **a** x **b** $n + 3$ **c** $\dfrac{m}{n}$ **d** $m - \dfrac{1}{n}$

3 **a** $10 < 74 < 100$

$\therefore \lg 10 < \lg 74 < \lg 100$

$\therefore 1 < \lg 74 < 2$

b ≈ 1.87

4 **a** ≈ 1.415 **b** ≈ 2.766 **c** ≈ 0.699 **d** ≈ 3.154

e ≈ -0.155 **f** ≈ 1.959 **g** ≈ -1.523 **h** no solution

5 **a** ≈ 1.6128 **b** $41 \approx 10^{1.6128}$

6 **a** $\approx 10^{0.6990}$ **b** $\approx 10^{1.6990}$ **c** $\approx 10^{2.6990}$
d $\approx 10^{3.6990}$ **e** $\approx 10^{-0.3010}$ **f** $\approx 10^{-1.3010}$
g $\approx 10^{1.5798}$ **h** $\approx 10^{2.5798}$ **i** $\approx 10^{3.5798}$
j $\approx 10^{0.5798}$ **k** $\approx 10^{-0.4202}$ **l** $\approx 10^{-1.4202}$
m $\approx 10^{-2.4202}$

7 **a** $\lg x$ is positive if x is greater than 1.
b $\lg x$ is negative if x is between 0 and 1.

8 A negative number cannot be written in the form 10^b where $b \in \mathbb{R}$, so its logarithm cannot be found.

9 **a** **i** $\lg 4 \approx 0.6021$ **ii** $\lg 40 \approx 1.6021$
b $\lg 40 = \lg(4 \times 10)$
$= \lg(10^{\lg 4} \times 10^1)$
$= \lg(10^{\lg 4 + 1})$
$= \lg 4 + 1$

10 **a** **i** $\lg 6 \approx 0.7782$ **ii** $\lg(0.006) \approx -2.2218$
b $\lg(0.006) = \lg(\frac{6}{1000})$
$= \lg\left(\frac{10^{\lg 6}}{10^3}\right)$
$= \lg(10^{\lg 6 - 3})$
$= \lg 6 - 3$

11 **a** $x = 1000$ **b** $x = 10$ **c** $x = 1$
d $x = 0.000\,01$ **e** $x = \sqrt[3]{10}$ **f** $x = \frac{1}{\sqrt[3]{10}}$
g $x = 10\,000\,000$ **h** $x = \frac{1}{\sqrt[4]{10}}$ **i** $x \approx 2.23$
j $x \approx 10\,600$ **k** $x \approx 0.119$ **l** $x \approx 0.001\,13$

12 **a** $x = 7$ **b** $x = 1$ **c** $x = \pm\frac{1}{10}$

EXERCISE 5B

1 **a** $10^3 = 1000$ **b** $3^2 = 9$ **c** $2^4 = 16$
d $7^0 = 1$ **e** $5^{-1} = \frac{1}{5}$ **f** $2^{\frac{1}{2}} = \sqrt{2}$
g $6^{-2} = \frac{1}{36}$ **h** $3^{-\frac{1}{2}} = \frac{1}{\sqrt{3}}$ **i** $4^{\frac{3}{2}} = 8$

2 **a** $\log_3 27 = 3$ **b** $\log_2 8 = 3$ **c** $\log_9 81 = 2$
d $\log_4 4 = 1$ **e** $\log_5 1 = 0$ **f** $\log_2(\frac{1}{8}) = -3$
g $\log_7(7\sqrt{7}) = \frac{3}{2}$ **h** $\log_{16} 8 = \frac{3}{4}$ **i** $\log_9(\frac{1}{3}) = -\frac{1}{2}$

3 **a** 5 **b** $\frac{1}{5}$

4 **a** 2 **b** 3 **c** -1 **d** $\frac{1}{2}$ **e** 2 **f** $-\frac{1}{2}$
g -5 **h** $\frac{1}{2}$ **i** 0 **j** 1 **k** $\frac{5}{2}$ **l** $\frac{1}{2}$
m $\frac{1}{3}$ **n** $-\frac{1}{2}$ **o** $\frac{2}{3}$ **p** $\frac{5}{2}$ **q** $-\frac{3}{2}$ **r** $-\frac{1}{2}$

5 **a** $x = 243$ **b** $x = \frac{1}{32}$ **c** $x = 16$ **d** $x = \frac{1}{3}$
e $x = 64$ **f** $x = \frac{1}{9}$ **g** $x = -\frac{3}{2}$ **h** $x = \frac{5}{3}$
i $x = \frac{1+\sqrt{5}}{2}$ **j** $x = -\frac{1}{5}$ **k** $x = 9$ or -7
l $x = 16$ or -11

6 **a** 3 **b** -2 **c** $\frac{1}{3}$ **d** $-\frac{1}{2}$

EXERCISE 5C

1 **a** $\lg 15$ **b** $\lg 14$ **c** $\lg 4$ **d** $\lg 9$
e $\lg 6$ **f** 2 **g** 3 **h** -1
i $\lg 30$ **j** $\lg(\frac{4}{5})$ **k** -3 **l** $\lg 90$

2 **a** $\lg 400$ **b** $\lg(\frac{3}{10})$ **c** $\log_3 15$
d $\log_2(\frac{7}{4})$ **e** $\lg 5000$ **f** $\lg(\frac{1}{125})$
g $\lg(m \times 10^k)$ **h** $\log_a(11a^2)$ **i** $\log_3(\frac{81}{20})$

3 **a** $\lg 8$ **b** $\lg(\frac{1}{25})$ **c** $\lg 80$ **d** $\lg 18$ **e** 2
f -2 **g** $\lg 300$ **h** $\lg 7$ **i** $\lg(\frac{1250}{9})$

4 **a** $\frac{3}{2}$ **b** $\frac{2}{3}$ **c** $\frac{3}{4}$ **d** $\frac{1}{8}$ **e** -3 **f** $-\frac{1}{3}$

5 **a** 2 **b** -1 **c** 1 **d** 2

6 **a** $4\lg x$ **b** $3\lg 2x$ **c** $-5\lg 3y$

7 **a** $y + z$ **b** $x + 2y$ **c** $x + z - y$
d $3z + \frac{1}{2}x$ **e** $\frac{1}{3}y - 2z$ **f** $x + \frac{1}{2}z - 4y$

8 **a** $q + r$ **b** $2p + r$ **c** $p + q + r$
d $q - 2p$ **e** $\frac{1}{2}r - 2p - q$ **f** $2q - 3p - 3r$

9 $\log_b Q = 3$

10 **a** $\log_t A + 3\log_t B = 15$, $2\log_t A - \log_t B = 9$
b $\log_t A = 6$, $\log_t B = 3$ **c** $\log_t(B^5\sqrt{A}) = 18$
d $B = t^3$

EXERCISE 5D.1

1 **a** $\lg y = x \lg 2$ **b** $\lg y = 3\lg x$
c $\lg M = 4\lg d$ **d** $\lg T = x\lg 5$
e $\lg y = \frac{1}{2}\lg x$ **f** $\lg y = \lg 7 + x\lg 3$
g $\lg S = \lg 9 - \lg t$ **h** $\lg M = 2 + x\lg 7$
i $\lg T = \lg 5 + \frac{1}{2}\lg d$ **j** $\lg F = 3 - \frac{1}{2}\lg n$
k $\lg S = \lg 200 + t\lg 2$ **l** $\lg y = \frac{1}{2}\lg 15 - \frac{1}{2}\lg x$

2 **a** $y = 7^x$ **b** $D = 2x$ **c** $F = \dfrac{5}{t}$ **d** $y = 6 \times 2^x$
e $P = \sqrt{x}$ **f** $N = \dfrac{1}{\sqrt[3]{p}}$ **g** $P = 10x^3$ **h** $y = \dfrac{10^x}{2}$
i $y = \dfrac{x^2}{10}$ **j** $T = 2k^5$ **k** $P = \dfrac{n^4}{9}$ **l** $y = 8 \times 16^x$

3 **a** $y = \dfrac{x^3}{2}$ **b** **i** $y = 4$ **ii** $y = 32$

4 **a** $y = 100(10^{\frac{1}{3}x})$ **b** **i** $y = 100$ **ii** $y = 1000$

5 **a** If there is a *power* relationship between y and x, for example $y = 5x^3$, then there is a *linear* relationship between $\lg y$ and $\lg x$.
b If there is an *exponential* relationship between y and x, for example $y = 4 \times 2^x$, then there is a *linear* relationship between $\lg y$ and x.

EXERCISE 5D.2

1 **a** $x = 25$ **b** $x = 67$ **c** $x = 20$ **d** $x = \frac{125}{64}$
e $x = 5$ **f** no solution **g** $x = \frac{9}{8}$ **h** no solution

2 **a** $x = 5$ **b** $x = 3$ or 6 **c** $x = 2$ or 4 **d** $x = 2$
e $x = 1$ **f** no solution **g** $x = 2$ **h** $x = 4$

3 **a** $x = 8$ **b** $x = 3$ **c** $x = 6$ **d** $x = 4$

4 **a** $x = 8$, $y = \frac{1}{2}$ **b** $x = 3$, $y = 5$ or $x = \frac{3}{2}$, $y = \frac{19}{2}$

EXERCISE 5E.1

1 **a** 3 **b** 5 **c** -4 **d** $\frac{1}{2}$ **e** k
f $2 + k$ **g** $a - b$ **h** $\frac{3}{2}$

2 **a** 5 **b** 8 **c** $\frac{1}{4}$ **d** $\frac{1}{9}$

3 **a** ≈ 2.303 **b** ≈ 4.248 **c** ≈ 0.693 **d** ≈ -1.204
e ≈ 6.685

4 There is no real value of x such that $e^x = -5$.

5 **a** $e^{3.4012}$ **b** $e^{1.9459}$ **c** $e^{4.5433}$ **d** $e^{-2.3026}$
e $e^{6.2146}$

6 **a** $x \approx 7.39$ **b** $x \approx 2.72$ **c** $x = 1$
d $x \approx 0.0498$ **e** $x \approx 0.002\,48$ **f** $x \approx 2.10$
g $x \approx 4.57$ **h** $x \approx 0.0813$

7 **a** $x \approx 6.72$ **b** $x \approx -0.211$ **c** no solution
d $x = -1$ or -3

8 **a** $k > -e^2$ **b** $k = -e^2$ **c** $k < -e^2$

9 $x = 1,\ y = 4$ or $x = 4,\ y = 1$

EXERCISE 5E.2

1 **a** $\ln 20$ **b** $\ln 5$ **c** $\ln 42$ **d** $\ln 7$
e $\ln(\frac{9}{2})$ **f** $\ln(\frac{6}{11})$ **g** $\ln 70$ **h** $\ln 6$
i 0 **j** $\ln(5e^2)$ **k** $\ln\left(\dfrac{8}{e}\right)$ **l** $\ln\left(\dfrac{e^4}{10}\right)$

2 **a** $\ln 125$ **b** $\ln(\frac{1}{16})$ **c** $\ln 5$ **d** $\ln 54$
e $\ln(\frac{16}{27})$ **f** $\ln 144$ **g** $\ln(\frac{81}{4})$ **h** $\ln 3$
i $\ln 8$ **j** $\ln(\frac{7}{25})$ **k** $\ln 4$ **l** $\ln 225$

3 **Hint:** In **d**, $\ln\left(\dfrac{e^2}{8}\right) = \ln e^2 - \ln 2^3$.

4 $\ln\left(\dfrac{3\sqrt{3}}{e^5}\right)$

5 **a** $D = ex$ **b** $F = \dfrac{e^2}{p}$ **c** $P = 5e^{2x}$
d $M = e^3y^2$ **e** $B = \frac{1}{4}e^{3t}$ **f** $N = \dfrac{1}{\sqrt[3]{g}}$
g $K = 3e^{5x}$ **h** $Q \approx 8.66x^3$ **i** $D \approx 0.518n^{0.4}$
j $T \approx \dfrac{4.85}{e^x}$

EXERCISE 5F

1 **a** $x \approx 4.64$ **b** $x \approx 2.29$ **c** $x \approx -0.325$
d $x \approx 3.68$ **e** $x \approx 5.68$ **f** $x \approx 6.34$

2 **a** $x \approx 1.43$ **b** $x \approx 1.56$ **c** $x \approx 3.44$
d $x \approx 5.82$ **e** $x \approx -1.34$ **f** $x \approx 2.37$
g $x \approx 0.275$ **h** $x \approx 1.81$ **i** $x \approx 9.64$

3 **a** $x = \ln 8$ **b** $x = \ln 50$ **c** $x = \ln 13$
d $x = 2\ln 12$ **e** $x = \ln 10 + 3$ **f** $x = \frac{1}{2}\ln 6$
g $x = \frac{1}{3}(\ln 20 + 1)$ **h** $x = \frac{1}{5}\ln 5$ **i** $x = 2\ln(\frac{6}{5})$

4 **a** $x = \frac{1}{2}\ln 300$ **b** $x \approx 2.85$

5 **a** $x = -\dfrac{\lg(0.03)}{\lg 2}$ **b** $x = \dfrac{10\lg\left(\frac{10}{3}\right)}{\lg 5}$ **c** $x = \dfrac{4\lg 8}{\lg 3}$

6 **b** $t \approx 6.93$ hours

7 **a** 50 g **b** $\approx 13\,200$ years

8 **a**

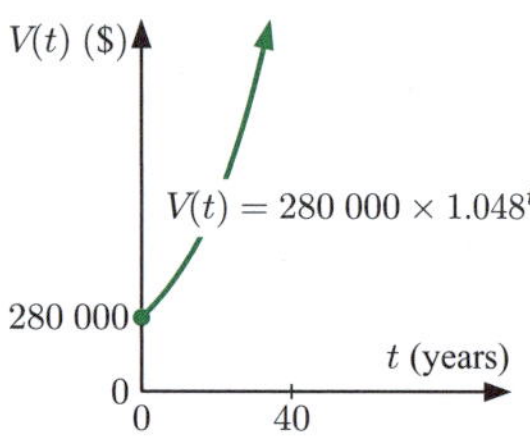

b ≈ 7.6 years

c $t = \dfrac{\lg\left(\frac{V}{280\,000}\right)}{\lg 1.048}$

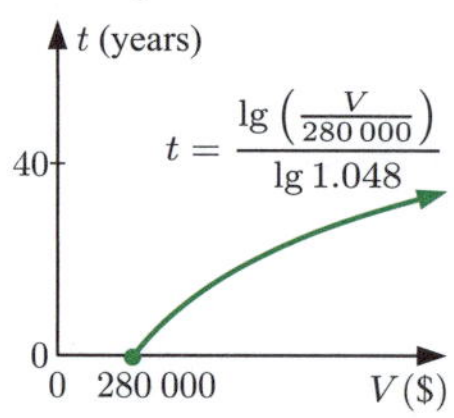

9 **a** $x = \ln 2$ **b** $x = 0$ **c** $x = \ln 2$ or $\ln 3$ **d** $x = 0$
e $x = \ln 4$ **f** $x = \ln\left(\frac{3+\sqrt{5}}{2}\right)$ or $\ln\left(\frac{3-\sqrt{5}}{2}\right)$

10 **a** $(\ln 2, 5)$ **b** $(\ln 3, 3)$ **c** $(0, 2)$ and $(\ln 5, -2)$
d $(\ln(\frac{1}{2}), 11)$ and $(\ln 3, 7\frac{2}{3})$

EXERCISE 5G

1 **a** ≈ 3.58 **b** ≈ 4.43 **c** ≈ -2.46 **d** ≈ 5.42

2 **a** $x \approx -4.29$ **b** $x \approx 3.87$ **c** $x \approx 0.139$

3 **a** $\log_9 26 = \frac{1}{2}\log_3 26$ **b** $\log_2 11 = 2\log_4 11$
c $\dfrac{6}{\log_7 25} = 3\log_5 7$

4 **Hint:** $\log_3 4 = \dfrac{\log_2 4}{\log_2 3}$ **5** $\log_2 7$ **6** $\log_a 5$

7 **a** $x = \sqrt[3]{50}$ **b** $x = \sqrt{13}$ **c** $x = 49$ **d** $x = 5$
e $x = 8$ **f** $x = 16$

8 **b** **i** $x = \frac{1}{9}$ or 9 **ii** $x = \frac{1}{2}$ or 32 **iii** $x = 2$ or 64

9 $x = \log_6 18$ **10** **b** $3\log_{14} 5$

EXERCISE 5H

1 **a** $x = -4$ **b** x-intercept -3, y-intercept 2
c Domain is $\{x : x > -4\}$, Range is $\{y : y \in \mathbb{R}\}$

2 **a** **i** $b = -1$ **ii** $a = 1$ **iii** $k = 3$
b Domain is $\{x : x > -1\}$, Range is $\{y : y \in \mathbb{R}\}$
c **i** $f(8) = 6$ **ii** $x \approx 0.442$

3 **a** Domain is $\{x : x > 3\}$, Range is $\{y : y \in \mathbb{R}\}$
b $x = 3$ **c** 4 **d** $x = 7$
e

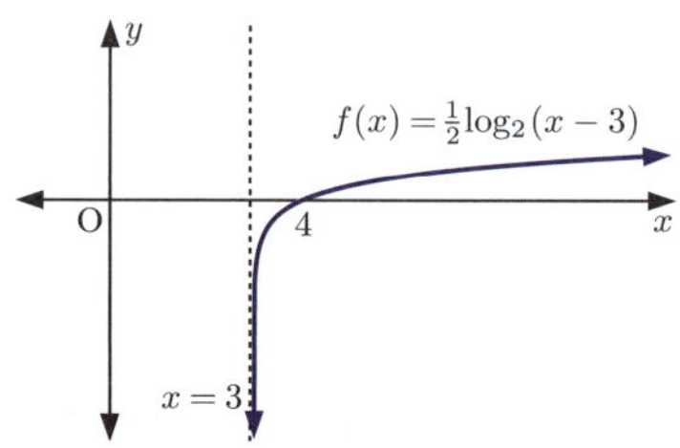

f $f^{-1}(x) = 2^{2x} + 3$

4 **a** Domain is $\{x : x > \frac{1}{2}\}$, Range is $\{y : y \in \mathbb{R}\}$
b $x = \frac{1}{2}$ **c** 1 **d** $x = 3$
e $f^{-1}(x) = \frac{1}{2}(5^x + 1)$
f

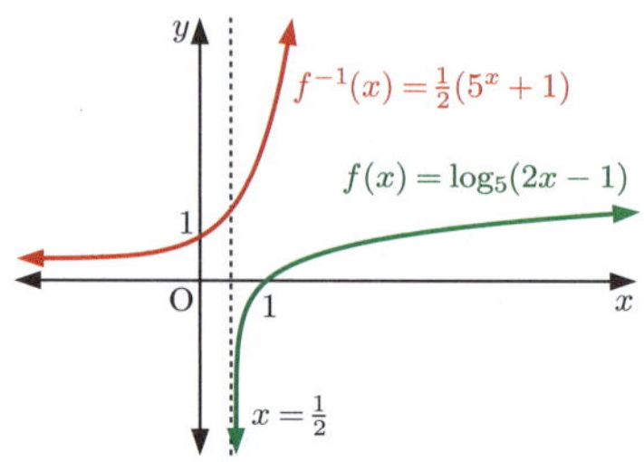

5 **a** $f^{-1}(x) = \log_3(x - 1)$
b f: Domain is $\{x : x \in \mathbb{R}\}$, Range is $\{y : y > 1\}$
f^{-1}: Domain is $\{x : x > 1\}$, Range is $\{y : y \in \mathbb{R}\}$

c

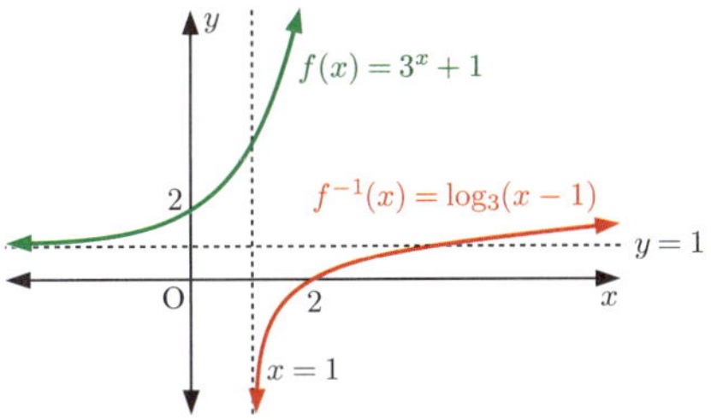

6 **a** **i** $f(5) = 3$ **ii** $f(x^2) = \log_2(x^2 + 3)$
iii $f(2x-1) = 1 + \log_2(x+1)$
b $\{x : x > -3\}$ **c** $x = \pm 5$

7 **a** $x = -2$ **b** P$(-1, 0)$ **c** $x = e^{\frac{2}{3}} - 2$
d Domain is $\{x : x > -2\}$, Range is $\{y : y \in \mathbb{R}\}$

8 **a** Domain is $\{x : x > 1\}$, Range is $\{y : y \in \mathbb{R}\}$
b $\frac{4}{3}$ **c** $f^{-1}(x) = \frac{1}{3}e^x + 1$
d

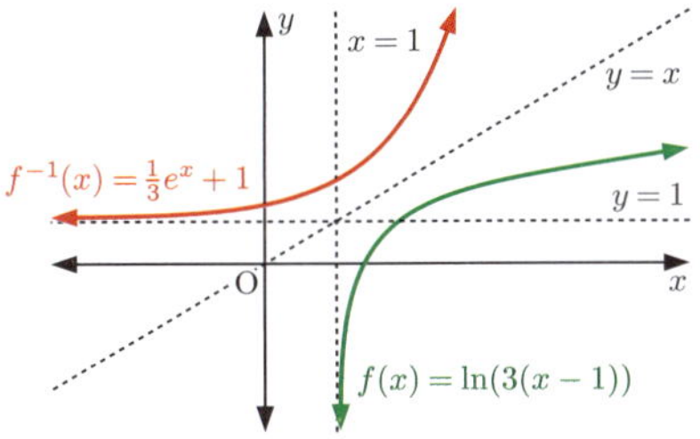

e Domain is $\{x : x \in \mathbb{R}\}$, Range is $\{y : y > 1\}$

9 **a** **B** **b** **D** **c** **A** **d** **C**

10 **a** Domain is $\{x : x > 2\}$, Range is $\{y : y \in \mathbb{R}\}$
b $x = 2$ **c** $\frac{5}{2}$
d

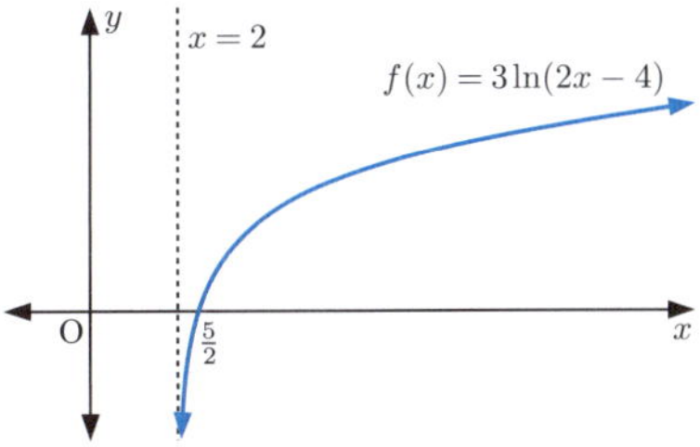

11 **a** Domain of $f(x)$ is $\{x : x > 1\}$
Range of $f(x)$ is $\{y : y \in \mathbb{R}\}$
Domain of $g(x)$ is $\{x : x > -5\}$
Range of $g(x)$ is $\{y : y \in \mathbb{R}\}$
b $f(x)$ has x-intercept 2, no y-intercept.
$g(x)$ has x-intercept -4, y-intercept $\ln 5$.
c $(4, \ln 9)$

12 **a** Domain is $\{x : x > 1\}$, Range is $\{y : y \in \mathbb{R}\}$
b $f(10) = \frac{1}{3}\ln 9 \approx 0.732$ **c** $x = e^6 + 1$
d $f^{-1}(x) = e^{3x} + 1$
Domain is $\{x : x \in \mathbb{R}\}$, Range is $\{y : y > 1\}$
e $(f \circ f^{-1})(x) = (f^{-1} \circ f)(x) = x$

13 **a** $f^{-1}(x) = \frac{1}{2}\ln x$
i $(f^{-1} \circ g)(x) = \frac{1}{2}\ln(2x-1)$
ii $(g \circ f)^{-1}(x) = \frac{1}{2}\ln\left(\frac{x+1}{2}\right)$
b $x = 13$

14 **a** $f(1) = \frac{10}{e}$, $g(6) = \ln 3$ **b** 4
c $fg(x) = \frac{10}{x-3}$ **d** $x = \ln 2$

15 **a** $\{x : x > -6\}$ **b** $f^{-1}(x) = e^x - 6$
c x-intercept -5, y-intercept $\ln 6$
d $x = e^3 + \ln 3 - 6 \approx 15.2$ **e** $x = -\frac{8}{3}$ or 3

REVIEW SET 5A

1 **a** 2 **b** $1 + k$ **c** $\frac{1}{3}$ **d** $-\frac{5}{2}$

2 **a** $\log_3(\frac{1}{81}) = -4$ **b** $\log_8 16 = \frac{4}{3}$

3 **a** $\lg 54$ **b** $\log_2 7$ **c** $\lg 8000$ **4** -1

5 **a** $\lg P = \lg 3 + x \lg 7$ **b** $\lg m = 3\lg n - \lg 5$

6 **a** $x = 3$ **b** $x = 5$

7 **Hint:** Use change of base rule.

8 **a** $T = \frac{x^2}{5}$ **b** $K = 3 \times 2^x$

9 **a** $5\ln 2$ **b** $3\ln 5$ **c** $6\ln 3$

10

Function	$y = \log_2 x$	$y = \ln(x+5)$
Domain	$x > 0$	$x > -5$
Range	$y \in \mathbb{R}$	$y \in \mathbb{R}$

11 **a** $2A + 2B$ **b** $A + 3B$ **c** $3A + \frac{1}{2}B$
d $4B - 2A$ **e** $3A - 2B$

12 **a** $x = 0$ or $\ln\left(\frac{2}{3}\right)$ **b** $x = e^2$

13 **a** $x \approx 2.46$ **b** $x \approx 1.88$ **14** $(\lg 8)^2$

15 **a** $\{y : y > 1\}$
b **i** $f^{-1}(x) = \ln\left(\frac{5}{x-1}\right)$ **ii** $f^{-1}(2) = \ln 5$
c $\{x : x > 1\}$ **d** $x = 6$
e

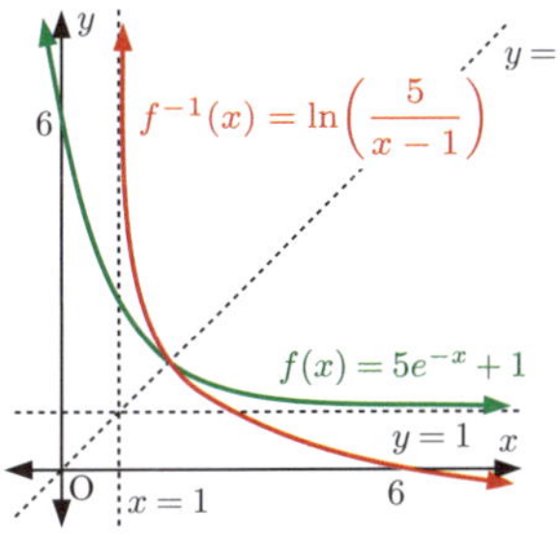

REVIEW SET 5B

1 **a** $10^{1.204}$ **b** $10^{-1.620}$ **c** $10^{3.863}$

2 **a** $x = \sqrt{10}$ **b** $x = \frac{1}{\sqrt{3}}$ **c** $x = \frac{1}{e^2} + 3$

3 **a** 6 **b** $3 - n$ **c** $t + \frac{1}{2}$

4 **a** $k \approx 3.25 \times 2^x$ **b** $Q = 5P^3$ **c** $A = 6 \times 2^x$

5 **a** $x = \frac{\lg 70}{\lg 3}$ **b** $x = 2$

6 -1 **7** $\log_8 30 = \frac{1}{3}\log_2 30$

8 **a** $x = 8$ **b** $x = 3$ **9** **a** 9 **b** $\ln 5$

10 **a** $\lg M = \lg 5 + x\lg 6$ **b** $\lg T = \lg 5 - \frac{1}{2}\lg l$
c $\lg G = \lg 4 - \lg c$

11 **a** $x = \ln 3$ **b** $x = \ln 3$ or $\ln 4$

12 **a** $x = -2$
b Domain is $\{x : x > -2\}$, Range is $\{y : y \in \mathbb{R}\}$
c -1 **d** $\ln 8$ **e** $x = e^2 - 2 \approx 5.389$

13 **a** Domain is $\{x : x > -2\}$, Range is $\{y : y \in \mathbb{R}\}$
b VA is $x = -2$, x-intercept is 7, y-intercept is ≈ -1.37
c $g^{-1}(x) = 3^{x+2} - 2$
Domain is $\{x : x \in \mathbb{R}\}$, Range is $\{y : y > -2\}$
d

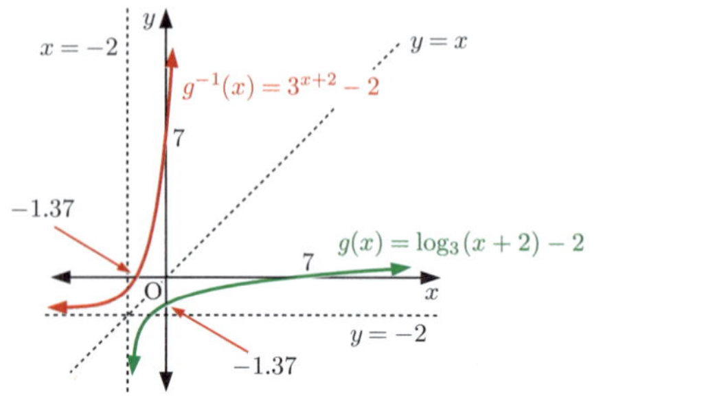

14 **a** $x = 5$ **b** $x = 32$ or $\frac{1}{32}$ **c** $x = 9$ or 81

15 **a** Domain is $\{x : x > 4\}$, Range is $\{y : y \in \mathbb{R}\}$
b x-intercept is 5, no y-intercept **c** $x = 4 + \sqrt{6}$
d $x = 0$

EXERCISE 6A

1 **a** polynomial **b** polynomial
c not a polynomial, one term has x as the index
d polynomial
e not a polynomial, the index in the second term is not an integer
f not a polynomial, the last term has a negative index

2 **a** **i** 2 **ii** 1 **iii** 4 **b** **i** 5 **ii** 2 **iii** 3
c **i** 6 **ii** -2 **iii** -7 **d** **i** 4 **ii** $\frac{1}{2}$ **iii** $-\frac{2}{3}$

EXERCISE 6B.1

1 **a** $3x^2 + 9x - 6$ **b** $-8x^3 + 4x + 16$
c $-x^2 - 3x + 2$ **d** $6x^3 - 3x - 12$
e $-2x^3 + x^2 + 4x + 2$ **f** $-2x^3 - x^2 - 2x + 6$

2 **a** **i** $3x^2 + 3x - 3$ **ii** $-x^2 - 3x + 5$
b **i** $3x^3 - 5x^2 + x + 7$ **ii** $-x^3 - 3x^2 + 3x - 5$
c **i** $-5x^4 + x^3 - 9x^2 - x - 4$
ii $-5x^4 - x^3 + 7x^2 + 3x + 4$
d **i** $2x^4 - 4x^3 + 2x + 5$ **ii** $2x^4 + 2x^3 + 12x - 11$

3 **a** $x^3 - 6x^2 + 2x - 13$ **b** $3x^3 - 4x^2 - x - 11$
c $-2x^3 + 6x^2 - x + 14$

EXERCISE 6B.2

1 **a** $3x^2 + 5x - 2$ **b** $2x^3 - 9x^2 + x + 12$
c $4x^5 - 21x^4 + 21x^3 - 2x^2 - 10x + 8$
d $-2x^7 - x^5 + 17x^4 + x^3 + 2x^2 - 35x$

2 **a** $3x^4 - 9x^3 + 12x - 3$ **b** $-10x^3 + 15x^2 - 30$
c $2x^4 - 9x^2 + 8x + 16$
d $-x^4 + 11x^3 - 12x^2 - 4x + 25$
e $2x^7 - 9x^6 + 9x^5 + 14x^4 - 32x^3 + 3x^2 + 24x - 6$
f $4x^6 - 12x^5 + 9x^4 + 24x^3 - 36x^2 + 36$

3 **a** $6x^3 + 3x^2 - 8x - 4$ **b** $-x^4 + 3x^3 - 2x^2 + 5x + 3$
c $-x^4 + 4x^3 + 3x^2 - 20x + 10$
d $-2x^5 + 11x^4 - 14x^3 - 2x^2 + x$

4 **a** leading coefficient is 2, constant term is -8
b coefficient of x is -2, degree is 4
c coefficient of x^2 is 2, constant term is 20
d degree is 5, leading coefficient is 6

EXERCISE 6C

1 **a** no **b** no **c** no **d** yes

2 **a** $-4, 2$ **b** $-8, 2$ **c** $-\frac{1}{3}, 2$ **d** $0, \pm 2$
e $0, \pm\sqrt{11}$ **f** $\pm 2, \pm\sqrt{2}$

3 **a** $x = -5$ or 3 **b** $x = -\frac{1}{2}$ or $\pm\sqrt{3}$
c $x = -3, \frac{1}{3}$, or 2 **d** $x = 0$ or $1 \pm \sqrt{3}$
e $x = 0$ or $\pm\sqrt{7}$ **f** $x = \pm\sqrt{2}$ or $\pm\sqrt{5}$

5 **a** $(2x+3)(x-5)$ **b** $x(x-7)(x-4)$
c $(x - 3 - \sqrt{6})(x - 3 + \sqrt{6})$
d $x(x + 1 + \sqrt{5})(x + 1 - \sqrt{5})$ **e** $x(3x-2)(2x+1)$
f $(x+1)(x-1)(x+\sqrt{5})(x-\sqrt{5})$

6 $P(\alpha) = 0$, $P(\beta) = 0$, $P(\gamma) = 0$

7 **a** $P(x) = ax(x-3)(x+4)$, $a \neq 0$
b $P(x) = a(x+1)(x-1)(x-5)$, $a \neq 0$
c $P(x) = a(x-3)(x^2 - 4x + 1)$, $a \neq 0$
d $P(x) = a(4x+1)(x^2 + 2x - 6)$, $a \neq 0$

8 **a** $P(x) = a(x^2 - 1)(x^2 - 2)$, $a \neq 0$
b $P(x) = a(x-2)(5x+1)(x^2 - 3)$, $a \neq 0$
c $P(x) = a(x+3)(4x-1)(x^2 - 2x - 1)$, $a \neq 0$
d $P(x) = a(x^2 - 4x - 1)(x^2 + 4x - 3)$, $a \neq 0$

EXERCISE 6D

1 **a** $a = 5$, $b = -2$ **b** $a = 4$, $b = 3$
c $a = 3$, $b = 8$ **d** $a = 2$, $b = 5$

2 **a** $a = 3$, $b = -5$ **b** $a = 2$, $b = 1$

3 **a** $a = 1$, $b = 6$, $c = 8$ **b** $a = 2$, $b = -1$, $c = 3$
c $a = 2$, $b = 5$, $c = -4$

4 $a = 1$, $b = 4$, $c = -3$, $k = 22$

5 **a** $a = 1$, $b = 6$, $c = -7$ **b** $(x+3)(x+7)(x-1)$

6 **a** $p = 2$, $q = 7$, $r = 5$ **b** $x = \frac{1}{2}, -1$, or $-\frac{5}{2}$

7 **a** $a = 3$, $b = -2$, $c = 1$
b $3x^3 + 10x^2 - 7x + 4 = (x+4)(3x^2 - 2x + 1)$
$3x^2 - 2x + 1$ has $\Delta = -8$
$\therefore$ the only real zero is -4.

8 **a** $(x+5)(x+2)(x-3)$ **b** $(x-2)^2(x+4)$
c $(x+4)(2x+3)(x-3)$ **d** $(3x-2)(2x-3)(x+4)$

9 $k = 27$, $(x+3)^3$

10 **a** $(2x+1)(x^2 - 3x + 5)$
b $x^2 - 3x + 5$ has $\Delta = 9 - 20 < 0$
$\therefore$ the only real zero is $-\frac{1}{2}$
$\therefore$ there is only one real linear factor.

11 $a = -2$, $b = 3$
$x = -3, -2, -1$, and 5 are the roots of the equation.

12 $m = 2$, $n = -1$, $(x+1)^2(x^2 - x + 1)$

13 $p = -4$, $q = 2$
$x^2 - 4x + 5$ has $\Delta = 16 - 20 < 0$
$x^2 + 2x + 2$ has $\Delta = 4 - 8 < 0$
$\therefore$ there are no real linear factors.

EXERCISE 6E

1 **a** quotient is $x+1$, remainder is 2
$x^2+5x+6=(x+1)(x+4)+2$
b quotient is $2x+1$, remainder is 3
$2x^2-3x+1=(2x+1)(x-2)+3$
c quotient is $x-4$, remainder is 14
$x^2-x+2=(x-4)(x+3)+14$
d quotient is x^2+x+3, remainder is 5
$x^3+3x^2+5x+11=(x^2+x+3)(x+2)+5$
e quotient is x^2+3x-4, remainder is -5
$x^3+2x^2-7x-1=(x^2+3x-4)(x-1)-5$
f quotient is x^2-2, remainder is 11
$x^3+4x^2-2x+3=(x^2-2)(x+4)+11$

2 **a** $x-1+\dfrac{2}{x-3}$ **b** $3x-2-\dfrac{3}{x+1}$
c $x+2+\dfrac{13}{x-2}$ **d** $x^2-5x-14-\dfrac{63}{x-4}$
e $x^2-2x+9-\dfrac{22}{x+2}$
f $2x^3-x^2+4x-10+\dfrac{26}{x+3}$

3 **a** $(x+3)(x+1)(x-2)$ **b** $(x-3)(x+4)^2$
c $(x+3)(3x+1)(2x-5)$

EXERCISE 6F

1 **a** 11 **b** -10 **c** 46 **d** 5
2 **a** $a=-2$ **b** $a=4$ **c** $a=5$
3 **a** $a=-4,\ b=7$ **b** $a=-5,\ b=6$
4 **a** $a=2,\ b=-5$ **b** 79
5 **a** $-2m^2+8m+6$ **b** $2-\sqrt{7}<m<2+\sqrt{7}$
6 **a** **i** $-10\frac{7}{8}$
ii $P(-\frac{1}{2})=-10\frac{7}{8}$ which is the same result as **a i**
b $\dfrac{P(x)}{2x+1}=Q(x)+\dfrac{R}{2x+1}$
$\therefore\ P(x)=Q(x)\,(2x+1)+R$
$\therefore\ P(-\frac{1}{2})=Q(-\frac{1}{2})(0)+R$
$\therefore\ R=P(-\frac{1}{2})$
c **i** 7 **ii** $\frac{4}{9}$ **iii** $-\frac{1}{4}$

EXERCISE 6G

1 **a** factor **b** not a factor **c** factor **d** not a factor
2 **a** $c=2$ **b** $c=-2$ **c** $b=3$
3 **a** $a=1,\ b=10$ **b** $p=\frac{12}{7},\ q=\frac{75}{7}$ **c** $m=2,\ n=7$
4 $a=3,\ b=2$ **5** $a=-1,\ b=-7$
6 **a** $k=-8$ **b** $P(x)=(x-3)(3x^2+x-2)$
c $x=-1,\ \frac{2}{3},\ 3$
7 **a** $f(3)=0$ **b** $f(x)=(x-3)(x^2+x-20)$
c $f(x)=(x-3)(x-4)(x+5)$
8 **a** $P(a)=0\ \therefore\ x-a$ is a factor of $P(x)$
b $(x-a)(x^2+ax+a^2)$
9 **a** $P(-a)=0\ \therefore\ x+a$ is a factor of $P(x)$
b $(x+a)(x^2-ax+a^2)$
10 $n=3$

EXERCISE 6H

1 **a** $x=1,\ 2,\ 3$ **b** $x=-1,\ 2$ {2 is a double root}
c $x=1,\ -1,\ -2$ **d** $x=-1,\ 3,\ 4$ **e** $x=-5,\ -4,\ 4$
f $x=-3,\ -5$ {-5 is a double root}
2 **a** $x=-3,\ -2,\ 3$ **b** $x=-3,\ 2,\ 5$ **c** $x=3$ or $\frac{5\pm\sqrt{57}}{2}$
3 **a** $x=-2,\ 2,\ 3$ **b** $x=-3,\ -2,\ 6$ **c** $x=-3,\ 4,\ 7$

EXERCISE 6I

1 **a**
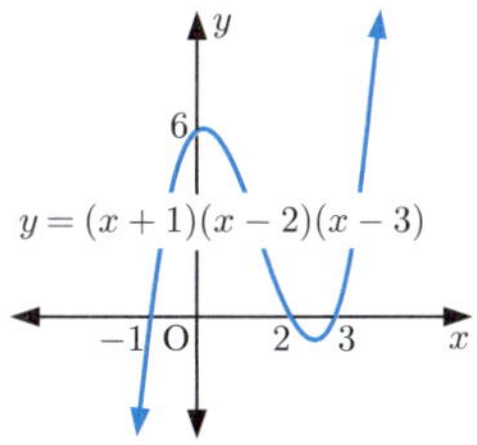

b
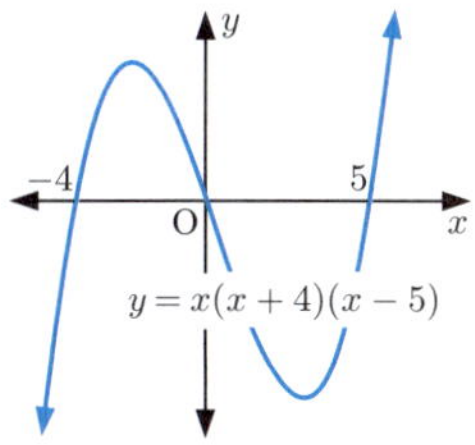

c
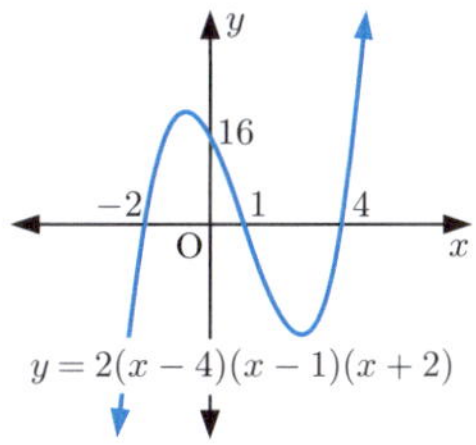

d
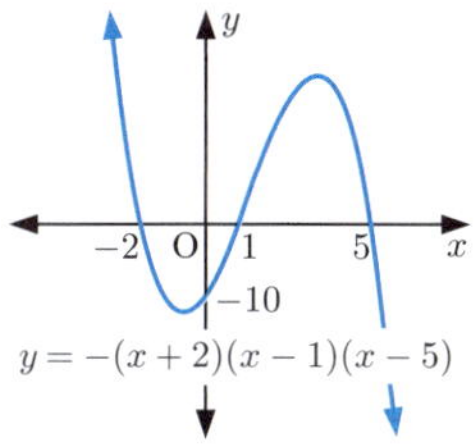

e
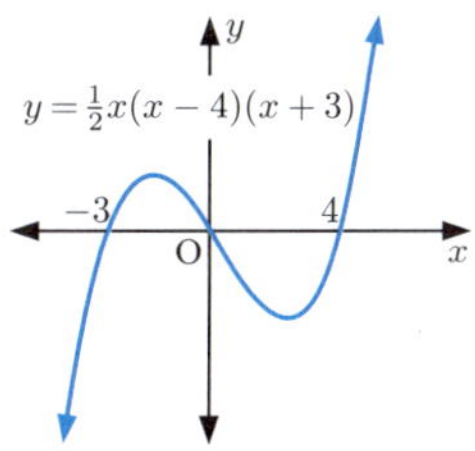

f
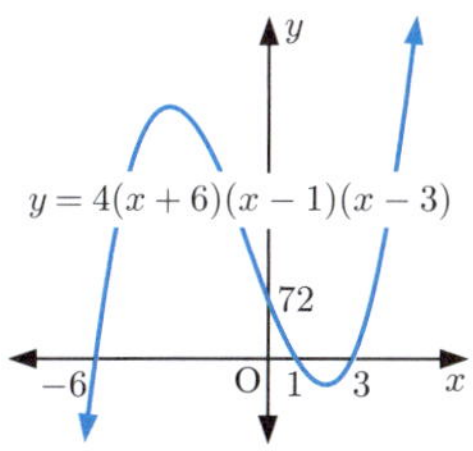

g
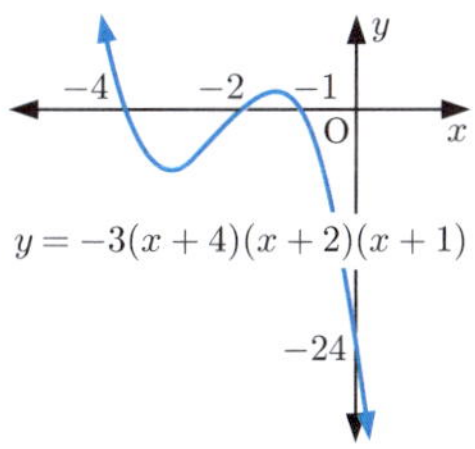

h
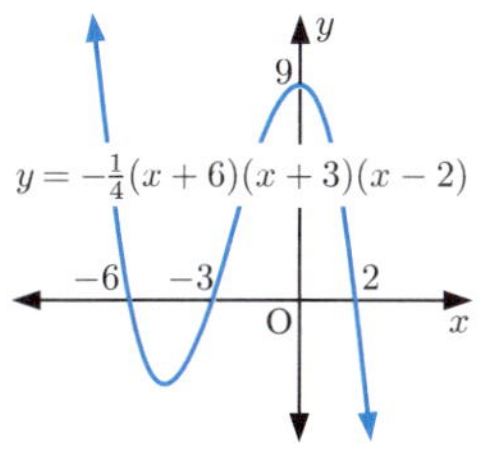

i
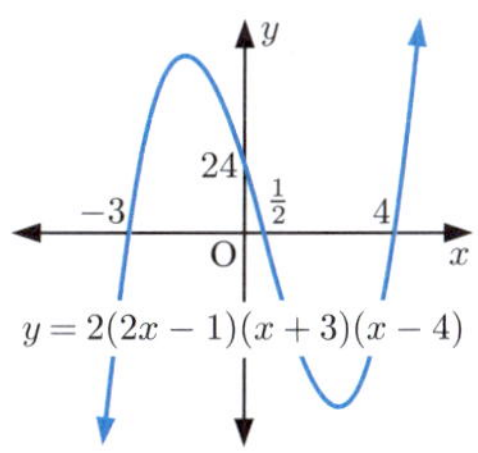

j
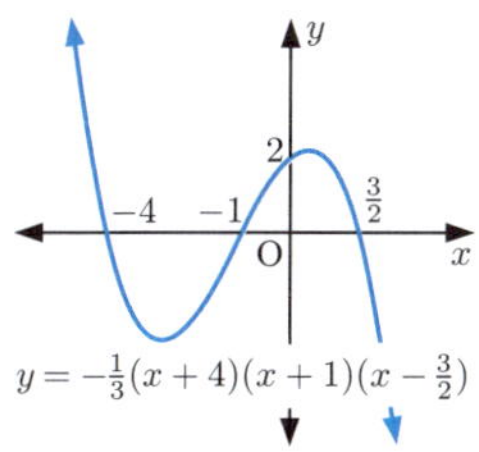

2 **a**
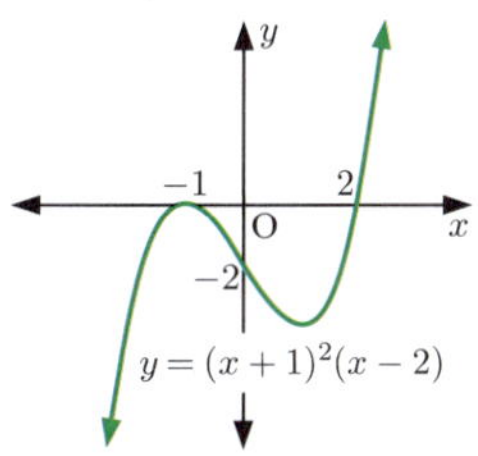

b
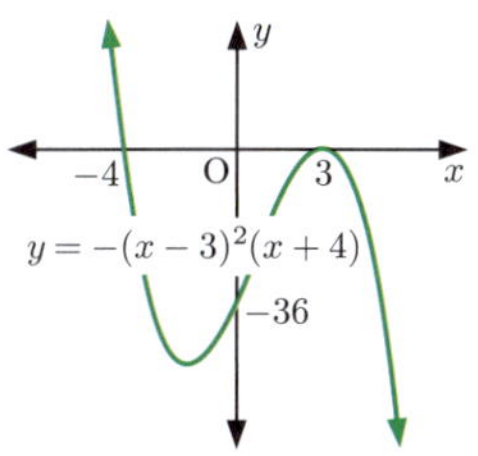

c

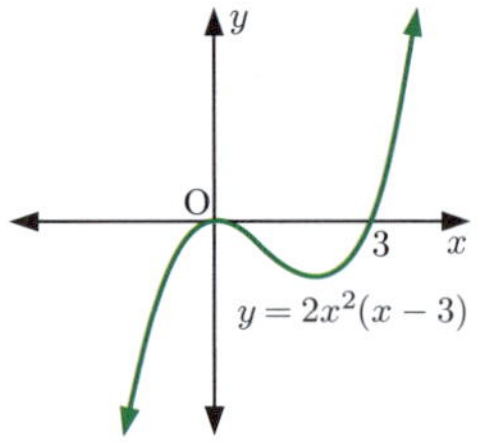

d

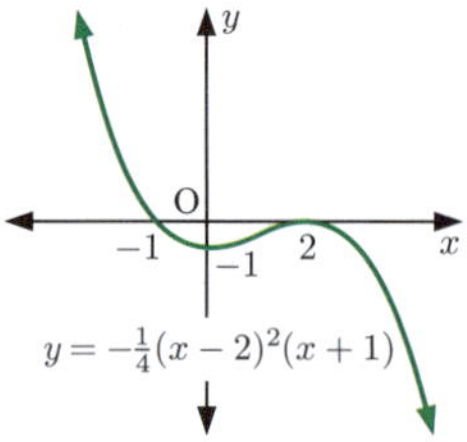

e

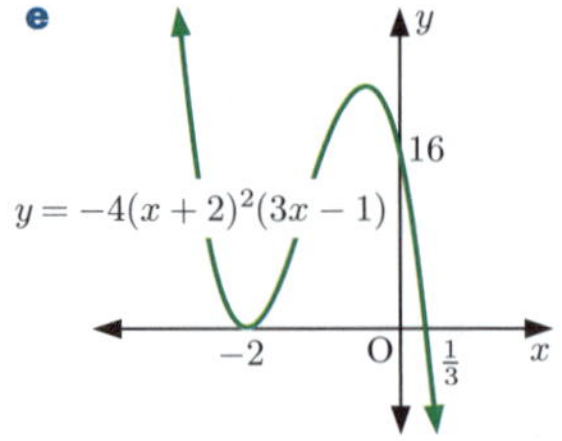

f

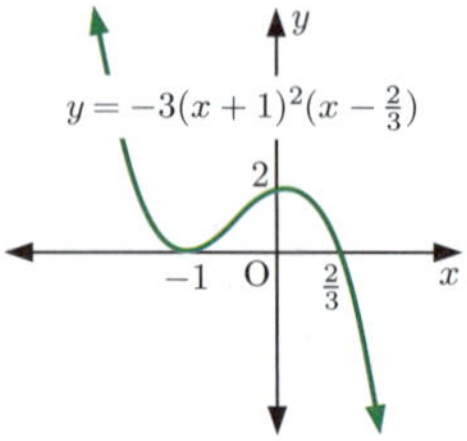

3 **a**

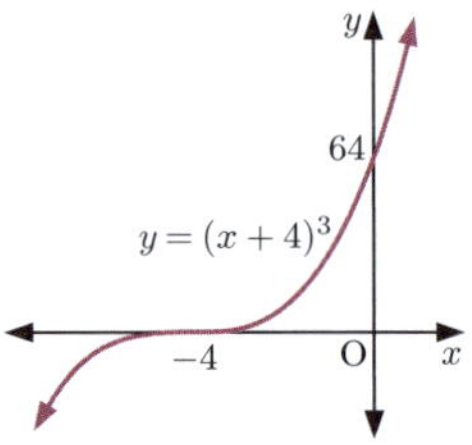

b

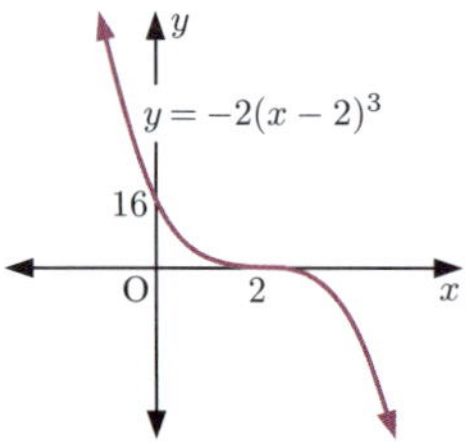

c

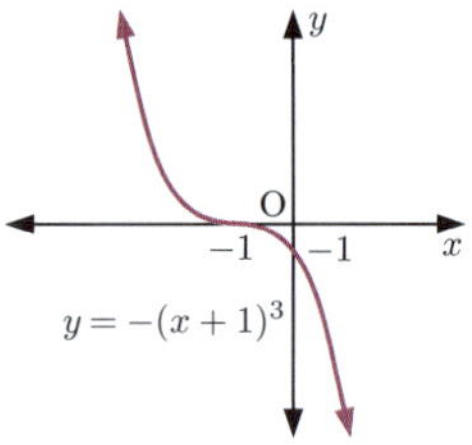

d

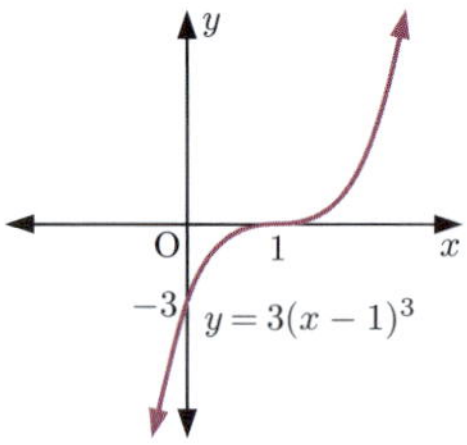

4 **a**

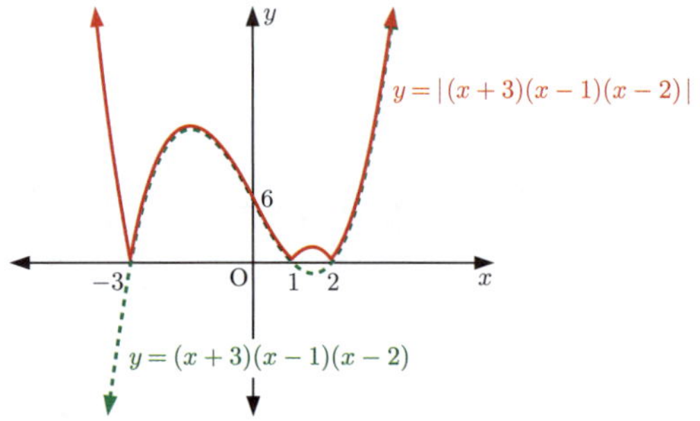

b

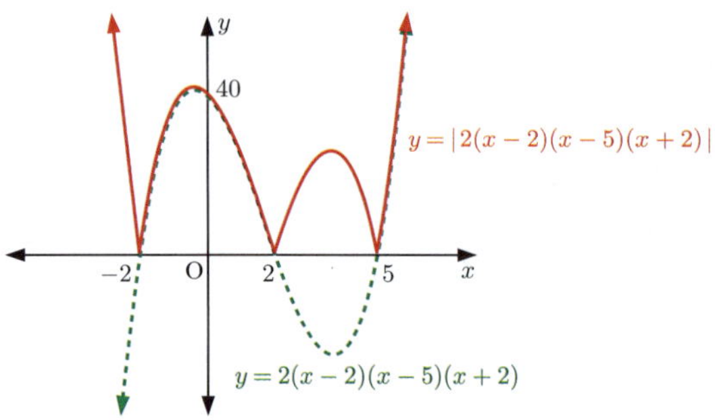

c

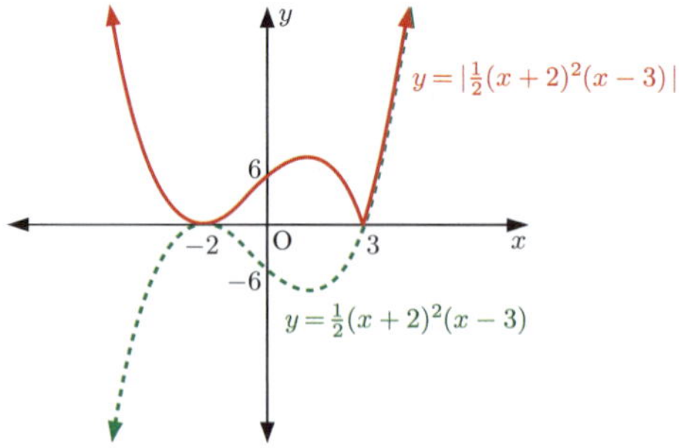

d

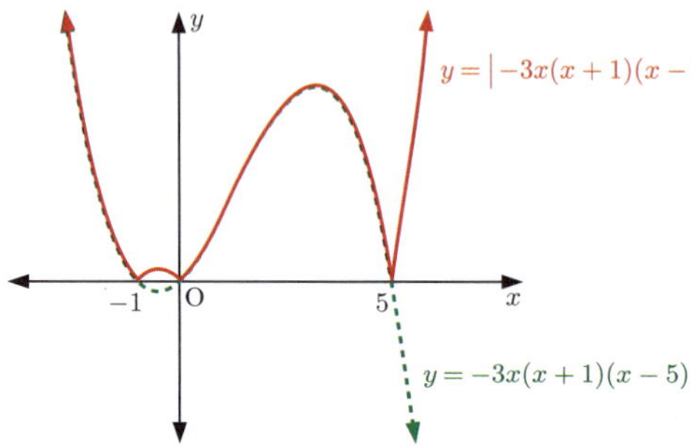

e

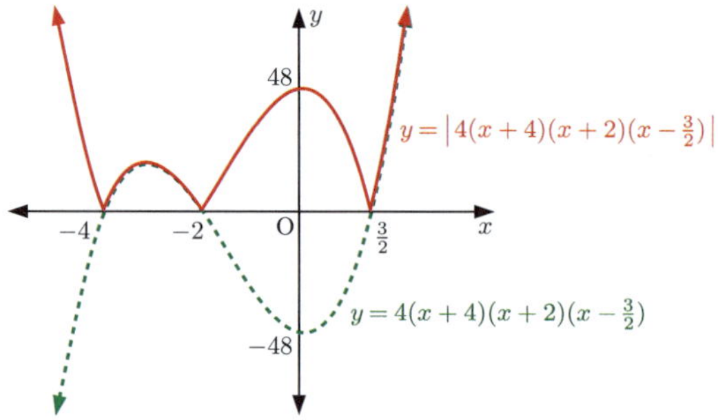

f

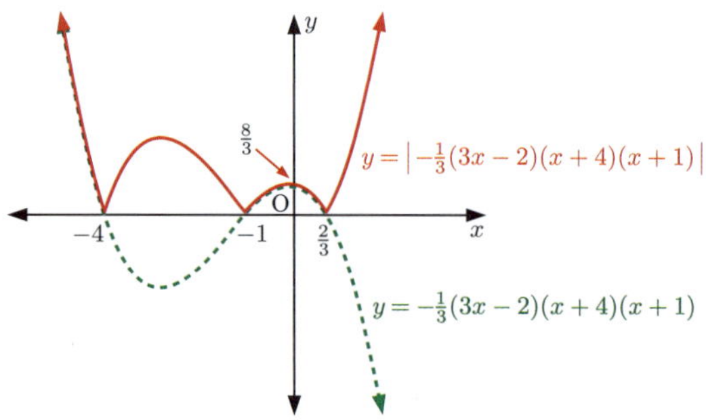

5 **a** $x < -1$ or $2 < x < 5$ **b** $-1 \leqslant x \leqslant 2$ or $x \geqslant 5$

6 **a**

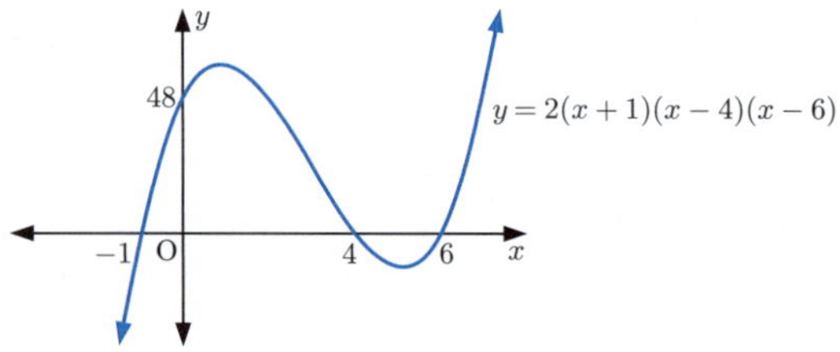

b $-1 < x < 4$ or $x > 6$

7 **a** $-1 \leqslant x \leqslant 3$ or $x \geqslant 5$ **b** $-3 < x < 2$ or $x > 6$

c $x < -4$ or $-\frac{1}{2} < x < 1$ **d** $x < -3$ or $-3 < x < \frac{3}{2}$

8 **a**

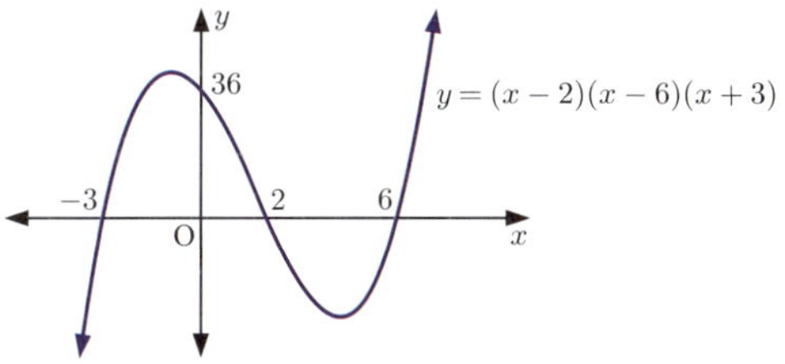

b $-3 < x < 2$ or $x > 6$

9 **a**

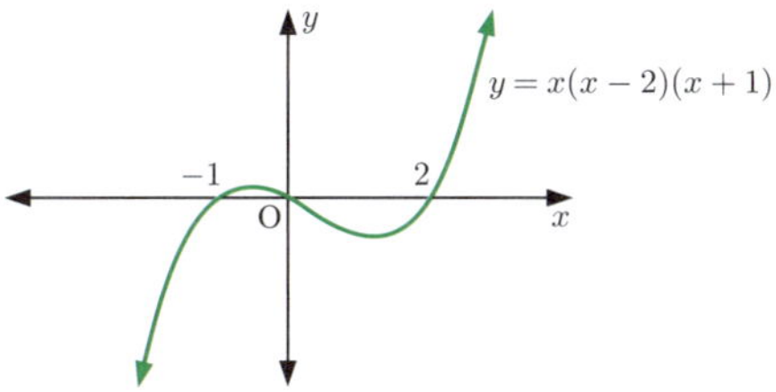

b $x = 1$ or $\pm\sqrt{2}$ **c** $x < -\sqrt{2}$ or $1 < x < \sqrt{2}$

10 **a**

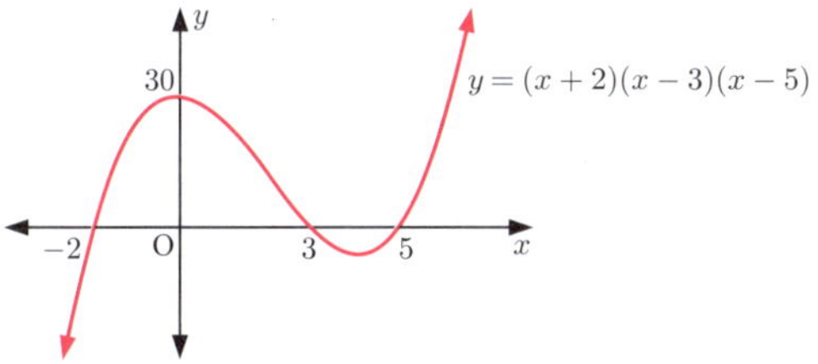

b $-2 < x < 3$ or $x > 5$ **c** $x = -1, 1$, or 6

d $x \leqslant -1$ or $1 \leqslant x \leqslant 6$

REVIEW SET 6A

1 **a** 5 **b** 3 **c** -4 **d** -1

2 **a** $8x^2 + 6x + 3$ **b** $7x^2 - 9x + 9$

c $15x^4 + 32x^3 + 29x - 4$

3 **a** quotient is $2x + 5$, remainder is 3

b quotient is $x^2 - 4x + 2$, remainder is -5

4 **a** no **b** yes **5** **a** $\frac{4}{3}$ and -2 **b** $-4 \pm \sqrt{5}$

6 **a** $(x + 4)(x - 7)$ **b** $(x + 1 + \sqrt{7})(x + 1 - \sqrt{7})$

7 $a = 3$, $b = 1$, $c = -3$, $k = -11$

8 **a** $a = 1$, $b = -2$, $c = 3$

b For $x^2 - 2x + 3$, $\Delta = -8$

$\therefore$ the only real root is $x = -3$.

9 **a** 1 **b** -53 **10** **a** not a factor **b** factor

11 $k = 6$ **12** $a = 4$, $b = -1$ **13** $c = 3$

14 **a** $a = -19$, $b = -20$ **b** $-5, -1$, and 4

15 **a** $f(-5) = 0$ **b** $f(x) = (x + 5)(x^2 - 4x + 3)$

c $f(x) = (x + 5)(x - 1)(x - 3)$

16 $x = -3, -1$, or 5

17 **a**

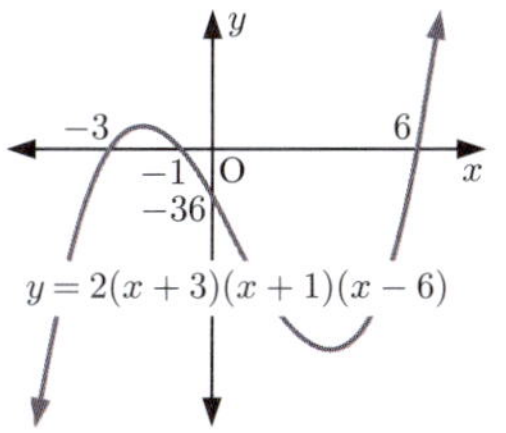

b

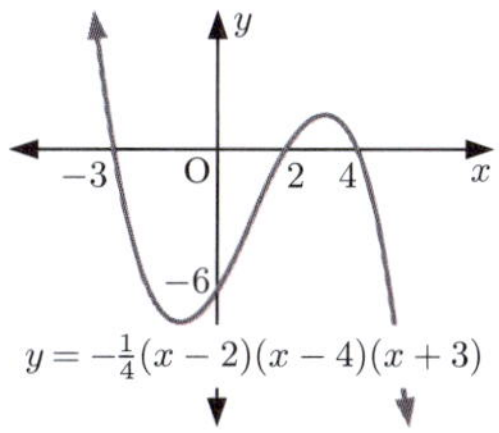

18 **a**

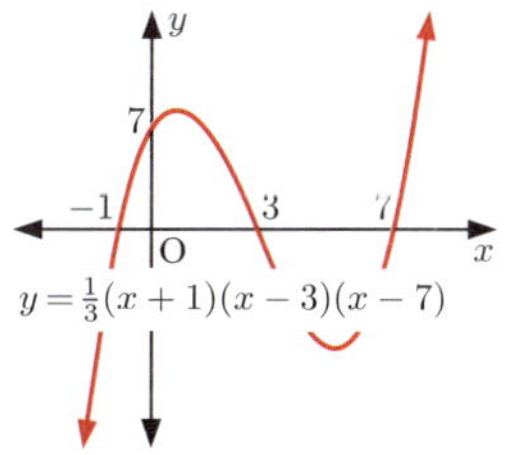

b $-1 < x < 3$ or $x > 7$

REVIEW SET 6B

1 coefficient -7, degree 4

2 **a** $x^3 + x^2 - 10x + 6$ **b** $x^3 - x^2 + 2x + 4$

c $x^5 - 6x^4 - 3x^3 + 29x^2 - 34x + 5$

3 **a** $x^2 - x + 1 + \dfrac{4}{x+1}$ **b** $x^3 - 2x^2 + x + 4 - \dfrac{6}{x-4}$

4 **a** 0, 4, and 5 **b** $x = \frac{3}{4}$ or $\frac{-1 \pm \sqrt{21}}{2}$

6 **a** $P(x) = a(4x - 1)(x^2 - 2x - 4)$, $a \neq 0$

b $P(x) = a(x + 4)(2x - 1)(x^2 - 4x - 3)$, $a \neq 0$

7 For $k = 3$, $b = 27$, $x = 3$ or -3.

For $k = -1$, $b = -5$, $x = -1$ or 5.

8 **a** -3 **b** -7 **9** **a** $a = 5$ **b** -12

10 **a** $3k^2 - 27k + 42$ **b** $k < -2$ or $k > 7$

11 **b** $(x - 2)(x^2 + 2x - 9)$ **c** 2, $-1 \pm \sqrt{10}$

12 $a = \frac{8}{7}$, $b = \frac{174}{7}$

13 **a** $a = -20$, $b = 12$ **b** $f(x) = (2x - 1)(x - 6)(x + 2)$

14 **a** $a = 6$ **b** $f(2) = 0$ **d** $f(x) = (x + 1)(x - 2)^3$

15 $x = -4, 2$, or 3

16 **a**

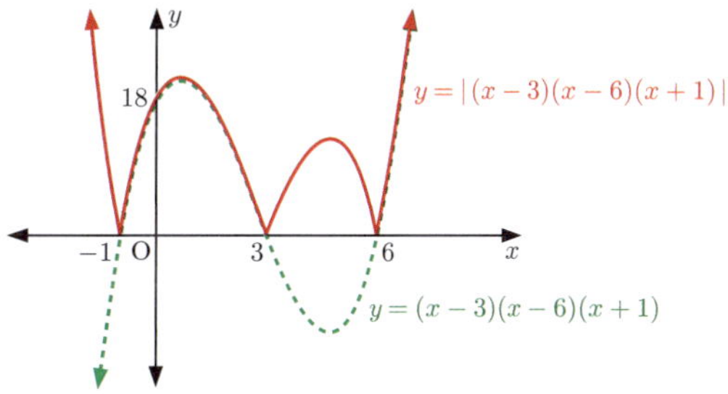

b

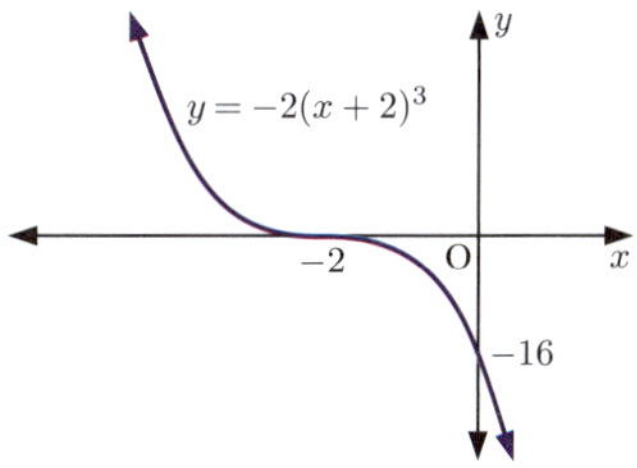

17 **a**

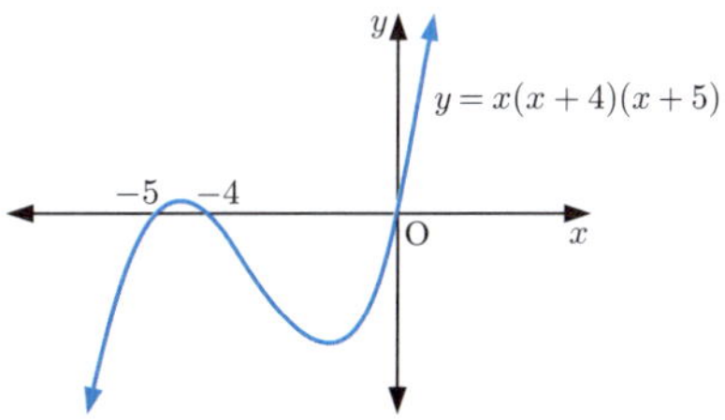

b $x < -5$ or $-4 < x < 0$ **c** $x = -6, -2$, or -1

d $-6 \leqslant x \leqslant -2$ or $x \geqslant -1$

EXERCISE 7A

1 **a** $y = \frac{1}{2}x^3 + 2$ **b** $y = 3\sqrt{x} - 1, \quad x \geqslant 0$
c $y = 3 - x^4$ **d** $y = \frac{1}{3} \times 2^x$
e $y = \dfrac{2}{x} + 1$ **f** $y = -\frac{3}{2} \times 3^x + 11$

2 **a** **i** $y = x^2 + 3x$ **ii** $y = 18$
b **i** $y = -\frac{1}{2}\sqrt{x} + \dfrac{10}{\sqrt{x}}, \quad x > 0$ **ii** $y = \frac{17\sqrt{3}}{6}$
c **i** $y = \dfrac{5}{3x} \times 2^x$ **ii** $y = \frac{40}{9}$
d **i** $y = 2x^3 - 9x$ **ii** $y = 27$
e **i** $y = \dfrac{1}{x^2} - \dfrac{12}{x} + 36$ **ii** $y = 32\frac{1}{9}$
f **i** $y = (x+2)^2 + 3$ **ii** $y = 28$

3 **a** $\lg y = 2x - 1$ **b** $y = \frac{1}{10} \times 10^{2x}$

4 $y = 1000 \times 10^{-\frac{3}{2}x}$

5 **a** $y = \frac{1}{10\,000} \times 10^x$ **b** $y = 10\,000 \times (\frac{1}{10})^x$
c $y = 5 \times 4^x$

6 **a** $y = 10 \times 10^{\frac{1}{3}x}$ **b** $y = 1000$

7 **a** **i** $y = e^3 \times e^{2x}$ **ii** $y \approx 20.1 \times 7.39^x$
b $y = e^7$ **c** $x \approx 1.35$

8 **a** $y \approx 148 \times 0.0498^x$ **b** $y \approx 54.6 \times 1.65^x$
c $y \approx 0.368 \times 1.28^x$ **d** $y \approx 7.39 \times 0.513^x$
e $y \approx 0.449 \times 3.32^x$ **f** $y \approx 1530 \times 0.0970^x$

9 **a** $A = 10, \quad b \approx 0.631$ **b** $y \approx 8.91$ **c** $x \approx 3.16$

10 **a** $A = e^2, \quad b = e^5$ **b** $y = e^{12}$ **c** $x = \frac{1}{25}$

11 **a** $\lg y = -\frac{1}{2}\lg x + 2$ **b** $y = \dfrac{100}{\sqrt{x}}$

12 **a** $y = x^{\frac{1}{4}}$ **b** $y = \dfrac{1000}{x}$ **c** $y = e^{\frac{3}{2}}x^2$

13 **a** $K = 7\sqrt{t}$ **b** $K = 21$ **14** **a** 3 **b** $\lg 4$

EXERCISE 7B

1 **a**

x^2	1	4	9	16
y	2	11	26	47

b

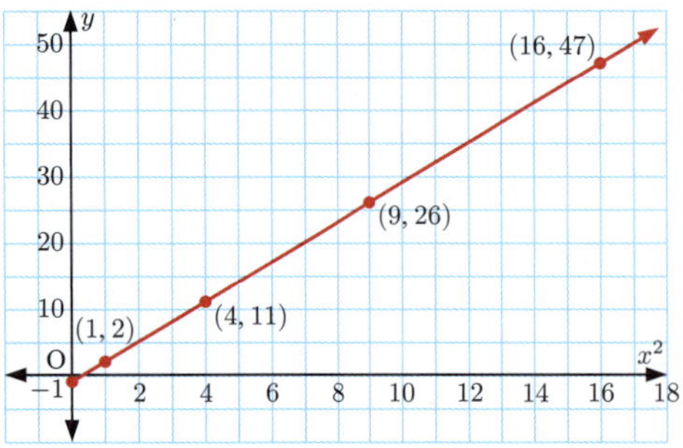

c $y = 3x^2 - 1$

2 **a**

x	1	2	3	4
$y\sqrt{x}$	9	14	19	24

b

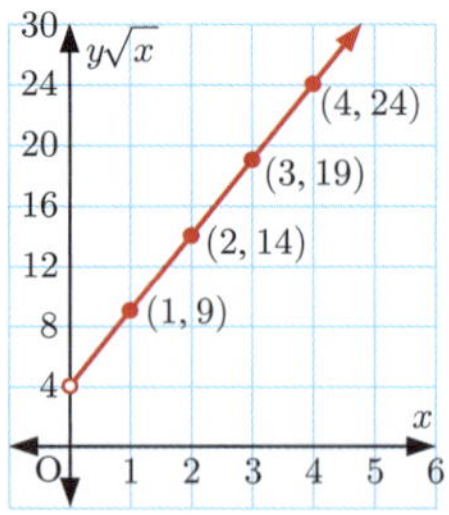

c $y = 5\sqrt{x} + \dfrac{4}{\sqrt{x}}, \quad x > 0$
d $y = 21$

3 **a**

$\dfrac{1}{x}$	1	0.5	0.33	0.2
xy	-1	0	0.33	0.6

b

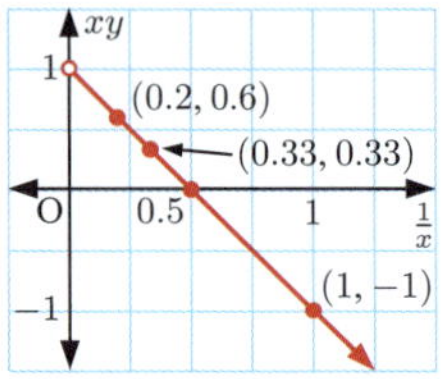

c $a = 1, \quad b = -2$
d $y = 0.08$
e $x = \frac{5}{2}$ or 10

4 **a**

$x\sqrt{x}$	2.83	8	14.70	22.63
$y\sqrt{x}$	7.41	10	13.35	17.31

b

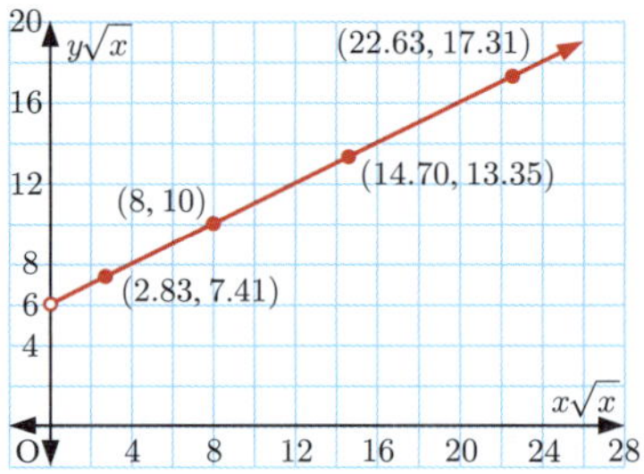

c $y = \frac{1}{2}x + \dfrac{6}{\sqrt{x}}, \quad x > 0$ **d** $y = 6.5$

5 **a**

t	1	2	3	4	5
$\lg M$	0.600	0.800	1	1.20	1.40

b

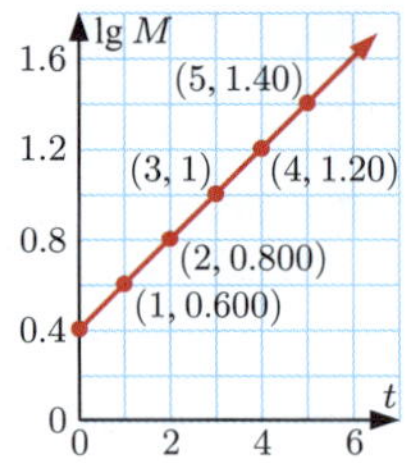

c $M \approx 2.51 \times 1.58^t$
d ≈ 2.51 g

6 **a** x^2
b

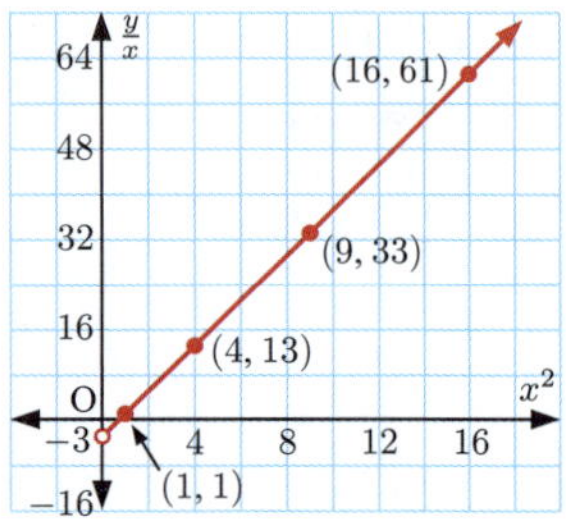

c $a = 4, \quad b = -3$
d $y = 485$

7 Plot xy against $\sqrt{x}$. $y = \frac{8}{x} - \frac{4}{\sqrt{x}}$ $\{a = 8,\ b = -4\}$

8 **a** $a \approx 4.90,\ b \approx 2.00$ **b** ≈ 44.1 m **c** ≈ 4.04 seconds

REVIEW SET 7A

1 **a** $y = \frac{1}{2}x^5 + 2$ **b** $y = -\frac{2}{3}x^2 + 4x$

2 **a** $y = \frac{3}{\sqrt{x}} - \frac{2}{x},\quad x > 0$ **b** $y = \frac{7}{9}$

3 **a** $\lg y = \frac{1}{2}\lg x + 1$ **b** $y = 10\sqrt{x}$

4 **a** $A = e^2,\quad b = \frac{3}{5}$ **b** $y = e^5$ **c** $x \approx 3.19$

5 **a**

x^3	1	8	27	64
xy	8	15	33.99	71

b

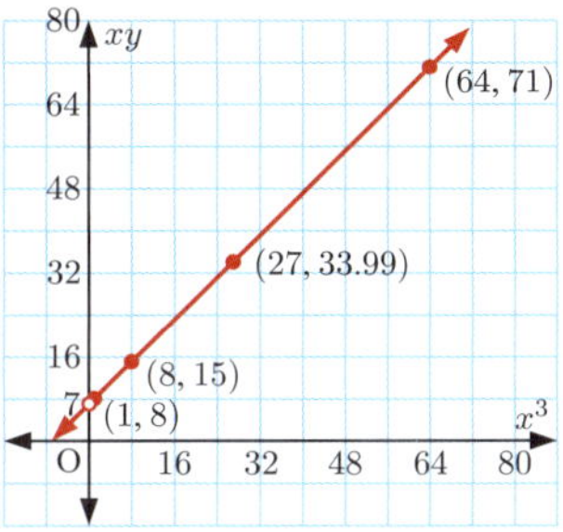

c $y = x^2 + \frac{7}{x}$

d $y = 50$

6 **a** Plot $\lg y$ against x.

$y = 100 \times (10^{-\frac{1}{3}})^x$ $\{a = 100,\ b = 10^{-\frac{1}{3}}\}$

b $y \approx 46.4$

REVIEW SET 7B

1 **a** $y = 5x - \frac{2}{x}$ **b** $y = 39\frac{3}{4}$

2 **a** $y \approx 4.48 \times 1.28^x$ **b** $y \approx 9.49$

3 **a**

$\sqrt{x}$	1	1.41	1.73	2
$\frac{y}{\sqrt{x}}$	2.5	3.74	4.69	5.5

b

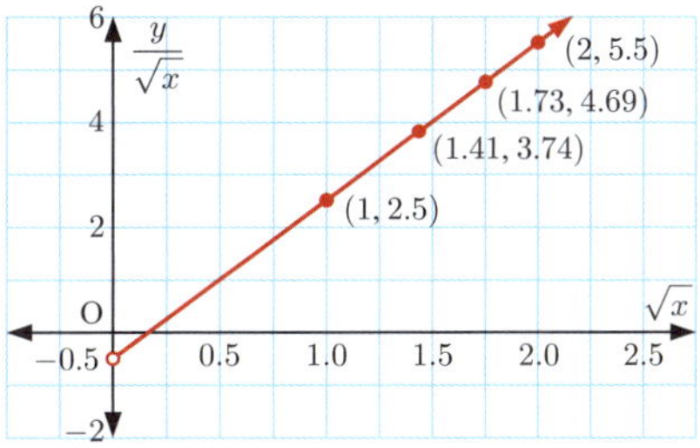

c $y = 3x - \frac{\sqrt{x}}{2}$

4 **a** $\frac{1}{x}$

b **i** $a = 6,\ b = 3$ **ii** $y = 11$ **iii** $x = 1$ or 4

5 **a**

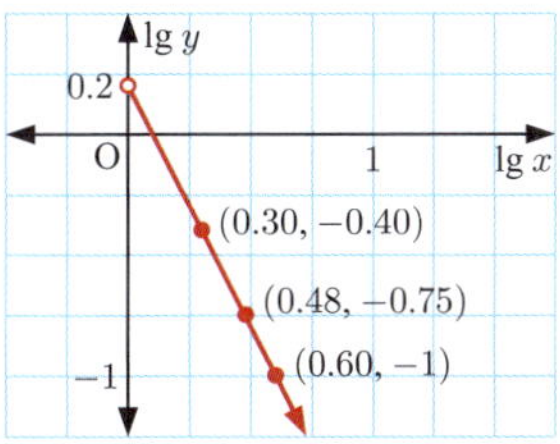

b $y = \frac{1.6}{x^2}$

c $y = \frac{2}{45}$

d $x \approx 1.79$

6 **a** Plot mP against m^3, $P = \frac{1}{2}m^2 - \frac{3}{m}$

b $P = 11.9$ **c** $m = -2, -1$, or 3

EXERCISE 8A

1 **a** π^c **b** $\frac{\pi}{2}^c$ **c** $\frac{\pi}{3}^c$ **d** $\frac{\pi}{9}^c$ **e** $\frac{\pi}{4}^c$
f $\frac{\pi}{18}^c$ **g** $\frac{\pi}{60}^c$ **h** $\frac{\pi}{5}^c$ **i** $2\pi^c$ **j** $4\pi^c$
k $\frac{3\pi}{2}^c$ **l** $\frac{5\pi}{6}^c$ **m** $\frac{2\pi}{5}^c$ **n** $\frac{2\pi}{3}^c$ **o** $\frac{5\pi}{9}^c$

2 **a** $\approx 0.375^c$ **b** $\approx 1.85^c$ **c** $\approx 5.28^c$ **d** $\approx 3.21^c$
e $\approx 4.15^c$

3 **a** $90°$ **b** $45°$ **c** $120°$ **d** $18°$ **e** $150°$
f $72°$ **g** $360°$ **h** $9°$ **i** $330°$ **j** $63°$

4 **a** $\approx 28.65°$ **b** $\approx 171.89°$ **c** $\approx 43.54°$
d $\approx 73.91°$ **e** $\approx 236.63°$

5 **a**

Degrees	0	45	90	135	180	225	270	315	360
Radians	0	$\frac{\pi}{4}$	$\frac{\pi}{2}$	$\frac{3\pi}{4}$	π	$\frac{5\pi}{4}$	$\frac{3\pi}{2}$	$\frac{7\pi}{4}$	2π

b

Deg.	0	30	60	90	120	150	180	210	240	270	300	330	360
Rad.	0	$\frac{\pi}{6}$	$\frac{\pi}{3}$	$\frac{\pi}{2}$	$\frac{2\pi}{3}$	$\frac{5\pi}{6}$	π	$\frac{7\pi}{6}$	$\frac{4\pi}{3}$	$\frac{3\pi}{2}$	$\frac{5\pi}{3}$	$\frac{11\pi}{6}$	2π

EXERCISE 8B

1 **a** 3 cm **b** 4.8 m **c** ≈ 8.95 cm

2 **a** 6.3 cm^2 **b** ≈ 26.2 m^2 **c** 40.5 cm^2

3 **a** 5.95 cm **b** 19.95 cm **c** 20.825 cm^2

4 **a** 1.2 radians **b** 15 cm^2

5 **a** $\theta = 0.8$ radians, area $= 40$ cm^2
b $\theta = 1.75$ radians, area $= 14$ cm^2
c $\theta = \frac{7}{9}$ radians, area $= 25.515$ m^2

6 **a** ≈ 6.11 m **b** ≈ 24.4 m^2

7 **a** ≈ 5.35 cm **b** ≈ 11.2 cm **c** ≈ 21.9 cm

8 **a** ≈ 15.3 cm^2 **b** ≈ 11.7 cm^2 **c** ≈ 3.53 cm^2

9 **a** **i** $\theta \approx 113°$ **ii** ≈ 1.97 radians
b ≈ 35.9 cm **c** ≈ 94.2 cm^2

10 **a** $P\widehat{Q}R = 30°$, $Q\widehat{P}R = 60°$
b perimeter ≈ 9.21 cm, area ≈ 1.29 cm^2

EXERCISE 8C

1 **a** **i** P(cos 56°, sin 56°), Q(cos 110°, sin 110°), R(cos 207°, sin 207°)
ii P(0.559, 0.829), Q(−0.342, 0.940), R(−0.891, −0.454)

b **i** P(cos 155°, sin 155°), Q(cos 235°, sin 235°), R(cos(−65°), sin(−65°))
ii P(−0.906, 0.423), Q(−0.574, −0.819), R(0.423, −0.906)

2 **a**

Quadrant	*Degree measure*	*Radian measure*	$\cos\theta$	$\sin\theta$	$\tan\theta$
1	$0° < \theta < 90°$	$0 < \theta < \frac{\pi}{2}$	+ve	+ve	+ve
2	$90° < \theta < 180°$	$\frac{\pi}{2} < \theta < \pi$	−ve	+ve	−ve
3	$180° < \theta < 270°$	$\pi < \theta < \frac{3\pi}{2}$	−ve	−ve	+ve
4	$270° < \theta < 360°$	$\frac{3\pi}{2} < \theta < 2\pi$	+ve	−ve	−ve

b **i** 1 and 4 **ii** 2 and 3 **iii** 3 **iv** 2

3

θ (degrees)	$0°$	$90°$	$180°$	$270°$	$360°$	$450°$
θ (radians)	0	$\frac{\pi}{2}$	π	$\frac{3\pi}{2}$	2π	$\frac{5\pi}{2}$
sine	0	1	0	-1	0	1
cosine	1	0	-1	0	1	0
tangent	0	undef	0	undef	0	undef

4 **a** **i** $\frac{1}{\sqrt{2}} \approx 0.707$ **ii** $\frac{\sqrt{3}}{2} \approx 0.866$ **iii** $\sqrt{3} \approx 1.73$
iv $\frac{1}{\sqrt{3}} \approx 0.577$

b

θ (degrees)	$30°$	$45°$	$60°$	$135°$	$150°$	$240°$	$315°$
θ (radians)	$\frac{\pi}{6}$	$\frac{\pi}{4}$	$\frac{\pi}{3}$	$\frac{3\pi}{4}$	$\frac{5\pi}{6}$	$\frac{4\pi}{3}$	$\frac{7\pi}{4}$
sine	$\frac{1}{2}$	$\frac{1}{\sqrt{2}}$	$\frac{\sqrt{3}}{2}$	$\frac{1}{\sqrt{2}}$	$\frac{1}{2}$	$-\frac{\sqrt{3}}{2}$	$-\frac{1}{\sqrt{2}}$
cosine	$\frac{\sqrt{3}}{2}$	$\frac{1}{\sqrt{2}}$	$\frac{1}{2}$	$-\frac{1}{\sqrt{2}}$	$-\frac{\sqrt{3}}{2}$	$-\frac{1}{2}$	$\frac{1}{\sqrt{2}}$
tangent	$\frac{1}{\sqrt{3}}$	1	$\sqrt{3}$	-1	$-\frac{1}{\sqrt{3}}$	$\sqrt{3}$	-1

5 **a** **i** ≈ 0.940 **ii** ≈ 0.940 **iii** ≈ 0.342 **iv** ≈ 0.342
v ≈ 0.766 **vi** ≈ 0.766 **vii** ≈ 0.574 **viii** ≈ 0.574

b $\sin(180° - \theta) = \sin\theta$

c $\sin\theta$ and $\sin(180° - \theta)$ have the same value, as P and Q have the same y-coordinate.

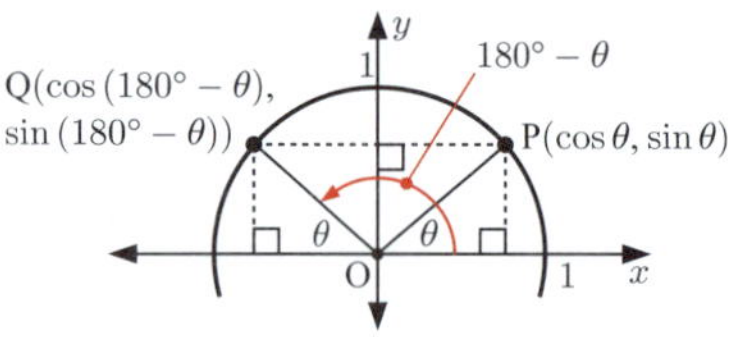

d **i** $170°$ **ii** $118°$ **iii** $\frac{2\pi}{3}$ **iv** $\frac{5\pi}{6}$

6 **a** **i** ≈ 0.766 **ii** ≈ -0.766 **iii** ≈ -0.985
iv ≈ 0.985 **v** ≈ 0.866 **vi** ≈ -0.866
vii ≈ 0.423 **viii** ≈ -0.423

b $\cos(180° - \theta) = -\cos\theta$

c $\cos(180° - \theta) = -\cos\theta$, as the x-coordinates of P and Q are negatives of each other.

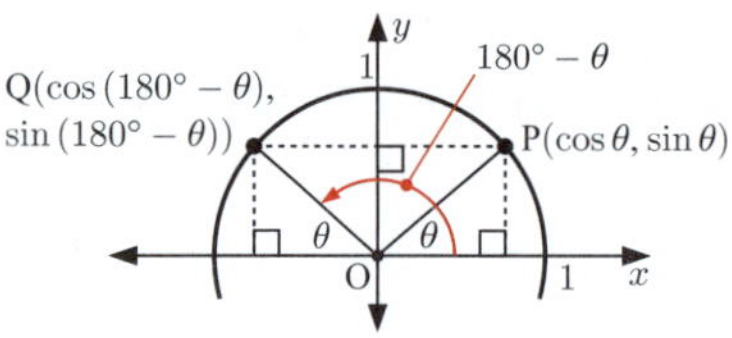

d **i** $160°$ **ii** $96°$ **iii** $\frac{4\pi}{5}$ **iv** $\frac{3\pi}{5}$

7 **a** ≈ 0.9511 **b** ≈ 0.5592 **c** ≈ -0.156
d ≈ 0.656 **e** ≈ 0.454 **f** ≈ -0.970

8 **a**

θ (radians)	$\sin\theta$	$\sin(-\theta)$	$\cos\theta$	$\cos(-\theta)$
0.62	≈ 0.581	≈ -0.581	≈ 0.814	≈ 0.814
1.403	≈ 0.986	≈ -0.986	≈ 0.167	≈ 0.167
4.283	≈ -0.909	≈ 0.909	≈ -0.416	≈ -0.416
5.901	≈ -0.373	≈ 0.373	≈ 0.928	≈ 0.928
-2.42	≈ -0.661	≈ 0.661	≈ -0.751	≈ -0.751

b $\sin(-\theta) = -\sin\theta$, $\cos(-\theta) = \cos\theta$

c P and Q have the same x-coordinates, and opposite y-coordinates.

d **i** $\cos(2\pi - \theta) = \cos(-\theta) = \cos\theta$
ii $\sin(2\pi - \theta) = \sin(-\theta) = -\sin\theta$

EXERCISE 8D

1

	a	b	c	d	e
$\sin\theta$	$\frac{1}{\sqrt{2}}$	1	$-\frac{1}{\sqrt{2}}$	-1	$-\frac{1}{\sqrt{2}}$
$\cos\theta$	$-\frac{1}{\sqrt{2}}$	0	$\frac{1}{\sqrt{2}}$	0	$\frac{1}{\sqrt{2}}$
$\tan\theta$	-1	undef	-1	undef	-1

2

	a	b	c	d	e
$\sin\beta$	$\frac{\sqrt{3}}{2}$	$-\frac{\sqrt{3}}{2}$	$\frac{1}{2}$	$-\frac{1}{2}$	$-\frac{\sqrt{3}}{2}$
$\cos\beta$	$\frac{1}{2}$	$-\frac{1}{2}$	$-\frac{\sqrt{3}}{2}$	$-\frac{\sqrt{3}}{2}$	$\frac{1}{2}$
$\tan\beta$	$\sqrt{3}$	$\sqrt{3}$	$-\frac{1}{\sqrt{3}}$	$\frac{1}{\sqrt{3}}$	$-\sqrt{3}$

3 **a** $\cos 150° = -\frac{\sqrt{3}}{2}$, $\sin 150° = \frac{1}{2}$, $\tan 150° = -\frac{1}{\sqrt{3}}$

b $\cos(-135°) = -\frac{1}{\sqrt{2}}$, $\sin(-135°) = -\frac{1}{\sqrt{2}}$, $\tan(-135°) = 1$

4 **a** $\cos 270° = 0$, $\sin 270° = -1$

b $\tan 270°$ is undefined

5 **a** $\frac{1}{4}$ **b** $\frac{3}{4}$ **c** 4 **d** $\frac{1}{4}$ **e** $-\frac{3}{2}$ **f** $\frac{7}{4}$
g $-\sqrt{3}$ **h** 3 **i** $-\frac{1}{2}$ **j** $\frac{11}{4}$ **k** -2 **l** $-\frac{2}{3}$

6 **a** $30°, 150°$ **b** $60°, 120°$ **c** $45°, 315°$
d $120°, 240°$ **e** $135°, 225°$ **f** $240°, 300°$

7 **a** $\frac{\pi}{4}, \frac{5\pi}{4}$ **b** $\frac{3\pi}{4}, \frac{7\pi}{4}$ **c** $\frac{\pi}{3}, \frac{4\pi}{3}$
d $0, \pi, 2\pi$ **e** $\frac{\pi}{6}, \frac{7\pi}{6}$ **f** $\frac{2\pi}{3}, \frac{5\pi}{3}$

8 **a** $\frac{\pi}{6}, \frac{11\pi}{6}, \frac{13\pi}{6}, \frac{23\pi}{6}$ **b** $\frac{7\pi}{6}, \frac{11\pi}{6}, \frac{19\pi}{6}, \frac{23\pi}{6}$ **c** $\frac{3\pi}{2}, \frac{7\pi}{2}$

9 **a** $\theta = k\pi, \quad k \in \mathbb{Z}$ **b** $\theta = \frac{\pi}{2} + k\pi, \quad k \in \mathbb{Z}$

10 **a** $\frac{\pi}{3}$ **b** $\frac{7\pi}{4}$ **c** $\frac{\pi}{2}$ **d** $\frac{7\pi}{6}$ **e** $\frac{4\pi}{3}$

EXERCISE 8E

1 **a** $\sin\theta = \pm\frac{\sqrt{3}}{2}$ **b** $\sin\theta = \pm\frac{\sqrt{5}}{3}$ **c** $\sin\theta = 0$
d $\sin\theta = \pm\frac{4}{5}$

2 **a** $\cos\theta = \pm\frac{2\sqrt{2}}{3}$ **b** $\cos\theta = \pm\frac{\sqrt{21}}{5}$ **c** $\cos\theta = \pm 1$
d $\cos\theta = 0$

3 **a** $\sin\theta = \frac{\sqrt{15}}{4}$ **b** $\cos\theta = -\frac{\sqrt{11}}{6}$ **c** $\cos\theta = \frac{3}{5}$
d $\sin\theta = -\frac{3\sqrt{5}}{7}$

4 **a** $\tan\theta = -\frac{2}{\sqrt{5}}$ **b** $\tan\theta = -\frac{\sqrt{21}}{2}$ **c** $\tan\theta = \frac{1}{2}$
d $\tan\theta = -\frac{\sqrt{33}}{4}$

5 **a** $\cos x = \frac{1}{\sqrt{5}}$, $\sin x = \frac{2}{\sqrt{5}}$ **b** $\cos x = -\frac{4}{5}$, $\sin x = \frac{3}{5}$

c $\cos x = -\frac{2}{\sqrt{7}}$, $\sin x = -\sqrt{\frac{3}{7}}$

d $\cos x = \frac{3}{\sqrt{10}}$, $\sin x = -\frac{1}{\sqrt{10}}$

6 **a** m is negative since $\tan\theta$ is negative when θ is in quadrant 2.

b $\cos\theta = -\dfrac{1}{\sqrt{m^2+1}}$, $\sin\theta = -\dfrac{m}{\sqrt{m^2+1}}$

EXERCISE 8F

1 **a** $\theta \approx 1.23$ or 5.05 **b** $\theta \approx 0.848$ or 2.29
c $\theta \approx 1.25$ or 4.39 **d** $\theta \approx 0.730$ or 2.41
e $\theta \approx 0.675$ or 3.82 **f** $\theta \approx 1.37$ or 4.91
g $\theta \approx 0.989$ or 4.13 **h** $\theta \approx 0.428$ or 5.86
i $\theta \approx 0.694$ or 2.45

2 **a** $\theta \approx 3.79$ or 5.64 **b** $\theta \approx 1.86$ or 4.42
c $\theta \approx 2.47$ or 5.61 **d** $\theta \approx 3.55$ or 5.87
e $\theta \approx 1.82$ or 4.96 **f** $\theta \approx 1.88$ or 4.41
g $\theta \approx 2.19$ or 5.33 **h** $\theta \approx 1.87$ or 4.42
i $\theta \approx 4.02$ or 5.41

3 **a** $\theta = 60°$ or $300°$ **b** $\theta \approx 44.4°$ or $136°$
c $\theta = 60°$ or $240°$ **d** $\theta \approx 9.59°$ or $170°$
e $\theta = 90°$ or $270°$ **f** $\theta = 240°$ or $300°$
g $\theta \approx 107°$ or $253°$ **h** $\theta \approx 115°$ or $295°$
i $\theta \approx 244°$ or $296°$

EXERCISE 8G

1 **a** $\frac{2}{\sqrt{3}}$ **b** $\sqrt{2}$ **c** $\frac{1}{\sqrt{3}}$ **d** 2 **e** -1
f $\sqrt{3}$ **g** $-\frac{2}{\sqrt{3}}$ **h** -1

2

	$\cos\theta$	$\sin\theta$	$\tan\theta$	$\sec\theta$	$\operatorname{cosec}\theta$	$\cot\theta$
a	$-\frac{1}{2}$	$\frac{\sqrt{3}}{2}$	$-\sqrt{3}$	-2	$\frac{2}{\sqrt{3}}$	$-\frac{1}{\sqrt{3}}$
b	$-\frac{1}{\sqrt{2}}$	$\frac{1}{\sqrt{2}}$	-1	$-\sqrt{2}$	$\sqrt{2}$	-1
c	$\frac{\sqrt{3}}{2}$	$-\frac{1}{2}$	$-\frac{1}{\sqrt{3}}$	$\frac{2}{\sqrt{3}}$	-2	$-\sqrt{3}$

3 **a** $\operatorname{cosec} x = \frac{5}{4}$, $\sec x = \frac{5}{3}$, $\cot x = \frac{3}{4}$
b $\operatorname{cosec} x = -\frac{4}{\sqrt{15}}$, $\sec x = 4$, $\cot x = -\frac{1}{\sqrt{15}}$

4 **a** $\sin\theta = -\frac{2\sqrt{2}}{3}$, $\tan\theta = -2\sqrt{2}$, $\operatorname{cosec}\theta = -\frac{3}{2\sqrt{2}}$,
$\sec\theta = 3$, $\cot\theta = -\frac{1}{2\sqrt{2}}$
b $\cos x = -\frac{\sqrt{7}}{4}$, $\tan x = \frac{3}{\sqrt{7}}$, $\operatorname{cosec} x = -\frac{4}{3}$,
$\sec x = -\frac{4}{\sqrt{7}}$, $\cot x = \frac{\sqrt{7}}{3}$
c $\sin x = \frac{\sqrt{5}}{3}$, $\cos x = \frac{2}{3}$, $\tan x = \frac{\sqrt{5}}{2}$, $\operatorname{cosec} x = \frac{3}{\sqrt{5}}$,
$\cot x = \frac{2}{\sqrt{5}}$
d $\sin\theta = \frac{1}{3}$, $\cos\theta = -\frac{2\sqrt{2}}{3}$, $\tan\theta = -\frac{1}{2\sqrt{2}}$,
$\sec\theta = -\frac{3}{2\sqrt{2}}$, $\cot\theta = -2\sqrt{2}$
e $\sin\beta = -\frac{2}{\sqrt{29}}$, $\cos\beta = -\frac{5}{\sqrt{29}}$, $\operatorname{cosec}\beta = -\frac{\sqrt{29}}{2}$,
$\sec\beta = -\frac{\sqrt{29}}{5}$, $\cot\beta = \frac{5}{2}$
f $\sin\theta = -\frac{1}{\sqrt{17}}$, $\cos\theta = -\frac{4}{\sqrt{17}}$, $\tan\theta = \frac{1}{4}$,
$\operatorname{cosec}\theta = -\sqrt{17}$, $\sec\theta = -\frac{\sqrt{17}}{4}$

5 **a** $60°, 300°$ **b** $60°, 120°$ **c** $150°, 330°$

6 **a** $\theta = k\pi,\ k \in \mathbb{Z}$ **b** $\theta = \frac{\pi}{2} + k\pi,\ k \in \mathbb{Z}$
c $\theta = \frac{\pi}{2} + k\pi,\ k \in \mathbb{Z}$ **d** $\theta = k\pi,\ k \in \mathbb{Z}$

REVIEW SET 8A

1 **a** $\frac{\pi}{10}^c$ **b** $\frac{5\pi}{18}^c$ **c** $3\pi^c$ **d** $\frac{10\pi}{9}^c$

2 **a** 4.8 cm **b** ≈ 18.8 m

3 **a** $\theta = \frac{7}{11}$ radians **b** 38.5 cm^2

4 **a** 18π cm^2 **b** $(12\pi + 6)$ cm

5 **a** 8 cm **b** $\theta \approx 1.85^c$ **c** ≈ 11.2 cm^2

6 **a** $57°$ **b** $\frac{\pi}{4}$ **c** $36°$

7

	$\sin\theta$	$\cos\theta$	$\tan\theta$
a	$\frac{\sqrt{3}}{2}$	$\frac{1}{2}$	$\sqrt{3}$
b	0	-1	0
c	$-\frac{1}{\sqrt{2}}$	$-\frac{1}{\sqrt{2}}$	1

8 **a** $\frac{2\pi}{3}$ **b** $\frac{7\pi}{6}$ **9** $\cos\theta = \pm\frac{\sqrt{55}}{8}$

10 **a** $\frac{3\sqrt{3}}{2}$ **b** -1 **c** 0

11 $\sin x = \frac{2\sqrt{2}}{3}$, $\tan x = -2\sqrt{2}$ **12** $\tan\theta = -2$

13 **a** $150°, 210°$ **b** $45°, 315°$ **c** $120°, 300°$

14 **a** $\theta \approx 1.37$ or 4.91 **b** $\theta \approx 0.337$ or 3.48
c $\theta \approx 0.290$ or 2.85

15 $\cos x = -\frac{\sqrt{15}}{4}$, $\tan x = \frac{1}{\sqrt{15}}$, $\operatorname{cosec} x = -4$,
$\sec x = -\frac{4}{\sqrt{15}}$, $\cot x = \sqrt{15}$

REVIEW SET 8B

1 **a** $40°$ **b** $126°$ **c** $\approx 47.0°$ **d** $\approx 111°$

2 **a** ≈ 9.42 cm^2 **b** 81 cm^2

3 **a** The arc length is less than the radius of the sector, so
$\theta = \frac{l}{r} < 1$.
b $\theta \approx 0.846$ **c** 71.5 m^2 **d** ≈ 8.23 m^2

4 **a** $5\sqrt{3}$ cm
b perimeter $= 10(\frac{\pi}{3} + \sqrt{3})$ cm, area $= 25(\sqrt{3} - \frac{\pi}{3})$ cm^2

5 **a** P$(\cos 140°, \sin 140°)$ **b** P$(-0.766, 0.643)$

6 $\cos\frac{\pi}{2} = 0$, $\sin\frac{\pi}{2} = 1$

8 **a** $\frac{9}{4}$ **b** $\frac{-\sqrt{3}-\sqrt{2}}{2}$ **c** $\sqrt{3} + 1$

9 **a** $\frac{2\pi}{3}, \frac{4\pi}{3}, \frac{8\pi}{3}, \frac{10\pi}{3}$ **b** $\frac{3\pi}{4}, \frac{7\pi}{4}, \frac{11\pi}{4}, \frac{15\pi}{4}$

10 $\sin\theta = \pm\frac{\sqrt{35}}{6}$ **11** $\cos\theta = -\frac{2}{\sqrt{5}}$, $\sin\theta = \frac{1}{\sqrt{5}}$

12 **a** $\theta \approx 3.45$ or 5.98 **b** $\theta \approx 2.60$ or 3.68
c $\theta \approx 3.00$ or 6.14

13 **a** 1 **b** -2 **c** $\frac{1}{2}$

14 $\sin\alpha = \frac{\sqrt{91}}{10}$, $\cos\alpha = -\frac{3}{10}$, $\tan\alpha = -\frac{\sqrt{91}}{3}$,
$\operatorname{cosec}\alpha = \frac{10}{\sqrt{91}}$, $\cot\alpha = -\frac{3}{\sqrt{91}}$

EXERCISE 9A

1 **a** periodic **b** periodic **c** not periodic
d periodic **e** not periodic **f** periodic

2 **a** 2 **b** 4

3 **a** **i** 5 **ii** $y = 3$ **iii** 2
b **i** 4 **ii** $y = 1$ **iii** 3
c **i** 6 **ii** $y = -\frac{5}{2}$ **iii** $\frac{7}{2}$
d **i** 2π **ii** $y = \frac{1}{2}$ **iii** $\frac{5}{2}$

4 **a** maximum $= 7$, minimum $= -1$
b **i** $f(9) = 4$ **ii** $f(21) = 2$

5 **a** q
b No, consider $f = \sin x$, $g = \sin 2\pi x$. f has period 2π, g has period 1, but $f + g$ is not periodic.

EXERCISE 9B

1 **a** 2 **b** $\frac{1}{2}$ **c** 4

2 **a** $y = 3$ **b** $y = 0$ **c** $y = -4$

3 **a** $\frac{2\pi}{3}$ **b** 2π **c** 6π

4 **a** $b = 3$ **b** $b = 5$ **c** $b = 6$ **d** $b = 4$ **e** $b = \frac{3}{2}$

5 a

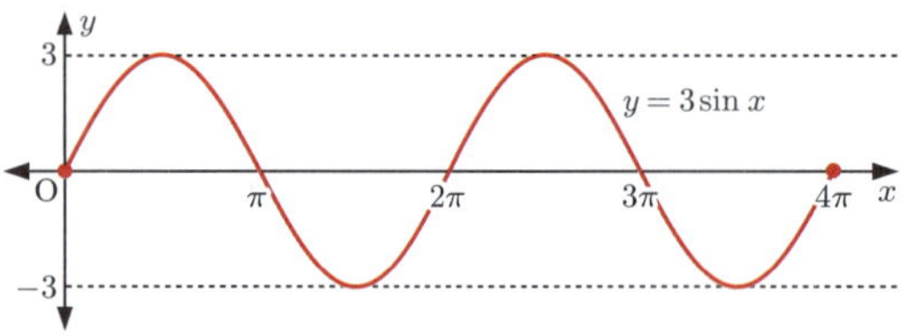

b

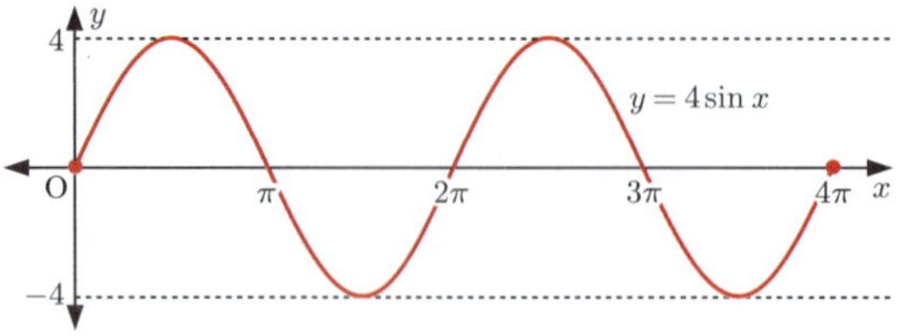

c

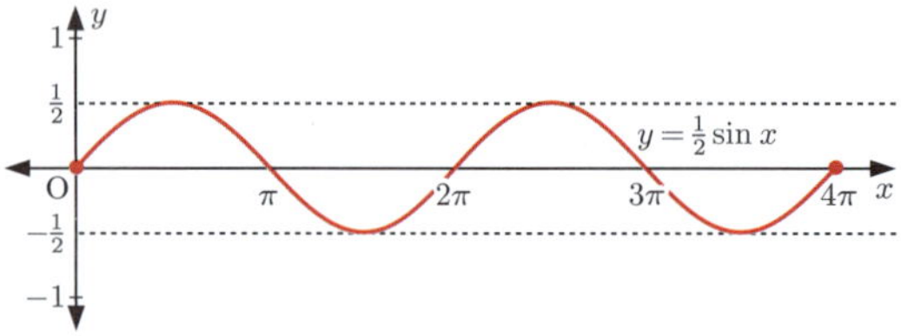

d

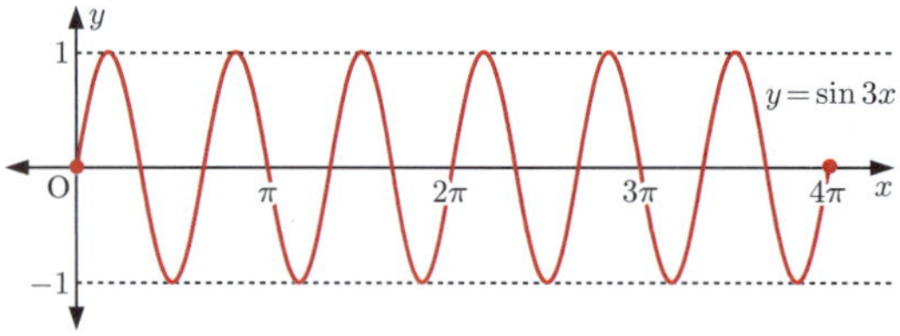

e

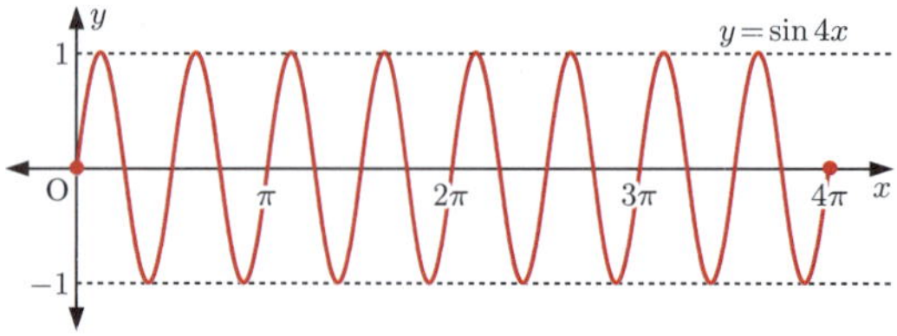

f

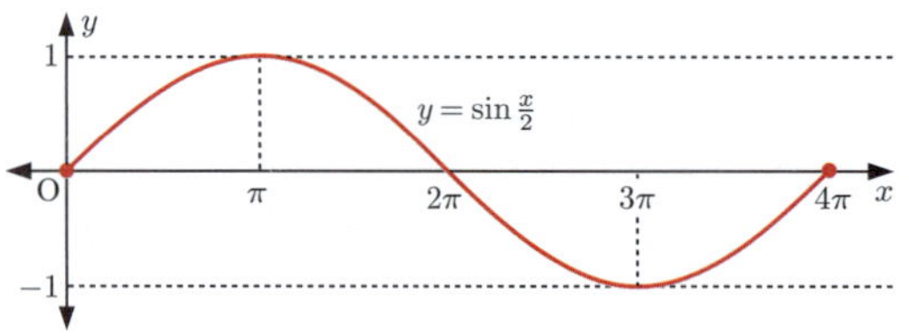

g

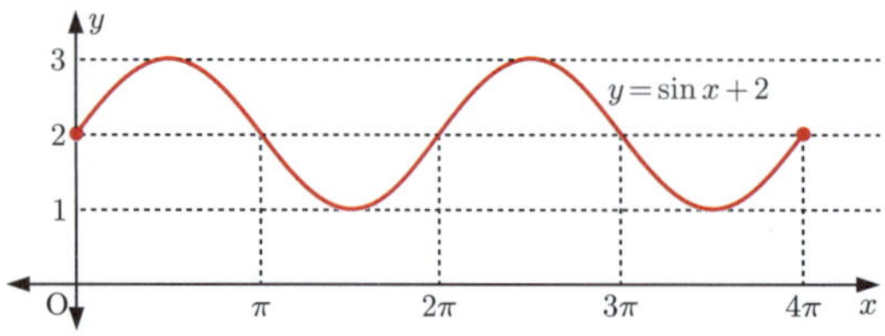

h

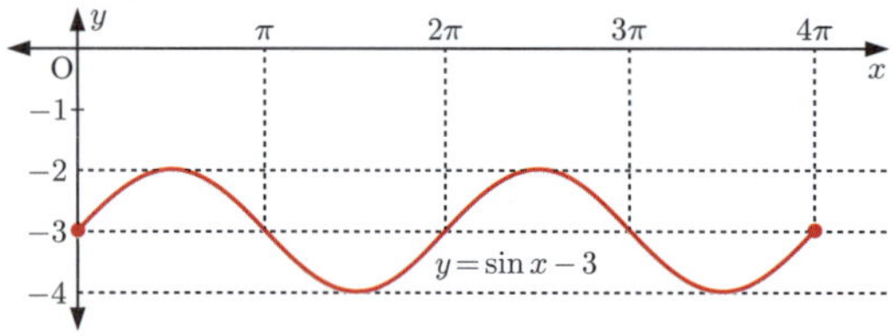

i

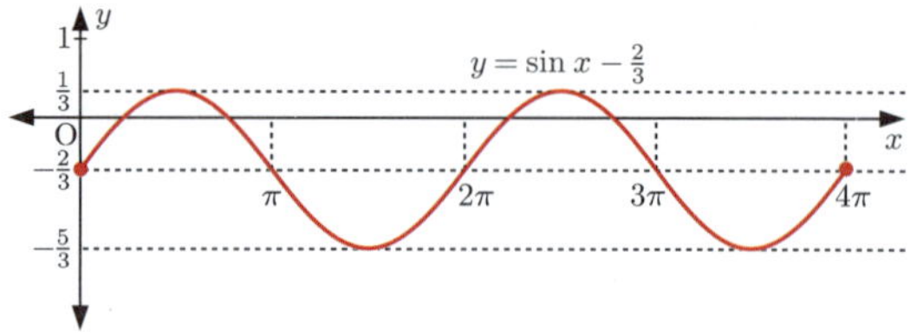

6 a

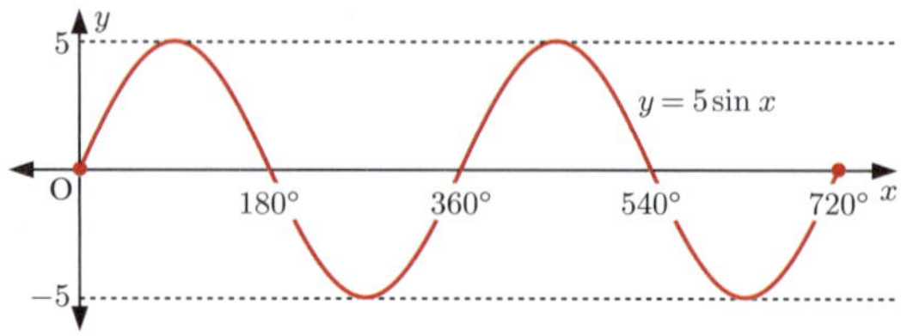

b

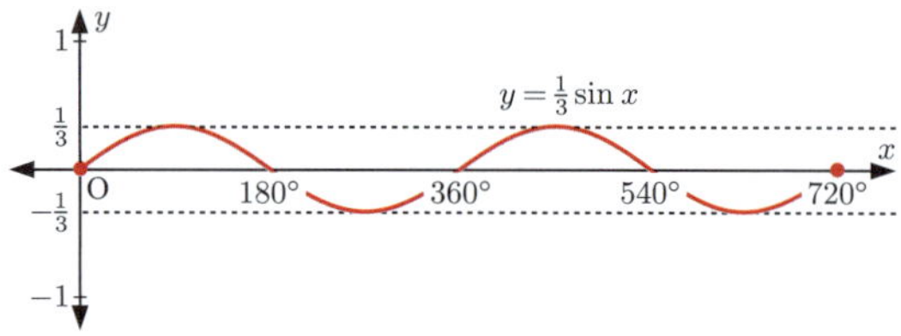

c

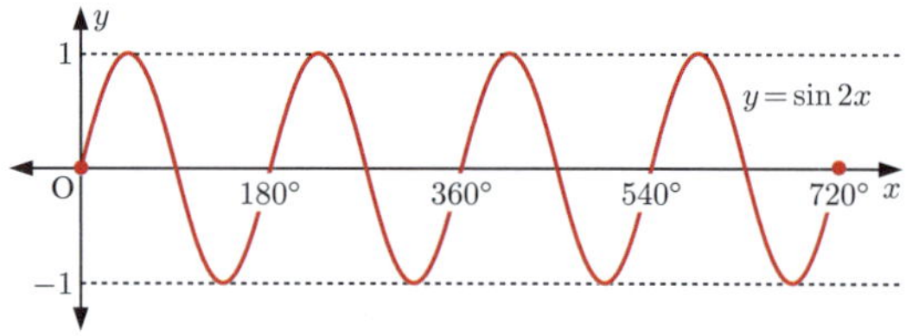

d

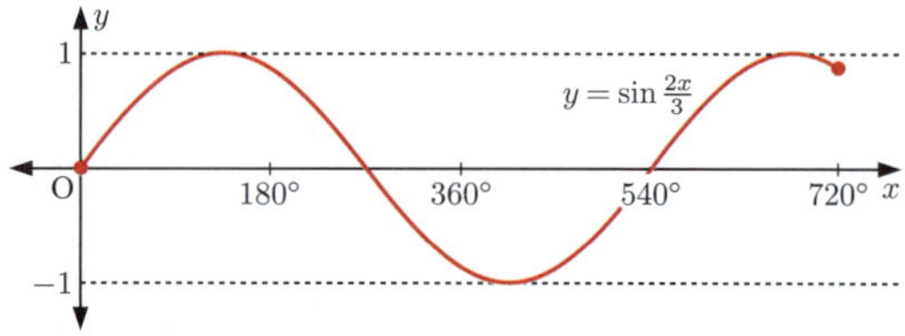

e

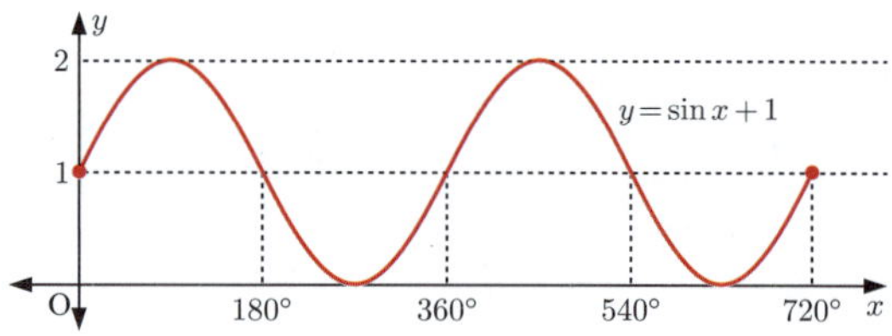

f

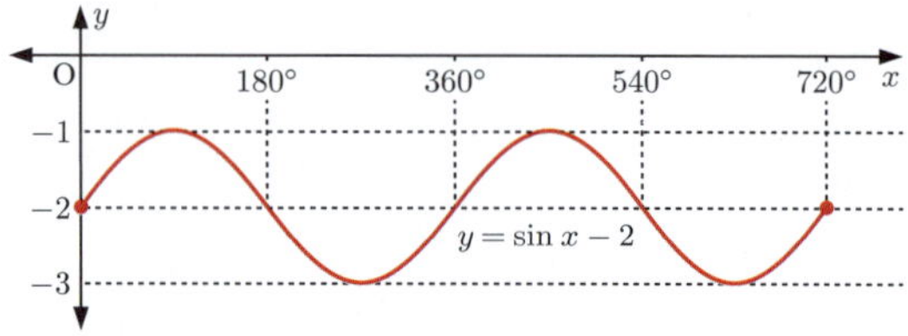

7 a

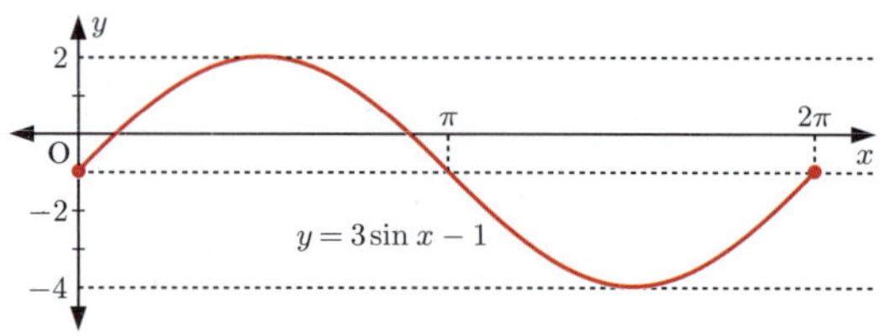

b

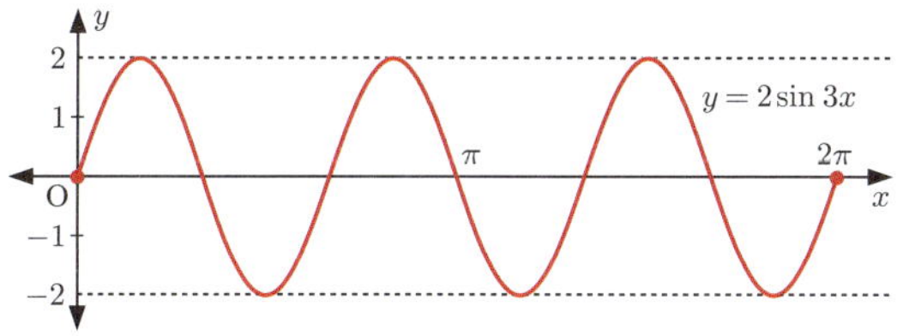

c

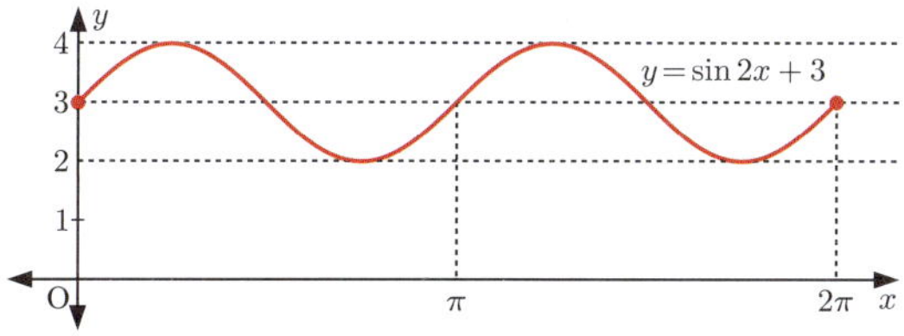

d

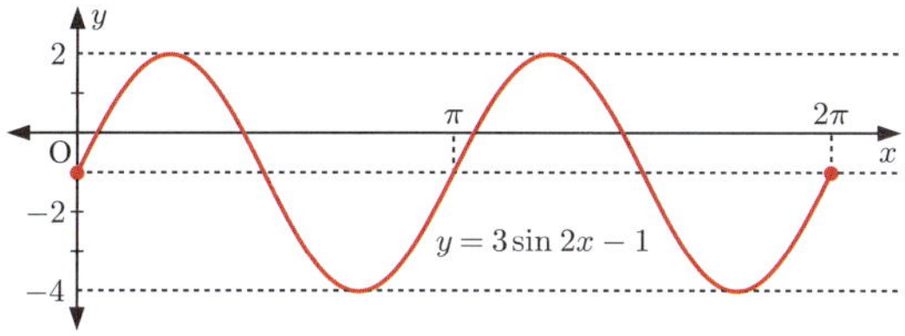

e

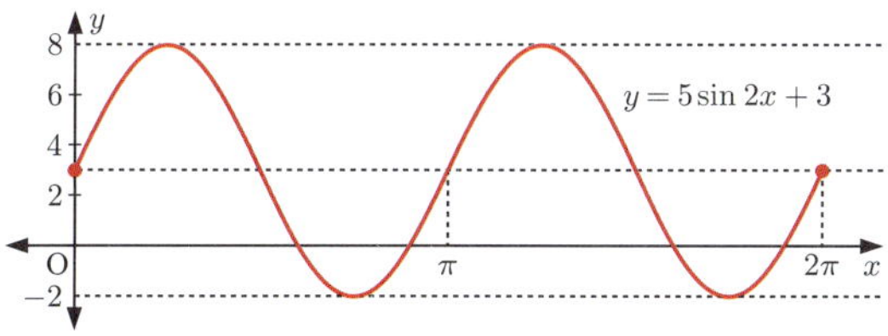

f

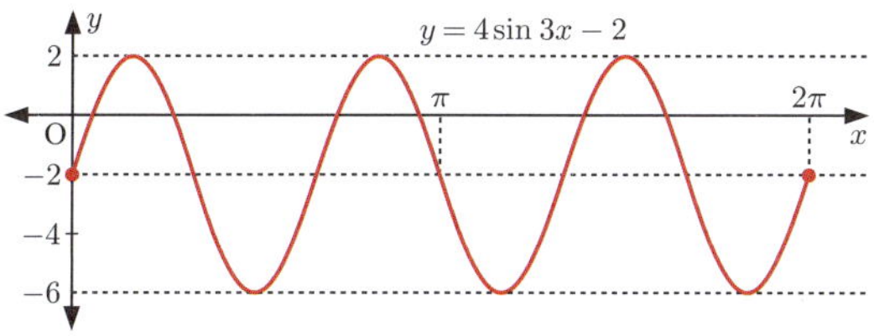

8 **a** $a = 3,\ b = 1,\ c = 0$ **b** $a = 2,\ b = 5,\ c = 6$
c $a = 5,\ b = 3,\ c = -2$ **d** $a = \frac{1}{4},\ b = \frac{5}{6},\ c = 7$

9 **a** $\{y : 3 \leqslant y \leqslant 5\}$ **b** $\{y : -4 \leqslant y \leqslant 2\}$
c $\{y : -7 \leqslant y \leqslant -3\}$

10 **a** $y = \sin x - 3$ **b** $y = 3\sin 3x + 2$ **c** $y = 3\sin 2x$

11 $m = 2,\ n = -3$

12 **a**

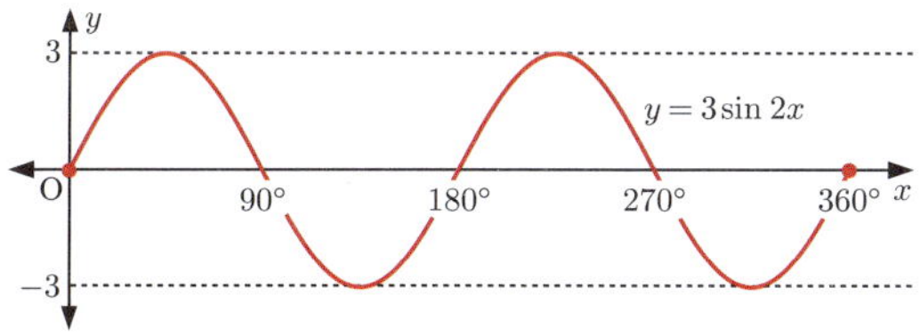

b

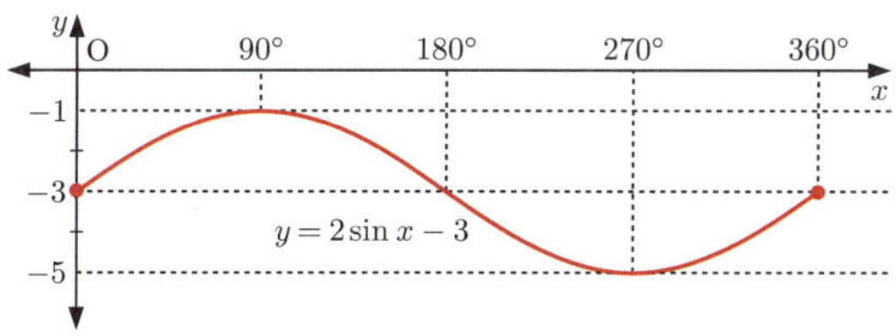

c

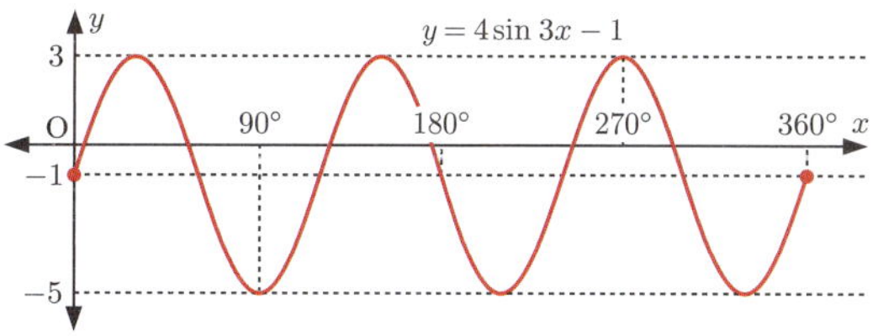

13 **a**

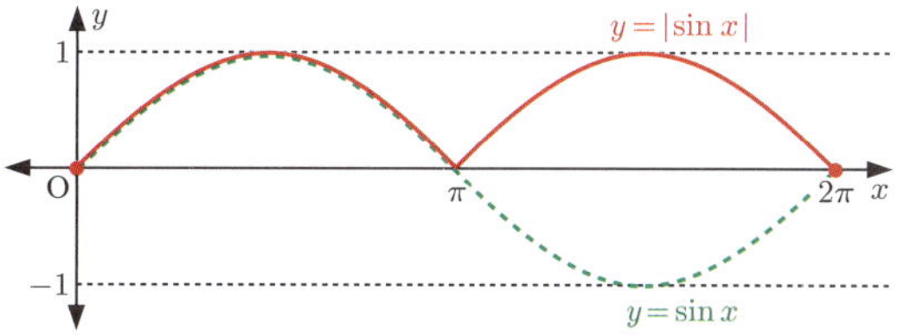

b

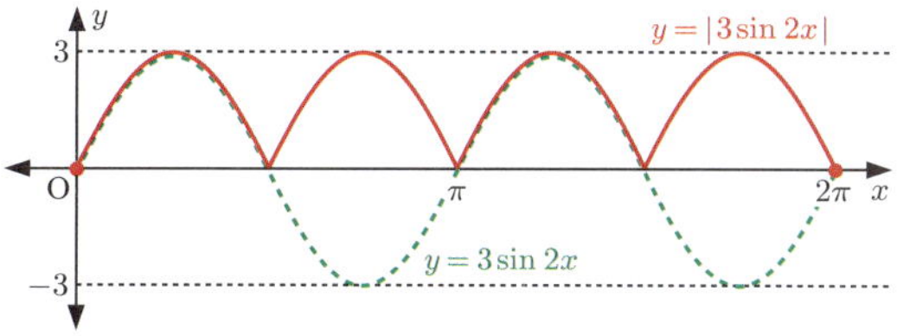

EXERCISE 9C

1 **a**

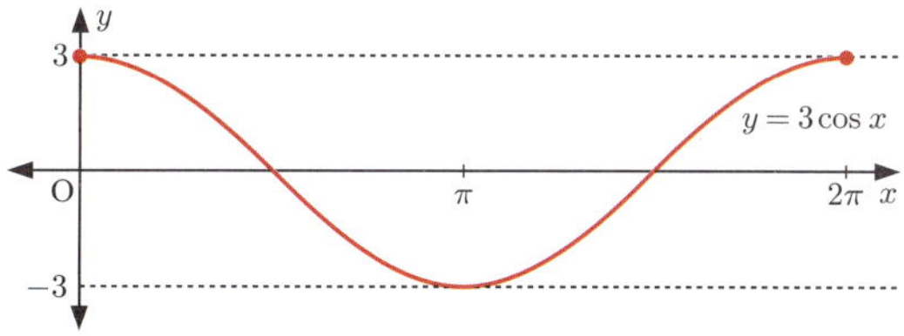

b

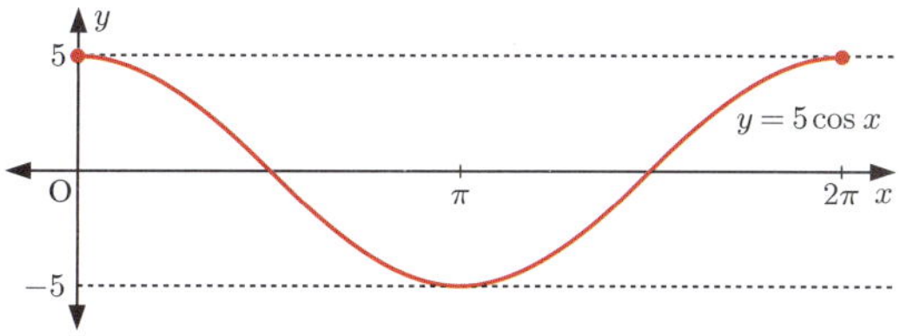

c

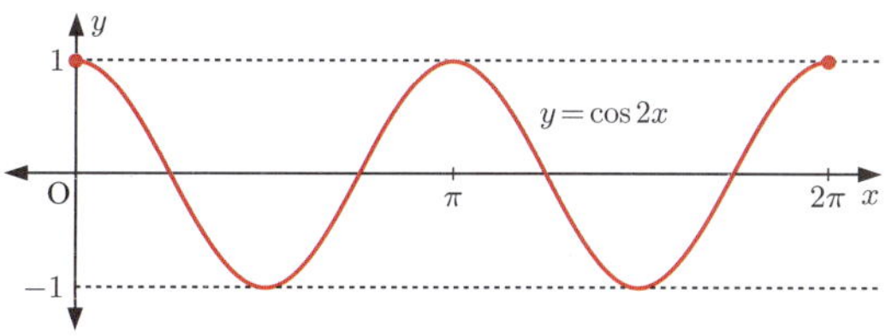

d

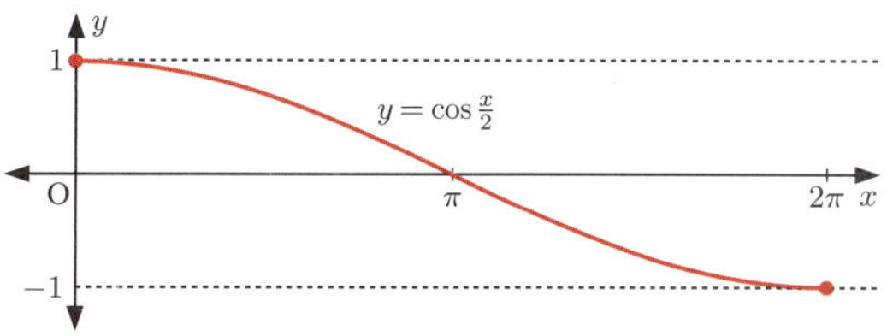

e

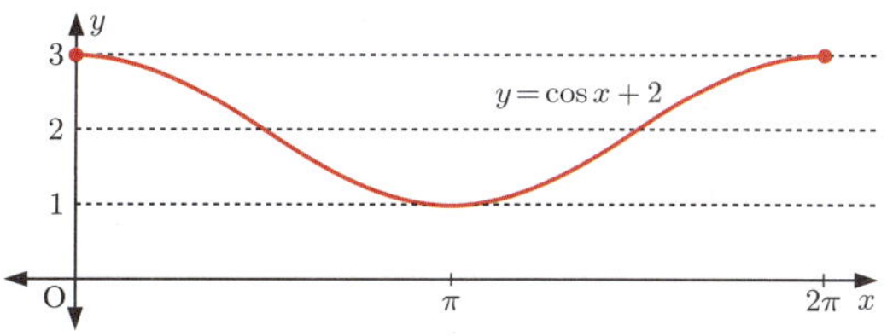

f

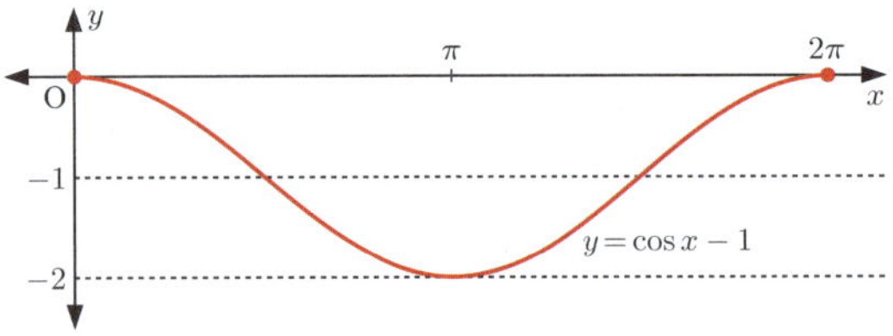

g

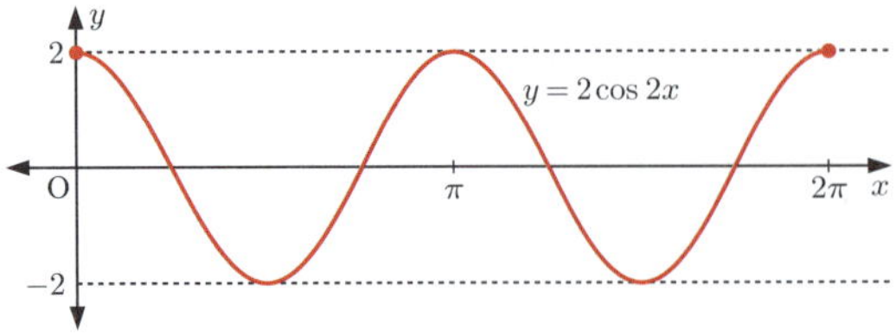

h

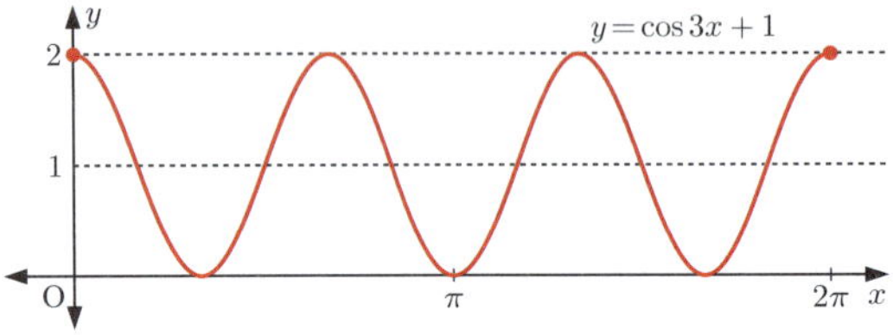

i

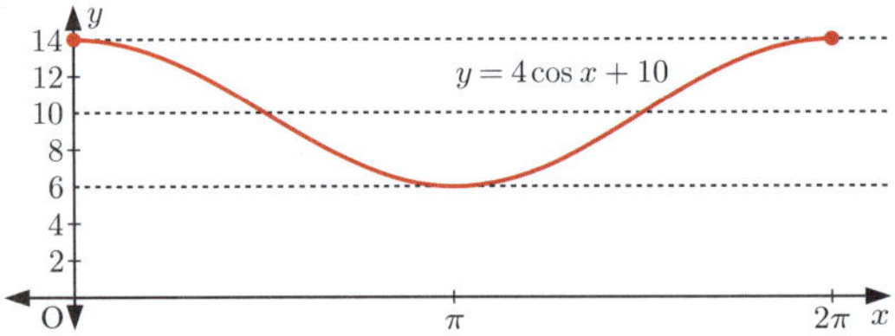

j

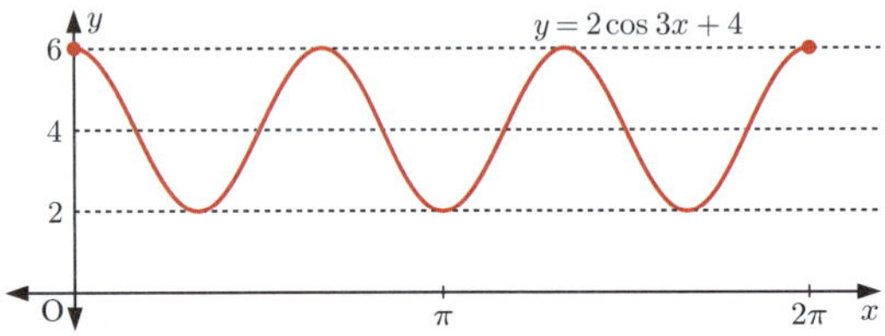

k

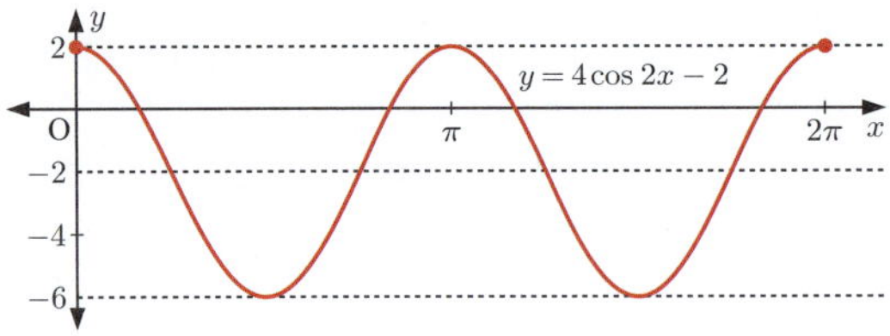

l

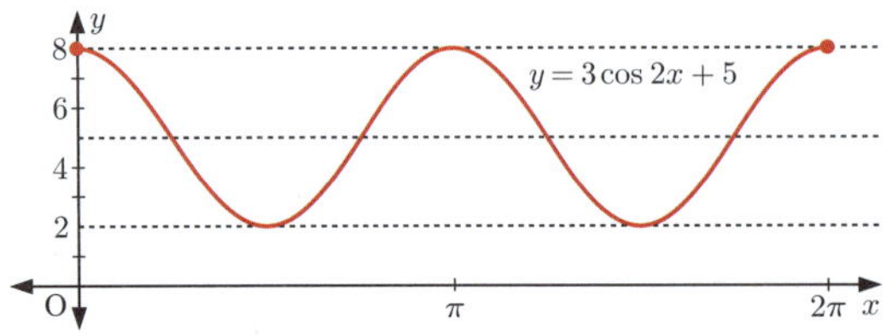

2 **a** $a = 4,\ b = 3,\ c = -1$ **b** $a = 3,\ b = 5,\ c = 3$
c $a = \frac{1}{6},\ b = \frac{3}{4},\ c = -4$

3 **a** maximum $= 5$, minimum $= -5$
b maximum $= 3$, minimum $= -1$
c maximum $= -4$, minimum $= -10$

4 **a** $y = 2\cos 2x$ **b** $y = \cos \frac{x}{2} + 2$

5 **a** $a = 5,\ b = \frac{4}{3},\ c = 1$

b

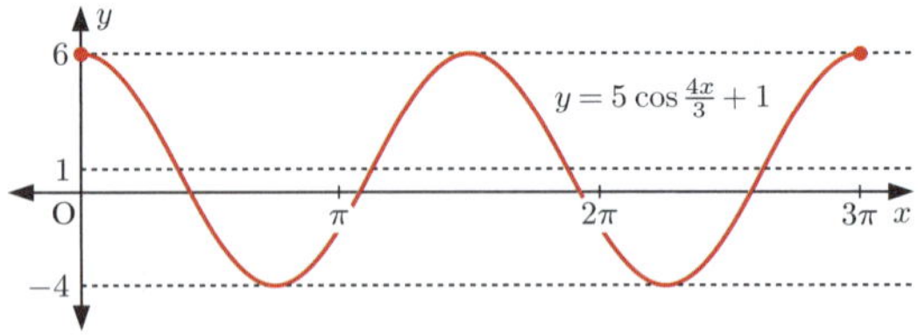

6 **a**

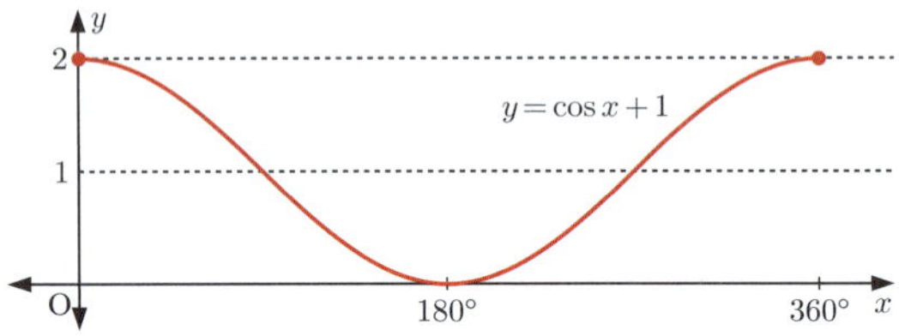

b

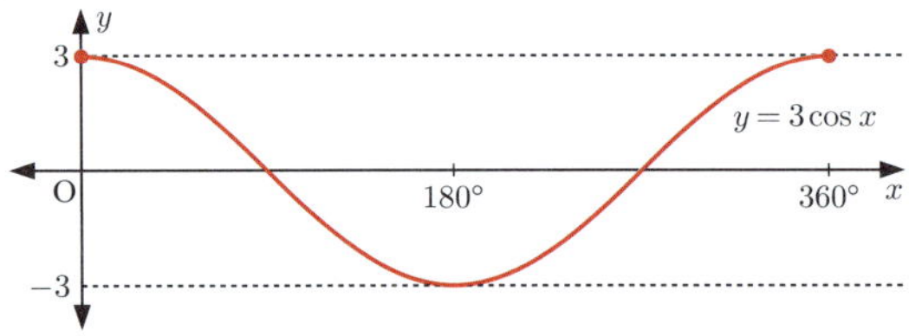

c

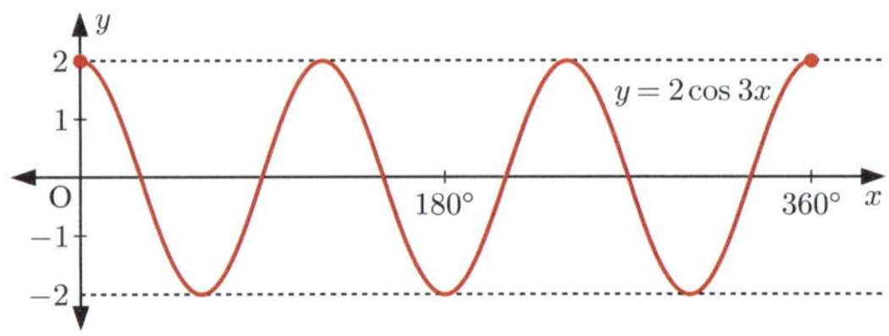

7 **a** $a = \frac{3}{2},\ b = 6,\ c = -\frac{3}{2}$

b

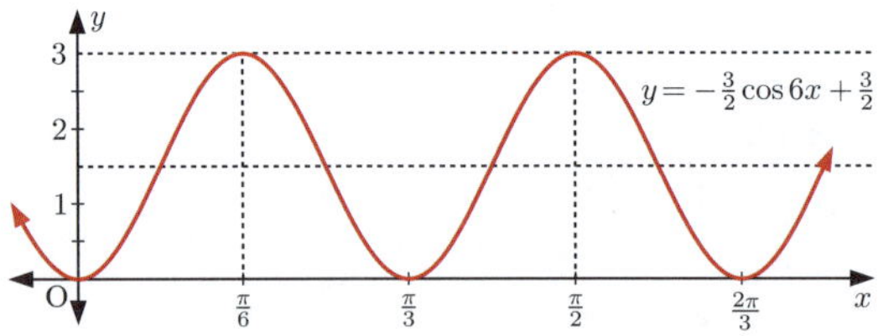

c $y = -\frac{3}{2}\cos 6x + \frac{3}{2}$

EXERCISE 9D

1 **a** $\frac{\pi}{2}$ **b** $\frac{\pi}{4}$ **c** 6π

2 **a**

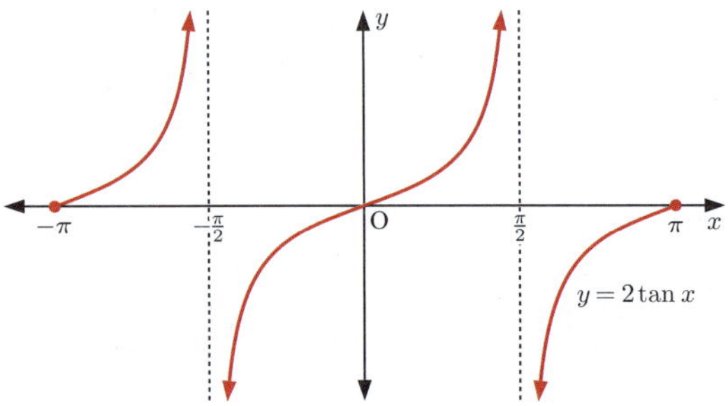

b

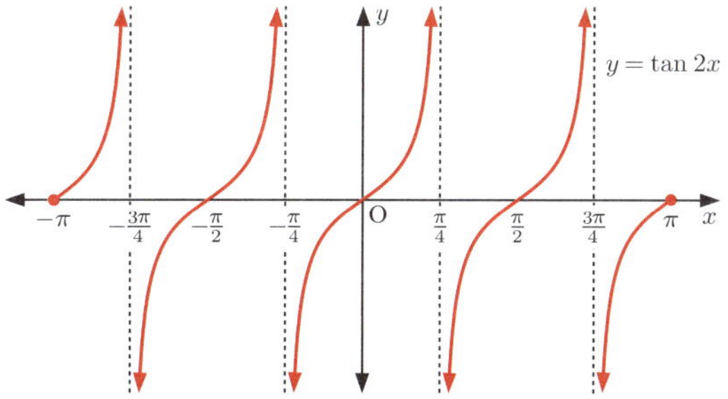

c

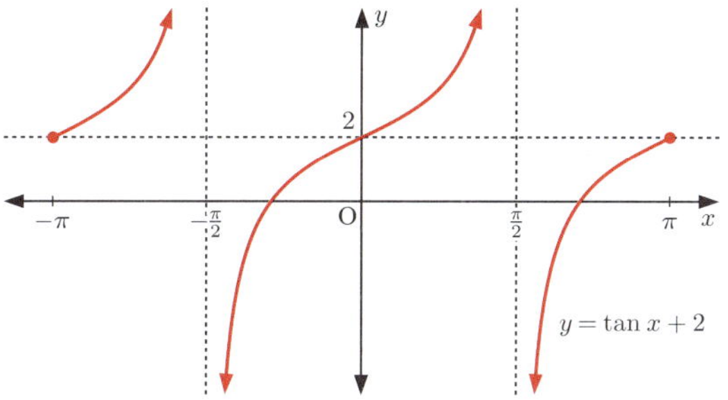

d

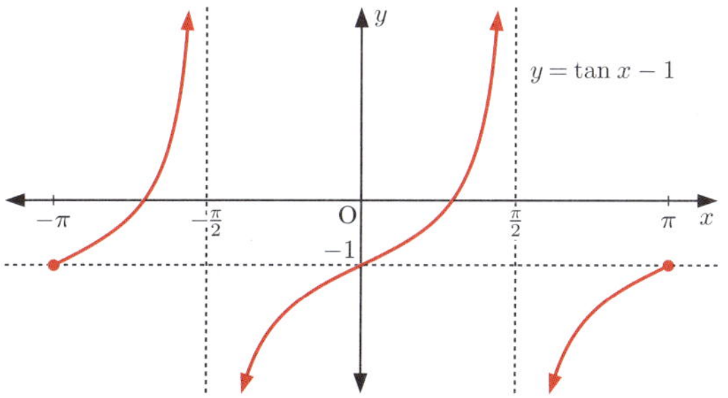

e

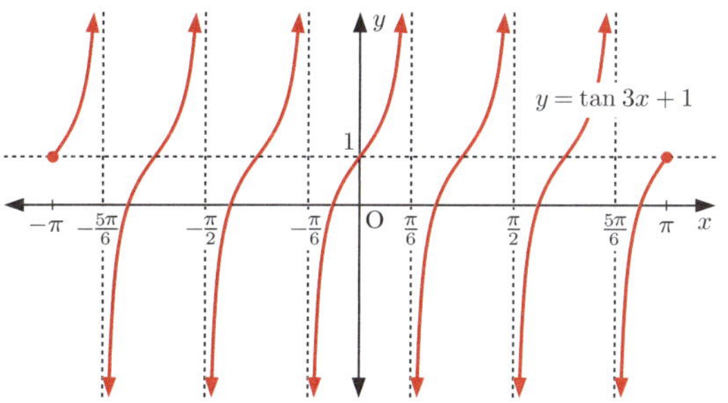

f

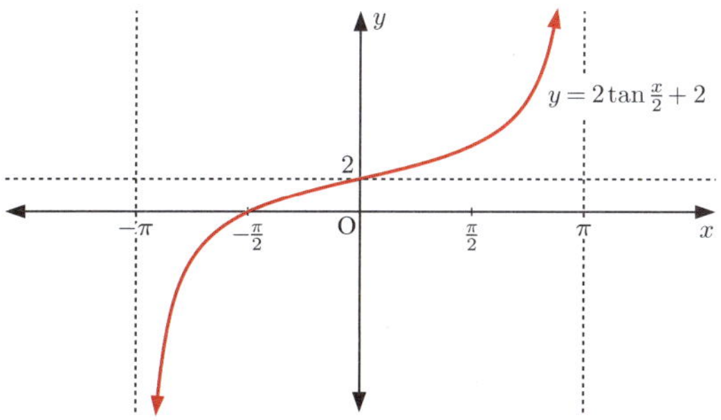

3 **a** $b = \frac{3}{2},\ c = 2$ **b** $b = 2,\ c = -3$

4 $m = \frac{1}{4},\ n = \frac{3}{4}$ **5** $a = 4,\ b = 3,\ c = 2$

EXERCISE 9E.1

1 **a** $x \approx 0.3, 2.8, 6.6, 9.1, 12.9$ **b** $x \approx 5.9, 9.8, 12.2$

2 **a** $x \approx 1.2, 5.1, 7.4$ **b** $x \approx 4.4, 8.2, 10.7$

3 **a** $x \approx 0.4, 1.2, 3.5, 4.3, 6.7, 7.5, 9.8, 10.6, 13.0, 13.7$
b $x \approx 1.7, 3.0, 4.9, 6.1, 8.0, 9.3, 11.1, 12.4, 14.3$

4 **a** $x \approx 0.3, 2.8, 3.5, 6.0, 6.6, 9.1, 9.7, 12.2, 12.9$
b $x \approx 0.9, 2.3, 4.0, 5.4, 7.2, 8.5, 10.3, 11.7, 13.5, 14.8$

5 **a** **i** ≈ 1.6 **ii** ≈ -1.1
b **i** $x \approx 1.1, 4.2, 7.4$ **ii** $x \approx 2.2, 5.3$

EXERCISE 9E.2

1 **a** $x = \frac{\pi}{3}, \frac{5\pi}{3}, \frac{7\pi}{3}$, or $\frac{11\pi}{3}$ **b** $x = \frac{5\pi}{4}, \frac{7\pi}{4}, \frac{13\pi}{4}$, or $\frac{15\pi}{4}$
c $x = \frac{\pi}{3}, \frac{4\pi}{3}, \frac{7\pi}{3}$, or $\frac{10\pi}{3}$

2 **a** $x = -\frac{5\pi}{3}, -\frac{4\pi}{3}, \frac{\pi}{3}$, or $\frac{2\pi}{3}$ **b** $x = -\frac{7\pi}{4}, -\frac{\pi}{4}, \frac{\pi}{4}$, or $\frac{7\pi}{4}$
c $x = -\frac{7\pi}{4}, -\frac{3\pi}{4}, \frac{\pi}{4}$, or $\frac{5\pi}{4}$

3 **a** $x = \frac{2\pi}{3}, \frac{4\pi}{3}$, or $\frac{8\pi}{3}$ **b** $x = \frac{\pi}{3}, \frac{2\pi}{3}, \frac{4\pi}{3}, \frac{5\pi}{3}, \frac{7\pi}{3}$, or $\frac{8\pi}{3}$

4 **a** $x = \frac{5\pi}{4}$ or $\frac{7\pi}{4}$ **b** $x = \frac{5\pi}{12}, \frac{7\pi}{12}, \frac{13\pi}{12}, \frac{5\pi}{4}$, or $\frac{7\pi}{4}$

5 **a** $x = 30°$ or $210°$ **b** $x = 60°$

6 **a** $x = \frac{\pi}{3}, \frac{2\pi}{3}, \frac{4\pi}{3}$, or $\frac{5\pi}{3}$ **b** $x = 60°$ or $300°$
c $x = \frac{2\pi}{9}, \frac{5\pi}{9}$, or $\frac{8\pi}{9}$ **d** $x = \frac{\pi}{2}, \frac{3\pi}{2}$, or $\frac{5\pi}{2}$
e $x = -160°, -80°, -40°, 40°, 80°$, or $160°$
f $x = \frac{\pi}{8}, \frac{5\pi}{8}, \frac{9\pi}{8}$, or $\frac{13\pi}{8}$

7 A$(\frac{\pi}{12}, 0)$, B$(\frac{11\pi}{12}, 0)$, C$(\frac{13\pi}{12}, 0)$, D$(\frac{23\pi}{12}, 0)$

8 **a**

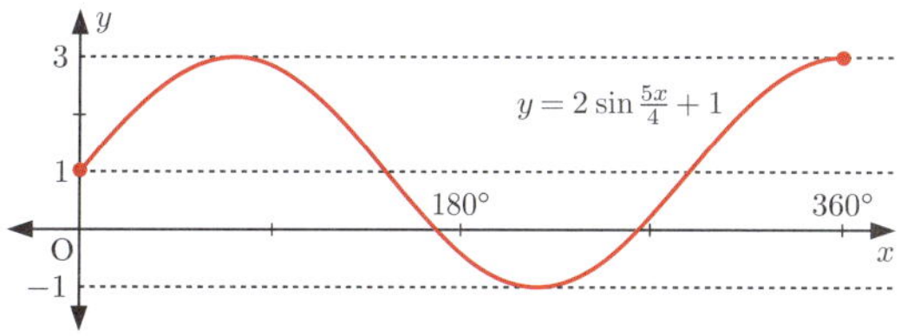

b $(168°, 0)$, $(264°, 0)$

9 **a** $x = \frac{5\pi}{12}$ or $\frac{23\pi}{12}$ **b** $x = \frac{\pi}{4}$ or $\frac{5\pi}{4}$
c $x = -\frac{5\pi}{3}, -\frac{\pi}{3}, \frac{\pi}{3}$, or $\frac{5\pi}{3}$ **d** $x = \frac{\pi}{12}$ or $\frac{11\pi}{12}$
e $x = \frac{4\pi}{9}$ **f** $x = -\frac{7\pi}{12}, -\frac{\pi}{12}, \frac{5\pi}{12}$, or $\frac{11\pi}{12}$

10 **a** $x = \frac{\pi}{4}$ or $\frac{5\pi}{4}$ **b** $x = \frac{3\pi}{4}$ or $\frac{7\pi}{4}$
c $x = \frac{\pi}{12}, \frac{5\pi}{12}, \frac{3\pi}{4}, \frac{13\pi}{12}, \frac{17\pi}{12}$, or $\frac{21\pi}{12}$
d $x = \frac{\pi}{6}, \frac{2\pi}{3}, \frac{7\pi}{6}$, or $\frac{5\pi}{3}$

11 **a** $\tan x = \pm\sqrt{3}$ **b** $x = \frac{\pi}{3}, \frac{2\pi}{3}, \frac{4\pi}{3}$, or $\frac{5\pi}{3}$

12 **a** $x = \frac{\pi}{6}, \frac{5\pi}{6}, \frac{7\pi}{6}$, or $\frac{11\pi}{6}$
b $x = 45°, 135°, 225°$, or $315°$
c $x = \frac{\pi}{4}, \frac{3\pi}{4}, \frac{5\pi}{4}, \frac{7\pi}{4}, \frac{9\pi}{4}, \frac{11\pi}{4}, \frac{13\pi}{4}$, or $\frac{15\pi}{4}$
d $x = -\frac{3\pi}{2}, -\frac{\pi}{2}, \frac{\pi}{2}$, or $\frac{3\pi}{2}$
e $x = 15°, 75°, 105°$, or $165°$
f $x = -\frac{2\pi}{3}, -\frac{\pi}{3}, \frac{\pi}{3}$, or $\frac{2\pi}{3}$ **g** $x = \frac{\pi}{3}, \frac{2\pi}{3}, \frac{4\pi}{3}$, or $\frac{5\pi}{3}$
h $x = \frac{\pi}{12}, \frac{\pi}{4}, \frac{5\pi}{12}, \frac{7\pi}{12}, \frac{3\pi}{4}$, or $\frac{11\pi}{12}$

13 **a** $\cos x = \frac{1}{2}$ **b** $x = \frac{\pi}{3}$ or $\frac{5\pi}{3}$

14 **a** $x = -\frac{5\pi}{6}$ or $-\frac{\pi}{6}$ **b** $x = -\frac{\pi}{4}$ or $\frac{\pi}{4}$
c $x = -\frac{5\pi}{6}$ or $\frac{\pi}{6}$ **d** $x = -\frac{\pi}{2}$ or $\frac{\pi}{2}$
e $x = -\frac{5\pi}{6}, -\frac{2\pi}{3}, \frac{\pi}{6}$, or $\frac{\pi}{3}$ **f** $x = -\frac{\pi}{12}$ or $\frac{11\pi}{12}$

15 **a** $x = \frac{\pi}{2}$ or $\frac{3\pi}{2}$ **b** $x = \frac{\pi}{3}, \frac{2\pi}{3}, \frac{4\pi}{3}$, or $\frac{5\pi}{3}$
c $x = \frac{\pi}{6}, \frac{5\pi}{6}, \frac{7\pi}{6}$, or $\frac{11\pi}{6}$

EXERCISE 9E.3

1 **a** $x \approx 0.93$ or 2.21 **b** $x \approx 1.32$ or 4.97
c $x \approx 0.61$ or 3.75 **d** $x \approx 1.80$ or 4.48
e $x \approx 3.87$ or 5.55 **f** $x \approx 2.16$ or 5.30

2 **a** $x \approx 0.18, 2.96, 6.46$, or 9.24
b $x \approx -5.39, -0.89, 0.89$, or 5.39
c $x \approx 2.60, 5.74$, or 8.88
d $x \approx 0.23, 1.34, 3.37$, or 4.48 **e** $x \approx -0.71$ or 2.43
f $x \approx 0.72, 1.38, 2.81, 3.47$, or 4.91
g $x \approx 2.30$ or 3.98 **h** $x \approx 1.50$ or 2.43

3 **a** $x \approx 199.5°$ or $340.5°$ **b** $x \approx 39.2°$ or $140.8°$
c $x \approx -162.9°, -102.9°, -42.9°, 17.1°, 77.1°$, or $137.1°$
d $x \approx 47.4°, 87.6°, 137.4°$, or $177.6°$
e $x \approx 46.2°, 73.8°$, or $166.2°$
f $x \approx 40.9°, 130.9°, 220.9°$, or $310.9°$

4 **a** $x \approx 0.34$ or 2.80 **b** $x \approx 1.77$ or 4.51
c $x \approx 0.46$ or 3.61 **d** $x \approx 0.67, 2.47, 3.81$, or 5.61
e $x \approx 2.50$ or 5.64 **f** $x \approx 0.86, 1.71, 4.01$, or 4.85

EXERCISE 9F.1

1 **a** $8\sin\theta$ **b** $5\tan\theta$ **c** $2\cos\theta$
d $-4\tan\theta$ **e** $4\sin^2\theta$ **f** $-3\cos^2\theta$

2 **a** 4 **b** -3 **c** 4 **d** $2\sin^2\theta$ **e** $5\cos^2 A$
f $-\cos^2\theta$ **g** $-4\sin^2 A$ **h** $\cos\theta$ **i** -1

3 **a** $4\tan\theta$ **b** $\tan^3\alpha$ **c** $\sec x$
d $4\sin x$ **e** $2\tan\theta$ **f** 1
g $\cot\theta$ **h** $\sec A$ **i** $5\sin x$

4 **a** $5\sec^2\theta$ **b** $4\operatorname{cosec}^2 A$ **c** 1
d $\tan^2\theta$ **e** $\tan^2\alpha$ **f** $\cos A$

EXERCISE 9F.2

1 **a** $(\cos\theta + \sin\theta)(\cos\theta - \sin\theta)$
b $(2 + \tan A)(2 - \tan A)$ **c** $(\sin\theta + 1)(\sin\theta - 1)$
d $\cos\theta(2\cos\theta + 1)$ **e** $\sin\alpha(\sin\alpha - 5)$
f $(2\sin\phi + \cos\phi)(2\sin\phi - \cos\phi)$
g $(\tan\theta + 1)(\tan\theta - 2)$ **h** $(\cos A + 1)(\cos A + 4)$
i $(\sin\beta + 2)(\sin\beta - 5)$ **j** $(2\cos\theta + 3)(\cos\theta - 2)$
k $(3\tan\phi - 2)(\tan\theta + 1)$ **l** $(3\sin A - 2)(2\sin A + 5)$

2 **a** $\cos\theta(\cos\theta - 1)$ **b** $(\sin\theta + 2)(\sin\theta - 1)$
c $(\tan\theta + 1)^2$

3 **a** $1 + \cos\theta$ **b** $\cos A - \sin A$ **c** $3 + \tan\phi$
d $\sin\theta + 1$ **e** $-\cos\phi - 2$ **f** $\tan\theta - 1$

5 $\dfrac{(3\cos x - 2)(\cos x + \sin x)}{\sin x}$

EXERCISE 9F.3

1 **b** $\theta = \frac{7\pi}{6}$ or $\frac{11\pi}{6}$ **2** **b** $\theta = 30°$ or $210°$

3 **b** $A = \frac{\pi}{6}, \frac{5\pi}{6}, \frac{7\pi}{6}, \frac{11\pi}{6}, \frac{13\pi}{6}, \frac{17\pi}{6}, \frac{19\pi}{6}$, or $\frac{23\pi}{6}$

4 **b** $\theta \approx 1.11$ or 4.25 **5** **b** $x = \frac{\pi}{6}, \frac{5\pi}{6}, \frac{7\pi}{6}$, or $\frac{11\pi}{6}$

6 **a** $\cos x(2\cos x - 1)$ **b** $x = \frac{\pi}{3}, \frac{\pi}{2}, \frac{3\pi}{2}$, or $\frac{5\pi}{3}$

7 **a** $(2\sin\theta - 1)(\sin\theta + 1)$ **b** $\theta = 30°, 150°$, or $270°$

8 **a** $x = 0, \frac{\pi}{4}, \pi, \frac{5\pi}{4}$, or 2π **b** $x = \frac{2\pi}{3}, \pi$, or $\frac{4\pi}{3}$
c $x = 0, \frac{\pi}{4}, \frac{\pi}{2}, \pi, \frac{5\pi}{4}, \frac{3\pi}{2}$, or 2π **d** $x = \frac{\pi}{3}$ or $\frac{5\pi}{3}$
e $x = \frac{3\pi}{2}$ **f** $x = \frac{3\pi}{4}, \frac{7\pi}{4}$ *or* $x \approx 1.25, 4.39$

9 **a** $\theta = 90°, 180°$, or $270°$ **b** $\theta = 270°$
c $\theta = 90°, 270°$ *or* $\theta \approx 131.8°, 228.2°$
d $\theta = 45°$ or $225°$ **e** $\theta = 45°, 105°, 165°, 225°$, or $285°$
f $\theta = 60°, 300°$ *or* $\theta \approx 109.5°, 250.5°$

10 **a** $x = \frac{\pi}{2}$ or $x \approx 3.48, 5.94$ **b** $x \approx 1.82$ or 4.46
c $x \approx 1.37$ or 4.91

11 **a** $(\tan^2 y - 3)(\tan^2 y - 1)$
b $y = \frac{\pi}{4}, \frac{\pi}{3}, \frac{2\pi}{3}, \frac{3\pi}{4}, \frac{5\pi}{4}, \frac{4\pi}{3}, \frac{5\pi}{3}, \frac{7\pi}{4}, \frac{9\pi}{4}, \frac{7\pi}{3}, \frac{8\pi}{3}, \frac{11\pi}{4}, \frac{13\pi}{4}, \frac{10\pi}{3}, \frac{11\pi}{3}$, or $\frac{15\pi}{4}$

12 **a** $(2\sin x + 1)(\sqrt{3}\sin x - \cos x) = 0$
b $x = 30°, 210°$, or $330°$

REVIEW SET 9A

1 **a** not periodic **b** periodic

2 **a** minimum = 2, maximum = 4
b minimum = −2, maximum = 2
c minimum = −3, maximum = 3
d minimum = −2, maximum = 0

3 **a** 2π **b** $\frac{\pi}{2}$ **c** π **d** $\frac{8\pi}{3}$

4 **a**

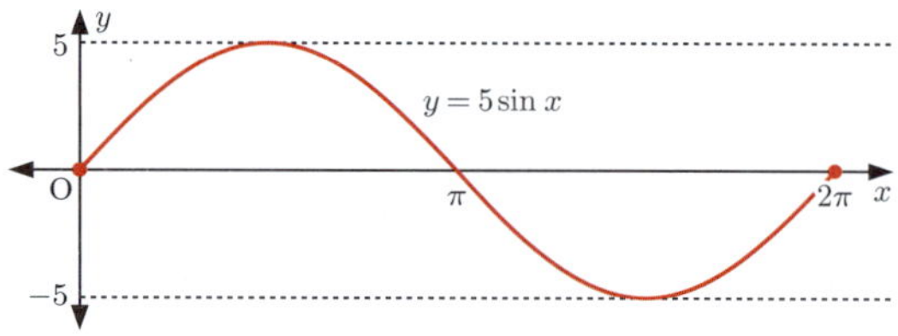

b

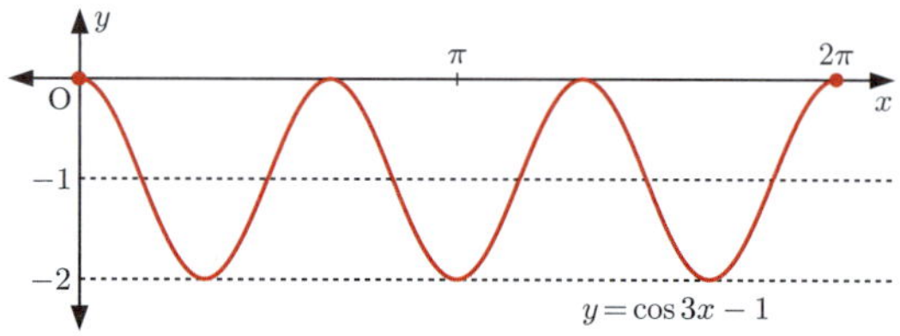

c

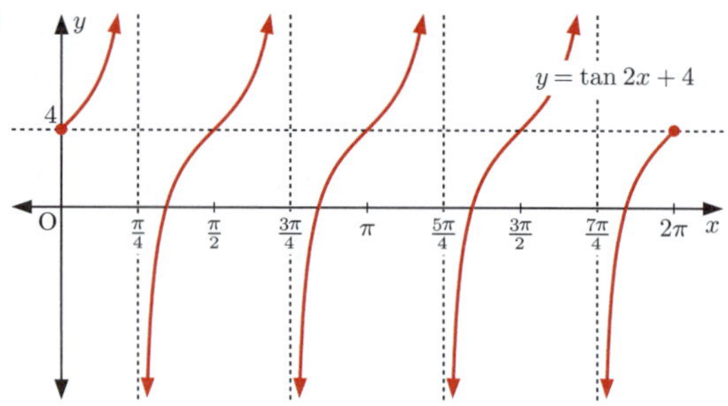

d

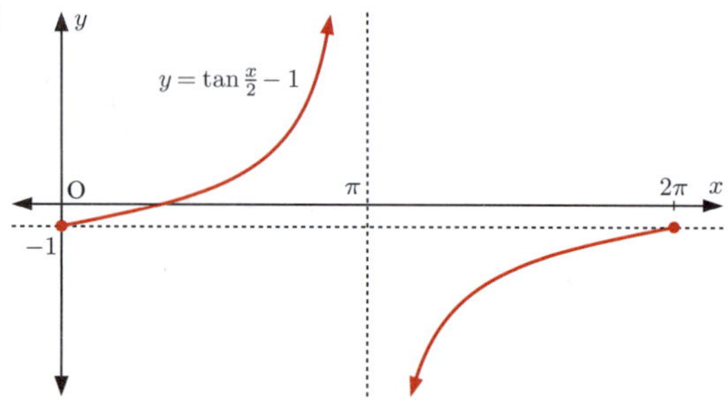

5

Function	Period	Amplitude
$y = 3\sin 2x + 1$	π	3
$y = \tan 2x$	$\frac{\pi}{2}$	undefined
$y = 2\cos 3x - 3$	$\frac{2\pi}{3}$	2

Function	Domain	Range
$y = 3\sin 2x + 1$	$x \in \mathbb{R}$	$-2 \leqslant y \leqslant 4$
$y = \tan 2x$	$x \neq \pm\frac{\pi}{4}, \pm\frac{3\pi}{4}, \ldots$	$y \in \mathbb{R}$
$y = 2\cos 3x - 3$	$x \in \mathbb{R}$	$-5 \leqslant y \leqslant -1$

6 $y = 4\cos 2x$

7

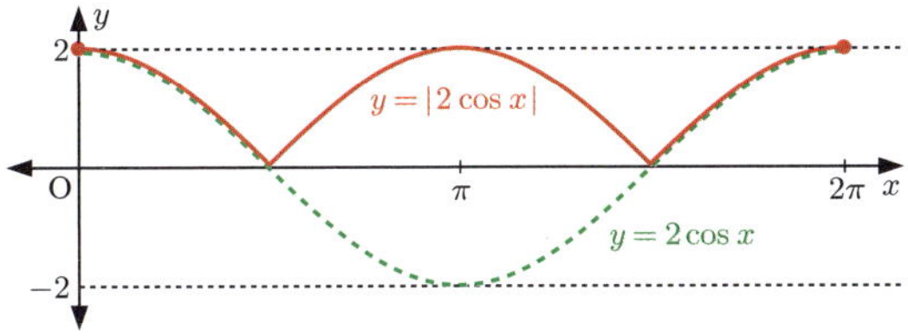

8 **a** $x = \frac{7\pi}{6}, \frac{11\pi}{6}, \frac{19\pi}{6}$, or $\frac{23\pi}{6}$
b $x = -\frac{7\pi}{4}, -\frac{5\pi}{4}, \frac{\pi}{4}$, or $\frac{3\pi}{4}$
c $x = \frac{4\pi}{9}, \frac{5\pi}{9}, \frac{10\pi}{9}, \frac{11\pi}{9}, \frac{16\pi}{9}$, or $\frac{17\pi}{9}$
d $x = \frac{\pi}{4}, \frac{7\pi}{4}, \frac{9\pi}{4}$, or $\frac{15\pi}{4}$

9 **a** $x = -\frac{2\pi}{3}, -\frac{\pi}{6}, \frac{\pi}{3}$, or $\frac{5\pi}{6}$ **b** $x = -\frac{7\pi}{12}$ or $-\frac{\pi}{12}$
c $x = -\frac{2\pi}{3}, -\frac{\pi}{3}, \frac{\pi}{3}$, or $\frac{2\pi}{3}$

10 **a** $x = 30°$ or $150°$ **b** $x = 135°$
c $x = 67.5°$ or $157.5°$

11 **a** $x \approx 0.80$ or 2.34 **b** $x \approx 2.18$ or 4.10
c $x \approx 1.21$ or 4.35

12 **a** $9\cos\theta$ **b** $2\sin\theta$ **c** $\cos\theta$

14 **b** $x = 0, \pi, 2\pi, 3\pi$, or 4π

15 **a** $(3 + \sin\theta)(3 - \sin\theta)$ **b** $(\cos\theta - 1)(\cos\theta - 2)$
c $(2\tan\theta - 1)(\tan\theta + 4)$

16 **a** $x = \frac{\pi}{6}, \frac{\pi}{2}$, or $\frac{5\pi}{6}$ **b** $x = \frac{\pi}{3}, \frac{2\pi}{3}, \frac{4\pi}{3}$, or $\frac{5\pi}{3}$
c $x = \frac{3\pi}{4}, \frac{7\pi}{4}$ *or* $x \approx 1.33, 4.47$

REVIEW SET 9B

1 **a** **i** 6 **ii** $y = 4$ **iii** 3
b **i** 4π **ii** $y = -\frac{3}{2}$ **iii** $\frac{9}{2}$

2 **a** $b = 6$ **b** $b = \frac{1}{2}$ **c** $b = \frac{\pi}{3}$

3 **a** minimum $= -5$, maximum $= -1$
b minimum $= -2$, maximum $= 4$
c minimum $= 5$, maximum $= 13$

4 **a**

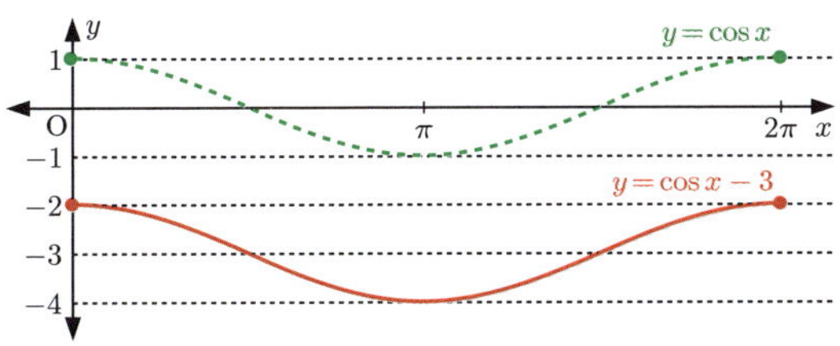

b

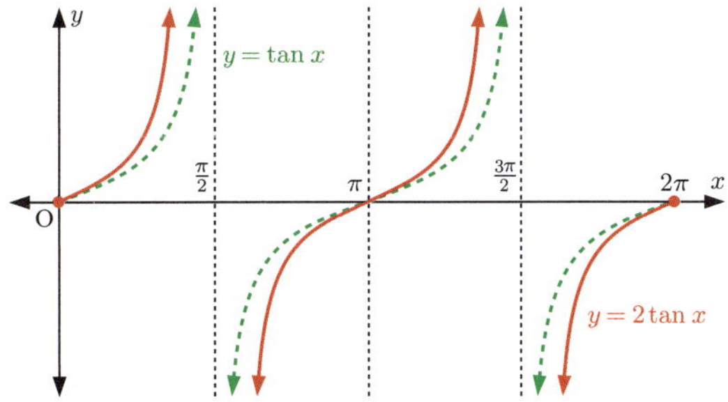

c

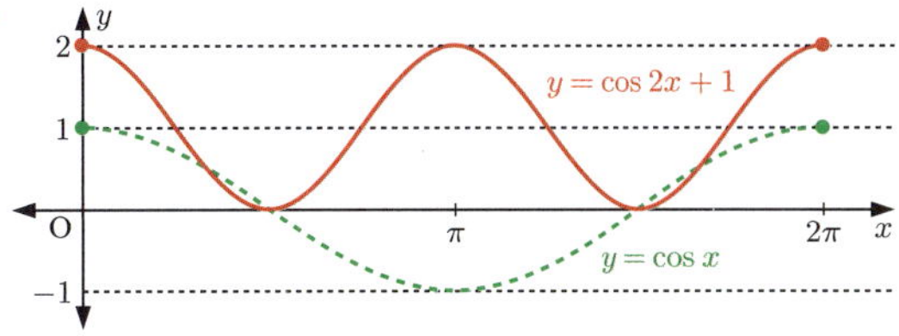

d

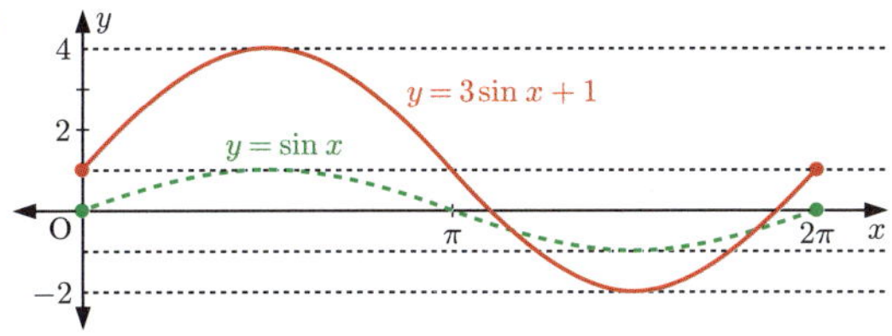

5 **a** $a = 2, \ b = 6, \ c = -2$
b

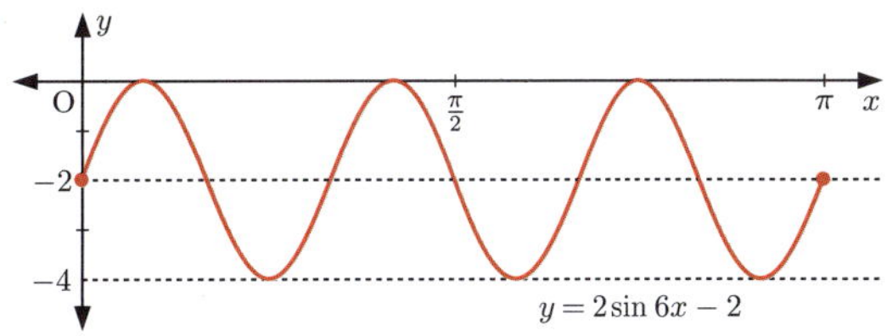

6 **a** $y = 2\tan x + 2$
b

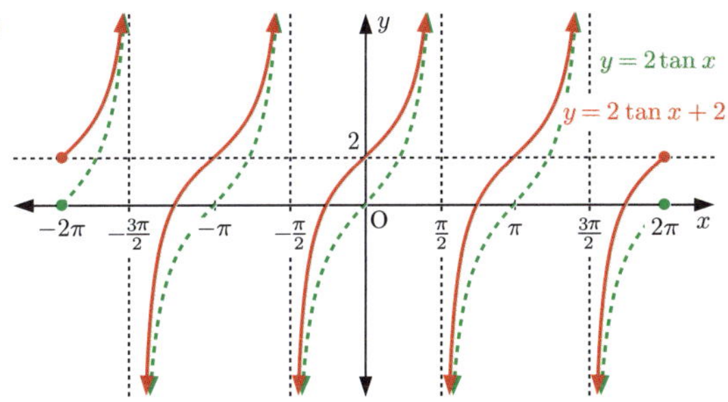

7 **a** $m = 3, \ n = -1$
b $(\frac{\pi}{18}, 0), (\frac{5\pi}{18}, 0), (\frac{13\pi}{18}, 0), (\frac{17\pi}{18}, 0), (\frac{25\pi}{18}, 0), (\frac{29\pi}{18}, 0)$

8 **a** $x \approx -6.1, -3.4$ **b** $x \approx 0.8$

9 **a** $x = \frac{4\pi}{3}$ or $\frac{5\pi}{3}$ **b** $x = \frac{\pi}{8}, \frac{7\pi}{8}, \frac{9\pi}{8}$, or $\frac{15\pi}{8}$
c $x = \frac{\pi}{6}$ or $\frac{7\pi}{6}$

10 **a** $x = -135°, -45°, 45°$, or $135°$
b $x = -135°, -45°, 45°$, or $135°$
c $x = -105°, -15°, 75°$, or $165°$

11 **a** $A \approx 66.4°$ or $293.6°$ **b** $A \approx 192.8°$ or $347.2°$
c $A \approx 7.0°, 97.0°, 187.0°$, or $277.0°$

12 **a** $6\cos^2\phi$ **b** $\text{cosec}\,\theta$ **c** $\tan^4\alpha$

13 b $\theta = \frac{\pi}{2}, \frac{3\pi}{2}, \frac{5\pi}{2}$, or $\frac{7\pi}{2}$

14 a $x = \frac{7\pi}{6}$ or $\frac{11\pi}{6}$ b $x = 0, \pi$, or 2π

c $x = \frac{3\pi}{4}, \frac{7\pi}{4}$ *or* $x \approx 0.983, 4.12$

15 a $y = \frac{\pi}{6}, \frac{5\pi}{6}, \frac{7\pi}{6}$, or $\frac{11\pi}{6}$

b $y \approx 0.85, 2.29, 3.87, 5.55$

16 a $(2\sin^2 x - 1)(4\sin^2 x - 3)$

b $x = 45°, 60°, 120°, 135°, 225°, 240°, 300°$, or $315°$

EXERCISE 10A

1 a 6 paths b 8 paths

2 a 12 routes b 36 routes

3 30 meal combinations 4 $26^5 = 11\,881\,376$ codes

5 30 ways 6 a 16 numbers b 12 numbers

7 $26^4 \times 10^3 = 456\,976\,000$ registration plates

EXERCISE 10B

1 a 5 paths b 7 paths c 11 paths d 16 paths

2 9 routes 3 a 3 paths b 4 paths

EXERCISE 10C

1 a 2 b 120 c 720 d 3 628 800

2 a 7 b 60 c 56 d 120 e $\frac{1}{20}$

f $\frac{1}{210}$ g 15 h 45

3 a $n+1$ b $\dfrac{n(n-1)}{2}$ c $\dfrac{(n+3)(n+2)(n+1)}{6}$

4 a $4!$ b $7!$ c $\dfrac{8!}{5!}$ d $\dfrac{15!}{11!}$ e $\dfrac{9!}{3!6!}$ f $\dfrac{13!}{4!9!}$

5 $\dfrac{10!}{6!} = 5040$ ways

EXERCISE 10D

1 a A, B b AB, BA

2 a P, Q, R, S

b PQ, PR, PS, QP, QR, QS, RP, RQ, RS, SP, SQ, SR

c PQR, PQS, PRQ, PRS, PSQ, PSR, QPR, QPS, QRP, QRS, QSP, QSR, RPQ, RPS, RQP, RQS, RSP, RSQ, SPQ, SPR, SQP, SQR, SRP, SRQ

d PQRS, PQSR, PRQS, PRSQ, PSQR, PSRQ, QPRS, QPSR, QRPS, QRSP, QSPR, QSRP, RPQS, RPSQ, RQPS, RQSP, RSPQ, RSQP, SPQR, SPRQ, SQPR, SQRP, SRPQ, SRQP

3 a 120 b 2520 c $\approx 1.01 \times 10^{19}$

4 72 ways 5 1680 ways

6 a 64 numbers b 24 numbers

7 a 240 ways b 3360 ways

8 a 720 ways b 120 ways c 48 ways

9 a 120 b 6 c 6

10 a 360 numbers b 60 numbers c 12 numbers

11 a 294 numbers b 108 numbers c 42 numbers

d 150 numbers

12 a 24 ways b 12 ways

13 a 5040 ways b 30 240 ways

14 a 24 ways b 6 ways c 18 ways d 12 ways

e 6 ways

15 a 720 ways b 144 ways c 72 ways d 144 ways

EXERCISE 10E

1 a 3 b 15 c 5 d 21

2 a PQ, PR, PS, PT, QR, QS, QT, RS, RT, ST

b $\binom{5}{2} = \dfrac{5 \times 4}{2 \times 1} = 10$ ✓

4 a $k = 2$ b $k = 5$ or 6

5 a permutation b combination c permutation

d combination

6 $\binom{15}{7} = 6435$ different teams

7 a $\binom{27}{5} = 80\,730$ different committees

b $\binom{26}{4} = 14\,950$ different committees

8 $\binom{5}{3} = 10$ different selections

9 $\binom{12}{6} = 924$ different teams

10 a $\binom{18}{11} = 31\,824$ different teams

b $\binom{16}{9} = 11\,440$ different teams

c $2\binom{16}{10} = 16\,016$ different teams

11 a $\binom{14}{3} = 364$ different committees

b i $\binom{6}{3} = 20$ different committees

ii $\binom{6}{2}\binom{8}{1} = 120$ different committees

iii $\binom{6}{1}\binom{8}{2} = 168$ different committees

iv $\binom{8}{3} = 56$ different committees

12 a $\binom{16}{4} = 1820$ ways b $\binom{7}{2}\binom{9}{2} = 756$ ways

c $\binom{7}{2}\binom{9}{2} + \binom{7}{1}\binom{9}{3} + \binom{9}{4} = 1470$ ways

13 $\binom{5}{1}\binom{7}{3}\binom{4}{2} = 1050$ ways

14 a $\binom{16}{5} = 4368$ ways b $\binom{10}{3}\binom{6}{2} = 1800$ ways

c $\binom{10}{5}\binom{6}{0} = 252$ ways

d $\binom{10}{3}\binom{6}{2} + \binom{10}{4}\binom{6}{1} + \binom{10}{5}\binom{6}{0} = 3312$ ways

15 a $\binom{15}{8} = 6435$ ways b $\binom{9}{5}\binom{6}{3} = 2520$ ways

c $\binom{6}{6}\binom{9}{2} = 36$ ways

d $\binom{9}{8}\binom{6}{0} + \binom{9}{7}\binom{6}{1} + \binom{9}{6}\binom{6}{2} + \binom{9}{5}\binom{6}{3} = 4005$ ways

16 a 2304 bows

b 1128 handshakes, the sum of the number of side lengths and diagonals of a 48 sided polygon.

c 4560 curtsies d 120 different orders

17 90 diagonals 18 $\binom{10}{5} = 252$ numbers

19 a $\binom{11}{3}\binom{8}{3}\binom{5}{4} = 46\,200$ ways

b $\binom{8}{1}\binom{7}{3} + \binom{8}{2}\binom{6}{2} = 700$ combinations

20 a $\binom{8}{4} = 70$ ways b $\binom{8}{2}\binom{6}{2}\binom{4}{2} = 2520$ ways

21 11 200 ways

EXERCISE 10F

1 a $x^3 + 6x^2 + 12x + 8$ b $x^3 - 3x^2 + 3x - 1$

c $1 + 6x + 12x^2 + 8x^3$ d $27x^3 - 54x^2 + 36x - 8$

e $x^3 + \frac{3}{2}x^2 + \frac{3}{4}x + \frac{1}{8}$ f $8x^3 - 12x + \dfrac{6}{x} - \dfrac{1}{x^3}$

2 **a** $x^4+12x^3+54x^2+108x+81$
b $x^4-4x^3+6x^2-4x+1$
c $16x^4+32x^3+24x^2+8x+1$
d $x^4-8x^3+24x^2-32x+16$
e $81x^4+432x^3+864x^2+768x+256$
f $x^4-8x^2+24-\frac{32}{x^2}+\frac{16}{x^4}$

3 **a** $x^5+5x^4+10x^3+10x^2+5x+1$
b $x^5-5x^4y+10x^3y^2-10x^2y^3+5xy^4-y^5$
c $32+\frac{80}{x}+\frac{80}{x^2}+\frac{40}{x^3}+\frac{10}{x^4}+\frac{1}{x^5}$
d $x^{10}-10x^7+40x^4-80x+\frac{80}{x^2}-\frac{32}{x^5}$

4 $64+160x^2+20x^4$

5 **a** 1 6 15 20 15 6 1
b **i** $x^6+6x^5+15x^4+20x^3+15x^2+6x+1$
ii $64-192x+240x^2-160x^3+60x^4-12x^5+x^6$
iii $x^6-6x^3+15-\frac{20}{x^3}+\frac{15}{x^6}-\frac{6}{x^9}+\frac{1}{x^{12}}$

6 $1+18x+135x^2+....$

7 **a** $10+6\sqrt{3}$ **b** $17+12\sqrt{2}$ **c** $76-44\sqrt{3}$

8 $\frac{59+34\sqrt{3}}{13}$

9 **a** $243-405x+270x^2-90x^3+15x^4-x^5$
b 238.976 910 149 9

10 **a** $a=3,\ b=-\cos x$
b $T_3=9\cos^2 x,\quad T_4=-\cos^3 x$

11 $3x^5+25x^4+80x^3+120x^2+80x+16$

12 **a** 80 **b** 540

EXERCISE 10G

1 **a** $(2x+5)^9=\binom{9}{0}(2x)^9 5^0+\binom{9}{1}(2x)^8 5^1+\binom{9}{2}(2x)^7 5^2+....+\binom{9}{8}(2x)^1 5^8+\binom{9}{9}(2x)^0 5^9$
b $(4x-3)^{13}=\binom{13}{0}(4x)^{13}(-3)^0+\binom{13}{1}(4x)^{12}(-3)^1+\binom{13}{2}(4x)^{11}(-3)^2+....+\binom{13}{12}(4x)^1(-3)^{12}+\binom{13}{13}(4x)^0(-3)^{13}$
c $\left(3x-\frac{2}{x}\right)^{16}$
$=\binom{16}{0}(3x)^{16}\left(-\frac{2}{x}\right)^0+\binom{16}{1}(3x)^{15}\left(-\frac{2}{x}\right)^1+\binom{16}{2}(3x)^{14}\left(-\frac{2}{x}\right)^2+....+\binom{16}{15}(3x)^1\left(-\frac{2}{x}\right)^{15}+\binom{16}{16}(3x)^0\left(-\frac{2}{x}\right)^{16}$

2 **a** $T_5=\binom{7}{4}(3x)^3 5^4$ **b** $T_8=\binom{11}{7}(x^2)^4(-2)^7$
c $T_7=\binom{16}{6}(x)^{10}\left(\frac{5}{x}\right)^6$
d $T_{13}=\binom{17}{12}\left(\frac{8}{x}\right)^5(-x^2)^{12}$

3 **a** **i** $T_3=\binom{7}{2}(2x)^5(-1)^2$ **ii** $T_5=\binom{7}{4}(2x)^3(-1)^4$
b **i** $T_3=672x^5$ **ii** $T_5=280x^3$

4 **a** $T_{r+1}=\binom{7}{r}x^{7-r}3^r$ **b** $\binom{7}{2}3^2=189$

5 **a** $\binom{12}{4}2^8 3^4=10\,264\,320$ **b** $\binom{12}{7}2^5 3^7=55\,427\,328$

6 **a** $\binom{10}{3}1^7(-3)^3=-3240$
b $\binom{10}{7}1^3(-3)^7=-262\,440$

7 **a** 144 **b** 5376 **c** 2304

8 **a** $T_{r+1}=\binom{8}{r}x^{8-r}a^r$ **b** $a=-2$

9 **a** $\binom{10}{5}5^5=787\,500$ **b** $\binom{15}{5}(-4)^5=-3\,075\,072$

10 **a** $\binom{10}{3}2^7(-5)^3=-1\,920\,000$ **b** $\binom{6}{5}5^1 4^5=30\,720$
c $\binom{6}{4}1^2 3^4=1215$ **d** $\binom{15}{5}1^{10}(-2)^5=-96\,096$

11 $k=5$ **12** $a=3$ **13** **b** $a=5,\ b=2$

14 **a** $\binom{6}{2}2^4(-3)^2+\binom{6}{1}2^5(-3)^1=1584$
b $3\binom{5}{2}(-2)^2+4\binom{5}{1}(-2)^1=80$
c $\binom{8}{3}3^5 2^3-4\binom{8}{2}3^6 2^2=-217\,728$

15 **a** $\binom{7}{4}3^3(-2)^4=15\,120$
b $\binom{7}{4}3^3(-2)^4+3\binom{7}{3}3^4(-2)^3=-52\,920$

16 **a** $\binom{8}{3}2^5(-5)^3-3\binom{8}{1}2^7(-5)^1=-208\,640$
b $\binom{6}{3}2^3-\binom{6}{4}2^4=-80$

17 $a=3,\ b=-2,\ c=57$ **18** $a=3$ **19** $n=8$

20 $n=6$ **21** $k=-3,\ n=6$

REVIEW SET 10A

1 48 paths

2 **a** 24 **b** 42 **c** 126

3 **a** 24 **b** 6 **c** 48

4 **a** 840 numbers **b** 360 numbers **c** 280 numbers

5 **a** JKL, JKM, JKN, JKO, JLM, JLN, JLO, JMN, JMO, JNO, KLM, KLN, KLO, KMN, KMO, KNO, LMN, LMO, LNO, MNO
b $\binom{6}{3}=\frac{6!}{3!3!}=20$ ✓

6 **a** $\binom{8}{3}=56$ triangles **b** 24 triangles

7 **a** $\binom{11}{4}=330$ teams **b** **i** 60 teams **ii** 215 teams

8 **a** $x^3+9x^2y+27xy^2+27y^3$
b $81-216x+216x^2-96x^3+16x^4$

9 **a** $a=x^{-\frac{3}{4}},\ b=\frac{3}{4}x^{\frac{9}{4}}$
b $\left(x^{-\frac{3}{4}}+\frac{3}{4}x^{\frac{9}{4}}\right)^4=x^{-3}+3+\frac{27}{8}x^3+\frac{27}{16}x^6+\frac{81}{256}x^9$

10 $-841+538\sqrt{3}$

11 **a** $\binom{6}{3}5^3=2500$ **b** $\binom{8}{5}(-3)^5=-13\,608$
c $2\binom{5}{2}(-3)^2+\binom{5}{1}(-3)=165$

12 **a** 60 **b** 240 **13** $k=\pm 2$

REVIEW SET 10B

1 6 paths **2** **a** $6!$ **b** $\frac{10!}{7!}$ **c** $\frac{12!}{8!4!}$

3 **a** 3024 **b** 840 **c** 42

4 **a** $\binom{15}{2}=105$ handshakes
b $\binom{30}{2}-2\binom{15}{2}=225$ handshakes

5 1440 ways **6** $\binom{9}{6}\binom{2}{1}\binom{6}{4}=2520$ ways

7 **a** $\binom{21}{5}=20\,349$ committees

b **i** $\binom{8}{5} = 56$ committees **ii** 14 140 committees

iii $2\binom{19}{4} = 7752$ committees

8 $(3-2x)^5 = 243 - 810x + 1080x^2 - ....$

9 $-103 + 74\sqrt{2}$

10 **a** $\binom{10}{3}2^7 3^3 = 414\,720$ **b** $\binom{10}{5}2^5 3^5 = 1\,959\,552$

c $\binom{10}{8}2^2 3^8 = 1\,180\,980$

11 $\binom{5}{2}2^3 1^2 - 3\binom{5}{1}2^4 1^1 = -160$ **12** $n = 7$

13 $k = -6, \ n = 5$

EXERCISE 11A.1

1 **a** Start with 7, and each term thereafter is 6 more than the previous term.

b $u_1 = 7, \ u_4 = 25$ **c** $u_5 = 31, \ u_6 = 37$

2 **a** Start with 5, and each term thereafter is 3 more than the previous term. The next two terms are $u_6 = 20$ and $u_7 = 23$.

b Start with 38, and each term thereafter is 4 less than the previous term. The next two terms are $u_6 = 18$ and $u_7 = 14$.

c Start with $\frac{1}{2}$, and each term thereafter is $1\frac{1}{2}$ more than the previous term. The next two terms are $u_6 = 8$ and $u_7 = 9\frac{1}{2}$.

d Start with 3, and each term thereafter is twice the previous term. The next two terms are $u_6 = 96$ and $u_7 = 192$.

e Start with 2, and each term thereafter is 5 times the previous term. The next two terms are $u_5 = 1250$ and $u_6 = 6250$.

f Start with 162, and each term thereafter is one third of the previous term. The next two terms are $u_5 = 2$ and $u_6 = \frac{2}{3}$.

3 **a** $u_6 = 25, \ u_7 = 36$ **b** $u_6 = 36, \ u_7 = 49$

c $u_6 = 125, \ u_7 = 216$ **d** $u_5 = \frac{1}{36}, \ u_6 = \frac{1}{49}$

e $u_6 = 16, \ u_7 = 22$ **f** $u_6 = 42, \ u_7 = 56$

4 **a** The first two terms are 1 and 1, and each term thereafter is the sum of the previous two terms.
The next three terms are $u_7 = 13, \ u_8 = 21, \ u_9 = 34$.

b The first two terms are 1 and 3, and each term thereafter is the sum of the previous two terms.
The next three terms are $u_6 = 18, \ u_7 = 29, \ u_8 = 47$.

c The nth term is the sum of the nth and $(n+1)$th prime numbers.
The next three terms are $u_7 = 36, \ u_8 = 42, \ u_9 = 52$.

EXERCISE 11A.2

1 **a** $u_1 = 6, \ u_2 = 13, \ u_3 = 22, \ u_4 = 33$

b $u_{10} = 141$ **c** $u_6 = 61$

2 **a** $u_1 = 7, \ u_2 = 9, \ u_3 = 11, \ u_4 = 13$

b $u_1 = 5, \ u_2 = 8, \ u_3 = 11, \ u_4 = 14$

c $u_1 = 3, \ u_2 = 0, \ u_3 = -3, \ u_4 = -6$

d $u_1 = 2, \ u_2 = 5, \ u_3 = 10, \ u_4 = 17$

e $u_1 = 5, \ u_2 = 12, \ u_3 = 31, \ u_4 = 68$

f $u_1 = 3, \ u_2 = 9, \ u_3 = 27, \ u_4 = 81$

g $u_1 = 10, \ u_2 = 20, \ u_3 = 40, \ u_4 = 80$

h $u_1 = 2, \ u_2 = 7, \ u_3 = 24, \ u_4 = 77$

i $u_1 = \frac{1}{3}, \ u_2 = \frac{8}{5}, \ u_3 = \frac{27}{7}, \ u_4 = \frac{64}{9}$

3 **a** $u_1 = 9, \ u_2 = 20, \ u_3 = 33, \ u_4 = 48$

b $u_1 = 9$, which is not prime, and $u_n = n(n+8)$, so n is always a factor of u_n.
$\therefore$ the sequence will not contain any primes.

c $u_{28} = 1008$

4 **a** $u_n = n^2$

b **i** $u_n = (n+1)^2$ **ii** $u_n = n^2 - 1$

iii $u_n = \dfrac{1}{n^2}$ **iv** $u_n = \dfrac{n}{(n+1)^2}$

EXERCISE 11B

1 **a** arithmetic, $d = 7$ **b** arithmetic, $d = 2\frac{1}{2}$

c arithmetic, $d = -4$ **d** not arithmetic

e arithmetic, $d = -3.9$ **f** not arithmetic

g not arithmetic **h** arithmetic, $d = \ln 3$

2 **a** $u_1 = 3, \ d = 4$ **b** $u_n = 4n - 1$

c $u_{60} = 239$ **d** $u_{126} = 503$

3 **a** $u_1 = 172, \ d = -7$ **b** $u_n = 179 - 7n$

c $u_{12} = 95$ **d** $u_{26} = -3$

4 **a** $u_{n+1} - u_n = \frac{5}{2}$ which is constant
$\therefore$ the sequence is arithmetic.

b $u_1 = 1, \ d = 2\frac{1}{2}$ **c** $u_{20} = 48\frac{1}{2}$ **d** $u_{41} = 101$

5 **a** $k = 5$ **b** $k = 3$ **c** $k = 4$ or -11

d $k = 3$ or $\frac{1}{2}$ **e** $k = -2$ **f** $k = 1$ or 2

g $k = 4$ **h** $k = 1$ **i** $k = 4$

6 **a** $k = -\frac{1}{2}$, 0, or 1

b For $k = -\frac{1}{2}$:

i $d = \frac{37}{8}$ **ii** $-\frac{11}{4}, \frac{15}{8}, \frac{13}{2}$ **iii** $u_{10} = 38\frac{7}{8}$

For $k = 0$:

i $d = 5$ **ii** $-3, 2, 7$ **iii** $u_{10} = 42$

For $k = 1$:

i $d = 5$ **ii** $-2, 3, 8$ **iii** $u_{10} = 43$

7 **b** $k^2 + 1 > 0, \ -2 - k^2 < 0, \ 3 > 0$
The first three terms alternate in sign, so there cannot be a common difference.
$\therefore$ the terms cannot be part of an arithmetic progression.

8 **a** $u_n = 3n + 20$ **b** $u_n = 53 - 7n$

c $u_n = \frac{29}{6}n - \frac{239}{6}$ **d** $u_n = -\frac{5}{2}n + 23$

9 $u_{30} = \frac{103}{7}$ **10** $u_{33} = 11$

11 **a** arithmetic, common difference $= d$

b arithmetic, common difference $= 5d$ **c** not arithmetic

d arithmetic, common difference $= 2d$ **e** not arithmetic

f arithmetic, common difference $= d^2$

12 **a** **i** $u_n = \sqrt{2} + n$ **ii** $u_n = (n-1)\sqrt{2}$

b No. The difference between two rational terms would be rational, which would imply that the common difference is also rational, and hence all terms would be rational.

EXERCISE 11C

1 **a** geometric, $r = 5$ **b** geometric, $r = -2$

c not geometric **d** geometric, $r = \frac{1}{10}$

e geometric, $r = -\frac{1}{4}$ **f** geometric, $r = \sqrt{7}$

g not geometric **h** geometric, $r = \frac{1}{2}$

2 **a** $u_1 = 5, \ r = 3$ **b** $u_n = 5 \times 3^{n-1}$

c $u_{10} = 98\,415$

3 **a** $u_1 = 320, \ r = -\frac{1}{2}$ **b** $u_n = 320 \times (-\frac{1}{2})^{n-1}$

c $u_7 = 5$ **d** $u_{10} = -\frac{5}{8}$

4 **a** $k = \pm 24$ **b** $k = \pm 7\sqrt{5}$ **c** $k = 11$
d $k = \frac{1}{4}$ or 10 **e** $k = 2$ **f** $k = 0$ or 4
g $k = 5$ **h** $k = \ln 2$ **i** $k = \frac{1}{27}$ or $3\sqrt{3}$

5 **a** $k = \frac{1}{3}$ or 5
b For $k = \frac{1}{3}$:
i $-\frac{8}{3}, \frac{4}{3}, -\frac{2}{3}$ **ii** $r = -\frac{1}{2}$ **iii** $u_{10} = \frac{1}{192}$
For $k = 5$:
i 2, 6, 18 **ii** $r = 3$ **iii** $u_{10} = 39\,366$

6 **a** **i** $p = \frac{17}{2}$ **ii** $p = 7$
b The terms of the geometric progression are all integers, while the terms of the arithmetic progression are all non-integers.

7 **a** $u_n = \frac{5}{2} \times 2^{n-1}$ **b** $u_n = 486 \times (\frac{1}{3})^{n-1}$
c $u_n = \frac{2}{5} \times 5^{n-1}$ **d** $u_n = 48 \times (\frac{1}{2})^{n-1}$
e $u_n = -\frac{1}{9} \times (-\frac{3}{2})^{n-1}$ **f** $u_n = -\frac{3}{5} \times (\frac{6}{7})^{n-1}$

8 **a** $u_1 = -160, \quad r = -\frac{3}{4}$ **b** $u_7 = -\frac{3645}{128}$

9 **a** $u_n = \sqrt{2} \times 2^n$ **b** $u_n = 2 \times (\sqrt{2})^{n-1}$

10 **a** geometric, common ratio $= r$
b neither geometric nor arithmetic
c arithmetic, common difference $= \ln r$
d geometric, common ratio $= r^3$
e geometric, common ratio $= r^2$
f geometric, common ratio $= r^4$

EXERCISE 11D

1 **a** $u_3 = 8$ **b** $S_3 = 16$ **c** $u_5 = 19$ **d** $S_5 = 46$

2 **a** **i** 5, 7, 9, 11 **ii** $S_4 = 32$
b **i** $-5, -2, 3, 10$ **ii** $S_4 = 6$
c **i** 7, 14, 28, 56 **ii** $S_4 = 105$
d **i** 3, 5, 11, 29 **ii** $S_4 = 48$

3 $u_6 = 12$

4 **a** $S_1 = \frac{1}{2}, \; S_2 = \frac{2}{3}, \; S_3 = \frac{3}{4}, \; S_4 = \frac{4}{5}$ **b** $\frac{100}{101}$

EXERCISE 11E

1 **a, b, c** 75 **2** $S_{12} = 450$ **3** $S_{20} = 220$

4 **a** 210 **b** 75 **c** 4075 **d** 6780
e -2280 **f** -2400 **g** 275 **h** 387.5

5 2500 **6** **a** 24 terms **b** 1476

7 **a** 775 **b** 1705 **c** 969 **d** 1040 **e** -345 **f** 306

8 **a** $S_n = \frac{n}{2}(2u_1 + (n-1)d)$ where $u_1 = 4, \; d = 4$
$= \frac{n}{2}(8 + 4(n-1))$ and so on
b 840

9 **a** \$65 **b** \$1350

10 **a** $\frac{n(n^2+1)}{2}$ **b** $\frac{3(3^2+1)}{2} = \frac{3(10)}{2} = 15$ ✓ **c** 260

11 **a** $S_n = 2n^2 + 3n$ **b** $n = 10$

12 **a** $m = \ln 2$ **b** $S_{20} = -55$ **13** $n = 11$

EXERCISE 11F.1

1 **a, b** 315

2 **a** 3280 **b** 4 882 812 **c** $\frac{1533}{32}$ **d** $63 + 63\sqrt{2}$
e -1364 **f** $\frac{1640}{27}$ **g** ≈ 52.2 **h** ≈ 12.8

3 **a** $u_1 = 2, \; r = 3$ **b** 59 048

4 **a** 10 terms **b** $S_{10} = 7161$

6 **a** $r = \pm\frac{2}{3}$ **b** For $r = \frac{2}{3}$: $S_8 = 700\frac{5}{9}$
For $r = -\frac{2}{3}$: $S_8 = -140\frac{1}{9}$

7 **a** $\approx$ \$130 000 **b** $\approx$ \$797 000

8 **a** **i** ≈ 791 mL **ii** ≈ 188 mL **iii** ≈ 45 mL
b The amount of water Doug drinks is a geometric sequence with $r = \frac{3}{4}$.
c **i** ≈ 9437 mL **ii** ≈ 9968 mL **iii** ≈ 9998 mL
d If Doug followed the formula, the amount would eventually become too small to measure, and too small to sustain him. (In theory Doug would never run out of water.)

EXERCISE 11F.2

1 **a** diverge **b** converge **c** diverge **d** converge

2 **a** **i** ≈ 23.44 **ii** ≈ 26.53 **iii** ≈ 26.99
b 27 **c** $S_\infty = \frac{9}{1-\frac{2}{3}} = 27$

3 **a** 32 **b** $\frac{5}{4}$ **c** 27 **d** 128 **e** $\frac{432}{7}$ **f** $\frac{2}{3}$

4 **a** $(8-4) + (2-1) + (\frac{1}{2} - \frac{1}{4}) + ...$
$= \quad 4 \quad + \quad 1 \quad + \quad \frac{1}{4} \quad + ...$
b $\frac{8}{1-(-\frac{1}{2})} = \frac{16}{3} = \frac{4}{1-\frac{1}{4}}$

5 **a** Start with $\frac{1}{3}$ unit^2 shaded blue. So $u_1 = \frac{1}{3}$.
Each step shades $\frac{1}{3}$ of the previous area, so $r = \frac{1}{3}$.
$\therefore \; S = \frac{1}{3} + \frac{1}{9} + \frac{1}{27} + ...$
b At each step, the same amount is coloured blue as is coloured red.
c If the process continues indefinitely, half the 1 unit^2 rectangle will be shaded blue and half will be shaded red. And from **a**, total blue area $= \frac{1}{3} + \frac{1}{9} + \frac{1}{27} + ...$
$\therefore \; \frac{1}{3} + \frac{1}{9} + \frac{1}{27} + ... = \frac{1}{2}$
d $S_\infty = \frac{\frac{1}{3}}{1-\frac{1}{3}} = \frac{1}{2}$

6 $10 + 5\sqrt{2}$

7 $12 + 3 + \frac{3}{4} + ..., \; u_1 = 12, \; r = \frac{1}{4}$
$4 + 3 + \frac{9}{4} + ..., \; u_1 = 4, \; r = \frac{3}{4}$

8 $S_\infty = \frac{e^5}{4(2+e)}$

9 $u_1 + u_2 + u_3 + u_4 + ...$ has common ratio r, $0 < r < 1$
a $u_1 - u_2 + u_3 - u_4 + ...$ has common ratio $-r$
$\therefore \; -1 < -r < 1$ and so the series is convergent.
$\sqrt{u_1} + \sqrt{u_2} + \sqrt{u_3} + \sqrt{u_4} + ...$ has common ratio $\sqrt{r}$
$\therefore \; -1 < \sqrt{r} < 1$ and so the series is convergent.
b $\frac{64}{3}$

REVIEW SET 11A

1 **a** $u_1 = 5, \; u_2 = 2, \; u_3 = -1, \; u_4 = -4$
b $u_1 = 0, \; u_2 = \frac{7}{2}, \; u_3 = \frac{26}{3}, \; u_4 = \frac{63}{4}$

2 **a** $u_1 = 5, \; d = 8$ **b** $u_n = 8n - 3$
c $u_{15} = 117$ **d** $u_{126} = 1005$

3 **a** $u_1 = 31, \; d = -4\frac{1}{2}$ **b** $u_{20} = -54\frac{1}{2}$

4 **a** geometric, $r = \frac{1}{\sqrt{5}}$ **b** geometric, $r = -4$
c not geometric **d** geometric, $r = -\frac{e}{2}$

5 **a** $k = 3$ **b** $k = -1$ or 7 **c** $k = \pm\sqrt{2}$ or $\pm 2\sqrt{2}$

6 **a** 1180 **b** -1410 **7** 968

8 **a** $1\,747\,625$ **b** ≈ 11.08

9 **a** $r = \frac{1}{3}$ or $-\frac{4}{3}$

b For $r = \frac{1}{3}$:
- **i** $S_5 = \frac{121}{18} \approx 6.72$
- **ii** $S_\infty = \frac{27}{4} = 6.75$

For $r = -\frac{4}{3}$:
- **i** $S_5 = \frac{181}{18} \approx 10.06$
- **ii** the series diverges

10 **a** $\frac{125}{4}$ **b** 16 **c** $\dfrac{18}{3-\sqrt{3}}$ **d** $-\frac{16}{3}\ln 3$

REVIEW SET 11B

1 **a** $u_1 = 4,\ u_2 = 10,\ u_3 = 18,\ u_4 = 28$

b $u_n = n(n+3)$, and the product of an even and an odd number is always even.
$\therefore$ all of the terms of the sequence are even.

c $u_{24} = 648$

2 **a** $u_n = 7n - 11$ **b** $u_n = -\frac{29}{8}n + \frac{47}{2}$

3 **a** $k = 5$ **b** $u_{25} = 238$ **c** $u_{51} = 498$

4 **a** $u_1 = 108,\ r = -\frac{2}{3}$ **b** $u_7 = \frac{256}{27}$ **c** $S_7 = \frac{1852}{27}$

6 **a** 1070 **b** 270 **c** -119 **d** $\frac{58\,025}{32} \approx 1813.3$

7 $S_6 = 15$ **8** $n = 10$

9 **a** $u_1 = \frac{250}{81},\ r = \frac{3}{5}$ **b** $S_\infty = \frac{625}{81}$

10 **a** $u_1 = m,\ r = \dfrac{\pi}{m}$ **b** $m < -\pi$ or $m > \pi$

c $\dfrac{16}{4-\pi} \approx 18.64$ **d** $m = 2\pi$

EXERCISE 12A

1 **a** $\begin{pmatrix} 5 \\ 3 \end{pmatrix} = 5\mathbf{i} + 3\mathbf{j}$ **b** $\begin{pmatrix} 4 \\ 0 \end{pmatrix} = 4\mathbf{i}$

c $\begin{pmatrix} -2 \\ 5 \end{pmatrix} = -2\mathbf{i} + 5\mathbf{j}$ **d** $\begin{pmatrix} 6 \\ -2 \end{pmatrix} = 6\mathbf{i} - 2\mathbf{j}$

e $\begin{pmatrix} 0 \\ -5 \end{pmatrix} = -5\mathbf{j}$ **f** $\begin{pmatrix} -4 \\ -4 \end{pmatrix} = -4\mathbf{i} - 4\mathbf{j}$

2 **a** $2\mathbf{i} + 5\mathbf{j}$

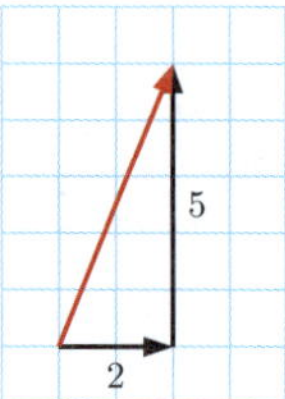

b $\mathbf{i} - 3\mathbf{j}$

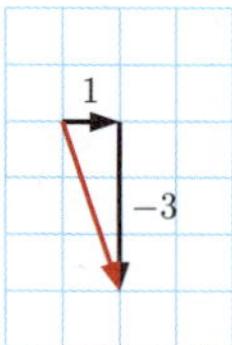

c $-5\mathbf{i}$

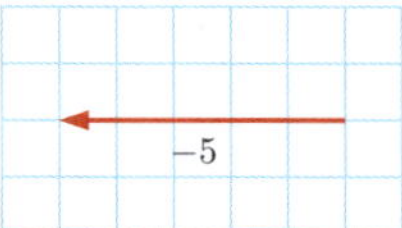

d $-3\mathbf{i} - 2\mathbf{j}$

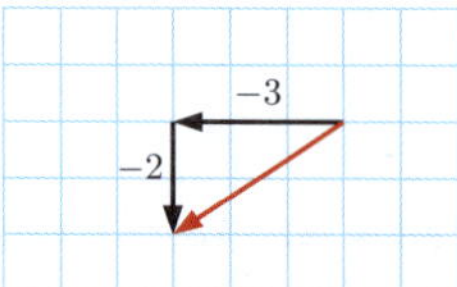

3 **a** $\begin{pmatrix} 3 \\ 4 \end{pmatrix}$

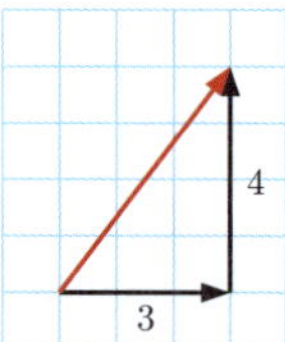

b $\begin{pmatrix} 2 \\ -1 \end{pmatrix}$

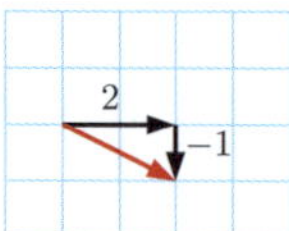

c $\begin{pmatrix} 0 \\ 3 \end{pmatrix}$

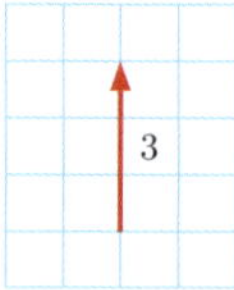

d $\begin{pmatrix} -5 \\ -4 \end{pmatrix}$

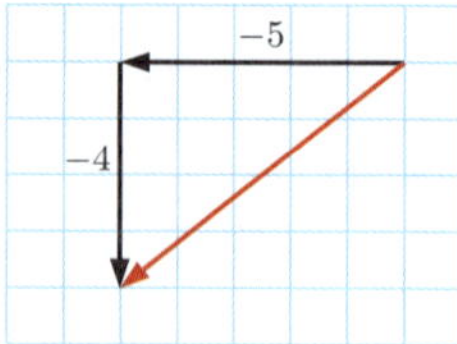

4 **a** **i** $\begin{pmatrix} 4 \\ 1 \end{pmatrix}$, $4\mathbf{i} + \mathbf{j}$ **ii** $\begin{pmatrix} -4 \\ -1 \end{pmatrix}$, $-4\mathbf{i} - \mathbf{j}$

iii $\begin{pmatrix} -1 \\ -5 \end{pmatrix}$, $-\mathbf{i} - 5\mathbf{j}$ **iv** $\begin{pmatrix} 2 \\ 0 \end{pmatrix}$, $2\mathbf{i}$

v $\begin{pmatrix} 3 \\ -4 \end{pmatrix}$, $3\mathbf{i} - 4\mathbf{j}$ **vi** $\begin{pmatrix} 4 \\ 1 \end{pmatrix}$, $4\mathbf{i} + \mathbf{j}$

b $\overrightarrow{AB}$ and $\overrightarrow{DE}$. They have the same magnitude and direction.

5 **a** $2\mathbf{i} - 3\mathbf{j}$ **b**

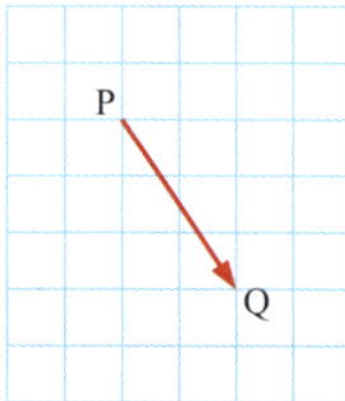

c $\overrightarrow{QP} = \begin{pmatrix} -2 \\ 3 \end{pmatrix}$

d No, $\overrightarrow{PQ} = -\overrightarrow{QP}$. They have the same magnitude but opposite direction.

6 **a** $k = 2$ **b** $k = -1$ **c** $k = 3$

7 **a** $a = 2,\ b = 5$ **b** $a = -1,\ b = -4$

c $a = 1,\ b = -2$ or $a = -\frac{11}{4},\ b = \frac{1}{2}$

EXERCISE 12B

1 **a** $\sqrt{5}$ units **b** 5 units **c** 3 units **d** $\sqrt{2}$ units

e $\sqrt{61}$ units

2 **a** $\sqrt{2}$ units **b** $\sqrt{13}$ units **c** $\sqrt{73}$ units **d** 5 units

e 10 units

3 **a** unit vector **b** not a unit vector **c** unit vector

d not a unit vector **e** unit vector

4 **a** $k = \pm 1$ **b** $k = \pm 1$ **c** $k = 0$

d $k = \pm\frac{2\sqrt{2}}{3}$ **e** $k = \pm\frac{\sqrt{21}}{5}$

5 $k = \pm 4$ **6** $p = \pm\sqrt{37}$

EXERCISE 12C

1 **a** $\mathbf{a} = \begin{pmatrix} 3 \\ 5 \end{pmatrix}$, $\mathbf{b} = \begin{pmatrix} 6 \\ -2 \end{pmatrix}$

b $\mathbf{a} + \mathbf{b} = \begin{pmatrix} 9 \\ 3 \end{pmatrix}$ $\mathbf{a} - \mathbf{b} = \begin{pmatrix} -3 \\ 7 \end{pmatrix}$

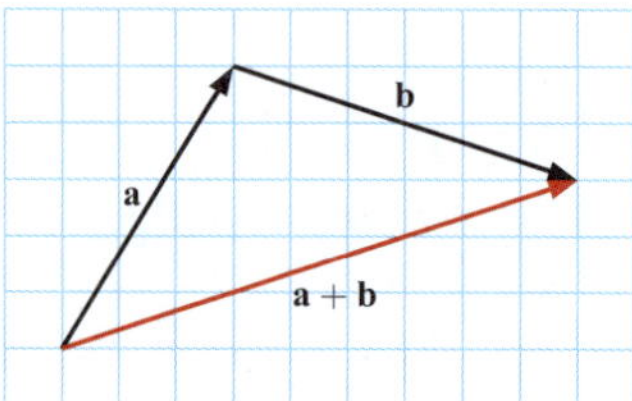

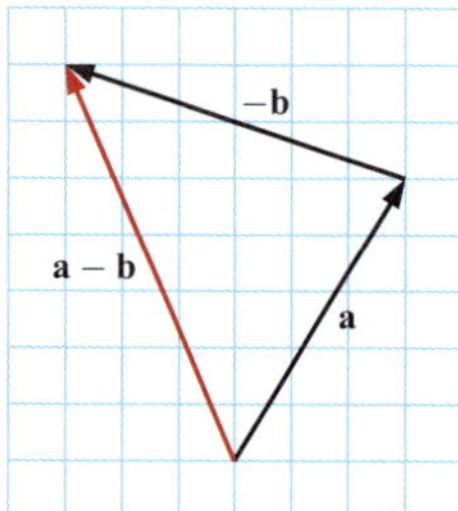

2 **a** $\begin{pmatrix}1\\8\end{pmatrix}$ **b** $\begin{pmatrix}1\\8\end{pmatrix}$ **c** $\begin{pmatrix}5\\1\end{pmatrix}$ **d** $\begin{pmatrix}5\\1\end{pmatrix}$
e $\begin{pmatrix}2\\3\end{pmatrix}$ **f** $\begin{pmatrix}2\\3\end{pmatrix}$ **g** $\begin{pmatrix}4\\6\end{pmatrix}$ **h** $\begin{pmatrix}4\\6\end{pmatrix}$

3 **a** $\begin{pmatrix}-4\\-4\end{pmatrix}$ **b** $\begin{pmatrix}4\\4\end{pmatrix}$ **c** $\begin{pmatrix}-3\\6\end{pmatrix}$ **d** $\begin{pmatrix}7\\-2\end{pmatrix}$
e $\begin{pmatrix}-8\\-4\end{pmatrix}$ **f** $\begin{pmatrix}0\\-4\end{pmatrix}$

4 **a** $\mathbf{v} = \begin{pmatrix}3\\2\end{pmatrix}$ **b** $2\mathbf{v} = \begin{pmatrix}6\\4\end{pmatrix}$, $-3\mathbf{v} = \begin{pmatrix}-9\\-6\end{pmatrix}$

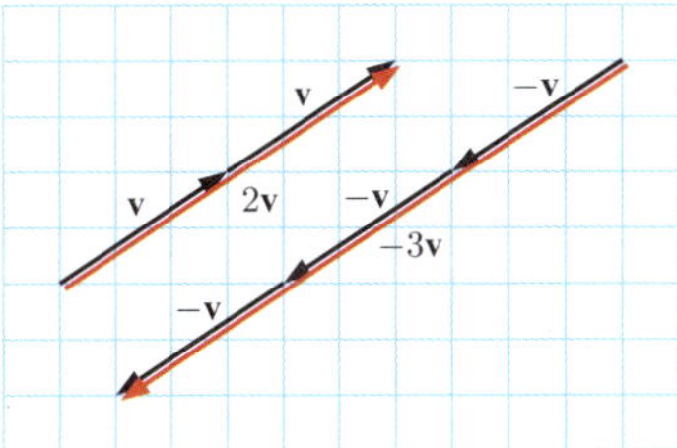

5 **a** $\begin{pmatrix}8\\4\end{pmatrix}$ **b** $\begin{pmatrix}-4\\-2\end{pmatrix}$ **c** $\begin{pmatrix}20\\10\end{pmatrix}$ **d** $\begin{pmatrix}-12\\-6\end{pmatrix}$
e $\begin{pmatrix}2\\1\end{pmatrix}$ **f** $\begin{pmatrix}-1\\-\frac{1}{2}\end{pmatrix}$

6 **a** $\begin{pmatrix}15\\-9\end{pmatrix}$ **b** $\begin{pmatrix}2\\-12\end{pmatrix}$ **c** $\begin{pmatrix}1\\3\end{pmatrix}$ **d** $\begin{pmatrix}12\\0\end{pmatrix}$
e $\begin{pmatrix}5\\24\end{pmatrix}$ **f** $\begin{pmatrix}-3\\-27\end{pmatrix}$ **g** $\begin{pmatrix}10\\-6\end{pmatrix}$ **h** $\begin{pmatrix}-24\frac{1}{3}\\17\end{pmatrix}$

7 **a**

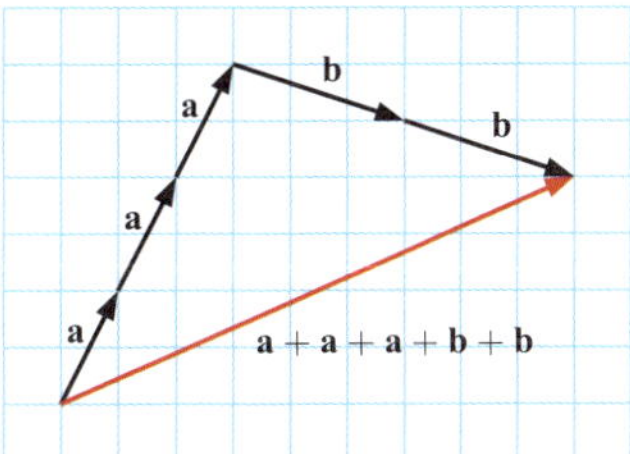

b

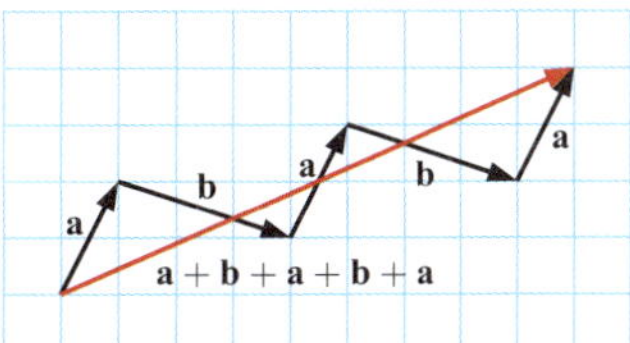

c

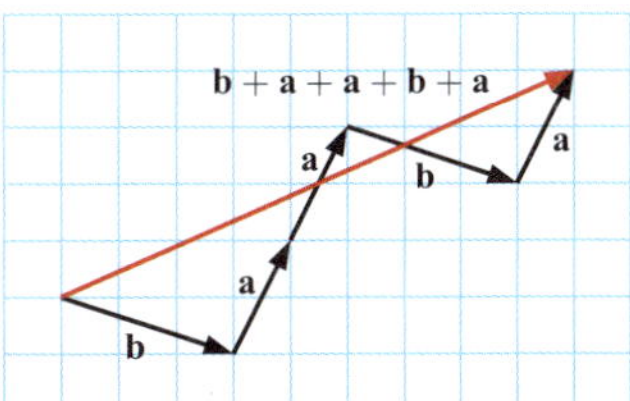

In each case, the result is $3\mathbf{a} + 2\mathbf{b} = \begin{pmatrix}9\\4\end{pmatrix}$.

8 **b**

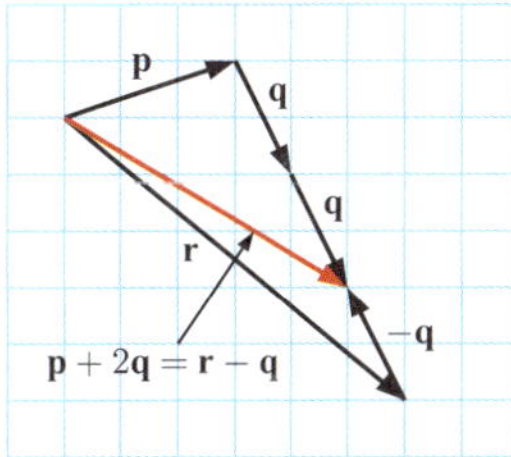

9 **a** 5 units **b** $\sqrt{37}$ units **c** $2\mathbf{i} + 10\mathbf{j}$
d $2\sqrt{26}$ units **e** $4\mathbf{i} - 2\mathbf{j}$ **f** $2\sqrt{5}$ units
g $\sqrt{53}$ units **h** $\sqrt{313}$ units

10 **a** $3\mathbf{i} + 2\mathbf{j}$ **b** $-\mathbf{i} + 9\mathbf{j}$ **c** $6\mathbf{i} - \mathbf{j}$
d $7\mathbf{j}$ **e** 2 units **f** $2\sqrt{10}$ units

11 **a** $k = \pm 5$ **b** $k = -3 \pm \sqrt{51}$

12 **a** $(1 + 4k)\mathbf{i} + (3 - k)\mathbf{j}$ **b** $k = 4$

13 $k_1 = 3,\ k_2 = 2$

EXERCISE 12D

1 **a** $\begin{pmatrix}5\\1\end{pmatrix}$ **b** $\begin{pmatrix}2\\-5\end{pmatrix}$ **c** $\begin{pmatrix}5\\-9\end{pmatrix}$ **d** $\begin{pmatrix}4\\-5\end{pmatrix}$
e $\begin{pmatrix}7\\1\end{pmatrix}$ **f** $\begin{pmatrix}6\\14\end{pmatrix}$

2 **a** $\begin{pmatrix}-1\\3\end{pmatrix}$ **b** $\begin{pmatrix}5\\-9\end{pmatrix}$ **c** $\begin{pmatrix}3\\1\end{pmatrix}$

3 **a** $\begin{pmatrix}2\\4\end{pmatrix}$ **b** $\begin{pmatrix}1\\-5\end{pmatrix}$ **c** $-5\mathbf{i} + 5\mathbf{j}$ **d** $3\mathbf{i} - 3\mathbf{j}$

4 **a** $4\mathbf{i} - 10\mathbf{j}$ **b** $2\sqrt{29}$ units **c** $-4\mathbf{i} + 10\mathbf{j}$ **d** $2\sqrt{29}$ units

5 **a** $-5\mathbf{i} + 3\mathbf{j}$ **b** $\frac{1}{2}\mathbf{i} - \frac{7}{2}\mathbf{j}$

6 **a** $\overrightarrow{AB} = \begin{pmatrix}k-2\\4\end{pmatrix}$, $|\overrightarrow{AB}| = \sqrt{(k-2)^2 + 16}$ units
b $k = 2 \pm \sqrt{33}$ **c**

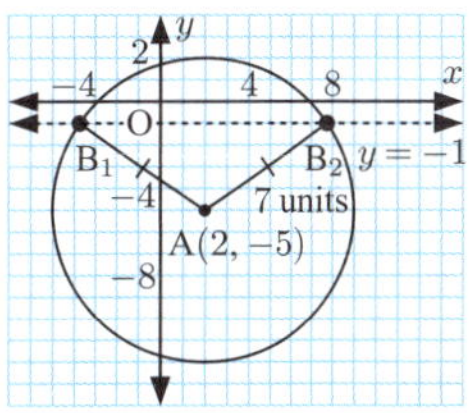

7 **a** **i** $\mathbf{b} - \mathbf{a}$ **ii** $2\mathbf{a} + 4\mathbf{b}$ **b** $\sqrt{445}$ units

8 **a** $\overrightarrow{AB} = \begin{pmatrix}4\\5\end{pmatrix}$, $\overrightarrow{AC} = \begin{pmatrix}5\\-4\end{pmatrix}$
b $-\overrightarrow{AB} + \overrightarrow{AC} = \overrightarrow{BA} + \overrightarrow{AC} = \overrightarrow{BC}$ **c** $\begin{pmatrix}1\\-9\end{pmatrix}$
d $\overrightarrow{BC} = \begin{pmatrix}2-1\\-3-6\end{pmatrix} = \begin{pmatrix}1\\-9\end{pmatrix}$ ✓

9 **a** $\begin{pmatrix}-3\\8\end{pmatrix}$ **b** $\begin{pmatrix}-1\\-2\end{pmatrix}$ **c** $\begin{pmatrix}1\\8\end{pmatrix}$

EXERCISE 12E.1

1 **a** parallel **b** not parallel **c** parallel
d parallel **e** not parallel **f** parallel

2 **a** $m = -12$ **b** $m = 6$ **c** $m = 20$
d $m = 4$ or -6

3 **a** $2\mathbf{v}$ **b** $-\frac{1}{3}\mathbf{v}$

4 **a** $\mathbf{v} = \begin{pmatrix} 5 \\ 5 \end{pmatrix}$, $\mathbf{a} = \begin{pmatrix} 2 \\ 2 \end{pmatrix}$, $\mathbf{b} = \begin{pmatrix} 1 \\ -1 \end{pmatrix}$, $\mathbf{c} = \begin{pmatrix} 5 \\ 5 \end{pmatrix}$,

$\mathbf{d} = \begin{pmatrix} -3 \\ -3 \end{pmatrix}$, $\mathbf{e} = \begin{pmatrix} 7 \\ 1 \end{pmatrix}$

b **i** **a**, **c**, and **d** **ii** **a** and **c** **iii** **c** and **e** **iv** **c**

5 **a** $\overrightarrow{PQ}$ is parallel and in the same direction as $\overrightarrow{RS}$, and 4 times its length.

b $\overrightarrow{AB}$ is parallel and the same length as $\overrightarrow{CD}$, and in the opposite direction.

c $\overrightarrow{KL}$ is parallel and in the same direction as $\overrightarrow{MN}$, and half its length.

6 **a** $\frac{1}{\sqrt{10}}\mathbf{i} + \frac{3}{\sqrt{10}}\mathbf{j}$ **b** $\frac{5}{\sqrt{29}}\mathbf{i} - \frac{2}{\sqrt{29}}\mathbf{j}$ **c** $-\frac{4}{\sqrt{17}}\mathbf{i} + \frac{1}{\sqrt{17}}\mathbf{j}$

7 **a** $3\mathbf{i} + 4\mathbf{j}$

b **i** $\frac{3}{5}\mathbf{i} + \frac{4}{5}\mathbf{j}$

ii $\frac{9}{5}\mathbf{i} + \frac{12}{5}\mathbf{j}$

iii $-6\mathbf{i} - 8\mathbf{j}$

c

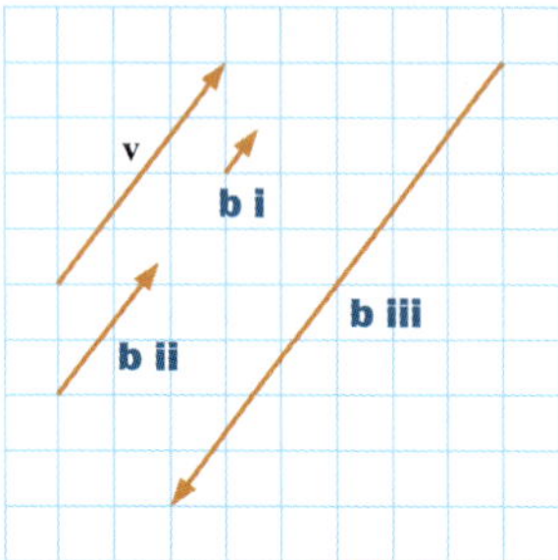

8 **a** $\frac{3}{\sqrt{34}}\mathbf{i} - \frac{5}{\sqrt{34}}\mathbf{j}$ **b** $\frac{15}{\sqrt{34}}\mathbf{i} - \frac{25}{\sqrt{34}}\mathbf{j}$

c $\frac{24}{\sqrt{34}}\mathbf{i} - \frac{40}{\sqrt{34}}\mathbf{j}$, $-\frac{24}{\sqrt{34}}\mathbf{i} + \frac{40}{\sqrt{34}}\mathbf{j}$

9 **a** $\begin{pmatrix} -\frac{6}{\sqrt{5}} \\ \frac{3}{\sqrt{5}} \end{pmatrix}$ **b** $\begin{pmatrix} \frac{24}{\sqrt{37}} \\ \frac{4}{\sqrt{37}} \end{pmatrix}$

10 **a** $-5\mathbf{i} + 4\mathbf{j}$ **b** $-\frac{5}{\sqrt{41}}\mathbf{i} + \frac{4}{\sqrt{41}}\mathbf{j}$

11 **a** $\overrightarrow{PR} = 2\mathbf{p} + \mathbf{q}$, $\overrightarrow{QS} = -4\mathbf{p} - 2\mathbf{q}$ **b** $\overrightarrow{QS} = -2\overrightarrow{PR}$

c opposite directions, since they are negative multiples of each other

12 **a** $\begin{pmatrix} \frac{12}{\sqrt{5}} \\ -\frac{6}{\sqrt{5}} \end{pmatrix}$ **b** $\begin{pmatrix} 5 + \frac{12}{\sqrt{5}} \\ 3 - \frac{6}{\sqrt{5}} \end{pmatrix}$ **c** $\left(5 + \frac{12}{\sqrt{5}},\ 3 - \frac{6}{\sqrt{5}}\right)$

EXERCISE 12E.2

1 **a** $\overrightarrow{AB} = 3\overrightarrow{BC}$, $\therefore$ A, B, and C are collinear.

b $\overrightarrow{PQ} = -\frac{3}{2}\overrightarrow{QR}$, $\therefore$ P, Q, and R are collinear.

c $\overrightarrow{XY} = \frac{5}{2}\overrightarrow{YZ}$, $\therefore$ X, Y, and Z are collinear.

2 $m = -6$

3 **a** $\overrightarrow{AB} = \mathbf{b} - \mathbf{a}$, $\overrightarrow{BC} = 4\mathbf{a} - 4\mathbf{b}$

b $\overrightarrow{AB} = -\frac{1}{4}\overrightarrow{BC}$, $\therefore$ A, B, and C are collinear.

4 **Hint:** Show that P, Q, and R are collinear.

EXERCISE 12F

1 **a** L_1 has direction vector $\begin{pmatrix} -5 \\ 1 \end{pmatrix}$,

L_2 has direction vector $\begin{pmatrix} -10 \\ 2 \end{pmatrix}$

b $\begin{pmatrix} -10 \\ 2 \end{pmatrix} = 2\begin{pmatrix} -5 \\ 1 \end{pmatrix}$, the direction vectors are a scale multiple of each other, so L_1 and L_2 are parallel.

2 **a** **i** $\begin{pmatrix} x \\ y \end{pmatrix} = \begin{pmatrix} 8 \\ 2 \end{pmatrix} + t\begin{pmatrix} 5 \\ -6 \end{pmatrix}$, $t \in \mathbb{R}$

ii $x = 8 + 5t$, $y = 2 - 6t$, $t \in \mathbb{R}$ **iii** $6x + 5y = 58$

b **i** $\begin{pmatrix} x \\ y \end{pmatrix} = \begin{pmatrix} 3 \\ 0 \end{pmatrix} + t\begin{pmatrix} -4 \\ 9 \end{pmatrix}$, $t \in \mathbb{R}$

ii $x = 3 - 4t$, $y = 9t$, $t \in \mathbb{R}$ **iii** $9x + 4y = 27$

c **i** $\begin{pmatrix} x \\ y \end{pmatrix} = \begin{pmatrix} -7 \\ -1 \end{pmatrix} + t\begin{pmatrix} 6 \\ 3 \end{pmatrix}$, $t \in \mathbb{R}$

ii $x = -7 + 6t$, $y = -1 + 3t$, $t \in \mathbb{R}$

iii $x - 2y = -5$

3 **a** **i** $(2, -1)$ **ii** $(8, 1)$ **iii** $(-7, -4)$ **b** $\frac{1}{3}$

4 **a** $x = 2 + 10t$, $y = 6 + 5t$, $t \in \mathbb{R}$

b

t	0	1	3	-1	-4
Point	(2, 6)	(12, 11)	(32, 21)	(−8, 1)	(−38, −14)

c

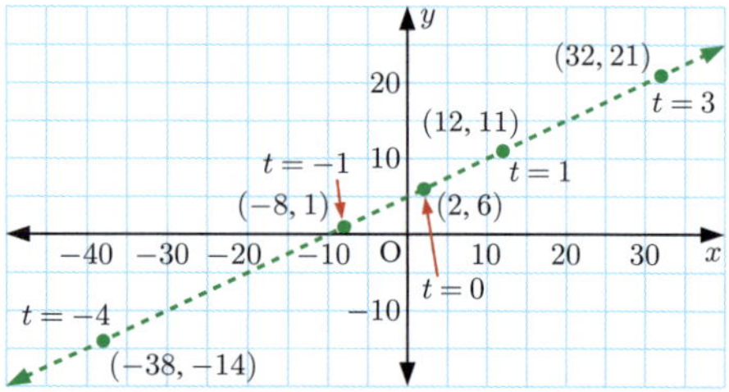

5 no **6** $k = 2$

7 **a** $(20, 0)$ **b** $(0, 10)$

8 **a** L_1: $2x + 3y = -2$

L_2: $2x + 3y = -2$

b L_1 and L_2 are the same line. **c** $s = 2 - 2t$

EXERCISE 12G

1 **a** **i** $(9, -7)$ **ii** $\begin{pmatrix} -2 \\ -8 \end{pmatrix}$ **iii** $2\sqrt{17}\ \text{m s}^{-1}$

b **i** $(2, -5)$ **ii** $\begin{pmatrix} -1 \\ 4 \end{pmatrix}$ **iii** $\sqrt{17}\ \text{m s}^{-1}$

c **i** $(-3, 4)$ **ii** $\begin{pmatrix} 10 \\ 1 \end{pmatrix}$ **iii** $\sqrt{101}\ \text{m s}^{-1}$

d **i** $(-5, 3)$ **ii** $\begin{pmatrix} 4 \\ -7 \end{pmatrix}$ **iii** $\sqrt{65}\ \text{m s}^{-1}$

2 **a** $(4, 1)$

b

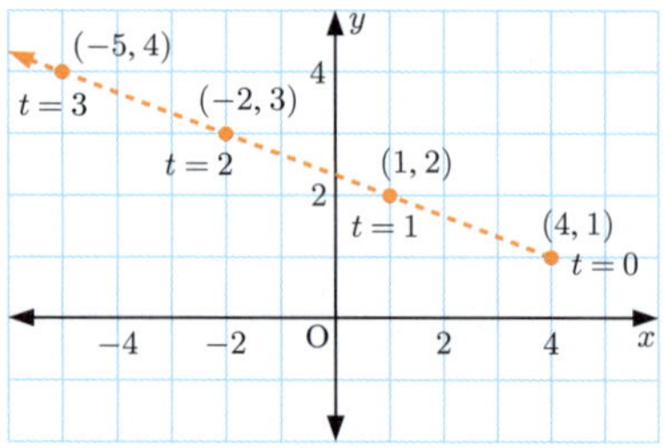

c $\sqrt{10}\ \text{m s}^{-1}$ **d** $(-20, 9)$ **e** $\begin{pmatrix} -\frac{21}{\sqrt{10}} \\ \frac{7}{\sqrt{10}} \end{pmatrix}$

3 **a** $\begin{pmatrix} -8 \\ 3 \end{pmatrix} + t\begin{pmatrix} 2 \\ 1 \end{pmatrix}$, $t \geqslant 0$ **b** $(-4, 5)$

c $\approx 063°$ **d** $t = 4$ minutes

4 **a** The bird's speed and direction can be written as the velocity vector $\begin{pmatrix}16\\12\end{pmatrix}$ km h^{-1}.

b $\approx 053.1°$ **c** $\begin{pmatrix}x\\y\end{pmatrix} = \begin{pmatrix}16\\12\end{pmatrix}t$ **d** 2 hours 30 minutes

5 **a** A is at $(-7, 3)$, B is at $(-10, -12)$

b For A, speed is $5\ \text{m s}^{-1}$. For B, speed is $\sqrt{29}\ \text{m s}^{-1}$.

c after 3 seconds **d** $(5, -6)$

6 **a** $p = 4$ **b** after 30 minutes **c** $\approx 323°$

d **i** $t = 2$ hours **ii** $(-2, 6)$ **iii** $q = 5$

7 **b** $\begin{pmatrix}5\cos 20°\\-5\sin 20°\end{pmatrix}$

c $(10 + 40\cos 40°,\ -20 + 40\sin 40°) \approx (40.6, 5.71)$

d **i** after ≈ 28 seconds **ii** $\approx (181, 124)$ **iii** $p \approx 172$

REVIEW SET 12A

1 **a** $\mathbf{a} = \begin{pmatrix}2\\5\end{pmatrix} = 2\mathbf{i} + 5\mathbf{j}$, $\mathbf{b} = \begin{pmatrix}-3\\2\end{pmatrix} = -3\mathbf{i} + 2\mathbf{j}$

b **i** $-\mathbf{i} + 7\mathbf{j}$ **ii** $7\mathbf{i} + 8\mathbf{j}$

2 **a** $\begin{pmatrix}3\\-1\end{pmatrix}$ **b**

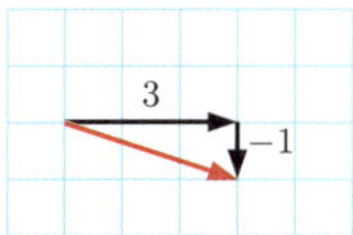

c $-3\mathbf{i} + \mathbf{j}$

d $\sqrt{10}$ units

3 **a** $k = \pm\frac{\sqrt{21}}{5}$ **b** $k = 0$ or -1

4 **a** $\begin{pmatrix}\frac{12}{\sqrt{17}}\\-\frac{3}{\sqrt{17}}\end{pmatrix}$ **b** $\begin{pmatrix}\frac{35}{\sqrt{34}}\\-\frac{21}{\sqrt{34}}\end{pmatrix}$

5 **a** $\begin{pmatrix}-2\\-10\end{pmatrix}$ **b** $\begin{pmatrix}4\\11\end{pmatrix}$ **c** $5\sqrt{2}$ units

6 **a** $\begin{pmatrix}2\\-3\end{pmatrix}$ **b** $\begin{pmatrix}-3\\-6\end{pmatrix}$ **c** $\sqrt{34}$ units

7 $m = -5$, $n = 2$ **8** $k = 3$

9 **a** $\begin{pmatrix}x\\y\end{pmatrix} = \begin{pmatrix}-6\\3\end{pmatrix} + t\begin{pmatrix}4\\-3\end{pmatrix}$, $t \in \mathbb{R}$

b $x = -6 + 4t$, $y = 3 - 3t$, $t \in \mathbb{R}$ **c** $3x + 4y = -6$

10 $m = 10$

11 **a** $\begin{pmatrix}70\\80\end{pmatrix} + t\begin{pmatrix}-2\\-4\end{pmatrix}$, $t \geqslant 0$ **b** $(40, 20)$

c $2\sqrt{5}\ \text{m s}^{-1}$ **d** **i** 48 seconds **ii** $96\sqrt{5}$ m

REVIEW SET 12B

1 **a** **i** $\begin{pmatrix}4\\0\end{pmatrix} = 4\mathbf{i}$ **ii** $\begin{pmatrix}-2\\-4\end{pmatrix} = -2\mathbf{i} - 4\mathbf{j}$

iii $\begin{pmatrix}-2\\4\end{pmatrix} = -2\mathbf{i} + 4\mathbf{j}$

b **ii** and **iii**, $|-2\mathbf{i} + 4\mathbf{j}| = |-2\mathbf{i} - 4\mathbf{j}| = 2\sqrt{5}$ units

2 **a** $a = 12$, $b = 7$

b $a = \frac{1}{2}$, $b = 1$ or $a = \frac{1}{18}$, $b = -\frac{1}{3}$

3 **a** $\sqrt{13}$ units **b** $\sqrt{10}$ units **c** $\sqrt{109}$ units

4 **a** $k = \pm 3\sqrt{2}$ **b** $k = -5 \pm 3\sqrt{15}$

5 $\begin{pmatrix}-7\\11\end{pmatrix}$ **6** **a** $\frac{1}{\sqrt{11}}\mathbf{v}$ **b** $\frac{2}{\sqrt{11}}\mathbf{v}$ **c** $-\frac{4}{\sqrt{11}}\mathbf{v}$

7 **a** $6\mathbf{i} + 6\mathbf{j}$ **b** $6\sqrt{2}$ units **c** $\frac{1}{\sqrt{2}}\mathbf{i} + \frac{1}{\sqrt{2}}\mathbf{j}$

8 $n = -7$

9 **a** $\overrightarrow{PQ} = \mathbf{q} - \mathbf{p}$, $\overrightarrow{QR} = 2\mathbf{q} - 2\mathbf{p}$

b P, Q, and R are collinear.

10 $\begin{pmatrix}x\\y\end{pmatrix} = \begin{pmatrix}0\\8\end{pmatrix} + t\begin{pmatrix}5\\4\end{pmatrix}$, $t \in \mathbb{R}$

11 **a** $k = -4$ **b** $\approx 214°$ **c** $t = 5$ seconds

d $t = 15$ seconds **e** $\sqrt{409}$ m

EXERCISE 13A

1 **a** 11 **b** 6 **c** 5 **d** -3 **e** 4 **f** 2

2 **a** 2 **b** -5 **c** 3 **d** 4 **e** -1 **f** 3

g -4 **h** -2 **i** 8

EXERCISE 13B

1 **a** -2 **b** $-\frac{1}{2}$ **2** **a** $100\ \text{km h}^{-1}$ **b** $50\ \text{km h}^{-1}$

3 **a**

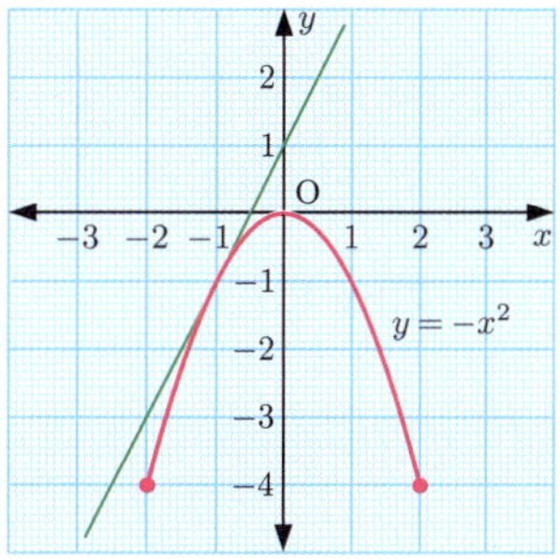

b 2

EXERCISE 13C

1 **a** $(2 + h)^2$

b gradient of FM $= \dfrac{(2+h)^2 - 4}{(2+h) - 2} = 4 + h$ for $h \neq 0$

c **i** 5 **ii** 4.5 **iii** 4.1 **iv** 4.01

d 4 **e** 4

2 **a** 6 **b** 5 **c** 5

3 **a** 4 **b** 12 **c** $-\frac{1}{3}$ **d** 0

4 **a** **i** 1 **ii** 2 **iii** 3 **b** a

EXERCISE 13D.1

1 $f'(1) = \frac{3}{2}$

2 $f'(3) = 1$, $f'(-1) = -7$, the gradient of the tangent to $y = f(x)$ is 1 at $x = 3$, and -7 at $x = -1$.

3 l_1 has gradient 2, l_2 has gradient -1, l_3 has gradient 7.

4 **a** P(2, 4) **b** -2 **c** **i** $(-1, 1)$ **ii** $(1, 5)$

EXERCISE 13D.2

1 **a** $f'(x) = 3$ **b** $f'(x) = 2x$ **c** $f'(x) = -1$

2 **a** $\dfrac{dy}{dx} = -2$ **b** $\dfrac{dy}{dx} = 2x + 4$ **c** $\dfrac{dy}{dx} = 3x^2$

3 **a** $f'(x) = 4x - 1$

b $f'(-1) = -5$, $f'(2) = 7$, the gradient of the tangent to $y = f(x)$ is -5 at $x = -1$, and 7 at $x = 2$.

4 **a** **i** $f'(x) = 0$ **ii** $f'(x) = 0$

b $f'(x) = 0$. The graph of $f(x) = c$ is a horizontal line, which has gradient 0.

5 **a** $f'(x) = 3x^2 - 1$ **b** 2

6 **a** $g'(x) = 2x$, $f'(x) = 10x$

b $f'(x) = \lim\limits_{h \to 0} \dfrac{cg(x+h) - cg(x)}{h}$

$= \lim\limits_{h \to 0} c\left(\dfrac{g(x+h) - g(x)}{h}\right)$

$= c\left(\lim\limits_{h \to 0} \dfrac{g(x+h) - g(x)}{h}\right) = cg'(x)$

7 **a** **i** $f'(x) = 4x$ **ii** $f'(x) = 9$ **iii** $f'(x) = 4x + 9$

b $f'(x) = \lim\limits_{h \to 0} \dfrac{(g(x+h) + h(x+h)) - (g(x) + h(x))}{h}$

$= \lim\limits_{h \to 0} \dfrac{(g(x+h) - g(x)) + (h(x+h) - h(x))}{h}$

$= \lim\limits_{h \to 0} \dfrac{g(x+h) - g(x)}{h} + \lim\limits_{h \to 0} \dfrac{h(x+h) - h(x)}{h}$

$= g'(x) + h'(x)$

REVIEW SET 13A

1 **a** 9 **b** $-\frac{1}{2}$ **c** -1 **2** **a** 3 **b** 0

3 **a** 1 **b** -4 **4** **a** 5 **b** 11 **5** $f'(1) = -\frac{2}{3}$

6 $f'(-3) = -5$, the gradient of the tangent to $y = f(x)$ at $x = -3$, is -5.

7 **a** $f'(x) = 5$ **b** $f'(x) = 2x - 2$

8 **a** $f'(x) = 6x - 4$ **b** 2

REVIEW SET 13B

1 **a** 3 **b** -4 **c** 2

2 **a**

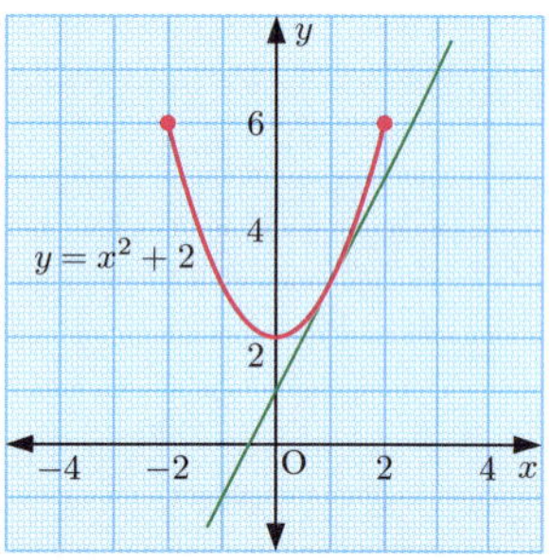

b 2

3 **a** $-\frac{1}{2}$ **b** 6 **4** $f'(2) = \frac{2}{5}$

5 l_1 has gradient 3, l_2 has gradient -1

6 **a** $\dfrac{dy}{dx} = -7$ **b** $\dfrac{dy}{dx} = 4x$

7 **a** $f'(x) = 3x^2 + 2$

b $f'(0) = 2$, $f'(2) = 14$, the gradient of the tangent to $y = f(x)$ is 2 at $x = 0$, and 14 at $x = 2$.

8 **a** $f'(x) = -4x - 4$ **b** 4 **c** $(2, -16)$

EXERCISE 14A

1 **a** $f'(x) = 4x^3$ **b** $f'(x) = 7x^6$ **c** $f'(x) = 10x^9$

d $f'(x) = 5$ **e** $f'(x) = 0$ **f** $f'(x) = 8x$

g $f'(x) = 18x^2$ **h** $f'(x) = -2$ **i** $f'(x) = -5x^9$

j $f'(x) = 2x - 3$ **k** $f'(x) = 2$

l $f'(x) = 6x - 1$ **m** $f'(x) = 3x^2 + 4x - 6$

n $f'(x) = 6x^2 - 8x - 5$ **o** $f'(x) = \frac{3}{2}x^2 - \frac{2}{3}x$

2 **a** $-\dfrac{1}{x^2}$ **b** $-\dfrac{3}{x^4}$ **c** $-\dfrac{7}{x^8}$ **d** $-\dfrac{4}{x^3}$

e $\dfrac{5}{x^2}$ **f** $-\dfrac{12}{x^5}$ **g** $1 - \dfrac{6}{x^2}$ **h** $-\dfrac{2}{x^3} + \dfrac{15}{x^6}$

i $2x + \dfrac{12}{x^4} - \dfrac{4}{x^5}$ **j** $-\dfrac{6}{x^2} + \dfrac{2}{x^3} - \dfrac{10}{x^6}$

k $-\dfrac{1}{7x^2}$ **l** $1 + \dfrac{4}{3x^3}$

3 **a** $f'(x) = 3x^{-\frac{1}{2}} = \dfrac{3}{\sqrt{x}}$ **b** $f'(x) = \frac{1}{4}x^{-\frac{3}{4}} = \dfrac{1}{4\sqrt[4]{x^3}}$

c $f'(x) = -x^{-\frac{3}{2}} = -\dfrac{1}{x\sqrt{x}}$

d $f'(x) = -\frac{1}{3}x^{-\frac{4}{3}} = -\dfrac{1}{3x\sqrt[3]{x}}$

e $f'(x) = 2x + 4x^{-\frac{1}{2}} = 2x + \dfrac{4}{\sqrt{x}}$

f $f'(x) = 7 - \frac{1}{4}x^{-\frac{1}{2}} = 7 - \dfrac{1}{4\sqrt{x}}$

g $f'(x) = 3x^{\frac{1}{2}} = 3\sqrt{x}$

h $f'(x) = -2x^{-2} + \frac{3}{2}x^{-\frac{5}{2}} = -\dfrac{2}{x^2} + \dfrac{3}{2x^2\sqrt{x}}$

i $f'(x) = \frac{1}{2}x^{-\frac{1}{2}} + 10x^{\frac{3}{2}} = \dfrac{1}{2\sqrt{x}} + 10x\sqrt{x}$

4 **a** $\dfrac{dy}{dx} = 2x + 5$ **b** $\dfrac{dy}{dx} = 2x - 6$ **c** $\dfrac{dy}{dx} = 1 - \dfrac{4}{x^2}$

d $\dfrac{dy}{dx} = \frac{3}{2}x^{-\frac{1}{2}} + x^{-\frac{3}{2}} = \dfrac{3}{2\sqrt{x}} + \dfrac{1}{x\sqrt{x}}$

e $\dfrac{dy}{dx} = 8x + 20$

f $\dfrac{dy}{dx} = \frac{1}{4}x^{-\frac{1}{2}} - \frac{1}{4}x^{-\frac{3}{2}} = \dfrac{1}{4\sqrt{x}} - \dfrac{1}{4x\sqrt{x}}$

5 **a** $10x - 4$ **b** $4x^{-\frac{1}{2}} - 3x^{-2} = \dfrac{4}{\sqrt{x}} - \dfrac{3}{x^2}$ **c** $1 + \dfrac{12}{x^3}$

6 **a** $\dfrac{dy}{dx} = 9x^2 - 5$ **b** $\dfrac{dy}{dt} = 7 + \dfrac{4}{t^3}$

c $\dfrac{du}{dx} = 2x^{-\frac{1}{2}} + x^{-2} = \dfrac{2}{\sqrt{x}} + \dfrac{1}{x^2}$

d $\dfrac{dP}{dt} = 2t + 8$ **e** $\dfrac{dT}{dx} = 2 + \dfrac{2}{x^2}$

f $\dfrac{dy}{du} = \frac{3}{2}u^{\frac{1}{2}} - 2u^{-\frac{1}{2}} - 2u^{-\frac{3}{2}} = \frac{3}{2}\sqrt{u} - \dfrac{2}{\sqrt{u}} - \dfrac{2}{u\sqrt{u}}$

7 **a** $f'(x) = 3x^2 + 4x - 3$ **b** $f(2) = 11$, $f'(2) = 17$

c For $f(x) = x^3 + 2x^2 - 3x + 1$, the gradient of the tangent at $(2, 11)$ is 17.

8 **a** -6 **b** $-\frac{3}{2}$ **c** 5 **d** 3 **e** $\frac{3}{4}$ **f** $-\frac{1}{4}$

9 **a** -3 **b** $\frac{3}{4}$ **c** 0

10 **a** $(-2, 1)$ **b** $(1, 1)$ **c** $(-1, -1)$ and $(\frac{1}{3}, -\frac{23}{27})$

d $(2, 3)$ and $(-2, -1)$ **e** $(0, 7)$ and $(4, -25)$

11 $(-2, 3)$ **12** $a = 4$, $b = -6$

13 **a** $m = -8$, $n = -11$ **b** $p(x) = (2x+1)^2(x-3)$

EXERCISE 14B.1

1 **a** $gf(x) = (4x+1)^3$ **b** $gf(x) = 4x^3 + 1$

c $gf(x) = \dfrac{1}{2 - x^2}$ **d** $gf(x) = 2 - \dfrac{1}{x^2}$

e $gf(x) = \sqrt{5x - 3}$ **f** $gf(x) = 5\sqrt{x} - 3$

2 **Note:** There may be other answers.

a $g(x) = x^2$, $f(x) = 6x - 1$

b $g(x)=\frac{2}{x},\quad f(x)=x^2-3$

c $g(x)=\sqrt{x},\quad f(x)=4x-7$

d $g(x)=\frac{1}{\sqrt{x}},\quad f(x)=8-x^2$

EXERCISE 14B.2

1 **a** $3u^4,\ u=2x-5$ **b** $5u^{-1},\ u=x^2-2$

c $u^{\frac{1}{2}},\ u=7x-1$ **d** $2u^{-3},\ u=8-x$

e $-4u^{-\frac{1}{2}},\ u=x+1$ **f** $u^{\frac{1}{3}},\ u=x^2+5x-2$

2 **a** $\frac{dy}{dx}=18x-12$ **b** $\frac{dy}{dx}=-\frac{2}{(2x-4)^2}$

c $\frac{dy}{dx}=10x(x^2-3)^4$ **d** $\frac{dy}{dx}=-\frac{40}{(5x+1)^3}$

e $\frac{dy}{dx}=\frac{2}{\sqrt{4x-5}}$ **f** $\frac{dy}{dx}=\frac{3x^2-7}{2\sqrt{x^3-7x}}$

g $\frac{dy}{dx}=-\frac{2(2x+1)}{(x^2+x-1)^2}$ **h** $\frac{dy}{dx}=\frac{1}{2(6-x)^{\frac{3}{2}}}$

i $\frac{dy}{dx}=\frac{2x+1}{3(x^2+x)^{\frac{2}{3}}}$

3 **a** $\frac{dy}{dx}=\frac{1}{\sqrt{2x+5}}$ **b** $\frac{1}{3}$

4 **a** 150 **b** $-\frac{1}{4}$ **c** -3 **d** $\frac{1}{2}$ **e** 2 **f** $\frac{3087}{16}$

5 $a=\frac{2}{3}$ or 2

6 **a** $k=4$ **b** $\frac{dy}{dx}=-\frac{2x}{(x^2-4)^2}$ **c** **i** $\frac{2}{9}$ **ii** $-\frac{6}{25}$

7 $a=8,\ b=3$

EXERCISE 14C

1 **a** $f'(x)=2x+3$ **b** $f'(x)=8x-4$

c $f'(x)=3x^2-4x$ **d** $f'(x)=\sqrt{x+1}+\frac{x}{2\sqrt{x+1}}$

e $f'(x)=3x^2+6x-1$ **f** $f'(x)=4x-5$

g $f'(x)=2x\sqrt{x-2}+\frac{x^2}{2\sqrt{x-2}}$

h $f'(x)=3x^2(2x+5)^2+4x^3(2x+5)$

i $f'(x)=5\sqrt{x^2-6}+\frac{5x^2}{\sqrt{x^2-6}}$

2 **a** $\frac{dy}{dx}=6x^2-3$ **b** $\frac{dy}{dx}=3(x-4)^3+9x(x-4)^2$

c $\frac{dy}{dx}=2x\sqrt{1-2x}-\frac{x^2}{\sqrt{1-2x}}$ **d** $\frac{dy}{dx}=11-8x$

e $\frac{dy}{dx}=2\sqrt{x+5}+\frac{x}{\sqrt{x+5}}$

f $\frac{dy}{dx}=\frac{(x-1)^3}{2\sqrt{x}}+3\sqrt{x}(x-1)^2$

g $\frac{dy}{dx}=2x(2-x)^3-3x^2(2-x)^2$

h $\frac{dy}{dx}=9x^2\sqrt{x+2}+\frac{3x^3}{2\sqrt{x+2}}$

i $\frac{dy}{dx}=\frac{(7-x^2)^3}{2\sqrt{x+1}}-6x\sqrt{x+1}(7-x^2)^2$

3 **a** 39 **b** $\frac{15}{2}$ **c** $\frac{105}{2\sqrt{2}}$ **d** -26

4 **a** $f'(x)=\frac{(2x-7)^2(14x-7)}{2\sqrt{x}}$ **b** $\frac{49}{4}$

c $x=\frac{1}{2}$ and $x=\frac{7}{2}$

5 $-\frac{8}{9}$

EXERCISE 14D

1 $\frac{dy}{dx}=\frac{2x^2+3}{x^2}$

Check: $y=2x-\frac{3}{x},\quad \frac{dy}{dx}=2+\frac{3}{x^2}=\frac{2x^2+3}{x^2}$

2 **a** $\frac{dy}{dx}=\frac{1}{(x+1)^2}$ **b** $\frac{dy}{dx}=\frac{6}{(3-x)^2}$

c $\frac{dy}{dx}=\frac{2x^2-6x}{(2x-3)^2}$ **d** $\frac{dy}{dx}=\frac{-x^2-10x-1}{(x^2-1)^2}$

e $\frac{dy}{dx}=\frac{2x^2+10x-5}{(2x^2+5)^2}$ **f** $\frac{dy}{dx}=\frac{2-x}{2\sqrt{x}(x+2)^2}$

g $\frac{dy}{dx}=\frac{3x-6\sqrt{x}}{2\sqrt{x}(\sqrt{x}-1)^2}$ **h** $\frac{dy}{dx}=\frac{1-9x}{2\sqrt{x}(3x+1)^3}$

i $\frac{dy}{dx}=\frac{4x+17}{2\sqrt{x+2}(x+2)}$

3 **a** $\frac{dy}{dx}=-\frac{7}{(2x-3)^2}$ **b** **i** -7 **ii** $-\frac{7}{9}$

4 **a** -3 **b** -12 **c** $-\frac{1}{2}$ **d** $-\frac{1}{16}$

5 **a** $k=4$ **b** 5 **6** **b** $(\frac{1}{2},4),\ (-2,-1)$

EXERCISE 14E

1 **a** $f'(x)=3e^x$ **b** $f'(x)=-e^x$ **c** $f'(x)=4e^{4x}$

d $f'(x)=2e^{2x}$ **e** $f'(x)=-\frac{1}{5}e^{-\frac{x}{5}}$

f $f'(x)=2xe^{x^2}$ **g** $f'(x)=2e^x-5e^{-x}$

h $f'(x)=2e^{2x+1}+\frac{e^{\frac{1}{x}}}{x^2}$ **i** $f'(x)=-56xe^{3-4x^2}$

2 **a** $2xe^x+x^2e^x$ **b** $2e^{4x}+8xe^{4x}$

c $5e^{-2x}-6xe^{-2x}$ **d** $\frac{e^{5x-1}(10x+1)}{2\sqrt{x}}$

e $\frac{xe^x-e^x}{x^2}$ **f** $\frac{3x^2-2x^3}{e^{2x}}$

g $\frac{4xe^x-2e^x+1}{2x\sqrt{x}}$ **h** $\frac{6-3e^x+e^{-x}}{(e^{-x}-1)^2}$

3 **a** $\frac{dy}{dx}=\frac{e^{2x}(4x+5)}{2\sqrt{x+1}}$ **b** **i** $\frac{5}{2}$ **ii** $\frac{17}{4}e^6$

4 **a** 1 **b** $9e$ **c** $-24\ln 2$ **d** -5

5 $k=-2$ **6** **a** $\frac{3}{16}e^2$ **b** $P(1,\sqrt{e})$

7 $k=\frac{1}{e}$, point of contact (e,e)

EXERCISE 14F

1 **a** $\frac{dy}{dx}=\frac{3}{x}$ **b** $\frac{dy}{dx}=-\frac{1}{2x}$ **c** $\frac{dy}{dx}=2-\frac{7}{x}$

d $\frac{dy}{dx}=\frac{5}{5x-1}$ **e** $\frac{dy}{dx}=\frac{-7}{2-7x}$

f $\frac{dy}{dx}=\frac{2x}{x^2+3}$ **g** $\frac{dy}{dx}=\frac{8x-1}{4x^2-x}$

h $\frac{dy}{dx}=\frac{2x-5}{x^2-5x+2}$ **i** $\frac{dy}{dx}=1$ **j** $\frac{dy}{dx}=\frac{1}{2x}$

k $\frac{dy}{dx}=5x^4+\frac{1}{2x}$ **l** $\frac{dy}{dx}=5e^{5x}+\frac{1}{x}$

2 **a** $f'(x) = \ln x + 1$ **b** $f'(x) = 6x\ln x + 3x$

c $f'(x) = e^x \ln x + \dfrac{e^x}{x}$ **d** $f'(x) = \dfrac{1 - 2\ln x}{x^3}$

e $f'(x) = 7\ln(3x+1) + \dfrac{21x}{3x+1}$

f $f'(x) = \dfrac{1 - \ln 5x}{x^2}$ **g** $f'(x) = \dfrac{6 - \ln(x^3)}{2x\sqrt{x}}$

h $f'(x) = \dfrac{3}{\sqrt{x}}\ln(5x^2 - 1) + \dfrac{60x\sqrt{x}}{5x^2 - 1}$

i $f'(x) = -\dfrac{2}{x(\ln x)^2}$

3 **a** $\dfrac{dy}{dx} = \dfrac{4}{x} - 1$ **b** $\dfrac{dy}{dx} = \dfrac{1}{x+3} - \dfrac{2}{2x-7}$

c $\dfrac{dy}{dx} = 2 + \dfrac{1}{2x}$ **d** $\dfrac{dy}{dx} = \dfrac{3}{2(3x-5)}$

e $f'(x) = \dfrac{2x}{x^2+1} - \dfrac{4}{4x-3}$ **f** $f'(x) = \dfrac{2}{x} + \dfrac{1}{1-x}$

g $f'(x) = \dfrac{12}{2x+5}$ **h** $\dfrac{dy}{dx} = \dfrac{2}{x} + \dfrac{1}{2(x-7)}$

i $\dfrac{dy}{dx} = 3 - \dfrac{4}{x}$ **j** $f'(x) = \dfrac{15}{3x-2} - \dfrac{1}{x+2}$

4 l_1 has gradient $-\frac{1}{2}$, l_2 has gradient $\frac{9}{22}$

5 **a** $\frac{2}{9}$ **b** 4 **c** $2\ln 3 + 2$ **d** $-\dfrac{\ln 5}{e^2}$

6 **a** $b = 2$ **b** $15 + \dfrac{2}{e^2}$

7 **a** $f'\left(\dfrac{1}{e}\right) = \dfrac{3}{2e}$

b The tangent at $x = \dfrac{1}{e}$ has a negative gradient

$\therefore\ f'\left(\dfrac{1}{e}\right) = \dfrac{3}{2e} > 0$ must be incorrect.

c Lee has incorrectly used the rule $\ln(a^n) = n\ln a$ in line 2 of his working by saying $(\ln x)^2 = 2\ln x$.
$\ln(x^2) = 2\ln x$ but $(\ln x)^2 \neq 2\ln x$ in general.

d $f'\left(\dfrac{1}{e}\right) = -\dfrac{3}{e}$

8 **a** $\{x : x < -3 \text{ or } x > \frac{1}{2}\}$ **b** $-\frac{7}{4}$

c $(-\frac{13}{2}, -\ln 4)$ and $(4, 0)$

EXERCISE 14G

1 **a** $\dfrac{dy}{dx} = 2\cos x$ **b** $\dfrac{dy}{dx} = -\sec^2 x$

c $\dfrac{dy}{dx} = -\frac{1}{3}\sin x$ **d** $\dfrac{dy}{dx} = 3\sec^2 3x$

e $\dfrac{dy}{dx} = 5\cos 5x$ **f** $\dfrac{dy}{dx} = -8\sin 4x$

g $\dfrac{dy}{dx} = 2\cos(2x-1)$ **h** $\dfrac{dy}{dx} = \sin(3-x)$

i $\dfrac{dy}{dx} = \frac{3}{2}\sec^2 \frac{x}{2}$ **j** $\dfrac{dy}{dx} = \cos x + \sin x$

k $\dfrac{dy}{dx} = -3\sin x - 2\sec^2 2x$

l $\dfrac{dy}{dx} = -\pi\sin\pi x - 2\cos(4-x)$

2 **a** $f'(x) = -2x\sin x^2$ **b** $f'(x) = \dfrac{\cos\sqrt{x}}{2\sqrt{x}}$

c $f'(x) = -\dfrac{\sec^2 \frac{1}{x}}{x^2}$ **d** $f'(x) = (2x-1)\cos(x^2 - x)$

e $f'(x) = -\dfrac{4\sin(\ln x)}{x}$

f $f'(x) = 6x^2\sec^2 x^3 + \cos(x+1)$

g $f'(x) = e^x\sec^2 e^x + 6\sin 2x$

h $f'(x) = -2\sin x\cos x$ **i** $f'(x) = \dfrac{\cos x}{2\sqrt{\sin x}}$

3 **a** $\sin x + x\cos x$ **b** $e^x\cos x - e^x\sin x$

c $2x\tan x + x^2\sec^2 x$ **d** $\cos^2 x - \sin^2 x$

e $\dfrac{-x\sin x - \cos x}{x^2}$ **f** $\dfrac{2x\sin x - x^2\cos x}{\sin^2 x}$

g $\dfrac{1}{1+\cos x}$ **h** $\dfrac{2x\sec^2 x - \tan x}{2x\sqrt{x}}$

i $-\sin x + \dfrac{\sin x}{x} + \ln x\cos x$

j $\dfrac{-2\cos 3x - 3\sin 3x}{e^{2x}}$

k $2\sec^2 x - \dfrac{2x\cos 2x - \sin 2x}{x^2}$ **l** $\dfrac{-1 - 2\cos x}{5\sin^2 x}$

4 **a** **i** A(0, 1) **ii** B$(\frac{\pi}{3}, -\frac{1}{2})$ **iii** C$(\frac{3\pi}{4}, 0)$

b $\dfrac{dy}{dx} = -2\sin 2x$ **c** **i** 0 **ii** $-\sqrt{3}$ **iii** 2

5 **a** 1 **b** 2 **c** $1 - \frac{\pi}{\sqrt{3}}$ **d** $-2\sqrt{3}$

EXERCISE 14H

1 **a** $\dfrac{dy}{dx} = 8x$ **b** y increases by ≈ 0.16

2 **a** $\dfrac{dy}{dx} = -\dfrac{12}{x^2}$

b **i** y increases by ≈ 0.04

ii x decreases by ≈ 0.0067

3 **a** $\dfrac{dy}{dx} = \dfrac{6}{(2-x)^2}$

b **i** y increases by ≈ 0.0612

ii x decreases by ≈ 0.0141

4 **a** $\dfrac{dy}{dx} = 3e^{3x}$ **b** y increases by ≈ 0.603 **c** $\dfrac{dy}{dx} = 3y$

d x decreases by ≈ 0.002

5 **a** y increases by ≈ 1.6

b using **a**, $2.02^5 \approx 33.6$
using technology, $2.02^5 \approx 33.63$

6 **a** From the graph, we see that y would initially decrease as x decreases from $\frac{\pi}{6}$.

b y decreases by ≈ 0.023 **c** $\sin\frac{19\pi}{120} \approx 0.477$

EXERCISE 14I

1 **a** $f'(x) = 8x^3 - 10x$ **b** $f''(x) = 24x^2 - 10$

2 **a** $\dfrac{dy}{dx} = \dfrac{3}{x} + 3x^2$ **b** $\dfrac{d^2y}{dx^2} = -\dfrac{3}{x^2} + 6x$

3 **a** $f''(x) = 8$ **b** $f''(x) = 20x^3 - \dfrac{4}{x^3}$

c $f''(x) = 24x - 6$ **d** $f''(x) = -\dfrac{1}{2x\sqrt{x}}$

e $f''(x) = -\dfrac{4}{x^2} - \dfrac{3}{4x^2\sqrt{x}}$ **f** $f''(x) = 168(2x-3)^5$

g $f''(x) = e^x$ **h** $f''(x) = -2\sin x$

i $f''(x) = 9e^{3x} + 4\cos 2x$

4 **a** $\dfrac{d^2y}{dx^2} = 4e^{2x}$ **b** $\dfrac{d^2y}{dx^2} = 4 + \dfrac{2}{(1-x)^3}$

c $\dfrac{d^2y}{dx^2} = \dfrac{2-2x^2}{(x^2+1)^2}$ **d** $\dfrac{d^2y}{dx^2} = 6e^x + 3xe^x$

e $\dfrac{d^2y}{dx^2} = \dfrac{12}{(2-x)^3}$ **f** $\dfrac{d^2y}{dx^2} = -\dfrac{14}{(x+2)^3} - 2$

g $\dfrac{d^2y}{dx^2} = 2\cos^2 x - 2\sin^2 x$

h $\dfrac{d^2y}{dx^2} = 10(x^2-1)^3(9x^2-1)$ **i** $\dfrac{d^2y}{dx^2} = \dfrac{2\sin x}{\cos^3 x}$

5 **a** $f(2) = -2$ **b** $f'(2) = -2$ **c** $f''(2) = 2$

6 **a** $f(\frac{\pi}{6}) = \frac{\pi\sqrt{3}}{12}$ **b** $f'(\frac{\pi}{6}) = \frac{\sqrt{3}}{2} + \frac{\pi}{6}$

c $f''(\frac{\pi}{6}) = 2 - \frac{\pi}{\sqrt{3}}$

7 **a** $f''(x) = \dfrac{2x^3 - 12x}{(x^2+2)^3}$ **b** $x = 0$ or $\pm\sqrt{6}$

8 **a** $\dfrac{d^2y}{dx^2} = \dfrac{1}{x^2}$ **b** $\dfrac{d^2y}{dx^2} = \dfrac{1}{x}$ **c** $\dfrac{d^2y}{dx^2} = \dfrac{2-2\ln x}{x^2}$

d $\dfrac{d^2y}{dx^2} = -\dfrac{1}{(x+2)^2} - \dfrac{1}{(x-5)^2}$

9 **a** $x = 0, \pi$, or 2π **b** $x = \frac{3\pi}{4}$ or $\frac{7\pi}{4}$ **c** $x = \frac{\pi}{2}$ or $\frac{3\pi}{2}$

10 $f(x) = 2x^2 - x + 4$

REVIEW SET 14A

1 **a** $f'(x) = 12x^3 - 4x$ **b** $f'(x) = -\dfrac{2}{x^2} + \dfrac{15}{x^4}$

c $f'(x) = \dfrac{4}{\sqrt{x}} + \dfrac{3}{2x\sqrt{x}}$

2 **a** $\dfrac{dy}{dx} = 30(5x-2)^5$ **b** $\dfrac{dy}{dx} = \dfrac{8}{(6-x)^3}$

c $\dfrac{dy}{dx} = \dfrac{2x-3}{2\sqrt{x^2-3x}}$

3 **a** $\frac{23}{6}$ **b** $\frac{3}{4}$

4 **a** $f'(x) = 3e^{3x-1}$ **b** $f'(x) = \dfrac{-5}{4-5x}$

c $f'(x) = (2x-1)\cos(x^2 - x)$

5 $-\frac{1}{4\sqrt{2}}$ **6** **a** $\dfrac{\sin x}{2\sqrt{x}} + \sqrt{x}\cos x$ **b** $\dfrac{1 - x\ln x}{xe^x}$

7 $a = 4,\ b = 3$ **8** **a** $k = -3$ **b** 4

9 **a** $\dfrac{dy}{dx} = \dfrac{-2}{1-2x} - \dfrac{2x}{x^2+3}$ **b** $\dfrac{dy}{dx} = \dfrac{3}{x} - \tan x$

10 **a** $f'(x) = \dfrac{e^x\cos x + e^x \sin x}{\cos^2 x}$

b $f'(0) = 1$, the gradient of the tangent at $x = 0$ is 1.

c $\left(\frac{3\pi}{4}, -\sqrt{2}e^{\frac{3\pi}{4}}\right)$, $\left(\frac{7\pi}{4}, \sqrt{2}e^{\frac{7\pi}{4}}\right)$

11 **a** $\dfrac{dy}{dx} = 4x^3$ **b** y increases by ≈ 2.16 **c** $3.02^4 \approx 83.16$

12 **a** $f''(x) = 14 - \dfrac{4}{x^3}$ **b** $f''(x) = e^{\sin x}(\cos^2 x - \sin x)$

REVIEW SET 14B

1 **a** $\dfrac{dy}{dx} = 15x^4 - 8x$ **b** $\dfrac{dy}{dx} = 2x - 8$ **c** $\dfrac{dy}{dx} = 5 - \dfrac{1}{x^2}$

2 **a** $\dfrac{dy}{dt} = \dfrac{1}{3\sqrt[3]{t^2}}$ **b** $\dfrac{dM}{dx} = -15(7-3x)^4$

3 $a = -6,\ b = -2$ **4** **a** -4 **b** $\ln 4 + \frac{3}{4}$

5 l_1 has gradient $\frac{1}{2}$, l_2 has gradient $\frac{1}{6}$

6 **a** $f'(x) = \dfrac{e^{2x}(12x+5)}{\sqrt{6x+1}}$ **b** $\dfrac{11e}{2}$

7 **a** $\dfrac{-6x}{(x^2-1)^2}$ **b** $4\cos 4x - \cos x + x\sin x$

c $\dfrac{60x}{6x^2-1}$ **d** $2\tan x \sec^2 x$

8 $k = \sqrt{2}$

9 **a** $f'(x) = \dfrac{1}{x} - \dfrac{1}{x-2}$ **b** $-\frac{1}{12}$

c $(-2, -\ln 2)$, $(4, \ln 2)$

10 **b** $(\frac{\pi}{4}, 1 + \frac{1}{2}\ln 2)$

11 **a** $\dfrac{dy}{dx} = \dfrac{5}{(1-x)^2}$ **b** **i** y increases by ≈ 0.15
ii x decreases by ≈ 0.05

12 **a** $\dfrac{d^2y}{dx^2} = \dfrac{-2e^x}{(e^x-2)^2}$ **b** $(\ln 4, \ln 2)$

EXERCISE 15A

1 **a** $y = 6x - 9$ **b** $y = 3x - 2$ **c** $y = 10x + 21$
d $y = -\frac{2}{3}x + 4$ **e** $y = \frac{2}{3}x + 6$ **f** $y = -x - 9$

2 **a** $y = 2x - 6$ **b** P$(0, -6)$, Q$(3, 0)$ **3** $6\frac{1}{4}$ units^2

4 **a** $x + 5y = 32$ **b** $x - 3y = -4$ **c** $y = 4x - 15$
d $2x - 5y = 12$ **e** $x + 6y = -12$ **f** $x + 15y = 604$

5 **a** -4 **b** $y = 6x - 16$ **c** $x + 6y = -22$

6 **a** $16x + 3y = 70$ **b** $(\frac{35}{8}, 0)$, $(0, \frac{70}{3})$

7 **a** $\left(\sqrt[3]{4}, \dfrac{5}{\sqrt[3]{2}}\right)$ **b** $y = 4x - \dfrac{3}{\sqrt[3]{2}}$

8 $b = -3,\ c = 1$

9 **a** $3x - 4y = -5$ **b** $y = -4x + 13$
c $3x - 4y = -2$ **d** $3x - 50y = -7$

10 **a** $2x + 3y = 6$ **b** $4x + 57y = 1038$
c $27x - 15y = 44$ **d** $y = 3x - 9$

11 $a = 6,\ b = 7$ **12** P$(6, 6)$

13 **a** $\{x : -3 \leqslant x \leqslant 3\}$

b **Hint:** Let P$(a, \sqrt{9-a^2})$ be a general point on the curve. Find, in terms of a, the equation of the normal at P.

c $y = \sqrt{9 - x^2}$ is the top half of a circle centred at the origin with radius 3. The normal to a point on the circle is a radius of the circle. The normal must therefore pass through the centre of the circle, which is the origin.

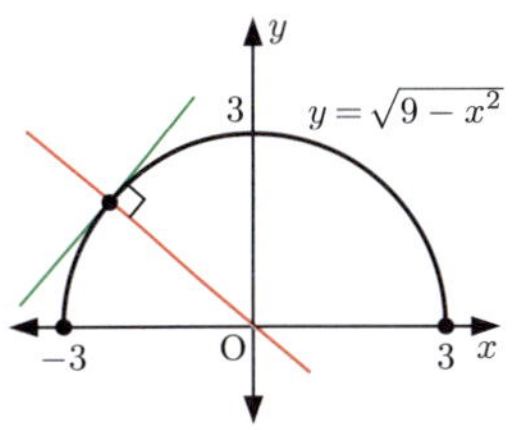

14 **a** $y = x - 1$ **b** $y = 8x - 8\ln 2 + 4$
c $5x - 4y = 20 - 8\ln 2$ **d** $x + e^2y = 4$
e $\sqrt{3}x + 2y = \frac{\pi}{\sqrt{3}} + 1$ **f** $6x + y = \pi$
g $2x - y = \frac{\pi}{2} + 1$ **h** $x - 2y = -1$

15 **a** $x + ey = 4 + e^2$ **b** $y = x - 3$
c $2x + 3\sqrt{3}y = \frac{\pi}{3}$ **d** $x - y\ln 2 = -(\ln 2)^2$

16 5 units^2 **17** $a = 2,\ b = 3$

18 **a** $2x + y = -1$ **b** $(-2, 3)$ **19** $(2, 14)$

20 **a** $x + y = -2$ **b** $(1, -3)$

21 **a** **i** $y = -\frac{2}{k^3}x + \frac{3}{k^2}$ **ii** $-\frac{1}{2}k$ **b** $k = 2^{\frac{5}{6}}$

22 $\approx 63.43°$

EXERCISE 15B

1 **a** $f'(x) = -3x^2 + 6x + 24$
$= -3(x+2)(x-4)$

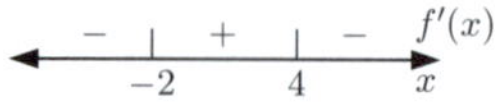

b increasing for $-2 \leqslant x \leqslant 4$, decreasing for $x \leqslant -2$ and $x \geqslant 4$

2 **a** increasing for $x \geqslant -1$, decreasing for $x \leqslant -1$

b increasing for $x \leqslant -1$ and $x \geqslant 1$, decreasing for $-1 \leqslant x \leqslant 1$

c increasing for $x \leqslant \frac{5}{2}$, decreasing for $x \geqslant \frac{5}{2}$

d decreasing for all $x \neq 0$, never increasing

e increasing for $0 < x \leqslant 6$, decreasing for $x < 0$ and $x \geqslant 6$

f increasing for $x \geqslant 0$, never decreasing

g increasing for $x \leqslant -\sqrt{5}$ and $x \geqslant \sqrt{5}$, decreasing for $-\sqrt{5} \leqslant x < 0$ and $0 < x \leqslant \sqrt{5}$

h increasing for $x \geqslant 0$, decreasing for $x \leqslant 0$

i increasing for $x \geqslant -1$, decreasing for $x \leqslant -1$

j decreasing for all $x \neq 3$, never increasing

3 **a** $f'(x) = -3x^2 + 4x - 3$

b $f'(x)$ has $a < 0$ and $\Delta = b^2 - 4ac = -20 < 0$
$\therefore$ $f'(x) < 0$ for all x.
$\therefore$ $f(x)$ is decreasing for all x.

EXERCISE 15C

1 **a** P is a minimum turning point, Q is a maximum turning point, and R is a stationary inflection.

b **i**

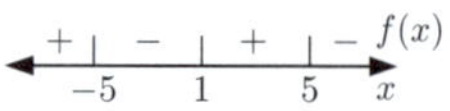

ii

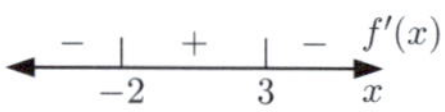

2 **a** $f'(x) = 3x^2 - 12$
$= 3(x+2)(x-2)$

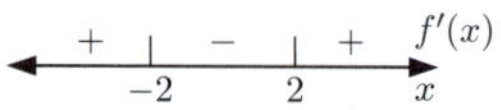

b increasing for $x \leqslant -2$ and $x \geqslant 2$, decreasing for $-2 \leqslant x \leqslant 2$

c maximum turning point $(-2, 17)$, minimum turning point $(2, -15)$

3 **a** minimum turning point $(-\frac{3}{2}, -\frac{9}{4})$

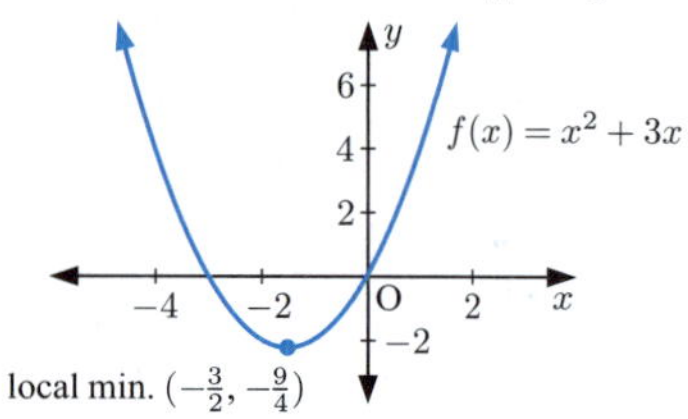

b stationary inflection $(0, -2)$

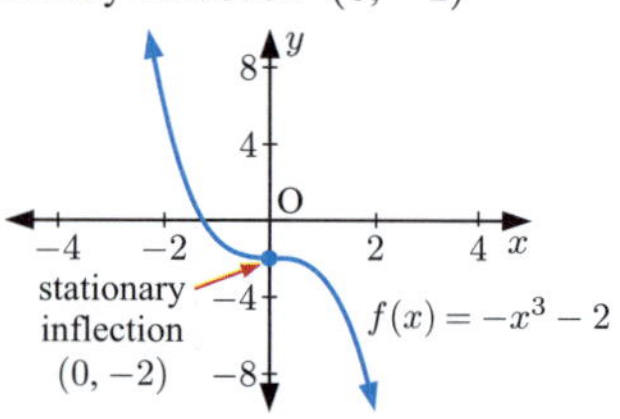

c maximum turning point $(2, 9)$, minimum turning point $(4, 5)$

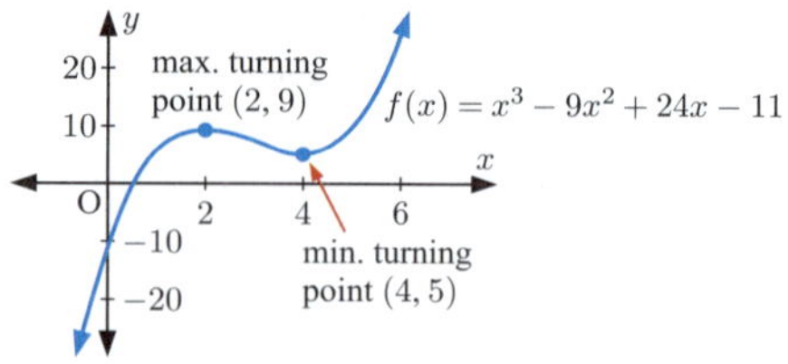

d minimum turning points $(-2, -13)$ and $(2, 13)$, maximum turning point $(0, 3)$

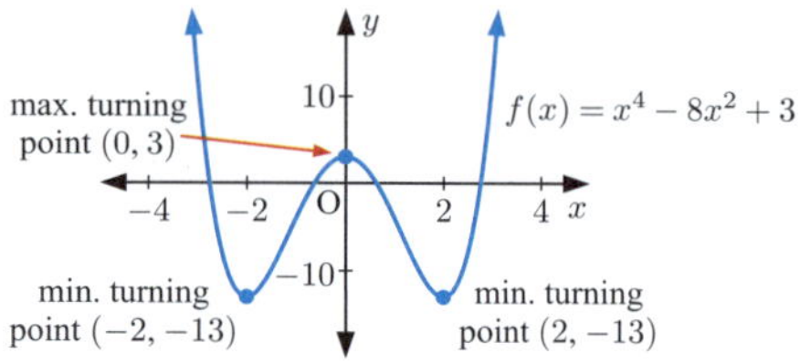

e minimum turning point $(1, -1)$

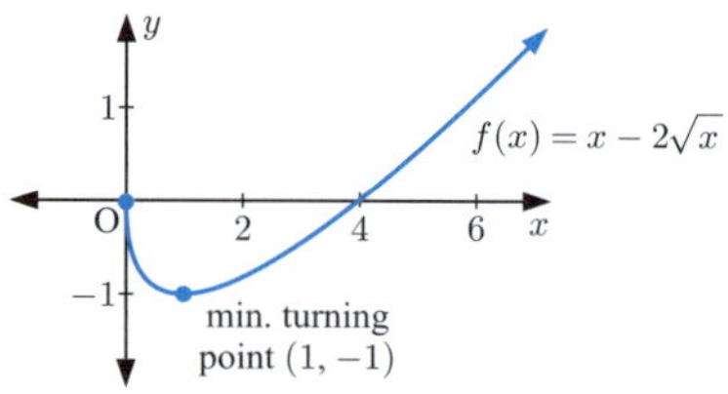

f minimum turning point $(1, -1)$

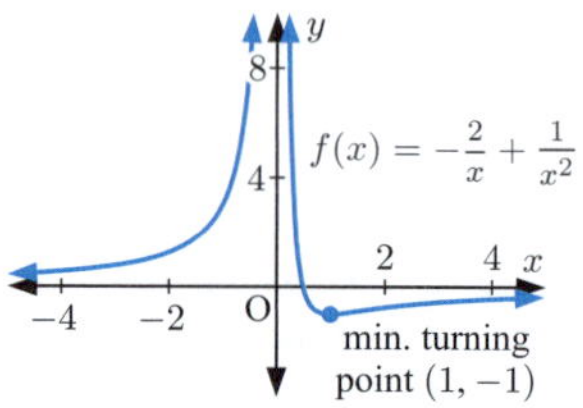

g minimum turning point $(0, -15)$, stationary inflection $(-3, 12)$

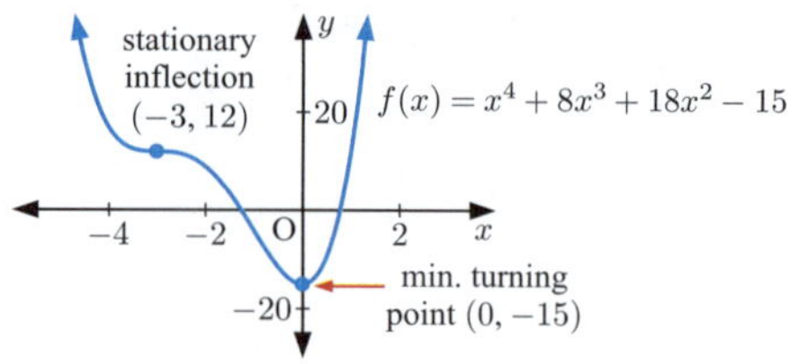

h minimum turning point $(2, 12)$

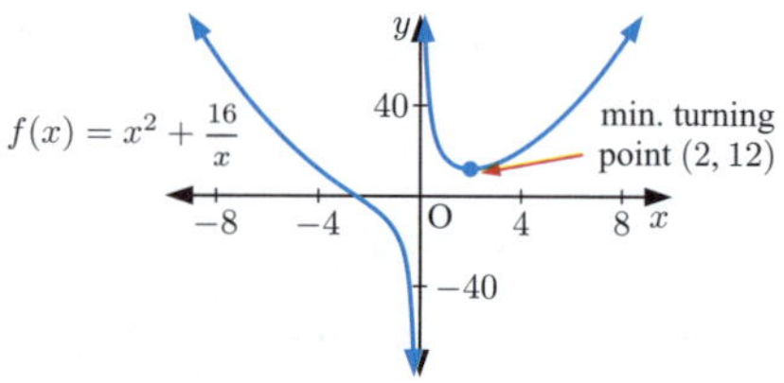

4 **a** $a = 4$ **b** $f''(x) = 6x + 8$

c $f''(-3) < 0$, so it is a maximum turning point

5 **a** $a = -4$, $b = 1$ **c** minimum turning point

6 **a** $A(-2, 0)$ **b** $B(0, 2)$ **c** $C(-1, e)$

7 **a** minimum turning point $\left(-\frac{1}{2}, -\frac{3}{2e}\right)$

b minimum turning point $(-1, -\frac{1}{2})$, maximum turning point $(1, \frac{1}{2})$

c minimum turning point $(-2, \ln 2)$

d minimum turning point $\left(\frac{1}{\sqrt{e}}, -\frac{1}{2e}\right)$

e stationary inflection $\left(-2, \frac{1}{2e^2}\right)$

f minimum turning point $\left(-\frac{5}{2}, -\frac{7\sqrt{35}}{4}\right)$, maximum turning point $(3, 2\sqrt{6})$

8 **a** $\frac{dy}{dx} = \frac{7}{(x+2)^2} - 7$, $\frac{d^2y}{dx^2} = -\frac{14}{(x+2)^3}$

b minimum turning point $(-3, 31)$, maximum turning point $(-1, 3)$

9 **a** $a = 6$, $b = 9$ **b** $\left(-5, \frac{4}{e^5}\right)$

10 **a** $f'(x) = 2\cos(x + \frac{\pi}{6})$ **b** $A(\frac{\pi}{3}, 2)$, $B(\frac{4\pi}{3}, -2)$

11 **a** maximum turning points $(0, 3)$ and $(2\pi, 3)$, minimum turning point $(\pi, -3)$

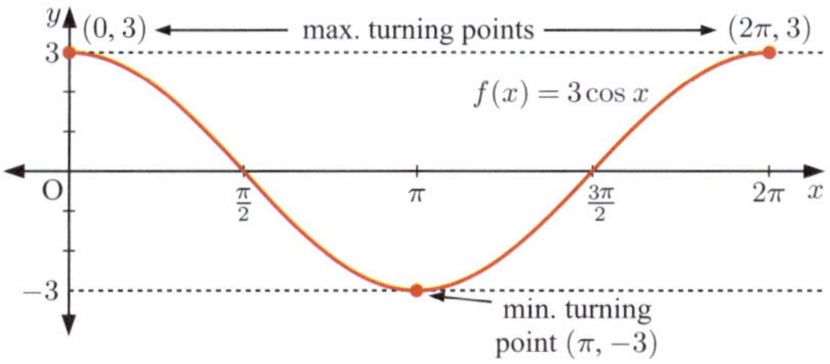

b maximum turning points $(\frac{\pi}{4}, 1)$ and $(\frac{5\pi}{4}, 1)$, minimum turning points $(\frac{3\pi}{4}, -1)$ and $(\frac{7\pi}{4}, -1)$

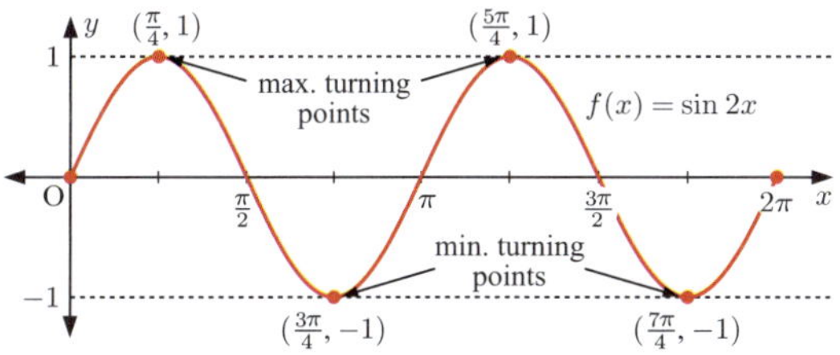

c maximum turning point $(\frac{\pi}{4}, 1)$, minimum turning point $(\frac{5\pi}{4}, -1)$

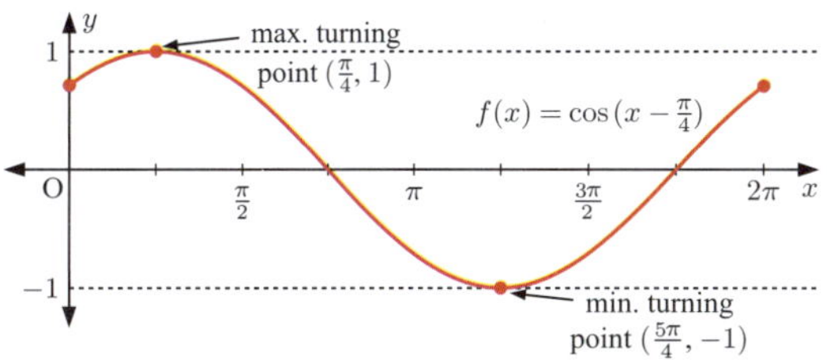

d maximum turning point $(\frac{\pi}{4}, \sqrt{2})$, minimum turning point $(\frac{5\pi}{4}, -\sqrt{2})$

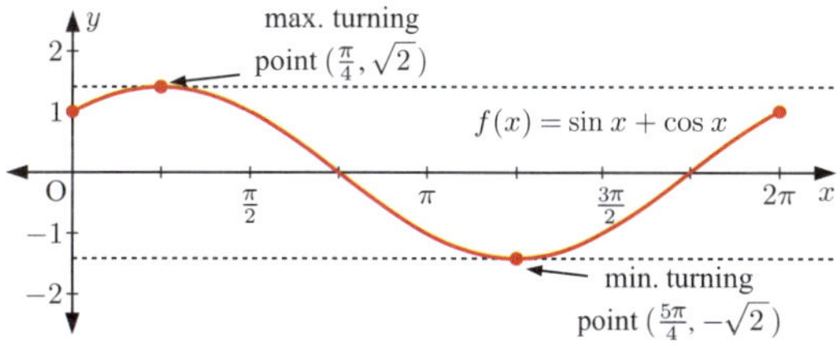

e maximum turning points $(0, 1)$, $(\pi, 1)$, and $(2\pi, 1)$, minimum turning points $(\frac{\pi}{2}, 0)$ and $(\frac{3\pi}{2}, 0)$

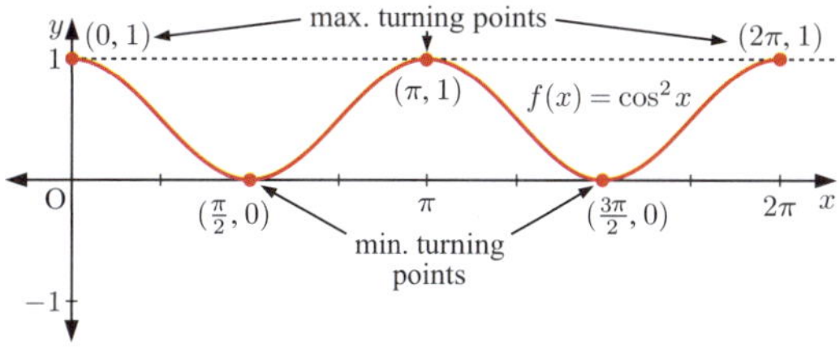

f maximum turning point $(\frac{\pi}{6}, 2)$, minimum turning point $(\frac{7\pi}{6}, -2)$

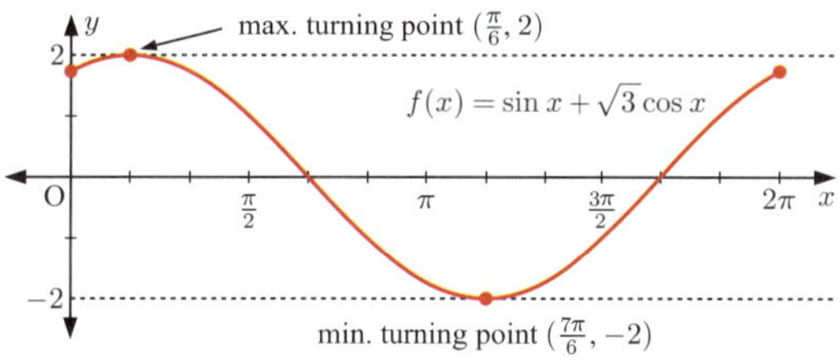

12 **Hint:** Show that $\frac{dy}{dx} \neq 0$.

13 **a** minimum turning point $(-\frac{2\pi}{3}, -\frac{2\pi}{3} - \sqrt{3})$, maximum turning point $(\frac{2\pi}{3}, \frac{2\pi}{3} + \sqrt{3})$

b minimum turning point $(-\frac{\pi}{2}, \frac{1}{3})$, maximum turning point $(\frac{\pi}{2}, 1)$

c minimum turning point $\left(-\frac{\pi}{4}, \sqrt{2}e^{-\frac{\pi}{4}}\right)$, maximum turning point $\left(\frac{3\pi}{4}, -\sqrt{2}e^{\frac{3\pi}{4}}\right)$

14 **a** $f'(x) = -2\cos x \sin x + \cos x$

b $(\frac{\pi}{6}, -\frac{3}{4})$, $(\frac{\pi}{2}, -1)$, $(\frac{5\pi}{6}, -\frac{3}{4})$, $(\frac{3\pi}{2}, -3)$

c $f''(x) = 2\sin^2 x - 2\cos^2 x - \sin x$

d maximum turning points $(\frac{\pi}{6}, -\frac{3}{4})$ and $(\frac{5\pi}{6}, -\frac{3}{4})$, minimum turning points $(\frac{\pi}{2}, -1)$ and $(\frac{3\pi}{2}, -3)$

15 $a = -6$, $b = 12$, $c = -5$

EXERCISE 15D.1

1 **a** **i** 3 m **ii** 5 m

b $v(t) = 2t - 3 \text{ m s}^{-1}$, $a(t) = 2 \text{ m s}^{-2}$

c $v(4) = 5 \text{ m s}^{-1}$, $a(4) = 2 \text{ m s}^{-2}$

2 **a** $v(t) = 3t^2 - 4t - 10 \text{ cm s}^{-1}$, $a(t) = 6t - 4 \text{ cm s}^{-2}$

b $a(2) = 8 \text{ cm s}^{-2}$ **c** $t = 3$ s **d** 11 cm s^{-1}

3 **a** 10 m **b** $\left(5 + \frac{1}{2\sqrt{5}}\right) \text{ m s}^{-1}$ **c** $t = 1$ s

4 **a** $1\ \text{m s}^{-1}$ **b** $t = 6\ln 2$ s **c** $e(e-2)^2\ \text{m s}^{-2}$

5 **a** $4\ \text{m s}^{-1}$ **b** $t = 4$ s
c **i** $\frac{\pi\sqrt{3}}{2}\ \text{m s}^{-2}$ **ii** $-\frac{\pi}{2}\ \text{m s}^{-2}$

EXERCISE 15D.2

1 **a** $v(t) = 2t - 6\ \text{cm s}^{-1}$, $a(t) = 2\ \text{cm s}^{-2}$

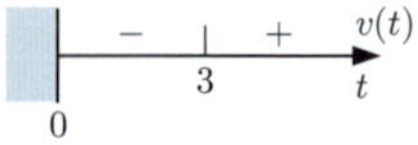

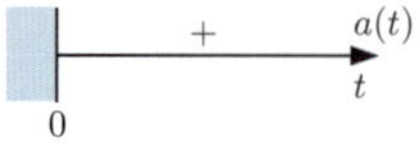

b $s(0) = 4$ cm, $v(0) = -6\ \text{cm s}^{-1}$, $a(0) = 2\ \text{cm s}^{-2}$
The object is initially 4 cm to the right of the origin and is moving to the left at $6\ \text{cm s}^{-1}$. It is accelerating at $2\ \text{cm s}^{-2}$ to the right.
c At $t = 3$, $s(3) = -5$, the object changes direction 5 cm to the left of the origin.
d

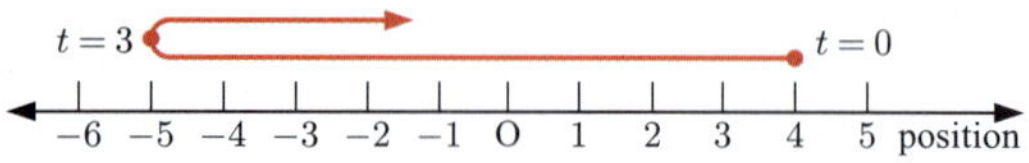

e $s(5) = -1$ cm **f** 13 cm

2 **a** $v(t) = 3t^2 - 10t + 7\ \text{m s}^{-1}$, $a(t) = 6t - 10\ \text{m s}^{-2}$

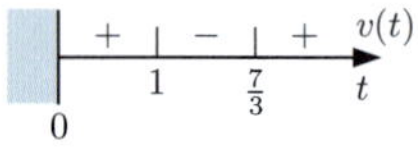

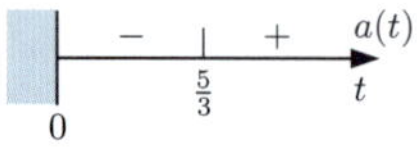

b $s(0) = 2$ m, $v(0) = 7\ \text{m s}^{-1}$, $a(0) = -10\ \text{m s}^{-2}$
The particle is initially 2 m to the right of the origin, moving to the right at $7\ \text{m s}^{-1}$, and accelerating at $10\ \text{m s}^{-2}$ to the left.
c **i** $1 \leqslant t \leqslant \frac{7}{3}$ **ii** $0 \leqslant t \leqslant 1$, $t \geqslant \frac{7}{3}$
iii $1 \leqslant t \leqslant \frac{5}{3}$, $t \geqslant \frac{7}{3}$ **iv** $0 \leqslant t \leqslant 1$, $\frac{5}{3} \leqslant t \leqslant \frac{7}{3}$
d $8\ \text{m s}^{-2}$ **e** 5 m to the right of the origin **f** 4 m

3 **a** $v(t) = -9.8t + 19.6\ \text{m s}^{-1}$, $a(t) = -9.8\ \text{m s}^{-2}$

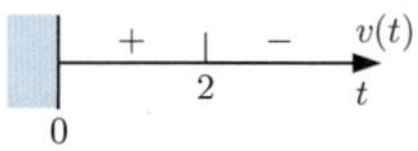

b $s(0) = 1.5$ m, $v(0) = 19.6\ \text{m s}^{-1}$
The ball is initially 1.5 m above the ground, moving upwards at $19.6\ \text{m s}^{-1}$.
c $t = 2$ s **d** 21.1 m **e** 24.5 m

5 **a** $v(0) = \ln 7\ \text{m s}^{-1}$ **b** $-1\ \text{m s}^{-2}$
c $t = 2$ s and $t = 3$ s **d** $2 \leqslant t \leqslant 3$
e $t = 3$ s and $t = 4$ s

6 **a** **i** $\approx 57.3\ \text{m s}^{-1}$ **ii** $\approx 93.8\ \text{m s}^{-1}$
b $a(t) = 70e^{-\frac{t}{5}} - 14te^{-\frac{t}{5}}\ \text{m s}^{-2}$
c 5 seconds, after which the velocity begins to decrease.

7 **a** $v(t) = 15\cos 3t\ \text{cm s}^{-1}$
b

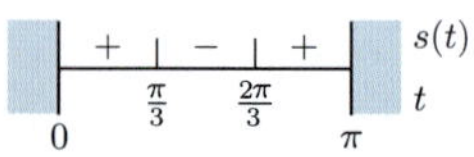

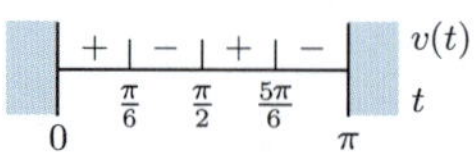

c **i** $0 \leqslant t \leqslant \frac{\pi}{3}$, $\frac{2\pi}{3} \leqslant t \leqslant \pi$ **ii** $\frac{\pi}{3} \leqslant t \leqslant \frac{2\pi}{3}$
iii $0 \leqslant t \leqslant \frac{\pi}{6}$, $\frac{\pi}{2} \leqslant t \leqslant \frac{5\pi}{6}$
iv $\frac{\pi}{6} \leqslant t \leqslant \frac{\pi}{2}$, $\frac{5\pi}{6} \leqslant t \leqslant \pi$
d $7.5\ \text{cm s}^{-1}$ **e** $-45\ \text{cm s}^{-2}$

8 **a** ≈ 1.54 m **b** $1.5\ \text{m s}^{-1}$ **c** $t = \frac{2\pi}{3}, \frac{4\pi}{3}$, or $\frac{8\pi}{3}$ s
d $t = \frac{7\pi}{6}$ s **e** ≈ 2.91 m

EXERCISE 15E

1 **a** **i** 2 m **ii** 4 m **b** $H'(t) = \dfrac{4}{(t+1)^2}$
c **i** 1 metre per year **ii** $\frac{1}{4}$ metres per year
d As $H'(t) = \dfrac{4}{(t+1)^2} > 0$ for all $t \geqslant 0$, the height of the giraffe is always increasing.

2 **a** **i** $\approx 553\,000$ dollars **ii** $\approx 642\,000$ dollars
b $\dfrac{dV}{dt} = 25\,000e^{\frac{t}{20}}$ dollars per year
c **i** increasing at $\approx 32\,100$ dollars per year
ii increasing at $\approx 41\,200$ dollars per year
d **i** $e^{\frac{t}{20}} > 0$ for all $t \geqslant 0$, so $\dfrac{dV}{dt} > 0$, the value of the house is always increasing.
ii $\dfrac{d^2V}{dt^2} = 1250e^{\frac{t}{20}} > 0$, the rate at which the house increases in value is always increasing.

3 **a** $\frac{1000}{18} \approx 55.6°$C
b $\dfrac{dT}{dx} = -\dfrac{2000x}{(x^2+2)^2}$ °C per metre
c **i** decreases at $\approx 111°$C per metre
ii decreases at $\approx 13.7°$C per metre
d The temperature always decreases as the distance from the campfire increases.

4 **a** **i** 12.5 m **ii** 32 m
b **i** increasing at 2 metres per m s^{-1}
ii increasing at 5 metres per m s^{-1}
c **i** $\dfrac{dD}{dv} = \frac{1}{4}v > 0$ for all $v > 0$, the stopping distance is always increasing as v increases.
ii $\dfrac{d^2D}{dv^2} = \frac{1}{4} > 0$, the rate at which the stopping distance increases is always increasing.

5 **a** **i** ≈ 30.4 cm **ii** ≈ 34.4 cm
b **i** increasing at $\frac{1}{12} \approx 0.0833$ cm per day
ii increasing at $\frac{1}{20} = 0.05$ cm per day
c **i** $\dfrac{dC}{dt} = \dfrac{10}{t+100} > 0$, for all $t \geqslant 0$, the head circumference of a baby is always increasing.
ii $\dfrac{d^2C}{dt^2} = -\dfrac{10}{(t+100)^2} < 0$ for all $t \geqslant 0$, the rate at which the circumference of a baby's head increases is always decreasing.

6 **a** **i** ≈ 126.5 cents per litre **ii** ≈ 131.7 cents per litre
b **i** increasing at ≈ 8.39 cents per litre per day
ii decreasing at ≈ 12.1 cents per litre per day

7 **a** $\dfrac{dP}{dt} = \dfrac{97\,500}{e^{1.3t}(1+10e^{-1.3t})^2}$
b **i** ≈ 1315 parrots per year **ii** ≈ 1365 parrots per year
c $(1+10e^{-1.3t})^2 > 0$, $e^{1.3t} > 0$
$\therefore \dfrac{dP}{dt} = \dfrac{97\,500}{e^{1.3t}(1+10e^{-1.3t})^2} > 0$
d $\dfrac{dP}{dt} \to 0$ as $t \to \infty$, the rate of population growth decreases over time.

8 **a** $A = \pi r^2$ m^2 **b** $\dfrac{dA}{dr} = 2\pi r$

c increasing at 10π m^2 per m

9 increasing at 20 cm^2 per radian

10 increasing at $36\sqrt{3}$ cm^2 per cm

11 increasing at ≈ 7.93 cm per radian

12 **a** $A = \frac{49}{2}(\theta - \sin\theta)$

b **i** increasing at ≈ 7.18 cm^2 per radian

ii increasing at ≈ 45.7 cm^2 per radian

iii increasing at 12.25 cm^2 per radian

EXERCISE 15F

1 **b** $x = 10$ **c** 10 m $\times$ 20 m

2 **a** $0 < x < 9$ **c** 3 cm $\times$ 3 cm

3 $\theta = 90°$ **4** ≈ 4.41 months old

5 **a** **Hint:** Show that the trapezium has height $h = \sqrt{16 - x^2}$.

b $x = 2$

6 **a** $\mathrm{P}(a, 9 - a^2)$ **b** $0 < a < 3$

c $A = 18a - 2a^3$ units^2 **d** $12\sqrt{3}$ units^2 when $a = \sqrt{3}$

7 $\mathrm{P}(\sqrt{3}, \sqrt{3})$

8 **a** $E'(t) = 5e^{-\frac{t}{4}} - \frac{5}{4}te^{-\frac{t}{4}}$

b **i** increasing by ≈ 2.92 units per minute

ii not increasing or decreasing

c at $t = 4$ minutes

9 **b** $x = 2\sqrt{\dfrac{15}{4+\pi}} \approx 2.90$

10 **b** $\dfrac{dA}{d\theta} = -400\sin(\theta - \frac{\pi}{2})\left[1 + \sin(\theta - \frac{\pi}{2})\right] + 400\cos^2(\theta - \frac{\pi}{2})$

c $\theta = \frac{2\pi}{3}$

11 **a** $y = 5 + \dfrac{65}{x}$ **b** $\sqrt{(x+13)^2 + \left(5 + \dfrac{65}{x}\right)^2}$

c ≈ 6.88 m **d** ≈ 24.6 m

EXERCISSE 15G

1 **a** $\dfrac{dy}{dt} = (6x^2 - 4)\,\dfrac{dx}{dt}$

b increasing at 6 units per second

2 **a** decreasing at 24 units per second

b increasing at $\frac{1}{30}$ units per second

3 **a** increasing at $\frac{4}{5}$ units per second

b increasing at $\dfrac{4}{e^2}$ units per second

4 **a** $V = x^3$

b **i** increasing at 6 m^3 per second

ii increasing at 96 m^3 per second

5 **a** $A = \pi r^2$

b **i** increasing at $\frac{2}{\pi}$ cm per second

ii increasing at $\frac{5}{6\pi}$ cm per second

6 decreasing at 0.9 cm s^{-1}

7 decreasing at 240π cm^2 per minute

8 **a** increasing at 0.3 cm s^{-1} **b** increasing at $\frac{5}{2\sqrt{41}}$ cm s^{-1}

9 **a** $x = \dfrac{8}{\cos\theta}$ **b** increasing at ≈ 0.0931 cm per minute

10 decreasing at ≈ 0.419 cm per minute

11 increasing at $\frac{1}{48}$ radians per second

12 **b** **i** increasing at ≈ 0.210 cm^2 per minute

ii increasing at $\frac{3\pi}{4}$ cm^2 per minute

13 **a** increasing at ≈ 0.262 cm s^{-1}

b increasing at ≈ 0.213 cm s^{-1}

14 **a** $r = \dfrac{30}{\theta + 2}$ cm

b **i** decreasing at ≈ 0.425 radians per second

ii decreasing at $\frac{12}{5}$ radians per second

REVIEW SET 15A

1 **a** $y = 3x + 1$ **b** $5x - 4y = -8$ **c** $x - 6y = -11$

d $y = 2x + 3$

2 **a** $y = 2ex - e$ **b** $x + 2ey = 1 + 2e^2$

3 $a = 2,\ b = 5$

4 **a** $f'(x) = 3x^2 - 6$

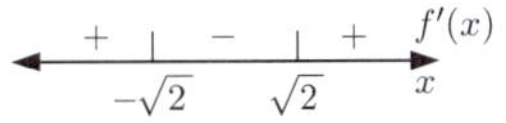

b $-\sqrt{2} \leqslant x \leqslant \sqrt{2}$

5 **a** increasing for $x \leqslant 3$, decreasing for $x \geqslant 3$

b increasing for $x \geqslant -1$, decreasing for $x \leqslant -1$

6 **a** $\mathrm{A}(0, \ln 3)$ **b** $\mathrm{B}(\frac{\pi}{2}, \ln 4)$ **c** $\mathrm{C}(\frac{3\pi}{2}, \ln 2)$

7 **a** maximum turning points $(-1, 5)$ and $(2, 32)$, minimum turning point $(0, 0)$

b

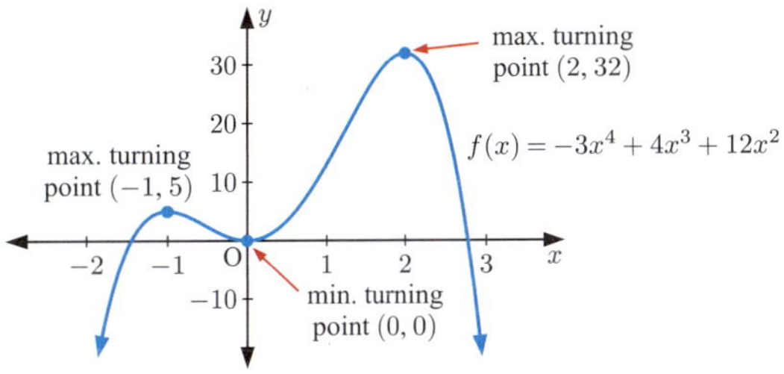

8 **a** $v(t) = 3t^2 - 6$ cm s^{-1} **b** 3 cm s^{-1} **c** $t = \sqrt{2}$ s

d 12 cm s^{-2}

9 **a** $v(t) = -9.8t + 24.5$ m s^{-1}

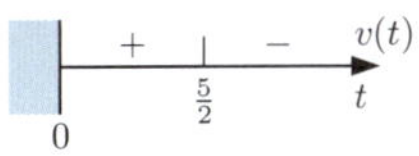

$a(t) = -9.8$ m s^{-2}

b $s(0) = 1$ m, $v(0) = 24.5$ m s^{-1}

c $\frac{253}{8} = 31.625$ m **d** $\frac{833}{20} = 41.65$ m

10 **a** **i** 5625 mL **ii** 2500 mL

b **i** decreasing at 375 mL per second

ii decreasing at 250 mL per second

11 increasing at $20\sqrt{2}\,\dfrac{d\theta}{dt}$ cm^2 per radian

12 $\mathrm{C}\left(\frac{\pi}{24}, \frac{\sqrt{3}}{2}\right)$ **13** decreasing at $\frac{2}{5}$ units per second

14 increasing at 16π cm^3 per second

REVIEW SET 15B

1 **a** $y = 2 - x$ **b** $x + 2y = 3e$ **2** 30 units^2

3 **a** $y = 4 - x$ **b** Q(2, 2)

4 **a** $f'(x) = 6x^2 - 10x + 6$
b $b^2 - 4ac = -44 < 0$
$f'(x)$ is a concave up parabola that does not cut the x-axis
$\therefore$ $f'(x) > 0$ for all x.

5 **a** $x < 0$ and $0 < x \leqslant \frac{1}{2}$ **b** $P(\frac{1}{2}, 3)$

6 **a** minimum turning point $(-\sqrt{3}, -6\sqrt{3})$,
maximum turning point $(\sqrt{3}, 6\sqrt{3})$
b minimum turning point $\left(-3, -\frac{27}{e^3}\right)$,
stationary inflection $(0, 0)$
c minimum turning point (e, e)

7 **a** $a = 3$ **b** minimum turning point $(-5, -\frac{1}{10})$

8 $P(-\pi, 2)$, $Q(\pi, 2)$

9 **a** $v(t) = 5e^{5t} + 3 \text{ cm s}^{-1}$
b $v(t) > 0$ for all t, so the object is never at rest.
c $25e^5 \text{ cm s}^{-2}$ **d** $(e^{10} + 5)$ cm

10 **b** **i** increasing at $\frac{2\pi}{5}$ cm^2 per minute
ii increasing at $\frac{4\pi}{15}$ cm^2 per minute

11 **b** $x \approx 6.18$ **12** $\theta \approx 1.11$

13 increasing at $3\sqrt{14}$ units per second

14 **a** **i** $P = \left(10 + \frac{10}{\cos\theta}\right)$ cm **ii** $A = 25\tan\theta$ cm^2
b **i** increasing at $\frac{5\pi}{27}$ cm per minute
ii increasing at $\frac{25\pi}{27}$ cm^2 per minute

EXERCISE 16A

1 **a** 12 **b** 10 **c** 28 **d** $\frac{35}{2}$

2 **a** a **b** $\frac{1}{2}a^2$

3 **a** Each strip extends above the graph of $y = x^3$. The sum of the areas of the strips includes the area between $y = x^3$ and the x-axis, and some area above $y = x^3$.
d $\dfrac{a^4}{4}$

4 **a** $\dfrac{a^5}{5}$ **b** $\dfrac{a^6}{6}$

EXERCISE 16B

1 **a** $F'(x) = x$ $\therefore$ $\displaystyle\int x\,dx = \frac{x^2}{2} + c$
b $F'(x) = x^4$ $\therefore$ $\displaystyle\int x^4\,dx = \frac{x^5}{5} + c$
c $F'(x) = 8x^7$ $\therefore$ $\displaystyle\int x^7\,dx = \frac{x^8}{8} + c$
d $F'(x) = -2x^{-3}$ $\therefore$ $\displaystyle\int x^{-3}\,dx = -\frac{x^{-2}}{2} + c$
e $F'(x) = \frac{1}{2}x^{-\frac{1}{2}}$ $\therefore$ $\displaystyle\int x^{-\frac{1}{2}}\,dx = 2x^{\frac{1}{2}} + c$
f $\displaystyle\int x^n\,dx = \frac{x^{n+1}}{n+1} + c$
g No, as we cannot divide by zero.

2 **a** $F'(x) = e^x$ $\therefore$ $\int e^x\,dx = e^x + c$
b $F'(x) = e^{2x}$ $\therefore$ $\int e^{2x}\,dx = \frac{1}{2}e^{2x} + c$
c $F'(x) = \frac{1}{3}e^{\frac{x}{3}}$ $\therefore$ $\int e^{\frac{x}{3}}\,dx = 3e^{\frac{x}{3}} + c$
d $\displaystyle\int e^{kx}\,dx = \frac{1}{k}e^{kx} + c$

3 **a** $F'(x) = 5$ $\therefore$ $\int 5\,dx = 5x + c$
b $\int k\,dx = kx + c$

4 **a** $F'(x) = x^6$
b **i** $\frac{1}{7}x^7 + c$ **ii** $\frac{2}{7}x^7 + c$ **iii** $\frac{5}{7}x^7 + c$
iv $-\frac{1}{7}x^7 + c$
c If k is a constant, then $\int k\,f(x)\,dx = k\,F(x) + c$ where $F(x)$ is the antiderivative of $f(x)$.

5 **a** $F'(x) = 6x + 8$ $\therefore$ $\int (6x + 8)\,dx = 3x^2 + 8x + c$
b **i** $3x^2 + c$ **ii** $8x + c$
c $\int (f(x) + g(x))\,dx = \int f(x)\,dx + \int g(x)\,dx$

6 **a** $\dfrac{dy}{dx} = -\dfrac{1}{2\sqrt{1-x}}$
b $\displaystyle\int \frac{1}{\sqrt{1-x}}\,dx = -2\sqrt{1-x} + c$

7 **a** $\dfrac{1}{\sqrt{2x+5}}$ **b** $\sqrt{2x+5} + c$

8 **a** **i** $\dfrac{1}{x}$ **ii** $\ln x + c$ **b** **i** $\dfrac{1}{x}$ **ii** $\ln(-x) + c$

9 **a** $\dfrac{dy}{dx} = \cos x$ **b** $\int \cos x\,dx = \sin x + c$

10 **a** $-\sin x$ **b** $\int \sin x\,dx = -\cos x + c$

11 **a** $\dfrac{dy}{dx} = \sec^2 x$ **b** $\int \sec^2 x\,dx = \tan x + c$

12 **a** $3\cos 3x$ **b** $\int \cos 3x\,dx = \frac{1}{3}\sin 3x + c$

EXERCISE 16C.1

1 **a** $\dfrac{x^{11}}{11} + c$ **b** $2x^4 + c$ **c** $-\frac{4}{9}x^9 + c$
d $-\dfrac{1}{3x^3} + c$ **e** $3x^2 - 2x + c$ **f** $x^3 + c$
g $6\ln|x| + c$ **h** $\dfrac{2x^3}{3} + x + c$
i $\dfrac{x^3}{3} + \dfrac{3x^2}{2} - 2x + c$ **j** $\dfrac{x^4}{4} - \dfrac{x^2}{2} + c$
k $\dfrac{5x^3}{3} - 2\ln|x| + c$ **l** $5x^2 - \dfrac{5}{x} + c$
m $\frac{10}{3}x^{\frac{3}{2}} + c$ **n** $-6\sqrt{x} + c$
o $\frac{4}{3}x^{\frac{3}{2}} + 7\ln|x| + c$ **p** $\dfrac{x^4}{2} + \dfrac{3x^2}{2} + \dfrac{4}{x} + c$
q $\dfrac{x^3}{3} + 2\sqrt{x} + c$ **r** $2x^{\frac{3}{2}} + \dfrac{1}{x^2} + c$

2 **a** $\dfrac{x^3}{3} - 2x^2 + 4x + c$ **b** $\dfrac{x^3}{3} - 3x^2 + c$
c $\dfrac{x^5}{5} - 2x^3 + 9x + c$ **d** $-\dfrac{9}{x} + 6x + \dfrac{x^3}{3} + c$
e $-\dfrac{16}{x} + 8\ln|x| + x + c$ **f** $2x^2 + 3x + c$

g $2x - \frac{5}{x} + c$ **h** $\frac{x^2}{2} + 2x - 3\ln|x| + c$

i $\frac{x^3}{3} + 2x^2 - \ln|x| + c$ **j** $4x^{\frac{3}{2}} + 8\sqrt{x} + c$

k $\frac{2}{5}x^{\frac{5}{2}} - 4\sqrt{x} + c$ **l** $18\sqrt{x} - 6\ln|x| - \frac{2}{\sqrt{x}} + c$

3 **a** $\frac{x^4}{4} + 2x^3 + 6x^2 + 8x + c$
b $\frac{27x^4}{4} + 9x^3 + \frac{9x^2}{2} + x + c$
c $8x - 6x^2 + 2x^3 - \frac{x^4}{4} + c$
d $\frac{x^5}{5} - x^4 + 2x^3 - 2x^2 + x + c$

4 **a** $2\sin x + c$ **b** $7e^x + c$
c $-5\cos x - e^x + c$ **d** $-\cos x + \sin x + c$
e $6\sin x - \frac{e^x}{4} + c$ **f** $2e^x + \frac{1}{3}\cos x + c$

5 **a** $f(x) = \frac{5x^4}{4} - x^2 + 6x + c$
b $f(x) = \frac{2}{7}x^{\frac{7}{2}} + \frac{2}{3x^3} + c$
c $f(x) = \frac{x^4}{4} - \sin x + c$
d $f(x) = 3x^2 - 4e^x + c$
e $f(x) = -\cos x - 4\ln|x| + c$
f $f(x) = 2e^x - 7\ln|x| + \frac{x^6}{6} + c$
g $f(x) = 2\sqrt{x} + 3\cos x + c$
h $f(x) = 3\sin x + \cos x + \frac{1}{5}e^x + c$

6 **a** $y = -\frac{2}{x^2} - 5x^2 + c$ **b** $y = \frac{x^2}{2} + 5\ln|x| + \frac{1}{x} + c$
c $y = \frac{2}{3}e^x - \frac{2}{9}x^{\frac{3}{2}} + c$ **d** $y = 6\sin x + 2\cos x + c$
e $y = \frac{2}{5}x^{\frac{5}{2}} + 6\sqrt{x} - 8e^x + c$
f $y = \frac{4x^3}{3} + 4x - \frac{1}{x} - \frac{1}{2}\sin x + c$

EXERCISE 16C.2

1 **a** $f(x) = 2x^2 - x + 5$ **b** $f(x) = \frac{x^2}{2} + 2x - 4$
c $f(x) = -\frac{6}{x} + 6$ **d** $f(x) = 2x^{\frac{3}{2}} - 11$

2 **a** $y = \frac{x^3}{3} - x^2 + \frac{2}{3}$ **b** $y = e^x + 3x + 3$
c $y = \frac{2}{3}x^{\frac{3}{2}} + 3\ln|x| + \frac{1}{3}$
d $y = -2\cos x - 6\sin x + 10 + \sqrt{3}$

3 $y = 10\sqrt{x} - 6$ **4** $y = \frac{3x^2}{2} - 4e^x + 2$

5 **a** $f(x) = x^3 - x^2 + 4x - 5$
b $f(x) = \frac{4}{3}x^{\frac{3}{2}} - \frac{3}{2}x^2 + 4x + \frac{1}{3}$
c $f(x) = -2\ln|x| - \frac{3}{x} + \frac{3}{e^2}$
d $f(x) = -\sin x + 3\cos x - 2x - 1$

6 $y = -\cos x - 5e^x + 6x - 4$

EXERCISE 16D

1 **a** $\frac{1}{10}(2x-1)^5 + c$ **b** $\frac{1}{20}(5x+2)^4 + c$
c $-\frac{1}{6}(3-x)^6 + c$ **d** $\frac{1}{15}(4+3x)^5 + c$
e $\frac{1}{12}(4x-3)^6 + c$ **f** $-\frac{1}{14}(5-2x)^7 + c$

2 **a** $\frac{-1}{9x+3} + c$ **b** $\frac{1}{9}(6x-1)^{\frac{3}{2}} + c$
c $-\frac{2}{5(5x-3)^2} + c$ **d** $-\frac{1}{6}(2-4x)^{\frac{3}{2}} + c$
e $\sqrt{2x+3} + c$ **f** $-\frac{3}{2}\sqrt{1-8x} + c$

3 **a** $\frac{1}{3}\ln|3x+5| + c$ **b** $\frac{1}{4}\ln|4x-1| + c$
c $-\frac{1}{2}\ln|7-2x| + c$ **d** $\frac{3}{8}\ln|8x-7| + c$
e $6\sqrt{x} - \ln|x+5| + c$
f $-\frac{1}{3}\ln|1-6x| + \frac{2}{3}(x+1)^{\frac{3}{2}} + c$

4 **a** $\frac{1}{3}e^{3x} + c$ **b** $\frac{1}{4}e^{4x-1} + c$ **c** $-\frac{1}{2}e^{9-2x} + c$
d $\frac{1}{2}e^{2x} - \frac{1}{4}e^{-4x} + c$ **e** $2e^{2x+1} + e^{-x} + c$
f $\frac{1}{2}e^{2x+2} - 10e^{x+1} + 25x + c$
g $\frac{1}{6}e^{6x-2} + \frac{27x^4}{4} - 18x^3 + 18x^2 - 8x + c$
h $\frac{1}{4}e^{4x} + 2x - \frac{1}{4}e^{-4x} + c$
i $-\frac{5}{3}e^{1-3x} - \frac{1}{2}\sqrt{4x+1} + c$

5 **a** $\frac{1}{4}\sin 4x + c$ **b** $-\frac{1}{3}\cos 3x + c$
c $\frac{1}{2}\sin\left(2x + \frac{\pi}{2}\right) + c$ **d** $\frac{2}{3}\sin 3x - \frac{1}{6}\cos 6x + c$
e $-5\sin\left(\frac{\pi}{2} - x\right) + c$
f $-\frac{1}{12}\cos 2x + \sin\left(x - \frac{\pi}{6}\right) + c$
g $-\frac{1}{5}\cos 5x - \frac{1}{3}x^3 + c$ **h** $\frac{1}{7}e^{7x} - \frac{3}{4}\sin 4x + c$
i $\frac{2}{3}\cos(\pi - 3x) - \frac{2}{3}(9-x)^{\frac{3}{2}} + c$

6 $f(x) = \frac{1}{15}(3x-1)^5 + \frac{1}{15}$ **7** $y = \frac{1}{2}\sqrt{4x-3} + \frac{7}{2}$

8 $y = \frac{3}{2}\ln|2x-6| + \ln 5$

9 **a** $y = 3x^2 - 2e^{2x} + 5$ **b** $y = 3 - 4x$

10 **a** $y = -2\cos 3x - 1$
b **i** $y = \frac{1}{6}x - \frac{\pi}{12} - 1$ **ii** $y = -\frac{1}{3\sqrt{3}}x + \frac{\pi}{27\sqrt{3}} - 2$

11 $y = \frac{1}{2}e^{-4x} - 4x + 4$

12 **a** $a = \frac{\pi}{2}$ **b** $y = -\frac{4}{\pi^2}\cos\frac{\pi x}{2} - \sin\left(x - \frac{\pi}{2}\right) + \frac{4}{\pi^2}$

13 $a = \frac{4k}{3},\ k \in \mathbb{Z}$ **14** $f(x) = \frac{1}{2}\sin\left(2x + \frac{\pi}{2}\right) + \frac{3}{2}$

EXERCISE 16E

1 **a** $3x^2 + c$ **b** 9 **2** **a** $2x^2 + 5x + c$ **b** 10

3 **a** 8 **b** $\frac{21}{2}$ **c** 16 **d** 12 **e** $-\frac{15}{2}$ **f** 1
g $\frac{64}{9}$ **h** $\frac{14}{3}$ **i** 8

4 **a** $\frac{8}{3}$ **b** $\frac{158}{3}$ **c** 3 **d** 0 **e** $-\frac{61}{2}$ **f** $\frac{256}{15}$

5 **a** $\ln\left(\frac{5}{2}\right)$ **b** $3\ln\left(\frac{2}{3}\right)$ **c** 12
d $2\ln\left(\frac{19}{11}\right)$ **e** $-\frac{5}{3}\ln 7$ **f** $\frac{7}{3}\ln\left(\frac{17}{2}\right)$

6 **a** $e - 1$ **b** $12 - 4e$ **c** $e^2 + 3$
d $\frac{1}{2}(e^6 - 1)$ **e** $3(e - \sqrt[3]{e})$ **f** $\frac{1}{5}(e^3 - e^{-12})$
g $\frac{1}{4}(e^7 - e^5)$ **h** $e - 2$ **i** $\frac{13}{2}$

7 **a** $\frac{1}{2}$ **b** 2 **c** 5 **d** $\frac{\pi^2}{8}-1$
e $-\frac{1}{3}$ **f** $\pi-1$ **g** $\frac{3}{2}$ **h** $-\frac{1}{6}$ **i** -2
j $\frac{1}{5}$ **k** $\frac{1}{2}-\frac{\sqrt{3}}{4}$ **l** $\frac{3\sqrt{3}}{4}-\frac{7}{12}\pi^{\frac{3}{2}}+\frac{3}{4}$

8 **a** $m=2$ **b** $m=\frac{1}{2}\ln 6$ **c** $m=\frac{2\pi}{3}$

10 **a** 14 **b** -9 **c** 0 **d** 20

11 **a** 18 **b** 8 **c** 2

12

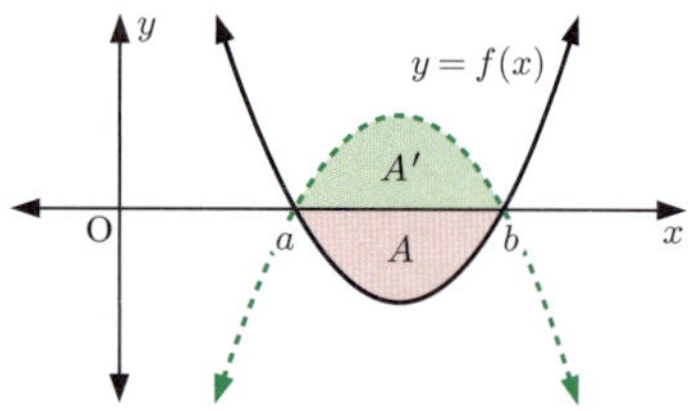

$A'=\int_a^b(-f(x))\,dx$
$=-\int_a^b f(x)\,dx$
$=A$

13 **a** 6 **b** 6 **c** 14 **d** 18

14 **a** $-\frac{9}{2}$ **b** $-9+\frac{9\pi}{4}$ **c** 0 **d** $-\frac{27}{2}+\frac{9\pi}{4}$

15 **b** $\frac{12}{5}$ **16** **b** 12

17 **a** $3x^2e^{x^3}$ **b** $\frac{1}{3}e^{x^3}+c$ **c** $a=\sqrt[3]{\ln 10}$

REVIEW SET 16A

1 16 **2** **a** $\frac{dy}{dx}=\frac{1}{\sqrt{2x-3}}$ **b** $\sqrt{2x-3}+c$

3 **a** $\frac{x^{10}}{10}+c$ **b** $2x^3+\frac{x^2}{2}-x+c$
c $2x^{\frac{3}{2}}+4\ln|x|+c$ **d** $-\frac{4}{x^2}-7x+c$
e $9x+12\ln|x|-\frac{4}{x}+c$ **f** $\frac{10}{3}x^{\frac{3}{2}}-8\sqrt{x}+c$

4 **a** $f(x)=x-3e^x+4$ **b** $f(x)=-2\cos x-\frac{1}{2}x^2-4$

5 **a** $\frac{1}{8}(2x+4)^4+c$ **b** $-\frac{1}{6}(7-x)^6+c$
c $\frac{2}{15}(5x-2)^{\frac{3}{2}}+c$ **d** $\frac{1}{4}\ln|4x-3|+c$
e $-\frac{1}{3}e^{6-3x}+c$ **f** $-\frac{2}{3}(2-x)^{\frac{3}{2}}+5e^{\frac{x}{5}}+c$

6 **a** $y=x-\frac{1}{3}\sin 3x-\frac{1}{3}$ **b** $y=x$

7 **a** $\frac{x^3}{3}+2x^2+c$ **b** $\frac{32}{3}$

8 **a** 16 **b** $\frac{38}{3}$ **c** $\frac{9}{2}$
d $\frac{1}{2}(e^3-e^2)$ **e** $9\ln 3-8$ **f** 0

9 **a** $m=-3$ **b** $m=\frac{14}{3}$

10 **a** 2π **b** -4 **c** $\pi-3$

11 **a** $\frac{4x}{2x^2+5}$ $\therefore \int\frac{x}{2x^2+5}\,dx=\frac{1}{4}\ln(2x^2+5)+c$
b $\frac{1}{4}\ln\left(\frac{37}{23}\right)$

REVIEW SET 16B

1 $\frac{21}{2}$ **2** **a** $f'(x)=\ln x$ **b** $x\ln x-x+c$

3 **a** $3x-\frac{x^6}{6}+c$ **b** $\frac{2}{5}x^{\frac{5}{2}}-\frac{2}{3}x^{\frac{3}{2}}+c$
c $5\sin x+c$ **d** $-2e^x+\cos x+c$
e $\frac{1}{4}\sin x+\frac{1}{x^3}+2\ln|x|+c$ **f** $\frac{1}{2}e^{2x}+2e^x+x+c$

4 **a** $y=\frac{5}{3}e^x-\frac{4}{9}x^{\frac{3}{2}}+c$ **b** $y=5\sin x+\frac{4}{3}\cos 3x+c$

5 **a** $f(x)=4\sqrt{x}+6x-3$
b $f(x)=-\sin x-e^x+4x+6$

6 **a** $\frac{1}{3(2-x)^3}+c$ **b** $-6\sqrt{2x-1}+c$
c $\frac{1}{2}e^{2x}+\frac{1}{3}e^{-3x}+c$ **d** $2\sin\frac{x}{2}-\frac{7x^3}{3}+c$
e $-\frac{1}{4}\cos 4x+\frac{5}{e^x}-\frac{3}{2}\ln|1-4x|+c$
f $\frac{1}{20}\sin\left(2x-\frac{\pi}{4}\right)-\frac{8}{x}+c$

7 **a** $y=12e^{\frac{x}{2}}-2x-12e$
b $y=\frac{x}{2-6e}+\frac{1}{3e-1}-4$

8 **a** $\frac{57}{2}$ **b** $\frac{25}{2}-\ln(\frac{4}{3})$ **c** $12(e-1)$
d 2 **e** $-\frac{10}{3}$ **f** 0

9 **a** -4 **b** 11 **c** 28 **10** **b** $1-\frac{1}{\sqrt{3}}$

EXERCISE 17A.1

1 **a** 10 units^2 **b** 9 units^2 **c** 8 units^2

2 **a** 15 units^2 **b** 8 units^2 **c** 2 units^2
d 6 units^2 **e** $4\frac{2}{3}$ units^2 **f** $2\ln 4$ units^2
g $5\frac{1}{3}$ units^2 **h** 4 units^2 **i** $\frac{1}{2}\ln 5$ units^2

3 **B**

4 **a** $\frac{1}{2}(e^8-e^2)$ units^2 **b** 1 unit^2
c $2\ln 3$ units^2 **d** $24\frac{2}{3}$ units^2

5 **a** $85\frac{1}{3}$ units^2 **b** $4\frac{1}{2}$ units^2 **c** $5\frac{1}{3}$ units^2
d $(4\ln 4-3)$ units^2

6 **a** $P(\pi, 1)$, $Q(2\pi, 0)$ **b** 2 units^2

7 **a** $P\left(-\frac{1}{\sqrt{3}}, \frac{2}{3\sqrt{3}}\right)$ **b** $\frac{5}{36}$ units^2

8 **a** $(\pi+\sqrt{3})$ units^2 **b** 3π units^2

9 **b** $\sqrt{2}$ units^2 **10** **b** $\frac{4}{e^2}$ units^2

11 **a** $k=3$ **b** $k=\frac{3}{2}$

12 **a** **i** $A(3, 0)$ **ii** $B(1, 4)$ **b** $6\frac{1}{6}$ units^2

13 $k=9-4\sqrt{2}$

EXERCISE 17A.2

1 $-\int_{-2}^{4}f(x)\,dx$

2 **a** 9 units^2 **b** $1\frac{1}{3}$ units^2 **c** $\left(4+\frac{1}{e}\right)$ units^2
d 1 unit^2 **e** $\left(\frac{\pi^2}{2}-2\right)$ units^2 **f** $(4-\ln 3)$ units^2

3 **a** $P(4, -8)$ **b** $26\frac{1}{5}$ units^2

4 **a** $A(\frac{\pi}{2}, 0)$, $B(\pi, 0)$ **b** 4 units^2

5 **a** $B(2, 0)$, $C(4, 0)$ **b** $21\frac{1}{12}$ units^2

6 **a** A_1-A_2 **b** A_1+A_2

7 **a** A_1 is larger: $\int_{-4}^{5}f(x)\,dx<0$, so the area below the x-axis on this interval must be larger than the area above the x-axis.
b 19 units^2

EXERCISE 17B

1 **a** $10\frac{2}{3}$ units^2 **b** $5\frac{1}{2}$ units^2

2 **a** $P(\frac{\pi}{6}, \frac{1}{2})$, $Q(\frac{5\pi}{6}, \frac{1}{2})$ **b** $(\sqrt{3} - \frac{\pi}{3})$ units^2

3 **a** $4\frac{1}{2}$ units^2 **b** 32 units^2

c $\left(1 - \dfrac{2}{e} + \dfrac{1}{e^2}\right)$ units^2 **d** $\left(\frac{\sqrt{3}}{2} + \frac{\pi}{3}\right)$ units^2

4 **a**

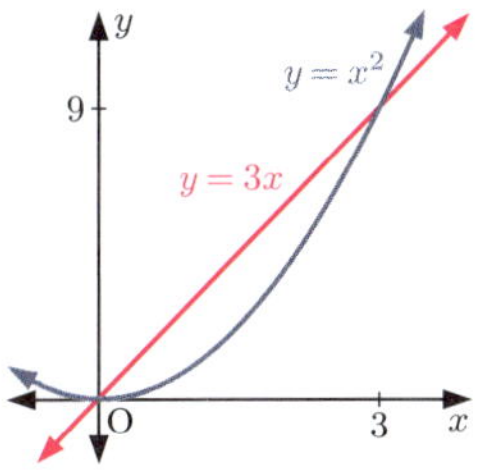

b $(0, 0)$ and $(3, 9)$
c $4\frac{1}{2}$ units^2

5 **a**

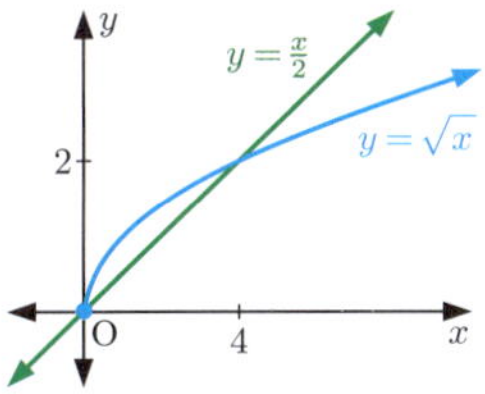

b $(0, 0)$ and $(4, 2)$
c $1\frac{1}{3}$ units^2

6 **a** $A(0, 4)$, $B(2, -8)$, $C(5, 4)$ **b** $21\frac{1}{12}$ units^2

7 $21\frac{1}{12}$ units^2 **8** **a** $k = \frac{1}{2}$ **b** $(3 - 2\ln 2)$ units^2

9 **b** $B(2, 1)$ **c** $6\frac{2}{3}$ units^2 **10** 6 units^2

11 **a** $P(-3, -4)$, $Q(1, 0)$ **b** $21\frac{1}{3}$ units^2

12 **b** $\approx 68.3\%$ **13** $5\frac{1}{3}$ units^2 **14** $\ln 3$ units^2

EXERCISE 17C.1

1 **a** 1000 m **b** 95 m

2 **a** $4\ \text{m s}^{-1}$ **b** **i** 75 m **ii** 125 m

3 **a** after 25 seconds **b** 262.5 m
c 112.5 m in the positive direction

4 **a**

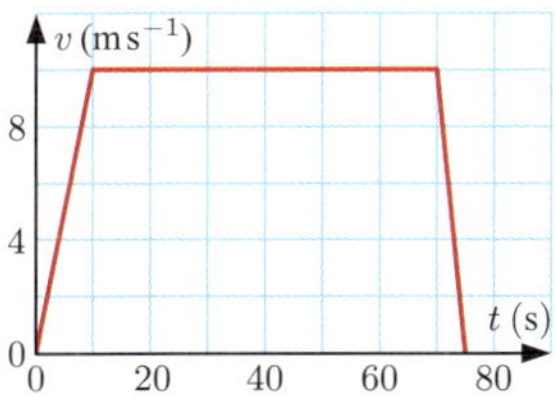

b 675 m

5 **a** **i**

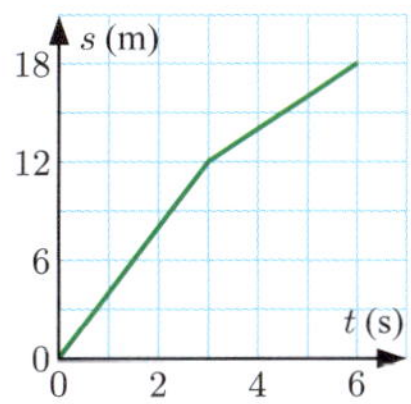

ii 18 m
iii 18 m in the positive direction

b **i**

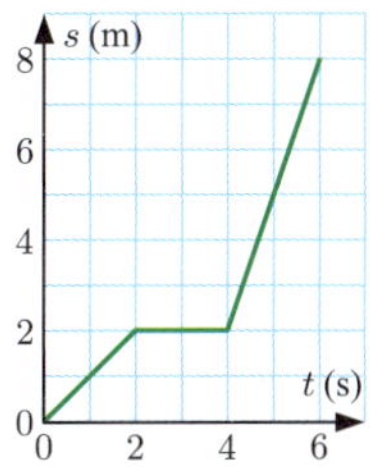

ii 8 m
iii 8 m in the positive direction

c **i**

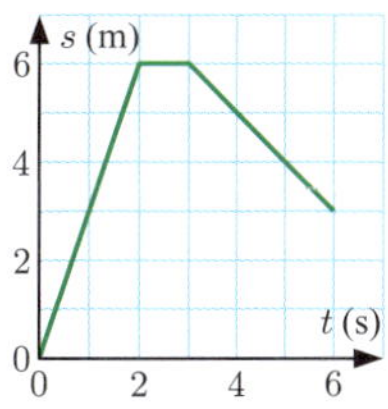

ii 9 m
iii 3 m in the positive direction

6 **a** $u = \frac{192}{19} \approx 10.1$

b gradient $= -\dfrac{u}{6} = -\frac{32}{19} \approx -1.68$;
between 10 and 14 seconds, the acceleration of the particle was $-\frac{32}{19} \approx -1.68\ \text{m s}^{-2}$.

7 **a**

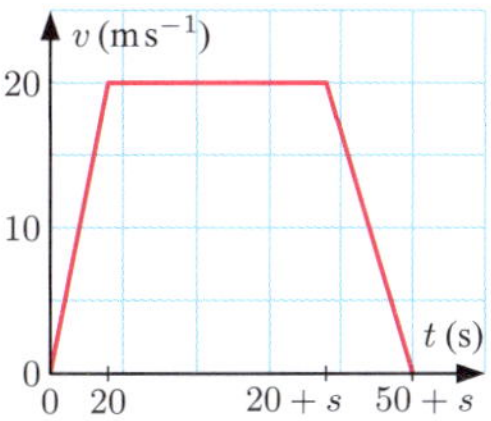

b $s = 75$

EXERCISE 17C.2

1 **a** $v(t) > 0$ for all $t \geqslant 0$ **b** $s(t) = t^2 + 3t + 4$ m
c 18 m

2 **a** $s(t) = 6t - t^2$ cm **b** 10 cm **c** 8 cm to the right of O

3 **a** $s(t) = -4.9t^2 + 29.4t + 1.4$ m **b** after 3 seconds
c 63.7 m

4 **a** $26\ \text{cm s}^{-1}$
b $v(2) = 0\ \text{cm s}^{-1}$, so the particle changes direction after 2 seconds.
c $\int_0^5 v(t)\,dt = -45\frac{5}{6}$ cm; the displacement of the particle after 5 seconds is $45\frac{5}{6}$ cm to the left of its initial position.
d $\int_0^2 v(t)\,dt - \int_2^5 v(t)\,dt = 71\frac{1}{6}$ cm; the total distance travelled in the first 5 seconds is $71\frac{1}{6}$ cm.

5 **a** $3\ \text{m s}^{-1}$ **b** $v(t) > 0$ for all $t \geqslant 0$
c **i** $s(t) = 4\ln|3t + 1| + 2$ m
ii $a(t) = -\dfrac{36}{(3t+1)^2}\ \text{m s}^{-2}$
d $4\ln 10$ m **e** $2\frac{1}{4}\ \text{m s}^{-2}$ to the left

6 **a** **i** $2\ \text{m s}^{-1}$ **ii** $6\ \text{m s}^{-1}$
b $s(t) = \frac{1}{3}(4t+1)^{\frac{3}{2}} - \frac{1}{3}$ m
c **i** 12 seconds **ii** $14\ \text{m s}^{-1}$

7 **a** $4\left(1 - \dfrac{1}{e}\right) \approx 2.53\ \text{m s}^{-1}$
b The kayaker's speed approaches $4\ \text{m s}^{-1}$.
c ≈ 224 m **d** $\approx 0.0821\ \text{m s}^{-2}$

8 **a** $1\ \text{cm s}^{-2}$ to the right **b** $v(t) = 5 + 3t - \frac{1}{2}t^2\ \text{cm s}^{-1}$
c **i** $9\ \text{cm s}^{-1}$ **ii** $3\ \text{cm s}^{-1}$

9 **a** **i** $v(t) = 2e^{2t} - 4\ \text{m s}^{-1}$ **ii** $s(t) = e^{2t} - 4t - 1$ m
b after $\frac{1}{2}\ln 2 \approx 0.347$ seconds **c** ≈ 3.16 m

10 **a** The pendulum is initially at the extreme positive position. This is the point where the pendulum changes direction. So the pendulum must be at rest at this point. That is, $v = 0$ when $t = 0$.

b $v(t) = -20\sin 4t$ cm s^{-1} **c** $s(t) = 5\cos 4t$ cm
d 20 cm s^{-1} **e** $\frac{\pi}{2} \approx 1.57$ seconds

REVIEW SET 17A

1 **a** 21 units^2 **b** $5\frac{1}{3}$ units^2 **c** $(e^2 - 1)$ units^2

2 **a** $A(\frac{\pi}{4}, \sqrt{2})$, $B(\frac{3\pi}{4}, 0)$ **b** $\sqrt{2}$ units^2

3 **a** **i** P(4, 20) **ii** Q(6, 0) **b** $61\frac{1}{3}$ units^2

4 **a** At A, x-intercept is $\frac{\pi}{6}$; at B, x-intercept is $\frac{\pi}{2}$.
b $\frac{2}{3}$ units^2

5 **a**

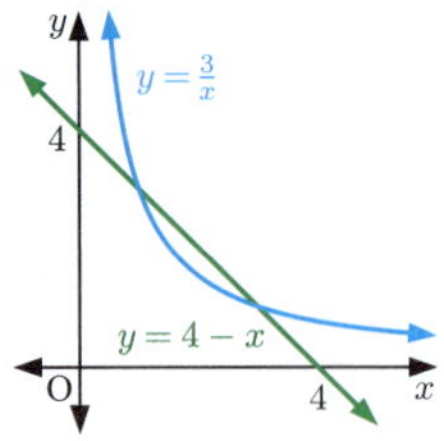

b (3, 1) and (1, 3)
c $(4 - 3\ln 3)$ units^2

6 $49\frac{1}{3}$ units^2 **7** **a** 750 m **b** 13.5 m

8 **a** $y = 3x - 4$ **b** 9 units^2

9 **a** $s(t) = \frac{1}{3}t^3 + 2t^2 - 32t$ m **b** $101\frac{1}{3}$ m
c after 3 seconds

10 **a** 36 m s^{-1} **b** $s(t) = \frac{27}{5}t^{\frac{5}{3}}$ m
c **i** 27 seconds **ii** 1312.2 m = 1.3122 km

REVIEW SET 17B

1 **a** P(3, 0) **b** 4 units^2 **2** **b** $\frac{12}{e}$ units^2

3 **a** $(5 - \frac{5}{3}\ln 4)$ units^2 **b** $\frac{3}{4}$ units^2

4 **a** B(2, −16), C(5, 5) **b** $78\frac{1}{12}$ units^2

5 **a**

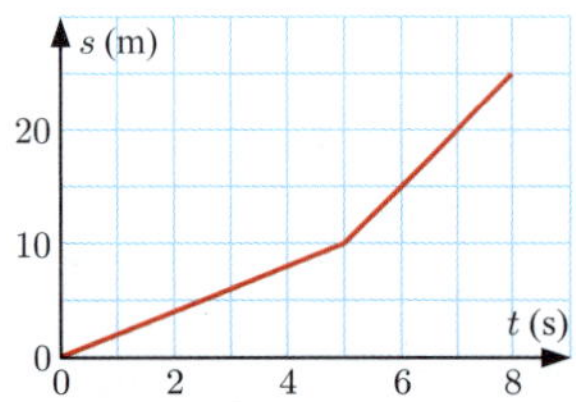

b 25 m

6 **a** 6 m s^{-1} **b** 270 m **c** 0.8 m s^{-2}

7 **a** $P(\frac{\pi}{6}, \sqrt{3})$, $Q(\frac{11\pi}{6}, \sqrt{3})$ **b** $(\frac{5\sqrt{3}}{3}\pi + 2)$ units^2

8 **a** P(ln 3, 0), Q(ln 6, 3) **b** $(2 + \ln 288)$ units^2

9 **a** **i** 9 cm s^{-1} **ii** 3 cm s^{-1} **b** after 4 seconds
c $s(t) = 12t - \frac{3}{2}t^2$ cm **d** 30 cm

10 **a** **i** $v(t) = 6\sqrt{t+1} - 12$ m s^{-1}
ii $s(t) = 4(t+1)^{\frac{3}{2}} - 12t - 2$ m
b after 3 seconds, 6 m to the left of the origin **c** 24 m

INDEX